Wild Plants
The Treasure of Natural Healers

T0133752

Editors

Mahendra Rai
Department of Biotechnology, SGB Amravati University
Amravati, Maharashtra, India

Shandesh Bhattarai
Nepal Academy of Science and Technology
Khumaltar, Lalitpur, Nepal

Chistiane M. Feitosa
Department of Chemistry, Federal University of Piaui
Petronio Portela Campus, Teresina
Brazil

CRC Press
Taylor & Francis Group
Boca Raton London New York

CRC Press is an imprint of the
Taylor & Francis Group, an **informa** business

A SCIENCE PUBLISHERS BOOK

CRC Press
Taylor & Francis Group
6000 Broken Sound Parkway NW, Suite 300
Boca Raton, FL 33487-2742

© 2021 by Taylor & Francis Group, LLC
CRC Press is an imprint of Taylor & Francis Group, an Informa business

No claim to original U.S. Government works

Version Date: 20200608

International Standard Book Number-13: 978-0-367-82087-9 (Hardback)

Library of Congress Cataloging-in-Publication Data

Names: Rai, Mahendra, editor.
Title: Wild plants : the treasure of natural healers / editors, Mahendra
  Rai, Department of Biotechnology, SGB Amravati University, Amravati,
  Maharashtra, India, Shandesh Bhattarai, Nepal Academy of Science and
  Technology, Khumaltar, Lalitpur, Nepal, Chistiane M. Feitosa, Department
  of Chemistry, Federal University of Piaui, Petronio Portela Campus,
  Teresina, Brazil.
Description: Boca Raton : CRC Press, [2020] | Includes bibliographical
  references and index.
Identifiers: LCCN 2020019435 | ISBN 9780367820879 (hardcover)
Subjects: LCSH: Medicinal plants. | Wild plants, Edible. |
  Ethnopharmacology.
Classification: LCC QK99.A1 W53 2020 | DDC 581.6/34--dc23
LC record available at https://lccn.loc.gov/2020019435

Visit the Taylor & Francis Web site at
http://www.taylorandfrancis.com

and the CRC Press Web site at
http://www.routledge.com

Foreword

Traditional medicine, both codified (e.g., Chinese medicine, Ayurveda, Unani) and non-codified, has become a global movement with rapidly growing economic importance. In many Asian countries, traditional medicine is widely used, even though western medicine is often readily available. The number of visits to providers of traditional medicine in the US now exceeds the number of visits to primary care physicians. Many medicinal plant species are easily available in online trade, often without correct scientific identification, and possible contamination, which creates large safety concerns. In developing countries, uncodified traditional medicine is often the only accessible and affordable treatment available. Doctors are mostly located in cities and other urban areas, and are therefore inaccessible to rural populations. In Africa, up to 80% of the population uses traditional medicine as the primary healthcare system. In Latin America, the WHO Regional Office for the Americas (AMRO/PAHO) reports that 71% of the population in Chile and 40% of the population in Colombia have used traditional medicine. In many Asian countries, traditional medicine is widely used, even though western medicine is often readily available. In Japan, 60–70% of allopathic doctors prescribe traditional medicines for their patients. In China, traditional medicine accounts for about 40% of all healthcare, and is used to treat roughly 200 million patients annually. The expenses for the use of traditional and complementary-alternative medicine are exponentially growing in many parts of the world. Traditional and complementary-alternative medicine is also gaining more and more respect by national governments and health providers. As one example, Peru's National Program in Complementary Medicine and the Pan American Health Organization recently compared complementary medicine to allopathic medicine in clinics and hospitals operating within the Peruvian Social Security System.

However, this globalization of traditional remedies, in particular from non-codified traditional pharmacopoeiae, leaves many questions unanswered: does the use of traditional medicine reflect major health issues? Some plants may have beneficial properties, while others can cause adverse reactions. Even when the herbal ingredients themselves have proven benefits and no known safety concerns, some of the administration methods may be harmful. Most importantly, how can safety concerns associated with traditional medicines and practices be identified, monitored, and communicated to users and other stakeholders, and how can the safety and sustainability of the global supply of medicinal plants be ensured?

The present volume addresses a variety of these crucial questions, while closely keeping focus on the tremendous wealth wild plants provide to the global community. The editors give a very well-placed introduction to the complicated and often polemically discussed value of medicinal species, and the need for their sustainable use, while the following chapters ensure the outlining of the potential danger of disappearance of important species, including the tremendous danger of replacement of species once the original sources have been depleted. The editors have managed to include a wide range of case studies in this impressive volume, giving up-to-date information from countries as diverse as Peru, Nepal, Bangladesh, to name a few, and incorporating both general information, as well as examples on fungal use and plants as snake bite antidotes. In the final section of this impressive volume, specific species are used to illustrate the tremendous potential of medicinal plant species, as well as the dangers of their use without consultation and supervision of a trained specialist.

The present volume comes at a very important time, when dedicated programs aiming to establish *in situ* collections of important species, detailed phytochemical profiles for each species, as well as the repatriation of traditional knowledge in local languages, under the guidelines of the Nagoya Protocol, are urgently needed. Plants and their associated bio-cultural knowledge play an essential role in the ecosystem services that support all life on Earth. There is a great urgency to address the vital importance of traditional knowledge about plants, their utility, management, and conservation. This unique, often ancient, and detailed knowledge is typically held and maintained by local and indigenous communities. The implementation of the "Nagoya Protocol on Access to Genetic Resources and the Fair and Equitable Sharing of Benefits Arising from their Utilization to the Convention on Biological Diversity" has brought a great boost for the rights of indigenous and local communities. Benefit sharing in this context also needs to not only include the repatriation of the new data gathered, in a language and form accessible to the traditional owners, but also the translation and repatriation of the results of previous studies conducted in the same indigenous or local community, if not already done by the original researchers. In addition, informants, should they so desire, must be allowed full participation as authors in all publications of a study, rather than simply being mentioned as a sideline in the acknowledgments.

Professor Rainer W. Bussmann

Former Director and
William L. Brown Curator of Economic Botany/Senior Curator
William L. Brown Center at Missouri Botanical Garden
Principal Scientist
Department of Ethnobotany
Institute of Botany
Ilia State University
Tbilisi, 0105, Georgia

Preface

In the earliest period of human history, people used wild plants and animals as their staple food and medicine. These resources are culturally acceptable, important, cheap, and easily reachable to local people across the globe, particularly in the remote areas and mountains. Aboriginals have developed their traditional knowledge regarding use of wild plants, its management, and sustainable conservation. Several useful plant-derived drugs were discovered as a result of scientific follow-up of well-known plants used in traditional medicine through the ethnopharmacological research. Many of today's major diseases, such as HIV and cancer are being treated with a group of potential products derived from biodiversity. Thus, modern medicine has also benefited from wild plants that were originally used as home herbal remedies.

The global diversity of wild plants is very high, but little is known about their food/medicinal significance. Wild plants offer great potential for the discovery of unique molecules and new sources of active compounds. Therefore, wild plants should be investigated by modern scientific techniques to establish their safety and efficacy and to determine their potential as a source of new drugs. In this context, the present book provides a comprehensive overview of the wild plants, their usage, status, and threats, as well as the growing interest in ethnopharmacology research. The book comprises of important issues, such as diversity of wild plants, with emphasis on medicinal and food plants, threats to wild plants, and traditional ethnobotanical knowledge of their uses in skin diseases, snake-bites, and in cosmeceuticals. Moreover, the ethnopharmacological relevance of wild plants in Latin America has been discussed.

This book will be useful for the researchers working in the areas of conservation biology, botany, ethnobiology, ethnopharmacology, national and international policymakers, etc. The principal aim of this book is to collect the dispersed global data about the wild plant resources and their usage. Some contributors from India, Nepal, Pakistan, Brazil, China, Bangladesh, etc., have provided important information regarding the importance of their wild plant resources. Hence, the proposed publication will be considered as the main source of information and will be entirely different from other related publications.

We take the opportunity to thank all the contributors for their generous cooperation and efforts in offering up-to-date chapters. Further, we express our sincere thanks to the publisher and the authors of the chapters, whose research work has been cited in the present book. We are also thankful to the entire team of CRC Press for their cooperation, efforts, timely help, and patience in the publication of this book. Finally, MKR and CMF thank CNPq (National Council for Scientific and Technological Development, Brazil) for financial support (Process number 403888/2018-2), and SB acknowledges NAST (Nepal Academy of Science and Technology), Kathmandu, Nepal for varied support.

Mahendra Rai, India
Shandesh Bhattarai, Nepal
Chistiane M. Feitosa, Brazil

Contents

Specific Plants and Ailments

General

1

Wild Plants as a Treasure of Natural Healers
The Need for Unlocking the Treasure

Mahendra Rai,[1,3] *Shandesh Bhattarai*[2],* and *Chistiane Mendes Feitosa*[3]

Introduction

Plants are primarily multicellular, mostly photosynthetic eukaryotes of the kingdom Plantae which are distributed worldwide. The evolution of plants has resulted from the earliest algal mats, through bryophytes, lycopods, ferns to gymnosperms, and angiosperms (www.encyclopedia.com). Plants in all of these groups continue to flourish in the environments in which they evolved and have some of the largest genomes among all organisms (Todd and Scott 2013). The largest plant genome (in terms of gene number) is that of *Triticum asestivum*, estimated to encode ≈ 94,000 genes (Brenchley et al. 2012), and thus nearly five times as many as the human genome. The first plant genome sequenced was that of *Arabidopsis thaliana*, which encodes about 25,500 genes (Arabidopsis Genome Initiative 2000). In terms of sheer DNA sequence, the smallest published genome is that of the *Utricularia gibba* at 82 Mb (28,500 genes) (Enrique et al. 2013), while the largest, from the *Picea abies*, extends over 19,600 Mb (encodes about 28,300 genes) (Nystedt et al. 2013).

The estimates for the number of described plant species in the world vary in literature (Groombridge and Jenkins 2002, Thorne 2002, Scotland and Wortley 2003, Chapman 2009, Funk et al. 2009, Chase et al. 2015, APG IV 2016, Christenhusz and Byng 2016, RBG Kew 2016, Shrestha et al. 2018). The updated publication revealed that the described and accepted number of plant species in the world to be about 374,000, of which nearly 308,312 are described and accepted, vascular plant species of which 295,383 are Angiosperms [monocots (74,273), eudicots (210,008), Gymnosperms (1,079), Ferns (10,560), and Lycopods (1,290)] (Christenhusz and Byng 2016). The estimated numbers of liverworts are 9,000 species (Crandall Stotler and Stotler 2000), 200–250 hornworts (Villarreal et al. 2010), 12,700 mosses (Crosby et al. 1999, Cox et al. 2010), about 44,000 algae (Guiry 2012), amounts to a total of about 374,000 (~374,262) plant species worldwide. These numbers differ from earlier estimates by Chapman (2009), which has substantially lower estimates, with 310,129 as the total number of plant species, of which 281,621 are vascular plants, but a higher estimate is by Pimm and Joppa (2015), which states that there are an estimated 450,000 species. The largest vascular plant families are Orchidaceae (*about* 28,000 species), followed by Asteraceae (*about* 24,700 species) (Funk et al. 2009, Chase et al. 2015, Shrestha et al. 2018).

---
[1] Sant Gadge Baba Amravati University, Amravati, Maharashtra, India.
[2] Nepal Academy of Science and Technology, Khumaltar, Lalitpur.
[3] Department of Chemistry, Federal University of Piaui, Petronio Portella Campus, Brazil.
* Corresponding author: shandeshbhattarai@gmail.com

Large parts of the world are still in need of additional biological expeditions (Christenhusz and Byng 2016). The key countries that yield the greatest numbers of new species are Australia, Brazil, China, and New Guinea, although much smaller African, American, Pacific, and Central and tropical Asian countries also contribute substantial numbers (Zhang et al. 2014), but the newest species are perhaps to be established in the world's biodiversity hotspots (Joppa et al. 2011). The top ten countries in the world with the highest number of vascular plants are as follows: Brazil (56,215), Colombia (51,229 species), China (32,200 species), Indonesia (29,375 species), Mexico (26,071 species), South Africa (23,420 species), Venezuela (21,073), USA (19,473 species), Ecuador (19,362 species), and India (18,664 species) (Groombridge and Jenkins 2002, Shrestha 2016, Shrestha et al. 2018).

The research on wild plants to identify the commercially valuable genetic and biochemical resources has been on for centuries, and is now well accepted. This review aimed to gather information about the treasures of wild plants and stress the necessity to unlock such treasures for the conservation of wild plant resources and traditional knowledge.

Wild Plants as Natural Heritage

Wild plants denote species that are neither cultivated nor domesticated, but available from the natural habitat (Beluhan and Ranogajec 2010). Commonly, wild plants are unnoticed (Scoones et al. 1992), which include herbs, shrubs, trees, and grasses that grow without human help and are an important part of nature's biodiversity. In the earliest period of human history, people used wild plants and animals as their primary food and medicine. These resources are culturally acceptable, cheap, and easily accessible to local people, particularly in remote regions (Bhattarai et al. 2006, 2009). Indigenous people have developed their traditional knowledge regarding wild plant use, conservation, and management.

The role of wild plants used for foods in peoples' diets and medicine should not be underrated. There has been a growing interest in studying the consumption of wild food plants for sustainable use and management (Pfoze et al. 2011). Wild plants offer a diversity of food needed to sustain a rich and healthy diversity of insects, birds, and animals. Some insects feed off of specific plants so that a loss of a wild plant in an area could lead to the loss of an insect (www.praying-nature.com). Of the Earth's half-million plant species, about 3,000 species have been used as crops, and only 150 species have been in large scale cultivation (Mohammed et al. 2008).

The difference between wild and domesticated species is not easy. Domestication is an extended and difficult process, and many plants are found in various stages of domestication as a result of human selection (Pegu et al. 2013). In most civilizations, the use of wild plants forms part of indigenous systems of knowledge and practice that have developed and accrued over generations (Slikkerveer 1994).

The global diversity of wild plants is very high but little is known about their food, medicinal, and other use-value. Wild plants are a potential source for the discovery of novel molecules and new bioactive compounds, mainly because of the environmental stress to which they are exposed (Cordell 2002, Balunas and Kinghorn 2005, Chin et al. 2006, Newman and Cragg 2007, Carvalho 2011, Atanasov et al. 2015). Many plant-derived drugs were revealed as a result of scientific follow-up of well-known plants used in ethnomedicine (Cordell 2002, Carvalho 2011), and main diseases, such as HIV and cancer, are being treated with an array of potential products derived from biodiversity (Balunas and Kinghorn 2005, Chin et al. 2006, Newman and Cragg 2007, Atanasov et al. 2015). Hence, modern medicine has also benefited from wild plants that were used as home herbal therapies (Cordell 2002, Balunas and Kinghorn 2005, Newman and Cragg 2007, Carvalho 2011).

Wild plants have been playing a major part as a source of foods and medicines, and have a vital socio-economic role through their use in fuelwoods, dyes, poisons, shelter, fibers, religious and ritual ceremonies. The use of wild plants in providing sources of income and livelihoods in rural areas is acknowledged globally (Jain 1963, Agyemang 1996, Moreno-Black et al. 1996, Sajeev and Sasidharan 1997, LaRochelle and Berkes 2003, Kar 2004, Sawain et al. 2007, Yesodharan et al. 2007, Patiri and Borah 2007, Kar and Borthakur 2007, 2008, Misra et al. 2008, Aryal et al. 2009, Kalaba et al. 2009, Giliba et al. 2010, Kutum et al. 2011, Legwaila et al. 2011, Sarmah and Arunachalam 2011, Singh and

Rawat 2011, Seal 2011, 2012, Dutta 2012). The sustainable harvest of wild plants is very important, as it can offer vital resources as well as generate income for local people.

The usage of plants is not restricted to documented literature because copious knowledge is still available in traditional daily life (Cakir 2017). Globally, wild plants in traditional medicines have been increasingly used by various communities due to their significant role in maintaining good health (Mahapatra et al. 2019). About 50,000 plants are believed to be used in traditional medicine, but the exact number of medicinally useful constituents residing inside these 50,000 plants is still unknown (Gewali 2008). Hence, wild plants are becoming a part of a new way of thinking because of their hidden medicinal and food values, and in managing health, healthy food, food safety, and slow food movements (Yeşil et al. 2019).

Today, an increasing number of plants used in traditional medicine are described to have diverse activities in infectious diseases (Mahapatra et al. 2019). A single plant as a medicine must have been selected after various hits and trials of experiments, of which useful species were treasured as medicines (Gewali 2008). Nowadays, we take an important medicine, aspirin, for the relief of the pain, but as early as in 400 BC, Hippocrates gave Greek women willow-leaf tea (the tea contained aspirin-like constituents) to relieve the pain of childbirth. Similarly, Quinine from Cinchona bark saved the life of many people from malaria and Reserpine from *Rauvolfia serpentina* (Figure 1.1) was instrumental in bringing the peace of mind as well as the relief of psychotic behaviors (Gewali 2008). Periwinkle constituents, vinblastine, and vincristine were the first effective drugs against different forms of cancer (Gewali 2008). *Taxus* species (e.g., Figure 1.2) has afforded taxol used to cure breast and ovary cancers. Thus, the list of successful drugs originated from wild plants is long (Gewali 2008), but even today, only a few facts regarding the medicinal usage of wild plants have been captured and used by the natural healers in natural healing practices and still, other facts are unexplored and hidden.

Figure 1.1 *Rauvolfia serpentina* (L.) Benth. ex Kurz.

Figure 1.2 *Taxus* species.

Natural Healers—The key to Unlocking the Treasure

Globally, natural healing systems are widely accepted. A natural healer is a person who prescribes natural medicine to the patient. Natural healers play an important role in continuing their indigenous medical knowledge and practices as their tradition and culture (Aryal et al. 2016). Traditional healers' knowledge is a community-based scheme of knowledge that has been developed, preserved, and maintained over generations by the communities through their continuous interactions, observations, and investigations with their surrounding environment (Pushpangadan and Nair 2005).

Natural healers (Figures 1.3a, b) have increasingly taken culture into account because traditional medicines are rich and rooted in their cultural heritage. Indigenous knowledge of medicine is generally transmitted orally over many generations through a community, family, and individuals. The elements of knowledge of traditional medicine may be known by many or may be collected and applied by those in a definite role of the healer (*Acharya and Anshu* 2008).

Figure 1.3a Tharu People/Fisherman from Kailali District of Nepal carrying wild plant (*Bauhinia* sp.) leaves for domestic use.

Figure 1.3b Tharu People from Bardiya District of Nepal collecting and hanging wild plant for home use.

The Need for Recognition of Natural Healers

The healing system is comprised of a wide range of medical beliefs, knowledge, and practices, including medical doctors (specialized in allopathic medicine), natural healers (long knowledge in natural medicine), and others. Natural healers have long been getting extensive public acceptance and playing a major role in meeting the health care needs of local people. In many parts of the world, the natural healing system has been recognized as the cheapest and most accessible health care means for the majority of the rural people. Healers' knowledge is powerfully influenced by several factors, and they make safe, efficacious, quality, and affordable traditional medicines available to the vast majority of the people.

In remote villages, where do people go for treatment when they are sick? Local people can cure their diseases with the help of modern or allopathic medicine, or with more traditional, alternative, or complementary medicine. The system of natural healing has been widely recognized, established, and well accepted in the remote villages in different regions of the world. Although modern allopathic medicine is known to be common, effective, fast curing, and frequently being improved through scientific research, many patients feel comfortable and find better results by using both the modern and complementary medicines and following both systems of medication. Thus, these days, in fact, complementary/alternative medical systems have increased in availability and scope, and are receiving more popularity.

The Potential Role of Natural Healers

Traditional or indigenous or folk medicine encompasses medical aspects of indigenous knowledge established over generations within various societies (WHO 2008, Carvalho 2011). In some Asian and African countries, up to 80% of the population relies on traditional medicine for their primary health care needs (WHO 2008, 2013). Traditional medicine is often the measured practice of alternative medicine. Traditional medicines comprise traditional European medicine, traditional Chinese medicine, traditional Korean medicine, traditional African medicine, Ayurveda, Siddha medicine, Unani, ancient Iranian Medicine, Iranian (Persian), Islamic medicine, Muti, and Ifá. Ethnobotany, ethnomedicine, ethnobiology, herbalism, and medical anthropology are some scientific disciplines that study traditional medicine (WHO 2008, 2013, www.en.wikipedia.org).

From a global perspective, the socio-cultural and economic concerns in modern medicine may limit its potentiality to provide satisfactory health care (Cordell 2002, Bhattarai 2009, Rios 2011). Due to the safety, effectiveness, and quality of biomedical care, traditional medicines, along with the role of natural healers, are widely accepted for diagnosis and treatment (Cordell 2002, Balunas and Kinghorn 2005, Rios 2011).

According to Mwu and Gbodossou (2000), three factors, i.e., their own beliefs, the success of their actions, and the beliefs of the community validate the role of the healer. Linking to this, Setswe (1999) divided the roles of natural healers in primary health care into two parts: (i) Protective and promotive health, (ii) Curative and psychosocial care, but here we have made a slight modification.

(i) Protective and promotive health

The defensive roles of traditional healers have been stressed in numerous programs commenced in different parts of South Africa (Setswe 1999). The national HIV/STD preventive program trained 1,510 healers in HIV/STD prevention in 1992, directing the kind of collaboration that healers could deliver in working with modern practitioners (Green et al. 1995). Abdool Karim et al. (1994) discovered potential preventive health roles that traditional healers could play concerning the AIDS epidemic. He further concluded that their role could go beyond education to actively influence the community's views and attitudes to risk-associated behaviors. Some traditional healers have been actively involved in growth monitoring, oral rehydration, breastfeeding, immunization, family planning, food supplementation, and female education in South Africa, where traditional healers function in a similar role to that of village health workers (Freeman and Motsei 1992). Immunization against witchcraft, forecasts of future events, declaration of secrets, and annual check-ups are some of the known preventive roles of traditional healers (Abdool Karim et al. 1994, Kale 1995).

(ii) Curative and psychosocial care

Traditional medicines are generally effective with ailments, such as diarrhea, cough, cold, headaches, other pains, swellings, and sedating patients (Freeman and Motsei 1992). The key role of the natural healers is in the realm of psychiatry, considering the methodology used in natural medicine and the fact that mental illness is a product of society (Kelly 1995). The success of traditional healers in treating psychological problems is well documented and often recognized (Hoff 1992). To treat the psychological problems, a big part of healers' practice is committed to counseling individuals whose problems are the consequences of quick social and economic changes in the community (Abdool Karim et al. 1994).

The promotion of family planning, prevention and treatment of childhood diarrhea through oral rehydration therapy, improved nutrition, safe water and sanitation, personal hygiene, recognizing and managing tuberculosis, leprosy, malnutrition, and basic first aid information are recognized as the roles of traditional healers in other developing countries (Hoff 1992).

Wild Plants as a Potential Source of Medicine

Wild plants are often prescribed by natural healers to cure diseases. The use of plants as a source of medicine has been the earliest practice and an important component of the health care system (Bhattarai et al. 2009, Patro 2016). Even today, people residing in both rural and urban areas depend on a diverse group of wild plant products (Bhattarai 2009, Rios 2011). An estimated 50,000–70,000 medicinal and aromatic species are harvested from the wild (www.traffic.org).

Herbal medicines can be more accessible than expensive biomedical treatments. Biomedical treatments are often unfriendly, corrupt, with treatments offering too many side effects, and in some cases, disappointing (Cordell 2002, Graz et al. 2011). Natural medicines are sold widely (Cunningham 1997) to treat diseases where modern medical facilities are limited and expensive. The belief in natural healers and healing systems, coupled with the decline of conservative medical treatments, has led to a search for natural medicines to treat various ailments (Cunningham 1997, Bhattarai et al. 2006).

Medicinal plants are widely used in non-industrialized societies because they are widely available and cheaper than modern medicines (Cordell 2002, Bhattarai 2009, Rios 2011). However, during the last decade, there has been a growing concern in traditional and alternative systems of medicine in industrialized countries (Setswe 1999, Cordell 2002, Carvalho 2011). According to the World Health Organization, the global market for herbal products is over USD 60 billion (Nirali and Shankar 2015). The annual global export value of the thousands of types of plants with suspected medicinal properties was estimated to be USD 2.2 billion in 2012 (www.traffic.org). In 2017, the potential global market for botanical extracts and medicines was estimated at several hundred billion dollars (Ahn 2017).

Bioprospecting for Therapeutic Plants

Bioprospecting is the discovery and commercialization procedure of novel products, comprising economically valuable species and genes from biological resources. These novel products can be useful in many fields, including pharmaceuticals, agriculture, bioremediation, and nanotechnology (Oli and Dhakal 2009, Beattie et al. 2011). Biodiversity prospecting provides economic value to ensure sustainable conservation of natural biodiversity in developing countries (Eisner 1989), which are rich in biocultural diversity.

The Himalayan medicinal plants offer great potential for the discovery of novel molecules and new sources of active compounds, mainly because of the environmental stress to which they are subjected (Jackson and Dewick 1984). Plants in harsh environmental conditions (e.g., freezing temperature, drought, defoliation, high-intensity light, etc.) have developed a morphological, chemical, and genetic modification for their success in respective habitats. At least a few out of a large number of ethnomedicinal plants may contain important phytochemicals that can be used for the treatment of serious diseases (Bhattarai 2009).

The variety of medicinal plants is very high, but little is known about the biochemical and pharmacological properties. About 1.5% of the flowering plants of the world have been screened for pharmaceutical compounds (Farnsworth 1990), but the detailed biochemical analysis is limited to only a few species having very high use values (Cordell 2002). Considering the negative effects of synthetic drugs, people are looking for natural remedies, which are safe and effective (Cordell 2002). In this respect, medicinal plants used in the traditional therapy could be the alternative source for the development of new therapeutic agents to combat the resistant organisms, and at present, a number of drugs derived from medicinal plants have been shown to have various biological activities (Kashman et al. 1992, Cordell 2002, Balunas and Kinghorn 2005). Therefore, plants with medicinal values should be investigated by modern scientific techniques to establish their safety and efficacy, and to determine their potential as a source of new drugs (Balunas and Kinghorn 2005, Carvalho 2011). Numerous scientific studies are going on to isolate potent phytochemicals for antimicrobial therapy. Many useful plant-derived drugs were discovered as a result of scientific follow-up of well-known plants used in traditional medicine (Cordell 2002, Balunas and Kinghorn 2005, Carvalho 2011).

Traditional medicine is the basis of healthcare to treat various infectious diseases. Although some infectious diseases have been oppressed by modern medicines, new diseases are constantly evolving (Cordell 2002). Thus, one of the fruitful approaches to overcome resistant microbes is to search for new anti-infective agents of plant origin (Farnsworth 1990). There is an enormous wealth of information on cheap and culturally accepted ethnopharmacology-based remedies (Balunas and Kinghorn 2005). Several medicinal plants have been used in traditional medicinal practices for centuries, but until now, scientists have not been able to capitalize on this herbal wealth adequately. Although exploration and preliminary screening of ethnomedicinal plants have been carried out for several species, biomedical research at chemical and molecular level warrants further research (Martin 1995, Balunas and Kinghorn 2005, Bhattarai et al. 2008a, b).

It is well known that the use of natural plant products in drugs is the only answer to the problem of healthcare of the huge human population in the future. Although the isolation and identification of novel natural products to be used in drugs are costly and time-consuming, they are safe and their sustainable supply can be ensured (Cordell 2002). These days, the plant resources and the associated indigenous knowledge are disappearing at an alarmingly rapid rate, and scientists are deeply distressed over these losses and have pledged themselves to find a way to arrest the destruction of biodiversity and indigenous knowledge.

Wild Plants for Food and Nutrition

Wild plants are non-cultivated and non-domesticated plants (Tardio et al. 2006), which play a significant role in the life of rural people. Millions of people in developing countries still depend on wild plants to meet their food requirements, especially during food crisis (FAO 2004, Balemie and Kebebew 2006, Pfoze et al. 2011). Food and nutrition safety is vital for a healthy and productive life. The method of food production and ingestion has shaped human society and the environment (Desor 2017). Traditionally, humans may have consumed more than 7,000 wild edible plants (Grivetti and Ogle 2000), but many such food resources and treasured plants are still to be explored (Mohan Ram 2000). The people living in the mountainous region have been facing large challenges in food and nutrition security. Although progress has been made in calorie intake, malnutrition remains a serious challenge (Rasul et al. 2019).

About 50% of the total population in the world suffers from malnutrition, but women and children suffer more (Rasul et al. 2019). The global population facing food and nutritional insecurity increased from 777 million in 2015 to 815 million in 2016 (FAO 2017). It was documented that wild edible plants raise the nutritional value of rural diets, incorporating micronutrients (vitamins and minerals), which are sometimes higher than those of domesticated varieties (Msuya et al. 2010). A large fraction of rural populations does not yield adequate food, and so meet their nutritional requirement by consuming various wild plants (Singh and Arora 1978, Bhattarai et al. 2009, Rasul et al. 2019). It was projected that approximately one billion people globally use wild food plants to supplement their diets (Shumsky et al. 2014). Thus, it is crucial to know the contribution of these plants to food and nutritional security in communities (Ojelel et al. 2019).

Various studies reported that wild edible plants are potential sources of nutrition, but in many cases, are described to be more nutritious than conventionally eaten crops (Grivetti and Ogle 2000). The diverse foods from plants provide nutritional diversity, and are also a source of major food during famine or scarcity (Hatloy et al. 1998, Balemie and Kebebew 2006). It has been reported that wild plants are equivalent, in terms of nutritive values, with commercial fruits, and thus they can be promoted as alternative sources of nutrition (Bajracharya 1980, Sundriyal and Sundriyal 2001).

Wild foods are components of diets and local economies from rural Africa (Ncube et al. 2016) to urban USA (McLain et al. 2014). In many regions, wild foods contribute considerably to household food security, dietary diversity, and nutritional safety (Kajembe et al. 2000), because they add diversity to the mostly starch-based, staple diets of households (Bharucha and Pretty 2010, Uusiku et al. 2010, Powell et al. 2011). In many regions, diets are in transition as a consequence of globalization and increasing market access (Pingali 2007, Damman et al. 2008, FAO 2010, Ncube et al. 2016). The former is connected to a decline in agrobiodiversity, dietary diversity, and knowledge of wild foods and local cultivars, whereas the latter brings exposure to and convenience of new foods that may be easily available (Van Vliet et al. 2015).

Herbal Formulations

Herbal formulations are important characteristics of designing medicines (Mohamed et al. 2010), which is a stable and acceptable structure formed following a particular formula (www.en.wikipedia.org), which is often used in a dosage form. Based on the method/route of administration, dosage forms come in several types, including many kinds of liquid, solid, and semisolid, but the common dosage forms include the pill, tablet/capsule, drink/syrup. The drug delivery route depends on the dosage form of the active compound. Various dosage forms may occur for a single particular drug, since different medical conditions can permit different routes of administration (www.bbc.co.uk).

The formulation techniques are crucial to confirm the quality, taste, safety, and stability of the drug (Sharma et al. 2016). The traditional medicine formulation contains plant material as its primary component (Seema 2014, Thillaivanan and Samraj 2014), encompassing medicinal plants, minerals, organic matter, etc. Herbal drugs constitute those traditional medicines that principally use medicinal plant preparations for healing (Pal and Shukla 2003). Drugs derived from plant origin are recognized to have less risk and low side effects and single plants can provide the exact fraction of all the constituents for various ailments (Williamson 2001, Inamdar et al. 2008, Dandagi et al. 2009, Palav and D'mello 2009).

The use of more than one herb in a medicinal preparation is polyherbal formulation, and the preparations are either as single herbs or as groups of herbs in composite formulae. Polyherbal formulations consist of several bioactive components that are responsible for synergistic activity, consequently enhancing the medicinal value. Each bioactive component of the polyherbal formulation is connected to each other and is very important (Musthaba et al. 2009, Thillaivanan and Samraj 2014). The prehistoric indigenous system of medicine mentions several single and compound drug preparations of plant origin to cure various disorders (Newman and Cragg 2007, Parasuraman et al. 2014). The Ayurvedic literature has also highlighted the concept of polyherbal formulation and mentioned combined extracts of plants rather than individual ones (Srivastava et al. 2013, Parasuraman et al. 2010). Ayurvedic herbals are prepared in several dosage forms, in which almost all of them are polyherbal formulations (Srivastava et al. 2013, Parasuraman et al. 2010).

With the emerging nanotechnology, there has been significant progress on the improvement of new herbal formulations, including polymeric nanoparticles, nanocapsules, liposomes, phytosomes, nanoemulsions, microsphere, transferosomes, and ethosomes. These formulations have been described as having numerous benefits over the traditional formulations (Sharma et al. 2016). Herbal formulations are being patented in recent years, but the number of patents being filed for plant origin is lesser in the past few years (Mohamed et al. 2010). The reason for lesser number of patents is that the herbal formulations need a novelty to obtain a patent, which requires scientific proof of their pharmacological or pharmacodynamics property against the disease for which they are anticipated. However, most of the companies which produce herbal formulations do not have the scientific evidence of their biological activity, and they follow the orally transmitted traditional knowledge (Mohamed et al. 2010, Musthaba et al. 2009).

Benefit-sharing, a Basic Need

The Rio Declaration and the Convention on Biological Diversity (1992) explained the rights of traditional people and local communities. Also, various treaties and national laws have been ratified worldwide to control the use of the intellectual property and to establish equitable benefit sharing (Laird 2002, Dutfield and Suthersanen 2008, Oli and Dhakal 2009). Benefit sharing is the sharing of the consumption of biological resources, community knowledge, technologies, innovations, or practices. It also means the sharing of all forms of compensation for the use of genetic resources, whether monetary or non-monetary (CBDa, b). Monetary benefits may be open payments, access fees, milestone payments, license fees, salaries and infrastructure research funding, joint schemes, and joint ownership of intellectual property rights, whereas, non-monetary profits may enclose the sharing of research results, scientific research collaboration, participation in product development, collaboration in education and training and technology transfer (CBDa, b, Oli and Dhakal 2009).

Benefit-sharing Arrangements with the Kani Tribe

The Kani is a tribal community living in the Thiruvananthapuram district of the Western Ghats of South India, Kerala. The Kani people consider the plant (*Trichopus zeylanicus*) to be a very essential medicine with brilliant healing properties. At first, the juice of pounded mass of the fresh tuberous root is mixed with an equal quantity of the juice of coconut kernel. The mixture is then boiled for some time to get a semisolid form after cooling is administered. About 10–15 ml semisolid medicine is taken twice a day for 15 to 30 days to cure all kinds of peptic ulcers and related afflictions. It is also recommended for stamina as a roborant, and blood purifier (Pushpangadan et al. 1990). The plant was first described by Joseph and Chandrasekharan of the Botanical Survey of India in 1978 (Pushpangadan et al. 1990). It is a perennial herbaceous plant with milky latex under the family Dioscoreaceae. The roots are moniliform, tuberous, highly aromatic, and 30 cm long in clusters. A single healthy plant yields upto 5 kg of fresh roots (Pushpangadan et al. 1990).

Based on the abovementioned indigenous knowledge of the Kani tribe, a benefit-sharing arrangement was prepared between the Tropical Botanical Garden and Research Institute (TBGRI) and the Kani tribe for the development of a drug called '*Jeevani*' using *Trichopus zeylanicus* (Anuradha 1998, Pushpangadan et al. 1990, 1998, Oli and Dhakal 2009). The healers of the Kani tribe usually transfer their medicinal knowledge of herbs from one generation to another. These herbal healers are known as plathis. The knowledge was communicated by three Kani tribal members to the scientists of TBGRI, who isolated twelve active compounds from *T. zeylanicus*, and developed the drug *Jeevani* and filed two patent applications on the drug (Pushpangadan et al. 1990, Oli and Dhakal 2009). The prepared drug *Jeevani* is a restorative, immuno-enhancing, anti-stress, and anti-fatigue agent. Later, this technology was licensed to Arya Vaidya Pharmacy, Ltd, which is an Indian company producing Ayurvedic herbal formulations. Moreover, in order to share the profits arising from the commercialization of the herbal drug, a trust fund was established. This experience has provided insight for developing benefit-sharing provisions in the National Biodiversity Policy and Macrolevel Action Strategy, as well as in biodiversity legislation (Oli and Dhakal 2009). The utilization of the fund with the participation of all pertinent stakeholders and the sustainable harvesting of the plant has generated some glitches which offer lessons on the role of intellectual property rights in benefit-sharing over medicinal plant and traditional medicinal knowledge (Pushpangadan et al. 1990, 1998, Oli and Dhakal 2009).

The Cases of Biopiracy

The traditional knowledge of tribal people being used by others for profit, without authorization or compensation to the tribals themselves is called biopiracy. There have been limited cases of biopiracy of traditional knowledge from underdeveloped biodiversity-rich countries, but developing countries, such as India, Brazil, and Malaysia also encountered numerous cases of biopiracy. Various foreign corporations achieved patents based on biological materials without acknowledging the source of their knowledge or sharing the benefits (Tripathi 2003). Some of these cases include patents obtained in other countries on Turmeric (Haldi), Bitter gourd (karela), Neem, Basmati rice, etc., but many of these patients were successfully opposed, and the patents got revoked.

Neem Patent as an example of Traditional knowledge and Patent issues

Traditional knowledge is sustained and passed on from elder to younger generation within a community. The invention should be protected using the patent system, but it is not easily protected by the current intellectual property system. Currently, the intellectual property system normally grants safety for a limited period to inventions and original works (www.wipo.int). The current international system for protecting intellectual property was shaped during the age of industrialization in the West, and developed subsequently in line with the perceived needs of technologically advanced societies. In the current approach, aboriginal people/local communities and administrations, mainly in developing countries, have claimed equal protection for traditional knowledge systems (www.wipo.int).

The Indian texts printed 2,000 years back described the usage of Neem by local communities in agriculture, human and veterinary medicine, toiletries, cosmetics, as well venerated in the culture, religions, and literature (Porter 2006, Pankaj et al. 2011, Singh et al. 2011). The Neem (*Azadirachta indica*) plant, with multiple uses, originates from the Indian subcontinent (i.e., India, Nepal, Pakistan, Bangladesh, Sri Lanka, and the Maldives), and now grows in more than 50 tropical and semi-tropical regions around the world (Singh et al. 2011, www.en.wikipedia.org).

Several powerful compounds, including the chemical *Azadirachtin* in the Neem seeds were found, which is used as an astringent. The barks, leaves, flowers, and seeds of Neem are used to treat a variety of diseases, ranging from leprosy to diabetes, skin disorders, and ulcers (DPR 2016, www.emedicinehealth.com), and the twigs are used as antiseptic toothbrushes. The pesticidal property of Neem was first described in India in 1928, but after 30 years, systematic research was started. After 1928, research on the Neem plant started globally, leading to isolation and identification of hundreds of the active compounds, from various parts (Keher et al. 1949, Devakumar et al. 1993, Brahmachari 2004, Akhila and Rani 1999, Biwas et al. 2002), with pesticidal, nematicidal, antifungal (Khan and Wassilew 1987, Iyer and Williamson 1991, Bhatnagar and McCromick 1988, Allameh et al. 2001, Mossini et al. 2004), antibacterial (Siddique et al. 1992), anti-inflammatory (Okapanyi and Ezeukwu 1981), antitumor, antiviral (Gogate and Marathe 1989, Rao et al. 1969, Badam et al. 1999, Parida et al. 1997, 2002), anti-ulcer (Febry et al. 1996, Garg et al. 1993, Srirupa et al. 2002, Chattopadhyay et al. 2004), antimalarial (Badam et al. 1987, Vasanth et al. 1990) antioxidant, antimutagenic, and anticarcinogenic activity (Hanachi et al. 2004, Baral and Chattopadhyay 2004, Sarkar et al. 2009), and found its applications in pesticide, medical, healthcare, and cosmetic industry (Murthy and Sirsi 1958, Bhargava ct al. 1970, Pillai and Santhakumari 1981, Fujiwara et al. 1982, Pant et al. 1986, Biswas et al. 2002, Pankaj et al. 2011).

Since the 1980s, many Neem-related processes and products have been patented. According to Rekhi (2006), 171 products of Neem have been patented in Japan (59), followed by USA (54), India (36), Germany (05), EPO (05), Great Britain (02), Austria, Belgium, Denmark, Ireland, France, Greece, etc. [PCT] (10). The patent for Neem was first filed by W.R. Grace and the Department of Agriculture, the USA in the European Patent Office. The Neem patent is a technique of controlling fungi on plants comprising of contacting the fungi with a Neem oil formulation. Legal disapproval has been filed by India against the grant of the patent (Singh et al. 2011). The legal opposition to this patent was lodged by the New Delhi-based Research Foundation for Science, Technology, and Ecology, in cooperation with the International Federation of Organic Agriculture Movements and Magda Aelvoet, former green Member of the European Parliament (MEP) (Singh et al. 2011, www.countercurrents.org). The opponents' submitted evidence of ancient Indian ayurvedic texts that have described the hydrophobic extracts of neem seeds were known and used for centuries in India, for the cure of skin diseases, and in protecting crops from infections. The EPO identified the lack of novelty, inventive step, and possibly form a relevant prior art and revoked the patent.

Conclusions

Research in the different ethnobotanical fields on wild plants has advanced rapidly in the past decade. Traditional ethnobotanical knowledge of wild plants is passed orally from elders to younger generations through the word of mouth. The younger generations obtain names of wild plants at home and study to identify the collected plants by supporting their parents in fields and forests. Further, they utilize their experience to collect wild plants for foods. Wild plants offer great potential for the discovery of novel molecules and new sources of bioactive compounds. Therefore, wild plants should be studied by modern scientific techniques to establish their safety and efficacy, and to determine their potential as a source of new drugs. The main intention of several scientists for this kind of bioprospecting research is to support the scientific and economic development of the country through the discovery of new herbal drugs. Examples are from some countries, such as camptothecin from *Camptotheca acuminata* (India), Artemisinin from *Artemisia annua* (China) confirm that plants from ethnomedicine provide drugs which give therapeutical progress worldwide. Additionally, new chemical structures could be used as lead structures or as pharmacological tools.

Acknowledgments

MKR and CMF thank CNPq (National Council for Scientific and Technological Development, Brazil) for financial support (Process number 403888/2018-2). SB thanks NAST (Nepal Academy of Science and Technology, Kathmandu, Nepal) for varied support.

References

Abdool Karim, S.S., Ziqubu-Page, T.T., Arendse, R. 1994. Bridging the gap: project report for the South African Medical Research Council. Sou. Afr. Med. J. Suppl. 84: 1–14.

Acharya, D., Anshu, S. 2008. Indigenous Herbal Medicines: Tribal Formulations and Traditional Herbal Practices. Jaipur: Aavishkar Publishers.

Agyemang, M.M.O. 1996. The Leaf Gatherers of Kwapanin, Ghana, Forest Participation. Series No. 1, IIED, London.

Ahn, K. 2017. The worldwide trend of using botanical drugs and strategies for developing global drugs. BMB Rep. 50(3): 111–116.

Akhila, A., Rani, K. 1999. Chemistry of the Neem Tree (*Azadirachta indica* A. Juss.). Fortschr. Chem. Org. Naturst. 78: 47–49.

Allameh, A., Razzaghi-Abyaneh, M., Abyaneh, M.R., Shams, M., Rezaee, M.B., Jaimand, K. 2001. Effect of Neem Leaf Extracts on production of Aflatoxins and Activities of Fatty Acid Synthetase, Isocitrate dehydrogenase and Glutathione S-transferase in *Aspergillus parasiticus*. Mycopath. 154: 79–84.

Anuradha, R.V. 1998. Sharing with the Kanis: A case study from Kerala, India. Secretariat of the Convention on Biological Diversity, Montreal.

APG IV. 2016. An update of the Angiosperm Phylogeny group classification for the orders and families of flowering plants: APG IV. Bot. J. Linn. Soc. 181: 1–20.

Arabidopsis Genome Initiative. 2000. Analysis of the genome sequence of the flowering plant *Arabidopsis thaliana*. Nat. 408(6814): 796–815.

Aryal, K.K., Dhimal, M., Pandey, A., Pandey, A.R., Dhungana, R., Khaniya, B.N., Mehta, R.K., Karki, K.B. 2016. Knowledge Diversity and Healing Practices of Traditional Medicine in Nepal. Kathmandu, Nepal: Nepal Health Research Council.

Aryal, K.P., Berg, A., Ogle, B. 2009. Uncultivated plants and livelihood support—A case study from Chepang people of Nepal. Ethno. Res. App. 7: 409–422.

Atanasov, A.G., Birgit, W., Eva-Maria, P.-W., Thomas, L., Christoph, W., Pavel, U., Veronika, T., Limei, W., Stefan, S., Elke, H.H., Judith, M.R., Daniela, S., Johannes, M.B., Valery, B., Marko, D.M., Brigitte, K., Rudolf, B., Verena, M.D., Hermann, S. 2015. Discovery and resupply of pharmacologically active plant-derived natural products: A review. Biotec. Adv. 33(8): 1582–1614.

Badam, L., Deolankar, R.P., Kulkarni, M.M., Nagsampgi, B.A., Wagh, U.V. 1987. *In vitro* antimalaria activity of neem leaf and seed extract. Indian J. Malariol. 24: 111–117.

Badam, L., Joshi, S.P., Bedekar, S.S. 1999. *In vitro* antiviral activity of neem (*Azadirachta indiabout* A. Juss) leaf extract against group B coxsackieviruses. J. Commun. Dis. 31(2): 79–90.

Bajracharya, D. 1980. Nutritive values of Nepalese edible wild fruits. *Z. Labansm.* Unters. Forsch. 171: 363–366.

Balemie, K., Kebebew, F. 2006. Ethnobotanical study of wild edible plants in Derashe and Kucha Districts, South Ethiopia. J. Ethnobio. Ethnomed. 2: 53.

Baral, R., Chattopadhyay, U. 2004. Neem (*Azadirachta indica*) Leaf mediated immune activation causes prophylactic growth inhibition of Murine Ehrlich Carcinoma and B16 Melanoma. Int. Immunopharmacol. 4: 355–366.

Beattie, A.J., Hay, M., Magnusson, B., NYS, R., Smeathers, J., Vincent, J.F.V. 2011. Ecology and bioprospecting. Austral. Ecol. 36(3): 341–356.

Beluhan, S., Ranogajec, A. 2010. Chemical composition and non-volatile components of Crotial wild edible mushrooms. F. Chem. 124: 1076–1082.

Bhargava, K.P., Gupta, M.B., Gupta, G.P., Mitra, C.R. 1970. Anti-inflammatory activity of saponins and other natural products. Ind. J. Med. Res. 58(6): 724–730.

Bharucha, Z., Pretty, Z. 2010. The roles and values of wild foods in Agricultural systems. Phil. Trans. R. Soc. B. 365: 2913–2926.

Bhatnagar, D., McCromick, S.P. 1988. The inhibitory effect of neem (*Azadirachta indica*) leaf extracts on aflatoxin synthesis in *Aspergillus parasiticus*. JAOCS 65: 1166–1168.

Bhattarai, S., Chaudhary, R.P., Taylor, R.S.L. 2006. Ethnomedicinal plants used by the people of Manang district, Central Nepal. J. Ethnobio. Ethnomed. 2: 41.

Bhattarai, S., Chaudhary, R.P., Taylor, R.S.L. 2008a. Screening of selected ethnomedicinal plants of Manang district, Central Nepal for antibacterial activity. Ethnobot. 20: 9–15.

Bhattarai, S., Chaudhary, R.P., Taylor, R.S.L. 2008b. Antibacterial activity of selected ethnomedicinal plants of Manang District, Central Nepal. J. Theor. Expt. Bio. 5(1 & 2): 01–09.

Bhattarai, S., Chaudhary, R.P., Taylor, R.S.L. 2009. Wild edible plants used by the people of Manang District, Central Nepal. Eco. Food Nut. 47: 1–20.

Bhattarai, S. 2009. Ethnobotany and Antibacterial Activity of Selected Medicinal Plants of Nepal Himalaya. Ph.D Dissertation (Central Department of Botany), Tribhuvan University, Kirtipur, Kathmandu, Nepal.

Biswas, K., Chattopadhyay, I., Banerjee, R.K., Bandyopadhyay, U. 2002. Biological Activities and Medicinal Properties of Neem (*Azadirachta indica*). Curr Sci. 82: 1336–1345.

Brahmachari, G. 2004. Neem—An omnipotent plant: a retrospection. Chem. Bio Chem. 5: 408–421.

Brenchley, R., Spannagl, M., Pfeifer, M., Barker, G.L.A., Amore, R.D., Allen, A.M., McKenzie, N., Kramer, M., Kerhornou, A., Bolser, D., Kay, S., Waite, D., Trick, M., Bancroft, I., Gu, Y., Huo, N., Luo, M.-C., Sehgal, S., Kianian, S., Gill, B., Anderson, O., Kersey, P., Dvorak, J., McCombie, R., Hall, A., Mayer, K.F.X., Edwards, K.J., Bevan, M.W., Hall. N. 2012. Analysis of the bread wheat genome using whole-genome shotgun sequencing. Nat. 491(7426): 705–710.

Cakir, E.A. 2017. Traditional knowledge of wild edible plants of Iğdır Province (East Anatolia, Turkey). Acta. Soc. Bot. Pol. 86(4): 3568.

Carvalho, J.C.T. 2011. Brazilian ethnomedicinal plants with anti-inflammatory action. pp. 76–114. *In*: Rai, M., Acharya, D., Rios, J.L. (eds.). Ethnomedicinal Plants: Revitalization of Traditional Knowledge of Herbs. Scientific Publishers, Enfield, New Hampshire.

CBD. 1992. The Convention on Biological Diversity. Montreal: Secretariat of the Convention on Biological Diversity. www.cbd.int/traditional/(accessed on March 2009).

CBD (no date b). Article 8(j): Traditional Knowledge, Innovations and Practices. Montreal: Secretariat of the Convention on Biological Diversity. www.cbd.int/traditional/(accessed on March 2009).

Chapman, A.D. 2009. Numbers of Living Species in Australia and the World, 2nd edition. A Report for the Australian Biological Resources.

Chase, M.W., Cameron, K.M., Freudenstein, J.V., Pridgeon, A.M., Salazar, G., Van den Berg, C., Schuiteman, A. 2015. An updated classification of Orchidaceae. Bot. J. Linn. Soc. 177: 151–174.

Chattopadhyay, I., Nandi, B., Chatterjee, R., Biswas, K., Bandyopadhyay, U., Banerjee, R.K. 2004. Mechanism of Antiulcer Effect of Neem (*Azadirachta indica*) Leaf Extract: Effect on H+-K+ ATPase, Oxidative Damage and Apoptosis. Inflammopharmacol. 12: 153–176.

Chin, Y.-W., Balunas, M.J., Chai, H.B., Kinghorn, A.D. 2006. Drug Discovery from Natural Sources. The AAPS J. 8(2): 240-253.

Christenhusz, M., Byng, J.W. 2016. The number of known plant species in the world and its annual increase. Phytot. 261(3): 201–217.

Convention on Biological Diversity. www.cbd.int/convention.

Cordell, G.A. 2002. Natural products in drug discovery-creating a new vision. Phytochem. Rev. 1: 261–271.

Cox, C.J., Goffinet, B., Wickett, N.J., Boles, S.B., Shaw, J. 2010. Moss diversity: a molecular phylogenetic analysis of genera. Phytot. 9: 175–195.

Crandall-Stotler, B., Stotler, R.E. 2000. Morphology and classification of the Marchantiophyta. pp. 21. *In*: Jonathan, A., Shaw & Bernard Goffinet (eds.). Bryophyte Biology. (Cambridge: Cambridge University Press).

Crosby, M.R., Magill, R.E., Allen, B., He, S. 1999. Checklist of Mosses. Missouri Botanical Gardens, St. Louis.

Cunningham, A.B. 1997. An Africa-wide overview of medicinal plant harvesting, conservation and healthcare. pp. 116–129. *In*: FAO Medicinal Plants for Forest Conservation and Healthcare, FAO, Rome, Italy.

Damman, S., Eide, W.B., Kunhlein, H.V. 2008. Indigenous peoples' nutrition transition in a right to food perspective. Food Pol. 33(2): 135–155.

Dandagi, P.M., Patil, M.B., Mastiholimath, V.S., Gadad, A.P., Dhumansure, R.H. 2009. Development and evaluation of hepatoprotective polyherbal formulation containing some indigenous medicinal plants. Ind. J. Pharm. Sci. 70: 265–268.

Desor, S. 2017. Ideas and initiatives towards an alternative food system in India. Kalpavriksh: Deccan Gymkhana, Pune 411004, India.

Devakumar, C., Dev, S., Randhawa, N.S., Parmar, B.S. 1993. Neem research and development. Soc. Pest. Sci., India.

DPR. 2016. Medicinal Plants of Nepal. Second Edition. Government of Nepal, Ministry of Forests and Soil Conservation, Department of Plant Resources, Thapathali, Kathmandu, Nepal.

Dutfield, G., Suthersanen, U. 2008. Global Intellectual Property Law. Edward Elgar Publishing; Cheltenham.

Dutta, U. 2012. Wild vegetables collected by the local communities from the Chirang Reserved Forest of BTAD, Assam. Int. J. Sci. Adv. Tec. 2(4): 116–126.

Eisner, T. 1989. Prospecting for nature's chemicals riches. Iss. Sci. Tec. 6(2): 31–34.

Enrique, I.-L., Eric, L., Gustavo, H.-G., Claudia, A.P.-T., Lorenzo, C.-P., Tien-Hao, C., Tianying, L., Andreanna, J.W., Maria, J.A.J., June, S., Araceli, F.-C., Mario, A.-V., Elsa, G.-C., Gustavo, A.-H., Stephan, C.S., Heinz, H., Andre, E.M., Sen, X., Michael, L., Araceli, O.-A., Sergio, A.C.-P., Maria de, J.O.-E., Jacob, I.C.-L., Todd, P.M., Todd, M., Douglas, B., Alfredo, H.-E., Victor, A.A., Luis, H.-E. 2013. Architecture and evolution of a minute plant genome. Nat. 498: 94–98.

FAO. 2004. The State of Food and Agriculture: Agricultural Biotechnology; Meeting the Needs of the Poor? Rome: FAO.

FAO. 2010. Sustainable Diets and Biodiversity, Directions and Solutions for Policy research and actions. pp. 309. *In*: Burlingame, B., Dernini, S. (eds.). Proceedings of the International Scientific Symposium. Biodiversity and Sustainable diets united against hunger, 3–5 November 2010, FAO Headquarters, Rome.

Farnsworth, N.R. 1990. The role of ethnopharmacology in drug development. *In*: Bioactive Compounds from Plants, 154, 2–21. CIBA Found. Symp. John Wiley and son Chichester.

Febry, W., Okema, P., Ansorg, R. 1996. Activity of East African Medicinal Plants against *Helicobacter pylori*. Chemot. 42: 315–317.

Food and Agriculture Organization of the United Nations. 2017. The state of food security and nutrition in the world 2017: building resilience for peace and food security. Rome: FAO.

Freeman, M., Motsei, M. 1992. Planning health care in South Africa—is there a role for traditional healers? Soc. Sci. Med. 34: 1183–1190.

Fujiwara, T., Takeda, T., Ogihara, Y., Shimizu, M., Nomura, T., Tomita, Y. 1982. Studies on the structure of polysaccharides from the bark of *Melia azadirachta*. Chem. Pharm. Bull. 30: 4025–4030.

Funk, V.A., Susanna, A., Stuessy, T.F., Bayer, R.J. (eds.). 2009. Systematics, evolution, and biogeography of Compositae. International Association for Plant Taxonomy, Vienna.

Garg, G.P., Nigma, S.K., Ogle, C.W. 1993. The gastric antiulcer effects of the leaves of the neem tree. Plan. Med. 59: 215–217.

Gewali, M.B. 2008. Aspects of Traditional Medicine in Nepal. Institute of Natural Medicine, University of Toyama, Toyama, Japan.

Giliba, R.A., Lupala, Z.J., Kayombo, C., Mwendwa, P. 2010. Non-Timer Forest Products and their contribution to poverty Alleviation and Forest Conservation in Mbulu and Babati districts—Tanzania. J. Hun. Ecol. 31(2): 73–78.

Gogate, S.S., Marathe, A.D. 1989. Antiviral Effects of Neem Leaf (*Azadirachta indica* Juss.). J. Res. Edu. Ind. Med. 8(1): 1–3.

Graz, B., Kitua, A.Y., Malebo, H.M. 2011. To what extent can traditional medicine contribute a complementary or alternative solution to malaria control programs? Mal. J. 10(1): S6–12.

Green, E.C., Zokwe, B., Dupree, D. 1995. The experience of an AIDS prevention program focused on South African traditional healers. Soc. Sci. Med. 40: 503–515.

Grivetti, L.E., Ogle, B.M. 2000. Value of traditional foods in meeting macro—and micronutrient needs: the wild plant connection. Nut. Res. Rev. 13: 31–46.

Groombridge, B., Jenkins, M.D. 2002. World Atlas of Biodiversity. Prepared by the UNEP World Conservation Monitoring Centre, Berkeley: University of California Press.

Hanachi, P., Fauziah, O., Peng, L., Wei, L.C., Nam, L.L., Tian, T.S. 2004. The effect of (*Azadirachta indica*) on distribution of antioxidant elements and glutathione S-transferase activity in liver of rats during hepatocarcinogenesis. Asia. Pac. J. Clin. Nutr. 13: S170.

Hatloy, A., Torheim, L.E., Oshaug, A. 1998. Food variety—a good indicator of nutritional adequacy of the diet? A case study from an urban area in Mali, West Africa about Europ. J. Cli. Nut. 52: 891–898.

Hoff, W. 1992. Traditional healers and community health. Wor. Hea. For. 13: 182–187.

https://www.traffic.org/news/traffic-launches-sustainable-wild-harvested-medicinal-plant-project-in-viet-nam. Traffic launches sustainable wild harvested medicinal plant project in Viet Nam; Published 9th April 2012. Accessed on June, 2019.

http://www.praying-nature.com/site_pages.php?section=Ecology+Matters%21&category_ref. Accessed on October 2019.

https://en.wikipedia.org/wiki/Azadirachta_indiabout Accessed on October 2019.

https://en.wikipedia.org/wiki/Pharmaceutical_formulation. Accessed on October 2019.

https://en.wikipedia.org/wiki/Traditional_medicine. Accessed on October 2019.

https://www.bbc.co.uk. Doctors 'missed' fatal overdoses. Accessed on 13 September 2019.

https://www.countercurrents.org/bhargava140709.htm. Accessed on October 2019.

https://www.encyclopedia.com/plants-and-animals/botany/botany-general/evolution-plants. Accessed on October 2019.

https://www.wipo.int/edocs/pubdocs/en/wipo_pub_941_2018-chapter2.pdf. Accessed on November 2019.

https://www.wipo.int/tk/en/. Accessed on November 2019.

Inamdar, N., Edalat, S., Kotwal, V.B., Pawar, S. 2008. Herbal drugs in milieu of modern drugs. Int. J. Green Pharm. 2: 2–8.

Iyer, S.R., Williamson, D. 1991. Efficacy of some plant extracts to inhibit the protease activity of *Trichophyton* species. Geob. 18: 3–6.

Jackson, D.E., Dewick, P.M. 1984. Aryltetralin lignans from *Podophyllum hexandrum* and *Podophyllum peltatum*. Phytochem. 23: 1147–1152.

Jain, S.K. 1963. Wild plant—foods of the tribals of BAstar (Madhya Pradesh). Bull. Bot. Sur. Ind. 30(2): 56–80.

Joppa, L.N., Roberts, D.L., Myers, N., Pimm, S.L. 2011. Biodiversity hotspots house most undiscovered plant species. Proc. Nat. Aabout Sci. USA 108: 13171–13176.

Kajembe, G.C., Mwenduwa, M.I., Mgoo, J.S., Ramadhani, H. 2000. Potentials of Non Wood Forest Products in Household Forest Security in Tanzania: The Role of Gender Based Local Knowledge. Gender, Biodiversity and Local Knowledge Systems (LinKS) to Strength Agricultural and Rural Development (GCP/RAF/338/NOR).

Kalaba, F.K., Chirwa, P.W., Prozesky, H. 2009. The contribution of indigenous fruit trees in sustaining rural livelihoods and conservation of natural resources. J. Hor. For. 1(1): 001–006.

Kale, R. 1995. Traditional healers in South Africa: a parallel health care system. Bri. Med. J. 310: 1182–1185.

Kar, A. 2004. Common wild vegetables of Aka tribe of Arunachal Pradesh. Ind. J. Tra. Kno. 3(3): 305–313.

Kar, A., Borthakur, S.K. 2007. Wild vegetables sold in local markets of Karbi Anglong, Assam. Ind. J. Tra. Kno. 6(1): 169–172.

Kar, A., Borthakur, S.K. 2008. Wild vegetables of Karbi Anlong district. Nat. Pro. Rad. 7(5): 448–460.

Kashman, Y., Gustafson, K.R., Fuller, R.W., Cardellina, II J.H., McMohan, J.B., Currens, M.J., Buckheit, Jr. R.W., Hughes, S.K., Cragg, G.M., Boyd, Mr. 1992. HIV inhibitory natural products. Part 7. The calanolides, a novel HIV-inhibitory class of coumarin derivatives from the tropical rainforest tree, *Calophyllum lanigerum*. J. Med. Chem. 35(15): 2735–2743.

Keher, N.D., Negi, S.S., Atal, C.K., Kapur, M. 1949. Cultivation and utilization of medicinal plants. Regional Research Laboratory, Jammu Tawi, India.

Kelly, J.C. 1995. Cooperation between traditional healers and medical personnel. Sou. Afr. Med. J. 85: 686.

Khan, M., Wassilew, S.W. 1987. Natural pesticides from the neem tree and other tropical Plants. GTZ, Eschborn, Germany.

Kutum, A., Sarmah, R., Hazarika, D. 2011. Ethnobotanical study of Mishing tribe living in fringe villages of Kaziranga National Park of Assam, India. Int. J. Fund. App. Lif. Sci. 1(4): 45–61.

Laird, S.A. (Ed.). 2002. Biodiversity and Traditional Knowledge. Earthscan Publications; London.

LaRochelle, S., Berkes, F. 2003. Traditional Ecological Knowledge and practice for wild plants: Biodiversity use by the Rara_mure in the Sierra Tarahumara, Mexico. Int. J. Sustain. Dev. Wor. Eco. 10: 361–375.

Legwaila, G.M., Mojeremane, W., Madisa, M.E., Murolotsi, R.M., Rampart, M. 2011. Potential of traditional food plants in rural household food security in Boswana. J. Hor. For. 3(6): 171–177.

Mahapatra, A.D., Bhowmik, P., Banerjee, A., Das, A., Ojha, D., Chattopadhyay, D. 2019. Ethnomedicinal wisdom: an approach for antiviral drug development. pp. 35–61. *In*: New Look to Phytomedicine, Advancements in Herbal Products as Novel Drug Leads; Khan MSA, Ahmad I, and Chattopadhyay D).

Martin, G.J. 1995. Ethnobotany: A Methods Manual. Chapman and Halls, London.

McLain, R.J., Hurley, P.T., Emery, M.R., Poe, M.R. 2014. Gathering "Wild" Food in the City: Rethinking the role of foraging in Urban ecosystem planning and management. Local Env. 19: 220–240.

Mwu, M., Gbodossou, E. 2000. Alternative Medicine: Nigeria The role of traditional medicine (PDF). The Lancet.

Misra, S., Maikhuri, R.K., Kala, C.P., Rao, K.S., Saxena, K.G. 2008. Wild leafy vegetables: a study of their subsistence dietetic support to the inhabitants of Nanda Devi Biosphere Reserve, India. J. Ethnobiol. Ethnomed. 4: 115.

Mohamed, M., Sanjula, B., Tanwir, M.D.A., Kamal, Y.T., Sayeed, A., Javed, A. 2010. Patented Herbal Formulations and their Therapeutic Applications. Rec. Pat. Dru. Del. Form. 4: 231–244.

Mohammed, S.A.S., Rana, M.J., Jehan, H.A.S., Wafa, A.E., Fatemah, A.K., Kifayeh, H.Q., Isra, S.K., Israa, M.S., Aseel, A.M., Buthainah, A.I., Hanan, M.H., Rasha, B.K., Samiah, M.A., Ghadah, M.S., Muna, A.A., Maha, M.H.A., Nehaya, A.S., Hebah, K.A., Hanadi, A.N. 2008. Traditional knowledge of wild edible plants used in Palestine (Northern West Bank): A comparative study. J. Ethnobiol. Ethnomed. 4: 13.

Mohan Ram, H.Y. 2000. Plant Resources of Indian Himalaya. 9th G.P. Pant Memorial Lecture, G B Pant Institute of Himalayan Development, Gangtok, Sikkim.

Moreno-Black, G., Somnasong, W.P., Thamathawan, S., Brozvosky, P. 1996. Non-Domesticated food resources in the marketplace and marketing system in Northeastern Thailand. J. Ethnobiol. 16(1): 99–117.

Mossini, S.A., de Oliveira, K.P., Kemmelmeier, C. 2004. Inhibition of Patulin production by *Penicillium expansum* cultured with Neem (*Azadirachta indica*) leaf extracts. J. Bas. Microbiol. 44: 106–113.

Msuya, T.S., Kideghesho, J.R., Mosha, T.C. 2010. Availability, preference, and consumption of indigenous forest foods in the Eastern Arc Mountains, Tanzania. Ecol. Food Nutr. 49(3): 208–27.

Murthy, S.P., Sirsi, M. 1958. Pharmacological studies on Melia *Azadirachta indiabout.* Ind. J. Phys. Pharm. 2: 387–396.

Musthaba, S.M., Baboota, S., Ahmed, S., Ahuja, A., Ali, J., Baboota, S. 2009. Nano approaches to enhance pharmacokinetic and pharmacodynamics activity of plant origin drugs. Curr. Nanosci. 5(3): 344–352.

Ncube, K., Shackleton, C., Swallow, B., Dassayanake, W. 2016. Impacts of HIV/AIDS on food consumption and wild food use in rural South Africa. Food Sec. 1–18. DOI 10.1007/s12571-016-0624-4.

Newman, D.J., Cragg, G.M. 2007. Natural Products as sources of new drugs over the last 25 Years. J. Nat. Prod. 70: 461–477.

Nirali, B.J., Shankar, M.B. 2015. Global market analysis of herbal drug formulations. Int. J. Ayu. Pharma. Chem. 4(1): 59–65.

Nystedt, B., Street, N.R., Zuccolo, A., Lin, Y.-C. 2013. The Norway spruce genome sequence and conifer genome evolution. Nat. 1–6. Doi: 10.1038/nature12211.

Ojelel, S., Mucunguzi, P., Katuura, E., Kakudidi, E.K., Namaganda, M., Kalema, J. 2019. Wild edible plants used by communities in and around selected forest reserves of Teso-Karamoja region, Uganda. J. Ethnobio. Ethnomed. 15: 3.

Okapanyi, S.N., Ezeukwu, G.C. 1981. Antiinflamatory and Anti-pyretic activities of *Azadirachta indiabout.* Pla. Med. 41: 34–39.

Oli, K.P., Dhakal, T.D. 2009. Access and Benefit Sharing from Genetic Resources and Associated Traditional Knowledge, Training of Trainers and Resource Manual, (Ed.) A. Beatrice Murray. International Centre for Integrated Mountain Development (ICIMOD), Kathmandu, Nepal.

Pal, K.S., Shukla, Y. 2003. Herbal Medicine: Current Status and the Future. Asi. Pac. J. Can. Prev. 4: 281–288.

Palav, Y.K., D'mello, P.M. 2009. Standardization of selected Indian medicinal herbal raw materials containing polyphenols as major phytoconstituents. Ind. J. Pharm. Sci. 68: 506–9.

Pankaj, S., Lokeshwar, T., Mukesh, B., Vishnu, B. 2011. Review on Neem (*Azadirachta indica*): Thousands Problems one solution. Int. Res. J. Pharm. 2(12): 97–102.

Pant, N., Garg, H.S., Madhusudanan, K.P., Bhakuni, D.S. 1986. Sulfurous compounds from *Azadirachta indica* leaves. Fito. 57: 302–304.

Parasuraman, S., Kumar, E.P., Kumar, A., Emerson, S.F. 2010. Anti-hyperlipidemic effect of triglize, a polyherbal formulation. Int. J. Pharm. Pharm. Sci. 2: 118–22.

Parasuraman, S., Thing, G.S., So, D.A. 2014. Polyherbal formulation: Concept of Ayurveda. Pharmacogn. Rev. 8(16): 73–80.

Parida, M.M., Pandya, G., Bhargava, R., Jana, A.M. 1997. Assessment of *in vitro* antiviral activity of certain indigenous plants against Polio Virus Type-3. Ind. J. Vir. 13(2): 101–115.

Parida, M.M., Upadhyay, C., Pandya, G., Jana, A.M. 2002. Inhibitory potential of Neem (*Azadirachta indica* Juss) leaves on Dengue Virus Type-2 replication. J. Ethnopharmacol. 79: 273–278.

Patiri, B., Borah, A. 2007. The Wild Edible Plants of Assam. Department of Forest and Environment, Government of Assam.

Patro, L. 2016. Medicinal Plants of India: with special reference to Odisha. IJARIIE 2(5): 2395–4396.

Pegu, R., Gogoi, J., Tamuli, A.K., Teron, R. 2013. Ethnobotanical study of wild edible plants in Poba Reserved Forest, Assam, India: multiple functions and implications for conservation. Res. J. Agri. Forestry Sci. 1: 1–10.

Pfoze1, N.L., Yogendra, K., Bekington. M. 2011. Survey and assessment of floral diversity on wild edible plants from Senapati district of Manipur, Northeast India. J. Bio. Env. Sci. 1(6): 50–62.

Pillai, N.R., Santhakumari, G. 1981. Anti-arthritic and anti-inflammatory activity actions of nimbidin. Pla. Med. 45: 59–63.

Pimm, S.L., Joppa, L.N. 2015. How many plant species are there, where are they, and at what rate are they going extinct? Ann. Misso. Bot. Gar. 100: 170–176.

Pingali, P.L. 2007. Westernization of Asian diets and the transformation of food systems: Implications for research and policy. Fod Pol. 32(3): 281–298.

Porter, A.H. 2006. Neem: India's tree of life. In 17 April, 2006. BBC News.

Powell, B., Hall, J., Johns, T. 2011. Forest cover, use and dietaryintake in the East Usambara Mountains, Tanzania International Forestry Review 13(3), online.

Pushpangadan, P., Rajasekhran, S., Ratheesh, K.P.K., Jawahar, C.R., Velayudhan, N.V., Lakshmi, N., Sarad, A.L. 1988. Arogyapacha (*Trichopus Zeylanicus* Gaertn.). The Ginseng of Kani Tribes of Agasthyar Hills (Kerala) for Eevergreen Health and Vitality. Anc. Sci. L. 7: 13–16.

Pushpangadan, P., Rajasekharan, A., Ratheeshkumar, P.K., Jawahar, C.R., Radhakrishnan, K., Nair, C.P.R., Amma, L.S., Aicrpe, A.V.B. 1990. Amrithapala (Janakia arayalpatra, Joseph & Chandrasekharan), A new drug from the Kani tribe of Kerala. Anc. Sci. L. IX(4): 212–214.

Pushpangadan, P., Nair, K.N. 2005. Value Addition and commercialization of Biodiversity and Associated Traditional Knowledge in the context of the Intellectual Property Regime. J. Intell. Pro. Rig. 10: 441–453.

Rao, A.R., Kumar, S.S., Paramasivam, T.B., Kamalkashi, S., Parashuram, A.R., Shantha, M. 1969. Study of Antiviral Activities of Tender Leaves of Margosa. Ind. J. Med. Res. 57: 495–498.

Rasul, G., Saboor, A., Tiwari, P.C., Hussain, A., Ghosh, N., Chettri, G.B. 2019. Food and Nutrition Security in the Hindu Kush Himalaya: Unique Challenges and Niche Opportunities. The Hindu Kush Himalaya assessment. pp. 301–338.

RBG Kew. 2016. The State of the World's Plants Report – 2016. Kew, Royal Botanic Gardens.

Rekhi, J.S. 2006. The patent system in India. Office of DC (SSI), Ministry of Industry, New Delhi.

Rios, J.L. 2011. Ethnomedicinal plants: progress and the future of drug development. pp. 1–24. *In*: Rai, M., Acharya, D. Rios, J.L. (eds.). Ethnomedicinal Plants: Revitalization of Traditional Knowledge of Herbs. Scientific Publishers, Enfield, New Hampshire.

Sajeev, K.K., Sasidharan, N. 1997. Ethnobotanical observations on the tribals of Chinnar Wildlife Sanctuary. Anc. Sci. Lif. XVI(4): 284–292.

Sarkar, K., Bose, A., Haque, E., Chakraborty, K., Chakraborty, T., Goswami, S., Ghosh, D., Baral, R. 2009. Induction of Type 1 Cytokines during Neem Leaf Glycoprotein Assisted Carcinoembryonic Antigen Vaccination is Associated with Nitric Oxide Production. Int. Immunopharmacol. 9(6): 753–760.

Sarmah, R., Arunachalam, A. 2011. Contribution of Non-Timer Forest Products (NTFPs) to livelihood economy of the people living in forest fringes in Changlang district of Arunachal Pradesh. Ind. J. Fund. App. Lif. Sci. 1(2): 157–169.

Sawain, J.T., Jeeva, S., Lynden, F.G. 2007. Wild edible plants of Meghalaya, North-east India. Nat. Pro. Rad. 6(5): 410–426.

Scoones, I., Melnyk, R., Pretty, J.N. 1992. The hidden harvest: wild foods in agricultural systems: a literature review and annotated bibliography. International Institute of Environment and Development London, pp. 256.

Scotland, R.W., Wortley, A.H. 2003. How many species of seed plants are there? Tax. 52: 101–104.

Seal, T. 2011. Evaluation of some wild edible plants from nutritional aspects used as vegetable in Meghalaya state of India. Wor. App. Sci. J. 12(8): 1282–1287.

Seal, T. 2012. Antioxidant activity of some wild edible plants of Meghalaya state India: A comparison using two solvent extraction systems. Int. J. Nut. Met. 4(3): 51–56.

Seema, A. 2014. Recent development of herbal formulation—a novel drug. Int. Ayur. Med. J. 2(6): 952–958.

Setswe, G. 1999. The role of traditional healers and primary health care in South Africa. Hea. Gesond. 4(2): 56–60.

Sharma, P., Verma, S., Misri, P. 2016. Global need for novel herbal drug formulations. Int. J. Pharma. Phytoche. Res. 8(9): 1535–154.

Shrestha, K.K. 2016. Global biodiversity and taxonomy initiatives in Nepal. pp. 177–223. *In*: Jha, P.K., Siwakoti, M., Rajbhandary, S. (eds.). Frontiers of Botany. Kathmandu: Central Department of Botany, Tribhuvan University.

Shrestha, K.K., Bhattarai, S., Bhandari, P. 2018. Handbook of Flowering Plants of Nepal (Volume 1. Gymnosperms and Angiosperms: Cycadaceae – Betulaceae). Scientific Publishers, Jodhpur, India, pp. 648.

Shumsky, S.A., Hickey, G.M., Pelletier, B., Johns, T. 2014. Understanding the contribution of wild edible plants to rural social-ecological resilience in semi-arid Kenya. Ecol. Soc. 19(4).

Siddique, S., Faizi, S., Siddique, B.S., Ghisuddin, F.S. 1992. Constituents of *Azadirachta indica*: Isolation and structure elucidation of a new antibacterial Tetranortriterpenoid, Mahmoodin, and a New Protolimonoid, Naheedin. J. Nat. Prod. 55: 303–310.

Singh, H.B., Arora, R.K. 1978. Wild Edible Plants of India. ICAR, New Delhi.

Singh, G., Rawat, G.S. 2011. Ethnomedicinal survey in Kedernath Wildlife Sanctuary in Western Himalaya, India. Ind. J. Fun. App. Lif. Sci. 1(1): 35–46.

Singh, O., Khanam, Z., Ahmad, J. 2011. Neem (*Azadirachta indica*) in context of Intellectual Property Rights (IPR). Rec. Res. Sci. Tech. 3(6): 80–84.

Slikkerveer, L. 1994. Indigenous agricultural knowledge systems in developing countries: a bibliography. Indigenous Knowledge Systems Research and Development Studies No. 1. Special issue: INDAKS project report 1 in collaboration with the European Commission DG XII. Leiden, the Netherlands, Leiden Ethnosystems and Development Programme (LEDP).

Srirupa, D., Jena, J., Patnaik, S.L., Mukherjee, D. 2002. The antiulcer effects of *Azadirachta indica* in Pyloric-Ligated rats. Ind. J. Pharmacol. 34: 145–146.

Srivastava, S., Lal, V.K., Pant, K.K. 2013. Polyherbal formulations based on Indian medicinal plants as antidiabetic phytotherapeutics. Phytopharmacol. 2: 1–15.

Sundriyal, M., Sundriyal, R.C. 2001. Wild edible plants of the Sikkim Himalaya: nutritive values of selected species. Eco. Bot. 55(3): 377–390.

Tardío, J., Pardo de Santayana, M., Morales, R. 2006. Ethnobotanical review of wild edible plants in Spain. Bot. J. Linn. Soc. 152: 27–71.

Thillaivanan, S., Samra, J.K. 2014. Challenges, Constraints and Opportunities in Herbal Medicines—A Review. Int. J. Her. Med. 2(1): 21–24.

Thorne, R.F. 2002. How many species of seed plants are there? Tax. 51: 511–512.

Todd, M.P., Scott, J. 2013. The First 50 Plant Genomes. The Plant Gen. 6(2): 1–7.

Tripathi, S.K. 2003. Intellectual property and genetic resource, traditional knowledge and folklore: International, Regional and National perspectives, trends and strategies. J. Intell. Prop. Rig. 8(6): 468–77.

Uusiku, N.P., Oelofse, A., Duodu, K.G., Bester, M.J. and Faber, M. 2010. Nutritional value of leafy vegetables of sub-Saharan Africa and their potential contribution to human health: a review. J. Food Compos. Analy. 23: 499–509.

van Vliet, N., Quiceno-Mesa, M.P., Cruz-Antia, D., Tellez, L., Martins, C., Haiden, E., Ribeiro Oliveira, M., Adams, C., Morsello, C., Valencia, L., Bonilla, T., Yagüe, B., Nasi, R. 2015. From fish and bushmeat to chicken nuggets: the nutrition transition in a continuum from rural to urban settings in the tri-frontier amazon region. Ethnobio. Conser. https://doi.org/10.15451/ec2015-7-4.6-1-12.

Vasanth, S., Gopal, R.H., Rao, R.H., Rao, R.B. 1990. Plant Antimalarial Agents. J. Sci. Ind. Res. 49: 68–77.

Villarreal, J.C., Cargill, D.C., Hagborg, A., Söderström, L., Renzaglia, K.S. 2010. A synthesis of hornwort diversity: patterns, causes and future work. Phytot. 9(1): 150–166.

WHO. 2008. Traditional Medicine: Definitions. World Health Organization.

WHO. 2013. WHO traditional medicine strategy: 2014–2023. The World Health Organization.

Williamson, E.M. 2001. Synergy and other interactions in phytomedicines. Phytomed. 8: 401–409.

Yeşil, Y., Çelik, M., Yılmaz, B. 2019. Wild edible plants in Yeşilli (Mardin-Turkey), a multicultural area. J. Ethnobio. Ethnomed. 15: 52.

Yesodharan, K., Sujana, K.A. 2007. Wild edible plants traditionally used by the tribes in Parambikam Wildlife Sanctuary, Kerala, India. Nat. Prod. Rad. 6(1): 74–80.

Zhang, Z.-Q., Christenhusz, M.J.M., Esser, H.-J., Chase, M.W., Vorontsova, M.S., Lindon, H., Monro, A., Lumbsch, H.T. 2014. The making of the world's largest journal in systematic botany. Phytot. 191(1): 1–9.

2

The Disappearance and Substitution of Native Medicinal Species

Nelida Soria

Introduction

Medicinal plants have been used since time immemorial to control human health problems, by preventing diseases, palliating symptoms, or even healing, and still today, it constitutes the first alternative for disease treatments in the populations of developing countries where there is a lack of effectiveness of health systems to respond to health problems in communities.

Thus, in the 1970s, and in order to achieve "Health for All in the year 2000", the World Health Organization, guided the research and application of Natural and Traditional Medicine (MNT), due the fact that this form of medicine is more natural, more innocuous, effective, with a rational cost, and accessible to large population groups. WHO supports member states in promoting the use of Traditional Medicines (TM) as a Primary Health Care (PHC), basing on ensuring the safety and quality of medicines and educating consumers to use them correctly (WHO 2002).

In addition, TM could be effective as a frontline treatment and prevention, for conditions, such as colds, diarrhea, stomach aches, light fevers. Its wide use is favored because it is firmly rooted in the belief system which is accepted culturally (Soria and Ramos 2015).

However, despite the fact that Medicinal Plants (PM) are used by the vast majority of the population, their incorporation into primary health care is still regulated due to the barriers presented by health systems, services, and personnel, and it is unusual to have traditional and allopathic medicine integrated in with the same service. Thus, many professionals of allopathic medicine, even in those countries with a strong history of TM, express great reservations and often serious disbelief about its benefits (WHO 2002).

In some countries, the barriers that have limited the use of TM by health personnel have been identified, and also an insufficient preparation of doctors and nurses to orient population toward TM use has been indicated. This could be due to the hegemonic medical model, which presents features such as scientific rationality (Rodríguez Ramos 2014). There are also countries in Latin America in which medicinal plants are part of primary health care, for example, Brazil, where since 1996 this practice is incorporated in the Unified Health System (SUS) (Borges et al. 2010).

The World Health Organization (WHO 2013) aims to promote the safety, effectiveness, and quality of traditional medicine by the broadening of the knowledge base and providing advice on

Faculty of Applied Sciences, National University of Pilar; Scientific Society of Paraguay; nsoria2000@yahoo.com

regulatory standards and quality assurance. Nevertheless, this objective may be difficult to achieve since the medicinal plants used by the communities are changing as environmental modifications occur, and a species currently used may not be the same as the one that the community used in the past, although it retains the same common name.

On the other hand, research carried out in recent years to identify the native species used by the communities demonstrates that the use of the plants depends on the ethnic groups that are studied, as well as the natural habitat near which these communities settle. Also, many of the studied ethnic groups have medicinal gardens where they cultivate species that they consider effective for some health conditions, and that in general are exotic species, such as *Aloe vera* (aloe), *Artemisia absinthium* (wormwood), *Mentha piperita* (mint), and others.

From a wide perspective, plant-based medicine can be considered, and it should be considered as the knowledge and the field of interaction between cultural resources, local practices and knowledge, natural resources, and the preservation of biodiversity. In other words, the users and their interactions with nature and the professionals of the healthcare team (Espinos Becerra et al. 2014).

However, in the Latin American context, the experiences applied from an intercultural approach have generally been characterized by a health treatment that is disconnected from the rest of the problems of the populations and communities. This approach is not linked to the social and economic structure, and as a result the ancestral medicine moves away from the official health system. The concept of intercultural health has been frequently used to identify strategies that take into consideration the ethnic-cultural variables of the indigenous population within the healthcare process, and not the variables of the population in general (Almeida Vera and Almeida Vera 2014).

Moreover, the evolution in the use of medicinal plants, from the consumption of fresh or dried herbs to processed products, has determined their use as raw materials in the production of phytopharmaceuticals, which are the remedies or drugs based on herbs, also referred to as botanical drugs or herbal medicine. Then, medicinal plants constitute the raw material for industrialization or semi-industrialization processes. This has an influence on the conservation of the species, as well as their replacement, especially because industrialization makes it necessary to have enough high-quality feedstock that is required by these industries.

In fact, depending on the industry that demands the raw material, whether pharmaceutical, phytotherapy, cosmetics, or essential oils, the quantity required, as well as the quality of the material, and of course the farming method can vary, affecting the conservation of these plants.

These factors can positively affect the conservation of medicinal plants, because generally, the raw material to be commercialized is obtained directly from their natural habitat, and it does not undergo domestication or cultivation processes. This happens mainly due to the lack of policies related to the conservation of medicinal plants as sustainable resources.

The mentioned dynamics of medicinal plants over time allows one to find species that are replaced by one another, different species that are known with the same common name, and vulnerable species prone to disappear due to the over-exploitation to which they are subjected, especially when the plant part that is used is underground (root, root bark, rhizome, tuber, bulb, others) (Soria and Basualdo 2015).

Factors Influencing the Disappearance of Medicinal Plant Species

When the factors that influence the disappearance of the plant species are analyzed, we find that the anthropic actions that produce habitat modification, transculturation, and plant trade constitute the actions that affect the survival of the medicinal species (Figure 2.1). Also, the evolution of the communities causes some medicinal species to stop being used, and/or being replaced by others, so, for example, species that are used to lose weight or to cure various types of cancer become fashionable, increasing the number of species classified as medicinal, with no scientific evidence to confirm the effectiveness of the species.

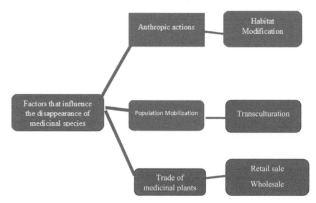

Figure 2.1: Factors that influence the disappearance of medicinal plant species.

Anthropic Actions

The modern society, which uses materials from nature for its development, usually does not consider the effects on biological diversity caused by its actions. Many times, the modifications that are carried out to build up or expand cities, roads, bridges do not take into account the necessity to have a territorial planning where all the important aspects for biodiversity, those that could allow the survival of a species, must be indicated.

In fact, factors such as climate, soil fertility, temperatures, or rainfall are ecological principles that influence the distribution of plant and animal populations on Earth. Today's society and its activities lead to important variations in these factors and as a consequence of this, certain populations are driven to the brink of extinction or led to the extinction of the whole species.

Medicinal plants do not escape the effects of these anthropic actions and many of them disappear due to habitat modification, and their properties are lost without even being known or studied. Habitat modification is one of the most important anthropic actions for the species conservation.

Habitat Modification

Habitat modification occurs due to changes in ecosystem, for example, increase in the size of cities, road constructions, bridges, change to agricultural/livestock land, and other modifications due to anthropic action. Due to this, there is a reduction in the size of the original ecosystems, that is to say, the habitat is fragmented, and this produces a progressive loss of the species that inhabit the site, and the plant populations are reduced as the surface size of the sites decreases. In this context, medicinal species may be affected, populations decline, genetic erosion takes place, and eventually the population disappears.

This loss of habitat can be regional, with the consequent reduction in the size of populations of medicinal species. As a consequence, the regional density of the species decreases (number of individuals per unit of area in the mentioned entire region), and this may result in a decrease in the capacity index to reduce point extinctions through the contribution of individuals from less altered areas.

Therefore, the modification of ecosystems and habitat produces:

- Increase of fragmented habitat. This trend progressively reduces the size of the populations maintained in each one of the single fragments, increasing the risk of reaching a threshold below which they are unfeasible, and causing genetic erosion.
- Increase in the distance between fragments, which makes difficult the exchange of individuals among isolated populations, as well as to recover, to recolonize, leading the species to the possibility of further extinction.

- Finally, there is an increase in the perimeter-area ratio and, consequently, greater exposure of fragmented habitat to multiple interferences from peripheral habitats, generically known as habitat matrix. Consequently, there is a growing edge effect that causes a deterioration of the quality of the habitat in regression, affecting the survival of the populations cliffed in the fragments (Group of Medicinal Plants Specialists 2006).

Indeed, despite the numerous efforts that have been made in recent years to call attention to the problems faced by nature as a result of deforestation and destruction of ecosystems, mainly due to anthropogenic actions, the loss of biological diversity with the consequent disappearance of species has been increasing, and degradation is increasingly linked to factors associated with common economic and social problems in developing countries.

An example is the fragmentation of habitat that occurs, for instance, with the disorderly growth of populations due to urban and rural development without planning processes, adding pressure on the available natural resources. In addition, overexploitation, livestock, industrialization, and road constructions and other road works have contributed significantly to the modification of ecosystems, accelerating the processes of wildlife extinction.

All this anthropic modification affects medicinal plants, which decrease in quantity and tend to disappear from the places where they were collected, so the population replaces those species with others that are more easily accessible to be obtained, and that are morphologically similar to the traditional species.

In this context, we must remember that the plants that are used for medicinal purposes are commercialized in almost all the countries of America, and they are obtained mainly from their natural habitat. In Mexico, more than 90% of the medicinal plants that are consumed come from wild populations with no kind of sustainable management being held. This situation is shared with the other countries of America, since to be sold they are extracted from their original areas, and sometimes they are over-exploited, putting their survival at risk (Gutiérrez Domínguez and Betancourt 2004, Basualdo et al. 2004).

The overexploitation for commercial use, either locally or internationally, especially when they constitute the raw material of some herbal medicine, in more industrialized processes of phytotherapeutic products, or in various industrial processes (food, cosmetics), causes problems in the conservation of species.

In addition, the vulnerability of the species is closely linked to the plant organ that is used as a medicinal product, so the disappearance of the medicinal species may be due to factors directly linked to its use, especially when the organ used as a medicinal product is the root, the rhizome, or part of them. In that case, for their use, the species must be totally extracted, endangering their survival (Basualdo et al. 1991, 1995, Soria and Basualdo 2015).

There are numerous examples that show us how species can struggle with pressures of the extensive use, especially when they enter a cycle that can be called the "fashion cycle", as happened with *Petiveria alliacea* and *Momordica charantia* that were the objects of an economic interest in Central America. These plants have applied a strong pressure to extractive activities on the natural populations, taking them to the risk of extinction and currently *Uncaria tomentosa*, which is the object of an intensive international trade that has propitiated domestication activities to satisfy the demand and to reach the correct management of these resources (Ocampo Sánchez and Valverde 2000).

In addition, species that can be called "fashionable" come into a trade as those that are used to lose weight, considered the most important condition of our time, and therefore the most frequent disease today. Slimming species become fashionable and disappear from trade after some time, as has happened with *Vernonia chamaedrys*, *Moringa oleifera*, among others (Basualdo and Soria 2014).

The implementation of conservation policies for species in general, and medicinal products in particular, in a first stage, involves a determination study of the degree of threat of each medicinal species, and the updating of inventories of medicinal plants is required for this purpose.

The cultivation of medicinal plants needed to satisfy the world market seems to be a difficult problem to solve due to various factors, such as:

- The lack of knowledge about the growth and reproduction requirements of most medicinal plant species.
- Little experience and research is dedicated to domestication; in general crops are expensive and relatively few species have large and reliable markets necessary to maintain these budgets.
- In many communities where collecting medicinal and aromatic plants is an important source of income, the availability of land for the cultivation of non-food products is limited (Group of Medicinal Plants Specialists 2006).

It can be said that the low production of medicinal species is due, among other factors, to the lack of knowledge about the production strategies and the lack of training in breeding management by producers and technicians who provide technical assistance to the public and private sector.

Population Movements

Migration processes contribute to the variation of plants of medicinal uses. When a human population moves from a place to another, the inhabitants move with all their things, including the medicinal plants that are part of their therapeutic arsenal, which somehow influences traditional knowledge, because the mixture of cultures occurs.

Transculturation

Transculturation is the process in which a social group progressively adopts the cultural practices of another social group over time. This process can occur due to the migration of communities from one place to another, or in border cities where the trade of medicinal plants occurs between countries (Chiappe 2015).

It is important to remember that the culture of the use of medicinal plants has evolved by itself and with the passage of time, plants that were not of traditional use or species that have similar characteristics to those originally known were incorporated into the pharmacopoeia, replacing one by the other. This happens because the commercialization is carried out by the common names. For example, we can mention that the plant known as "katuava" is used as an energizer and aphrodisiac; various species were identified by this common name, such as *Anemopaegma arvense* (Bignoniaceae), *Psidium cinereum* var. *paraguariensis* (Myrtaceae), *Trichilia catigua* (Meliaceae), and *Erythroxylum vaccinifolium* (Erytroxylaceae), all of which are species from different genus and families (Degen et al. 2005).

All this dynamics of knowledge gave rise to what is currently called "Urban Ethnobotany", which demonstrates how the traditional practices and plants used by migrants in a given area are used not only by them, but also by the local population where they live. These plants are incorporated into a new urban environment, with their own plant medicines, as well as the knowledge associated with them and the ways of use (Hurrel et al. 2016).

These multicultural contexts can be considered complex systems in which the interrelationship between immigrants and the local population becomes very dynamic. These systems take place in a situation of change when immigrants try to recreate their cultural heritage as best they can in the new pluricultural context using the original natural resources of the area where they come from, as well as products derived from them, and they absorb the culture where they reside, increasing their knowledge about the use of medicinal plants.

In this context, it can be said that immigrants in interaction with the local population are creating a cultural change, where botanical knowledge linked to group identity is mixed with local communities. This expands the "plant-based remedies" in the new residence site (Ladio and Albuquerque 2014).

In general, the use attributed to the species is empirical, based on personal experiences, which are then enhanced by its use within the community. Thus, many chronic conditions, such as diabetes and hypertension, which are frequent diseases in our times, are frequently treated with numerous plant species that are not used for these conditions in other communities. As a consequence, there is an increase in the number of species in the "Herbal Pharmacopoeia" of the countries (Basualdo et al. 2003, Soria and Basualdo 2015).

Considering that the species are often used without scientific studies, and only based on experience, it is common that different plant species, in gender and family, are used for the same condition, which could show a lack of therapeutic response of the plant to proposed use. However, there may be intrinsic and extrinsic reasons for the species, such as the type of soil where it grows or the time of collection, which could modify its therapeutic response.

In some cases, more than 20 species can be used to treat the same condition, as mentioned by Degen de Arrúa and González (2014), citing 37 species used as anti-inflammatories in communities in Paraguay. Puentes (2016) includes 115 medicinal vascular plants species marketed as antidiabetics in the Buenos Aires-La Plata conurbation, Argentina, as a demonstration of the variety of species that are used to fight the same disease, and many of them are not sufficiently studied to support their indicated use.

In some countries, such as Paraguay, the use of fresh plants for medicinal purposes is still common. Plants are used in cold water maceration, in a typical drink called "terere", which consists of placing chopped up yerba mate, *Ilex paraguariensis*, in a bowl, and then cold water is poured along with the fresh herbs, which is considered to be "refreshing", and have a diuretic effect. This is drunk using a special straw, and these herbs are often used as a refresher and preventive to avoid diseases (Basualdo et al. 2004).

The disappearance of species and the substitution of some of them for others may be related to the lack of information from those who collect species from their natural habitat, or to premeditated acts of malice, considering that the raw material may be chopped, and not be obvious to the unaided eye.

The conservation of medicinal species, and consequently, their sustainability over time, is based primarily on determining the degree of threat in order to ensure their survival, since the use of medicinal plants constitutes an ancestral tradition with the risk of disappearing, if measures that allow its conservation through sustainable management are not implemented.

The low production rate at the commercial level is due, among other factors, to the lack of information about the production techniques and the lack of training in crop management by producers and technicians who provide technical assistance from the public and private sector (Fretes 2008).

Variation in the Number of Medicinal Plant Species

Given all the above, it is demonstrated that the dynamics in the use of medicinal plants, and the variation in use over time, can be maintained or changed. When we analyze the species used as medicinal in the different countries of America, it is shown that the number of medicinal plants varies, and in general, is related to the diversity of species cited for the country. For example, in Mexico, it is estimated that there are 4,500 medicinal species, in Ecuador 2,900, in Colombia 2,600 plants, in Argentina 1,529 species, while in Paraguay the number reaches 266 (Basualdo et al. 2004, Trillo et al. 2011, Ortega-Cala et al. 2019) (Table 2.1).

Of the plants mentioned as medicinal by the different groups, there are very few studies that can guarantee their safe and effective use. Even so, the number of species used for medicinal purposes is increasing in the different countries of America, for example, Basualdo et al. (2003, 2004) indicated that in the capital and metropolitan areas of Paraguay, 266 species are marketed with medicinal purposes that are used to combat, prevent, or cure 57 conditions, while Pin et al. (2009) mention about 300 species used for medicinal purposes for the same area, representing a 15% increase in 5 years between publications.

Table 2.1: Biodiversity Relationship of Medicinal Plants/Plants in American countries.

Country	Estimated Number of Vascular Species	Estimated Number of Medicinal Species
Mexico	30,000	4,500
Colombia	22,840	2,660
Ecuador	15,306	3,118
Peru	18,652	3,408
Bolivia	15,345	3,000
Argentina	15,000	1,529
Brazil	46,716	5,000
Paraguay	8,500	266

The variation in the amount of number of plants that are used as medicinal, may be due to factors, such as the interconnection that exists between the different communities since the borders between the countries have almost disappeared today, so, when there is movement of populations, the inhabitants move with all their things including the medicinal plants that are part of their therapeutic arsenal, which in some way influences traditional knowledge, because a transculturation takes place, that is, a mixture of culture.

On the other hand, the irrational use of some species that leads to the overexploitation and degradation of the natural ecosystems where these resources grow, contributes to the disappearance of the medicinal species that are used in the different places. All this demonstrates the importance of elaborating inventories of medicinal plants in rural, indigenous, and urban communities, as well as developing programs for the rescue and conservation of traditional knowledge (Lastres et al. 2015).

The process of transculturation mentioned in this paper can also have an effect on the knowledge of medicinal species, since young generations will only remember the species currently used in their community, and no longer remember the plants from the places where their ancestors previously lived. Plus, as the transmission is oral, it is very likely that their parents' knowledge is forgotten. Thereby, the knowledge related to native medicinal species disappears, hence underlining the importance of registering and inventorying the species used in the communities.

Furthermore, the population movements take into consideration other species that were unknown to the populations that receive migrants. Thereby, it makes visible and disseminates various resources of plant origin that, without these migratory processes, would not become part of the repertoire of local medicinal plants (Acosta et al. 2018).

In addition, as the collection is carried out in the natural habitat in which the plant grows, these species become adulterated or replaced, either intentionally or due to lack of control, especially when the morphological characters are similar, as occurs with the species *Thitonia rotundifolia* that replace the species *Jungia floribunda*. Both of these species belong to the Asteraceae Family, which is used as a hypoglycemic and is known by its common name "yaguareté po" (Table 2.2).

That is, as medicinal plants are collected from wild populations, they can become contaminated with other species or parts of plants, causing an incorrect identification. Contamination may be accidental or intentional adulteration may occur. These circumstances may adversely affect the safety of the products.

Another aspect that contributes to the disappearance and/or replacement of medicinal species is the confusion of species, that is, the replacement of the species that were originally used in the communities. This can happen because plants are collected and marketed using their common name, while their botanical identification is unknown.

Apart from that, there are species that morphologically resemble each other, and the communities group them together, giving them the same common name or similar common names. These species share some characters, so they end up forming a group of species that are traded for the same uses.

Table 2.2: Medicinal species that are marketed with the same popular name, of similar or different use.

Common name	Species	Family	Habitat	Plant part	Use
Acaryso	*Hydrocotile bonariensis* Lam.				
	Hydrocotile leucocephala Cham. & Schltdl.	Apiaceae	Wet ground	Entire plant	Skin. External use
Árnica del campo	*Arnica montana* L.	Asteraceae	Countryside	Flower, Roots	Painkiller for joints, rheumatism, and muscles. External use
	Aldama linearifolia (Chodat) E.E. Schill. & Panero	Asteraceae	Countryside	Flower, Roots	Enfermedades renales. Internal use
Caña brava	*Costus arabicus* L.	Zingiberaceae			
	Hedychium coronarium J.Köning	Zingiberaceae	Swamps	Rhizome	Lithiasis. Anti-sifilitic
Calabacita	*Momordica charantia* L.	Cucurbitaceae	Side road, ruderal	Leaf	Antidiabetic
Cangorosa	*Maytenus ilicifolia* Mart.	Celastraceae	Countryside, Forest	Root, Bark of the root	Anticancerous
	Salacia pittieriana A.C. Sm.	Celastraceae	Countryside	Root, Bark of the root	Anticancerous
Cola de caballo (Horsetail)	*Equisetum arvense* L.	Equisetaceae	Wet ground	Aerial part	Diuretic
	Equisetum giganteum L.	Equisetaceae	Wet ground	Aerial part	
jaguarete ka'a	*Baccharis crispa* Spreng.	Asteraceae	Countryside	Entire plant	
	Baccharis trimera (Less.) DC.	Asteraceae	Countryside	Entire plant	Digestive
	Baccharis microcephala (Less.) DC.	Asteraceae	Countryside	Entire plant	
katuava	*Anemopaegma arvense* (Vell.) Stellfeld ex J.F. Souza *Psidium cinereum* var. *paraguariensis*	Bignoniaceae	Countryside	Rhizome	
	Trichilia catigua A. Juss.	Myrtaceae	Countryside	Leaf	Energizer
	Erytroxylum vaccinifolium Mart.	Meliaceae	Forest	Leaf	
		Erytroxylaceae	Forest	Leaf	
kambara	*Gochnatia polymorpha* (Less.) Cabrera	Asteraceae	Countryside	Leaf	Antitussive Expectorant
	Buddleja madagascariensis Lam	Scrophulariaceae	Farmed	Leaf	Antitussive Expectorant
Malva blanca (White mallow)	*Sida cordifolia* L.	Malvaceae	Countryside	Aerial part	Antitussive
	Walteria albicans Turcz.	Malvaceae	Countryside	Aerial part	

Table 2.2 contd. ...

...Table 2.2 contd.

Common name	Species	Family	Habitat	Plant part	Use
Mburucuja	*Passiflora cincinnata* Mast.	Passifloraceae	Forest	Flower, Fruit, seed	Tranquilizer. Sedative
	Passiflora alata Dryand.	Passifloraceae	Forest	Flower, Fruit, seed	
Ñandypa	*Genipa americana* L.	Rubiaceae	Forest	Leaf	Lowers cholesterol,
	Sorocea bonplandii (Baill.) W.C.Burger, Lanj. & Wess Beer.	Moraceae	Forest	Leaf	Slimmer
Palo azul (Blue stick)	*Cyclolepis genistoides* Don	Asteraceae	Xerophytic vegetation	Leaf	Antidiabetic
	Eysenhardtia polystachya (Ortega) Sarg.	Fabaceae	Dry forest	Leaf	Detoxifying, diuretic
Moringa	*Moringa oleifera* L.	Moringaceae	Farmed	Entire plant	Especies de moda. Slimming
Typycha	*Vernonia chamaedrys* Less.	Asteraceae	Countryside	Aerial part	
Suruvina	*Couepia grandiflora* (Mart. & Zucc.) Benth.	Chrysobalanaceae	Country side	Bark of the root	Antidiabetic
Tajuja	*Ceratosanthes* sp.	Cucurbitaceae	Side Road	Roots	Abortive.
	Cayaponia espelina Cogn.	Cucurbitaceae			Antirheumatic. Against Hepatitis A.
Toro rati	*Acanthospermum hispidum* DC.	Asteraceae	Roadside	Whole plant	Tonsillitis. Pharyngitis
	Acicarpha tribuloides Juss.	Calyceraceae	Roadside	Whole plant	
Uña de gato (cat nail)	*Uncaria tomentosa* (Willd.) DC.	Rubiaceae	Forest	Bark, root	Antidiabetic
	Dolichandra unguis-cati (L.) L.G. Lohmann	Bignoniaceae	Forest	Bark	Antidiabetic
Urusu katii	*Trixis nobilis* (Vell.) Katinas	Asteraceae	Countryside	Root	Antiparasitic
	Trixis pallida Less.	Asteraceae		Root	
Yaguareté po	*Jungia floribunda* Less.	Asteraceae	Water stream shores	Leaf	Antidiabetic
	Tithonia diversifolia (Hemsl.) A. Gray	Asteraceae	Farmed	Leaf	
Yerba de lucero	*Pluchea sagittalis* (Lam.) Cabrera	Asteraceae	Forest	Aerial part	Digestive. Indigestion
	Hyptis brevipes Poit.	Lamiaceae	Forest	Aerial part	

These groups do not remain static or constant over time, but some species stop being used, while others are incorporated into the group. It is essential to develop descriptive and comparative studies that account for this dynamic in the use of medicinal plants (Pochettino et al. 2008).

Medicinal Plants Trade

As it was told, in the world, 17 mega-diverse countries have been identified, from which eight are in Latin America—Bolivia, Brazil, Colombia, Costa Rica, Ecuador, Mexico, Peru, and Venezuela. From the existing plant species on the planet, less than 10% have been scientifically evaluated for therapeutic purposes. And the estimates indicate that about 15,000 medicinal plants are already endangered (PAHO 2019).

In America, it is estimated that about 23,000 plant species are used as medicinal, mostly in traditional medicine systems. Most species do not have scientific studies that can guarantee their effectiveness and safety.

The trade of medicinal plants, then, has two characteristics: the small quantity trade or retail trade, and the collection in large quantities for sale as raw materials for industry or for export.

It is important to mention that, from the number of species that are used, only a relatively small number is used in a significant volume as a raw material in the pharmaceutical, food, and/or cosmetic industries, although the vast majority comes from their natural habitat.

Also, it is important to mention that the use and commercialization of medicinal plants is stimulated by the growing demands of industries, such as agribusiness companies, which trade medicinal species as bulk sale products, such as yerba mate in countries, such as Paraguay, Argentina, Brazil, increasing the growing demand of the industry for medicinal raw materials.

The trade of medicinal plants in the different countries of America includes native, cultivated, and imported species. The preponderance of exotic species is very common in South America, and it can include up to 216 species from Europe, Asia, North America, Africa, and the Pacific (Giraldo et al. 2009, Zambrano et al. 2015).

In addition, the foreign trade of herbs is stimulated by studies that show that medicinal herbs produce less side-effects in contrast to synthetic drugs.

Retail Trade

Medicinal plants trade in the various countries of America is carried out in local markets, and fresh or dried plants are offered. In almost all countries, it is considered a marginal economic activity, although some studies show that they move a significant amount of money every year. Despite this, market information is generally not collected in official statistics, and little attention is paid to its ecological, economic, and cultural impact (Giraldo et al. 2009).

Retail merchants that are represented by market stalls do not have a record in terms of volumes of product sold, even though prices are relatively homogeneous within sales centers. Mostly fresh herbs are sold and when they are not fully sold, they are dried and packaged for sale as a dry product (USAID 2010).

Wholesale Trade

According to data from the International Trade Center (2001), the world market for medicinal plants has grown tremendously, and offers good development prospects for exports. It is estimated that sales of medicinal herbs increased from USD 12,500 million in 1994 to USD 30,000 million in 2000, representing a 5% to 15% annual growth rate depending on the region. It is considered that the segment of herbal food supplements has been registering an even greater annual growth, estimating that over time, the increase reaches 50% (Cañigueral et al. 2003).

Thus, it shows that half of the plants used in the world market are used in human food, and the rest in cosmetics and in the pharmaceutical industry. The productive countries are mainly developing countries and the products are aimed to developed countries, mainly the United States of America, Japan, European countries such as Germany, United Kingdom, France, and Spain (Cañigueral et al. 2003).

This growing global demand for medicinal plants has generated sustained and sometimes uncontrolled illegal trade of plant materials that are irregularly extracted, especially from developing countries, whose biodiversity has been highly affected by the indiscriminate collection of wild species threatened with extinction (Ocampo and Mora 2010).

As an example of this, we have the case of the medicinal plants trade carried out by Paraguay, one of the less biodiverse countries in the America in terms of plants, where according to the data provided by the National Plant Quality Service (SENAVE 2017), 1,849.1 tons of medicinal species were exported in 2017 (Table 2.3). Among the marketed species, the blue stick is mentioned—*Cyclolepis genistoides* is included in the list of endangered species in the country (MADES, Resolution 470/2019), and it is not clear if that amount came from cropping for export. The list includes introduced species that are obtained from crops and native species which are not cultivated and that appear to come from their natural habitat.

Although, in developed countries, and especially in Europe and the United States, the market for medicinal herbs is very regulated and its access is very difficult, for developing countries whose products are not subject to strict control procedures applied by the pharmaceutical industry, the trade with these plants continues to increase.

The growing demand for phytotherapy and natural drugs, both local cultures and their biological resources, will become increasingly vulnerable to the ongoing pressure of market economies.

In some cases, provisioning is becoming critical, as indicated by the increasing distances that collectors are required to go to collect medicinal plants. This would be demonstrating the overexploitation of the species, as occurs for example with the horsetail *Equisetum arvense*, *E. giganteum*, or with the yaguareté kaa *Baccharis trimera*, *B. crispa*. For these plants, the gatherers mention that they must walk long distances to find the species.

Developing countries are origin centers of more than two two-thirds of the world's plant species—of which at least 35,000 have potential medicinal value. According to the United Nations Environment Program, the estimated value of pharmaceutical materials in the southern hemisphere derived from medicinal species would range from USD 35,000 to 47,000 million.

Table 2.3: Species exported by Paraguay (2017).

Popular name	Species	Origin	Conservation Status
Amba'y	*Cecopia pachystachya* Trec.	Natural habitat	Not evaluated
Horsetail	*Equisetum arvense* L.	Natural habitat	Vulnerable
Jaguareté ka'a	*Baccharis trimera* Less.	Natural habitat - Cropped	Vulnerable
Blue stick	*Cyclolepis genistoides* Don.	Natural habitat	Endangered
Ox limb	*Bahuinia* sp.	Natural habitat - Cropped	Not evaluated
Cedar grass	*Cymbopogon citratus* L.	Cropped	Not evaluated
Cedar Paraguay	*Aloysia triphylla* L´Herit.	Cropped	Not evaluated
Moringa	*Moringa oleifera* L.	Cropped	Not evaluated

Source: SENAVE 2017

Conclusion

As noted, there is a close relationship between medicinal plants and conservation of natural resources. In fact, it can be affirmed that the "extinction" of medicinal plants is not only a problem of characteristics of the species, but also a cultural, economic, and political problem. This was already stated in the Declaration of Chiang Mai (1988), which served as a starting tool for the analysis of the conservation of medicinal species and which we cite here literally:

LA CHIANG MAI DECLARATION: "Save lives, saving plants"

"We, health professionals and plant conservation specialists who have come together for the first time at the WHO/IUCN/WWF International Consultation on Conservation of Medicinal Plants, held in Chiang Mai, 21–26 March 1988, do hereby reaffirm our commitment to the collective goal of "Health for All by the Year 2000" through the primary health care approach, and to the principles of conservation and sustainable development in the World Conservation Strategy:

- Recognize that medicinal plants are essential in primary health care, both in self-medication and in national health services;
- Are alarmed at the consequences of the loss of plant diversity around the world;
- View with great concern the fact that many of the plants that provide traditional and modern drugs are threatened;
- Draw the attention of the United Nations, its agencies and the Member States, other international agencies, their members, and non-governmental organizations to:

  - The vital importance of medicinal plants in health care;
  - The increasing and unacceptable loss of these medicinal plants due to habitat destruction and unsustainable harvesting practices;
  - The fact that the plant resources in one country are often of critical importance to other countries;
  - The significant economic value of the medicinal plants used today and the great potential of the plant kingdom to provide new drugs;
  - The continuing disruption and loss of indigenous cultures, which often hold the key to finding a new medicinal plant that may benefit the global community;
  - The urgent need for international cooperation and coordination to establish programs for the conservation of medicinal plants to ensure that adequate quantities are available for future generations.

We, the members of the Chiang Mai International Consultation, hereby call on all people to commit themselves to "Save plants, to save lives"."

Chiang Mai, (Thailand) -WHO, IUCN, WWF- March 26, 1988

Despite the time elapsed, and after all the efforts made worldwide, the disappearance of native medicinal species remains a big problem without a solution.

References

Acosta, M.E., Ladio, A.H., Vignale, N.D. 2018. Bolivian migrant herbalist in a context Argentine Northwest. Bol. Latinoam. Caribe Plant Med. Aromat. 17(2): 217–237.

Almeida Vera, L., Almeida Vera, L. 2014. Foundation of the Ecuadorian intercultural management model in primary health care. MEDISAN 18(8): 1201–1214.

Basualdo, I., Zardini, E., Ortíz, M. 1991. Medicinal Plants of Paraguay: Underground Organs. Econ. Bot. 45(1): 86–96.

Basualdo, I., Zardini, E, Ortíz, M. 1995. Medicinal Plants of Paraguay: Underground Organs II. Econ. Bot. 49(4): 380–394.

Basualdo, I., Soria, N., Ortíz, M., Degen, R. 2003. Medicinal use of plants commercialized in the markets of Asuncion and greater Asunción. Rev. Soc. Cient. Parag. 14: 5–22.

Basualdo, I., Soria, N., Ortíz, M., Degen, R. 2004. Medicinal Plants Marketed in Markets of Asunción and Gran Asunción (Part I). Rojasiana 6(1): 95–114.

Basualdo, I., Soria, N. 2014. Medicinal plants marketed in the municipal market of the city of Pilar, Dpto. Ñeembucu, Paraguay. Dominguezia 30(2): 47–53.

Borges, A.M., Ceolin, T., Barbieri, R.L., Heck, R.M. 2010. The insertion of medicinal plants in nursing practice: a growing challenge. Global Nursing 18: 1–8.

Cañigueral, S., Dellacassa, E., Bandoni, A.L. 2003. Medicinal Plants and Phytotherapy: Dependency Indicators or Development Factors? Acta Farm. Bonaerense 22(3): 265–78.

Chiappe, C.M. 2015. Transculturation or Aculture? Conceptual nuances In Juan Van Kessel and Alejandro Lipschutz?. Rev. Ciencias Sociales. Vol. (35): 47–57.

Degen, R. Basualdo, I., Soria, N. 2005. Marketing and conservation of plant species medicinal products of Paraguay. Rev. Fitoterapia 4(2): 129–137.

Degen de Arrúa, R., González, Y. 2014. Plants used as anti-inflammatory in Paraguayan folk medicine. Bol. Latinoam. Caribe Plant. Med. Aromat. 13(3): 213–231.

Espinos Becerra, N., Chaparro Chaparro, J.A., Chaparro Chaparro, N.Y. 2014. Local Knowledge about the use and management of natural resources from El Consuelo's Moor. Rev. Cult. Cient. 47–55.

Fretes, F. 2008. Plantas Medicinales y Aromáticas una Alternativa de Producción Comercial. 80 pp.

Giraldo, D., Baquero, E., Bermúnez, A., Oliveira-Miranda, M.A. 2009. Medicinal plant trade characterization in popular markets of Caracas, Venezuela. Acta Bot. Venez. V. 32(2): 267–301.

Group of Medicinal Plants Specialists. 2006. International Standard for the Sustainable Wild Collection of Medicinal and Aromatic Plants—ISSC-MAP. Working Draft (June 2006). Steering Group for the Development of an International Standard for the Sustainable Wild Collection of Medicinal and Aromatic Plants.

Gutiérrez Domínguez, M.A., Betancourt Aguilar, Y. 2004. The market of medicinal plants in Mexico, current situation and development prospects. http://www.herbotecnia.com.ar/c- public-003.html, accessed 10/05/2019.

Hurrell, J.A., Puentes, J., Arenas, P.M. 2016. Ethnobotanical studies in Buenos Aires-La Plata conurbation, Argentina: Medicinal plant products introduced by Paraguayan immigrants. Bonplandia, 25(1): 43–52.

International Trade Center (ITC). 2001. www.intracent.org/itc/market-info-tool, accessed 12/04/2019.

Ladio, A.H., Albuquerque, U.P. 2014. The concept of hybridization and its contribution to urban ethnobiology. Ethnobio. Conserv. 3: 6.

Lastres, M., Ruiz-Zapata, Th., Castro, M., Torrecilla, P., Lapp, M., Hernández-Chong, L., Muñoz, D. 2015. Knowledge and use of the Medicinal Plants of the Valle De La Cruz Community, Aragua State. Pittieria 39: 59–89.

MADES. (Ministry of Environment of Paraguay). 2019. Update of native species of endangered flora in Paraguay. Resolution 470/2019.

Ocampo Sánchez, R.A., Valverde, R. 2000. Manual of cultivation and conservation of medicinal plants. 1 a. ed. - San Jose; Costa Rica. 148 p.

Ocampo, R., Mora, G. 2010. (PDF) Las Plantas Medicinales de América Latina como Materia Prima ¿Cuál es, o debería ser su papel? Available from: https://www.researchgate.net/publication/49598517_Las_Plantas_Medicinales_d e_A [accessed Oct 09 2019].

Ortega-Cala, L.l., Monroy-Ortiz, C., Monroy-Martínez, R., Colín-Bahena, O. Flores-Franco, G., Luna-Cavazos, M., Monroy-Ortiz, R. 2019. Medicinal plants used for diseases of the digestive system in Tetela del Volcán, Estado de Morelos, Mexico. Bol. Latinoam. Caribe Plant. Med. Aromat. 18(6): 106–129.

PAHO. Pan American Health Organization. Situation of medicinal plants in Peru. Meeting report of the group of experts in medicinal plants. Lima: OPS; 2019. 13p.

Pin, A., González, G., Marín, G., Céspedes, G., Cretton, S., Christen, P., Roguet, D. 2009. Medicinal Plants of the Asunción Botanical Garden. Asunción, Paraguay. 441 pp.

Pochettino, M.L., Arenas, P., Sánchez, D., Correa, R. 2008. Traditional botanical knowledge, commercial circulation and consumption of medicinal plants in an urban area of Argentina. Bol Latinoam. Caribe Plant. Med. Aromat. 7(3): 141–148.

Puentes, J. 2016. Medicinal plants and derived products commercialized as antidiabetics in Buenos Aires-La Plata conurbation, Argentina]. Bol. Latinoam. Caribe Plant. Med. Aromat. 15(6): 373–397.

SENAVE. National Service of Vegetable and Seed Quality and Health. 2017. Statistic yearbook. 1–48 p. http://web.senave.gov.py:8081/docs/informes/ANUARIO%20ESTADISTICO%20 SENAVE %202018.pdf (accedido, 10/10/2019).

Soria, N., Basualdo, I. 2015. Plant genetic resources Conservation of medicinal species in Paraguay (Part I). Dominguezia Vol. 31(1): 41–47.

Soria, N., Ramos, P. 2015. Use of medicinal plants in primary health care in Paraguay: some considerations for safe and effective use. Mem. Inst. Investig. Cienc. Salud 13(2): 8–17.

Rodríguez Ramos, R. 2014. Naturopathic Medicine and Primary Health Care. Cuban experience Available in: http/:www.uva.org.Ar/cuba .

Trillo, C., Arias Toledo, B., Colantonio, S. 2011. A Review of the Ethnomedicine in Argentina: The construction of the discipline and perspectives for the future. Bonplandia 20(2): 405–417.

USAID. Agencia del Gobierno de los Estados Unidos para el Desarrollo Internacional. 2010 Medicinal and Aromatic Plants and Alternative of Commercial Production. 58 p.

WHO-IUCN-WWF. 1988. Guidelines on the conservation of medicinal plant. The Chiang Mai Declaration, 32 p.

WHO. 2002. Traditional Medicine Strategy: 2002–2005. Geneve, Suiza. 74 p.

WHO. 2013. Estrategia de la OMS sobre medicina tradicional 2014–2023. 74 p. Geneve, Suiza. https://www.who.int/traditional-complementary-integrative-medicine/en/.

Zambrano, L.F., Buenaño, M.P., Mancera, N.J., Jiménez, E. 2015. Estudio etnobotánico de plantas medicinales utilizadas por los habitantes del área rural de la Parroquia San Carlos, Quevedo, Ecuador. Rev Univ. Ssalud. 17(1): 97–111.

Specific Countries

3

Wild Plants of Northern Peru
Traditions, Scientific Knowledge, and Innovation

Fidel A. Torres-Guevara,[1,2] *Mayar L. Ganoza-Yupanqui,*[*,2,3]
Luz A. Suárez-Rebaza,[3] *Gonzalo R. Malca-García*[4] *and Rainer W. Bussmann*[5]

Introduction

The knowledge and the use of native plants by native healers of rural agrarian societies located in the moorland and cloud forests of the Northern Peruvian Andes constitute intangible and tangible treasures for sustainable development of these highly diverse ecosystems. Such development is a social opportunity in the hands of farming and peasant communities, and may become viable if a network or innovation system is created with other public and private agents of these territories in order to articulate regional, national, and international economic advantages. In this way, the role of rural Andean communities will be to sustainably establish the primary link between science and innovation chains for an economy based on the exploitation and conservation of wild plants, and the richness they possess and know well.

The participative investigation of wild plant species to put a value on their diversity, incorporates the local traditional knowledge about the use and conservation of plants, with Western scientific knowledge about phytochemistry (phytochemical discovery) in an intercultural dialogue, in order to verify, standardize, and expand the former, then make it available to the wider society as an expression of mutual interest between academia and rural society. This implies a new relationship between the academic investigator(s) and the traditional investigator(s), getting away from the investigator-informant relationship, to one as co-authors.

Academic scientific knowledge has permitted corroborating and amplifying not only the chemical properties of culturally important plants, but also proving the validity and efficacy of traditional practices for efficient extraction of important plant-based active principles, and the important influence

[1] The Mountain Institute INC, Vargas Machuca 408 Urb. San Antonio, Lima, Perú.

[2] Asociación para la Ciencia e Innovación Agraria de la Red Norte-AgroRed Norte, Mz O Lote 20 Urb. Los Cocos del Chipe, Piura, Perú.

[3] Departamento de Farmacología, Facultad de Farmacia y Bioquímica, Universidad Nacional de Trujillo, Av. Juan Pablo II S/N, Ciudad Universitaria, Trujillo, Perú.

[4] UIC/NIH Center for Botanical Dietary Supplements Research, Department of Medicinal Chemistry and Pharmacognosy, College of Pharmacy, University of Illinois at Chicago, 833 S. Wood St., Chicago, IL 60612, USA.

[5] Department of Ethnobotany, Institute of Botany, Ilia State University, Tbilisi, 0105, Georgia.

* Corresponding author: mganoza@unitru.edu.pe

of growing location and harvest time on the biochemical proportions of wild plans, due to the influence of microclimates, microhabitats, and annual cycles in the cloud forest and Páramo ecosystems.

The economic use of knowledge generated through a participative investigation based on traditional knowledge requires that any benefits are also collective property of the society involved in the investigation. This focus, with the idea to improve livelihoods and well-being of the participating community, implies it is a new form of organization, implying the necessity for organizational innovations, such as the conversion of natural or formal local organizations into legal entities in order to be able to access funding instruments for sustainable environmental projects.

Thus, the community organizations participating in the investigation of wild plants become the owners of the results, through a change from their current role as sole providers of primary material and knowledge about plant use, to providers of crude drugs, essential oils, liquid extracts, or high quality natural products needed for phytochemical analyses of toxicity, pharmacological activity, concentration of phenolic compounds, antioxidant activity, antibacterial activity, and identification of the most important bioactive compounds (e.g., flavonoids, tanins).

In order to allow an intensive use of studied species with potential for further development, it is necessary to establish propagation and production protocols to increase and guarantee the availability of prima material with added value. This can be achieved through three alternative efforts: (1) Cultivation in agro-ecologic fields or gardens through the uses of seeds or cuttings; (2) expansion of the wild population of the species, especially if it is as common and abundant, through community controlled, sustainable wild-collection; (3) in case of species that cannot be propagated through seeds due to difficult germination, through the establishment of *in vitro* protocols to establish clonal publications of the species.

The challenge of innovation based on the great wealth that wild plant biodiversity represents, lies in the reality of a social system dominated by the exclusion of rural agrarian societies with low connectivity, due to a lack of public services, e.g., road infrastructure, transport, education, and information, both with regard to quality and quantity. This context of gaps in spacial and social connectivity leads to a significant increase in terms of the cost of the cooperation between academics or technicians and the community organizations that hold the natural resources and traditional knowledge for the development of markets for innovations based on the diversity of wild plants with high potential. In order to overcome this challenge, both public and private actors need to engage in a process of governance and policy change in order to establish a system of innovation that will allow a change in local development policies.

Environment of Wild Plants of the Northern Andes

Moorland and Cloud Forests of the Northern Peruvian Andes

The moorland and cloud forest ecosystem complex of the Northern Peruvian Andes is located between 4°43'48"/5°50'S and 79°36'/79°24"W at 1,500 to 3,700 m above sea level. It occupies a surface area of about 120,000 ha (Perú 2015). These ecosystems are known as primary source of diversity for many groups of plants. In Peru, there are about 17,000 flowering plants and gymnosperms, with more than 8,000 endemics (approximately 47%). In the moorland-cloud forest of the Piura, Cajamarca, and Amazonas regions, there are more than 715 endemic species, which represent about 20% of the endemism of the entire country, in less than 8% of the national land area. Among these species, there are least 11 endemic genera within five flowering-plant families registered in northern Peru (Sagástegui et al. 1999). These ecosystems not only represent richness per se, but also, because of their vegetation type, they constitute hydric catchments and regulation areas for the fluvial basins in northern Peru, where the plains depend exclusively on water from the moorlands and cloud forests. Therefore, territorial agreements are required for their conservation. The best strategy is creating innovative territorial systems oriented toward the conservation of the biodiversity and hydric capacity of these ecosystems (Figure 3.1) (Gomez-Peralta et al. 2008, ANA 2017, Lindsay 2019).

Figure 3.1: Moorland and cloud forests of the Northern Peruvian Andes (Ayabaca and Huancabamba): 2,700 y 3,500 masl.

Andean Moorlands

The Andean moorlands of Piura are located between 3,000 and 7,000 masl, and occupy a surface area of 66,300 hectares (More et al. 2013). It has been calculated that 46,184 hectares of them are in relatively pristine condition (Recharte et al. 2015). The moorland ecosystem possesses the greatest tropical mountain flower diversity in the world, with approximately 1,400 non-vascular, and 3,400 vascular plant species (Sánchez 2012, Hofstede et al. 2014). The importance of this flora lies not only in being a source of resources for health and nutrition for rural communities, but also because it is a layer of vegetative material that facilitates catchment, filtration, and distribution of water. As a consequence, agrarian societies that live within these ecosystems have a great environmental responsibility to maintain the biodiversity and the hydric regulation capacity of the watersheds where they live. Peruvian moorlands located in the Piura and Cajamarca regions are complex, and the study of their floristic composition is still emerging. According to Sánchez (2012), the physiognomy of the moorlands is very similar to the Peruvian high plateaus (*jalca*), but the vegetation communities possess their own distinct species composition: ecotonal communities of the woods and grassy scrubland (*pajonal graminoso*) with the species: *Hypericum laricifolium*, *Brachyotum* spp., *Pernettya prostrata*, and *Podocarpus oleifolius*. Communities of the grassy scrubland are made up of vegetation of the moorlands proper: *Calamagrostis* spp., *Agrostis* spp., *Stipa* spp., *Paspalum bonplandianum*, *Neurolepis aristata*. Accompanying pteridophytes include *Huperzia* spp., *Lycopodium* spp., *Jamesonia* spp., *Niphidium* spp., and *Lophosoria* spp. Angiosperms, dicots, as well as small herbaceous shrubs are also found within the pasture landscape.

Besides the environmental services of protecting the hydric capacity of basins and providing sustenance for endemic animals in danger of extinction, such as the spectacled bear and tapir, the vegetation of the moorlands also offers a special vegetative diversity useful for human medicine and nutrition. This diversity contains a collection of bioactive molecules that act as pharmacological, cosmeceutical, and nutraceutical industrial feedstocks, as well as compounds that act as nutricosmetics, perfumes, fragrances, flavors, scents, bifunctional food, biocides, repellents, and natural pigments (Carhuapoma Yance 2011).

Andean Cloud Forests

The landscape of this type of forest is characterized by persistent humidity and precipitation. This dense, steep-sloped woodland is associated with a large amount of shrubby plants and epiphytes, such as mosses, ferns, orchids, bromeliads. Numerous thin waterfalls run through the hillsides. The tree stratum is not very high, but it is very tangled, and includes tree ferns up to 10 m high.

The majority of tropical mountain cloud forests are considered highly fragile ecosystems because they play important hydrological and ecological roles, and they are becoming one of the most threatened ecosystems due to rapid human settlement and the small amount of land area the forests cover. Many institutions and decision-making bodies are still not aware of the serious consequences of the disappearance of these woodlands, whose deforestation could trigger catastrophic erosive consequences. The cloud forests of the western Andean slopes of Northern Peru and Southern Ecuador are habitats of high phytodiversity with a high endemism index. Indeed, these cloud forests have a larger number of endemic species than humid tropical forests; because of this, urgent measures are required for their study and protection (Amanzo et al. 2003, Weigend et al. 2006, Peña 2015).

In addition to providing protection from erosion and their function as hydrological regulators, there are many other arguments in favor of protecting, investigating, managing appropriately, and providing information about the value and potential of mountain cloud forests to society in general, and in particular to the populations that depend on them (Llerena et al. 2010). The biodiversity of cloud forests is impressive compared to the lowland forest. However, the cloud forests have barely been studied and relatively little is known about them. Located on the slope of the Amazon and the Pacific-facing mountain slopes between 1,500 and 2,000 meters above sea level, the cloud forests catch sufficient humidity to sustain a high plant density.

Cloud forests in Peru occupy about 10% of the national territory and are found in 11 regions of the country: Amazonas, Ayacucho, Cajamarca, Cusco, Huánuco, Junín, Madre de Dios, Pasco, Puno, San Martín, and Ucayali. The cloud forests are inhabited by 1.8 million people (Perú 2015). In northern Peru, the area surrounding the zone of the Huancabamba Depression (Piura, Cajamarca, and Amazonas Regions) has been widely recognized as the primary source of diversity in many plant groups (Sagástegui et al. 1999, 2003). In the deflection zone, there are at least 715 endemic species that represent approximately 10% of the endemism in the entire country, but in less than 8% of its area. Also, there are 126 species concentrated in Ayabaca and Huancabamba. At the altitude range of 1,500–3,000 masl, where the cloud forests are distributed, conditions of humidity, temperature, and pressure vary significantly with altitude, thus generating a gradient of environmental changes that are important in the adaptation of species (Young et al. 2012).

In Huancabamba Province, the cloud forest surrounding the Blanco River, a headwater of the Chinchipe river, is dominated by the "romerillo" (*Podocarpus oleifolius*), the only native tropical forest conifer of South America, and species such as "mountain cedar" (*Cedrela lilloi*), Grossulariaceae, Juglandaceae, Myrtaceae, Lauraceae, and Moraceae, and Chilean myrtle, which are typical species of mountain forests. Tree ferns of the genus *Cyathea* spp. and epiphytic bromeliads (*Tillandsia* spp.), as well as *Puya* spp., indicate the high humidity of the environment. The herbaceous vegetation is dominated by Asteraceae, Labiatae, Polemoniaceae (*Cantua* sp.), or Araceae (*Anturium* sp.). These herbs are accompanied by others with high potential for nutritional or medicinal use, such as wild species of *Physalis* spp., *Solanum caripense*, *Solanum quitoense*, Cucurbitaceae, and wild tomatoes (*Lycopersicon* spp.), as well as diverse types of bromeliads, a preferred food of the spectacled bear, and diverse species of orchids flowering in different months of the year (Torres Guevara 2006). These cloud forests are intermediate spaces between the lowland forests of the Amazon and the moorlands, with important functions for regional ecological dynamism, and provide high-potential economic opportunities.

Traditional Knowledge: Ethnobotany in the Systematization and Documentation of Native Collective Knowledge

Interculturality in Ethnobotanical Research. Botanical Culture of the High Andean Communities

Culture consists in the expression of beliefs, stories, knowledge, language, and festivals. It is the feeling of belonging to the world around us: its mountains, lakes, and forests, according to where one lives, thinking that we form a part of the same world with both cultivated and wild plants and with domestic and mounted animals. Beliefs, feelings, and emotions that are shared and passed on comprise the culture of a village (Monroe Morante and Arenas Barchi 2003). The construction of the emotions about what has sense, meaning, and priority on what we think and do, represents our beliefs to decide on each moment and aspect of life. This construction is made from established communication with those with whom we frequently interact and form our social circle: this is what we call culture.

Communication among people about plants and their uses, even if these uses are different for the same plant or same for different plants, allows for the organization and prioritization of their reality. In this way, daily experiences with plants of their environment become understandable, meaningful, and imaginable. This social relationship between children and adults causes the progressive formation of a mental model of reality that is similar to the models of the child's parents and community. This model makes possible a framework or system of ideas and concepts that allows for the sharing of experiences and makes them feel like they inhabit the same world. However, no communication is made without encoding emotions. People also communicate and learn by observing the emotions of others; the seeds of these emotions are planted during childhood, when certain words are associated with an emotion (for example "healthy"). If there are different situations when the child is relieved from or cured of a disease with plants used by their parent, and the child hears the word "healthy", this word unites the whole village in those cases where the word "healthy" generates the same sensation (relief and cure), and it is also associated with the plants that healed them. All of this constitutes the underpinnings of a health culture based on available medicinal plants (Samaja 2000, Castilla del Pino 2002, Pardo de Santayana et al. 2003, Barrett 2018).

For communities living within the jalca, an ecosystem similar to the moorland, the significance of the value of plants is associated with myths their ancestors left them as an inheritance, and also with the protection and life generation power of the mountains. The plants of the jalca have a greater medicinal, nutritious, or reproductive power than plants of lower altitudes that have similar properties, because they thrive and reproduce in the extreme conditions present at high altitudes. Therefore, some transmit their strength, health, and fertility to people who know how to use and respect them, neither removing too much nor too often. This is due to the force transmitted by the mountains, to which they pay respect, worship, and offer payment before entering their summits, jalcas, or moorlands. (Walter 2017).

Two types of knowledge are integrated during participative research of wild plants: ethnobotany (native or cultural knowledge of the use and preservation of herbal resources) and botanical discovery (western scientific knowledge, e.g., plant phytochemistry). The first one provides groundwork for scientific hypotheses, and the second one verifies, expands, and standardizes the first one in order to share it universally. This is a key concept in terms of applied interculturality. Research that involves peasant families and scientists requires the effort to maintain reciprocity between traditional and scientific knowledge. This type of participative research, results not only in the creation of new knowledge, but it is also the beginning of a new practice of use of wild plants for new innovations. As rural societies are involved, the understanding of the culture of the mountains of northern Peru is crucial to communicate and manage the results of innovation, in the fulfillment of reciprocal interests between local population and scientists.

The current prestige of many Peruvian plant species as high-quality foods available worldwide (maca, quinoa, native potatoes, kiwicha, camu-camu, and others) is a result of the scientific verification of their nutritional and medicinal properties. This research was initially based on hypotheses drawn from the knowledge of inhabitants of Andean and Amazonian rural communities about the use, practice, and diversity of these foods. However, this has happened in an asymmetric and informal environment between Peruvian traditional knowledge and Western scientific knowledge, which has not benefited the country, much less the traditional owners of the knowledge base at the bottom of the research and innovation chain (Hersch-Martínez 2002, Bussmann and Sharon 2015). Therefore, traditional and scientific knowledge requires an innovative institutional shift that better promotes a national agenda of technological scientific innovation based on biodiversity.

In the new economy of natural knowledge, the entrance of rural communities into the processing of herbal products that gives added value to the vegetation they know and use, can activate the use of the products and services offered by biodiversity, such as genes endemic to the bioeconomy, species with hydric functions, and climate change indicator species (Abramovay 2013).

Changes in the Ethnobotanical Approach, from Traditional Individual to Organizational Knowledge

Culturally, the traditional healer is the intellectual heir who remembers and represents the richness of collective traditional knowledge of medicinal plants present within a rural society. Their mastery over the diversity of plants for therapeutic use has been acquired as part of the construction of knowledge in the society. However, in order to generate wealth in community well-being, government investments to value collective traditional knowledge through scientific research should focus on the registry and systematization of knowledge, from the traditional healer to the communal organization in the society that the healer belongs to. This means that when registering the knowledge regarding plants of a territory, the knowledge of the healer should be integrated as part of the collective knowledge of that community instead of being considered as an exclusive reference. In this way, the knowledge of community members can complement and corroborate one another.

Benefiting from the richness and knowledge of biodiversity for processes of social inclusion through scientific knowledge that revalues, adds value, and innovates from this diversity is a collective enterprise of wealth generation, provided that these innovations are aimed at the conservation of biodiversity. This requires a collective agreement and new institutional frameworks that can only be assumed by the community organizations as operational socio-economic units of this process.

In order for scientific institutions to engage in discovery to provide added value to systematization of traditional knowledge, it is necessary that community organizations acquire joint roles in the research and innovation chain. The integration of research of traditional knowledge of medicinal plants and, e.g., phytochemical research to find bioactive molecules generates new knowledge that may be used to reevaluate traditional knowledge, as well as the development of new lines of phytochemical research oriented using native biodiversity to meet the demands of the health and bio-commerce industries. This equal relationship between traditional and scientific knowledge to establish a competitive process highlights the need for intercultural interaction between communal organizations and academic entities that establish common interest agreements based on their particular purposes (Horák 2015, Granda et al. 2015). This process requires two investments not commonly considered: (1) The investment necessary for permanent intercultural communication; and (2) The investment in the trust necessary to establish fair and reciprocal contracts between communal organizations and scientific entities.

The traditional scientific approach aimed at the registration of local knowledge, in which rural experts are only individual informants, without faces or names as owners of knowledge, does not contribute to the participation of the communities. The communities wind up missing out on the

Figure 3.2: Community organizations associated with participatory research (a and b).

integration of scientific knowledge to improve and increase the added value of offered resources in the value chain of which they can be co-authors. The intercultural approach described here diverges from the one that assumes that rural societies can solely be suppliers of raw material and non-valuable knowledge to the scientific knowledge. The modern approach encourages development of innovations that use native species as well as scientific research programs that generate new knowledge about them. Although the generation of the new knowledge had started with traditional knowledge before being transformed and investigated, the communities that contributed to it were not included in the process as receptors of the resulting benefits. The non-inclusive behavior of certain industries and research entities establishes their innovation and research agendas as rationale for extraction of natural resources and knowledge without any interest in reciprocity to the communities that own those resources. Such behavior may be alien or create conflict with the interests and expectations of the communities of these regions and the country (Hersch-Martínez 2002).

The growing Peruvian demand for medicinal plants and natural products because of their acceptance and use in traditional medicine and the expansion of complementary medicine (Villar López et al. 2016) has led to an increase in the demand for quality products by the consumer. This, in turn, has repercussions in product process quality, which creates a need to generate new knowledge to add value to the products in each segment of the production chain. The quality requirement in each link of the supply chain, including processing and marketing of medicinal plants, will encourage supply chain stakeholders to form new organizational arrangements, collective agreements, and organize research. New spaces for knowledge exchange will be required to establish priorities, strategies, division of responsibilities, and management of science and technological and institutional innovation of medicinal plants (Figure 3.2).

Rural Societies Demanding Scientific Systematization of their Native Knowledge to Assess their Wealth

The incipient knowledge and scientific use of plant species of the moorlands, which are rich in compounds of high therapeutic and nutritional interest, weakens the valorization and sustainable management of these ecosystems, which blocks their normalization, under uses their economic potential, and prevents the formulation of a defined policy of development based on biodiversity to face the challenges of climate change.

In the last decades, Andean communities of the moorlands and cloud forests have notably evolved as a result of having their ecosystems threatened by large-scale metal mining operations in their territory, including changes in hydric services, and provided and used bioservices. Due to this, community members have begun to express demands for agroforestry knowledge and innovations in order to conserve the natural resources of the moorland and cloud forests of their territory, on which they depend for their life and culture. In 2012, they began to create formalized legal organizations to take advantage of financing from the Peruvian government to promote economically sustainable environmental projects based on business plans; this constitutes the first actions of innovation oriented toward the conservation of moorland and cloud forests.

The participation of community organizations with co-financing of innovation projects based on demand represents a change in behavior in which passive participation of community organizations as a beneficiary of goods and services is replaced by competitive participation of the organization to co-finance the goods and services necessary to enhance the richness of wild plant diversity through participatory research.

In 2014, the organizations of moorland communities that implement innovation and research projects with an innovation horizon evaluated the limitations that isolated enterprises have for sustainability, as well as the low incidence of individual action on local governments to achieve better conditions or facilities in the provision of support services to assess their natural resources. In response to this, the Platform for Coordination of the Management of the Moorlands was created as the first attempt to integrate the particular objectives of different organizations linked by the common interest to conserve the biodiversity of wild plants of the moorland and cloud forests.

Scientific Knowledge

Appropriate Botanical Discovery of Plants According to Local and Cultural Conditions

Communal organizations of the moorlands and cloud forests that participate in scientific investigation designate which community members act as guides in the identification of species of interest, both for their taxonomic determination and for analysis of the active substances that they possess. This creates scientific knowledge about the identity of communally known species, as well as verifies and potentially discovers new therapeutic uses. This may increase the potential for use and possible processing.

The selection of the representatives of the communal organization is carried out by consensus of assemblies, and is based on the approval of the participatory role in the research. It requires an agreement between the parties (scientific entities and community organizations), as well as a mutual interest based on intercultural learning and communication, in which traditional and scientific knowledge establish a relationship of mutual respect, which must be revealed in the mutual satisfaction of the achieved results.

With the participation of knowledgeable people in collection expeditions, the community not only contributes with collective knowledge about the use of plants (useful structure, mode of use, condition that controls, and preparation), but also with information regarding its location, abundance, time of phenological development, access routes, meteorological conditions, social conditions, plant habits (if they can or not be domesticated), propagation method, varieties of useful species, their distribution in the moorland, and/or cloud forest at different altitudes. Together, this constitutes the ecophysiological knowledge of these species of plants (Table 3.1).

One important aspect for the sharing of traditional knowledge through scientific knowledge is the identity of plants with therapeutic properties, so that the expected results may be repeated consistently. Clear identity of plant species removes the danger of using similar species or varieties which may produce inconsistent results. Accurate identification is particularly important because different species are registered with the same name and have different uses, or the same species has different names and is used for the same purpose from one community to another within the same territory. Also, the source or ecological niche of a species can change the physiology of plants according to altitude, especially in those who grow in both ecosystems. This has implications in use

Table 3.1: Ethnobotanical description of moorland species and cloud forest of the Andeans from Piura prioritized by local experts.

Local name of plant	Scientific name	Type of plant	Ecosystem	Part used	Age of the plant	Quantity to use (g)	Form of preparation	Time of preparation	Form of consumption	Dosis	Effect of the plant
"hierba del toro"	*Cuphea cilliata* Ruiz & Pav.	Tree	Cloud forests	Fruit	Adult		Crude	Direct	Toasted	Not applicable	Child nutrition
"ushpa"	*Vaccinium floribundum* Kunth	Bush	Moorland	Fruit	Adult		Crude/ Macerated	Direct 15 days	Fresh/ Drink	Not applicable/ Half glass	Nutritional/Respiratory conditions
"zarcilleja"	*Brachyotum angustifolium* Wurdack	Herb	Moorland	Whole plant	In bloom	10 g/l	Raw juice or infusion	30 min in infusion	Drink	One cup 50 ml a day	Antibiotic, depurative
"pega-pega"	*Acaena ovalifolia* Ruiz & Pav.	Herb	Moorland	Leaves	In bloom	30 g	Dry powder in hot water	10 min	Drink	Half cup of 50 ml. Twice daily.	Infection control
"payana"	*Bejaria resinosa* Mutis ex L. f.	Bush	Moorland	Leaves	Adult	20 g/l	Infusion	15 min	Drink	Half cup of 50 ml. Twice daily.	Anti-inflammatory, female matrix infection control
"lanche colorado"	*Myrcianthes rhopaloides* (Kunth) McVaugh	Tree	Cloud forest	Leaves	Adult	100 g/l	Infusion	20 min	Drink	One cup 50 ml daily.	Colds
"lanche chiquito"	*Myrcianthes myrsinoides* (Kunth) Grifo	Tree	Cloud forest	Leaves	Adult	100 g/l	Infusion	15 min	Drink	Water time	Digestive drink, regulates the body
"chupicaure"	*Muehlenbeckia hastulata* (Sm.) I.M. Johnst.	Herb	Cloud forest	Whole plant	In bloom	100 g	Lotion	20 min	Rub	Not applicable	Skin infections/chickenpox control

and processing because the same species of plant can vary its concentration of desired secondary metabolites depending on the environment.

Although interviews contribute to the analysis and organization of traditional knowledge, the words used to transmit this knowledge have a local and culturally specific character which may not necessarily coincide with the meaning of the researcher's questions. In this regard, the importance of intercultural communication is highlighted. The cultural exchange of knowledge lies within the agreement between both partners on the meaning of the concepts used for the uses and values assigned to the plants according to cultural traditions of the communities.

The standardization and delimitation of these similarities and differences, the use and dissemination of the results of the studies in an interdisciplinary line of action are the aspects that make it possible to overcome the local character of the traditional knowledge. Thus, in order to standardize and share it, traditional knowledge can be reassessed by using the species with a new added value based on the conservation and sustainable use of the diversity of the ecosystem to which they belong.

Phytochemical analyses have corroborated and expanded not only the knowledge of chemical properties of the selected plants, but have confirmed the efficacy of traditional practices in the extraction of the active chemicals of medicinal interest. Due to this, scientific standardization of these procedures allows peasant organizations to provide value-added products made on-site. Additionally, phytochemical analysis has shown the important influence microclimates have on the biochemistry of wild plants, where chemical content varies based on geography, temperature, humidity, and radiation even within the same ecosystem. The latitudinal-longitudinal difference includes combinations of these factors, such as relief changes with difference in exposure to solar radiation, wind speed, distribution, and intensity of rainfall and relief. These effects were verified in the significant difference that the same species show when grown in the moorland or cloud forests of Ayabaca or Huancabamba. It was found that the concentration of phenolic compounds and the mean oxidant inhibitory concentration and antioxidant activity was higher in Huancabamba despite having similar altitude and soil conditions as Ayabaca (Tables 3.2–3.4).

As an example, an important contribution of phytochemical studies to traditional knowledge has been the analysis of properties of two species of the same genus with similar names (*Myrcianthes myrsinoides* "lanche chiquito" y *Myrcianthes rhopaloides* "lanche colorado"). Both are used intensively for the same purposes: as a digestive and flu medicine. It was demonstrated that *M. rhopaloides* contains greater amount of total phenolic compounds, higher free radical inhibitory

Table 3.2: Phenolic compounds (PC) of moorland (3,000 to 3,700 masl) and cloud forest (1,900 to 2,900 masl) species of Huancabamba and Ayabaca, expressed in Gallic acid equivalent.

Extract	PC (mg GAE/g dried plant sample)							
	Muehlembeckia hastulata		*Cuphea cilliata*		*Acaena ovalifolia*		*Vaccinium floribundum*	
	Hbba	Ayabaca	Hbba	Ayabaca	Hbba	Ayabaca	Hbba	Ayabaca
Ethanol 96% (A)	53.150	6.678	45.912	19.812	65.462	3.193	82.745	15.715
Ethanol 70% (B)	62.449	10.286	84.085	25.727	105.881	5.901	109.853	31.389
Ethanol 45% (C)	78.709	8.490	92.547	43.676	102.996	6.004	99.033	58.774
Infusion (D)	53.355	9.152	67.450	36.806	87.599	1.549	88.155	33.986
Decoction (E)	73.562	6.606	67.597	41.376	24.959	1.864	119.801	41.588
HSD[a] ($P < 0.01$)[b]	5.947 AD CE	1.727 AE BCD	6.436 DE	2.951 CE	7.715 BC	1.024 BC DE	9.003 AD	2.693 BD

a: Express sublineing does not make statistical difference; b: Minimum significant difference

Table 3.3: Antioxidant activity (AA) of moorland (3,000 to 3,700 masl) and cloud forest (1,900 to 2,900 masl) species of Huancabamba and Ayabaca, expressed in Quercetin equivalent.

Extract	AA (mg Quercetin/g dried species sample)							
	Muehlembeckia hastulata		*Cuphea cilliata*		*Acaena ovalifolia*		*Vaccinium floribundum*	
	Hbba	Ayabaca	Hbba	Ayabaca	Hbba	Ayabaca	Hbba	Ayabaca
Ethanol 96% (A)	54.852	3.025	24.160	13.680	61.795	0.643	56.920	14.836
Ethanol 70% (B)	44.808	3.507	76.945	26.498	117.977	2.975	81.885	14.706
Ethanol 45% (C)	74.351	3.624	101.302	50.279	101.105	4.651	66.719	30.897
Infusion (D)	26.392	13.995	56.231	30.536	86.710	0.064	50.618	7.734
Decoction (E)	49.190	8.629	59.284	35.072	4.924	0.649	63.289	0.849
HSD[a] ($P < 0.01$)[b]	15.569 ABE	0.809 ABC	13.011 DE	7.156 BD DE	9.942	0.372 AE	12.004 ACDE	3.721 AB

a: Express sublineing does not make statistical difference; b: Minimum significant difference

Table 3.4: Inhibitory concentration (IC_{50}) of moorland (3,000 to 3,700 masl) and cloud forest (1,900 to 2,900 masl) species of Huancabamba and Ayabaca, expressed in Gallic acid equivalent.

Extract	IC_{50} (mg GAE/ml of extract)							
	Muehlembeckia hastulata		*Cuphea cilliata*		*Acaena ovalifolia*		*Vaccinium floribundum*	
	Hbba	Ayabaca	Hbba	Ayabaca	Hbba	Ayabaca	Hbba	Ayabaca
Ethanol 96%	0.148	0.340	0.125	0.290	0.344	0.440	0.158	0.200
Ethanol 70%	0.142	0.390	0.114	0.200	0.126	0.290	0.143	0.330
Ethanol 45%	0.130	0.330	0.133	0.180	0.105	0.250	0.142	0.290
Infusion	0.179	0.150	0.128	0.220	0.139	0.650	0.201	0.490
Decoction	0.164	0.170	0.128	0.220	0.047	0.450	0.201	0.420

Table 3.5: Comparative analysis of two genera of *Myrcianthes* (*M. myrsinoides* "lanche chiquito" and *M. rhopaloides* "lanche colorado") in concentration of phenolic compounds (PC), inhibitory concentration (IC_{50}), and antioxidant activity (AA).

Extract	IC_{50} (mg GAE/ml)		PC (mg GAE/g)		AA (mg Quercetin/g)	
	Mm	*Mr*	*Mm*	*Mr*	*Mm*	*Mr*
Ethanol 96% (A)	0.130	0.200	32.244	27.930	59.477	32.438
Etahnol 70% (B)	0.250	0.170	28.968	30.136	23.242	44.478
Etahnol 45% (C)	0.270	0.170	33.148	44.235	24.217	55.952
Infusion (D)	0.210	0.420	40.492	57.581	40.279	27.169
Decoction (E)	0.180	0.170	38.161	59.745	48.154	74.046
HSD[a] ($P < 0.01$)[b]	0.051	0.018 BCE	4.837 ABC DE	2.737 AB DE	3.464 BC	3.870

a: Express sublineing does not make statistical difference; b: Minimum significant difference
Mm: *Myrcianthes myrsinoides*; *Mr*: *Myrcianthes rhopaloides*

Table 3.6: Bactericidal and bacteriostatic activity of *Myrcianthes myrsinoides* "lanche chiquito" and *Myrcianthes rhopaloides* "lanche colorado" of Ayabaca (2,800 masl).

Extract	Minimum Bactericidal Concentration (MBC)/Minimum Inhibitory Concentration (MIC)									
	S. aureus ATCC 25923		*B. subtilis* ATCC 6633		*E. coli* ATCC 25922		*P. aeruginosa*		MRSA ATCC 43300	
	Mm	*Mr*	*Mm*	*Mr*	*Mm*	*Mr*	*Mm*	*Mr*	*Mm*	*Mr*
Decoction	b	b	b	B	b	b	b	b	b	B
Infusion	b	b	b	b	b	B	b	B	b	B
Ethanol 45%	b	b	B	B	b	b	b	B	b	B
Ethanol 70%	b	b	b	B	b	b	b	B	b	B
Ethanol 96%	b	b	B	B	b	B	b	B	b	B

b: bactericide; B: bacteriostatic; *S. aureus*: *Staphylococus aureus*; *B. subtilis*: *Bacillus subtilis*; *E. coli*: *Escherichia coli*; *P. aeruginosa*: *Pseudomonas aeruginosa*
Mm: *Myrcianthes myrsinoides*; *Mr*: *Myrcianthes rhopaloides*

means efficiency and better antioxidant activity. However, *M. myrsinoides* showed better bactericidal effect against Gram-negative bacteria, such as *Escherichia coli* (Tables 3.5 and 3.6).

Chromatograms of phenolic compounds may be used in fingerprinting to identify species and verify drug quality with certainty with the species that is being used. This represents one of the most important aspects of quality of natural products for complementary medicine and biobusiness (Figures 3.3 and 3.4).

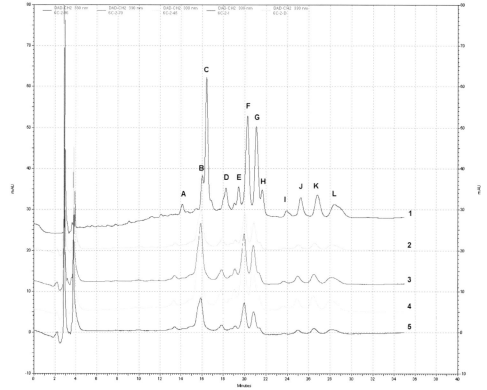

Figure 3.3: Identification of the retention times of the characteristic peaks of the chromatograms of the phenolic compounds purified with Amberlite XAD7HP of the extracts of *Myrcianthes myrsinoides* "lanche chiquito" (leaf) of Ayabaca by HPLC at 330 nm, ethanol 96% (1), ethanol 70% (3), ethanol 45% (4), infusion (5), and decoction (2).

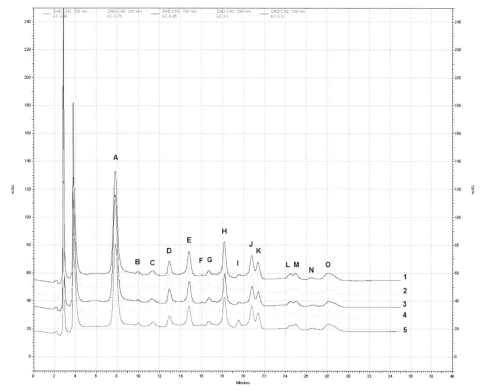

Figure 3.4: Identification of the retention times of the characteristic peaks of the chromatograms of the phenolic compounds purified with Amberlite XAD7HP of the extract *Myrcianthes rhopaloides* "lanche colorado" (leaf) of Ayabaca by HPLC at 330 nm, ethanol 96% (5), ethanol 70% (4), ethanol 45% (3), infusion (2), and decoction (1).

Phytochemistry of Medicinal Plants

The available electronic literature on the whole medicinal plants were collected using database searches, including Scopus, Google Scholar, Pubmed, web of Science, and Scifinder. The searches were limited to peer-reviewed English journals with the exception of books, and a few articles in foreign languages which were included.

Interestingly, no data was found about the phytochemistry and biological activity of *Cuphea cilliata*, *Brachyotum* spp., *Acaena ovalifolia*, and *Myrcianthes rhopaloides*.

Another example is *Vaccinium floribundum*, commonly known as "mortiño or ushpa", a wild shrub native to the Andes region of South America. Its berries are widely consumed in Ecuador and Peru as fresh fruit or processed products, such as juice and jam (Schreckinger et al. 2010b). In addition to the nutritional value, local communities use the extracts of this plant to treat various medical conditions, including diabetes and inflammation (Schreckinger et al. 2010b, a). This medicinal plant is rich in vitamins, polyphenolic, and anthocyanin compounds (Prencipe et al. 2014, Kumar et al. 2019).

However, there is only one study of the phytochemistry, describing tentative identification by LC-MS, as well as the identification of hydroxycinnamic acid, flavonoids, anthocyanidin (anthocyanin and proanthocyanidin), and cyclohexanecarboxylic acid using electrospray ionization tándem mass spectrosmetry (Figure 3.5) (Esquivel-Alvarado et al. 2019, Vasco et al. 2009). Fruits can also be used for wine preparation with antioxidant and antimicrobial properties due to the high content of phenolic compounds (Ortiz et al. 2013, Llivisaca et al. 2018). It has also been reported that fruits of *V. floribundum* Kunth showed a higher protective effect on human dermal fibroblasts compared to *Rubus glaucus* Benth (Alarcón-Barrera et al. 2018).

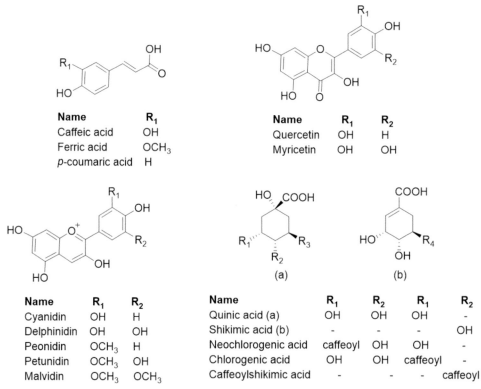

Figure 3.5: Chemical structures proposed by Vasco et al. (2009).

Bejaria resinosa is known as "pena de cerro" in Ecuador, and is a medicinal plant species used traditionally by the Saraguro ethnic group in Ecuador to treat nervous system, swollen wounds, and inflammations of genitals, liver, and cancer (Suárez et al. 2015). Ursonic acid was reported as the principal active component responsible for the *in vitro* cytotoxicity on tumor cells (Suárez et al. 2015). Unfortunately, a poor phytochemical and pharmacological investigation has been performed until now to identify other active components.

Myrcianthes myrsinoides (Myrtaceae) essential oil was analyzed for phytochemical and antibacterial bio-assays. Performing gas chromatography and gas chromatography-mass spectrometry, it was possible to determine 28 compounds, such as *p*-terpinen-4-ol, *o*-cymene, spathulenol, and caryophyllene oxide (Araujo et al. 2017). The essential oil was reported to exhibit important effects against *B. cereus*, *B. subtilis*, and *S. epidermidis* (Araujo et al. 2017).

Muehlenbeckia hastulata, commonly known as quilo mollaca and voqui, has been used for a long time in Peru, Argentina, and Chile as a diuretic, hypotensive, antihemorragic, sedative, and for reumatism treatment and burns. The chemical composition of the aerial part shows the presence of tannins, flavonoids, and anthraquinones (Figure 3.6) (Erazo et al. 2002, Mellado et al. 2012, 2013). The crude extract of quilo reported antioxidant activity on 2,2-diphenyl-1-pycrylhydrazil (DPPH) assay (Mellado et al. 2012). Interestingly, crude extract was evaluated *in vitro* for the biological activity against influenza virus proliferation in MDCK cells. On the other hand, three active compounds responsible for the biological activity, such as pheophorbide A, hepericin, and protohypericin were determined (Figures 3.7–3.14) (Yasuda et al. 2010).

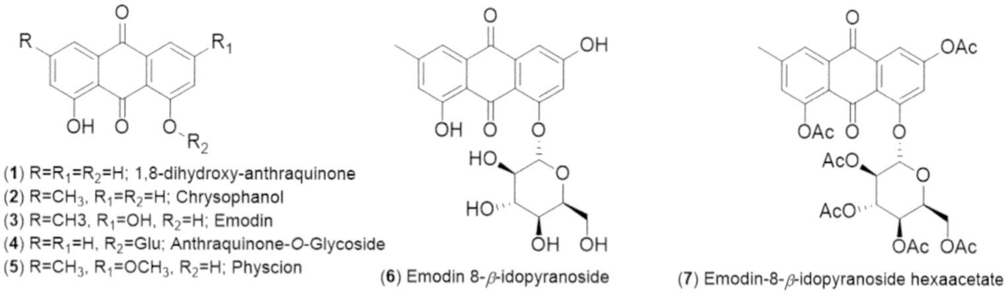

(1) R=R₁=R₂=H; 1,8-dihydroxy-anthraquinone
(2) R=CH₃, R₁=R₂=H; Chrysophanol
(3) R=CH3, R₁=OH, R₂=H; Emodin
(4) R=R₁=H, R₂=Glu; Anthraquinone-O-Glycoside
(5) R=CH₃, R₁=OCH₃, R₂=H; Physcion

(6) Emodin 8-β-idopyranoside

(7) Emodin-8-β-idopyranoside hexaacetate

Figure 3.6: Compounds isolated from *Muehlenbeckia hastulata* (J.E. Sm) Johnst (Mellado et al. 2013).

Figure 3.7: *Bejaria resinosa* "payana".

Figure 3.8: *Brachyotum angustifolium* "zarcilleja".

Figure 3.9: *Acaena ovalifolia* "pega-pega".

Figure 3.10: *Myrcianthes myrsinoides* "lanche chiquito".

Figure 3.11: *Myrcianthes rhopaloides* "lanche colorado".

Figure 3.12: *Muehlenbeckia hastulata* "chupicaure".

Figure 3.13: *Cuphea ciliata* "hierba del toro".

Figure 3.14: *Vaccinium floribundum* "ushpa".

Innovation

Transaction costs in Linking Traditional and Scientific Knowledge for Innovation

The wealth of native knowledge of wild plants of the moorlands and cloud forests of northern Peru may generate healthy rural societies that possess and conserve their knowledge, provided that conditions for the establishment of a territorial innovation system oriented to their conservation are created. It is the network formed by individuals, organizations, and companies for the creation, dissemination, and concerted use of knowledge for economic use in new products, processes, or organizational forms under the conditions of an institutional and policy that affect their behavior and performance (CEPAL 2004, Espinoza 2004, World Bank 2006, Rosas et al. 2014). An innovation system is comprised of agents of rural societies of the moorland and cloud forests that interact with agents of urban scientific and technological institutions.

Two agents that simplify building this innovation system can be identified. First are the agricultural producers who have already integrated themselves into an economy that does not treat them with equity, but have inherited natural and intellectual advantages, but may or may not know of the opportunities existing in urban institutions, or how to express their demands. Second are technologists and scientists who are also seeking to integrate into the economy based on the specialized knowledge they have developed; but who do not know the opportunities existing in rural societies, or to whom they can present their offers; but they also have need for resources and knowledge to create new innovations and research. Also, community organizations demand new technological and scientific knowledge, as well as offer natural resources and traditional knowledge about them. Technological and scientific institutions offer and generate specialized knowledge, but in turn need new resources and traditional knowledge to continue investigating. Both agents, one from the countryside and other from the city, have needs and offers to exchange. Both are needed to implement a mutually beneficial innovation system that helps all stakeholders involved.

The present situation of these two main agents needed for the construction of an innovation system in the moorland and cloud forests is that they have distinct problems that increase their costs in linking traditional knowledge and scientific innovation. Rural communities suffer from a lack of quality public services, such as good transportation options, education, and information. On the other hand, scientists, technologists, and project formulators are located in cities, both physically and culturally far away from the rural Andean societies, and have difficulty finding the proper channels to contact them and understand their needs. This hinders their ability to offer scientific or technological knowledge consistent with the needs of the rural community. Additionally, these spatial and social connectivity gaps mean that the transaction costs between academics or technologists and the peasant organizations possessing the natural resources and knowledge to which they seek to add value are significantly high. These investments include not only significant investment in travel to difficult to reach areas, but also include investment of a non-monetary nature accrued over time and with the intervention of local allies, which generates trust of communal organizations. Investments such as these need to be in place before the transaction of reciprocal interests as suppliers and demanders of goods and services can take place.

Rural Innovation Based on Biodiversity: Vegetable Resources and High-Quality Natural Product Prepared in situ at High Biodiversity Zones

Plant product discovery that explores the future possibilities of plants, based on historical and/or present evidence that has its source in traditional knowledge, does so through scientific research of chemical compounds, genetics, ecological functions, or principles of action contained in plants with potential utility (pharmaceutical, medical, cosmetic, perfumer, industrial gastronomy, etc.). This constitutes an area of opportunities for conservation, and economic, sustainable, and fair use of biodiversity. Within an appropriate legal context, it will benefit all parties, from businesses that risk capital investment to biological resources and knowledge providers who are usually poor towns or communities from mainly tropical or subtropical countries, such as Peru (Pastor and Sigüeñas 2008, Carhuapoma Yance 2011). However, if rural community farmers get involved at the beginning of the research by contributing what they know, they expect as a return a tangible benefit instead of a theoretical product from scientific knowledge, in terms of secondary metabolites from the studied species. Instead, they expect an operative alternative for use of those results, such as feasible practices and techniques for processing and transformation to give the species that they own an *in situ* added value.

Local Experts in the in situ Detection of Key Phytoconstituents to Select Promising Species

In order to integrate community experts in the research process of wild plants, the use of portable field equipment (prepared and tested by Mayar Ganoza and Fidel Torres) for *in situ* identification of highly important phytoconstituents (phenols, flavonoids, and tannins) in plants used in traditional medicine permits the orientation and optimization of selecting promising species with less uncertainty, reducing collection costs, identification of taxonomic details, stabilization, transport, and analysis.

Although the *in situ* and rapid detection of medically important species by measurement of three groups of metabolites provides information for making a decision, it does not mean that species that do not have significant quantities of these compounds should be considered useless, as they may contain other important metabolites. However, due to the high cost of field collection and phytochemical analysis, prioritization by rapid field detection is justified for the preliminary stages of wild plant prospection, which also increases the capacities of local experts in the knowledge of their plant diversity (Figures 3.15 and 3.16).

In the emerging natural products industry, very few businesses have implemented quality control products for raw materials in terms of agricultural and agroecological management, organic certification, post-harvest selection, and management. This is because most of what is used comes from intensive extraction of whole plants from their natural systems (Pastor and Sigüeñas 2008, Nolasco Cruz 2016). The innovation chain of natural products with possible links to biocommerce

Figure 3.15: Community-oriented botanical collection.

Figure 3.16: On-site test of the presence of bioactive substances.

find its major limitation in the initial step: sourcing of appropriate vegetable raw materials containing bioactive substances of interest (taxonomic identity, origin, physiological status, cleanliness, and stability). This limitation sets up an informal system with deficient quality from the beginning, and expands to the following steps of the value chain. Also, this situation does not promote the need of doing research or generation of scientific-technological knowledge to improve the quality of the natural products that are offered (Vila 2009).

This situation is explained by the gap between natural products companies located in the cities who demand vegetable raw materials and high-quality natural products, but do not know what the rural communities can offer them. Also, productive rural organizations offer their knowledge and species with the potential required; however, they are unaware of the existence of demand. The transaction costs to meet both rural farmer and companies are high in order to express their needs and reciprocal offers in order to create shared interest chains that increase their competitiveness.

Currently, there is an unsatisfied demand for high quality natural products with added value in Peru. The complementary medicine system of the Peruvian Ministry of Health is the principal consumer, as a consequence of the increasing development of the complementary systems nationwide from 5 to 36 regional centers (oral communication from Dr Luis Fernandez, Head of Complementary Medicine Center, Trujillo-Peru). On the other hand, products offered by the natural herbals products industry do not provide guarantee of their origin, identity, and quality of the vegetable raw material used. Therefore, there is a great opportunity for the sustainable use of biodiversity based on innovation.

Quality of Plant or Natural Products from Rural Communities

Quality is a requirement for the supply of therapeutic plant and natural products. This means that rural communal organizations possessing traditional knowledge on how to use medicinal plants within their territories need to integrate this specialized knowledge already available for the improvement of these goods, or they must generate knowledge to add value to those already available. Therefore, communal organizations who participate in wild plant research and prospecting should appropriate the scientific results by changing their current role of suppliers of raw materials and as informants of the cultural use of wild plants to suppliers of raw drugs, essential oils, fluid extracts, or high quality natural products based on toxic-pharmacological and phytochemical analyses, concentration of phenolic compounds, analysis of antioxidant ability, studies of antibacterial activity, and identification of the most important bioactive substance (flavonoids, tannins).

The quality of an improved good is determined as the safety of the reliable characteristics it possesses at the moment of its use or consumption. In the case of phytopreparations, its quality depends on the quality of the vegetable resource or raw drug (Figures 3.17–3.19).

Agroecological Management Strategy of Promising Species Selected by Ethnobotanical and Phytochemical Studies by Organizations that Contribute with Traditional Knowledge. Sustainability from the Needs of EsSalud

Ethnobotanical field studies result in the identification of species of interest to the communities because of their value, significance, and use. The identified species are then subjected to phytochemical analysis to measure their profile of secondary metabolites, level of toxicity, content of phenolic compounds, antioxidant activity, and antibacterial activity. These results determine whether or not the species is promising for innovation. However, in the approach of this type of research, the possibilities of innovation of a species are conditioned by the guarantee of its conservation. That is to say, the knowledge of the use and phytochemical richness of a species is not enough to start commercialization of a product if its conservation is not guaranteed. This means that it is not the extraction of plant resources from their environment that is the consequence of their use for innovation; but its *ex situ* propagation or controlled population management depending on its abundance, which does not degrade the ecosystem to which it belongs.

From a grouping of 60 moorland and cloud forest species studied based on traditional knowledge and phytochemical analyses, seven showed promise to be considered for additional innovation: "payana" (*Bejaria resinosa*), "bull grass" (*Cuphea cilliata*), "paste-paste" (*Acaena ovalifolia*), "ushpa" (*Vaccinium floribundum*), "chupicaure" (*Muehlenbeckia hastulata*), "zarcilleja chiquita" (*Brachyotum angustifolium*), and "lanche" (*Myrcianthes myrsinoides*). For these species, controlled propagation is the next phase to increase the availability of raw materials that will be given added value. For the purposes of controlled management, there are three alternatives: (1) Its cultivation in plots or agroecological gardens through its sexual seeds or vegetative structures, such as "paste-paste", "bull grass", and "chupicaure"; (2) Promotion of the natural expansion of the population of the species if it is abundant, through measures of social control of the community aimed at preventing the intervention of these ecosystems by the agricultural or livestock expansion of families of the community; in this way, controlled, planned, and focused extractions can be carried out, as in the cases of "payana" and "ushpa"; (3) In the case that a species has difficult to germinate seeds and vegetative propagation is not viable, an *in vitro* propagation system is used to produce a clonal population of the species, such as the "lanche".

The north coast of Peru, being an arid region that depends completely on the water sourced from moorland and cloud forests, promotes reforestation programs aimed at maintaining the water capacity of the basins. Three appropriate species for this purpose are "lanche" which is a tree species, and "ushpa" and "payana" as shrub species, in which in all three cases play an important role in the ground cover for the collection, filtration, retention, and distribution of rainfall water (2,500 mm annual average); becoming a profitable way of conserving biodiversity and the hydrological cycle of the basins. This is possible due to collective agreements that can only be achieved by community organizations with the capacity to make the decision to convert the wealth of their wild plants into an economic strategy for the conservation of environmental services and biodiversity.

Institutionalization of Research and Innovation of Biodiversity from Peasant Organizations as a Local Development Policy

The sustainability of science and innovation based on the diversity of wild plants known and used by peasant societies around the moorland and cloud forests has possibilities of sustainability to the

Figure 3.17: Phytopreparations module in Totora (2,600 masl) in the surroundings of the moorland and cloud forests.

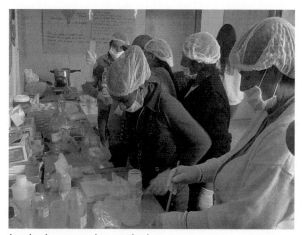

Figure 3.18: Phytopreparations by the community organization.

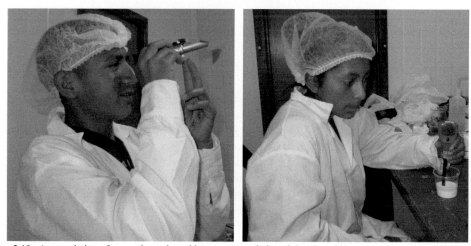

Figure 3.19: Appropriation of research conducted by young people in training (a and b).

extent that in a collective effort of the economic and social stakeholders from these ecosystems, they are able to build a territorial system of innovation. It is fostered as a local development policy, based on the management of their particular interests integrated by common objectives, which allows them to significantly increase their efficiency, management efficacy, and governance of a territorial system of innovation, framed as a local development policy.

The new rules of the game or institutional systems needed for the establishment of an innovation system supported by the biodiversity integrate research, production, and commercialization chains. It requires a territorial emphasis that recognizes the farming rural organizations at the origin of the chains in order to guarantee the results obtained. On one hand, the scientific research chain demands systematized specialized native knowledge as sources of scientific hypotheses, as well as high quality vegetable resources. On the other hand, the innovation chain of natural products based on endemic wild plants requires formal communal organizations that offer raw drugs and high-quality natural products with added value.

The encounter of suppliers and consumers of goods, services, and knowledge regarding wild plants for the generation of reciprocal benefits should be oriented towards the conservation of natural resources, as well as a need for an institutionality or rules that establish incentives and equal access to resources and services for an advantageous economical linkage that is possible by the establishment of a territorial innovation system (Roseboom et al. 2006, Glave and Jaramillo 2007). Institutional change can only be created with difficulty by those who exercise power to maintain inequality and exclusion. That is, a structural change from the top down is not possible, but it is from the bottom up. This implies organizations that manage to scale their governance capacities. Governance innovation using the approach of territorial innovation systems' proposals can be imposed by force. This political power depends on the organizational capacity of governance through collective management units, such as dialogue platforms (Rendón Schneir 2010).

In the summer of 2017, one of the most important expressions of climate change in northern Peru took place as extreme precipitation that resulted in a flood disaster. After this, the Peruvian state initiated a process of regional reconstruction, which includes the priority project "Integral Treatment for the Reduction of Vulnerability to Flooding and Water Scarcity in the Chira Piura Watershed" prepared by the Resources Council of the Chira-Piura Basin of the National Water Authority. This project involves, as a principal component, the conservation and protection of moorland and cloud forests that regulate the hydric cycle of where watersheds receive the main volume of precipitation annually. This represents an opportunity for an advantageous and sustainable economic articulation for communal organizations of the moorlands and cloud forests.

With its new contingent, the communal organizations of the moorlands are expanding and diversifying with new organizational arrangements (Cooperativa Territorial de los Páramos) as an institutional innovation to advantageously manage their biodiversity conservation projects and their wild plants.

The Territorial Multiservice Co-op of the Moorlands constitutes an advanced platform of business nature with economic interests in the development of the moorland territory. It has political influence to promote its conservation interests and principles in the different local governmental offices that are associated with its organizations. Its further development towards a new institutional innovation is creating a Territorial Commonwealth of the Moorlands. This commonwealth will be the strategy for the establishment of a territorial innovative system of the moorlands and cloud forests, based on the richness of vegetable diversity and native knowledge, and such innovations are sustained by the conservation of these ecosystems. The innovation process must inevitably be supported by the institutional structure of the desert plains that absolutely depend on the water supply of the moorlands and cloud forests.

Both science and technology build new social realities, but also destroy or erode others by making contact with them. This has happened with the native or traditional knowledge of the wild plants of the Andean societies of northern Peru, which has been severely weakened and eroded by schools and universities that devalue its significance. This discrimination is effectively implemented through public health and education programs, which import food to combat malnutrition using funds generated by the export of natural resources. Biodiversity is not seen as something that can feed the poor. This fallacious formula is based on the fact that wild and cultivated plant diversity is dismissed in dietary surveys, laboratory analyses, by public health agencies, and those who determine nutritional status, and from any other platform from which decisions and policy formulation are taken (Johns 2004).

In countries such as Peru, converting the wealth of knowledge of wild plants into a collective wealth of rural agrarian societies to which healers belong, means the construction of a new social reality, based on a new paradigm of knowledge sharing based on intercultural communication. It is given that innovation systems or networks are not just meetings of stakeholders, but the spaces of interaction for the exchange of knowledge that develop learning capacities by the principle that knowledge is the central engine of economic growth (Kuramoto 2007, Pomareda 2015).

Conclusions

The economically sustainable use of wild plants in the Páramos and cloud forests depends on the establishment of a territorial innovation system, which can integrate in strategic alliances involving rural agricultural organizations, scientific institutions, government institutions and private industries to systematize traditional collective knowledge, as well as plant diversity in an intercultural approach, as a first step in the prospective scientific research chain, orienting competition for innovations of complementary medicine system such as public health area and natural product industry. Phytochemical analyses have not only corroborated to, and expanded, the knowledge about selected plants, but also confirmed the efficiency and efficacy of traditional practices in the extraction of active ingredients of medicinal interest. They have also verified the influence of environmental parameters, e.g., on the synthesis of plant phenolic compounds, or the antioxidant capacity of species, and by means of chromatograms of the phenolic compounds, determined fingerprints that can verify the identity of species as, as important means for providing indicators for the quality of natural products in complementary medicine and bio-commerce.

References

Abramovay, R. 2013. Más allá de la economía verde: Temas Grupo Editorial.
Alarcón-Barrera, K.S., Armijos-Montesinos, D.S., García-Tenesaca, M., Iturralde, G., Jaramilo-Vivanco, T., Granda-Albuja, M.G., Giampierie, F., Alvarez-Suarez, J.M. 2018. Wild Andean blackberry (*Rubus glaucus* Benth) and Andean blueberry (*Vaccinium floribundum* Kunth) from the Highlands of Ecuador: Nutritional composition and protective effect on human dermal fibroblasts against cytotoxic oxidative damage. J Berry Res. 8(3): 223–236.
Amanzo, J., Acosta, R., Aguilar, C., Eckhardt, K., Baldeon, S., Pequeño, T. 2003. Evaluacion biologica rapida del Santuario Nacional Tabaconas - Namballe y Zonas Aledañas: Perú.
ANA, Autoridad Nacional del Agua. 2017. Tratamiento integral para la reducción de la vulnerabilidad frente a inundaciones y escasez hídrica en la cuenca Chira Piura: propuesta. Consejo de Recursos Hídricos de Cuenca Chira Piura.
Araujo, L., Rondón, M., Morillo, A., Páez, E., Rojas-Fermín, L. 2017. Antimicrobial activity of the essential oil of *Myrcianthes myrcinoides* (Kunth) Grifo (Myrtaceae) collected in the Venezuelan Andes. PhOL 201: 200–204.
Barrett, L.F. 2018. La vida secreta del cerebro: Como se construyen las emociones. Ediciones Paidós, Barcelona España, 552 p.
Bussmann, R.W., Sharon, D. 2015. Medicinal Plants of the Andes and the Amazon: The Magic and Medicinal Flora of Northern Peru. Graficart Srl. Perú.
Carhuapoma Yance, M. 2011. Plantas aromáticas nativas del Perú: biocomercio de fragancias, sabores y fitocosméticos. Consejo Nacional de Ciencia, Tecnología e Innovación Tecnológica, Lima Perú, 238 p.
Castilla del Pino, C. 2002. Teoría de los sentimientos: Tusquets Editores.
CEPAL. 2004. Políticas para promover la innovación y el desarrollo tecnológico. *In*: Desarrollo Productivo en Economías Abiertas.
Erazo, S., Muñoz, O., García, R., Lemus, I., Backhouse, N., Negrete, R., San Feliciano, A., Delporte, C. 2002. Constituents and biological activities from *Muehlenbeckia hastulata*. Z Naturforsch C 57(9-10): 801–4.

Espinoza, H. 2004. ¿Inversión en Investigación y Desarrollo para generar Competitividad?: Un Análisis de sus Efectos y Determinantes a Nivel de Empresas Manufactureras – Perú 1998: Informe Final. Centro de Estudios para el Desarrollo y la Participación (CEDEP).

Esquivel-Alvarado, D., Muñoz-Arrieta, R., Alfaro-Viquez, E., Madrigal-Carballo, S., Krueger, C.G., Reed, J.D. 2019. Composition of Anthocyanins and Proanthocyanidins in Three Tropical *Vaccinium* Species from Costa Rica. J Agric Food Chem. (In Press 2019).

Glave, M., Jaramillo, M. 2007. Perú: Instituciones y Desarrollo. Avances y agenda de Investigación (301–349). En: Investigación, políticas y Desarrollo en el Perú. : Lima; GRADE.

Gomez-Peralta, D., Oberbauer, S.F., McClain, M.E., Philippi, T.E. 2008. Rainfall and cloud-water interception in tropical montane forests in the eastern Andes of Central Peru. Forest Ecol Manag. 255(3): 1315–1325.

Granda, L., Rosero, M.G., Rosero, A. 2015. Plantas medicinales de la región Andina tropical. Quinoa (Chenopodium quinoa Willd.) y coca (*Erythroxylum* sp.), tesoros milenarios para tratamiento medicinal (72–92). En M. Horák (Ed). Etnobotánica y Fitoterapia en América. Brno, República Checa: Universidad de Mendel en Brno.

Hersch-Martínez, P. 2002. La doble subordinación de la etnobotánica latinoamericana en el descubrimiento y desarrollo de medicamentos: algunas perspectivas. Etnobiología 2(1): 103–119.

Hofstede, R., Calles, J., López, V., Polanco, R., Torres, F., Ulloa, J., Vásquez, A., Cerra, M. 2014. Los Páramos Andinos. ¿Qué sabemos? Estado del conocimiento sobre el impacto del cambio climático en el ecosistema páramo: UICN, Quito. Ecuador.

Horák, M. 2015. Etnobotánica y Fitoterapia en América. Brno, República Checa: Universidad de Mendel en Brno.

Johns, T. 2004. Especies subutilizadas y nuevos retos para la salud global. *LEISA* 20(1):9-10.

Kumar, B., Vizuete, K.S., Sharma, V., Debut, A., Cumbal, L. 2019. Ecofriendly synthesis of monodispersed silver nanoparticles using Andean Mortiño berry as reductant and its photocatalytic activity. *Vacuum* 160: 272–278.

Kuramoto, J. 2007. Sistemas de Innovación Tecnológica (105–133). *In*: Investigación, políticas y Desarrollo en el Perú. : Lima, GRADE.

Lindsay, A. 2019. Investing upstream: Watershed protection in Piura, Peru (9–27). Vol. 96: *In*: Environmental Science & Policy.

Llerena, C., Cruz-Burga, Z., Durt, É., Marcelo-Peña, J., Martínez, K., Ocaña, J. 2010. Gestión ambiental de un ecosistema frágil. Los bosques nublados de San Ignacio, Cajamarca, cuenca del río Chinchipe: Lima, Perú: Soluciones Prácticas.

Llivisaca, S., Manzano, P., Ruales, J., Flores, J., Mendoza, J., Peralta, E., Cevallos-Cevallos, J.M. 2018. Chemical, antimicrobial, and molecular characterization of mortiño (*Vaccinium floribundum* Kunth) fruits and leaves. Food Sci Nutr. 6(4): 934–942.

Mellado, M., Madrid, A., Jara, C., Espinoza, L. 2012. Antioxidant effects of *Muehlenbeckia hastulata* J. (Polygonaceae) extracts. J. Chil. Chem. Soc. 57(3): 1301–1304.

Mellado, M., Madrid, A., Peña-Cortes, H., Lopez, R., Jara, C., Espinoza, L. 2013. Antioxidant activity of anthraquinones isolated from leaves of Muehlenbeckia hastulata (J.E. SM.) JOHNST. (Polygonaceae). J. Chil. Chem. Soc. 58(2): 1767–1770.

More, A., Viñas, P., De Bievre, B., Acosta, L., Ochoa, B. Establecimiento de un sistema de monitoreo hidrológico de páramo andino como base para la determinación de medidas de adaptación al cambio climático. CONDESAN, NCI. 2013]. Available from https://www.slideshare.net/InfoAndina/establecimiento-de-un-sistema-de-monitoreo-hidrolgico-del-pramo-andino-como-base-para-la-determinacin-de-medidas-de-adaptacin-al-cambio-climtico.

Monroe Morante, J., Arenas Barchi, F. 2003. ¿Somos Iguales? Un aporte para el diálogo sobre la identidad cultural en las escuelas de la Sierra del Perú. Coordinadora de Ciencia y Tecnología en los Andes. Lima, Perú.

Nolasco Cruz, E.E. 2016. Tendencias actuales de las plantas medicinales producidas en el Perú. Tesis Ingeniero Industrial. Facultad de Ciencias Agropecuarias, Facultad de Ciencias Agropecuarias, Universidad Nacional de Trujillo. Perú.

Ortiz, J., Marin-Arroyo, M.R., Noriega-Dominguez, M.J., Navarro, M., Arozarena. 2013. Color, phenolics, and antioxidant activity of blackberry (*Rubus glaucus* Benth.), blueberry (*Vaccinium floribundum* Kunth.), and apple wines from Ecuador. J. Food Sci. 78(7): C985–93.

Pardo de Santayana, M., Gómez Pellón, E. 2003. Ethnobotany: traditional management of plants and cultural heritage. Anales Jard Bot. Madrid 60(1): 171–182.

Pastor, S., Sigüeñas, M. 2008. Bioprospección en el Perú. Edición: Isabel Lapeña – Sociedad Peruana de Derecho Ambiental. Lima – Perú.

Peña, G. 2015. Composición y diversidad arbórea en un área del bosque Chinchiquilla, San Ignacio – Cajamarca, Perú. Arnaldoa 22(1): 139–154.

Perú. Ministerio del Ambiente. 2015. Mapa nacional de cobertura vegetal: memoria descriptiva: Ministerio del Ambiente. Dirección General de Evaluación, Valoración y Financiamiento del Patrimonio Natural - Lima: MINAM.

Pomareda, C. 2015. La Agricultura y la Economía Rural en el Perú. 200 ed, Agro Enfoque.

Prencipe, F.P., Bruni, R., Guerrini, A., Rossi, D., Benvenuti, S., Pellati, F. 2014. Metabolite profiling of polyphenols in *Vaccinium berries* and determination of their chemopreventive properties. J. Pharm. Biomed. Anal. 89: 257–267.

Recharte, J., Torres, F. 2015. Donde la Amazonia contempla al Pacífico (199-209). En: LA AMAZONIA. Sílabas del agua, el hombre y la naturaleza: Banco Crédito del Perú. Lima, Perú.

Rendón Schneir, E. 2010. La Gestión Pública de la Innovación Agraria en el Perú: Antecedentes y Perspectivas. Escuala de Postgrado de la UPC, cuadernos de Investigacion EPG 11.

Rosas, A., Tostes, M., Torres, F. 2014. Rol de los Fondos Competitivos en la Gestión del Sistema Nacional de Innovación: el caso de INCAGRO en el norte del Perú 2005–2010. En M. Tostes (Coordinadora), Experiencias de Innovación en el Agro del Norte Peruano: innovación, cadenas productivas y asociatividad. Lima, Perú: EXCEDESA.

Roseboom, J., McMahon, M., Akanayaque, I. 2006. La innovación institucional en los sistemas de investigación y extensión agrícolas en América Latina y el Caribe: Banco Mundial. LEDEL SAC; Perú.

Sagástegui, A., Sánchez, I., Zapata, M., Dillon, M. 2003. Diversidad Florística del Norte del Perú; tomo II: Fondo Editorial de la Universidad Privada Antenor Orrego; Trujillo, Perú.

Sagástegui, A., Dillon, M.O., Sánchez, I., Leiva, S., Lezama, P. 1999. Diversidad Florística del Norte de Perú: Graficart, Trujillo.

Samaja, J. 2000. Aportes de la metodología a la reflexión epistemológica. En: Díaz, Esther. 2000. La posciencia: el conocimiento científico en las postrimerías de la modernidad: 1° ed. Buenos Aires, Editorial Biblos.

Sánchez, I. 2012. La diversidad biológica en Cajamarca: visión étnico-cultural y potencialidades: Gobierno Regional de Cajamarca.

Schreckinger, M.E., Lotton, J., Lila, M.A., de Mejia, E.G. 2010a. Berries from South America: a comprehensive review on chemistry, health potential, and commercialization. J. Med. Food 13(2): 233–46.

Schreckinger, M.E., Wang, J., Yousef, G., Lila, M.A., Gonzalez de Mejia, E. 2010b. Antioxidant capacity and *in vitro* inhibition of adipogenesis and inflammation by phenolic extracts of *Vaccinium floribundum* and *Aristotelia chilensis*. J. Agric. Food Chem. 58(16): 8966–76.

Suárez, A.I., Armijos, C., Andrade, J.M., Quisatagsi, E.V., Cuenca, M., Cuenca-Camacho, S., Bailon-Moscoso, N. 2015. The cytotoxic principle of Bejaria resinosa from Ecuador. JPP 4(3): 268–272.

Torres Guevara, F. 2006. Escenario de Riesgo para el Agua y la Biodiversidad: Pretensión de minería metálica en las cuencas del norte del Perú (Piura): Coordinadora Rural, CEPESER, PIDECAFE, CEPICAFE, Junta de Usuarios del Distrito de Riego de San Lorenzo, Proyecto BINACIONAL CATAMAYO-CHIRA.

Vasco, C., Riihinen, K., Ruales, J., Kamal-Eldin, A. 2009. Chemical composition and phenolic compound profile of mortino (*Vaccinium floribundum* Kunth). J. Agric. Food Chem. 57(18): 8274–81.

Vila, G. 2009. Análisis del uso de plantas medicinales en mercados de abastos del distrito de Ventanilla-Callao, 2007 Facultad de Farmacia y Bioquímica, Universidad Nacional Mayor de San Marcos, Lima-Perú.

Villar López, M., Ballinas Sueldo, Y., Soto Franco, J.N., Medina Tejada, N. 2016. Conocimiento, aceptación y uso de la medicina Tradicional, alternativa y/o complementaria por Médicos del seguro social de salud. RPMI. 1(1): 13–18.

Walter, D. 2017. Algunos aportes a la etnobotánica en la Cordillera Blanca (Sierra de Ancash). Indiana 341: 149–176.

Weigend, M., Dostert, N., Rodriguez-Rodriguez, E.F. 2006. Bosques relictos de los Andes peruanos. *In*: Centrales, M. Moraes, R., Ollgaard, B., Kvist, L.P., Borchsenius, F., Balslev, H. (eds.). Botanica Economica de los Andes: Universidad Mayor de San Andres, La Paz.

World Bank. 2006. Enhancing Agricultural Innovation: How to Go Beyond the Strengthening of Research Systems: World Bank Publications.

Yasuda, T., Yamaki, M., Iimura, A., Shimotai, Y., Shimizu, K., Noshita, T., Funayama, S. 2010. Anti-influenza virus principles from *Muehlenbeckia hastulata*. J. Nat. Med. 64(2): 206–11.

Young, B., Young, K., Josee, C. 2012. Vulnerabilidad de los ecosistemas de los andes tropicales al cambio climático (195–208). En: Herzog, S., Martinez, R., Jorgensen, P., Tiesen, H. (eds.). Cambio climático y biodiversidad en los Andes Tropicales: Instituto Interamericano para la investigación del cambio global.

4

Ethnic Uses of Plant species Among Magar People in Nepal

Shanta Budha-Magar

Introduction

Ethnic communities depend on plant and plant products for their primary healthcare and to fulfill their daily basic needs (Coburn 1984, Ghimire et al. 2000). They are vastly known for the use of plant resources found around them (Pásková 2017). Furthermore, the use of such medicinal plants in primary healthcare is growing in urban areas too. There has been increasing plant-based medication in Europe and America as well (Eloff 1998). Traditional knowledge owned by indigenous people linked with market value chain of the resources has a major impact on the use of medicinal plants and is effective against different ailments (Salim et al. 2019). Today, thousands of species used by ethnic groups have been confirmed with potential medicinal properties in pharmaceutical and drug industries. Many drugs have been successfully discovered based on indigenous knowledge (Fabricant and Farnsworth 2001, Brijesh et al. 2009, Taye et al. 2011). Thus, the traditionally used medicinal plants by ethnic groups have been recognized as complementary in scientific and pharmaceutical studies (Taylor et al. 2001, Eisenberg et al. 2011, Sagar 2014).

However, the traditional knowledge of plant use has been decreasing due to the reluctance of the young generation for the use of locally available resources and the unwillingness of elders to share their knowledge (Reyes-García et al. 2014, Sujarwo et al. 2014, Ianni et al. 2015). So, the documentation of indigenous knowledge on how local plant resources are utilized by the ethnic groups or communities is vital to save the indigenous knowledge of any country (Rao 2006, Luczaj et al. 2012). In this regard, an ethnobotanical study, which is the understanding of knowledge system by using both the anthropological and botanical approaches, is used in documentation of indigenous knowledge (Ford 1978, Davis 1995). Not only this, the plants consumed by local people or groups of people in a particular region and cultural context are necessary to be determined (Martin 1995). Now, there has been a snowballing interest to improve the traditional compilation-style of ethnobotanical studies (Hoffman and Gallaher 2007). One of the key issues to these studies is the relative importance of plant taxa to different human groups by elaborating indices of use values (Tardio and Santayana 2008). The consensus values are important to show the relationship between the use of the plant resources and knowledge of indigenous people about it (Singh et al. 2012).

Nepal, a multicultural country endowed with about 60 ethnic groups within two broad categories distributed along contour lines, paralleling preferred climates, and crops having specific cultural norms

Central Department of Botany, Tribhuvan University, Kirtipur, Kathmandu; shantabmgr@gmail.com and values (Manandhar 2002), has diverse cultural contexts regarding natural resources. The ethnic groups have their own way to understand and interpret nature. Various studies related to dependency of ethnic groups on plant resources in different communities have been previously done in Nepal (Manandhar 2002, Rajbhandari 2001, Shrestha et al. 2004, Baral and Kurmi 2006, Luitel et al. 2014). However, studies on the Magar community about their relationship with plant and plant products are rare, although they completely depend on these resources. There are some studies regarding ethnic uses of plant species in Magar community (Sapkota 2008, 2010, Ale et al. 2009, Acharya 2012, Thapa 2012, Singh and Hamal 2013, Malla et al. 2015, Singh et al. 2018).

Ethnobotany of Magar People in Nepal

Magar, one of the main ethnic groups in Nepal, entered the country around 1100 B.C. They live mainly in Jajarkot, Rukum, Rolpa, Myagdi, Baglung, Pyunthan, Palpa, Parbat, Gorkha, and Tanahu in western and central Nepal, and Sindhuli and Udaypur toward eastern Nepal. They have subsequently migrated to most parts of the country. This tribe is genetically isolated because they marry among their community. They have Mongol features, medium build, white in complexion, oval or round face, black straight hair, and razor cut eyes. This generally describes the physique of Magar, and by nature, they are gentle, honest, brave, charming, and happy people. They are very jovial people who love to sing and dance. They have numerous kinds of dances, as well as tribal games that they frequently play. The Magar has their own language which is rooted in the Tibeto-Burman family and the script called Akkha Lipi. Magar language is old and native spoken, used by Magar community (Manandhar 2002).

Based on the language, customs, and geographical distribution, the Magars are divided into Barha Magarat Magar, Atha Magarat Magar, High mountain Magars Chhantyal, and other Magars (Manandhar 2002). However, these four groups do not differ in their original traditions and other social affairs. Within the various Magar communities, there are also different clans: there are more than 700 sub thars (family name), such as Budha, Roka, Pun, Jhankri, Thapa, Rana, Aale, Benglasi, Gharti, Gurungachan, Thumsing, etc. The total population of Magar in the country is 1,887,733, among which the number of Magar-Kham speaking people is 27,113 (male—12,934 and female—14,179) (CBS 2011).

Magar has a typical dress, where male Magar wears a short tunic, shirt, and vest. Female Magar wears a *cholo* (blouse with long sleeves), a short wrap instead of a full-length sari (skirt) and a *patuka* (sash wrapped around the waist) as a belt. The women are fond of wearing a lot of ornaments, especially during festivals. They live in small egg-shaped houses thatched with grass in western Nepal (Manandhar 2002). They also build two-story houses roofed with grass thatching, wooden planks, or slates. They use locally available materials for house building. They paint their houses white or red or grey depending on the availability of natural paints in their surrounding area (Manandhar 2002).

Thus, Magar is one of the ethnic communities who are scattered all over the rugged terrain of the country (CBS 2011), and totally depends on the natural resources for their daily basic needs, and is rich in ethnomedicinal knowledge (Manandhar 2002, Singh et al. 2012, Acharya 2012). According to Singh et al. 2018, Magars are efficient in the use of various medicinal plants for their healthcare. They have acquired this knowledge from their long-term experiences and from their ancestors. He reported that Magars of Palpa use 70 ethnomedicinal plant species in different ways. According to Malla et al. (2015), Magar and Majhi are rich in ethnomedicinal knowledge. A total of 132 species were reported to be used by Magar and Majhi in Palpa district. Similarly, Acharya (2012) revealed 161 species in the treatment of various ailments, ranging from gastro-intestinal to headaches and fevers, respiratory tract related problems to dermatological problems, snake bites to ophthalmic and cuts and wounds. Thapa (2012) has reported 75 species to treat 39 different ailments by Magars in Parbat district. Out of a total of 221 plant species in and around Seti Hydropower Project, 43% of the plant species were ethnobotanically important and used by Magar ethnic community in food, medicine, and timber (Uprety et al. 2011). In east Nepal, they use medicinal plant species to treat particular diseases/disorders (Oli et al. 2005).

However, the Magar-Kham community, one of the major Magar communities in Western Nepal with a unique lifestyle, has not been explored concerning their knowledge on plant and plant products, except about their culture for instance about the language, the shamanism, and social status by few international scholars (Molnar 1981, 1982a, b, Watters 2002, Noonan 2003, Hatlebakk 2009, De Sales 2010, 2017). Hence, this paper aims to fulfill this gap. This paper deals with an assessment of the use pattern of plant species and documentation of the ethnic uses of plant species among the Magar-Kham.

Ethnic Uses of Plant species Among Kham Magar in Thabang: A Case Study

Study Site

The ethnobotanical data collection was conducted in Thabang, which lies in Thabang Rural Municipality to the northern part of Rolpa district in Province 5, Nepal. Thabang village was the main center of the civil war of the last ten years of Maoists' revolution and was known as the capital of Maoists (Gellner 2007). Geographically, Rolpa lies between 28.8^0 to 28.38^0 N latitudes and 82.10^0 to 83.90^0 E longitudes, with an elevation range of 701-3639 masl. (DFO 2018). It covers an area of 187,150 sq. meters, which is surrounded by five districts- Baglung and Pyuthan to the East, Salyan to the West, Rukum to the North, and Dang to the South (Figure 4.1).

Socio-demographic Structure

There are several ethnic groups in Rolpa, but of the total population 224,506 ((male- 103,100 and female- 121406)) of the district, Magar (43.78%) are the dominant inhabitants, followed by Chhetris,

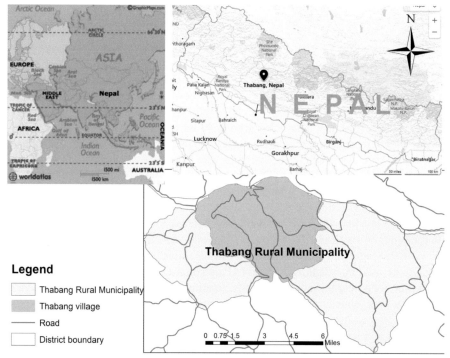

Figure 4.1: Map of Study Area.

Dalits, Newars, Gurungs, and others (CBS 2011). Thabang RM is inhabited by ethnic groups Kham Magar, Kami, Damai, Gurung, and Chhetri (CBS 2011). The total population of the study site of Thabang village is 4,841, including 2,251 males and 2,391 females, with a total of 937 households (DPR 2015), which are populated by only Kham Magar, Gurungs, Dalit, and Thakuri. In Thabang, the majority of people- around 2,414 (out of 5,922 in the district), speak Magar Kham language dialect, whereas Dalits (Bishwokarma and Nepali) speak Khas Nepali (locally *Khasanta*), an Indo-Aryan language.

Regarding Magar, there are five subtribes who entered Nepal from different directions around 2300 BC (Thapa 2014). Of the five subtribes, Kham Magar living in Athâra magarât region (Manandhar 2002) use Kham (Tibeto-Burman) dialect. They have their own lifestyle and typical culture (TAAN 2013). They have their own specific festival calendars, different ways of covering haystacks (by means of goatskin or with straw roof), and the different itineraries they follow with their flocks (De Sales 2000). Male Kham Magar people wear a short tunic (locally *sur*), shirt (locally *Chubandi*), vest, and gada (locally *khadi*). Female Kham Magar people wear *cholo* (blouse with long sleeves; locally *khyo*), *guniyo* (locally *bhey*), a patuka (sash wrapped around the waist; locally *waanfo*), and gada (locally *sur*). They believe in nature rather than in gods and goddesses. They don't have specific deities. They worship stones, trees, water, and directions. Some of them are adopting Hindu religion and a few are Christians.

Agriculture (animal husbandry and crop production) is the main livelihood strategy adopted by Kham Magar. More than 80% of the total households are involved directly or indirectly in traditional agriculture and animal husbandry (CBS 2011). Local farmers rear cattle, goat, sheep, pig, and poultry for their subsistence. Besides these, non-timber forest products (NTFPs), including medicinal and aromatic plants are the next source of income for local people. About 14% people depend on resources from the forest (DFO 2018). The trend of young generation going to foreign countries to earn is also a common activity.

Data Collection and Analysis

The field study was conducted in two visits in September 2014 and April 2015. The procedure of requiring Prior Informed Consent (PIC) as established by CBD was followed before recording the indigenous knowledge of people and also before accessing the natural resources (plant vegetation) around them. Their rights on their knowledge and natural resources were not violated. Basic ethnobotanical information of useful plants regarding their local name, part(s) use, purpose, and mode of use were collected through organizing focus group (*n* = 5) discussions in Phuntibang village. There were 20–30 people, including traditional faith healers, village heads, and other key informants from different sex and age. Secondly, for personal interviews, a total of 50 (30 male and 20 female, including five *Jhankris* (four male and one female), the traditional healers) forest dwellers, women, herders who were involved in collection of useful plants, were interviewed, applying semi-structured questionnaires. Interviews were conducted in the places where respondents felt the most comfortable. The questionnaire covered plant taxa, their distribution, major use categories, part(s) used, mode of collection, trade status, and strategy for mainstay. Thirdly, the local people were asked to select knowledgeable villagers to participate in the field. Based on local suggestions, two local people participated in the field, during which time different plant specimens cited as useful by local people were collected, and detailed ethnobotanical information was recorded.

Various kinds of recent literature reviews on the flora of Nepal and ethnobotany (e.g., Rajbhandari 2001, Lama et al. 2001, Manandhar 2002, Baral and Kurmi 2006, Bhattarai et al. 2006, Dutta 2007, DPR 2007, Ghimire et al. 2008, Rajbhandary 2013) were consulted to compare plant uses.

The number of species in different categories, such as species composition, and use categories, were analyzed in Microsoft Excel 2010, which includes the graphical presentation of useful species based on their use categories, parts used, and the number of use categories for each plant species. Here, the useful plant species were analyzed under two categories: (i) useful only in the study area, and (ii) useful in the study area and other parts of the country.

Species Composition

A total of 175 plant species, belonging to 73 families and 144 genera were identified as ethnobotanically useful among Kham Magar. Of the total, 25 species cited were reported ethnobotanically new to Nepal (Table 4.1), and the other 150 species were found useful both in the study area and elsewhere, in other parts of the country. Out of 25 ethnobotanically new species in the list of useful flora of Nepal, 10 are medicinal, one poisonous, and 14 species have other uses (Table 4.1). Three species (*Aconitum gammiei*, *Aconitum ferox*, and *Santalum album*), not found in the study area, were taken from other places. The plant species were grouped into 11 use categories; the majority (47.42%) of species were medicines, followed by food (42.85%), fuelwood (26.85%), fodder (21.71%), and social and religious use (18.85%) (Figure 4.2). Female respondents hold good knowledge in overall (36.5) and food (15.5) species, where male respondents are efficient in medicinal (11.5) and other categories (11). Majority of useful species were herbs ($n = 83$, 47.42%), followed by tree (25.71%, $n = 45$), shrub (21.14%, $n = 37$), and climbers (6.28%, n = 11). Regarding parts use, eight different parts were used. The majority of the species were harvested for shoots (48.57%), followed by leaves (46.85%), roots (27.42%), and fruits (26.28%) (Figure 4.3). Some of the species (9.14%) were cherished for their whole plant parts.

The medicinal plants were mostly used in the form of juice, followed by chewing, paste, decoction, and powder (Figure 4.4). Food plants were either eaten raw or cooked depending on their types. Raw food included fruits (45.33%), shoot tenders (4%), and seeds (5.33%). Cooked food included vegetables (30.67%), condiments and pickles (4% each), and oil and fermented food (2.67% each) (Figure 4.5).

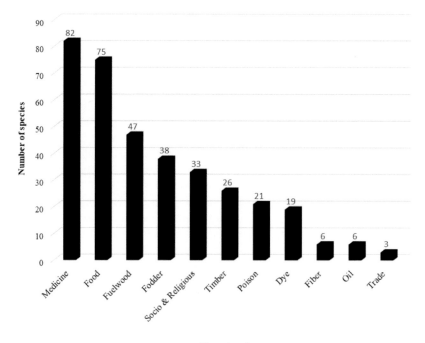

Figure 4.2: Plant use categories based on field surveys.

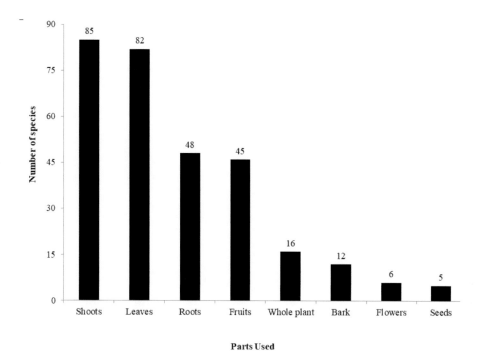

Figure 4.3: Plant parts used based on a field survey.

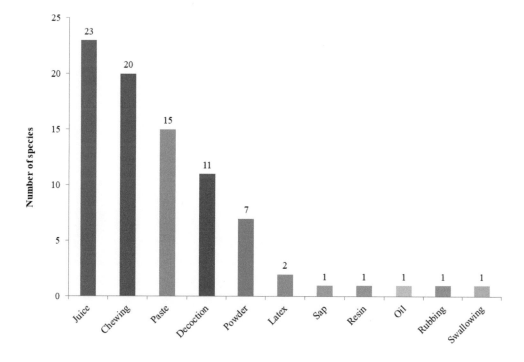

Figure 4.4: Number of plant species used in different modes for medicinal purposes.

Table 4.1: Plant species used by Kham Magar community in western Nepal.

S.N.	Botanical names of plant species	Family	Local name (Magar)	Parts use	Uses	
					Present findings	Comparison to previous findings (Literature review)
1	*Abies spectabilis* (D. Don) Mirb.	Pinaceae	Bhum	Wd, Cn	Cone extract used as dye, Whole flag pole (*Marla*) during religious festival "Bhume"	Wd: Fuel, Timber (2); Tr: Social, Religious (8); Cn: Dye (4); Lf: Medicine (2, 5, 9)
2	*Acer caudatum* Wall. *	Sapindaceae	Rijhyausai	Wd	Plants used as fuelwood and timber	
3	*Aconitum ferox* Wall. ex Seringe	Ranunculaceae	Chari bikh	Rt, St, Lf	Root as poison during hunting	Rt, St: Poison (1, 2, 4, 5, 9)
4	*Aconitum gammiei* Stapf	Ranunculaceae	Bikhama	Rt	Root in heartburn (gastritis) and stomachache (*Khayopiyo*)	Rt: Medicine (1, 2, 5)
5	*Aconitum spicatum* (Brühl) Stapf	Ranunculaceae	Ning	Rt, Sh, Lf	Leaf juice in cuts and wounds	Wp: Medicine (4, 6); Rt: Poison (2), Trade, Medicine (3, 7); Tu: Poisonous (5, 9), Commercial (8)
6	*Acorus calamus* L.	Zingiberaceae	Baja	Rt	Rhizome is used during cough or sore throat, scabies in dog	Rh: Medicine (1, 2, 4, 5, 9)
7	*Aesculus indica* (Colebr. ex Cambess.) Hook.	Sapindaceae	Pangar	Sh, Fr	Seed paste used in Mumps (*Baangle*)	Br, Lf: Medicine, Fodder (1); Sd: Food (1, 2, 4, 9) Medicine (5), Sh, Lf: Food (4); Wd: Material (4)
8	*Agave americana* L.	Asparagaceae	Siundi	Lf	Leaf paste is in joint-ache (*Chhaare*)	Wp: Food, Medicine, Fiber, Material, Poison (1, 4, 5); Rt: Medicine (5)
9	*Ageratina adenophora* (Spreng.) R. King and H. Rob.	Asteraceae	Maobaadi jangal	Lf	Leaf in bleedings and sick chicken	Lf: Medicine (1, 2, 9)
10	*Albizia julibrissin* Durazz.	Fabaceae	Bakar	Wd	Wood is used fuelwood	Br, Fl: Medicine (5); Br, Wd: Medicine, Fuel, Furniture (2); Br, Lf: Medicine (1)
11	*Allium sativum* L.	Amaryllidaceae	No:	Rt	Bulbs used as anti-poison	Bu: Medicine (1,5), Food (2)
12	*Allium tuberosum* Rottler ex Spreng	Amaryllidaceae	Haddi jhor	Rt	Bulb paste in bone fracture	Lf, Sd: Medicine (1)
13	*Allium wallichii* Kunth	Amaryllidaceae	No:ngan	Rt, Lf	Young leaves as vegetable or condiment	Lf: Food (2, 3); Wp: Food (1, 8); Bu, Lf: Medicine (9)
14	*Alnus nepalensis* D.Don	Betulaceae	Jhaar	Wd, Br	Wood for fuelwood and timber. Bark extract to dye clothes	Wd: Furniture, Timber (2); Br, Lf: Medicine (1); Br: Medicine (1, 9); Rt, Lf: Medicine (5)

No.	Scientific name	Family	Local name	Parts used	Uses	Uses codes
15	*Ampelocissus rugosa* (Wall.) Planch.	Vitaceae	Jiprang	Lt, Fr	Fruits food. Sap to treat conjunctivitis and latex is used to treat warts (*Jojhai*)	St: Medicine (2)
16	*Anemonastrum polyanthes* (D. Don) Holub Synonym: *Anemone polyanthes* D. Don*	Ranunculaceae	Rathabiratha	Fl	Flowers ornamental	
17	*Arisaema griffithii* Schott	Araceae	Dhokaya	Lf	Fermented leaves as *gundruk* (*gundru*)	Rh: Medicine (3), Rh, Lf: Food (3); Lf, Cm: Medicine, Food (8)
18	*Arisaema jacquemontii* Blume **	Araceae	Bhinu	Rt, Sh, Lf, Sd	Seeds in worm infestation (*Chanpakira*)	Wp: Medicine (1, 5); Lf, Cm: Food (2); Tu: Medicine (7)
19	*Artemisia dubia* Wall. ex Bess.	Asteraceae	Paati	Sh, Lf	Fresh leaf juice in wounds. Dried leaves as incense	Lf: Medicine (2, 9); Lf, Fl: Social, Religious (1)
20	*Asparagus racemosus* Willd.	Asparagaceae	Kurilla	Ts	Tender shoots as vegetable	Sh: Food (2, 4, 9); Rt: Medicine (1, 6)
21	*Begonia picta* Sm.	Begoniaceae	Kumkum	Fr	Fruits in toothache	Wp: Medicine, Food (2, 10)
22	*Berberis aristata* DC.	Berberidaceae	Jhethe chyantro	St, Br, Fr	Stem bark in dye. Ripened fruits wine preparation	Fr, St, Rt, Br: Medicine (1, 2, 3, 4, 5, 6, 9), Food (2, 3, 4, 8), dye (2, 8)
23	*Berberis asiatica* Roxb. ex DC.	Berberidaceae	Kattike chyantro	Rt, Sh, Br, Fr, Fl	Root extract in jaundice (*Pihile*) and diarrhea (*fu*). Stem bark extract in dye.	Fr, St, Rt, Br: Medicine, Food (8), Dye (1, 2, 8, 9)
24	*Bergenia ciliata* (Haw.) Sternb.	Saxifragaceae	Hangawo	Rt	Rhizome juice in typhoid (*Ghamjoro*), diarrhea, dysentery, vomiting, headache and menstruation disorder	Wp, Fl, Rh, Lf: Medicine (2, 4, 5, 7, 9); Rh, Sd, Wp: Medicine (1, 8); Commercial (3); Rh, Rt: Medicine (4); Rt: Medicine (5)
25	*Betula alnoides* Buch.-Ham. ex D. Don	Betulaceae	Puum	Wd	Woods as timber	Wd, Br, Lf: Timber, Medicine, Fodder (2) Wp, Br: Medicine (1, 5, 9)
26	*Betula utilis* D. Don	Betulaceae	Bhuja:	Lf	Leaves as fodder	Br, Wd: Medicine, Food, Material, Incense (2); St, Br: Medicine (1, 5); Sh, Br ,Rn, Wd: Medicine, Fuel, Social, Commercial, Material (3); Rt, Lf, Fl, Fr: Medicine (1, 9); Fl, Sd, Wd: Medicine, Food, Social (3); Br, Rn: Medicine (5); Wd, Rn: Material, Food
27	*Brucea javanica* (L.) Merr.	Simaroubaceae	Bhakimla/ Kharga	Fr	Ripe fruits pickled, to cure diarrhea	Fr: Medicine (5); Lf, Fr: Dye, Medicine, Food (2, 4, 9)

Table 4.1 contd. ...

...Table 4.1 contd.

S.N.	Botanical names of plant species	Family	Local name (Magar)	Parts use	Uses Present findings	Comparison to previous findings (Literature review)
28	*Campylotropis speciosa* (Schindl.) Schindl.	Fabaceae	Sangkhina	Rt	Root juice to cure diarrhea	Fl: Food (2, 9)
29	*Cannabis sativa* L.	Cannabaceae	Bhango	Br, Lf, Sd	Bark fiber make rough clothes (*Bhangaura*), sacks, bags, etc. Resinous, young leaves and inflorescence to treat pneumonia and fever. Seeds in pickle	Br, Lf, Sd: Fiber, Medicine, Food (1, 2, 4, 5, 9)
30	*Capsella bursa-pastoris* (L.) Medik.	Brassicaceae	Chaangan	St, Lf	Tender leaves vegetable	Wp, Lf: Medicine, Food (2); Wp: Medicine (1); Rt: Medicine (3); Lf: Food (8)
31	*Cenchrus flaccidus* (Griseb.) Morrone Synonym: *Pennisetum flaccidum* Griseb	Poaceae	Kajabhu	Wp	The plant as fodder	Wp: Fodder (2)
32	*Prunus cerasoides* D. Don Synonym: *Cerasus cerasoides* (Buch.-Ham. ex D. Don) S.Ya. Sokolov	Rosaceae	Paiya	St,Br	Barks decoction in backaches (*Pireko*), swellings, and dye	Wp, Br, Lf, Sd: Material, Fodder, Medicine (2); Br, Lf, Sd: Medicine (5); Br: Medicine (9)
33	*Cheilanthus dalhausiae* Hook.	Pteridaceae	Dumni sinka	Lf	Leaf paste in cuts and wounds	Rh; Medicine (2); Wp, Rh, Lf: Medicine (1,9)
34	*Chenopodium album* L.	Amaranthaceae	Bithu	St, Lf	Tender leaves as vegetable	Lf, Ts: Food (1, 2, 8, 9); Wp: Medicine (5); Lf, Fl, Sd: Medicine (1); Wh, Rt, Sd: Medicine (2, 9)
35	*Chrysopogon gryllus* (L.) Trin.	Poaceae	Syopal	Lf	Leaves as fodder	Wp, Rt: Fodder, Medicine (2)
35	*Cinnamomum glanduliferum* (Wall.) Nees.	Lauraceae	Malegiri	Wd, Fr	Woods in religiously function. Dry seeds as spice and fish poison	Wd, Lf: Material (1)
37	*Cirsium verutum* (D. Don) Spreng.	Asteraceae	Jhyankaal	Rt, Sh	Root juice in fever (*Joro*), Root to cure weeping disease. Stem culm eaten fresh	Rt: Medicine (1, 2), Food (2); Rt, Sh: Food (9)
38	*Clematis terniflora* DC.*	Ranunculaceae	Abijale	Sh, Lf, Br	Plant paste in gout, itching	
39	*Coriaria nepalensis* Wall.	Coriariaceae	Ghumil	Fr, Sd	Woods as fuel	Wd, Lf, Br, Fr: Material, Poison, Medicine, Food (2); Fr: Food (9), Poison (1); Lf: Poison (5)

No.	Scientific name	Family	Local name	Part	Uses	Uses (refs)
40	*Cornus capitata* Wall.	Cornaceae	Phuli	Wd, Fr	Ripe fruits eaten, brew to distill alcohol. Fruits in headache	Wd, Fr: Fuel, Food (2)
41	*Corylus ferox* Wall.	Betulaceae	Ghyamo naa	Fr	Ripe fruits eaten	Fr: Food (2)
42	*Cotoneaster frigidus* Wall. ex Lindl.	Rosaceae	Kaali monjyar	Wd	Wood as sticks	Sd: Medicine (2); Wd: Material (8)
43	*Cotoneaster microphyllus* Wall. ex Lindl.	Rosaceae	Saapithala	Wd, Fr	Ripe fruits eaten	Fr, Wp, Lf: Food, Material, Incense (2); Fr: Medicine, Food (3)
44	*Curcuma angustifolia* Roxb.	Zingiberaceae	Kachur	Rt	Root paste bone fracture	Rt: Medicine (2, 9)
45	*Cynodon dactylon* (L.) Pers.	Poaceae	Dubaa	Lf	Whole plant in social and cultural rituals	Wp, Lf: Medicine, Fodder, Religious (2)
46	*Cyperus cyperoides* (L.) Kuntze*	Cyperaceae	Tolaagaathaa	Rt	Roots in gastritis (*Gando*)	Rt
47	*Dactylorhiza hatagirea* (D. Don) Soó	Orchidaceae	Paanchwaule	Rt	Root in stomachache, headache, and typhoid	Rt, Lf: Medicine (2, 9); Tu: Medicine (1, 7), Tu: Medicine, Food, Commercial (3, 8)
48	*Daphne bholua* Buch.-Ham. ex D. Don	Thymelaeaceae	Barruwaa/Ratuwaai charo	Rt, Br	Root juice in fever. Bark for fiber	Rt, Br, Lf: Medicine, Material, Poison (2, 4); Rt: Medicine (9); Rt, Br: DPR (5) Br: Material, Medicine, Poison (3)
49	*Debregeasia longifolia* (Burm. f.) Wedd.	Urticaceae	Tooshaaraa	Fr	Ripe fruits eaten	Wd, Br, Lf, Fr: Fuel, Material, Fodder, Food (2); Lf, Br: Medicine, Material (9)
50	*Delphinium himalayae* Munz	Ranunculaceae	Nirmasii	Rt	Root in headache	Rt: Medicine (1); Tu: Medicine (1, 3, 8), Poison (1), Commercial (3, 8)
51	*Delphinium grandiflorum* L. *	Ranunculaceae	Atis	Rt	Root in stomach pain, typhoid, gout	
52	*Delphinium vestitum* Wall. ex Royle	Ranunculaceae	Mangraamhul	Rt, Lf	Root anthelmintic, during delivery for cattle	Rt: Medicine (1, 2, 9), Poison (2)
53	*Desmodium elegans* DC.	Fabaceae	Til daalaa	Lf	Leaves as fodder	Rt, Br, Lf: Medicine, Food, (1); Rt: Medicine (1, 9); Lf, St: Medicine, Food, Fuel, Material (3); Rt, Br: Medicine (5); St: Material (8)
54	*Dipsacus inermis* Wall.	Dipsacaceae	Mipa/Taukejhaar	Lf	Leaf juice in burns and cuts	Wp: Fodder (2); Lf, Td: Food (8)
55	*Drymaria cordata* (L.) Willd. ex & Schult.	Caryophyllaceae	Nayaa jungle	Sh, Lf	Plant juice in wounds and itching	Wp, Lf, Ts: Medicine, Food (2)

Table 4.1 contd. ...

...Table 4.1 contd.

S.N.	Botanical names of plant species	Family	Local name (Magar)	Parts use	Uses — Present findings	Comparison to previous findings (Literature review)
56	*Elaeagnus parvifolia* Wall. ex Royle	Elaeagnaceae	Dhakari	Lf, Fr	Leaves as fodder. Ripe fruits eaten	Fr: Medicine, Food (2); Fl: Medicine (1,3); Lf: Food, Fodder (3)
57	*Equisetum arvense* L.*	Equisetaceae	Mikrop	Rt, Sh	Roots and stems in fever	
58	*Eriobotrya elliptica* Lindl. *	Rosaceae	Mahaakhyaa	Lf	Leaves in social and religious rituals. Plant as fodder	
59	*Eriocapitella rivularis* (Buch. Ham. ex DC.) Chr.stenh. & Byng Synonym: *Anemone rivularis* Buch.-Ham. ex DC.	Ranunculaceae	Dhaskim	Lf	Leaf juice as anti-leech	Wp: Medicine (1, 2, 5, 9); Sd: Food (2, 9); Fr, Sd: Medicine (7)
60	*Eriocapitella vitifolia* (Buch.-Ham. ex DC.) Nakai Syn. *Anemone vitifolia* Buch.-Ham. ex DC.	Ranunculaceae	Kapaso	Fr	Powdered achene in cuts and wounds	Rt, Lf, Fr: Medicine (1, 2, 9)
61	*Eriophorum comosum* (Wall.) Nees	Poaceae	Syambhun	Sh	The plant as fiber	Wp: Fodder (2)
62	*Eulaliopsis binata* (Retz.) C.E. Hubb.	Poaceae	Babya	Lf	Plant in religious ceremony	Wp: Material (2)
63	*Euonymus lucidus* D.Don*	Celastraceae	Gunamohar	Wd, Lf	The plant as fodder and fuelwood	
64	*Eurya acuminata* DC.	Theaceae	Khoraasane	Wd, Lf	Woods as fuelwood. Leaves as fodder	Wd, Lf: Fuel, Fodder (2)
65	*Fagopyrum dibotrys* (D. Don) H. Hara	Polygonaceae	Baan bhade	Rt	Root in diarrhea	Lf, Ts: Food (2,9)
66	*Ficus neriifolia* Sm.	Moraceae	Dudeula	Wd, Lf, Fr	Woods as fuelwood. Plant as fodder and figs eaten	Wp, Lt, Fr: Fodder, Medicine, Food (2)
67	*Ficus religiosa* L.	Moraceae	Peepal	Wp	Stems fuelwood. Planted as hedge	Wp, Br, Lf, Lt, Fr: Religious, Medicine, Dye, Fodder (2)
68	*Ficus sarmentosa* Buch. ex J.E. Sm.	Moraceae	Bidu	Lf	Leaves as fodder	Sh, Lf, Fr: Fodder, Food (2). Br, Sp, Fr: Medicine (9)
69	*Flemingia strobilifera* (L.) W.T. Aiton	Fabaceae	Bhuisnankhinaa	Rt	Root in diarrhea and dysentery	Rt, Br, Fr: Medicine Food (2), Rt, Br: Medicine (1, 9)

No.	Scientific name	Family	Local name	Part used	Uses	Coded uses
70	*Fragaria nubicola* Lindl.	Rosaceae	Jhompaasaii	Fr	Ripe fruits eaten	Wp, Fr: Medicine, Food (2, 3, 7, 9); Lf, Fl, Fr: Medicine (1), Fr: Food (8)
71	*Fraxinus floribunda* Wall.	Oleaceae	Rhyankhuli	Sh, Lf, Br	Bark extract as dye	St, Rn: Medicine (1), Wd, Sh, Br: Material, Medicine (2)
72	*Galinsoga parviflora* Cav.	Asteraceae	Raawande	Sh, Lf	Plant as fodder	Wp: Medicine (1), Poison (2), Ts, Lf: Food, Medicine (9)
73	*Galium asperuloides* Edgew.	Rubiaceae	Khasare	Lf	Plant in cuts and wounds	Wp: Medicine (1)
74	*Gaultheria nummarioides* D.Don**	Ericaceae	Kaasai	Fr	Ripe fruits eaten and as dye	Fr: Food (2); Lf, Fr: Medicine, Food (1, 9),
75	*Geranium procurrens* Yeo*	Geraniaceae	Chhaapaa	Sh, Lf	Root in wounds (*Piriu*)	
76	*Girardinia diversifolia* (Link) Friis	Urticaceae	Puwa	Br	Bark as threads, ropes, and rough clothes	Lf, Br: Food, Medicine, Fiber (2, 4, 8, 9)
77	*Gonostegia hirta* (Blume) Miq.	Urticaceae	Barmitina	Rt	Root in bowls (*Pilo*). Roots eaten	Wp, Rt: Medicine, Food, Material (1, 2, 9)
78	*Hedera nepalensis* K. Koch	Araliaceae	Piplepaatte	Lf	Plant as fodder	Wp: Medicine, Fodder (2); Lf, Fr: Medicine; Lf, Fl: Medicine, Social (3) Lf: Medicine (9); Lf, Fr: Medicine (1)
79	*Hedychium coronarium* J. Koenig*	Zingiberaceae	Tunti	Rt	Root as anti-allergic, fever	
80	*Heracleum candicans* var. *obtusifolium* (Wall. ex DC.) F.T. Pu & M.F. Watson	Apiaceae	Tee	Rt, Sh	Roots in stomach pain. Tender shoots as vegetable	Rt, Fr: Medicine (3, 7)
81	*Holboellia latifolia* Wall.	Lardizabalaceae	Banbaalu	Fr	Ripe fruits eaten	Fr: Food (2, 9)
82	*Ilex dipyrena* Wall.	Aquifoliaceae	Syaru	Sh, Lf	Woods for fuelwood and leaves as fodder	Wd, Fr: Fuel, Food (2, 9)
83	*Impatiens urticifolia* Wall.*	Balsaminaceae	Banbhaango	Fr	Seeds eaten raw	Sd: Food (2)
84	*Imperata cylindrica* (L.) P. Beauv.	Cyperaceae	Siru	Lf	Plant for fodder and in worms and diarrhea in animals	Wp, Rt: Fodder, Medicine (1, 2)
85	*Jasminum humile* L.**	Oleaceae	Phaadulla	Fr, Fl	Ripe fruits as dye	Rt, Fl: Medicine (1, 2, 3); Rt, Br: Medicine (1, 3); Br: Medicine (9)
86	*Juglans regia* L.	Juglandaceae	Khaasai	Sh, Br, Sd	Bark in toothache and as dye and fish poison	Br, Lf, Sd, Wd: Medicine (2, 9), Br, Lf: Medicine (1); Br, Nt: Medicine (7), Poison, Food, Commercial (3); Fr, Sd: food (8), Br, Lf, Fr: Medicine, Food, Dye (4)

Table 4.1 contd. ...

...Table 4.1 contd.

S.N.	Botanical names of plant species	Family	Local name (Magar)	Parts use	Uses	
					Present findings	**Comparison to previous findings (Literature review)**
87	*Juniperus indica* Bertol.	Pinaceae	Dhupa	Lf	Leaves as incense	Wd, Lf: Incense (2, 9), Material (8); Lf, Fr, Wd: Medicine, Social, Material, Fuel, Commercial (3)
88	*Leibnitzia nepalensis* (Kunze) Kitam.	Asteraceae	Jhula	Sh, Lf	Root in stomachache. Leaf in wounds and as tinder	St, Lf, Fr: Medicine (3)
89	*Leucosceptrum canum* Sm.	Lamiaceae	Phusare	Rt	Root in wounds	Lf, Fr: Fodder, Food (2)
90	*Lilium nepalense* D. Don	Liliaceae	Gaa	Rt	Bulbs as food	Bu: Food (1), Medicine (2)
91	*Lindenbergia muraria* (Roxburgh ex D. Don) Brühl*	Orobanchaceae	Garichan	Rt, Sh, Lf	Plant in cuts, wounds, and burns	
92	*Lindera neesiana* (Wall. ex Nees) Kurz	Lauraceae	Tipa	Fr	Fruits antidotes, in flatulence	Rt, Br, Lf, Fr, Sd: Medicine (15); Fr, Sd: Medicine (2, 9)
93	*Lindera pulcherrima* (Nees) Hook. f.	Lauraceae	Phusare	Wd, Lf	Woods as fuelwood. The plant for fodder	Wd, Lf: Fuel, Fodder (2)
94	*Lyonia ovalifolia* (Wall.) Drude	Ericaceae	Sirwaan	Wd, Lf	Woods as fuelwood and timber	Wd, Lf: Medicine, Poison, Fuel (2); Lf, Bd, Rt: Poison, Medicine (1, 5, 9); Wp. Rt. Lf: Medicine, Poison (4)
95	*Machilus duthiei* King ex Hook. fil. Synonym: *Persea duthiei* (King) Kosterm.	Lauraceae	Jyang	Lf	Leaves fodder	Sh, Lf: Medicine (1, 9)
96	*Maharanga emodi* (Wall.) A. DC. **	Boraginaceae	Mahaarangi	Rt	Root extracts as dye	Rt: Medicine (2, 3)
97	*Mahonia napaulensis* DC.	Berberidaceae	Maadale chyantro	Rt,Fr	Ripe fruits eaten and as dye. Root bark as dye	Br, Fr: Medicine, Dye, Food (1, 2, 4, 9)
98	*Maianthemum purpureum* (Wall.) LaFrankie	Asparagaceae	Jyabir	Lf	Tender shoots and leaves vegetable. Roots in trade	Lf, Ts: Food (2)
99	*Marsdenia lucida* Edgew. ex Madden	Apocynaceae	Moor ralaa	Sh, Lf	Tender shoots and leaves are poisonous to cattle	Wp: Fencing (2)
100	*Mentha spicata* L.		Baasmati	Rt	Leaves as pickle	Lf: Medicine (1); Lf, Sd: Medicine, Food (2)

No.	Scientific name	Family	Local name	Parts used	Uses	Use details
101	*Morina longifolia* Wall. ex DC.	Capprifoliaceae	Jhyankaatu	Rt	Root or latex in fever	Wp: Incense (2); St, Lf, Fl: Medicine (1)
102	*Morus alba* L.	Moraceae	Hoi/Toot	Fr	Ripe fruits eaten	Fr: Food (2, 4, 8)
103	*Myriactis nepalensis* Less.	Asteraceae	Lese kura	Lf	Plant in cuts, wounds and burns	Fr: Food (2)
104	*Myrsine semiserrata* Wall.	Primulaceae	Makya	Lf	Leaves as fodder	Wd, Lf: Fuel, Fodder (2)
105	*Neohymenopogon parasiticus* (Wall.) Bennet	Rubiaceae		Lf	Dry leaves used by Witch-doctors	Fr: Medicine (1, 2, 9)
106	*Ophioglossum costatum* R. Br. Synonym: *Ophioglossum pedunculosum* Desv. *	Ophioglossaceae	Bargulaa	Lf	Fronds as vegetable	
107	*Oreocnide frutescens* (Thunb.) Miq.	Urticaceae	Sarghil	Lf	Tender shoots as vegetable	Lf: Medicine (1); Rt, Ts, Lf, Br: Medicine, Food, Material (2)
108	*Paris polyphylla* Sm.	Melanthiaceae	Satawaa	Rt	Root in cuts, wounds, and antidote	Rh: Medicine (1, 2, 3, 4, 8, 9); Commercial (3);
109	*Persicaria nepalensis* (Meisn.) Miyabe	Polygonaceae	Ratane	Sh, Lf	Plant as fodder	Rt: Medicine (1); Lf, Ts, Rt, Wp: Food, Medicine, Poison (2)
110	*Phytolacca latbenia* (Moq.) H. Walter	Phytolaccaceae	Jargo	Rt. Sh	Root in gastritis. Tender leaves and shoot as vegetable	Lf, Ts: Food (2); Rt: Medicine (9); Rt, Fr: Medicine, Condiment (1); Rt: Medicine (5)
111	*Pinus roxburghii* Sarg.	Pinaceae	Dang	Wd	Woods as fuelwood and timber	Wd, Lf, Rn, Sd: Timber, Medicine, Food (2); Wd, Rn: Timber, Material, Medicine, Trade (1, 4, 5); Rn, Sd; Medicine, Food (9)
112	*Pinus wallichiana* A.B. Jacks.	Pinaceae	Dhupi	Wd, Rn	Woods as fuelwood and timber. The resin in skin fracture	Rn: Medicine (1, 5), Wd, Rn: Material, Commercial (2, 3); Rn: Medicine (1), Wd: Religious (8)
113	*Plantago asiatica* subsp. *erosa* (Wall.) Z.Y. Li	Plantaginaceae	Hate/ Gandowasa/ Aantkatawa	Rt	Root in gastritis	Rt: Medicine (2)
114	*Pleione humilis* (Sm.) D. Don	Plantaginaceae	Ghabhyato	Rt	Bulb in bone fracture	Bu: Medicine (1, 2)
115	*Polygonatum verticillatum* (L.) All.	Asparagaceae	Rukan	Lf	Tender leaves and shoots as vegetable	Tu, Lf, Sh: Medicine, Food, (2, 9), Rt: Medicine, Food (1, 3); Lf: Food (8)
116	*Polygonum milletii* (H. Lév.) H. Lév.	Polygonaceae	Paat wa:/Laapse	Rt	Root in dysentery and diarrhea, on cuts and wounds	Sd: Food (2)

Table 4.1 contd. ...

...Table 4.1 contd.

S.N.	Botanical names of plant species	Family	Local name (Magar)	Parts use	Uses	
					Present findings	Comparison to previous findings (Literature review)
117	*Potentilla lineata* Trevir.	Rosaceae	Baan mulaa	Rt, Sh	Root in gastritis	Rt: Medicine (1, 9)
118	*Prinsepia utilis* Royle	Rosaceae	Kaaikiram	Sh, Sd	Seeds to extract oil	Sd, Fr, Lf: Medicine, Food, Fodder (2); Sd: Food (8), Medicine (5, 9)
119	*Prunella vulgaris* L.	Lamiaceae	Dhaakar	Lf	Plant in backaches	Wp: Medicine (2)
120	*Prunus cornuta* (Wall. ex Royle) Steud. *Synonym: Padus cornuta* (Wall. ex Royle) Carrière	Rosaceae	Gong rikureli	Fr	Ripe fruits eaten	Fr: Food (2)
121	*Prunus napaulensis* (Ser. ex DC.) Steud. *Synonym: Padus napaulensis* (Ser.) C.K. Schneid.	Rosaceae	Rikureli/Chhitu	Wd, Lf, Fr	Woods as fuelwood and timber. Ripe fruits as dye	Fr, Wd, Lf: Food, Material, Poison (2); Fr: Food (8)
122	*Pyracantha crenulata* (D. Don) M. Roem.	Rosaceae	Ghangaaru	Fr	Ripe fruits eaten	Wp, Fr: Hedge, Medicine, Food (2)
123	*Pyrus pashia* Buch.-Ham. ex D. Don	Rosaceae	Mihel	Sh, Fr	Woods as fuelwood. Ripe fruits eaten	Fr, Lf: Food, Medicine, Fodder (2, 9); Wd, Fr: Materials, Medicine (4)
124	*Quercus lanata* Sm.	Fagaceae	Mising	Sh, Lf, Fr	Plant as fodder. Woods as fuelwood and timber. Bark extract in sprains. Resin in soothing body aches	Wd, Br, Rn, Lf: Fuelwood, Material, Medicine, Fodder (2); Br, Rn, Ct: Medicine (1)
125	*Quercus leucotrichophora* A. Camus *Synonym: Quercus oblongata* D. Don	Fagaceae	Saipaa	Wd, Lf	The plant as fodder. Woods as fuelwood. Acorns for pigs	Wd, Br, Lf, Fr: Fuel, Material, Fodder, Food (2)
126	*Quercus mespilifolia* Wall. ex A.DC. *Synonym: Quercus mespilifoloides* A. Camus*	Fagaceae	Sari	Wd, Lf, Fr	Woods as fuelwood and timber. Plants for fodder. Fruits for pigs	
127	*Quercus semecarpifolia* Sm.	Fagaceae	Kar	Wd, Lf	Plant as fodder. Woods as fuelwood and timber. Bark extract in sprains. Resin in soothing body aches	Wd, Br, Lf, Rn, Sp: Timber, Fuelwood, Medicine, Food (2)

#	Scientific name	Family	Local name	Parts	Uses	Coded uses
128	*Rheum australe* D. Don	Polygonaceae	Chhikum	Rt, Sh, Lf	Root in stomach pain and as dye. Petioles as pickle	Rh: Medicine (1, 2, 3, 6), Dye (2, 8), Trade (8); Pt: Food (3, 8)
129	*Rhododendron arboreum* Sm.	Ericaceae	Sarwaai	Wd, Lf Fr	Flower in fish bone plugins in throat. Woods as fuelwood and timber	Wd, Br, Lf, Fl: Material, Medicine, Food, Poison (2); Br, Lf, Fl: Medicine, Poison (4, 5, 9)
130	*Rhododendron barbatum* Wall. ex G. Don*	Ericaceae	Kanwaai	Wd	Woods as fuelwood and timber	
131	*Rhododendron campanulatum* D. Don	Ericaceae	Chemalaa	Sh, Lf, Fl	Flower in scabies (*Dhaada*). Leaf as rat poison. The plant as fuelwood	Lf, Sh; Medicine, Fuel (2); Fl: Medicine (1), Wd: Material (8)
132	*Rhus succedanea* L.	Anacardiaceae	Bhalaayo	Wd, Lf	Woods as fuelwood. Bark and leaves as corrosive and vesicant	Lf, Fr: Allergy, Food (2, 4)
133	*Ribes glaciale* Wall.	Grossulariaceae	Rijhyaunsai	Fr	Ripe fruits eaten	Fr: Food (2)
134	*Ricinus communis* L.	Euphorbiaceae	Aareda	Sd	Seed in vegetable ghee	Rt, Br, Lf, Fl, Sd, Ct: Medicine, Poison (2)
135	*Rosa brunonii* Lindl.	Rosaceae ·	Dhankila	Sh	Plant as fuelwood	
136	*Rosa macrophylla* Lindl.	Rosaceae	Bhaarmase	Rt	Root in heartburn and stomach pain	Fl, Lf: Medicine, Food, Fodder (2, 3); Fr: Medicine (7)
137	*Rubia manjith* Roxb. ex Fleming	Rubiaceae	Khasare	Lf	Leaf on cuts and wounds	Rt, St: Medicine, Dye (2); Lf: Medicine (1); Rt: Dye (8)
138	*Rubus ellipticus* Sm.	Rosaceae	Angselu	Fr	Ripe fruits eaten	Wp: Food, Medicine, Fodder (2, 9); Fr: Food (8)
139	*Rubus hoffmeisterianus* Kunth & Bouché*	Rosaceae	Zoosai	Fr, Lf	Fermented leaf in Marasmus and in-appetite. Ripe fruits eaten	
140	*Rubus nepalensis* (Hook. f.) Kuntze	Rosaceae	Naample	Fr	Ripe fruits eaten	Fr: Food (2, 9)
141	*Rumex hastatus* D. Don	Polygonaceae	Kaapu	Rt, Sh, Lf, Fr	Roots, tender shoots and leaves as condiment in fish and as pickle	Wp, Lf, Ts: Medicine, Food, Fodder (2)
142	*Rumex nepalensis* Spreng.	Polygonaceae	Theulaa	Rt, Lf	Root in gastritis and diarrhea	Wp: Medicine, Food (1, 2, 3)
143	*Sabia campanulata* Wall. *	Sabiaceae	Kaalilara	Sh	Shoot tenders as vegetable	
144	*Salix babylonica* L.	Salicaceae	Bainsa	Sh, Lf	Woods used as fuelwood. Leaves as fodder	Br: Medicine (1, 5); Wd: Material (2)
145	*Salix sikkimensis* Andersson*	Salicaceae	Kyang	Sh, Lf	Woods as fuelwood and timber. Leaves as fodder	

Table 4.1 contd. ...

...Table 4.1 contd.

S.N.	Botanical names of plant species	Family	Local name (Magar)	Parts use	Uses	
					Present findings	Comparison to previous findings (Literature review)
146	*Santalum album* L.	Santalaceae	Chandan	Sh	Woods in religious functions	Wd: Medicine (1, 5)
147	*Satyrium nepalense* D. Don	Orchidaceae	Sirki	Rt	Roots in stomach pain	Rt: Medicine (1, 11); Bu, Lf: Food (2)
148	*Schisandra grandiflora* (Wall.) Hook. f. & Thomson	Schisandraceae	Ringhul	Fr	Ripe fruits are edible	Fr: Food (2)
149	*Scurrula elata* (Edgew.) Danser	Loranthaceae	Jokhaare	Fr	Ripe fruits edible	Fr: Food (2)
150	*Selaginella biformis* A. Braun ex Kuhn	Selaginellaceae	Pa	Lf	Plant in cuts and wounds	Lf, Rh: Medicine (2)
151	*Selinum wallichianum* (DC.) Raizada & Saxena	Apiaceae	Surkun	Lf	Stem pith eaten. Tender leaves as vegetable	Rt, Lf, Ts: Medicine, Food (2); Rh, St: Medicine, Food (9)
152	*Setaria viridis* (L.) P. Beauv. *	Poaceae	Ghundebanso/Nawang	Sh, Lf	The plant as fodder	
153	*Smilax aspera* L.	Smilacaceae	Daangru	Sh	Tender shoots and leaves as vegetable pickle	Lf, Ts, Fr: Food, Medicine (2)
154	*Solena heterophylla* Lour.	Cucurbitaceae	Bidumba	Rt, Fr	Plant juice in stomachaches. Fruits eaten	Wp, Fr: Medicine, Fodder, Food (2)
155	*Sorbus cuspidata* (Spach) Hedl. Synonym: *Sorbus vestita* (Wall. ex G. Don) Lodd.	Rosaceae	Porou naa	Wd, Lf, Fr	Woods as fuelwood. Leaves as fodder. Ripe fruits eaten	
156	*Stellaria vestita* Kurtz	Caryophyllaceae	Armaale	Lf	Tender shoots leaf as vegetable. Leaf on cuts wounds	Wp: Medicine (1, 2); Ts, Lf: Food (2)
157	*Strobilanthes lachenensis* C.B. Clarke*	Acanthaceae	Angaari	Sh, Lf	Poisonous to cattle	
158	*Swertia chirayta* (Roxb.) Karst.	Gentianaceae	Runka	Rt, Sh, Lf, Br, Fr, Fl	Plant in cold, cough, headache and jaundice	Wp: Medicine (1, 2, 3, 4, 5, 9)
159	*Taraxacum parvulum* DC. *	Asteraceae	Dhaalmundraa	Rt, Sh, Lf	Leaf and latex in cuts and wounds. Roots in trade. Dried plant as tea	
160	*Taxus contorta* Griff.**	Pinaceae	Jham chettri	Sh,	Woods as fuelwood and timber. Stem in cancer. Leaves as incense	Wd, Lf: Incense, Medicine (2), Religious (8)
161	*Thalictrum chelidonii* DC. *	Ranunculaceae	Dhongare	Sh, Lf	Tender shoots and leaves as vegetable	
162	*Thamnocalamus spathiflorus* (Trin.) Munro	Poaceae	Salma	Sh, Lf	Culms in social and religious ceremony. Shoot tender vegetable. Leaves as fodder	Cu, Lf: Material, Fodder (2)

	Scientific name	Family	Local name	Parts used	Uses	Previous uses
163	*Themeda arundinacea* (Roxb.) A. Camus	Poaceae	Pusai	Sh, Lf	Plants for fodder	Cu: Construction (9)
164	*Toona sinensis* (A. Juss.) M. Roem. *	Meliaceae	Tooni	Sh, Lf	Plant as woods. Leaves in social and religious rituals	
165	*Tragopogon gracilis* D. Don*	Asteraceae	Sorno	Rt, Sh, Lf	Root and latex in cuts and wounds (*Chhedar*). Shoots eaten	
166	*Typhonium diversifolium* Wall. ex Schott	Araceae	Tin-chyo	Lf	Young leaves as vegetable or fermented (*Gundru*)	Lf: Food (2, 3), Medicine (3); Cm: Food (8)
167	*Urtica dioica* L.	Urticaceae	Nganti	Lf	Tender shoots and leaves as vegetable	Rt, Lf: Medicine, Food (2, 9); Wp: Medicine, Food (3), Lf, Ts: Food (8)
168	*Valeriana hardwickii* Wall.	Capprifoliaceae	Somaayaa	Rt	Root in trade, headache	Lf, Rh: Medicine, Food, Religious (3); Rh: Incense, Medicine (8)
169	*Verbascum thapsus* L.	Scrophulariaceae	Yume	Rt	Root in diarrhea, dyspepsia, and fever	Wp, Rt: Medicine, Poison (1, 2, 9); Lf, Rt, St, FL; Medicine, Poison (5); Lf, St, Fl: Medicine (7)
170	*Viburnum cotinifolium* D. Don	Capprifoliaceae	Huirong	Sh	Branches as walking sticks	Br, Fr: Medicine (8), Fr: Food (9)
171	*Viburnum cylindricum* Buch. -Ham. ex D. Don *	Capprifoliaceae	Munumchornii	Wd, Br, Lf, Fr	Woods as fuelwood. Leaves as fodder. Bark extract and fruits as dye	
172	*Viburnum erubescens* Wall.	Capprifoliaceae	Hyanbur	Wd, Fr	Woods as fuelwood. Ripe fruits are edible	Lf, Sd: Medicine (1); Rt, Fr: Medicine, Food (2); Lf, Fr, Sd: Medicine, Food (9)
173	*Viburnum mullaha* Buch. -Ham. ex D. Don	Capprifoliaceae	Bataapsaii	Wd, Fr	Woods as fuelwood. Ripe fruits eaten and fruits as dye	Fr: Medicine (1), Food (9); Wd, Lf, Fr: Material, Medicine, Food, Dye (2)
174	*Viscum album* L.	Santalaceae	Jokhaare	Fr	Ripe fruits eaten	Wp: Medicine (1, 2, 4, 5, 9), Food (2, 9)
175	*Zanthoxylum armatum* DC.	Rutaceae	Tinbur	Fr, Sd	Seeds in cold, stomach disorder, and poison. Fruits as pickle	Rt, Br, Lf, Fr, Sd: Medicine, Food (1, 2, 4, 5, 9)

Parts used: Rt: Root, Wd: Wood, St: Stem, Sh: Shoot, Br: Bark, Lf: Leaf, Fr: Fruit, Fl: Flower, Sd: Seed, Cu: Culm, Tu: Tuber, Rh: Rhizome, Rn: Resin, Ts: Tender shoot, Ct: Cotyledon, Wp: Whole Plant, Tr: Tree, Bu: Bulb, Lt: latex.

Use categories: Dt: Dye and Tanning, Fb: fiber, Fd: Food, Fo: Fodder, FW: Fuelwood, Md: Medicine, Po: Poison, SR: Social and Religious, Tm: Timber, Tr: Trade, Ol: Oil.

Previous findings based on: 1-Baral and Kurmi (2006), 2-Manandhar (2002), 3- Ghimire et al. (2008), 4- Dutta (2007), 5-DPR (2007), 6-Bhattarai et al. (2006), 7-Lama et al. (2001), 8-Gautam (2012), 9-Rajbhandari (2001), 10-Rajbhmadary (2013).

*New use flora for Nepal, ** Documented different uses in previous literatures.

Nomenclature based on: http://www.efloras.org/flora_page.aspx?flora_id=110, http://www.theplantlist.org/, and http://www.catalogueoflife.org/col/search/all/key/Valeriana+hardwickii+/fossil/1/match/1

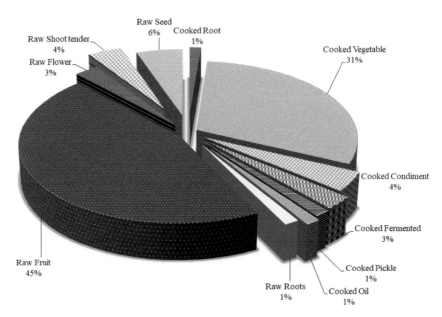

Figure 4.5: Percentage of plant species in different forms used for food purposes.

Informant Consensus Factor (Fic), Use Frequency (UF), and Use Value (UV)

The level of informant agreement was high (Fic = 1) for eight ailment categories and total consensus (Fic = 1) was obtained for dandruff, cancer, frightening, warts, pneumonia, jaundice, anesthetic, skin fracture, and eye problem (Table 4.2). Heartburn, vomiting, asthma, itching, joint ache, mumps, bone fracture, and the delivery problem showed a low level of consensus. Burns, cuts, wounds, and cancer ailments were very significant, since each of the 19 species in total was used against these ailments, followed by diarrhea and stomachache (16 species each), fever and gastritis (10 species each).

Based on use frequency (UF), the most frequently used species of the study area were *Taxus contorta* (0.86), *Juglans* regia (0.78), *Quercus semecarpifolia* (0.76), *Quercus lanata* (0.76), *Abies spectabilis* (0.64), *Lyonia ovalifolia* (0.62), *Quercus mespilifolioides* (0.58), *Berberis aristata* (0.56), *Cornus capitata* (0.52), *Fragaria nubicola* (0.52), and *Holboellia latifolia* (0.52). The most frequently used medicinal species include *Swertia chirayta* (0.38), *Paris polyphylla* and *Anemone vitifolia* (0.34 each), *Myriactis nepalensis* (0.24), and *Bergenia ciliata* (0.22). Similarly, *Cornus capitata* and *Polygonatum verticillatum* (0.38 each), *Holboellia latifolia* (0.36), *Fragaria nubicola* (0.28), *Arisaema griffithii* (0.26), and *Corylus ferox* (0.24) were among the most frequently used food species.

The most useful plant species in terms of overall use value considering all use categories were *Berberis aristata* (UV = 2.82), *Aesculus indica* (2.75), *Cerasus cerasoides* (2.53), *Berberis asiatica* (2.42), *Ficus neriifolia* (2.08), *Pinus wallichiana* (2.06), and *Aconitum ferox* (2.00). In terms of medicinal use value, *Bergenia ciliata* (UV = 3.29), *Swertia chirayta* (2.08), *Aconitum ferox*, *Aconitum gammeie*, and *Acorus calamus* (UV for each 2.00), *Delphinium vestitum* (1.93), and *Dactylorhiza hatagirea* (1.67) were highly important, with the highest use values for medicinal use category. Considering food use value, *Polygonatum verticillatum* (0.30 each), *Arisaema jacquemontii* and *Lindera neesiana* (0.04 each), *Asparagus racemosus*, *Rumex nepalensis*, *Tragopogon gracilis*, and *Rhododendron arboreum* (0.03 each) were highly important with highest use values.

Table 4.2: Informant consensus factor (Fic) for different ailment categories.

Use category (Local name)	Use reports (Nur)	Number of taxa (Nt)	Fic	Botanical Name of Plant species
Dandruffs (*Ghurul*)	3	1	1	*Urtica dioca*
Cancer	2	1	1	*Tsuga dumosa*
Vomiting (*Wakya*)	3	2	0.5	*Bergenia ciliata, Swertia chirayta*
Diarrhea (*Phui*)	38	16	0.595	*Berberis asiatica, Bergenia ciliata, Daphne bholua, Ageratina adenophora, Fagopyrum dibotrys, Imperata cylindrica, Campylotropis speciosa, Potentilla lineata, Brucea javanica, Polygonum milletii, Rumex nepalensis,*
Frightening (*Sato*)	2	1	1	*Equisetum arvense*
Warts (*Jojhai*)	5	1	1	*Ampelocissus rugosa*
Gastritis (*Gãndo*)	31	10	0.7	*Bergenia ciliata, Myriactis nepalensis, Fagopyrum dibotrys, Cyperus cyperoides, Plantago asiatica, Potentilla lineata, Rubus hoffmeisterianus, Rumex nepalensis, Zanthoxylum armatum, Phytolacca latbenia*
Typhoid (*Ghãnjoro*)	4	4	0	*Aconitum gammiei, Dactylorhiza hatagirea, Swertia chirayta, Delphinium grandiflorum*
Pneumonia (*Sardi*)	3	1	1	*Cannabis sativa*
Jaundice (*Pahenle*)	4	1	1	*Swertia chirayta*
Joint ache (*Chhare ghãsi*)	11	8	0.3	*Anemone rivularis, Arisaema jacquemontii, Cannabis sativa, Clematis terniflora, Agave americana, Swertia chirayta, Curcuma angustifolia*
Allergy (*Jaabe*)	5	4	0.25	*Aconitum gammiei, Acorus calamus, Rumex nepalensis, Hedychium coronarium*
Burns, Cuts, and Wounds	52	19	0.647	*Aconitum spicatum, Polygonum milleti, Delphinium vestitum, Dipsacus inermis, Myriactis nepalensis, Drymaria cordata, Ageratina adenophora, Lindenbergia muraria, Paris polyphylla, Pinus wallichiana, Galium asperuloides, Rubia manjih, Taraxacum parvulum, Leibnitzia nepalensis, Eriocapitella vitifolia, Selaginella biformis, Stellaria vestita, Cheilanthus dalhausiae,*
Fever (*Joro*)	12	10	0.182	*Aconitum gammiei, Bergenia ciliata, Cirsium verutum, Cynodon dactylon, Daphne bholua, Paris polyphylla, Taraxacum parvulum, Verbascum thapsus, Morina longifolia, Hedychium coronarium*
Stomach pain	30	19	0.379	*Aconitum gammiei, Aconitum spicatum, Solena heterophylla, Bergenia ciliata, Dactylorhiza hatagirea, Daphne bholua, Delphinium vestitum, Fagopyrum dibotrys, Paris polyphylla, Potentilla lineata, Quercus semecarpifolia, Rheum australe, Rhododendron campanulatum, Brucea javanica, Satyrium nepalense, Urtica dioica, Zanthoxylum armatum, Leibnitzia nepalensis, Rosa macrophylla, Delphinium grandiflorum*
Scabies (*Khaira*)	3	3	0	*Acorus calamus, Rhododendron campanulatum, Rumex nepalensis*

Table 4.2 contd. ...

...Table 4.2 contd.

Use category (Local name)	Use reports (Nur)	Number of taxa (Nt)	Fic	Botanical Name of Plant species
Mumps (*Baangale*)	4	3	0.333	*Verbascum Thapsus, Gonostegia hirta, Aesculus indica*
Cough (*Rughaa*)	26	6	0.8	*Acorus calamus, Bergenia ciliata, Lindera neesiana, Rhododendron arboreum, Swertia chirayta, Zanthoxylum armatum*
Bone fracture	4	3	0.333	*Pleione humilis, Ageratina adenophora, Aconitum gammiei*
Skin problem	2	1	1	*Pinus wallichiana*
Headache	4	2	0.667	*Dactylorhiza hatagirea, Swertia chirayta*
Eye problem	5	1	1	*Ampelocissus rugosa*
Asthma	3	2	0.5	*Cotoneaster microphyllus, Rheum australe*
Poison	24	8	0.696	*Aconitum gammiei, Allium sativum, Asparagus racemosus, Lindera neesiana, Solena heterophylla, Delphinium vestitum, Paris polyphylla, Zanthoxylum armatum*
Delivery	5	4	0.25	*Lindera neesiana, Delphinium vestitum, Paris polyphylla, Zanthoxylum armatum*
Itching	3	2	0.5	*Acorus calamus, Clematis terniflora*
Toothache	2	2	0	*Juglans regia, Begonia picta*

Diversity of Ethnic Uses of Plant species in Kham Magar

Kham Magar of Rolpa district is highly knowledgeable about the use of plant species. The diverse wild plant resources, including medicinal, food, fiber, and dye yielding plants found in the region, and the remoteness of the region allows them to understand and use them. Specifically, the older generation was highly knowledgeable about the diverse use of the species. The use of a few species, such as *Juglans regia*, *Fraxinus floribunda*, and *Maharanga emodi* as dye was reported from two elders who were in their nineties. The very first case study conducted in Kham Magar community documented and provided the importance of plants for the fulfillment of their daily needs in the region. A total 175 species were identified as ethnobotanically useful species based on empirical ethnobotanical study. Twenty five new species have been added in the list of useful flora of Nepal from this study, of which 10 are medicinal, 1 is poisonous, and 14 species have other uses (Table 4.1). The findings of this study support the possibility of recording new information on ethnobotanical importance through extensive ethnobotanical in Kham Magar community. Most of the wild plant species were used for food and medicine, as reported in other parts of Nepal (Kunwar and Bussmann 2008, Gautam 2012, Uprety et al. 2012), as well as elsewhere in the world (Rossato et al. 1999).

However, Kham Magar people are mainly dependent on agriculture and animal husbandry, and they use plant resources in different ways in their daily basic needs. This might not only be due to remoteness, insufficient basic infrastructures (transportation, health facilities), and poverty, but also due to traditional practice, which has been more culturally acceptable from their forefathers (Chaudhary 1998). Of the 50 interviewees, the majority of them were females ($n = 30$), and the rest of the 20 were males by gender, and 5 were *Jhankries*, the traditional healers. Some of the respondents were reluctant to share their experiences. This may be owing to continuous repression by the so-called dominant classes in the past (Pásková 2017). Considering the medicinal plant species, elders and *Jhankries* were more knowledgeable in comparison to other general local people and the young generation. Regarding gender, Kham Magar women hold good knowledge of food plants, but were less knowledgeable on medicine, fiber, and dye yielding plant species. This is because women are mainly involved in household activities, such as collection of fodder and cattle grazing (Uprety et al. 2012), rather than on treatment using medicinal plants.

The majority of the plant species were used in medicine, followed by food, fuelwood, and fodder. This result was different from similar studies in different parts of the country. For instance, Gautam (2012) reported the majority of species used in food, and Thapa (2015) found the majority of species used in fodder. The use of high portion of plant in medicinal may be because Kham Magar have strong dependency on natural remedies for their primary healthcare. Moreover, this may be because of strong cultural beliefs and limited interactions with the outer world (Manandhar 2002). Kham Magar people are selective in the use of plant parts. They use achenes of *Anemone vitifolia* as an antioxidant and sap of *Ampelocissus rugosa* in cataract. They commonly use herbs for their shoots, which was different from the previous studies (Shrestha and Dillion 2003, Gautam 2012, Thapa 2015). This may be because herbs are more abundant and easier to collect, process, and transport. Furthermore, shoots contain a high concentration of bioactive compounds (Luitel et al. 2014).

Kham Magars are proficient in formulation of medicinal plants for use. They formulate plant parts mostly as juice, which is followed by chewing, paste, decoction, and powder (Kunwar et al. 2013). In some cases, sap, latex, and resins of the plants are also used as ointment. Medicinal plant parts were mostly formulated as powder (Thapa 2015). Some plants are formulated in multiple preparation methods based on the knowledge provided by predecessors (Uprety et al. 2010, Bhattarai et al. 2006). The food plants were eaten mostly raw than cooked. Mainly fruits, seeds, and shoot tenders were eaten raw.

Cultural Valuation of Plant species by Kham Magars

Kham Magars are efficient in the valuation of plant species in different categories. Three quantitative techniques—Informant Consensus Factor (Fic), Use Frequency (UF), and Use Value (UV) have been used to analyze the usefulness of the ethnobotanical species among Kham Magar (Philips and Gentry 1993, Rossato et al. 1999, Tardio and Santayana 2008). Fic value was determined in order to know the agreement among the informants of the study area for the use of plants to treat certain ailment categories. The Fic value in eight ailment categories was found to be '1', indicating a high level of informant agreement compared to similar studies conducted in Nepal Himalaya (Kunwar et al. 2010). The plant species having high value of Fic were supposed to be efficient in treating particular ailments. In this study, *Urtica dioica* with Fic = 1 may be sufficient in the treatment of dandruff. Similarly, *Ampelocissus rugosa* is useful in treatment of warts, *Taxus contorta* in cancer, and *Swertia chirayita* in jaundice. So, the plant species that have high consensus values are socio-economically significant and important for pharmacological research to elucidate the chemical compound responsible for the antibacterial activity of plants (Canales et al. 2005).

The UF of overall useful plant species was high compared to medicinal and food plant species. This is because the UF value depends on the number of informants who cite plant species for its effectiveness and easy accessibility (Tardio and Santayana 2008). The medicinal plant species were mostly cited by *Jhankries* for the particular sub-use category, so the UF was less compared to overall useful species. Here, *Taxus contorta* with the highest UF value was the most frequently used species, as this species was cited by most of the informants ($n = 43$) because of its effectiveness. Similarly, the UV was also high for the overall species compared to medicinal and food species. This is because for medicinal plants, the number of use report for a species was less. The UV of food species was lowest compared to overall and medicinal species. This may be due to the low use reports for each species, because most of the species were cited for only single-use category (Tardio and Santayana 2008). Thus, the highest important (Fic, UF, and UV) values shown by useful plants indicate that these species are highly preferred by local people, which might be due to easy accessibility or easy to harvest and high phytochemical constituents.

Conclusion

Documentation of ethnic knowledge is important before they set with elders. Many elders in Magar community have passed away before they could transmit their knowledge to the next generation. The case study conducted in Kham Magar documented 175 ethnically used plant species, of which 25 species (nine medicinal, one poisonous, and 15 with other uses) were new to use flora of the country. There is further more possibility of recording new uses through extensive ethnobotanical study in adjoining areas among Magar Kham. Magar Kham is highly knowledgeable specifically in the use of medicinal plants, so a high diversity of medicinal species has been reported. The quantitative ethnobotanical analysis revealed that 47 species were used for various ailment categories. Species such as *Taxus contorta*, *Juglans regia*, *Quercus semecarpifolia*, *Quercus lanata*, *Abies spectabilis*, and *Lyonia ovalifolia* were frequently used species, and species such as *Aconitum gammiei*, *Swertia chirayta*, *Acorus calamus*, *Bergenia ciliata*, and *Rheum austale* were the most valuable species based on use value. The highest important (Fic, UF, and UV) values shown by medicinal plants indicate that these species are highly preferred by local people, which might be due to high phytochemical constituents. Many species were harvested by Kham Magar for their livelihood. Continuous collection of these species from the forest may lead to the extinction of species in the local area.

Acknowledgments

I am deeply grateful to all informants, who generously shared their time and knowledge. Special thanks are due to my Masters thesis supervisor, Dr. Suresh Kumar Ghimire, Professor in Central Department of Botany, Kathmandu Nepal, for his supervision of this research work and Prabin Bhandari, a Ph.D. scholar in Chinese Academy of Science, China for his support throughout the research, including fieldwork. I am also obliged to my family for continuously supporting me.

References

Acharya, R. 2012. Ethnobotanical study of medicinal plants of Resunga Hill used by Magar community of Badagaun VDC, Gulmi district, Nepal. Sci. World. 10(10): 54–65.

Ale, R., Raskoti, B.B., Shrestha, K. 2009. Ethnobotanical knowledge of Magar community in Siluwa VDC, Palpa district, Nepal. J. Nat. Hist. Mus. 24(1): 58–71.

Baral, S.R., Kurmi, P.P. 2006. A Compendium of Medicinal Plants in Nepal. Mrs. Rachana Sharma, Kathmandu Nepal.

Bhattarai, S., Chaudhary, R.P., Taylor, R.S.L. 2006. Ethnomedicinal plants used by the people of Manang district, central Nepal. J. Ethnobiol. Ethnomed. 2: 41.

Brijesh, S., Daswani, P., Tetali, P., Antia, N., Birdi, T. 2009. Studies on the anti-diarrhoeal activity of *Aegle marmelos* unripe fruit: Validating its traditional usage. BMC Complemen. Altern. Med. 9(1): 47.

Canales, M., Hernandez, T., Caballero, J., Romo de Vivar, A., Avila, G., Duran, A., Lira, R. 2005. Informant consensus factor and antibacterial activity of the medicinal plants used by the people of San Rafael Coxcatlan, Puebla, Mexico. J. Ethnopharmacol. 97: 429–439.

Catalogue of life. 2019. Indexing the known species of World. Published in HYPERLINK "http://www.catalogueoflife. org" www.catalogueoflife.org. (Accessed on 1 April 2020).

CBS [Central Bureau of Statistics] 2011. Nepal Census 2011. District profiles (Demography). Central Bureau of Statistics (CBS), National Planning Commission Secretariat, Kathmandu, Nepal. (Demography). https://data. humdata.org/dataset/nepal-census-2011-district-profiles-demography assessed on July 10, 2019.

Chaudhary, R.P. 1998. Biodiversity in Nepal: Status and Conservation. Tecrss Books, Bangkok, Thailand.

Coburn, B. 1984. Some native medicinal plants of western Gurung. Kailash 11(1-2): 55–88.

Davis, E.W. 1995. Ethnobotany: an old practice, a new discipline. pp. 40–51. *In*: Schultts, R.E., Reis, S.V. (eds). Ethnobotany: Evolution of a Discipline. Dioscoriodes Press, Oregon.

De Sales, A. 2000. The Kham Magar country, Nepal: between ethnic claims and Maoism. Europ. Bul. Himal. Res. 19: 41–71.

De Sales, A. 2010. The biography of a Magar communist: In Varieties of activist experience: Civil. Society in South Asia, SAGE Publications India: 17–45.

De Sales, A. 2017. From ancestral conflicts to local empowerment: two narratives from a Nepalese community. In Windows into a Revolution, Routledge 185–206.

DFO [District Forest Office], 2018. Annual Report of Community Forest Development Program, Government of Nepal Ministry of Forests and Soil conservation Department of Forest, District Forest Office, Rolpa.

DPR [Department of Plant Resources], 2007. Medicinal Plants of Nepal. Department of Plant Resources, Ministry of Forest and Soil Conservation, Government of Nepal. Thapathali, Kathmandu, Nepal.

DPR [District Profile Rolpa]. 2015. Village Development Committee wise Population distribution. http://ddcrolpa. gov.np/en/district-profile/ accessed on August 30, 2019.

Dutta, I.C. 2007. Non-Timber Forest Products of Nepal: Identification, Classification, Ethnic uses and Cultivation. Hillside Press (P.) Ltd., Kathmandu, Nepal.

Eflora. 2019. Annotated Checklist of Flowering Plants of Nepal published in HYPERLINK "http://www.efloras.org" www.efloras.org. (Accessed on 1 April 2020).

Eisenberg, D.M., Harris, E.S.J., Littlefield, B.A., Cao, S., Craycroft, J.A., Scholten, R., Zhao, Z., Chen, H., Liu, Y., Kaptchuk, T., Hahn, W. Wang, X., Roberts, T., Shamu, C.E., Clardy, J. 2011. Developing a library of authenticated Traditional Chinese Medicinal (TCM) plants for systematic biological evaluation—rationale, methods and preliminary results from a Sino-American collaboration. Fitoterapia 82(1): 17–33.

Eloff, J.N. 1998. A sensitive and quick microplate method to determine the minimal inhibitory concentration of plant extracts for bacteria. Plant. Medica. 64(08): 711–713.

Fabricant, D.S., Farnsworth, N.R. 2001. The value of plants used in traditional medicine for drug discovery. Envron. Health Perspect. 109(suppl 1): 69–75.

Ford, R.L. 1978. The nature and status of ethnobotany. *In*: Ford, R.L. (ed.). Anthropological Papers. Museum of Anthropology, University of Michigan, USA.

Gautam, R.K. 2012. Diversity of useful plant species in Humla Karnali Basin, Northwest Nepal. M. Sc. thesis in Plant Systematics and Biodiversity, Central Department of Botany Tribhuvan University.

Gellner, D.N. 2007. Democracy in Nepal: four models. In Seminar (Vol. 576). Seminar Publications.

Ghimire, S.K., Shrestha, A.K., Shrestha, K.K., Jha, P.K. 2000. Plant resources use and human impact around Royal Bardiya National Park, Nepal. J. Nat. Hist. Mus. 19: 3–26.

Ghimire, S.K., Pyakurel, D., Nepal, B., Sapkota, I.B., Prajuli, R.R., Oli, B. 2008. A Manual of NTFs of Nepal Himalyal. WWF Nepal, Kathmandu, Nepal.

Hatlebakk, M. 2009. Explaining Maoist control and level of civil conflict in Nepal. Chr. Michelsen Institute.

Hoffman, B., Gallaher, T. 2007. Importance Indices in Ethnobotany. Ethnobotany Research and Applications 5: 201–218.

Ianni, E., Geneletti, D., Ciolli, M. 2015. Revitalizing traditional ecological knowledge: A study in an alpine rural community. Environ. Manag. 56(1): 144–156.

Kunwar, R.M., Bussmann, R.W. 2008. Ethnobotany in the Nepal Himalaya. J. Ethnobiol. Ethnomed., 4(24): 1746–1769.

Kunwar, R.M., Shrestha, K.P., Bussmann, R.W. 2010. Traditional herbal medicine in Far-west Nepal: a pharmacological appraisal. J. Ethnobiol. Ethnomed. 6(1): 35.

Kunwar, R.M., Mahat, L. Acharya, R.P., Bussmann, R.W. 2013. Medicinal plants, traditional medicine, markets and management in far-west Nepal. J. Ethnobiol. Ethnomed 9(1): 24 http://www.ethnobiomed.com/content/9/1/24.

Lama, Y.C., S.K. Ghimire and Aumeeruddy-Thomas, Y. 2001. Medicinal Plants of Dolpo: Amchis' Knowledge and Conservation. WWF Nepal Program, Kathmandu.

Luczaj, L., Pieroni, A., Tardío, J., Pardo-de-Santayana, M., Sõukand, R., Svanberg, I., Kalle, R. 2012. Wild food plant use in 21 st century Europe, the disapperance of old traditions and the search for new cuisines involving wild edibles. Act. Societ. Bot. Polon. 81(4).

Luitel, D.R., Rokaya, M.N., Timisina, B., Munzbergova, Z. 2014. Medicinal plants used by the Tamang community in the Makawanpur district of central Nepal. J. Ethnobiol. Ethnomed. 10(1): 5. http://www.ethnobiomed.com/content /10/1/5.

Malla, B., Gauchan, D.P., Chhetri, R.B. 2015. An ethnobotanical study of medicinal plants used by ethnic people in Parbat district of western Nepal. J. Ethnopharmacol. 165: 103–117.

Manandhar, N.P. 2002. Plants and People of Nepal. Timber Press, Portland, Oregon.

Martin, G.J. 1995. Ethnobotany: A methods Manual. Chapman and Hall, London.

Molnar, A.M. 1981. The Kham Magar Women of Thabang. Status of Women in Nepal, Volume 2, Part 2, CEDA Centre for Economic Development Administration, Kathmandu: Tribhuvan University.

Molnar, A.M. 1982a. Flexibility and Option: A Study of the Dynamics of Women's Participation Among the Kham Magar of Nepal: in Dissertation Abstracts in Himalaya the journal of Association for Nepal and Himalayan Studies 2(2): 35–36.

Molnar, A.M. 1982b. women and politics: case of the Kham Magar of western Nepal. Amer. Ethnol. 9(3): 485–502. DOI: 10.1525/ae.1982.9.3.02a00030.

Noonan, M. 2003. Recent language contact in the Nepal Himalaya. Linguistics, Nepal, 1–24. https://doi.org/10.11588/xarep.00000189

Oli, B.R., Ghimire, S.K., Bhuju, D.R. 2005. Ethnographic validity and use values of plants locally utilized in the Churiya of east Nepal: A quantitative approach to ethnobotany. Bot. Orient. 5: 40–47.

Pásková, M. 2017. Local and indigenous knowledge regarding the land use and use of other natural resources in the aspiring Rio Coco geopark. In IOP Conference Series: Earth and Environmental Science 95(5): 052018. IOP Publishing.

Phillips, O., Gentry, A.H. 1993. The useful plants of Tambopata, Peru: I. Statistical hypotheses tests with a new quantitative technique. Eco. Bot. 47: 15–32.

Rao, S.S. 2006. Indigenous knowledge organization: An Indian scenario. Int. J. Inform. Man. 26(3): 224–233.

Rajbhandari, K.R. 2001. Ethnobotany of Nepal. Kathmandu: Ethnobotanical Society of Nepal (ESON). 189.

Rajbhandary, S. 2013. Traditional Uses of Begonia species (Begoniaceae) in Nepal. J. Nat. Hist. Mus. 27: 25–34.

Reyes-García, V., Paneque-Gálvez, J., Luz, A.C., Gueze, M., Macía, M.J., Orta-Martínez, M., Pino, J. 2014. Cultural change and traditional ecological knowledge. An empirical analysis from the Tsimane'in the Bolivian Amazon. Hum. Org. 73(2): 162.

Rossato, S.C., Leitão-Filho, H.F., Begossi, A. 1999. Ethnobotany of Caiçaras of the Atlantic forest coast (Brazil). Eco. Bot. 53: 387–395.

Sagar, P.K. 2014. Adulteration and substitution in endangered, ASU herbal medicinal plants of India, their legal status, scientific screening of active phytochemical constituents. I. J. Pharm. Sci. Res. 5(9): 4023.

Salim, M.A., Ranjitkar, S., Hart, R., Khan, T., Ali, S., Kiran, C., Xu, J. 2019. Regional trade of medicinal plants has facilitated the retention of traditional knowledge: case study in Gilgit-Baltistan Pakistan. J. Ethnobiol. Ethnomed. 15(1): 6.

Sapkota, P.P. 2008. Ethno-ecological Observation of Magar of Bukini, Baglung, Western, Nepal. Dha. J. Soc. Ant. 2: 227–252.

Sapkota, P.P. 2010. the Ritual use of jhakro in Magar Community. Dha. J. Soc. Ant., 4: 223–234.

Shrestha, P.M., Dillion, S.S. 2003. Medicinal plant diversity and use in the highlands of Dolakha district, Nepal. J. Ethnopharm. 86: 81–96.

Shrestha, K.K., Rajbhandary, S., Tiwari, N.N., Poudel, R.C., Uprety, Y. 2004. Ethnobotany in Nepal: Review and perspectives. Kathmandu, Nepal: WWF Nepal Program.

Singh, A.G., Kumar, A., Tewari, D.D. 2012. An ethnobotanical survey of medicinal plants used in Terai forest of western Nepal. J. Ethnobiol. Ethnomed. 8: 1–19.

Singh, A.G., Hamal, J.P. 2013. Traditional phytotherapy of some medicinal plants used by Tharu and Magar communities of Western Nepal, against dermatological disorders. Sci. Worl. 11(11): 81–89.

Singh, A.G., Kumar, A., Tewari, D.D., Bharati, K.A. 2018. New ethnomedicinal claims from Magar community of Palpa district, Nepal. NISCAIR-CSIR, India.

Sujarwo, W., Arinasa, I.B.K., Salomone, F., Caneva, G., Fattorini, S. 2014. Cultural erosion of Balinese indigenous knowledge of food and nutraceutical plants. Eco. Bot. 68(4): 426–437.

TAAN [Trekking Agencies' Association Nepal], 2013. Jaljala Trek the Hidden Treasurer of Rolpa. Imag. Nep. Bi-Mon. Tra. Mag. 25(5): 22–26.

Tardio, J., Santayana, M.P.D. 2008. Cultural Importance Indices: A Comparativce Analysis Based on the Useful Wild Plants of Southern Cantabrai (Northern Spain). Eco. Bot. 62(1): 24–39.

Taye, B., Giday, M., Animut, A., Seid, J. 2011. Antibacterial activities of selected medicinal plants in traditional treatment of human wounds in Ethiopia. Asi. Pac. J. Trop. Biomed. 1(5): 370–375.

Taylor, J.L.S., Rabe, T., McGaw, L.J., Jäger, A.K., Van Staden, J. 2001. Towards the scientific validation of traditional medicinal plants. Plan. Grow. Reg. 34(1): 23–37.

Thapa, S. 2012. Medico-ethnobotany of Magar community in Salija VDC of Parbat district, central Nepal. Our nat. 10(1): 176–190.

Thapa, G.S. 2014. Magarati Sanskirti: A Historical Review. Magar Pragati Samuha, Thapathali, Kathmandu.

Thapa, C. 2015. Diversity of useful plant species along an elevation gradient in Chungsa Valley, Humla, Northwest Nepal. M.Sc. thesis in Plant Systematics and Biodiversity, Central Department of Botany Tribhuvan University.

The Plant List. 2019. Version 1.1. Published on the Internet; http://www.theplantlist.org/ (accessed on 1 April 2020).

Uprety, Y., Asselin, H., Boon, E.K., Yadav, S., Shrestha, K.K. 2010. Indigenous use and bio-efficacy of medicinal plants in the Rasuwa District, Central Nepal. J. Ethnobiol. Ethnomed. 6: 1–6.

Uprety, Y., Poudel, R.C., Asselin, H., Boon, E. 2011. Plant biodiversity and ethnobotany inside the projected impact area of the Upper Seti Hydropower Project, Western Nepal. Env. Dev. Sus., 13(3): 463–492.

Uprety, Y., Paudel, R.C., Shrestha, K.K., Rajbhandary, S. Tiwari, N.N, Shrestha, U.B., Asselin, H. 2012. Diversity of use and local knowledge of wild edible plant resources in Nepal. J. Ethnobiol. Ethnomed. 8: 1–1.

Watters, D.E. 2002. A Grammar of Kham. Cambridge University Press.

5

Some Plants Used as Phytomedicine by Tribal Healers of Chittagong Hill Tracts, Bangladesh

Khoshnur Jannat,[1] Rownak Jahan,[1] Taufiq Rahman,[2] Md Shahadat Hossan,[3] Nasrin Akter Shova,[1] Maidul Islam[1] and Mohammed Rahmatullah[1,]*

Introduction

Ethnobotany implies the study of interactions between plants and people, the study comprising of gathering and documentation of the traditional knowledge and culture of the people in their use of plants for food, clothing, and medicine. Food and medicine are closely related, so much so that Hippocrates said "let food be thy medicine and let medicine be thy food" (Smith 2004). It is a fact that a lot of plants consumed by early humans and even modern humans have dietary, preventive, and therapeutic values. For instance, garlic has food value as a spice, the cloves are cardioprotective, and eating garlic can reduce cholesterol in hypercholesterolemic subjects (Das et al. 2016). A more selective term in ethnobotany is ethnomedicine—understood by most scientists as traditional medicine based on plants and animals, practiced by various ethnic groups, and whose traditional medicine(s) differs from allopathic medicine. Ethnomedicine is increasingly gaining importance in recent years; plants have always formed a source of new drugs, and it is being more and more appreciated that instead of synthetic chemistry, plants can be rich sources of more drugs in the future.

Traditional medicine has different names in different countries, but all have originated from the medicinal practices of indigenous people. In China, traditional Chinese medicine (TCM) dates back thousands of years ago; in India, Ayurveda developed around 4000 BC; Unani medicine developed in Greece around 2,500 years ago, and then through Arabs took root in India and some other Muslim countries; Kampo or traditional Japanese medicine came to Japan through Korea possibly in the 5th or 6th century AD; traditional medicines practiced by Australian aborigines and African tribal shamans have unknown time periods of origin; Russian herbal medicine developed in the 10th century (Yuan et al. 2016). All of these traditional medicinal systems are basically phytotherapeutic in nature. Human

[1] Department of Biotechnology and Genetic Engineering, University of Development Alternative, Lalmatia, Dhaka-1207, Bangladesh.

[2] Department of Pharmacology, University of Cambridge, Tennis Court Road, CB2 1PD, UK.

[3] School of Pharmacy, University of Nottingham, Nottingham NG7 2RD, UK.

* Corresponding author: rahamatm@hotmail.com

beings, since their very advent either by trial and error, or through fortuitous circumstances, or even by watching animals, learned to use plants as medicines. Wild chimpanzees in Uganda are known to ingest medicinal plants during various sicknesses (Krief et al. 2005); the same has been observed for wooly spider monkeys in Brazil (Petroni et al. 2017). Interestingly, at least a good proportion of the medicinal plants believed to be ingested by chimpanzees or wooly spider monkeys were also used by humans residing in or near the same forests for what appeared to be similar therapeutic purposes.

It can then be safely said that plants from time immemorial had formed the mainstay of human medicinal uses (besides satisfying nutritional needs). It is also self-evident that before the advent of the presently codified traditional medicinal systems and allopathic medicine, tribal medicinal practitioners were the repositories of medicinal knowledge, which in the absence of writing, were orally transmitted from generation to generation—a practice still present in recent times. Although allopathic medicine is in vogue for only about two centuries, a number of allopathic drugs have been discovered from observing and documenting the medicinal practices of indigenous people (Gilani and Rahman 2005). Major plant drugs for which no synthetic ones are available include vinblastine, reserpine, quinine, pilocarpine, cocaine, morphine, codeine, and artemisinin, to name only a few (Kumar et al. 1997). More plant-originating drugs are likely to enter the market or have done so, which include romidepsin, curcumin, ternatolide, camptothecin, and lovastatin (Yuan et al. 2016).

Considering that the modern age is seeing the emergence of new diseases (Ebola, MERS. Nipah, Hanta, bird flu to name a few) and emergence of drug-resistant vectors, the modern world is in desperate need for new drugs that can combat these twin menaces. There are also other considerations, such as modern synthetic drugs may give adverse effects and are not affordable to people of developing countries. Just as an example, cisplatin, which is a chemotherapy medication used against a number of cancers, is nephrotoxic (Nasri 2013); on the other hand, garlic which has anti-cancer potential (Petrovic et al. 2018), has renoprotective effect (Nasri 2012). Other factors also necessitate the discovery of efficacious drugs, such as increase in non-contagious diseases, such as cardiovascular disorders, diabetes, cancer, and neurodegenerative disorders.

There are about 391,000 vascular plants known to science, of which 369,000 are flowering plants, according to Planeta.com. Since all plants produce dozens to hundreds of secondary metabolites, which have pharmacological activities ranging from toxic to having therapeutic values, plants may be our main hope in preventing and curing diseases. However, instead of moving in the dark, the scientific first step would be to monitor the therapeutic uses of plants by traditional medicinal practitioners, preferably practitioners belonging to the indigenous people, for they are the persons who, for generations, have treated diseases with plants. This review will attempt to cover some medicinal plants used by the tribal people of Bangladesh. Bangladesh has perhaps over 100 large and small tribes within its borders (Chakma and Maitrot 2016), and it is a longstanding debate among the anthropologists about the number of tribes, and whether the tribes are indigenous or recent arrivals. We shall use the two terms interchangeably for the documented arrival of the most recent tribes (mainly tea plantation workers, who are essentially fragmented tribes brought to work in the tea estates by the then British rulers, dating around 1850).

Research on Medicinal Plants of Chittagong Hill Tracts

The list of plants used by the various tribal people and as presented here is not a complete list. Although we had been conducting surveys among various tribes since 2009 (Mia et al. 2009, Hossan et al. 2009, Shahidullah et al. 2009), we believe that we have surveyed only about 20% of the tribes, and that too not completely, since we did not visit all the communities of any given tribe. Tribes can be spread out over large and remote areas. For instance, the Garo tribe can be found in the districts of Tangail, Jamalpur, Sherpur, Mymensingh, Netrakona, Sunamganj, Sylhet, and Gazipur (Figure 5.1). The Chakmas, who form the largest ethnic group in Bangladesh, can be seen in the

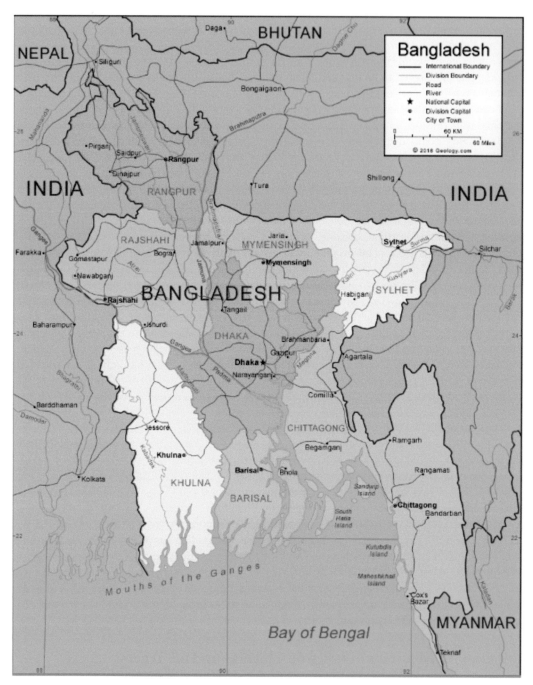

Figure 5.1: District map of Bangladesh.

districts of Chittagong, Cox's Bazaar, Khagrachhari, and Rangamati, as well as the states of Arunachal, Mizoram, and Tripura in neighboring India. The phytotherapeutic practices of tribal healers can vary widely; often two tribal or folk medicine healers in adjoining villages (or paras) will use different plant(s) to treat the same ailment (Mollik et al. 2010).

The various medicinal plants of tribes residing in the Chittagong Hill Tracts region in the southeast portion of Bangladesh are shown in Table 5.1. This region comprises of four districts, namely Bandarban, Cox's Bazaar, Khagrachhari, and Rangamati (Figure 5.1). Cumulatively, the region is inhabited by several large and small ethnic groups (tribes), including the Chakmas, the largest ethnic group in Bangladesh. The whole region essentially comprises of low hills covered with dense forests in parts. In the last decade or two, the primary forest is mostly gone, possibly due to the shifting slash-and-burn (jhum) cultivation methods practiced by the various ethnic groups, and also an influx of mainstream population, who have totally cleared forest areas and have prepared the ground for year-round cultivation. A sizeable chunk of once forested area has also been taken over by commercial planters, who cultivate crops such as tobacco, banana, cashew nut, and various other exotic fruits, such as dragon fruit, oranges, peaches, and durian—the fruits being previously never known to be cultivated in Bangladesh. The tribal population cannot use the forest area as before. Even 30 years back, in their slash-and-burn cultivation, a small section of the forest would be burned and cultivated for 2–3 years. The forest part would then be left to regenerate for the next 20 years at least. As a consequence, primary forest (trees) would come back and be able to grow for a sufficient amount of time to bear fruits and seeds, and so attract insects, birds, and fauna.

Secondary forests have now largely taken the place of primary forests (Figure 5.2). The forest area now mostly comprises of herbs, shrubs, and occasionally non-indigenous trees, such as *Tectona grandis*, planted for its wood. The primary forest area now comprises of around 15–20% of the total forest area (Ahammad and Stacey 2016). However, the remaining primary and secondary forests contain over 3,000 floral species, a large proportion of which are medicinally used by the various tribes.

A number of facts need to be borne in mind regarding the present review. First, the data presented here on the medicinal plants of the various tribes has been collected by the authors and various other researchers in actual field surveys; the details can be found in the references. No other work has been consulted beyond what has been surveyed by the principal author's (MR) team(s) at various periods since 2008. Second, this list (Table 5.1) is not a comprehensive list for a variety of reasons, such as not visiting all the tribes, not visiting all the residential communities (paras) of any given tribe, for selection of medicinal plants to treat a certain disease may vary widely between herbal practitioners of the same tribe, and not listing plants mentioned by other authors. Moreover, ethnomedicinal surveys are basically still at its infancy in Bangladesh, and much ground needs to be covered to get a comprehensive idea about the tribal medicinal plants. On the other hand, it is of utmost importance that such surveys be conducted in the quickest possible manner, for not only are forests dwindling because of human habitat pressure, but also that tribal knowledge and culture is being forgotten with the influx of the culture, habits, and education system(s) of the mainstream Bengali speaking population and foreign missionaries. It is the practical experience of the authors that young tribal people do not

Figure 5.2: A section of forest land in Bandarban district, Bangladesh.

Table 5.1: Plants used for therapeutic purposes by various tribes of Bangladesh (not a comprehensive list).

Plant Name	English name	Family	Local name	Parts used	Diseases treated	Tribe	References
Acanthus ilicifolius L.	Holy mangrove	Acanthaceae	Fereng-jubang	Root	Sex stimulant, rheumatic pain, cloudy urination	Marma	Rahmatullah et al. 2009
Andrographis paniculata (Burm.f.) Wall. Nees.	Creat	Acanthaceae	Chirata	Whole plant, leaf, stem	Fever, pain, malaria, diabetes, stomachic, tonic, alterative, helminthiasis, cholagogue, general debility, dysentery, certain forms of dyspepsia, liver complaints mainly of children, flatulence, diarrhea in children, spleen complaints, colic, strangulation of intestine, constipation, diarrhea, cholera, phthisis, jaundice	Chakma	Tasannun et al. 2015, Malek et al. 2014
Dicliptera bupleuroides Nees	Not known	Acanthaceae	Kala jaro	Leaf, stem	Fever, headache, wounds, sores between fingers	Chakma	Malek et al. 2014
Hemigraphis hirta (Vahl) T. Anderson	Hairy Hemographis	Acanthaceae	Kakra giluk shak	Whole plant	Headache	Chakma	Malek et al. 2014
Justicia adhatoda L.	Malabar nut tree	Acanthaceae	Bashok, Ludi bashok	Leaf	Coughs, mucus, hemorrhoids, rheumatic pain, breathing problems, asthma, helminthiasis, diarrhea, constipation	Chakma, Pankho, Marma	Malek et al. 2014, Sarker et al. 2013, Esha et al. 2012, Afroz et al. 2013, Rahmatullah et al. 2009
Justicia gendarussa Burm.	Gendarussa	Acanthaceae	Kala jaro gach	Leaf	Constipation, asthma	Chakma	Malek et al. 2014, Afroz et al. 2013
Lepidagathis incurva Buch.-Ham. ex D.Don	Curved Lepidagathis	Acanthaceae	Aarae uri nolakkher	Leaf, bark, root	Skin cancer	Chakma	Esha et al. 2012
Thunbergia grandiflora (Roxb. ex Rottler) Roxb.	Blue Thunbergia	Acanthaceae	Del lota, Jong-haileng	Stem. leaf	Conjunctivitis, eye problem, inflammation, cuts and wounds, astringent	Chakma, Bawm	Malek et al. 2014, Hossan et al. 2014
Acorus calamus L.	Sweet Flag	Acoraceae	Paan Raja, Boch	Leaf, root	Constipation, alopecia, asthma, mental disorders	Chakma, Pankho, Tonchongya	Tasannun et al. 2015, Sarker et al. 2013, Wahab et al. 2013, Rashid et al. 2012

Scientific name	Common name	Family	Local name	Part used	Uses	Ethnic group	References
Acrostichum aureum L.	Golden leather fern	Adiantaceae	Mou-chai-pang	Leaf	To increase physical strength, cloudy urination in women, sex stimulant	Marma	Rahmatullah et al. 2009
Adiantum lunulatum Burm.f.	Walking maiden hair fern	Adiantaceae	Arthichum	Wholeplant	Swollen eye, conjunctivitis	Bawm	Hossan et al. 2014
Adiantum philippense L.	Maiden hair fern	Adiantaceae	Goyali lota, Bandor thala, Kijau-pai-bang	Leaf, root	Fever, dysentery, crying in feverish children, sex stimulant	Chakma, Marma	Tasannun et al. 2015, Rahmatullah et al. 2009
Dracaena spicata Roxb.	Dracaena	Agavaceae	Boang-khola-paing-da	Leaf	Long-term fever, coughs, and mucus in nose	Marma	Rahmatullah et al. 2009
Sansevieria roxburghiana Schultes & Schultes f.	Indian bowstring hemp	Agavaceae	Lankh-hi-pang	Leaf	Leucorrhea, abscess	Chakma, Marma	Tasannun et al. 2015, Rahmatullah et al. 2009
Achyranthes aspera L.	Prickly chaff-flower	Amaranthaceae	Chirchiri, Obalenga	Leaf, whole plant, root	Abscess, having trouble during urination, passing of blood with urine, jaundice, respiratory problems	Pankho, Tonchongya, Marma	Sarker et al. 2013, Wahab et al. 2013, Rahmatullah et al. 2009
Aerva sanguinolenta L.	Kapok bush	Amaranthaceae	Lal pata	Leaf	Insect or snake bite	Pankho, Tonchongya	Sarker et al. 2013, Wahab et al. 2013
Amaranthus spinosus L.	Prickly amaranth	Amaranthaceae	Kanta marich, Kanta notay	Root, leaf	Excessive bleeding during menstruation, fever, diarrhea, malaria	Pankho, Tonchongya, Marma	Sarker et al. 2013, Wahab et al. 2013, Afroz et al. 2013
Celosia argentea L.	Silver cockscomb	Amaranthaceae	Dhupful gach, Kheyang marek	Leaf, stem, flower	Irregular menstruation, impotency, skin infections	Chakma	Malek et al. 2014, Esha et al. 2012
Celosia cristata L.	Crested cockscomb	Amaranthaceae	Morogful, Kromopa	Branch, leaf, flower, root	Leucorrhea, menstrual irregularity, uterus enlargement	Chakma, Marma	Afroz et al. 2013
Cyathula prostrata (L.) Blume	Small prickly chaff flower	Amaranthaceae	Uvo langera, Aarihuri nolahar	Leaf, root, whole plant	Itching, expectorant, emetic, demulcent, vulnerary	Chakma	Tasannun et al. 2015, Malek et al. 2014
Crinum asiaticum L.	Poison bulb	Amaryllidaceae	Hobaron, Shada roshun	Stem, leaf, fruit	Jaundice, Coughs, abscess on leg	Chakma, Pankho	Malek et al. 2014, Sarker et al. 2013
Gomphrena celosioides Mart.	Globe amaranth	Amaranthaceae	Ranga ameleya	Leaf, flower	Paralysis	Chakma	Afroz et al. 2013
Pancratium maritimum L.	Sea Daffodil	Amaryllidaceae	Khobarun	Leaf	Tumor	Chakma	Tasannun et al. 2015

Table 5.1 contd. ...

...*Table 5.1 contd.*

Plant Name	English name	Family	Local name	Parts used	Diseases treated	Tribe	References
Leea macrophylla Roxb.	Hathikana	Ampelidaceae	Bakaraj pata gach	Leaf, stem, root	Chest pain, back pain	Chakma	Malek et al. 2014
Anacardium occidentale L.	Cashew	Anacardiaceae	Gasnak	Fruit	Skin rash, skin disorder, alopecia, and helminthiasis	Bawm	Hossan et al. 2014
Polyalthia longifolia (Sonn.) Thwaites (PL)	False Ashok	Annonaceae	Debdaru	Bark	Burning sensations, fever, and diabetes	Chakma	Esha et al. 2012
Centella asiatica (L.) Urb.	Asian pennywort	Apiaceae	Thankuni	Leaf	Dysentery and other stomach disorders	Tripura	Afroz et al. 2013
Alstonia scholaris (L.) R.Br.	Devil tree	Apocynaceae	Sechsena gach, Chalai-bang, Nariath-ku-o	Leaf, stem, root, bark	Lack of milk in mother following childbirth, cold sores, fevers, and diabetes, typhoid fever	Chakma, Marma, Bawm	Esha et al. 2012, Rahmatullah et al. 2009, Hossan et al. 2014
Calotropis gigantea R.Br.	Giant Milkweed	Apocynaceae	Akondo, Mru-na	Leaf	Pain, rheumatism, and joint pain	Marma, Bawm	Afroz et al. 2013, Hossan et al. 2014
Catharanthus roseus L.	Madagascar periwinkle	Apocynaceae	Badam boot, Nayan tara	Leaf, bark, stem, root	Ear infection, hypertension, helminthiasis, and passing of blood with urine	Chakma, Tripura	Tasannun et al. 2015, Esha et al. 2012, Afroz et al. 2013
Holarrhena antidysenterica (Roxb. ex Fleming) Wall.	Conessi tree	Apocynaceae	Kurok gach	Bark	Dysentery	Pankho	Sarker et al. 2013
Ichnocarpus frutescens R.Br.	Black creeper	Apocynaceae	Monri chocha	Leaf	Chicken pox	Chakma	Malek et al. 2014
Rauwolfia serpentina (L.) Benth. ex Kurz.	Serpentine	Apocynaceae	Sursang, Churmang	Leaf, root, whole plant	High blood pressure, hypertension, and stomach pain	Chakma, Pankho, Tonchongya, Marma, Tripura	Tasannun et al. 2015, Esha et al. 2012, Sarker et al. 2013, Wahab et al. 2013, Afroz et al. 2013
Tabernaemontana corymbosa Wall.	Pinwheel flower	Apocynaceae	Sisaida	Root	Fever and flatulence	Bawm	Hossan et al. 2014

Scientific name	Common name	Family	Local name	Part used	Ailments/Uses	Tribe	References
Tabernaemontana divaricata (L.) R.Br. ex Roem. & Schult.	Pinwheel flower	Apocynaceae	Kathal khatya, Chanle-pang	Leaf stalk, leaf, root, fruit	Any disease in newborn infants, redness in eyes, conjunctivitis, ulcer, and breathing problems	Chakma, Marma	Malek et al. 2014, Esha et al. 2012, Rahmatullah et al. 2009
Thevetia peruviana (Pers.) K. Schum.	Yellow Oleander	Apocynaceae	Korobi, Khungkrom	Fruit	Hair loss	Tripura	Afroz et al. 2013
Aglaonema hookerianum Schott.	White dragon's head	Araceae	Gach pettai	Leaf, stem	Rheumatism	Chakma	Malek et al. 2014
Alocasia cucullata (Lour.) G. Don.	Chinese taro	Araceae	Bilae kochu	Leaf	Infections from being cut by thorns, and snake bite	Chakma	Esha et al. 2012
Areca catechu L.	Betelnut palm	Arecaceae	Supari gach	Root (east-sided)	Dysentery in children aged between 1–6 months	Marma	Afroz et al. 2013
Cocos nucifera L.	Coconut	Arecaceae	Daba	Fruit, root	Gonorrhea, syphilis, sedative, alopecia, flu, inflammation	Bawm	Hossan et al. 2014
Syngonium podophyllum Schott.	African evergreen	Araceae	Patabahar	Leaf	Influenza and pneumonia	Chakma	Tasannun et al. 2015
Typhonium trilobatum (L.) Schott.	Bengal arum	Araceae	Ghet Kochu, Nirbich	Leaf, Tuber	Snake bite	Chakma	Tasannun et al. 2015
Aristolochia tagala Cham.	Dutchman's pipe	Aristolochiaceae	Horinkan shak lota	Leaf, root	Fever, bloating, and fever	Chakma	Malek et al. 2014
Calotropis gigantea (L.) Ait.f.	Bowstring hemp	Asclepiadaceae	Akon shak	Leaf	Whitish discharge in urine, hypertension, helminthiasis (hook worm)	Chakma	Esha et al. 2012
Aloe vera (L.) Burm.f.	Indian Aloe	Asphodelaceae	Ghritokumari	Fleshy pulp of leaf	To cool head, heart problems, cholera, diarrhea, and physical weakness	Tripura	Afroz et al. 2013
Ageratum conyzoides L.	Goat weed	Asteraceae	Moni muchaher, Mogojo gach	Leaf	Insomnia, external cuts, and wounds	Chakma	Malek et al. 2014, Afroz et al. 2013
Blumea clarkei Hook.f.	Not known	Asteraceae	Tora gach	Leaf	Fracture and jaundice	Chakma	Malek et al. 2014
Blumea lacera (Burm.f.) DC.	Malay blumea	Asteraceae	Chichaknu	Leaf	Acidity, gastrointestinal troubles, flatulence, and stomach discomfort	Bawm	Hossan et al. 2014
Blumea membranacea Wallich ex de Candolle	Not known	Asteraceae	Kalo ambosh	Leaf	Crying in children with squirming	Chakma	Malek et al. 2014

Table 5.1 contd. ...

...*Table 5.1 contd.*

Plant Name	English name	Family	Local name	Parts used	Diseases treated	Tribe	References
Calendula officinalis L.	Pot Marigold	Asteraceae	Gada ful	Leaf	To stop bleeding from external cuts and wounds	Tripura	Afroz et al. 2013
Cirsium arvense (L.) Scop.	Creeping Thistle	Asteraceae	Bhosh mola	Leaf	Scorpion or other poisonous insect bite	Chakma	Malek et al. 2014
Chromolaena odorata (L.) R. M. King & H.Rob.	Siam weed	Asteraceae	Mojakkher, Khut toring	Young leaf at top of stem, root	Bleeding from external cuts and wounds, abscess with pain.	Chakma, Pankho, Tonchongya	Esha et al. 2012, Sarker et al. 2013, Wahab et al. 2013, Rashid et al. 2012
Elephantopus scaber L.	Elephant foot	Asteraceae	Rambohok, Pau-ma-fang	Leaf, root.	Inflammation, edema, and astringent, stomach pains, during gastric ulcer	Bawm, Marma	Hossan et al. 2014, Rahmatullah et al. 2009
Eupatorium odoratum L.	Christmas Bush	Asteraceae	Kaingja-pongja	Leaf	Stop bleeding, stimulate clot formation	Marma	Rahmatullah et al. 2009
Gynura nepalensis DC.	Mollucan spinach	Asteraceae	Sidereh beshak	Leaf	Stomach tumor	Chakma	Malek et al. 2014
Mikania cordata Burm.f.	Bitter vine	Asteraceae	Japai-nueh, Bache-a	Leaf	Bleeding from external cuts and wounds, stop bleeding; stimulate clot formation, emollient, gripes (sharp pains in the bowels), itch, wound	Tonchongya, Marma, Bawm	Rashid et al. 2012, Rahmatullah et al. 2009, Hossan et al. 2014
Spilanthes calva DC.	Toothache plant	Asteraceae	Japan-ankhasha, Aankhasa	Flower	Energizer, toothache	Bawm	Hossan et al. 2014
Spilanthes paniculata Wall. ex DC.	Not known	Asteraceae	Oshun shak	Leaf	Insomnia, oral lesion	Tonchongya	Rashid et al. 2012
Synedrella nodiflora Gaertn.	Cinderella weed	Asteraceae	Lung-ankhasha	Whole plant	Helminthiasis	Bawm	Hossan et al. 2014
Tagetes erecta L.	Marigold	Asteraceae	Ganda gach	Leaf	Hemorrhoids	Chakma	Esha et al. 2012
Vernonia cinerea L.	Purple Fleabane	Asteraceae	Dondo Urphong	Leaf, root	Rheumatic pain, if somebody is afraid or possessed by genies or evil spirits, feeling afraid, being possessed by "genies" or "ghosts"	Chakma, Pankho, Tonchongya	Tasannun et al. 2015, Sarker et al. 2013, Wahab et al. 2013
Basella alba L.	Malabar spinach	Basellaceae	Tharbak	Leaf	Dysentery, diarrhea, febrifuge, ulcer, edema, fever, wounds, nutritive	Bawm	Hossan et al. 2014

Scientific name	Common name	Family	Local name	Part used	Uses	Tribe	References
Begonia barbata Wall. ex A.De.	Wax Begonia	Begoniaceae	Shiltedoi, Kukthur (white)	Leaf, stem, wholeplant	Pain in the urinary tract while urinating, diarrhea in children, menstrual difficulties, irregular menstruation	Pankho, Tonchongya, Bawm	Sarker et.al. 2013, Wahab et.al. 2013, Hossan et al. 2014
Begonia silhetensis (A. DC.) C.B. Clarke)	Not known	Begoniaceae	Kukthur	Leaf, stem, whole plant	Diarrhea in children, irregular menstruation, dysmennorhea, headache	Bawm	Hossan et al. 2014
Heliotropium indicum L.	Devil weed	Boraginaceae	Hatichora	Leaf, root	Snake bite	Tripura	Afroz et al. 2013
Oroxylum indicum Vent.	Indian trumpet flower	Bignoniaceae	Kanai dinga, Khona gach	Bark, leaf, root	Jaundice, rheumatic fever, sudden unconsciousness, epilepsy, skin disorders, sexual stimulant, diarrhea, fever, wound	Chakma, Pankho, Marma, Bawm	Tasannun et al. 2015, Sarker et al. 2013, Rahmatullah et al. 2009, Hossan et al. 2014
Bombax ceiba L.	Cotton tree	Bombacaceae	Kumpangkung	Root	Aphrodisiac, premature ejaculation, helminthiasis	Bawm	Hossan et al. 2014
Ananas comosus (L.) Merr.	Pineapple	Bromeliaceae	Naindra-bang, Lothi	Leaf, fruit	Pneumonia, asthma, respiratory problems, helminthiasis, nutritive	Marma Bawm	Rahmatullah et al. 2009, Hossan et al. 2014
Opuntia dillenii (Ker Gawl) Haw.	Prickly pear	Cactaceae	Rengkung	Plant sap	Joint pain, arthritis	Bawm	Hossan et al. 2014
Carica papaya L.	Papaya	Caricaceae	Kukia, Kamkor	Latex, leaf, fruit	Hard stool, constipation, piles, ringworm	Tripura, Bawm	Afroz et al. 2013, Hossan et al. 2014
Casuarina equisetifolia L.	Beach Casuarina	Casuarinaceae	Pailong-pang	Root	To maintain healthy teeth	Marma	Rahmatullah et al. 2009
Anogeissus acuminata Wall.ex C.B.Clarke	Axle wood	Combretaceae	Sai-ki-bang, Sung-chubu	Bark, leaf	Toothache, loosening of tooth, lesions within the mouth or around the tooth, fever	Marma, Bawm	Rahmatullah et al. 2009, Hossan et al. 2014
Terminalia arjuna (Roxb.) Wight & Arn.	Arjun	Combretaceae	Arjun	Bark	Dysentery, flatulency, sex stimulant, paralysis	Chakma	Malek et al. 2014, Afroz et al. 2013
Terminalia bellirica Roxb.	Belliric Myrobalan	Combretaceae	Bohera, Bora gach	Fruit, bark, leaf	Anemia, asthma, gray hair, abscess, burning sensations on skin, hemorrhoids	Chakma	Tasannun et al. 2015, Esha et al. 2012

Table 5.1 contd. ...

...Table 5.1 contd.

Plant Name	English name	Family	Local name	Parts used	Diseases treated	Tribe	References
Terminalia chebula (Gaertn.) Retz.	Chebulic Myrobalan	Combretaceae	Uttel gach, Uoal	Fruits, leaf	Against all kinds of diseases, hemorrhoids, diabetes	Chakma	Malek et al. 2014, Esha et al. 2012
Argyreia nervosa (Burm.f.) Bojer.	Elephant Creeper	Convolvulaceae	Achar gach	Leaf, bark	Body ache	Chakma	Afroz et al. 2013
Ipomoea aquatica Forssk.	Water spinach	Convolvulaceae	Hormoma shak	Leaf	Constipation	Chakma	Esha et al. 2012
Ipomoea triloba L.	Aiea morning glory	Convolvulaceae	Del lodi	Leaf	Facial distortion	Chakma	Esha et al. 2012
Merremia umbellata (L.) Hallier f.	Hog vine	Convolvulaceae	Demra gach	Leaf	To stop bleeding from external cuts and wounds	Pankho	Sarker et al. 2013
Costus speciosus (J. Koenig) Sm.	Cane-reed	Costaceae	Ranga bishoma, Ketoki	Leaf, stem, rhizome	Hernia, hydrocele, ear pains, formation of pus in ear, eczema or itches around the nails, diarrhea, infertility, food	Chakma, Marma, Bawm	Tasannun et al. 2015, Esha et al. 2012, Rahmatullah et al. 2009, Hossan et al. 2014
Bryophyllum pinnatum (Lam.) Oken	Air plant	Crassulaceae	Rokia-pang-bang, Nasirkhaw	Leaf, whole plant	Pneumonia, cough, kidney or gall bladder stone, high blood pressure, cholera,constipation, diabetes, stomach or kidney stones, stone in urinary tract, muscle pain scabies, boils, rheumatism, nail inflammation, paronychia	Chakma, Marma, Bawm	Tasannun et al. 2015, Malek et al. 2014, Esha et al. 2012, Afroz et al. 2013, Rahmatullah et al. 2009, Hossan et al. 2014
Brassica juncea (L.) Czem.	Mustard greens	Cruciferae	Shorisha	Seed	See *Acorus calamus.*	Tonchongya	Rashid et al. 2012
Coccinia grandis (L.) Voigt	Ivy gourd	Cucurbitaceae	Hela hujur, Nichu-bang	Root, leaf, fruit	Diabetes, pain, frequent urination, menstrual problems like burning sensations during menstruation	Chakma, Marma	Tasannun et al. 2015, Hossan et al. 2014, Esha et al. 2012, Rahmatullah et al. 2009
Hodgsonia macrocarpa Cogn.	Kapayang	Cucurbitaceae	Keha-pang	Fruit	Fevers, malaria	Marma	Rahmatullah et al. 2009
Momordica charantia L.	Bitter melon	Cucurbitaceae	Tita pullo shak	Leaf	Diabetes, frequent urination	Chakma	Esha et al. 2012

Scientific name	Common name	Family	Local name	Uses	Part used	Tribe	References
Momordica cymbalaria Fenzl ex Naudin	Not known	Cucurbitaceae	Khedatol	Snake bite, piles	Leaf	Chakma	Tasannun et al. 2015
Trichosanthes tricuspidata Lour.	Redball snakegourd	Cucurbitaceae	Aak-um	Itch, scabies	Leaf	Bawm	Hossan et al. 2014
Thuja orientalis L.	Northern white-cedar	Cupressaceae	Farthing	Joint pain, edema, astringent	Young stem	Bawm	Hossan et al. 2014
Cuscuta reflexa (Roxb.)	Dodder	Cuscutaceae	Fayng, Jigro-bang	Jaundice, sexual stimulant, aphrodisiac, jaundice, liver disease, uterus pain, liver pain	Stem	Chakma, Tripura, Marma, Bawm	Afroz et al. 2013, Rahmatullah et al. 2009, Hossan et al. 2014
Cycas revoluta Thunb.	Sago palm	Cycadaceae	Moniraj ful	Stomach ache, vomiting, diarrhea, eye diseases.	Leaf, flower	Chakma	Afroz et al. 2013
Cyperus laxus Lam.	Not known	Cyperaceae	Kol	Urinary tract infection, urinary blockage, stone in urinary tract, irregular urination, burning sensation during urination	Leaf, root	Bawm	Hossan et al. 2014
Dillenia indica L.	Elephant Apple	Dilleniaceae	Debru-bang	Stimulate appetite, scabies.	Leaf, fruit	Marma	Rahmatullah et al. 2009
Diplazium esculentum (Retz.) Sw.	Linguda	Dryopteridaceae	Dheki shak	Abscess, overdose of any medicine	Root	Pankho, Tonchongya	Sarker et al. 2013, Rashid et al. 2012
Dryopteris filix-max (L.) Schott.	Male woodfern	Dryopteridaceae	Kraing-ha, Makokji	Increase physical strength, headache, sedative, destroy adverse effects of any other medication	Leaf	Marma, Bawm	Rahmatullah et al. 2009, Hossan et al. 2014
Tectaria heterosora (Baker) Ching	Wild Fern	Dryopteridaceae	Baidya nath	Diarrhea in infant	Root	Chakma	Esha et al. 2012
Antidesma roxburghii Wall. ex Tul.	Not known	Euphorbiaceae	Chung chungi prayjanga, Chung chunga fejang	Rheumatic pain, stomach pain, waist pain, paralysis of hand or leg	Leaf, stem, root, bark	Chakma	Malek et al. 2014, Esha et al. 2012
Baliospermum montanum Muell. Arg.	Red physic nut	Euphorbiaceae	Shovon pal	Joint pain, stomach tumor	Leaf	Chakma	Malek et al. 2014
Cnesmone javanica Bl.	Not known	Euphorbiaceae	Chagol chotta	Abdominal tumor	Leaf	Chakma	Malek et al. 2014

Table 5.1 contd. ....

...Table 5.1 contd.

Plant Name	English name	Family	Local name	Parts used	Diseases treated	Tribe	References
Codiaeum variegatum (L.) A.Juss.	Garden croton	Euphorbiaceae	Boangkhela-paingda	Leaf	Fevers, coughs, cold	Marma	Rahmatullah et al. 2009
Gelonium multiflorum A.Juss.	Not known	Euphorbiaceae	Aam kurut (Bandor kola)	Root	Throat pain	Pankho, Tonchongya	Sarker et al. 2013, Wahab et al. 2013
Jatropha curcas L.	Poison Nut	Euphorbiaceae	Khegoon gach	Bark	Irregular menstruation	Chakma	Esha et al. 2012
Mallotus philippinensis Muell. Arg.	Monkey Face Tree	Euphorbiaceae	Sholok jhara	Leaf	Rheumatism	Chakma	Malek et al. 2014
Pedilanthus tithymaloides (L.) Poit.	Jew's Slipper	Euphorbiaceae	Borokhud	Leaf	Pneumonia, influenza	Chakma	Tasannun et al. 2015
Ricinus communis L.	Castor bean	Euphorbiaceae	Ranga veron gach, Te-udol	Young stem with leaves, new leaf, leaf	Vomiting in children, blood dysentery, sexual disorders in men	Pankho, Tonchongya, Chakma	Sarker et al. 2013, Wahab et al. 2013, Afroz et al. 2013
Abrus precatorius L.	Rosary pea	Fabaceae	Kawz	Leaf	Stone in kidney, urethra, urinary bladder	Chakma	Afroz et al. 2013
Acacia farnesiana (L.) Willd.	Sweet acacia	Fabaceae	Ketna keshor, Aao-wia-pang	Bark, root	Insect bite, dog bite (rabies), fever, crying in children	Tripura, Marma	Afroz et al. 2013, Rahmatullah et al. 2009
Caesalpinia nuga (L.) W.T. Aiton	Not known	Fabaceae	Krong-khai-bang	Leaf, fruit	Anintoxicant, skin disorders	Marma	Rahmatullah et al. 2009
Cajanus cajan (L.) Millsp.	Pigeon pea	Fabaceae	Orolchoi, Kokleng	Leaf, fruit	Typhoid, pneumonia, jaundice, stomatitis, snakebite, bronchitis, coughs, hemorrhoids	Chakma, Tonchongya, Bawm	Afroz et al. 2013, Rashid et al. 2012, Hossan et al. 2014
Canavalia gladiata (Jacq.) DC.	Sword bean	Fabaceae	Mogno bichi	Seed pulp	Measles	Pankho, Tonchongya	Sarker et al. 2013, Wahab et al. 2013
Cassia alata L.	Ringworm Shrub	Fabaceae	Jowlong pata, Chakunda	Leaf	Skin disorder, eczema, skin infections, piles, ringworm, stomach pain due to bloating or ingigesion	Pankho, Tonchongya, Marma, Bawm, Chakma	Sarker et al. 2013, Wahab et al.2013, Afroz et al. 2013, Rashid et al. 2012, Rahmatullah et al. 2009, Hossan et al. 2014, Esha et al.2012

Scientific name	Common name	Family	Local name	Part used	Uses	Tribe	References
Cassia fistula L.	Golden shower	Fabaceae	Sitolsua, Nafi-keda-pang	Black seeds within fruits, fruit, bark	Constipation in children, fevers, to stimulate appetite	Pankho, Tonchongya, Marma	Sarker et al. 2013, Wahab et al. 2013, Rahmatullah et al. 2009
Clitoria ternatea L.	Bluebellvine	Fabaceae	Ungeful gach, Koai-khi-bang	Bark, root, leaf	Pneumonia, infections of genital organs, boils, itches in children	Chakma, Marma	Malek et al. 2014, Rahmatullah et al. 2009
Crotalaria pallida Aiton	Striped rattlepod	Fabaceae	Iji gach	Leaf	Kala azar	Chakma	Malek et al. 2014
Derris elliptica (Wallich) Benth.	Tuba root	Fabaceae	Mahaga	Root	Constipation, to cleanse bowel	Tonchongya	Rashid et al. 2012
Desmodium alata L.	Not known	Fabaceae	Ublangra	Leaf	Snake bite	Tonchongya	Rashid et al. 2012
Desmodium macrophyllum Desv.	Not known	Fabaceae	Tongkhto-chibang-khrung-pang	Leaf	Stomach acidity, stomach aches, abnormal heart palpitations	Marma	Rahmatullah et al. 2009
Desmodium triquetrum (L.) DC.	Trefle Gros	Fabaceae	Lori pata kher, Turgi modon	Leaf	Snake bite, dysentery, joint pain, sex stimulant.	Chakma	Tasannun et al. 2015, Malek et al. 2014
Erythrina variegata L.	Indian coral tree	Fabaceae	Kasai-pang, Pai-che-o	Bark, root	Helminthiasis	Marma, Bawm	Rahmatullah et al. 2009, Hossan et al. 2014
Flemingia congesta Roxb. ex W.T. Aiton	Not known	Fabaceae	Gach archanga	Leaf	Rheumatism	Chakma	Malek et al. 2014
Mimosa pudica L.	Action plant	Fabaceae	Lojjaboti, Lajori	Root, leaf, flower	Passing of blood during urination, burning sensations in urinary tract, wounds, labor pain	Chakma, Bawm	Esha et al. 2012, Afroz et al. 2013, Hossan et al. 2014
Moghania macrophylla (Willd.) Kuntze	Not known	Fabaceae	Lungmul-turpa	Leaf	Itching due to contact with poisonous caterpillars, or due to poisonous ant bites	Bawm	Hossan et al. 2014
Saraca indica L.	Asoka-tree	Fabaceae	Ker-shaye-a	Leaf	Chicken pox, small pox	Bawm	Hossan et al. 2014
Senna sophera (L.) Roxb.	Not known	Fabaceae	Jhunjhuni, Shot rahong	Root, leaf	Stomach pain, burning sensations during urination, leucorrhea, irregular urination or urinary blockage, burning sensations in urinary tract	Chakma, Bawm	Malek et al. 2014, Hossan et al. 2014

Table 5.1 contd. ....

...Table 5.1 contd.

Plant Name	English name	Family	Local name	Parts used	Diseases treated	Tribe	References
Senna tora (L.) Roxb.	Sickle pod	Fabaceae	Ijibiji gach, Histacin gach	Leaf	Sleeplessness, leechbite, sleeping problem in female	Chakma, Marma	Esha et al. 2012, Afroz et al. 2013
Uraria crinita (L.) Desvaux ex Candolle	Not known	Fabaceae	Billeh lengur	Root	Diarrhea or dysentery in children	Chakma	Malek et al. 2014
Elaeocarpus robustus Roxb.	Ceylon olive	Elaeocarpaceae	Jolpaithing	Root	Gastric acidity (symptoms: heartburn, chest pain, stomach pain, gas formation in stomach)	Bawm	Hossan et al. 2014
Swertia chirata (Roxb. ex Fleming) H. Karst.	Chiretta	Gentianaceae	Chirota	Fruit	Gastrointestinal disorders, helminthiasis	Chakma	Esha et al. 2012
Dicranopteris linearis (Burman f.) Underwood	Old world forked fern	Gleicheniaceae	Horang veher	Leaf	Blood clotting on bones or muscle	Chakma	Esha et al. 2012
Helminthostachys zeylanica L.	Kamraj	Helminthostachyaceae	Paing-jem	Leaf, root	Throat pain, pain in the larynx, sore throat, epistaxis (nose bleed)	Bawm	Hossan et al. 2014
Curculigo Latifolia Dryand.	Palm grass	Hypoxidaceae	Meloni pata	Leaf	Cancer, piles, snake bite	Chakma	Tasannun et al. 2015
Curculigo orchioides Gaertn.	Golden eye-grass	Hypoxidaceae	Dubo meloni, Jongli peyaz	Leaf, stem, rhizome	Snake bite, hydrocele, astringent, bleeding due to deep external wounds	Chakma, Bawm	Tasannun et al. 2015, Malek et al. 2014, Hossan et al. 2014
Curculigo recurvata W.T. Aiton	Not known	Hypoxidaceae	Meloni gach, Nathial	Root, rhizome	Allergy, aphrodisiac	Pankho, Tonchongya, Bawm	Sarker et al. 2013, Wahab et al. 2013, Hossan et al. 2014
Molineria capitulata (Lour.) Herb.	Palm grass	Hypoxidaceae	Dhubo melloni	Leaf	Rheumatism	Chakma	Malek et al. 2014
Eleutherine palmifolia (L.) Merr.	Not known	Iridaceae	Jharpo peyaz	Root	Jaundice in babies	Chakma	Malek et al. 2014
Eleutherine plicata Herb.	Tears of the Virgin	Iridaceae	Tong-krai-choi, Chikra-choi	Root	Difficulties in urination, elephantitis	Marma	Rahmatullah et al. 2009
Anisomeles indica (L.) Kuntze	Catmint	Lamiaceae	Horin ching	Leaf	Colic, dyspepsia, fever in children arising from teething	Chakma	Malek et al. 2014
Ajuga macrosperma Wall. ex Benth	Ground pine	Lamiaceae	Ram-thlung	Root	Stomachache, wounds	Bawm	Hossan et al. 2014

Scientific name	Common name	Family	Local name	Part used	Uses	Ethnic group	References
Gomphostemma crinitum Wall.	Not known	Lamiaceae	Nykia	Leaf	Pain, swelling, pain due to trauma	Bawm	Hossan et al. 2014
Hyptis capitata Jacq.	False ironwort	Lamiaceae	Chitra baishak	Leaf	Snake bite	Chakma	Tasannun et al. 2015
Hyptis suaveolens (L.) Poit.	Pignut	Lamiaceae	Tunga dana, Chong gadana	Stem, seed	Stomachache, acidity, ulcer, diabetes, jaundice, burningsensations during urination, feeling of excessive fullness of stomach, gastrointestinal disorders, premature ejaculation	Pankho, Chakma, Tonchongya, Bawm	Sarker et al. 2013, Afroz et al. 2013, Rashid et al. 2012, Hossan et al. 2014
Leucas aspera (Willd.) Link.	Tamba	Lamiaceae	Gousha khongor, Paing-sung-pang	Leaf, flower	Lesions on the tongue, pain due to hemorrhoids, lesions/infections within nostril	Chakma, Marma	Esha et al. 2012
Ocimum americanum L.	American basil	Lamiaceae	Sabrang, Vipena	Leaf	See *Acorus calamus*, itching, scabies, skin infections on hands or legs	Tonchongya, Bawm	Rashid et al. 2012, Hossan et al. 2014
Ocimum basilicum L.	Great basil	Lamiaceae	Sabarung gach, Jeth sabarang	Leaf, bark	Coughs, respiratory difficulties, fever, diabetes, skin diseases, if infant does not drink milk or cries incessantly, asthma, cold, chest pain in children	Chakma, Pankho, Tonchongya, Tripura	Esha et al. 2012, Sarker et al. 2013, Wahab et al. 2013, Afroz et al. 2013
Ocimum gratissimum L.	Clove basil	Lamiaceae	Ram tulshi	Leaf	Sexual weakness in males	Marma	Afroz et al. 2013
Ocimum tenuiflorum L.	Holy Basil	Lamiaceae	Khargi shukchand, Nikunta pata	Leaf	Coughs, respiratory difficulties, if infant does not drink milk or cries incessantly, sexual stimulant (males)	Chakma, Pankho, Tonchongya, Tripura	Esha et al. 2012, Sarker et al. 2013, Wahab et al. 2013, Afroz et al. 2013
Premna scandens Roxb.	Dusky Fire	Lamiaceae	Aan-orai-na	Leaf	Energizer, anorexia	Bawm	Hossan et al. 2014
Vitex agnus-castus L.	Chaste tree	Lamiaceae	Syamula	Young leaf	Cataract	Tonchongya	Wahab et al. 2013
Cinnamomum camphora (L.) J. Presl.	Camphorwood	Lauraceae	Korpur gach	Bark	Stomach pain, food poisoning	Tonchongya	Rashid et al. 2012

Table 5.1 contd. ....

...Table 5.1 contd.

Plant Name	English name	Family	Local name	Parts used	Diseases treated	Tribe	References
Cinnamomum tamala (Buch.-Ham.) Nees & Eberm. Synonym: *Cinnamomum obtusifolium* Roxb. ex Nees.	Indian Bay Leaf	Lauraceae	Tejpata, Shimkung	Leaf, bark	Infections on skin, stomach upset, feeling of excessive stomach fullness	Chakma, Bawm	Esha et al. 2012, Hossan et al. 2014
Dehaasia kurzii King ex Hook.f.	Not known	Lauraceae	Shigerae shik	Stem	If children bring out their tongue too often	Chakma	Malek et al. 2014
Litsea glutinosa (Lour.) C.B. Robinson	Common Tallow Laurel	Lauraceae	Khara jora gach, Mewa pata	Stem, leaf	Headache, stomach pain, respiratory difficulties due to mucus	Chakma	Malek et al. 2014, Esha et al. 2012
Litsea monopetala (Roxb.) Pers.	Not known	Lauraceae	Shurjo pata, Boro kukurchita	Leaf	Tumor	Chakma	Tasannun et al. 2015
Leea umbraculifera C.B. Clarke	Not known	Leeaceae	Aash ura gach	Leaf	Abscess, infections arising out from wounds due to being hit with a sharp iron utensil	Chakma	Esha et al. 2012
Allium cepa L.	Onion	Liliaceae	Peyaz	Bulb	See *Acacia farnesiana*	Tripura	Afroz et al. 2013
Asparagus racemosus Willd.	Buttermilk Root	Liliaceae	Choti chora	Fruit, root	Swelling or enlargement of testicles, abscess, skin disease, pain in genital regions	Pankho, Chakma	Sarker et al. 2013, Malek et al. 2014
Lygodium flexuosum (L.) Sw.	Maidenhair creeper	Lygodiaceae	Katto jug, Makla-pang	Leaf, root	Burns, rheumatism, fever and convulsion in children, sore throat, throat pain, inflammation	Chakma, Marma, Bawm	Malek et al. 2014, Rahmatullah et al. 2009, Hossan et al. 2014
Lagerstroemia speciosa (L.) Pers.	Giant crepe-myrtle	Lythraceae	Jarul	Leaf	Labor pain and related conditions	Chakma	Afroz et al. 2013
Abelmoschus moschatus Medikus	Musk okra	Malvaceae	Khunae gach	Leaf	Headache	Chakma	Malek et al. 2014
Abroma augusta L.f.	Devil's cotton	Malvaceae	Gach Chula	Leaf, bark, root	Irregular menstruation	Chakma	Eshaet al. 2012
Abutilon indicum (L.) Sweet	Indian mallow	Malvaceae	Flur-bang	Root	Diarrhea and other gastrointestinal disorders in both human and cattle	Marma	Rahmatullah et al. 2009

Scientific name	Common name	Family	Local name	Part used	Ailments/Uses	Tribe	References
Gossypium arboreum L.	Tree cotton	Malvaceae	Jom tula	Leaf	Epilepsy	Marma	Afroz et al. 2013
Hibiscus rosa sinensis L.	Rose mallow	Malvaceae	Honduby, Chuila-bai-pang	Flower, Leaf	Diarrhea, infections on palm of hand, bacterial skin infection (cellulitis), cataract, abortifacient	Chakma, Marma, Marma, Bawm	Esha et al. 2012, Afroz et al. 2013, Rahmatullah et al. 2009, Hossan et al. 2014
Hibiscus sabdariffa L.	Roselle	Malvaceae	Jarbo beroj, Kunae ful	Leaf, root	Rheumatism	Chakma	Malek et al. 2014
Pterospermum semisagittatu Buch.-Ham. ex. Roxb.	Bayur tree	Malvaceae	Noah-labai-pang	Leaf	Poisonous insect bites	Marma	Rahmatullah et al. 2009
Sida cordifolia L.	Flannel weed	Malvaceae	Khangra gilukonak	Leaf	Enlargement of uterus	Chakma	Afroz et al. 2013
Sida rhombifolia L.	Arrow leaf sida	Malvaceae	Bilbili gach, Boi uli pata	Leaf	Scabies, eczema, abscess	Chakma, Tonchongya	Esha et al. 2012, Rashid et al. 2012
Urena lobata L.	Caesar weed	Malvaceae	Fow-fi-i, Sujugmonglap	Flower, leaf, root	Chapped lips, skin lesions, urinary tract disorders, fever, cold sore, aphthae	Marma, Bawm	Rahmatullah et al. 2009, Hossan et al. 2014
Angiopteris evecta (J. R. Forst.) Hoffm.	Giant fern	Marattiaceae	Adib	Leaf	Joint pain	Chakma	Esha et al. 2012
Maranta arundinacea L.	Arrow-root	Marantaceae	Arraroot	Stem	Kidney stone	Chakma	Tasannun et al. 2015
Melastoma malabathricum L.	Pink lasiandra	Melastomataceae	Moha purti, Tong	Leaf, root	Red color of urine, burning sensations during urination, jaundice	Chakma, Bawm	Esha et al. 2012, Hossan et al. 2014
Azadirachta indica A. Juss.	Neem	Meliaceae	Neem pata	Young leaf, root, bark, stem	Fever, pain, to prevent tooth infections, diabetes, itch, skin problem, scabies	Pankho, Bawm	Sarker et al. 2013, Hossan et al. 2014
Anamirta cocculus (L.) Wight & Arn.	Levant nut	Menispermaceae	Ludi chibang	Leaf	Spots on new-born baby's skin	Chakma	Esha et al. 2012
Campylus sinensis Lour.	Chinese tinospora	Menispermaceae	Hoiccholodi	Leaf	Fever	Chakma	Tasannun et al. 2015
Cyclea barbata Miers	Green grass jelly	Menispermaceae	Bokpinem	Leaf	Skin infections in humans and animals, dermatitis, allergy, tetanus, throat sore	Bawm	Hossan et al. 2014
Parabaenus agittata Miers	Not known	Menispermaceae	Horin kan	Leaf	Snake bite	Chakma	Tasannun et al. 2015

Table 5.1 contd. ....

...Table 5.1 contd.

Plant Name	English name	Family	Local name	Parts used	Diseases treated	Tribe	References
Pericampylus glaucus (Lam.) Merill.	Broad-leaved moonseed	Menispermaceae	Patal pur	Root	Constipation in children	Pankho, Tonchongya	Sarker et al. 2013, Wahab et al. 2013
Stephania glabra Miers	Tape vine	Menispermaceae	Thanda manik	Leaf, fruit	Rheumatism, stomach pain	Chakma	Malek et al. 2014
Stephania japonica (Thunb.) Miers	Snake vine	Menispermaceae	Thanda alu, Koang-khawri	Leaf, root	Stomach pain, menstrual pain, pneumonia, cold, coughs, fever in children.	Chakma Bawm	Tasannun et al. 2015, Esha et al. 2012, Hossan et al. 2014
Tinospora cordifolia (Willd.) Miers ex. Hook. f. & Thoms.	Not known	Menispermaceae	Gulchi lota	Leaf, stem	Jaundice	Chakma	Afroz et al. 2013
Acacia catechu (L.f.) Willd.	Wadalee gum	Mimosaceae	Khuamui	Leaf, fruit	Diarrhea, cold sore, chronic dysentery	Bawm	Hossan et al. 2014
Mimosa pudica L.	Shame plant	Mimosaceae	Shra-pang	Leaf	Eczema, scabies, abscesses	Marma	Rahmatullah et al. 2009
Ficus hirta Vahl.	Hairy fig	Moraceae	Thammang gach	Leaf, root	Insanity, mental disorders, memory loss	Chakma	Esha et al. 2012
Ficus hispida L.f.	Hairy fig	Moraceae	Debida sura gach, Joana gach	Fruit, seed	Diabetes, hookworm	Chakma Tonchongya	Esha et al. 2012, Rashid et al. 2012
Ficus religiosa L.	Sacred fig	Moraceae	Ashwoth	Fruit	Hypertension	Chakma	Esha et al. 2012
Streblus asper Lour.	Siamee rough bush	Moraceae	Sharur gach	Leaf	To increase lactation in nursing mothers	Pankho	Sarker et al. 2013.
Ardisia solanacea Roxb.	Shoebutton ardisia	Myrsinaceae	Boro cholla	Leaf, stem	Rheumatism	Chakma	Malek et al. 2014
Eucalyptus citriodora Hook.	Lemon-scented gum	Myrtaceae	Gaster-epil	Leaf, root	Neuralgia (pain in the nerves), wounds, cold, inflammation	Bawm	Hossan et al. 2014
Psidium guajava L.	Guava	Myrtaceae	Goian, Koijem	Leaf, bark, fruit	Flatulence, gastrointestinal disorders, nutritive, toothache, gingivitis, scabie	Chakma, Bawm	Esha et al. 2012, Hossan et al. 2014
Syzygium cumini (L.) Skeels	Java plum	Myrtaceae	Chabri-shae-bang	Seed	Diabetes, urinary problems	Marma	Rahmatullah et al. 2009

Scientific name	Common name	Family	Local name	Part used	Ailments/Uses	Tribe	References
Boerhavia repens L.	Spreading hogweed	Nyctaginaceae	Punonama shak	Leaf	Edema, inflammation	Chakma	Afroz et al. 2013
Nymphaea nouchali Burm.f.	Blue waterlily	Nymphaeaceae	Kra-pang, Rilipar	Root, stem	Men having urination difficulties, biliary disorders, menstrual problems, diabetes, nutritive	Marma, Bawm	Rahmatullah et al. 2009, Hossan et al. 2014
Nyctanthes arbor-tristis L.	Coral jasmine	Oleaceae	Shefali phul, Shing guri phool gach	Leaf	Fever, skin diseases	Pankho, Chakma	Sarker et al. 2013, Esha et al. 2012
Helminthostachys zeylanica L.	Not known	Ophioglossaceae	Somacchi	Leaf, stem	Piles	Chakma	Tasannun et al. 2015
Cymbidium aloifolium (L.) Sw.	Aloe-leafed Cymbidium	Orchidaceae	Surimas	Leaf, whole plant, root, seed	Fever, tetanus, chest pain, cuts, injury, lesions	Chakma	Tasannun et al. 2015
Pandanus foetidus Roxb.	Screwpine	Pandanaceae	Ramlethi	Leaf, root	Joint pain	Bawm	Hossan et al. 2014
Phyllanthus amarus Schumach. & Thonn.	Black catnip	Phyllanthaceae	Baugari bhangaher	Whole plant	Burning sensations	Chakma	Malek et al. 2014
Phyllanthus emblica Gaertn.	Indian gooseberry	Phyllanthaceae	Sosha-ban, Hadamala gach	Leaf, fruit	Stimulate appetite, hemorrhoids, gastrointestinal disorders, ulcer, gastric pain, anemia	Marma, Chakma, Pankho	Rahmatullah et al. 2009, Esha et al. 2012, Sarker et al. 2013
Phyllanthus niruri L. English: Carry-me-seed, Egg woman, Stonebreaker, Seed-under-leaf		Phyllanthaceae	Bauli banga her	Leaf	Skin rash, skin diseases.	Chakma	Esha et al. 2012.
Peperomia pellucida L.	Shining bush plant	Piperaceae	Roha	Wholeplant	Snake bite, gastrointestinal problems, skin problems, boils, eczema	Bawm	Hossan et al. 2014
Piper betle L.	Betel leaf	Piperaceae	Paan	Leaf, petiole	Antiseptic, Also see *Acacia farnesiana*	Tripura	Afroz et al. 2013
Plantago ovata Forssk.	Spogel seeds	Plantaginaceae	Isabgul, Tumka	Seed	See *Cajanus cajan*. Also see *Melastoma* sp.	Tonchongya	Rashid et al. 2012
Plumbago indica L.	Fire plant	Plumbaginaceae	Agunitita	Leaf	Chest pain	Chakma	Tasannun et al. 2015

Table 5.1 contd. ....

...Table 5.1 contd.

Plant Name	English name	Family	Local name	Parts used	Diseases treated	Tribe	References
Bambusa bambos (L.) Voss.	Giant thorny bamboo,	Poaceae	Khai-wang-wah, Medi-wah	Leaf, root	Rheumatic pain, eczema, cough, leprosy	Marma	Rahmatullah et al. 2009
Chrysopogon aciculatus (Retz.) Trin.	Love grass	Poaceae	Ramthek	Root	Astringent, wound	Bawm	Hossan et al. 2014
Digitaria setigera Roth ex Roemer & J.A. Schultes	Bristly crabgrass	Poaceae	Chao mongra	Root	Spondylosis	Tripura	Afroz et al. 2013
Thysanolaena maxima (Roxb.) Kuntze	Tigergrass	Poaceae	Shuinda gach	Root	Insect bite	Tonchongya	Rashid et al. 2012
Polygonum chinensis L.	Chinese knotweed	Polygonaceae	Mon ijadar, Mone jojada	Leaf	Paralysis of hand or leg, wasting away of hands or legs	Chakma	Esha et al. 2012
Polygonum hydropiper L.	Marsh pepper	Polygonaceae	Mra-che-bang	Leaf, root	Eczema, scabies, anthelmintic	Marma	Rahmatullah et al. 2009
Drynaria quercifolia (L.) J. Sm.	Oak leaf fern	Polypodiaceae	Baiddonath Pata, Bokpinem	Whole plant, root, leaf	See *Cymbidium aloifolium*. Skin infections in humans, animals, dermatitis.	Chakma Bawm	Tasamun et al. 2015, Hossan et al. 2014
Cheilanthes belangeri (Bory in Belang.) C. Chr.	Silver Fern	Pteridaceae	Ching fuchi, Sil fushi	Leaf	Headache, feeing of hotness in head	Chakma	Esha et al. 2012
Pteris vittata L.	Chinese brake	Pteridaceae	Dingky shak	Root	Itch, scabies, any other type of skin disorders.	Tripura	Afroz et al. 2013
Ziziphus mauritiana Lam.	Desert apple	Rhamnaceae	Kul, Mrai-ra	Leaf, bark, fruit, root	Fever, flatulence, diarrhea, nutritive, tumor (external swelling without any known cause)	Chakma, Bawm	Esha et al. 2012, Hossan et al. 2014
Rubus moluccanus L.	Molucca raspberry	Rosaceae	Handa shoal	Leaf, stem	Rheumatism	Chakma	Malek et al. 2014
Adina cordifolia (Roxb.) Hook. f. ex Brandis	Saffron Teak	Rubiaceae	Pang-kha-bang	Leaf	Eye disorders like conjunctivitis	Marma	Rahmatullah et al. 2009
Hedyotis scandens Roxb.	Indian Madder	Rubiaceae	Rema-pang	Leaf	Itches, scabies, eczema	Marma	Rahmatullah et al. 2009
Hedyotis thomsonii Hook.f.	Not known	Rubiaceae	Gou o jhil her	Leaf	Excessive itching in the eyes	Chakma	Esha et al. 2012

Scientific name	Common name	Family	Local name	Ailment/Use	Parts used	Tribe	References
Hedyotis verticillata (L.) Lam.	Mallow	Rubiaceae	Boithita	Bursting of abscess followed by oozing of pus and reddish colored substance	Leaf	Chakma	Esha et al. 2012
Hymenodictyon orixense (Roxb.) Mabberley	Bridal couch tree	Rubiaceae	Dela gamari	Hemorrhoids	Top of stem	Chakma	Esha et al. 2012
Ixora athroantha Bremek.	Jungle flame	Rubiaceae	Ludi choulla, Ludi choilla	Diarrhea	Bark	Chakma	Esha et al. 2012
Ixora pavetta Andr.	Torchwood tree	Rubiaceae	Bath jora ful	Allergy	Stem	Chakma	Malek et al. 2014
Ixora parviflora Vahl.	Torchwood Ixora	Rubiaceae	Tualthu	Gastrointestinal troubles, diarrhea, scabies	Root	Bawm	Hossan et al. 2014
Morinda persicifolia Buch.-Ham.	Not known	Rubiaceae	Chui-tili-bang	Jaundice	Root	Marma	Rahmatullah et al. 2009
Mussaenda glabrata Hutch. ex Gamble	Dhobi Tree	Rubiaceae	Metoni	Headache	Leaf	Pankho, Tonchongya	Sarker et al. 2013, Wahab et al. 2013
Mussaenda roxburghii Hook. f.	East Himalayan Mussaenda	Rubiaceae	Rani thak, Hala garjan	Burning sensations in hands or legs, rheumatism, abscess, rheumatism, malaria, hemolytic anemia.	Leaf	Chakma, Bawm	Malek et al. 2014, Hossan et al. 2014, Esha et al. 2012
Paederia foetida L.	Chinese fever vine	Rubiaceae	Gondo madok, Gondho batali	Rheumatic pain, burning sensations during urination, rheumatic fever	Leaf	Chakma, Pankho, Tonchongya	Esha et al. 2012, Sarker et al. 2013, Wahab et al. 2013
Psychotria calocarpa Kurz.	Not known	Rubiaceae	Shudoma	Paralysis of hands or legs	Leaf	Chakma	Esha et al. 2012
Aegle marmelos (L.) Corr.	Stone apple	Rutaceae	Urik phang, Shifal	Jaundice, indigestion, gastrointestinal disorders like flatulence, constipation, diarrohea, dysentery, stomach disorders, stomachache, stress, sedative	Leaf, fruit, root	Chakma, Tripura, Marma, Bawm	Esha et al. 2012, Afroz et al. 2013, Rahmatullah et al. 2009, Hossan et al. 2014
Citrus limonum Risso	Lemon	Rutaceae	Khra-pang	Stimulates appetite fever, skin disorders, hair loss, vomiting tendency, lesions within the mouth	Fruit	Marma	Rahmatullah et al. 2009
Micromelum minutum Wight & Arn.	Java brucea	Rutaceae	Chadi uraccha	Tumors	Leaf, root	Chakma	Malek et al. 2014

Table 5.1 contd. ...

...Table 5.1 contd.

Plant Name	English name	Family	Local name	Parts used	Diseases treated	Tribe	References
Flacourtia jangomas (Lour.) Raeus.	Runeala plum	Salicaceae	Hada annol	Leaf	Rheumatism, to improve health	Chakma	Malek et al. 2014
Santalum album L.	Sandal wood	Santalaceae	Shet chondon	Wood	To remove scar marks or marks due to burns, skin diseases	Chakma	Esha et al. 2012
Allophylus cobbe (L.) Raeuschel	Not know	Sapindaceae	Jendra ma	Leaf	Pain in hand or leg	Chakma	Esha et al. 2012
Litchi chinensis Sonn.	Lychee	Sapindaceae	Lisuthing	Fruit,root	Nutritive, diarrhea, hiccups orchitis (inflammation of testis)	Bawm	Hossan et al. 2014
Cardiospermum halicacabum L.	Heart pea	Sapindaceae	Keda foshka	Leaf	Chicken pox	Pankho, Tonchongya	Sarker et al. 2013, Wahab et al. 2013
Mimusops elengi L.	Spanish cherry	Sapotaceae	Bokul	Leaf, bark	Skin wounds, skin infections, vitiligo	Chakma, Tripura	Esha et al. 2012, Afroz et al. 2013
Scoparia dulcis L.	Licorice weed	Scrophulariaceae	Aadam fuchi, Mikram-boi-pang	Lef, fruit, root	Physical weakness, dysentery, swelling of fingers, pain in chin or throat, tonsillitis, throat cancer, facialredness, eczema, skin diseases, spermatorrhea, snake bite, insect bite, antidote to poisoning	Chakma, Pankho, Marma, Bawm	Malek et al. 2014, Esha et al. 2012, Sarker et al. 2013, Rahmatullah et al. 2009, Hossan et.al. 2014
Smilax macrophylla Roxb.	Kumarika	Smilacaceae	Wisisong	Leaf, root	Toothache, skin disorder, rheumatism, joint pain	Bawm	Hossan et al. 2014
Smilax zeylanica L.	Kumarika	Smilacaceae	Kumujja loti, Gumujjej lodi	Leaf	Skin cancer, skin infections, udoramoy (diarrhea)	Chakma, Tonchongya	Esha et al. 2012, Rashid et al. 2012
Datura metel L.	Devil's-trumpet	Solanaceae	Kalo dhutra, Dhutura phool	Leaf, flower, root	Snake bite, anesthetic purposes, to stop bleeding from external wounds, neck ache, asthma	Chakma, Tripura	Tasannun et al. 2015, Esha et al. 2012, Afroz et al. 2013
Physalis micrantha Link.	Native gooseberry	Solanaceae	Pitting gulagach	Leaf	Flatulency in cows or buffaloes, urinary problem	Chakma	Malek et al. 2014
Solanum torvum Sw.	Devil's fig	Solanaceae	Khing khabang, Tita baegun	Leaf	Piles, gastric ulcer, hydrocele	Marma	Afroz et al. 2013
Solanum xanthocarpum Schrad. & Wendl.	Yellow berried nightshade	Solanaceae	Changkhai, Hati baegun	Fruit	Pinworm (small thread-like worm infesting human intestine and rectum)	Marma	Afroz et al. 2013

Scientific name	English name	Family	Local name	Part used	Diseases/Uses	Tribe	References
Sterculia villosa Roxb.	Elephant rope tree	Sterculiaceae	Sam being	Bark	Diarrhea	Pankho	Sarker et al. 2013
Aquilaria agallocha Roxb.	Agarwood	Thymeliaceae	Akod	Leaf	Coughs, mucus, rheumatic pain	Chakma	Esha et al. 2012
Grewia paniculata Roxb. ex DC.	Microcos	Tiliaceae	Ashar gach, Achat	Leaf, flower	Gastric troubles, bone fracture, indigestion, eczema, itch, small pox, typhoid fever, dysentery, syphilitic ulceration of the mouth	Chakma, Tripura	Esha et al. 2012, Afroz et al. 2013
Clerodendrum indicum L.	Glory bower	Verbenaceae	Bamonhati, Pilae shak	Leaf, root	Epilepsy, sudden bouts of unconsciousness, stomach pain, diabetes, obesity, hypertension, abscess	Chakma	Tasannun et al. 2015, Malek et al. 2014
Clerodendrum infortunatum L. English:	Hill glory bower	Verbenaceae	Beth gach	Leaf	Stomach ache	Pankho	Sarker et al. 2013
Clerodendrum serratum L.	Blue-flowered glory tree	Verbenaceae	Risente	Leaf, whole plant	Umbilical sore	Bawm	Hossan et al. 2014
Clerodendrum viscosum Vent.	Hill glory bower	Verbenaceae	Begh gach, Sujjara,	Leaf, root	Frequent urination, diabetes, malaria fever, any type of stomach pain, snake bite	Chakma, Tonchongya, Marma, Bawm	Esha et al. 2012, Rashid et al. 2012, Rahmatullah et al. 2009, Hossan et al. 2014
Vitex agnus L.	Monk's pepper	Verbenaceae	Samalu	New leaf	Cataract	Pankho	Sarker et al. 2013
Vitex negundo L.	Five leaved chaste tree	Verbenaceae	Shada tulshi, Moru-bang	Leaf, root, seed	If infant does not drink milk or cries incessantly, fever, hearing problem, headache, malaria, rheumatic pains, joint pains	Pankho, Tripura, Marma	Sarker et al. 2013, Afroz et al. 2013, Rahmatullah et al. 2009
Cissus javana DC.	Snake bitters	Vitaceae	Ajongma	Leaf	Jaundice, rheumatism	Chakma	Malek et al. 2014
Cissus quadrangularis L.	Adamant Creeper	Vitaceae	Harjora lota	Stem, leaves, young shoot	Bone fracture, laxative, tonic, analgesic, piles, tumor, loss of appetite, constipation, complaints of back and spine, otorrhea, epistaxis, scurvy, irregular menstruation, asthma, dyspepsia, bowel complaints	Chakma	Tasannun et al. 2015
Alpinia conchigera Griff.	Greater galangal	Zingiberaceae	Khet ranga	Rhizome	Dysentery, abdominal pain, stomach upset, gastric pain	Chakma	Malek et al. 2014

Table 5.1 contd. ...

...Table 5.1 contd.

Plant Name	English name	Family	Local name	Parts used	Diseases treated	Tribe	References
Alpinia nigra (Gaertn.) B. L. Burtt.	Bamboo leaf galangal	Zingiberaceae	Choia-bang	Root	Loss of sensation in hands and legs	Marma	Rahmatullah et al. 2009
Curcuma caesia Roxb.	Black turmeric	Zingiberaceae	Kalo holud	Rhizome	Bloating, menstrual disorders	Chakma	Malek et al. 2014
Curcuma longa L.	Turmeric	Zingiberaceae	Holud	Rhizome	Hypertension, abscess	Chakma	Esha et al. 2012
Curcuma zedoaria (Christm.) Roscoe	Zedoary	Zingiberaceae	Ranga holla	Stem	Jaundice	Chakma	Esha et al. 2012
Kaempferia galanga L.	Aromatic ginger	Zingiberaceae	Komla gach	Leaf	Bloating	Chakma	Malek et al. 2014
Zingiber montanum (J. Koenig) Link ex A. Dietr.	Cassumnar ginger	Zingiberaceae	Mone ada, Meni ada	Leaf	Swelling of joints, rheumatic pain	Chakma	Esha et al. 2012
Zingiber officinale Roscoe	Ginger	Zingiberaceae	Tumreng	Rhizome	Gastric acidity, gastrointestinal troubles, stomach upset	Bawm	Hossan et al. 2014

Figure 5.3: A Mogh (Marma) tribal healer in Bandarban district, Bangladesh.

believe in their traditional treatments with plants any more, but prefer going to allopathic doctors. This was borne out by a visit to a Mogh (Marma) tribal healer in Bandarban (Figure 5.3), who could give us the names of only ten plants, even though the healer, by his own admission, was practicing for over 20 years (Rahmatullah et al. 2019).

Medicinal Plants of Chittagong Hill Tracts for Drug Discovery

A number of plants can prove very useful and interesting to the scientists. For instance, malaria is a mosquito-transmitted life-threatening disease against which allopathic medicine practically has only one treatment option left, which is treatment with artemisinin-based combination therapies (World Health Organization 2015). However, the causative agent of malaria, *Plasmodium falciparum*, besides developing resistance against other antimalarial drugs, such as quinine earlier, has now started to develop resistance against artemisinin (Blasco et al. 2017). So, the world is in desperate need of effective antimalarial drugs. The various anti-malarial plants of the tribes of Chittagong Hill Tracts (CHT) include *Andrographis paniculata, Amaranthus spinosus, Hodgsonia macrocarpa, Mussaenda corymbosa, Clerodendrum viscosum*, and *Vitex negundo* (Table 5.1). Interestingly, *A. paniculata, A. spinosus* and *C. viscosum* have been shown to possess antimalarial (*Plasmodium* inhibitory) activity (Rehman et al. 1999, Mishra et al. 2009, Tiningsih et al. 2012, Goswami et al. 1998), while *V. negundo* has strong mosquito larvicidal properties (Raj et al. 2009). Thus, the potential for an anti-malarial novel drug from these plants is strong; the other two plants *H. macrocarpa* and *M. corymbosa* are of interest because their anti-malarial effects, if any, are yet to be scientifically studied.

Andrographis paniculata is known to contain a number of bio-active compounds, such as andrographolide, 14-deoxy-11-oxo-andrographolide, andrographosterol, andrographosterin, stigmasterol, and α-sitosterol (Bharati et al. 2011). The active anti-malarial compound has been reported to be andrographolide (Mishra et al. 2011). Betacyanins present in *Amaranthus spinosus* have been shown to possess antiplasmodial activity (Hilou et al. 2006). *Clerodendrum viscosum* is also rich in antioxidants and other bio-active compounds; these compounds include gallic acid, β-sitosterol, quercetin, oleanolic acid, clerodinin A, β-cubebene, viscosene, apigenin, and clerodolone (Kekuda et al. 2019). Apigenin is known to inhibit the growth of *Plasmodium berghei* (Amiri et al. 2018).

The CHT tribal practitioners used five plants against cancer, namely *Lepidagathis incurva* against skin cancer, *Carica papaya* and *Curculigo latifolia* against cancer (form of cancer not mentioned by the practitioners), *Scoparia dulcis* against throat and skin cancer, and *Smilax zeylanica* against skin cancer. Cytotoxic activities have been reported for *L. incurva* (Charoenchai et al. 2010). Aqueous extract of *C. papaya* leaves has been shown to inhibit proliferation of human breast cancer cells MCF-7 (Nisa et al. 2017). Although *C. latifolia* is yet to be reported for any anticancer potential, fractions of a related species *Curculigo orchioides* have reported anticancerous potential on cancer cell lines

HepG2, HeLa, and MCF-7 (Hejazi et al. 2018). The active principle of the plant against the metastasis of B16F10 melanoma cells was identified as curculigoside (Murali and Kuttan 2016). Benzoxazinoids from *Scoparia dulcis* (sweet broomweed) reportedly showed anti-proliferative activity against the DU-145 human prostate cancer cell line (Wu et al. 2012). In AGS human gastric adenocarcinoma cells, scopadulciol, isolated from *Scoparia dulcis*, has been found to induce β-Catenin degradation and overcome tumor necrosis factor-related apoptosis ligand resistance (Fuentes et al. 2015). Although no anticancer effect has been reported for *S. zeylanica*, other *Smilax* species reportedly possess anticancer properties (Fu et al. 2017, She et al. 2017). The structures of some bio-active plant constituents are shown in Figure 5.4, and some plant pictures are shown in Figure 5.5.

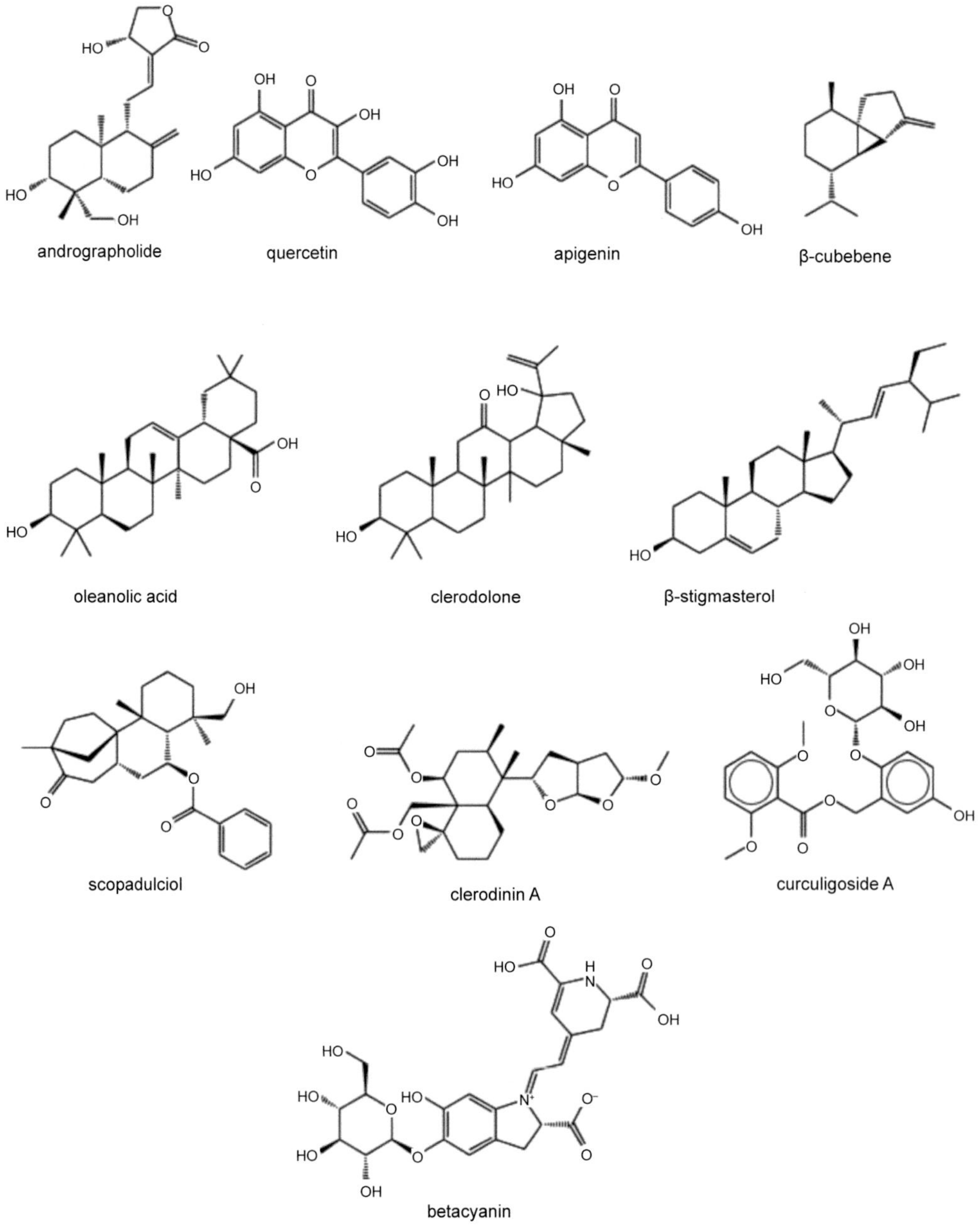

Figure 5.4: Some bio-active constituents isolated in various plants used by tribal healers of the Chittagong Hill Tract, Bangladesh.

Figure 5.5: Some medicinal plants used by the tribals of CHT—(a) *Acanthus illicifolius*, (b) *Calotropis gigantea*, (c) *Cuscuta reflexa*, (d) *Lygodium flexuosum*, (e) *Scoparia dulcis*, (f) *Vitex negundo*.

Taken together, the medicinal practitioners of various tribes in CHT area use several dozens of plants to treat a variety of diseases, including both common diseases, such as gastrointestinal disorders (diarrhea, constipation), and difficult to treat diseases, such as malaria, cancer, hypertension, and diabetes. Even successful treatment of common gastrointestinal disorders with plants can be beneficial to the tribal people. Tribal people's incomes are at or below the poverty level for the most part. As such, treatment with readily available and affordable plants can reduce medical costs. Moreover, the plants can prove of value to scientists looking for newer drugs, as enteric microorganisms are also getting resistant to existing drugs. Multi-drug resistant Gram-negative enteric bacteria have been reported from China (Liu et al. 2013). Such antibiotic resistance in enteric bacteria has been reported from other parts of the world also (Mutuku 2017). Thus, the plants of the tribal practitioners used against diarrhea and dysentery have potential for novel drug discoveries against antibiotic-resistant enteric microorganisms. At the same time, new and more effective drugs against the other diseases treated by the tribal practitioners with plants can aid conservation efforts, raise incomes of tribal people, and decrease medical costs, while at the same time offering better drugs.

Conclusion

Plants have always been a source of new drugs, and many drugs derived from plants are now being used in allopathic medicine. In recent times, because of emergence of new diseases and drug-resistant vectors, and also because of adverse effects of allopathic drugs, scientists are turning their attention to the plant kingdom for new and more efficacious drug discoveries. In this context, the various medicinal plants used by tribal people throughout the world, including the Chittagong Hill Tracts region of Bangladesh, can prove beneficial for new drug discoveries and alleviation of diseases which are hard to cure at present.

References

Afroz, S.S., Sen, U.S., Islam, M.J., Morshed, M.T., Bhuiyan, M.S.A., Ahmed, I., Haque, M.E., Rahmatullah, M. 2013. Ethnomedicinal plants of various tribal and folk medicinal practitioners of six localities of Rangamati and Khagrachari districts in Bangladesh. Am.-Eur. J. Sustain Agric 7(4): 240–250.

Ahammad, R., Stacey, N. 2016. Forest and agrarian change in the Chittagong Hill Tracts region of Bangladesh. In: Agrarian change in tropical landscapes. Center for International Forestry Research, Bogor, Indonesia [DOI: 10.17528/cifor/005867].

Amiri, M., Nourian, A., Khoshkam, M., Ramazani, A. 2018. Apigenin inhibits growth of the *Plasmodium berghei* and disrupts some metabolic pathways in mice. Phytother. Res. 32(6): DOI: 10.1002/ptr.6113.

Bharati, B.D., Sharma, P.K., Kumar, N., Dudhe, R., Bansal, V. 2011. Pharmacological activity of *Andrographis paniculata*: A brief review. Pharmacologyonline 2: 1–10.

Blasco, B., Leroy, D., Fidock, D.A. 2017. Antimalarial drug resistance: linking *Plasmodium falciparum* parasite biology to the clinic. Nat. Med. 23(8): 917–928.

Chakma, N., Maitrot, M. 2016. How ethnic minorities became poor and stay poor in Bangladesh: a qualitative enquiry. Working Paper 34, EEP/Shiree, Baridhara, Dhaka, Bangladesh.

Charoenchai, P., Vajrodaya, S., Somprasong, W., Mahidol, C., Ruchirawat, S., Kittakoop, P. 2010. Antiplasmodial, cytotoxic, radical scavenging and antioxidant activities of Thai plants in the family Acanthaceae. Planta Med. 76(16): 1940–1943.

Das, R., Biswas, S., Banerjee, E.R. 2016. Nutraceutical-prophylactic and therapeutic role of functional food in health. J. Nutr. Food Sci. 6(4): 527, DOI: 10.4172/2155-9600.1000527.

Esha, R.T., Chowdhury, M.R., Adhikary, S., Haque, K.M.A., Acharjee, M., Nurunnabi, M., Khatun, Z., Lee, Y.-K., Rahmatullah, M. 2012. Medicinal plants used by tribal medicinal practitioners of three clans of the Chakma tribe residing in Rangamati district, Bangladesh. Am.-Eur. J. Sustain Agric 6(2): 74–84.

Fu, S., Yang, Y., Liu, D., Luo, Y., Ye, X., Liu, Y., Chen, X., Wang, S., Wu, H., Wang, Y., Hu, Q., You, P. 2017. Flavonoids and Tannins from *Smilax china* L. Rhizome Induce Apoptosis Via Mitochondrial Pathway and MDM2-p53 Signaling in Human Lung Adenocarcinoma Cells. Am. J. Chin. Med. 45(2): 369–384.

Fuentes, R.G., Toume, K., Arai, M.A., Sadhu, S.K., Ahmed, F., Ishibashi, M. 2015. Scopadulciol, Isolated from *Scoparia dulcis*, induces β-catenin degradation and overcomes tumor necrosis factor-related apoptosis ligand resistance in ags human gastric adenocarcinoma cells. J. Nat. Prod. 78(4): 864–872.

Gilani, A.H., Rahman, A.U. 2005. Trends in ethnopharmacology. J. Ethnopharmacol. 100(1-2): 43–49.

Goswami, A., Dixit, V.K., Srivastava, B.K. 1998. Anti-malarial activity of aqueous extract of *Clerodendrum infortunatum*. Bionature 48: 45–48.

Hejazi, I.I., Khanam, R., Mehdi, S.H., Bhat, A.R., Rizvi, M.M.A., Thakur, S.C., Athar, F. 2018. Antioxidative and anti-proliferative potential of *Curculigo orchioides* Gaertn in oxidative stress induced cytotoxicity: *In vitro, ex vivo* and *in silico* studies. Food Chem. Toxicol. 115: 244–259.

Hilou, A., Nacoulmaa, O.G., Guiguemde, T.R. 2006. *In vivo* antimalarial activities of extracts from *Amaranthus spinosus* L. and *Boerhaavia erecta* L. in mice. J. Ethnopharmacol. 103(2): 236–240.

Hossan, M.S., Hanif, A., Khan, M., Bari, S., Jahan, R., Rahmatullah, M. 2009. Ethnobotanical survey of the Tripura tribe of Bangladesh. Am.-Eur. J. Sustain Agric, 3(2): 253–261.

Hossan, S., Hanif, A., Jahan, R., Rahmatullah, M. 2014. Ethnomedicinal plants of the bawm tribal community of Rowangchhari in Bandarban district of Bangladesh. J. Altern. Complement Med. 20(8): 581–589.

Kekuda, T.R.P., Shree, V.S.D., Noorain, G.K.S., Sahana, B.K., Raghavendra, H.L. 2019. Ethnobotanical uses, phytochemistry and pharmacological activities of *Clerodendrum infortunatum* L. (Lamiaceae): A review. J. Drug Deliv. Ther. 9(2): 547–559.

Krief, S., Hladik, C.M., Haxaire, C. 2005. Ethnomedicinal and bioactive properties of plants ingested by wild chimpanzees in Uganda. J. Ethnopharmacol. 101(1-3): 1–15.

Kumar, S., Shukla, Y.N., Lavania, U.C., Sharma, A., Singh, A.K. 1997. Medicinal and Aromatic Plants: Prospects for India. J. Med. Aromat. Plant Sci. 19(2): 361–365.

Liu, Y., Yang, Y., Zhao, F., Fan, X., Zhong, W., Qiao, D., Cao, Y. 2013. Multi-drug resistant Gram-negative enteric bacteria isolated from flies at Chengdu Airport, China. Southeast Asian J. Trop. Med. Public Health 44(6): 988–996.

Malek, I., Mia, N., Mustary, M.E., Hossain, M.J., Sathi, S.M., Parvez, M.J., Ahmed, M., Chakma, S., Islam, S., Billah, M.M., Rahmatullah, M. 2014. Medicinal plants of the Chakma community of Rangapanir Chara area of Khagraachari district, Bangladesh. Am.-Eur. J. Sustain Agric. 8(5): 59–64.

Mia, M.M., Kadir, M.F., Hossan, M.S., Rahmatullah, M. 2009. Medicinal plants of the Garo tribe inhabiting the Madhupur forest region of Bangladesh. Am.-Eur. J. Sustain. Agric. 3(2): 165–171.

Mishra, K., Dash, A.P., Swain, B.K., Dey, N. 2009. Anti-malarial activities of *Andrographis paniculata* and *Hedyotis corymbosa* extracts and their combination with curcumin. Malaria J. 8: 26. DOI: 10.1186/1475-2875-8-26.

Mishra, K., Dash, A.P., Dey, N. 2011. Andrographolide: A novel antimalarial diterpene lactone compound from *Andrographis paniculata* and its interaction with curcumin and artesunate. J. Trop. Med.: Article ID 579518.

Mollik, A.H., Hossan, M.S., Paul, A.K., Taufiq-Ur-Rahman, M., Jahan, R., Rahmatullah, M. 2010. Folk medicinal healers in three districts of Bangladesh and inquiry as to mode of selection of medicinal plants. Ethnobot. Res. Appl. 8: 195–218.

Murali, V.P., Kuttan, G. 2016. Curculigoside augments cell-mediated immune responses in metastatic tumor-bearing animals. Immunopharm. Immunot. 38(4): 264–269.

Nasri, H. 2012. Renoprotective effects of garlic. J. Renal. Inj. Prev. 2(1): 27–28.

Nasri, H. 2013. Cisplatin therapy and the problem of gender-related nephrotoxicity. J. Nephropharmacol. 2(2): 13–14.

Nisa, F.Z., Astuti, M., Murdiati, A., Mubarika Haryana, S. 2017. Anti-proliferation and apoptosis induction of aqueous leaf extract of *Carica papaya* L. on human breast cancer cells MCF-7. Pak J. Biol. Sci. 20(1): 36–41.

Petroni, L.M., Huffman, M.A., Rodrigues, E. 2017. Medicinal plants in the diet of wooly spider monkeys (*Brachyteles arachnoides*, E. Geoffroy, 1806)—a bio-rational for the search of new medicines for human use? Rev Bras Farmacogn 27: 135–142.

Rahmatullah, M., Hossan, M.S., Hanif, A., Roy, P., Jahan, R., Khan, M., Chowdhury, M.H., Rahman, T. 2009. Ethnomedicinal applications of plants by the traditional healers of the Marma tribe of Naikhongchhari, Bandarban district, Bangladesh. Adv. Nat. Appl. Sci. 3(3): 392–401.

Raj, P.V., Chandrasekhar, H.R., Dhanaraj, S.A., Vijayan, P., Nitesh, K., Subrahmanyam, V.M., Rao, J.V. 2009. Mosquito larvicidal activity of *Vitex negundo*. Pharmacologyonline 2: 975–990.

Rashid, M.M., Rafique, F.B., Debnath, N., Rahman, A., Zerin, S.Z., Harun-ar-Rashid, Islam, M.A., Khatun, Z., Rahmatullah, M. 2012. Medicinal plants and formulations of a community of the Tonchongya tribe in Bandarban district of Bangladesh. Am.-Eur. J. Sustain. Agric. 6(4): 292–298.

Rehman, N.N., Furuta, T., Kojima, S., Takane, K., Mohd, M.A. 1999. Antimalarial activity of extracts of Malaysian medicinal plants. J. Ethnopharmacol. 64(3): 249–254.

Sarker, M.N., Mahin, A.A., Munira, S., Akter, S., Parvin, S., Malek, I., Hossain, S., Rahmatullah, M. 2013. Ethnomedicinal plants of the Pankho community of Bilaichari Union in Rangamati district, Bangladesh. Am.-Eur. J. Sustain. Agric. 7(2): 114–120.

Shahidullah, M., Al-Mujahidee, M., Uddin, S.M.N., Hossan, M.S., Hanif, A., Bari, S., Rahmatullah, M. 2009. Medicinal plants of the Santal tribe residing in Rajshahi district, Bangladesh. Am.-Eur. J. Sustain. Agric. 3(2): 220–226.

She, T., Feng, J., Lian, S., Li, R., Zhao, C., Song, G., Luo, J., Dawuti, R., Cai, S., Qu, L., Shou, C. 2017. Sarsaparilla (*Smilax Glabra* Rhizome) Extract activates redox-dependent ATM/ATR pathway to inhibit cancer cell growth by S phase arrest, apoptosis, and autophagy. Nutr. Cancer 69(8): 1281–1289.

Smith, R. 2004. "Let food be thy medicine…". BMJ Clin. Res. 328(7433): DOI: 10.1136/bmj.328.7433.0-g.

Tiningsih, S., Ridwan, R., Prijanti, A.R., Sadikin, M., Freisleben, H.-J. 2012. Schizonticidal effect of a combination of *Amaranthus spinosus* L. and *Andrographis paniculata* Burm.f./Nees extracts in *Plasmodium berghei*-infected mice. Med. J. Indonesia 21(2): 66–70.

Tasannun, I., Ruba, F.A., Bhuiyan, B.U., Hossain, K.M., Khondokar, J., Malek, I., Bashar, A.B.M.A., Rahmatullah, M. 2015. Indigenous medicinal practices: medicinal plants of Chakma tribal medicinal practitioners in Rangamati district. Am.-Eur. J. Sustain. Agric. 9(5): 28–35.

Wahab, A., Roy, S., Habib, A., Bhuiyan, M.R.A., Roy, P., Khan, M.G.S., Azad, A.K., Rahmatullah, M. 2013. Ethnomedicinal wisdom of a Tonchongya tribal healer practicing in Rangamati district, Bangladesh. Am.-Eur. J. Sustain. Agric. 7(3): 227–234.

World Health Organization. 2015. Guidelines for the treatment of malaria, 3rd Edn. ISBN 978 92 4 154912 7.

Wu, W.H., Chen, T.Y., Lu, R.W., Chen, S.T., Chang, C.C. 2012. Benzoxazinoids from *Scoparia dulcis* (sweet broomweed) with antiproliferative activity against the DU-145 human prostate cancer cell line. Phytochem 83: 110–115.

Yuan, H., Ma, Q., Ye, L., Piao, G. 2016. The traditional medicine and modern medicine from natural products. Molecules 21: 559, DOI: 10.3390/molecules21050559.

6

Argentinian Wild Plants as Controllers of Fruits Phytopathogenic Fungi

Trends and Perspectives

María Inés Stegmayer,[1] *Norma Hortensia Álvarez,*[1] *María Alejandra Favaro,*[1]
Laura Noemí Fernandez,[1] *María Eugenia Carrizo,*[1] *Andrea Guadalupe
Reutemann*[1] and *Marcos Gabriel Derita*[1,2,]*

Introduction

Plant diseases caused by phytopathogenic fungi are responsible for economic losses arising mainly from crop yield reduction, but also resulting from diminished product quality and safety; sometimes they also represent a risk for human and animal health due to food contamination and the accumulation of toxic residues in the environment. Due to market globalization and climate change, the problem is growing at an accelerated pace (Pergomet et al. 2018).

Physical methods, some inorganic salts (Qin et al. 2010), and synthetic biocides that include sanitizing products (Mari et al. 2004) are among the main alternative strategies to diminish the threat posed by phytopathogens. However, they are complemented by emerging non-conventional approaches, such as biological control through the application of antagonistic microorganisms and biochemical control through natural antimicrobial substances (Grayer and Kokubun 2001). Treatments based on X-rays and radio frequency irradiation (Neven and Drake 2000), cold storage and cool/hot water have been tested, but their implementation requires large equipment, and sometimes exposure conditions may damage the sensory quality of fruits, including their firmness (Spadoni et al. 2013).

In addition, the use of biological agents, including *Saccharomyces cerevisiae*, *Aureobasidium pullulans*, *Bacillus subtilis* CPA-8, and *Metschnikowia fructicola*, among many others, have also been explored (Mari et al. 2012). However, the choice of the microorganism must be judicious, and to date, these types of biocontrol tools lack the capacity for eradication, and their effectiveness has been limited and variable.

[1] ICiAgro Litoral, Universidad Nacional del Litoral, CONICET, Facultad de Ciencias Agrarias, Kreder 2805, Esperanza, 3080HOF, Argentina.
[2] Farmacognosia, Facultad de Ciencias Bioquímicas y Farmacéuticas, Universidad Nacional de Rosario, Suipacha 531, S2002LRK, Rosario, Argentina.
* Corresponding author: mgderita@hotmail.com

Since regulations on the use of new and existing fungicides are becoming more and more stringent, it urges to identify and develop new chemical entities with antifungal activity. Nontoxic chemicals have emerged as promising alternatives to synthetic fungicides, as they provide effective protection against post-harvest spoilage. Different naturally occurring compounds (Fu et al. 2017), semisynthetic derivatives (Zhang et al. 2015), chitosan-based formulations (Novaes Azevedo et al. 2014), and plant products (Gatto et al. 2016), including extracts or essential oils (Mohammadi and Aminifard 2012, Di liberto et al. 2019) have been reported as part of this strategy.

The present chapter attempts to follow this line of thought and provide information about the reported use of wild plant species acting as antifungals, as well as some preliminary results based on *in vitro* studies performed in our laboratory.

Statistical Data of Argentinean Main Fruit Productions Along the Last 20 Years

Fruit and vegetable production in Argentina is carried out in almost the whole territory due to its climate diversity. The commercial production that provides to the main urban consumption centers is located in certain regions which offer competitive commercial advantages given by agro-climatic conditions, infrastructure (irrigation, roads), technology, and availability of supplies and services (technical assistance and qualified workforce) (Sordo et al. 2017).

Argentinean production data of oranges, strawberries, and peaches were analyzed according to FAO (Food and Agriculture Organization of the United Nations), Comtrade (United Nations Comtrade Database-International Trade Statistic-Import/export Data), and the MCBA (Central Market of Buenos Aires). The parameters studied were production (tons), harvested area (hectare), major producing states/provinces, and varieties used from 1997 to 2017. The performance obtained for orange production (tons) indicates an upward behavior with a crop area (hectare) that remains more or less constant, showing a maximum between 2005 and 2010 (Figure 6.1a). Entre Ríos, Buenos Aires, Santa Fe, and Tucumán provinces are the ones with the highest production represented by the major varieties Lane late, Salustiana, Valencia late, and Sanguinelli.

Regarding strawberry production and cultivated area, although an exponential behavior is observed worldwide, at the national level the trend is linear upward (Figure 6.1b). The provinces of Tucumán and Santa Fe turn out to be the ones with the highest production. Varieties Camarosa, Camino real, Cabrillo, and Sensación are the main ones that abound in the Central Market of Buenos Aires. Argentina exports mostly industrialized strawberry, and to a lesser extent, fresh fruit.

With respect to peaches, world production follows an exponential trend and harvested area, while in our country these parameters show a constant behavior except for 2014 and 2017, when production fell due to adverse weather conditions (Figure 6.1c). Regions with the highest production of peaches in Argentina are—north of Buenos Aires, south of Santa Fe, and Mendoza province, all of which are affected by a fungal pest (*Monilinia fructicola*) that prevents export to the EU. The main varieties commercialized by Argentina are: Flavor crest, Rich lady, and Red globe, among others.

Important Fruit Diseases caused by Phytopathogenic Fungi

Penicillium digitatum (Pers.) Sacc, *Botrytis cinerea* (Pers.: Fr.), and *Monilinia fructicola* (G. Wint.) Honey are three of the main phytopathogenic fungi that affect Argentine production of citrus, strawberries, and peaches in fields as well as during harvest and post-harvest stage. The main characteristics, disease cycle, epidemiology, and management of these pathogens will be reviewed below.

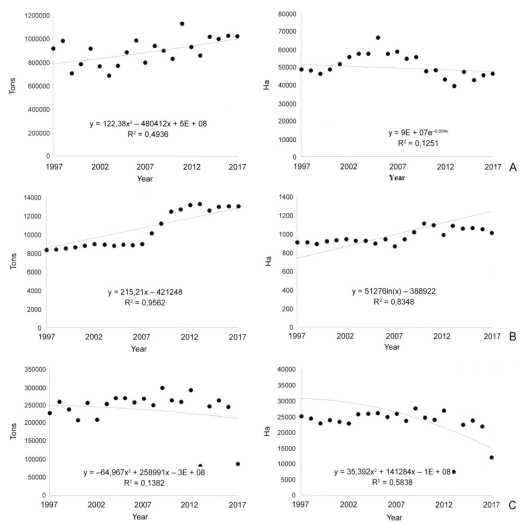

Figure 6.1: Production of oranges (A), strawberries (B) and peaches (C) in tons (left) and crop area data in hectare (right) between 1997 and 2017 in Argentina.

Citrus Green mold

Green mold rot disease caused by *P. digitatum* is the most important post-harvest disease of citrus fruit worldwide. In conditions conducive for disease, losses may reach 90% (Whiteside et al. 1988, Macarisin et al. 2007).

   P. digitatum is a necrotrophic wound parasite that requires a preexisting injury to penetrate the fruit peel. Inoculum sources for infection are present in rotten fruit on the ground of the orchard, in the packing house, or storage room. Infection occurs when masses of conidia are airborne, disseminated, and contact mature fruits which are highly susceptible. The optimum temperature is between 20–25°C (Whiteside et al. 1988, Palou et al. 2008, Kellerman et al. 2016). The infection and sporulation cycle can be repeated many times through the season (Whiteside et al. 1988). Symptoms appear as a soft, watery, and decolorized spot which rapidly enlarges due to the production of pathogenic hydrolytic enzymes. White mycelium accompanied by a mass of green spores (Figure 6.2a) appears on the fruit rind surface (Whiteside et al. 1988, Papoutsis et al. 2019).

Figure 6.2: Symptoms caused by (a) *Penicillium digitatum* in orange, (b) *Botrytis cinerea* in strawberry, and (c) *Monilinia fructicola* in peach.

Management of green molds is currently based on an integration of measures, such as minimizing fruit injury, sanitary practices, and fungicide treatments (Whiteside et al. 1988, Sukorini et al. 2013). The high number of pre- and post-harvest application of chemical fungicides has caused the development of *P. digitatum* resistant strains to several chemical groups (Papoutsis et al. 2019). In Argentina, the continuous use of the fungicides thiabendazol, imazalil, and pyrimethanil in citrus packing houses has led to the development of resistant *P. digitatum* isolates (Panozzo et al. 2018). Therefore, the requirement for alternative control strategies is increasing. The control of green mold without the application of chemical fungicides has been recently reviewed (Papoutsis et al. 2019). Among these non-chemical treatments, natural compounds, irradiations, hot water treatments, salts, and biocontrol agents constitute promising strategies for *P. digitatum* management (Sukorini et al. 2013, Papoutsis et al. 2019).

Strawberry Gray mold

B. cinerea (teleomorph *Botryotinia fuckeliana* (de Bary) Whetzel) causes serious diseases in more than 200 crop species worldwide (Williamson et al. 2007). Between them, gray mold is among the most devastating diseases affecting strawberry fruits worldwide (Hu et al. 2018). Losses for fruit rot in field may reach or exceed 50%, and significant losses during post-harvest and storage might occur (Mertely et al. 2002). Gray mold is between the most important post-harvest diseases in strawberry producing regions of Argentina (Murillo et al. 2016). In this crop, *B. cinerea* also affect leaves, petioles, flower buds, petals, and stems (Maas 1998).

Primary inoculum for flower infections is produced from overwintering sclerotia and plant debris in the form of conidia (Williamson et al. 2007). Infection is favored by high humidity, prolonged surface wetness, and moderate temperatures (15 to 25°C), but the pathogen develops well at a wide range of temperatures between 0 and 30°C (Mertely et al. 2002, Romanazzi and Feliziani 2014). After penetration, *B. cinerea* might remain latent until fruit maturity, when rot occurs and subsequent sporulation will provide inoculums for secondary infections (Williamson et al. 2007, Romanazzi and Feliziani 2014). Infections can also remain latent until storage, when the conditions of humidity and low temperatures slow down host defenses and favor disease development (Romanazzi and Feliziani 2014). Soft rot is most frequently found on the calyx end or on sides of fruits touching other rotten fruits. The fungus sporulates in the presence of free water, covering lesions with a gray mass of conidia and conidiophores (Maas 1987, Mertely et al. 2002, Williamson et al. 2007) (Figure 6.2b).

Management of *B. cinerea* constitutes a challenge because of the great diversity present in the pathogen, the ability to survive in diverse hosts for extended periods, and the variety of inoculum sources which increase adaptation in the field (Williamson et al. 2007, Hu et al. 2018). Thus, integrated disease management, including cultural practices, biological control, and chemical control based on methods for predicting the risk of disease are strongly recommended. However, traditional treatments are applied based on calendar schedules, so optimum spray intervals are rarely determined, and numerous field application are carried out from flowering until harvest (Mertely et al. 2002, Romanazzi and Feliziani 2014). Thus, development of resistance to multiple chemical types of fungicides has been reported worldwide between *B. cinerea* isolates (Williamson et al. 2007, Romanazzi and Feliziani 2014, Hu et al. 2018). The more times the fungicide is applied, higher is the risk of resistance arising in *B. cinerea*. Consequently, mixed spray programs have been devised, ideally with each spray chosen from a different fungicide group, to reduce the risk of substantial field resistance arising (Williamson et al. 2007).

Peach Brown-rot

M. fructicola is the causal agent of brown rot, a destructive pathogen on stone fruits worldwide, which also causes blossom blight and twig cankers (Mondino 2014, Dowling et al. 2019). In peach, *M. fructicola* is responsible for the main fruit losses during the growing season and in post-harvest. The reduction in yield is estimated to be between 20 and 80% in years conducive for disease (Hrustić et al. 2018).

In Argentina, the pathogen has been reported in almost all the producing regions, although the most affected region is the Pampean zone, where weather conditions favor the progress of disease (Mitidieri and Castillo 2014, Rista and Favaro 2014). Even though *M. fructicola* has recently been found to be present in many European countries, it is still considered a quarantine pest in the EPPO region (EPPO 2016).

M. fructicola is a polycyclic pathogen which survives mainly in mummified fruits and cankers on twigs, serving as a source of primary inoculum to infect blossoms, buds, and young shoots. The conidia developed in cankers are of epidemiological significance as secondary inoculum source for fruit rot (Mondino 2014, Dowling et al. 2019). The optimum temperature for fruit infection is 25°C, at which more than 79% of fruits could be infected under a wetting period of a minimum of 12 hours (h). However, if the wetting period is prolonged to 24 hours, infection can occur at a wider range of temperatures (Gell et al. 2008). Fruits are more susceptible to infection during ripening, since the fruit changes its color, and the presence of wounds facilitates penetration (Mondino 2014). Symptoms in fruits manifest as brown, firm rot, which rapidly enlarges, and after a few days post infection, the sign of the disease is visualized as a whitish to creamy powdery sporulation. Affected fruits might fall or remain mummified in plants (Figure 6.2c).

The management of the disease in Argentina integrates application of fungicides with cultural measures, such as the removal of fruit mummies and the pruning of twigs with cankers to reduce inoculum levels. At present, management of *M. fructicola* with fungicides constitutes a challenge for several reasons (Mondino 2014). A high number of applications are required to protect flowers and fruits, taking into account the long susceptibility period for the infection. Fungicides, such as dithiocarbamates, which present low possibility to generate resistance, have long waiting periods, which make application at fruit maturity difficult. Other fungicides with shorter waiting periods show high risk of resistance build-up. Fungicide resistance between isolates of *M. fruticola* has been widely reported for several groups of fungicides, such as quinone outside inhibitor, dicarboxamides, benzimidazoles, and demethylation inhibitors (Tran et al. 2019). Furthermore, carbendazim-resistant isolates have been found in producing areas of Argentina (Mitidieri and Castillo 2014).

For all those mentioned above, exploring the use of wilds plants in general to control fruit health constitutes an important field for interdisciplinary scientific research.

In vitro Evaluation of 10 Wild Species that Grow in the Central Region of Argentina for their Antifungal activities against *B. cinerea* and *M. fructicola*

In order to select plants for antifungal activity tests, it is necessary to take into account three important aspects: (1) if these plant species have proven to be antimicrobial against human pathogens; (2) if they are easy to find in nature; and (3) if they are available in good quantity (Petenatti et al. 2008). The antifungal activities of the plant species described below were evaluated against the fruit pathogens *B. cinerea* and *M. fructicola*. All of them are native or naturalized Argentinian species; they are widely distributed in the central region of the country, and they have potential antifungal activities. The scientific names of the plant and their botanical families are detailed in Table 6.1.

Table 6.1: Plant species evaluated against fruit pathogens *Botrytis cinerea* and *Monilinia fructicola*.

Plant Scientific Name	Family
Dysphania ambrosioides (L.) Mosyakin & Clemants	Amaranthaceae
Austroeupatorium inulifolium (Kunth) R.M. King & H. Rob.	Asteraceae
Wedelia glauca (Ortega) Hoffm. ex Hicken	Asteraceae
Rapistrum rugosum (L.) All.	Brassicaceae
Cyperus rotundus L.	Cyperaceae
Fimbristylis dichotoma (L.) Vahl	Cyperaceae
Schoenoplectus americanus (Pers.) Volkart ex Schinz & R. Keller	Cyperaceae
Schoenoplectus californicus (C.A.Mey.) Soják	Cyperaceae
Fumaria officinalis L.	Papaveraceae
Lantana camara L.	Verbenaceae

Plant species and their Activities

Within the Amaranthaceae family, one of the most recognized medicinal species for its properties is *Dysphania ambrosioides* "paico" (Figure 6.3a-c), a perennial herb that reaches 1.5 m high (Giusti 1997). Its fragrance comes from its leaves, which together with the fruits, are used in infusions or decoctions for digestive, anthelmintic, stimulant, and sudorific purposes (Kliks 1985, Eyssartier et al. 2009, Navone et al. 2014). In the Argentinian northwest area, it is used to treat "empacho" and parasites in children (Campos-Navarro and Scarpa 2013). In other regions of South America, it is also used for other medicinal purposes (Alonso and Desmarchelier 2005), such as treatment of skin diseases. Extracts in different concentration demonstrated an inhibition of *Aspergillus flavus*, *A. glaucus*, *A. niger*, *A. ochraceous*, *Colletotrichum gloeosporioides*, *C. musae*, *Fusarium oxysporum*, and *F. semitectum*, all of which are pathogens of post-harvest cultures (Jardim et al. 2008).

Asteraceae is another interesting family that includes species with potential antifungal value. This family is widely represented in Argentina, and some of its species are aggressive weeds in cultivated fields (Daehler 1998). In our work, two species of Asteraceae were studied—*Austroeupatorium inulifolium* and *Wedelia glauca*. The first one (Figure 6.3d-f), is a subshrub which is 1–2 m high, that grows up to 1,300 masl, in sandy soils, and humid fields. Their leaves are used as a cardiac stimulant, laxative, and anticoagulant (Dominguez 1924, Caius 1941, Arenas and Moreno Azorero

Figure 6.3: (a-c) *Dysphania ambrosioides,* (d-f) *Austroeupatorium inulifolium,* and (g-i) *Wedelia glauca.*

1977, Martínez Crovetto 1981). It is one of the ten plants most used in empirical medicine in rural areas of the Colombian Andes (Grande-Tovar et al. 2016). The essential oil and extracts obtained from this species have shown biological activities, including anti-inflammatory, insecticides, and antibacterial against *Staphylococcus aureus*, *Escherichia coli*, *Pseudomonas aeruginosa*, and *Bacillus subtilis* (Sanabria-Galindo et al. 1998, Álvarez et al. 2005). *In vitro* antifungal assays performed with the essential oil of *A. inulifolium* inhibited up to 70% of the growth of *Penicillium brevicompactum* and *Fusarium oxysporum* (Grande-Tovar et al. 2016). On the other hand, *W. glauca* (Figure 6.3g-i) is a rhizomatous perennial herb, up to 80 cm tall, and widely distributed in Argentina. Several species of this genus are used as digestive herbs all over the world (Rahman 2013). Hepatoprotective properties, antipyretic-analgesic, bactericidal, and molluscicidal activities are attributed to it (Li et al. 2007, Gastón et al. 2008). The acetone extracts obtained from the stem are effective against bacteria of the genera *Proteus* sp. and *Streptococcus* sp. It also causes inhibition in fungal cultures of the genera *Aspergillus* and *Candida* (De Arias et al. 1995).

Cyperaceae is another family widely represented in Argentina that includes aggressive weeds in crop fields. Moreover, many members of Cyperaceae form extensive populations in alluvial plains, streams, and ditches (Figure 6.4a). Cyperaceae, as many grass-like plants, have a high ability to reproduce vegetatively by rhizomes and stolons (Figures 6.4b and c) (Vrijdaghs 2006). Among their best known weeds are *Cyperus rotundus* and *Fimbristylis dichotoma*. *C. rotundus* (Figures 6.4c and d) is used for diarrhea, diabetes, pyresis, inflammation, malaria, stomach pains, and intestinal disorders (Peerzada et al. 2015). It synthesized bioactive components against numerous microorganisms (Adeniyi et al. 2014). Ethyl acetate extracts from its rhizomes have been cited as highly effective antifungal substances (Singh et al. 2011). For these characteristics, it is considered that *C. rotundus* has a high potential value to be used in the ecological control of plant diseases. Species of *Fimbristylis* (Figures 6.4e and f) are used in folk medicine to treat conditions, such as eczema, burns, diarrhea, intestinal infections, infestations, and fever, among other disorders (Simpson and Inglis 2001). However, unlike *C. rotundus*, its action against fungi has not been proven (Islam et al. 2011, Ismail and Siddique 2012, Kadam et al. 2018). On the other hand, sedges, such as *Schoenoplectus americanus* and *Schoenoplectus californicus* are important components of natural wetlands, where they form extensive "reeds" (Figure 6.4a). *Schoenoplectus* sp. has been proven as a phytoremediator in sediments contaminated with zinc, and in the purification of industrial waste as a constituent of wetlands (Poach et al. 2003, Arreghini et al. 2006, Thullen et al. 2008). The antimicrobial action has not been evaluated—neither in human fungi nor against phytopathogens.

The Brassicaceae family plays an important role in human nutrition and has several representatives in the Argentine flora (Jahangir et al. 2009). One of them is *Rapistrum rugosum* "nabo", an annual herb with glabrous or pubescent ellipsoid fruit that allow distinguishing it easily in the field (Figures 6.4g and h). The family which it belongs to is recognized for the production of glucosinolates (Holtz and Williamson 2004). These metabolites are responsible for the biocidal activity reported by several authors (Dubuis et al. 2005). Lazzeri and Mancini (2001), simulating this species as green manure, suppressed *Pythium* sp., and also induced an increase in total soil microbial activity.

Many times, plant species owe their name to some related medicinal property that they perform. *Fumaria officinalis*, from the Latin *fumus* which means "smoke" and *officinalis* that denotes uses in medicine (Figure 6.4i), is a very polymorphous herb included in the Papaveraceae family. It is consumed in infusion, tincture or syrups as sedative, depurative, hypotensive, antiasthmatic, choleretic/cholagogue, and spasmolytic (Alonso 1998, Rombi and Robert 1998). Its pharmacological action is situated in the regulation of choleresis (Del Vitto et al. 1998). It has shown activity against pathogens, such as *Staphylococcus aureus* and *Cladosporium herbarum* (Sengul et al. 2009).

The plant fragrance is another criterion that suggests the presence of secondary metabolites which could be used for medicinal purposes. *Lantana camara* (Figure 6.4j), which belongs to Verbenaceae family, offers a great amount of polycyclic triterpenoids in its aerial parts (Randrianalijaona et al. 2005, Ganjewala et al. 2009, Begum et al. 2010), whose antifungal and antibacterial effects against *P. aeruginosa*, *A. niger*, *F. solani*, and *C. albicans* have been demonstrated (Deena and Thoppil 2000). Naz and Bano (2013) found that the methanolic extracts of its leaves inhibited the growth of *A. fumigatus* and *A. flavus* to 71% and 66%, respectively. It also showed a larvicidal effect against *Aedes aegypti* and *Culex quinquefasciatus* (Kumar and Maneemegalai 2008).

Plant material

Plants were collected, mostly from farms and sides of roads in areas surrounding the Litoral region of Argentina, between March 2016 and February 2017. Each vegetal material was identified by AGR and MIS (co-authors of the present work), and a voucher specimen was deposited at the Herbarium of the FCA-UNL "Arturo Ragonese" (SF Herbarium), Kreder 2805-(3080HOF)-Esperanza, Argentina.

Figure 6.4: Species of (a-c) Cyperaceae family, (c and d) *Cyperus rotundus*, (e and f) *Fimbristylis dichotoma*, (g and h) *Rapistrum rugosum* "nabo", (i) *Lantana camara*, and (j) *Fumaria officinalis*.

After collection, plants were dried in a suitable environment, and different parts (leaves, flowers, fruits, seeds, bark, or the whole plant) were separated according to the extracts that had to be prepared.

For extract preparation, air-dried products of different aerial parts of each species (100 g) were powdered and successively macerated (3 × 24 hours each) with hexane, acetone, and methanol, using mechanical stirring to obtain the corresponding extracts, after filtration and evaporation.

Antifungal assays

Microorganisms and media: Monosporic strains of each fungus were obtained from fruits that presented the correspondent symptom and were morphologically characterized by the Micology Reference Center (CEREMIC, Rosario, Argentina) and the National Institute of Agricultural Technology (INTA, San Pedro, Argentina). Strains of *B. cinerea* CCC-100 and *M. fructicola* INTA-SP345 were grown on Potato-Dextrose-Agar (PDA) medium using petri dishes for 7 days at 15–20°C (as needed for the growth of each one), and sub-cultured every 15 days to prevent pleomorphic transformations. The inoculum of spore suspensions were obtained according to the CLSI reported procedures and adjusted to 1×10^4 Colony Forming Units (CFU)/mL (CLSI 2008).

Susceptibility tests: Diffusion tests were carried out using 9 cm in diameter sterile petri dishes provided with 4 divisions, so that each experiment was considered by quadruplicate. Extract solutions were prepared at a concentration of 50 mg/mL in DMSO, and once dissolved, 400 μL of this stock solution was taken and diluted in 20 mL of molten PDA culture medium. After vigorously shaking and before the mixture was solidified, 5 mL was poured into each of the 4 compartments of the petri dishes and cooled down. A conidia concentration between 10^4 and 10^5 CFU/mL was inoculated inside a well located in the center of each compartment once the medium containing 1000 ppm of each extract solidified. A negative control was performed using the commercial antifungal Carbendazim® and a positive one (growth control) employing the solvent DMSO without plant extract. Once the mycelium of the control plates completely covered the surface of the medium (in approximately 7 days), the measurements of the mycelium diameter developed in each plate treated with each plant extract were carried out by scanning the plates for later reading and analysis with ImageJ® software. The differences in the mean percentage of fungal growth in the presence of each extract were compared to positive and negative controls by statistical analysis with 95% confidence interval (CI).

Antifungal evaluation against B. cinerea: Among all hexane extracts of the 10 species evaluated (Figure 6.5), the most active ones turned out to be *D. ambrosioides* (aerial parts) and *W. glauca* (leaves), showing no significant differences with respect to the negative control, that is to say that they inhibited 100% of the fungal growth. Aerial parts of *C. rotundus* and *R. rugosum*, flowers of *F. dichotoma* and *R. rugosum*, and rhizomes of two species of *Schoenoplectus* were evaluated, and they were moderately active, showing significant differences between both positive and negative controls. Finally, *A. inulifolium* (flowers and leaves), *F. officinalis* (arial parts), and *L. camara* (leaves) resulted to be inactive with no (or small) significant differences with respect to the control growth.

Regarding acetone extracts (Figure 6.6), the most active ones were *W. glauca* (leaves), *S. californicus* (rhizomes), *R. rugosum* (flowers), and *F. officinalis* (aerial parts), showing no fungal growth (without significant differences respect to the negative control). Rhizomes of *S. americanus* and aerial parts of *R. rugosum* displayed moderate activity with significant differences between both controls. The rest of the extracts proved to be barely active or inactive without significant difference on their growth with respect to the positive control.

Methanolic extracts demonstrated to be less active than hexane and acetone ones against *B. cinerea* (Figure 6.7). Among them, *A. inulifolium* (flowers and leaves) and *R. rugosum* (flowers) showed no significant differences with respect to the inhibition control. *C. rotundus*, *S. californicus*, *R. rugosum* (aerial parts), and *F. officinalis* allowed the fungal development in approximately 50% of the cases, suggesting a moderate antifungal activity.

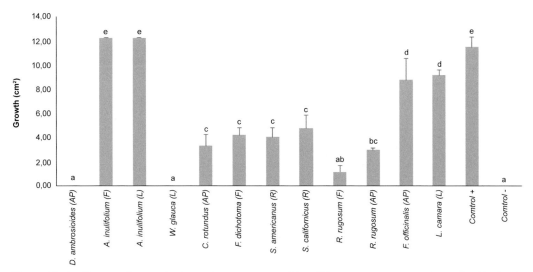

Figure 6.5: Antifungal activity evaluation of hexane extracts of 10 wild plant species against *B. cinerea*. The Y axis corresponds to fungal mycelium growth (cm²) in each compartment of the petri plate. The X axis corresponds to different samples evaluated. Control +: growth control; Control -: inhibition control using the commercial antifungal Carbendazim. Different letters mean statistically significant differences.

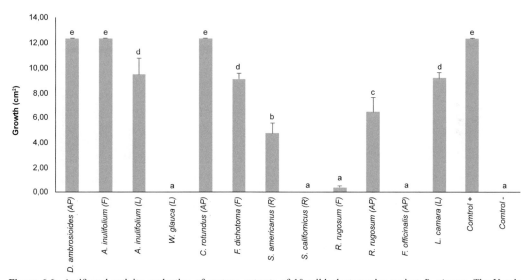

Figure 6.6: Antifungal activity evaluation of acetone extracts of 10 wild plant species against *B. cinerea*. The Y axis corresponds to fungal mycelium growth (cm²) in each compartment of the petri plate. The X axis corresponds to different samples evaluated. Control + = growth control; Control - = inhibition control using the commercial antifungal Carbendazim. Different letters mean statistically significant differences.

Antifungal activity evaluation against M. fructicola: Among all hexane extracts of the 10 species evaluated (Figure 6.8), the most active ones turned out to be *W. glauca* (leaves) and rhizomes of *S. californicus* showing no significant differences with respect to the negative control, that is to say that they inhibited 100% of the fungal growth. Hexane extracts of *R. rugosum* (flowers and aerial parts) were active, allowing a minimum fungal growth (without significant difference respect to the inhibition control). On the other hand, *D. ambrosioides*, *A. inulifolium*, and *F. officinalis* were moderately active, showing significant differences between both the positive and negative controls. Finally, *C. rotundus*, *F. dichotoma*, *S. americanus*, and *L. camara* resulted to be inactive with no significant differences with respect to the control growth.

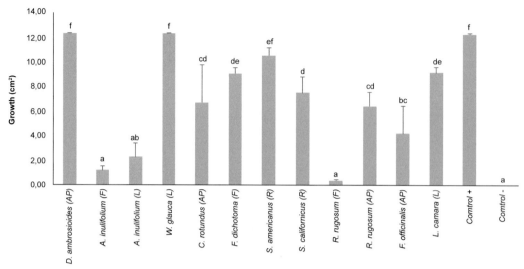

Figure 6.7: Antifungal activity evaluation of methanolic extracts of 10 wild plant species against *B. cinerea*. The Y axis corresponds to fungal mycelium growth (cm²) in each compartment of the petri plate. The X axis corresponds to different samples evaluated. Control +: growth control; Control -: inhibition control using the commercial antifungal Carbendazim. Different letters mean statistically significant differences.

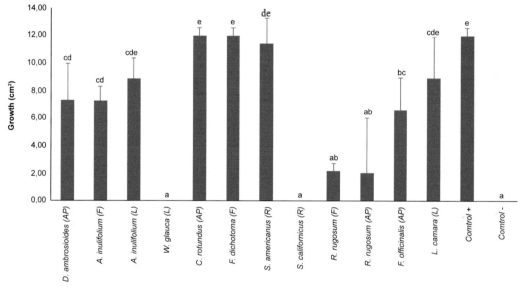

Figure 6.8: Antifungal activity evaluation of hexane extracts of 10 wild plant species against *M. fructicola*. The Y axis corresponds to fungal mycelium growth (cm²) in each compartment of the petri plate. The X axis corresponds to different samples evaluated. Control +: growth control; Control -: inhibition control using the commercial antifungal Carbendazim. Different letters mean statistically significant differences.

Regarding acetone extracts (Figure 6.9), the most active ones were *A. inulifolium* (leaves), *S. californicus* (rhizomes), and both extracts of *R. rugosum* (flowers and aerial parts), showing no fungal growth (without significant differences respect to the negative control) or an incipient fungal development. Acetone extracts of *F. dichotoma* and *F. officinalis* displayed moderate activity, with significant differences between both controls. The rest of the extracts proved to be barely active or inactive without significant difference in their growth with respect to the positive control.

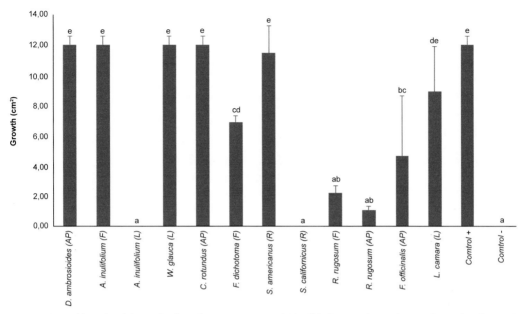

Figure 6.9: Antifungal activity evaluation of acetone extracts of 10 wild plant species against *M. fructicola*. The Y axis corresponds to fungal mycelium growth (cm^2) in each compartment of the petri plate. The X axis corresponds to different samples evaluated. Control +: growth control; Control -: inhibition control using the commercial antifungal Carbendazim. Different letters mean statistically significant differences.

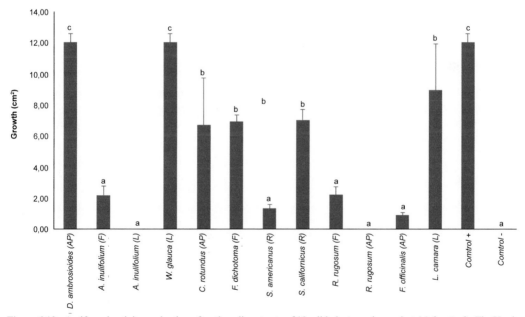

Figure 6.10: Antifungal activity evaluation of methanolic extracts of 10 wild plant species against *M. fructicola*. The Y axis corresponds to fungal mycelium growth (cm^2) in each compartment of the petri plate. The X axis corresponds to different samples evaluated. Control +: growth control; Control -: inhibition control using the commercial antifungal Carbendazim. Different letters mean statistically significant differences.

Methanolic extracts (Figure 6.10) could be classified into three categories taking into account their activities against *M. fructicola*: (1) extracts that inhibited the fungal growth as strong as the negative control carbendazim (*A. inulifolium*, *S. americanus*, *R. rugosum*, and *F. officinalis*); (2) extracts which displayed a moderate antifungal activity showing significant differences between both controls (*C. rotundus*, *F. dichotoma*, *S. californicus*, and *L. camara*); (3) inactive extracts showing no significant differences with respect to the control growth (*D. ambrosioides* and *W. glauca*).

Conclusions

In this chapter, it has been demonstrated that certain wild plants that grow in the central zone of Argentina could be used as potential antifungal agents against phytopathogenic fungi that affect the health of strawberries and bone fruits. Particularly, isolated monosporic strains of *B. cinerea* and *M. fructicola* were submitted to *in vitro* assays in order to explore their susceptibility to hexane, acetone, and methanolic extracts of 10 wild plant species. Some extract type of each species tested to be active, to a greater or lesser extent, against some of the pathogens evaluated. Since the different types of extracts are constituted by different kinds of secondary metabolites present in the plant, they can vary in the antifungal potency. Thus, *D. ambrosioides* and *W. glauca* hexane extract; *W. glauca*, *S. californicus*, *R. rugosum* flowers, and *F. officinalis* acetone extracts; and *A. inulifolium* and *R. rugosum* flowers methanolic extracts, inhibited 100% of the *in vitro* growth of *B. cinerea* under experimental conditions. On the other hand, *W. glauca* and *S. californicus* hexane extracts; *A. inulifolium* and *S. californicus* acetone extracts; and *A. inulifolium*, *S. americanus*, *R. rugosum* and *F. officinalis* methanolic extracts, inhibited 100% of the *in vitro* growth of *M. fructicola* under experimental conditions.

Future efforts will be focused in trying to determine the chemical composition of the active extracts, as well as their application procedures in field, or during harvest and post-harvest stages of strawberries and peaches.

Acknowledgments

Authors gratefully acknowledge Consejo Nacional de Investigaciones Científicas y Técnicas (CONICET) and Agencia Nacional de Promoción Científica y Tecnológica (ANPCyT) for financial support (PIP N° 2015-0524, PICT N° 2015-2259). MIS, NHA, and LNF are also thankful to CONICET for their fellowships.

References

Adeniyi, T.A., Adeonipekun, P.A., Omotayo, A.E. 2014. Investigating the phytochemicals and antimicrobial properties of three sedge (Cyperaceae) species. Not. Sci. Biol. 6(3): 276–281.

Alonso, J.R. 1998. Tratado de fitomedicina: bases clínicas y farmacológicas. Isis Ediciones. Buenos Aires.

Alonso, J., Desmarchelier, C. 2005. Plantas medicinales autóctonas de la Argentina. Lola. Buenos Aires.

Álvarez, M.E., Isaza, L.G., Echeverry, H.M. 2005. Efecto antibacteriano in vitro de *Austroeupatorium inulifolium* H.B.K. (Salvia amarga) y *Ludwigia polygonoides* H.B.K. (Clavo de laguna). Biosalud 4: 46–55.

Arenas, P., Moreno Azorero, R. 1977. Plants used as means of abortion, contraception, sterilization and fecundation by Paraguayan indigenous people. Econ. Bot. 31: 298–301.

Arreghini, S., de Cabo, L., de Iorio, A.F. 2006. Phytoremediation of two types of sediment contaminated with Zn by *Schoenoplectus americanus*. Int. J. Phytoremediat. 8(3): 223–232.

Begum, S., Zehra, S.Q., Ayub, A., Siddiqui, B.S. 2010. A new 28-noroleanane triterpenoid from the aerial parts of *Lantana camara* Linn. Nat. Prod. Res. 24(13): 1227–1234.

Caius, J.F. 1941. The medicinal and poisonous composites of India. J. Bombay Nat. Hist. Soc. 41: 607–645; 838–873.

Campos-Navarro, R., Scarpa, G.F. 2013. The cultural-bound disease "empacho" in Argentina. A comprehensive botanico-historical and ethnopharmacological review. J. Ethnopharmacol. 148(2): 349–360.

Clinical and Laboratory Standards Institute (CLSI). 2008. Reference method for broth dilution antifungal susceptibility testing for filamentous fungi (M38 A2), 2nd ed., Vol. 28. USA: Wayne, p. 1–35.

Daehler, C. 1998. The taxonomic distribution of invasive angiosperm plants: ecological insights and comparison to agricultural weeds. Biol. Conserv. 84(2): 167–180.

De Arias, A.R., Ferro, E., Inchausti, A., Ascurra, M., Acosta, N., Rodriguez, E., Fournet, A. 1995. Mutagenicity, insecticidal and trypanocidal activity of some Paraguayan Asteraceae. J. Ethnopharmacol. 45(1): 35–41.

Deena, M.J., Thoppil, J.E. 2000. Antimicrobial activity of the essential oil of *Lantana camara*. Fitoterapia 71(4): 453–455.

Del Vitto, L.A., Petenatti, E.M., Petenatti, M.E. 1998. Recursos herbolarios de San Luis (Argentina). Segunda parte: plantas exóticas cultivadas, adventicias y/o naturalizadas. Multequina online. Available in http://www.redalyc.org/articulo.oa?id=42800704. ISSN 0327-9375.

Di Liberto, M., Stegmayer, M.I., Svetaz, L., Derita, M. 2019. Evaluation of Argentinean medicinal plants and isolation of their bioactive compounds as an alternative for the control of postharvest fruits phytopathogenic fungi. Br. J. Pharmacognosy. In press. https://doi.org/10.1016/j.bjp.2019.05.007.

Dominguez, J.A. 1924. Sinopsis de la materia médica argentina. Rev. Med. Latino-Americ. 9(109): 1–26.

Dowling, M.E., Bridges, W.C., Cox, B.M., Sroka, T., Wilson, J.R., Schnabel, G. 2019. Preservation of *Monilinia fructicola* genotype diversity within fungal cankers. Plant Dis. 103(3): 526–530.

Dubuis, P.H., Marazzi, C., Städler, E., Mauch, F. 2005. Sulphur deficiency causes a reduction in antimicrobial potential and leads to increased disease susceptibility of oilseed rape. J. Phytopathol. 153(1): 27–36.

EPPO A2 List of pests recommended for regulation as quarantine pests: https://www.eppo.int/QUARANTINE/listA2.htm. (Accessed date: 19.06.2019).

Eyssartier, C., Ladio, A.H., Lozada, M. 2009. Uso de plantas medicinales cultivadas en una comunidad semi-rural de la estepa patagónica. Bol. latinoam. Caribe plantas med. aromát. 8(2). Available http://www.redalyc.org/articulo.oa?id=85611769004. ISSN 0717-7917.

Fu, W., Tian, G., Pei, Q., Ge, X., Tian, P. 2017. Evaluation of berberine as a natural compound to inhibit peach brown rot pathogen *Monilinia fructicola*. Crop Prot. 91: 20–26.

Ganjewala, D., Sam, S., Khan, K.H. 2009. Biochemical compositions and antibacterial activities of *Lantana camara* plants with yellow, lavender, red and white flowers. Eurasia J. Biosci. 3(10): 69–77.

Gastón, S., Bendersky, D., Barbera, P. 2008. Plantas tóxicas de la provincia de Corrientes. Sitio argentino de producción animal. Serie técnica n° 43.

Gatto, M.A., Sergio, L., Ippolito, A., Di Venere, D. 2016. Phenolic extracts from wild edible plants to control postharvest diseases of sweet cherry fruit. Postharvest Biol. Tec. 120: 180–187.

Gell, I., de Cal, A., Torres, R., Usall, J., Melgarejo, P. 2008. Relationship between the incidences of latent infections caused by *Monilinia* spp. and the incidence of brown rot of peach fruit: Factors affecting latent infection. Eur. J. Plant Pathol. 121: 487–498.

Giusti, L. 1997. Chenopodiaceae. Fl Fanerog Argent. 40: 1–52. Avaliable in: http://www.floraargentina.edu.ar/wp-content/uploads/2019/01/40-CHENOPODIACEAE.pdf.

Grande-Tovar, C.D., Chaves-Lopez, C., Viuda-Martos, M., Serio, A., Delgado-Ospina, J., Perez-Alvarez, J.A., Paparella, A. 2016. Sub-lethal concentrations of Colombian *Austroeupatorium inulifolium* (HBK) essential oil and its effect on fungal growth and the production of enzymes. Ind. Crops Prod. 87: 315–323.

Grayer, R., Kokubun, T. 2001. Plant-fungal interactions: the search for phytoalexins and other antifungal compounds from higher plants. Phytochemistry 56: 253–263.

Holst, B., Williamson, G. 2004. A critical review of the bioavailability of glucosinolates and related compounds. Nat. Prod. Rep. 21(3): 425–447.

Hrustić, J., Mihajlović, M., Grahovac, M., Delibašić, G., Tanović, B. 2018. Fungicide sensitivity, growth rate, aggressiveness and frost hardiness of Monilinia fructicola and Monilinia laxa isolates. Eur. J. Plant Pathol. 151: 389–400.

Hu, M.-J., Dowling, M.E., Schnabel, G. 2018. Genotypic and phenotypic variations in *Botrytis* spp. Isolates from single strawberry flowers. Plant Dis. 102: 179–184.

Islam, M., Barua, J., Karon, B., Noor, M. 2011. Antimicrobial, cytotoxic and antidiarrhoeal activity of *Fimbristylis aphylla* L. Int. J. Green Pharm. 5: 135–137.

Ismail, B.S., Siddique, A.B. 2012. Allelopathic inhibition by *Fimbristylis miliacea* on the growth of the rice plants. Adv. Environ. Bio. 30(2): 2423–2428.

Jahangir, M., Kim, H.K., Choi, Y.H., Verpoorte, R. 2009. Health-affecting compounds in Brassicaceae. Compr. Rev. Food Sci. Food Saf. 8(2): 31–43.

Jardim, C.M., Jham, G.N., Dhingra, O.D., Freire, M.M. 2008. Composition and antifungal activity of the essential oil of the Brazilian *Chenopodium ambrosioides* L. J. Chem. Ecol. 34(9): 1213–1218.

Kadam, S.K., Chandanshive, V.V., Rane, N.R., Patil, S.M., Gholave, A.R., Khandare, R.V., Govindwar, S.P. 2018. Phytobeds with *Fimbristylis dichotoma* and *Ammannia baccifera* for treatment of real textile effluent: An *in situ* treatment, anatomical studies and toxicity evaluation. Int. J. Environ. Agric. Res. 160: 1–11.

Kellerman, M., Joubert, J., Erasmus, A., Fourie, P.H. 2016. The effect of temperature, exposure time and pH on imazalil residue loading and green mould control on citrus through dip application. Postharvest Biol. Technol. 121: 159–164.

Kliks, M.M. 1985. Studies on the traditional herbal anthelmintic *Chenopodium ambrosioides* L.: ethnopharmacological evaluation and clinical field trials. Soc. Sci. Med. 21(8): 879–886.

Kumar, M.S., Maneemegalai, S. 2008. Evaluation of larvicidal effect of *Lantana camara* Linn against mosquito species *Aedes aegypti* and *Culex quinquefasciatus*. Adv. Biol. Res. 2(3): 39–43.

Lazzeri, L., Manici, L.M. 2001. Allelopathic effect of glucosinolate-containing plant green manure on *Pythium* sp. and total fungal population in soil. HortScience 36(7): 1283–1289.

Li, X., Dong, M., Liu, Y., Shi, Q.W., Kiyota, H. 2007. Structures and biological properties of the chemical constituents from the genus *Wedelia*. Chem. Biodivers 4(5): 823–836.

Maas, J.L. 1998. Compendium of strawberry diseases. Second edition. St. Paul, Minn., USA: APS Press, 98 pp.

Macarisin, D., Cohen, L., Eick, A., Rafael, G., Belausov, E., Wisniewski, M., Droby, S. 2007. *Penicillium digitatum* suppresses production of hydrogen peroxide in host tissue during infection of citrus fruit. Phytopathology 97: 1491–1500.

Mari, M., Gregori, R., Donati, I. 2004. Postharvest control of *Monilinia laxa* and *Rhizopus stolonifer* in stone fruit by peracetic acid. Postharvest Biol. Tec. 33: 319–325.

Mari, M., Martini, C., Guidarelli, M., Neri, F. 2012. Postharvest biocontrol of *Monilinia laxa*, *Monilinia fructicola* and *Monilinia fructigena* on stone fruit by two *Aureobasidium pullulans* strains. Biol. Control 60: 132–140.

Martínez Crovetto, R. 1981. Las plantas utilizadas en medicina popular en el noroeste de Corrientes. Instituto Miguel Lillo. Miscelánea 69: 7–139.

Mertely, J.C., MacKenzie, S.J., Legard, D.E. 2002. Timing of fungicide applications for *Botrytis cinerea* based on development stage of strawberry flowers and fruit. Plant Dis. 86: 1019–1024.

Mitidieri, M., Castillo, J.A. 2014. Manejo de la podredumbre morena (*Monilinia fructicola* y *M. laxa*) en huertos frutales de Uruguay, Chile, Bolivia, Brasil y Argentina. Argentina: CYTED, 87 pp.

Mohammadi, S., Aminifard, M.H. 2012. Effect of essential oils on postharvest decay and some quality factors of peach (*Prunus persica* var. Redhaven). J. Biol. Environ. Sci. 6: 147–153.

Mondino, P. 2014. Sintomatología, etiología, características epidemiológicas de la enfermedad. pp. 35–41. *In*: Mitidieri, M., Castillo, J.A. (eds). Manejo de la podredumbre morena (*Monilinia fructicola* y *M. laxa*) en huertos frutales de Uruguay, Chile, Bolivia, Brasil y Argentina. Argentina: CYTED.

Murillo, B.G., Guerrero, E.E.F., Zapata, S.R. 2016. Manejo ecológico en frutilla aplicando Trichoderma sp. como promotor de crecimiento y controlador biológico de *Botrytis cinerea*. AVERMA 20: 9–37.

Navone, G.T., Zonta, M., Gamboa, M. 2014. Fitoterapia Mbyá-Guaraní en el control de las parasitosis intestinales: Un estudio exploratorio con *Chenopodium ambrosioides* L. var *anthelminticum* en cinco comunidades de Misiones, Argentina. Polibotánica (37): 135–151.

Naz, R., Bano, A. 2013. Phytochemical screening, antioxidants and antimicrobial potential of *Lantana camara* in different solvents. Asian Pac. J. Trop Dis. 3(6): 480–486.

Neven, G., Drake, S.R. 2000. Comparison of alternative postharvest quarantine treatments for sweet cherries. Postharvest Biol. Tec. 20: 107–114.

Novaes Azevedo, A., Ribeiro Buarque, P., Oliveira Cruz, E.M., Fitzgerald Blank, A., Lins de Aquino Santana, L. 2014. Response surface methodology for optimization of edible chitosan coating formulations incorporating essential oil against several foodborne pathogenic bacteria. Food Control 43: 1–9.

Palou, L., Smilanick, J., Droby, S. 2008. Alternatives to conventional fungicides for the control of citrus postharvest green and blue molds. Stewart Postharvest Review 4: 1–16.

Papoutsis, K., Mathioudakis, M.M., Hasperué, J.H., Ziogas, V. 2019. Non-chemical treatments for preventing the postharvest fungal rotting of citrus caused by *Penicillium digitatum* (green mold) and *Penicillium italicum* (blue mold). Trends in Food Sci. & Tech. 86: 479–491.

Peerzada, A.M., Ali, H.H., Naeem, M., Latif, M., Bukhari, A.H., Tanveer, A. 2015. *Cyperus rotundus* L.: Traditional uses, phytochemistry, and pharmacological activities. J. Ethnopharmacol. 174: 540–60.

Pergomet, J., Di Liberto, M., Derita, M., Bracca, A., Kaufman, T. 2018. Activity of the pterophyllins 2 and 4 against postharvest fruit pathogenic fungi. Comparison with a synthetic analog and related intermediates. Fitoterapia 125: 98–105.

Petenatti, E., Gette, M., Derita, M., Petenatti, M., Solís, C., Zuljan, F., Zacchino, S. 2008. Importance of the ethnomedical information for the detection of antifungal properties in plant extracts from the Argentine flora.

In: South American Medicinal Plants as Potential Source of Bioactive Compounds. Transworld Research Network, Kerala 15–38.

Poach, M.E., Hunt, P.G., Vanotti, M.B., Stone, K.C., Matheny, T.A., Johnson, M.H., Sadler, E.J. 2003. Improved nitrogen treatment by constructed wetlands receiving partially nitrified liquid swine manure. Ecol. Eng. 20(2): 183–197.

Qin, G., Zong, Y., Chen, Q., Hua, D., Tian, S. 2010. Inhibitory effect of boron against *Botrytis cinerea* on table grapes and its possible mechanisms of action. Int. J. Food Microbiol. 138: 145–150.

Rahman, A.H.M. 2013. An ethnobotanical investigation on Asteraceae family at Rajshahi, Bangladesh. JBAMSR. 2(5): 133–141.

Randrianalijaona, J.A., Ramanoelina, P.A., Rasoarahona, J.R., Gaydou, E.M. 2005. Seasonal and chemotype influences on the chemical composition of *Lantana camara* L.: Essential oils from Madagascar. Anal. Chem. Acta. 545(1): 46–52.

Rista, L.M., Favaro, M.A. 2014. Manejo de Enfermedades. pp. 205–223. *In*: Gariglio, N., Bouzo, C., Travadelo, M. (eds.). Cultivos frutales y ornamentales para zonas templado–cálidas. Experiencias en la zona central de Santa Fe. Argentina: Ediciones UNL.

Romanazzi, G., Feliziani, E. 2014. Chapter 4: *Botrytis cinerea* (Gray Mold). pp. 131–146. *In*: Bautista-Baños, S. (ed.). Postharvest Decay. Control Strategies. USA: Academic Press.

Rombi, M., Robert, D. 1998. 100 plantes médicinales: composition, mode d'action et intérêt thérapeutique. Editions Romart 279 pp.

Sanabria-Galindo, A., Mendoza-Ruiz, A., Moreno, A.L. 1998. Actividad antimicrobiana *in vitro* de angiospermas colombianas. Rev. Colomb. Cienc. Quim. Farm. 27(1): 47–51.

Sengul, M., Yildiz, H., Gungor, N., Cetin, B., Eser, Z., Ercisli, S. 2009. Total phenolic content, antioxidant and antimicrobial activities of some medicinal plants. Pak. J. Pharm. Sci. 22(1): 102–106.

Simpson, D.A., Inglis, C.A. 2001. Cyperaceae of Economic, Ethnobotanical and Horticultural Importance: A Checklist. Kew Bull. 56: 257–360.

Singh, A., Mauryab, S., Singhc, R., Singh, U.P. 2011. Antifungal efficacy of some ethyl acetate extract fractions of *Cyperus rotundus* rhizomes against spore germination of some fungi. Arch. of Phytopathol. and Plant Protect. 44(20): 2004–2011.

Sordo, M.H., Travadelo, M., Pernuzzi, C. 2017. Strawberry crop evolution in province of Santa Fe (Argentina) in the last 50 years. Hort Arg. 36(90): 13–24.

Spadoni, A., Neri, F., Bertolini, P., Mari, M. 2013. Control of *Monilinia* rots on fruit naturally infected by hot water treatment in commercial trials. Postharvest Biol. Tec. 86: 280–284.

Sukorini, H., Sangchote, S., Khewkhom, N. 2013. Control of postharvest green mold of citrus fruit with yeasts, medicinal plants, and their combination. Postharvest Biol. Technol. 79: 24–31.

Thullen, J.S., Nelson, S.M., Cade, B.S., Sartoris, J.J. 2008. Descomposición de macrófitos en un humedal construido dominado por amoníaco de flujo superficial: tasas asociadas con variables ambientales y bióticas. Ingeniería ecológica. 32(3): 281–290.

Tran, T.T., Li, H., Nguyen, D.Q., Jones, M.G.K., Sivasithamparam, K., Wylie, S.J. 2019. *Monilinia fructicola* and *Monilinia laxa* isolates from stone fruit orchards sprayed with fungicides. Eur. J. Plant Pathol. 153: 985–999.

Vrijdaghs, A. 2006. A floral ontogenetic approach to homology questions in non-mapanioid Cyperaceae. Institut voor Plantkunde en Microbiologie, Laboratorium voor Plantensystematiek. Doctoral Thesis.

Williamson, B., Tudzynski, B., Tudzynski, P., Van Kan, J.A.L. 2007. Botrytis cinerea: the cause of grey mould disease. Mol. Plant Pathol. 8(5): 561–580.

Whiteside, J.O., Garnsey, S.M., Trimmer, L.W. 1988. Compendium of Citrus Diseases. St. Paul, Minn., USA: APS Press, 88 p.

Zhang, Y., Zeng, L., Yang, J., Zheng, X., Yu, T. 2015. 6-Benzylaminopurine inhibits growth of *Monilinia fructicola* and induces defense-related mechanism in peach fruit. Food Chem. 187: 210–217.

7

Plants from Brazil Used Against Snake Bites
Oleanolic and Ursolic Acids as Antiophidian Against *Bothrops jararacussu* venom

Jocimar de Souza,[1] *Bruna Stramandinoli Deamatis,*[1] *Fernanda Mayumi Ishii,*[1]
Ingrid Francine Araújo de Oliveira,[2] *Gustavo Rodrigues Toledo Piza,*[3]
Jorge Amaral Filho,[1] *Edson Hideaki Yoshida,*[1] *José Carlos Cogo,*[4]
Angela Faustino Jozala,[2] *Denise Grotto,*[3] *Rauldenis Almeida Fonseca Santos*[5]
and Yoko Oshima-Franco[1,*]

Introduction

There are approximately 2,900 snake species in the world, of which only 410 are considered poisonous (Cardoso et al. 2003). In Brazil, it is possible to find approximately 69 of these poisonous species, distributed between two families (Viperidae and Elapidae) and belonging mainly to the genera *Bothrops*, *Crotalus*, *Lachesis*, and *Micrurus* (Pinho and Pereira 2001, Carrasco et al. 2012).

Due to the large number of accidents and the severity of human events in tropical countries, snakebite accidents are considered a public health problem (Pinho and Pereira 2001). The clinical condition developed by the victim is very varied, depending on the amount of poison inoculated, the place of the bite, the age and physical condition of the victim, and especially the time between the accident and medical care (Borges et al. 1999).

Around 2.5 million cases per year are reported worldwide, with a mortality rate between 4 and 10% (Chippaux and Goyffon 1998). In Brazil, these accidents are around 20,000 per year, with a fatality rate of 0.4% (Pinho and Pereira 2001, Araújo et al. 2003). However, it is important to remember that

[1] Laboratory of Research in Neuropharmacology and Multidisciplinary (Lapenm), University of Sorocaba, Sorocaba, São Paulo, Brazil.
[2] Laboratory of Industrial Microbiology and Fermentation Process (Laminfe), University of Sorocaba, Sorocaba, São Paulo, Brazil.
[3] Laboratory of Toxicological Research (Lapetox), University of Sorocaba, Sorocaba, São Paulo, Brazil.
[4] Bioengineering and Biomedical Engineering Programs, Technological and Scientific Institute, Brazil University, São Paulo, Brazil.
[5] Federal Institute of Education, Science and Technology of Rondonia, Federal Institute of Rondônia, Calama, Rondônia, Brazil.
* Corresponding author: yoko.franco@prof.uniso.br

the data is not homogeneous. For example, in Brazil, the state of Paraná had a death rate of 51% of the 54.737 cases recorded by snakebite accident (Moura and Mourão 2012).

The toxins present in the venom of snakes have numerous properties, and act by different mechanisms—in paralysis, death, and digestion of prey, besides being a defense component against predators (Mebs 2002). They are constituted by a cocktail of chemical compounds mostly formed by enzymatic action proteins (Gutiérrez and Lomonte 1995, Gutiérrez 2002, Mebs 2002).

Poisoning by these snakes causes disorders that affect blood clotting, induce hemorrhage, edema, and local necrosis (Gutiérrez and Lomonte 1989, 1995, Gutiérrez 2002, Gutiérrez et al. 2007). The medical practice officially used to treat ophthalmic poisoning as the administration of serum therapy, which neutralizes venom toxins (Gutiérrez 2002). However, recovery from tissue damage rarely occurs (Cardoso et al. 2003, Gutiérrez et al. 1998), and the delay in patient care, even with serum therapy, may be insufficient, especially whet it occurs in rural areas far from medical care. In addition, traditional serum therapy may lead to anaphylactic reactions and hypersensitivity caused by whey proteins.

Thus, other complementary practices can be used, especially with the purpose of neutralizing local tissue damage, such as the use of medicinal plants (Cardoso et al. 2003, Da Silva et al. 2007).

In different parts of the world, medicinal plants have been providing metabolites capable of inhibiting the action of toxins present in snake venoms, and the ehnopharmacological survey of species popularly used to treat snakebites grows over time (Nakagawa et al. 1982, Mors 1991, Martz 1992, Houghton and Osibogun 1993, Mors et al. 2000, Pereira et al. 1996, Otero et al. 2000a, Soares et al. 2004, 2005), including investigating the real inhibition efficiency in *in vitro* and *in vivo* tests using isolated extracts and metabolites (Houghton and Osibogun 1993, Da Silva et al. 1997, Pereira et al. 1996, Daros et al. 1996, Mors et al. 2000, Januário et al. 2004, Soares et al. 2004, 2005, Ticli et al. 2005, Dey and De 2012).

Countries with high plant biodiversity stand out for the investigation of plants with antiophidic metabolites, and there has been an increase in scientific production associated with plants from the American continent, especially Central and South America (Otero et al. 2000a, Gutiérrez 2002, Cardoso et al. 2003, Soares et al. 2004, 2005).

In Brazil, one of the milestones for the investigation of plants capable of neutralizing toxins from snake venom occurred in 1882, with the investigation by Nakagawa and collaborators of the inhibitory capacity of the "specific person" infusion, a species of panacea used by the indigenous population of Amazon rainforest and absorbed by the Jesuits in the colonial period, used to combat different types of animal poisons—snakes, spiders, bees, scorpions, etc. (Pereira et al. 1996). From this infusion, cabenegrin pterocarpanes A-I and A-II were isolated, which showed proven antiophidic activity (Nakagawa et al. 1982).

The plant species present in the infusion was popularly known as "head of black", however, many species have this popular name, making the botanical identification impossible at the time. Subsequently, the same pterocarpanes were found in the *Harpalicia brasiliana* species, called the "snake root" (Da Silva et al. 1997), also used as an infusion to combat snakebite accidents by the local population, thus being one of the species present in "specific person".

Another Brazilian milestone in the fight against snakebite accidents occurred through the creation of epidemiological surveillance programs, such as Butantan Institute, Ezequiel Dias Foundation, and the Vital Brazil Institute, in the early 20th century, specialized in the development of serum therapy (Fan and Monteiro 2018).

Several scientific studies have been recording plant species popularly used to treat snakebite accidents. For example, Mors (1991) and Mors et al. (2000) listed 578 plant species popularly used to combat snake bites, Martz (1992) listed 11 species, and Soares et al. (2005) listed 850 species. In India, the main references are concentrated in the fight against venom of species of the genera *Naja*, *Daboia*, and *Ophiophagus* (Santhosh et al. 2013), where more than 520 plants have been registered for this purpose (Upasani et al. 2017). In Bangladesh, articles cite about 116 plants (Kadir et al. 2015), and in Sri Lanka there are reports of 341 plant species (Dharmadasa et al. 2015).

On the African continent, more than 100 plant species used for antiophidic purposes were recorded, distributed among the following countries—Mali, Democratic Republic of Congo, South Africa (Molander et al. 2014), and Kenya (Owuor and Kisangau 2006). In Colombia, 70 species have been listed, widely studied from a phytopharacologic point of view (Otero et al. 2000a, b, c, d). In Central America, about 260 plant species are described (Giovannini and Howes 2017).

In Brazil, different plants have already been registered and tested as snake venom inhibitors, described in different scientific articles, such as Soares et al. (2004), who report the use of 56 plants, Nishijima et al. (2009) report five plants, De Paula et al. (2010) report 12 plants, and Moura et al. (2015) described reports of 24 plants used by the community of Pará (state with the highest incidence of snakebite accidents in Brazil), and four species (*Bellucia dichotoma, Connarus favosus, Plathymenia reticulata, Philodendron megalophyllum*) showed 100% inhibition of induced hemorrhage by *Bothrops* poison. The number of plants used for this purpose is so large that digital databases have already been created in order to correlate the plant species with antiophidic tests (Amui et al. 2011). In addition, many other plant species are cited. However, they have never been evaluated pharmacologically (Moura et al. 2015, Indriunas and Aoyama 2018). Table 7.1 summarizes Brazilian native plants with proven action as antiophidians.

Although extracts of plant species have their capacity to neutralize proven ophidian toxins (Mors 1991, Houghton and Osibogun 1993, Otero et al. 2000a, b, c, d), fewer scientific studies report the chemical compounds responsible for inhibitory capacity (Giovannini and Howes 2017, Mors et al. 2000, Soares et al. 2005) (Table 7.2).

Plants containing triterpenes have antiophidic use described in several articles (Soares et al. 2005, Pereira et al. 1996, Daros et al. 1996, Mors et al. 2000), especially Lupane skeleton triterpenes, such as betulin, lupeol, and betulinic acid (Bernard et al. 2001, Ferraz et al. 2012, 2015).

Due to the known venom inhibition activity of *Bothrops* species, the study of triterpenic acids becomes a good alternative to obtain new ophidian venom inhibiting agents, such as oleanolic ($C_{30}H_{48}O_3$) and ursolic ($C_{30}H_{48}O_3$) acids (Figure 7.1), which naturally coexist (Liu 1995, Yin and Chan 2007) in food and medicinal plants (Table 7.3), as the free acid form or as aglycone for triterpenoid saponins (Price et al. 1987, Mahato et al. 1988, Wang and Jiang 1992), and have been reported to be beneficial and have notable therapeutic effects (Zhang et al. 2013a).

The pharmacological effects of oleanolic and ursolic acids comprise a vast list of properties. Wózniak et al. (2015) and Ayeleso et al. (2017) described the properties for ursolic and oleanolic acids, respectively, as being antibacterial, antimicrobial, anticancer/antitumour, antidiabetic, antihypertensive, anti-inflammatory, antioxidant, antiparasitic, and hepatoprotective, for both triterpenes. Besides, the properties attributed to ursolic acid also include protective effects on heart, brain, skeletal muscles, bones, and other organs (skin, kidney, and lung), although some studies point out negative effects of administration of this compound (Wózniak et al. 2015). Plants and derivatives have been assayed against snake venoms and other venomous animals, since complementary alternatives to antivenom therapy are desirable mainly by lack of this product in many communities (Knudsen and Laustsen 2018).

Oleanolic and Ursolic Acids as Antiophidian against *Bothrops jararacussu* venom: A Case Study

In this study, a new property of both triterpenes, isolatedly, was studied against the neurotoxic and myotoxic effects of *Bothrops jararacussu* snake venom, a snake of medical interest, on mouse phrenic nerve preparations, using a traditional myographic technique. Among the mechanisms by which snake venoms lead to toxic effects are the oxidative stress (Sunitha et al. 2015), and classic biomarkers were evaluated from bath samples of pharmacological assays. For the first time, the redox biomarkers were evaluated from the neuromuscular preparation's bath. The antimicrobial and citotoxicological profile of both acids were also evaluated.

Table 7.1: Brazilian native plants with proven action against snakebites.

Plant scientific name	Family	Parts used	Reference(s)
Aegiphila panamensis Moldenke	Verbenaceae	Leaves and Barks	Otero et al. 2000d
Aegiphila salutaris Kunth	Verbenaceae	-	Mors 1991
Allamanda catartica L.	Apocynaceae	Leaves and Barks	Otero et al. 2000d
Alternanthera brasiliana (L.) Kuntze	Amaranthaceae	Flowers	Félix-Silva et al. 2017, Moura et al. 2015
Anacardium occidentale L.	Anacardiaceae	Barks, Fruits, Leaves, and Roots	Ushanandini et al. 2009
Anacardium humile Martius	Anacardiaceae	Barks	Costa 2009
Aniba fragrans Ducke	Lauraceae	Barks	Moura et al. 2015
Aniba parviflora (Meisn.) Mez	Lauraceae	Leaves	Félix-Silva et al. 2017, Moura et al. 2015
Annona furfuracea A.St. Hil.	Annonaceae	-	Mors 1991
Annona montana Macfad.	Annonaceae	Leaves	Félix-Silva et al. 2017
Apuleia leiocarpa (Vogel) J.F. Macbr.	Fabaceae	Roots	Houghton and Osibogun 1993
Arisaema psittacus E. Barnes	Araceae	Roots	Breitbach et al. 2013
Aristolochia antihysterica Mart. ex Duch *is the synonym of Aristolochia triangularis* Cham.	Aristolochiaceae	-	Mors 1991
Aristolochia barbata Jacq.	Aristolochiaceae	Roots	Houghton and Osibogun 1993
Aristolochia birostris Duch.	Aristolochiaceae	Whole plant	França et al. 2005
Aristolochia cymbifera L.	Aristolochiaceae	Roots and Leaves	Da Silva et al. 2017
Aristolochia pilosa Kunth	Aristolochiaceae	Roots	Otero et al. 2000d
Aristolochia sprucei Mast.	Aristolochiaceae	-	Rodríguez 2010
Aristolochia theriaca Mart.	Aristolochiaceae	Roots	Houghton and Osibogun 1993
Aristolochia trilobata L.	Aristolochiaceae	Latex	Houghton and Osibogun 1993
Baccharis trimera Less.	Asteraceae	Aerial parts	Januário et al. 2004, Bernard et al. 2001
Bellucia dichotoma Cogn.	Melastomataceae	Barks	Moura et al. 2015
Bixa orellana L.	Bixaceae	Leaves and Branches	Otero et al. 2000d
Blutaparon portulacoides (A.St.-Hil.) Mears	Amaranthaceae	-	Carvalho et al. 2013
Bredemeyera floribunda Willd.	Polygalaceae	Roots	Houghton and Osibogun 1993, Soares et al. 2005
Brosimum guianensis (Aubl.) Huber	Moraceae	Leaves	Da Silva et al. 2017
Brunfelsia uniflora D. Don	Solanaceaae	Leaves	Houghton and Osibogun 1993
Bryonia bonariensis Miller	Cucurbitaceae	Leaves	Indriunas and Aoyama 2018
Buddleja brasiliensis Jacq.	Buddlejaceae	Roots	Mors et al. 2000
Bursera simaruba L. Sarg.	Burseraceae	Leaves	Mors 1991
Byrsonima crassa Nied.	Malpighiaceae	Leaves	Nishijima et al. 2009

Table 7.1 contd. ...

...Table 7.1 contd.

Plant scientific name	Family	Parts used	Reference(s)
Byrsonima crassifolia (L.) Kunth	Malpighiaceae	Leaves and Barks	Félix-Silva et al. 2017
Caesalpinia bonduc (L.) Roxb.	Fabaceae	Seeds	Mors et al. 2000
Casearia gossypiosperma Briq.	Salicaceae	Leaves	Camargo et al. 2010
Casearia mariquitensis Kunth	Salicaceae	Leaves	Izidoro et al. 2003
Casearia sylvestris Sw.	Salicaceae	Leaves	Mors et al. 2000, De Paula et al. 2010, Carvalho et al. 2013
Chiococca brachiata R. & Pav.	Rubiaceae	Roots	Houghton and Osibogun 1993
Chiococca racemosa **var.** *jacquiniana* Griseb.	Rubiaceae	Roots	Mors et al. 2000, Indriunas and Aoyama 2018
Cissampelos glaberrima A. St.-Hil.	Menispermaceae	Roots	Mors et al. 2000
Cissampelos pareira L.	Menispermaceae	Whole plant	Félix-Silva et al. 2017
Clibadium sylvestre (Aubl.) Baill.	Asteraceae	Whole plant	Otero et al. 2000d
Clusia fluminensis Guttiferae Juss.	Clusiaceae	Fruits	Da Silva et al. 2017
*Cocos coronata Mart.	Arecaceae	Barks	Félix-Silva et al. 2017
Combretum leprosum Mart.	Combretaceae	Roots	Da Silva et al. 2017
Connarus favosus Planch.	Connaraceae	Barks	Félix-Silva et al. 2017, Moura et al. 2015
Cordia verbenacea DC.	Boraginaceae	Leaves	Soares et al. 2005, Ticli et al. 2005
Costus guanaiensis Rusby	Costaceae	Barks	Otero et al. 2000d
Costus lasius Loes.	Costaceae	Leaves and Bark	Otero et al. 2000d
Crateva tapia L.	Capparaceae	Leaves	Félix-Silva et al. 2017, Moura et al. 2015
Croton urucurana Baill.	Euphorbiaceae	Barks	Da Silva et al. 2017
Cynophalla flexuosa L.	Capparaceae	Barks	Félix-Silva et al. 2017
Davilla elliptica St. Hill.	Dilleniaceae	Leaves	Nishijima et al. 2009
Derris amazonica Killip	Fabaceae	Roots	Félix-Silva et al. 2017
*Derris floribunda (Benth.) Ducke	Fabaceae	Roots	Félix-Silva et al. 2017
Derris sericea (Poir.) Ducke	Fabaceae	Roots	Mors et al. 2000
Derris urucu (Killip & Sm.) J.F.Macbr.	Fabaceae	Roots	Soares et al. 2005
Dipteryx alata Vogel	Fabaceae	Seeds	Ferraz et al. 2012
Dorstenia brasiliensis Lam.	Moraceae	Roots	Soares et al. 2005, Houghton and Osibogun 1993
Dracontium polyphyllum L.	Araceae	Leaves	Mors et al. 2000
Eclipta prostrata (L.) L.	Asteraceae	Aerial parts	Soares et al. 2005
Erechtites valerianifolia (Wolf) DC.	Asteraceae	Leaves and Barks	Otero et al. 2000d
Eupatorium ayapana Vent.	Asteraceae	Aerial parts	Indriunas and Aoyama 2018
Eupatorium triplinerve Vahl	Asteraceae	Leaves	Mors et al. 2000
Euphorbia hirta L.	Euphorbiaceae	Whole plant	Félix-Silva et al. 2017
Euterpe edulis Mart.	Arecaceae	Latex	Félix-Silva et al. 2017

Table 7.1 contd. ...

...Table 7.1 contd.

Plant scientific name	Family	Parts used	Reference(s)
Euterpe oleracea Mart.	Arecaceae	Fruits	Félix-Silva et al. 2017, Moura et al. 2015
Ficus nymphaeifolia Mill.	Moraceae	Leaves	Mors et al. 2000, Otero et al. 2000d
Gomphrena officinalis Mart.	Amaranthaceae	-	Mors 1991
Handroanthus barbatus (E.Mey.) Mattos	Bignoniaceae	Leaves	Félix-Silva et al. 2017, Moura et al. 2015
Harpalyce brasiliana Benth.	Fabaceae	Roots	Da Silva et al. 1997
Heterothalamus psiadioides Less.	Asteraceae	Whole plant	Mors et al. 2000
Humirianthera ampla (Miers) Baehni	Icacinaceae	Roots	Da Silva et al. 2017
Hypericum brasiliense Choisy	Hypericaceae	Roots	Carvalho et al. 2013
Hyptis capitata Jacq.	Lamiaceae	Leaves and Barks	Otero et al. 2000d
Ipomoea cairica (L.) Sweet	Convolvulaceae	Leaves and Barks	Otero et al. 2000d
Irlbachia alata (Aubl.) Maas	Gesneriaceae	Leaves	Otero et al. 2000d
Jatropha elliptica (Pohl) Oken	Euphorbiaceae	Leaves	De Paula et al. 2010
Jatropha gossypiifolia L.	Euphorbiaceae	Leaves	Da Silva et al. 2017
Jatropha ribifolia (Pohl) Baill.	Euphorbiaceae	Latex	Félix-Silva et al. 2017
Justicia pectoralis Jacq.	Acanthaceae	Leaves	Moura et al. 2015
Justicia secunda Vahl	Acanthaceae	Whole plant	Otero et al. 2000d
Kalanchoe brasiliensis Larrañaga	Crassulaceae	Leaves	Moura et al. 2015
Libidibia ferrea (Mart. ex Tul.) L.P.Queiroz	Fabaceae	Seeds	Félix-Silva et al. 2017, Moura et al. 2015
Lindernia difusa (L.) Wettst.	Scrophulariaceae	Whole plant	Otero et al. 2000d
Macfadyena unguis-cati (L.) A.H.Gentry	Bignoniaceae	Whole plant	Otero et al. 2000d
Machaerium eriocarpum Benth.	Fabaceae	Resin	Houghton and Osibogun 1993
Machaerium ferox (Mart. ex Benth.) Ducke	Fabaceae	Leaves	Félix-Silva et al. 2017, Moura et al. 2015
Mandevilla illustris (Vell.) Woodson var.	Apocynaceae	Roots and Leaves	Biondo et al. 2004
Mandevilla velutina K. Schum.	Apocynaceae	Roots	Carvalho et al. 2013, Soares et al. 2005, De Paula et al. 2010
Manihot esculenta Crantz	Euphorbiaceae	Roots	Félix-Silva et al. 2017
Marsypianthes chamaedrys (Vahl) Kuntze	Lamiaceae	Leaves	Moura et al. 2015
Marsypianthes hyptoides Mart. ex Benth.	Lamiaceae	Whole plant	Houghton and Osibogun 1993
Miconia albicans (Sw.)	Melastomataceae	Stems	De Paula et al. 2010, Félix-Silva et al. 2017
Miconia fallax DC.	Melastomataceae	Stems	De Paula et al. 2010, Félix-Silva et al. 2017
Miconia sellowiana Naudin	Melastomataceae	Stems	De Paula et al. 2010
Mikania cordifolia (L.f.) Willd.	Asteraceae	Whole plant	Houghton and Osibogun 1993

Table 7.1 contd. ...

...Table 7.1 contd.

Plant scientific name	Family	Parts used	Reference(s)
Mikania glomerata Spreng.	Asteraceae	Leaves	Mors et al. 2000, De Paula et al. 2010, Carvalho et al. 2013, Soares et al. 2005
*Mikania laevigata Sch.Bip. ex Baker	Asteraceae	Leaves	Collaço et al. 2012
Mikania opifera Mart.	Asteraceae	Aerial parts	Indriunas and Aoyama 2018
Mimosa pigra L.	Fabaceae	Roots	Houghton and Osibogun 1993
Mimosa pudica L.	Fabaceae	Roots	Houghton and Osibogun 1993
Momordica charantia L.	Cucurbitaceae	Aerial parts and Barks	Otero et al. 2000d
*Mouriri pusa Gardn.	Melastomataceae	Leaves	Nishijima et al. 2009
Ocimum micranthum Willd.	Lamiaceae	Aerial parts and Barks	Otero et al. 2000d
Passiflora quadrangulares L.	Passifloraceae	Aerial parts	Otero et al. 2000d
Peltodon radicans Phol.	Lamiaceae	Leaves, Stems, and Flowers	Costa et al. 2008
Pentaclethra macroloba (Willd.) Kuntze	Fabaceae	Barks	Carvalho et al. 2013, Soares et al. 2005, Da Silva et al. 2007
Periandra mediterranea (Vell.) Taub.	Fabaceae	Roots	Soares et al. 2005, Houghton and Osibogun 1993
*Periandra pujalu Taub.	Fabaceae	Roots	Houghton and Osibogun 1993
Philodendron megalophyllum Schott	Araceae	Vines	Moura et al. 2015
Philodendron tripartitum (Jacq.) Schott	Araceae	Aerial parts	Otero et al. 2000d
Phyllanthus acuminatus Vahl	Euphorbiaceae	Aerial parts	Otero et al. 2000d
*Phyllanthus klotzschianus Muell.-Arg.	Euphorbiaceae	Leaves	Houghton and Osibogun 1993
Piper arboreum Aubl.	Piperaceae	Whole plant	Otero et al. 2000d
*Piper caldense C.DC.	Piperaceae	Whole plant	Soares et al. 2005
Piper hispidum C.DC.	Piperaceae	Whole plant	Otero et al. 2000d
Piper marginatum Jacq.	Piperaceae	Whole plant	Otero et al. 2000d
Piper multiplinervium C.DC.	Piperaceae	Whole plant	Otero et al. 2000d
Piper peltatum L.	Piperaceae	Aerial parts and Barks	Houghton and Osibogun 1993
Piper reticulatum L.	Piperaceae	Aerial parts and Barks	Otero et al. 2000d
Plathymenia reticulata Benth.	Fabaccae	Barks	Moura et al. 2015
Polygala paniculata L.	Polygalaceae	Roots	Félix-Silva et al. 2017
Polygala spectabilis DC.	Polygalaceae	Roots	Félix-Silva et al. 2017
Portulaca pilosa L.	Portulacaceae	Leaves	Félix-Silva et al. 2017

Table 7.1 contd. ...

...Table 7.1 contd.

Plant scientific name	Family	Parts used	Reference(s)
Pothomorphe umbellata Miq.	Piperaceae	Whole plant	Mors et al. 2000
Prestonia coalita (Vell.) Woods.	Apocynaceae	Vines	Mors et al. 2000
Psychotria ipecacuanh (Brot.) Stokes	Rubiaceae	Aerial parts and Barks	Otero et al. 2000d
Psychotria poeppigiana Müll. Arg.	Rubiaceae	Fruits	Otero et al. 2000d
Quassia amara L.	Simaroubaceae	Whole plant	Otero et al. 2000d
Renealmia alpinia (Rottb.) Maas	Zingiberaceae	Roots	Otero et al. 2000d
Sapindus saponaria L.	Sapindaceae	Aerial parts	Da Silva et al. 2017
Schizolobium parahyba (Vell.) Blake	Fabaceae	Leaves	Carvalho et al. 2013
Scoparia dulcis L.	Plantaginaceae	Whole plant	Otero et al. 2000d
Senna reticulata Willd.	Fabaceae	Whole plant	Félix-Silva et al. 2017
Serjania erecta Radlk.	Sapindaceae	Leaves and Barks	Da Silva et al. 2017
Sida acuta Burm.f.	Malvaceae	Whole plant	Otero et al. 2000d
Simaba cedron Planch.	Simaroubaceae	Whole plant	Otero et al. 2000d
Simarouba versicolor A. St.-Hil.	Simaroubaceae	Seeds	Houghton and Osibogun 1993
Siparuna thecaphora (Poepp. & Endl.) A. DC	Monimiaceae	Aerial parts and Barks	Otero et al. 2000d
Solanum campaniforme Roem. & Schult.	Solanaceae	Leaves	Da Silva et al. 2017
Solanum nudum Dunal	Solanaceae	Aerial parts and Barks	Otero et al. 2000d
Stachytarpheta dichotoma Vahl.	Verbenaceae	Whole plant	Houghton and Osibogun 1993
*Staurostigma luschnathianum (Schott) K. Koch	Araceae	-	Mors 1991
Struthanthus orbicularis (Kunth) Eichler	Loranthaceae	Leaves and Branches	Mors et al. 2000
Strychnos pseudoquina St. Hil.	Loganiaceae	-	Nishijima et al. 2009
*Strychnos xinguensis Krukoff	Loganiaceae	Barks	Otero et al. 2000d
Stryphnodendron adstringens (Mart.)	Fabaceae	Barks	De Paula et al. 2010
Tabernaemontana catharinensis A. DC.	Apocynaceae	Roots and Barks	Carvalho et al. 2013
Tibouchina stenocarpa (Schrank & Mart. ex DC.) Cogn.	Melastomataceae	Roots	De Paula et al. 2010
Torresea cearensis Fr. Allem.	Fabaceae	Barks and Seeds	Félix-Silva et al. 2017
Tournefortia cuspidata A. DC.	Boraginaceae	Leaves and Barks	Otero et al. 2000d
Tradescantia geniculata Jacq.	Commelinaceae	-	Mors 1991
Wilbrandia ebracteata Cogn.	Cucurbitaceae	Roots	Houghton and Osibogun 1993
Xiphidium caeruleum Aubl.	Haemodoraceae	Whole plant	Otero et al. 2000d

*: endemic species from Brazil

Table 7.2: Chemical compounds isolated from plants with anti-snake venom activity found in the literature.

Plant scientific name	Family	Isolated compounds	Reference(s)
Apuleia leiocarpa (Vogel) J.F. Macbr.	Fabaceae	Terpenoids: β-Amirin, Apuleína Phytosteroids: Sitosterol, Stigmasterol Flavonoids	Houghton and Osibogun 1993
Aristolochia trilobata L.	Aristolochiaceae	Terpenoids Alkaloid: Aristolochic acid	Houghton and Osibogun 1993
*Aristolochia birostris Duch.	Aristolochiaceae	Terpenoids, Lignoids, Antraquinone, and Vanillin Alkaloid: Aristolochic acid	França et al. 2005 Félix-Silva et al. 2017
Aristolochia cymbiferae L.	Aristolochiaceae	Polyphenols Terpenoids Alkaloid: Aristolochic acid	Da Silva et al. 2017, Soares et al. 2005
Baccharis trimera Less.	Asteraceae	Terpenoids: Clerodane	Januário et al. 2004, Bernard et al. 2001, Soares et al. 2005, Dey and De 2012
Bredemeyera floribunda Willd.	Polygalaceae	Terpenoids: Bredemeyerosides B and D	Houghton and Osibogun 1993, Soares et al. 2005
Bursera simaruba L. Sarg.	Burseraceae	Flavonoid and Others	Mors 1991
Casearia gossypiosperma Briquet	Salicaceae	Quercetin	Soares-Silva et al. 2014
Casearia sylvestris Sw.	Salicaceae	Fitosteroid: Sitosterol, Stigmasterol Rutin	Mors et al. 2000, Cintra-Francischinelli et al. 2008b
Cabeça de Negro	-	Pterocarpans: Cabenegrines A-I and A-II	Nakagawa et al. 1982
Clusia fluminensis Guttiferae Juss.	Clusiaceae	Fitosteroid: Lanosterol Benzophenone: Clusianone	Da Silva et al. 2017
Combretum leprosum Mart.	Combretaceae	Terpenoid: Arjunolic acid	Da Silva et al. 2017
Cordia verbenacea DC.	Boraginaceae	Polyphenols: Rosmarinic acid e Chlorogenic acid	Soares et al. 2005, Ticli et al. 2005
Derris sericea (Poir.) Ducke	Fabaceae	Chalcone: Derricidin	Mors et al. 2000, Soares et al. 2005
Derris urucu (Killip & Sm.) J.F. Macbr.	Fabaceae	2,5-dihydroxymethyl-3,4-dihydroxypyrrolidine	Soares et al. 2005
Dipteryx alata Vogel	Fabaceae	Isoflavone, Lupane triterpenoids, Betulin, Phemolic acids	Ferraz et al. 2012, 2014, 2015, Yoshida et al. 2015
Dorstenia brasiliensis Lam.	Moraceae	Furanocoumarin: bergapten Monoterpenoid: dorstenin	Mors et al. 2000, Soares et al. 2005, Houghton and Osibogun 1993
Eclipta prostrata (L.) L.	Asteraceae	Coumestan: Wedelolactone Fitosteroid: Sitosterol, Stigmasterol Others: D-mannitol Demethylwedelolactone	Soares et al. 2005

Table 7.2 contd. ...

...Table 7.2 contd.

Plant scientific name	Family	Isolated compounds	Reference(s)
Euphorbia hirta L.	Euphorbiaceae	Flavonoid: Quercetin-3-O-α-rhamnoside Polyphenols	Félix-Silva et al. 2017
Eupatorium triplinerve Vahl	Asteraceae	Coumarins: Herniarin and Ayapin	Mors et al. 2000
Harpalyce brasiliana Benth.	Fabaceae	Pterocarpans: Edunol	Da Silva et al. 1997
**Mandevilla velutina* K. Schum.*	Apocynaceae	Steroids	Carvalho et al. 2013, Soares et al. 2005, De Paula et al. 2010
Mikania glomerata Spreng.	Asteraceae	Coumarin	Mors et al. 2000, De Paula et al. 2010, Carvalho et al. 2013
Mimosa pudica L.	Fabaceae	Fitosteroid: Sitosterol, Stigmasterol Others: D-mannitol	Houghton and Osibogun 1993, Soares et al. 2005
Pentaclethra macroloba (Willd.) Kuntze	Fabaceae	Macrolobinas Terpenoids Fitosteroids	Carvalho et al. 2013, Soares et al. 2005, Da Silva et al. 2007
Periandra mediterranea (Vell.) Taub.	Fabaceae	Fitosteroid: Sitosterol, Stigmasterol Triterpenes	Soares et al. 2005, Houghton and Osibogun 1993
**Phyllanthus klotzschianus* Muell.-Arg.*	Euphorbiaceae	Flavonoids: Quercetin, Rutin	Houghton and Osibogun 1993, Soares et al. 2005
**Piper caldense* C.DC.*	Piperaceae	N-methylaristolactam: Caldensin	Soares et al. 2005
Sapindus saponaria L.	Sapindaceae	Flavonoids	Da Silva et al. 2017, Soares et al. 2005
Serjania erecta Radlk.	Sapindaceae	Flavonoids, steroids, tannins, and catechins	Da Silva et al. 2017
Solanum campaniforme Roem. & Schult.	Solanaceae	Steroidal Alkaloids	Da Silva et al. 2017
Tabernamontana catharinensis A. DC.	Apocynaceae	Alkaloid: 12-methoxy-4-methyl-voachalotine	Carvalho et al. 2013

*: endemic species from Brazil

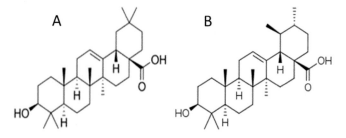

Figure 7.1: Chemical structure of oleanólic (A, 3β-hydroxyolean-12-en-28-oic acid) and ursolic (B, 3β-hydroxyurs-12-en-28-oic acid) acids.

Table 7.3: Oleanolic and ursolic acids concurrently found in plants.

Plant Scientific name	Family	Reference(s)
Sambucus nigra L.	Adoxaceae	Gleńsk et al. 2017
Plumeria obtusa L.	Apocynaceae	Alvarado et al. 2018
Baccharis uncinella DC.	Asteraceae	Zalewski et al. 2011, Passero et al. 2011
Cichorium endivia L. *Cynara cornigera* Lindl.		Hegazy et al. 2015
Helichrysum picardii Boiss. & Reuter		Santos Rosa et al. 2007
Silphium perfoliatum L. *Silphium trifoliatum* L. *Silphium integrifolium* Michx.		Kowalski 2007
Ilex paraguariensis A. St.	Aquifoliaceae	Puangpraphant and Mejia 2009
Arrabidaea triplinervia (Mart. ex DC.)	Bignomiaceae	Leite et al. 2006
Distictella elongata (Vahl) Urb.		Simões et al. 2011
Mansoa hirsuta DC.		Pereira et al. 2017
Radermachera boniana Dop.		Truong et al. 2011
Pterocephalus hookeri C.B.	Caprifoliaceae	Yang et al. 2007a
Hippocratea excelsa Kunth	Celastraceae	Cáceres-Castilho et al. 2008
Cornus officinalis Siebold & Zuccarini	Cornaceae	Huang et al. 2018
Davilla rugosa	Dilleniaceae	David et al. 2006, Gerardo and Aymard 2007, Fraga and Stehmann 2010
Diospyros kaki L.	Ebenaceae	Zhou et al. 2012
Hippophae rhamnoides L.	Elaeagnaceae	Zheng et al. 2009
Calluna vulgaris (L.) Hull.	Ericaceae	Garcia-Risco et al. 2015, Szakiel et al. 2013
Gaultheria procumbens L.		Michel et al. 2017
Vaccinium myrtillus L.		Szakiel et al. 2012
Vaccinium vitis-idaea L.		Szakiel and Mroczek 2007
Homonoia riparia Lour.	Euphorbiaceae	Yang et al. 2007b
Moussonia deppeana (Schltdl. & Cham.) Hanst	Gesneriaceae	Gutiérrez-Rebolledo et al. 2016
Cyclocarya paliurus Batalin	Juglandaceae	Lin et al. 2016
Eriope blanchetii L.	Lamiaceae	e Silva et al. 2012
Eriope latifolia (Mart. ex Benth.) Harley		Santos et al. 2011
Lepechinia caulescens (Ortega)		Aguirre-Crespo et al. 2006
Lycopus lucidus Turcz. ex Benth.		Lee et al. 2006
Ocimum basilicum L.		Marzouk 2009
Origanum vulgare L.		Nowak et al. 2013
Prunella vulgaris L.		Yang et al. 2016 Zhu et al. 2014
Rosmarinus officinalis L. var. *subtomentosus* Maire & Weiller		Nowak et al. 2013
Salvia L.		Kalaycioğlu et al. 2018

Table 7.3 contd. ...

...Table 7.3 contd.

Plant Scientific name	Family	Reference(s)
Salvia chinensis Benth.		Jiang and Zou 2014
Salvia chrysophylla Stapf.		Çulhaoğlu et al. 2013
Salvia filipes Benth.		Maldonado et al. 2016
Salvia lachnostachys Benth.		Erbano et al. 2012
Salvia viridis L.		Rungsimakan and Rowan 2014
Satureja parvifolia (Phil.) Epling		van Baren et al. 2006
Thymus mastichina L. subsp. *donyanae* R.		Gordo et al. 2012
Lecythis pisonis Cambess.	Lecythidaceae	Brandão et al. 2013, Silva et al. 2012
Cladocolea micrantha (Eichler) Kuijt	Loranthaceae	Guimarães et al. 2012
Punica granatum L.	Lythraceae	Katz et al. 2017
Miconia albicans		Vasconcelos et al. 2006
Miconia ligustroides DC.	Melastomataceae	Cunha et al. 2010
Miconia sp Ruiz & Pav.		Scalon Cunha et al. 2007
Eucalyptus globulus Labill.		Domingues et al. 2012
Syzygium aromaticum (L.) Merr. & L.M. Perry	Myrtaceae	Nowak et al. 2013
Ugni molinae Turcz.		Aguirre et al. 2006
Olea europeae L.	Oleaceae	Olmo-Garcia et al. 2016, Giménez et al. 2015, Allouche et al. 2009, Saimaru et al. 2007
Lopezia racemosa Cav.	Onagraceae	Moreno-Anzúrez et al. 2017
Paeonia lactiflora Pall.	Paeoniaceae	Zhou et al. 2011
Linaria alpina Mill.	Plantaginaceae	Venditti et al. 2015
Plantago major L.		Stenholm et al. 2013
Punica granatum L.	Punicaceae	Salah et al. 2014, Banihani et al. 2013
Ziziphus jujuba Mill.	Rhamnaceae	Zhang et al. 2011
Rosa canina L.		Saaby and Nielsen 2012, Wenzig et al. 2008
Chaenomeles sinensis (Dum. Cours.) Koehne		Miao et al. 2016
Chaenomeles speciosa (Sweet) Nakai		
Crataegus pinnatifida var. major N.E. Br.		Yang et al. 2004
Cydonia oblonga Mill.		Lorenz et al. 2008
Eriobotrya japonica Lindl.	Rosaceae	Cao et al. 2016, Li et al. 2015, Shi et al. 2014, Kikuchi et al. 2011, Lu et al. 2009
Malus domestica Borkh.		Andre et al. 2016 Brendolise et al. 2011
Prunus avium L.		Peschel et al. 2007
Prunus dulcis (Mill.) D.A. Webb		Amico et al. 2006
Prunus mume (Siebold) Siebold & Zucc.		Hattori et al. 2013

Table 7.3 contd. ...

...Table 7.3 contd.

Plant Scientific name	Family	Reference(s)
Duroia macrophylla Huber	Rubiaceae	Martins et al. 2013
Fadogia tetraquetra var. *tetraquetra*		Mulholland et al. 2011
Hedyotis corymbosa L.		Yang et al. 2013
Hedyotis diffusa Willd.		
Keetia leucantha Bridson		Beaufay et al. 2017, Bero et al. 2013
Mitracarpus scaber Zucc.		Gbaguidi et al. 2005
Oldenlandia diffusa (Willd.) Roxd. var. *polygonoides* Hook. f.		Gu et al. 2012
Psychotria viridis Ruiz & Pav.		Soares et al. 2017
Vitellaria paradoxa C.F.Gaertn	Sapotaceae	Catteau et al. 2017
Solanum lycopersicum L.	Solanaceae	Kalogeropoulos et al. 2012
Lantana camara L.	Verbenaceae	Srivastava et al. 2011

Experimental

Venom

Bothrops jararacussu venom was collected manually from adult specimens in Serpentario of the Center for Nature Studies (CNS). They were fed with white Swiss mice every two weeks. The venom was certified by Dr. José Carlos Cogo (Universidade Brasil, SP, Brazil), lyophilized, and stored at –4°C until use.

Phytochemicals

Oleanolic and ursolic acids were purchased commercially from Sigma®. The major pharmacological disadvantage of these phytocompounds is their poor water solubility (Oprean et al. 2016). Thus, the acids were solubilized using polyethylene glycol 400 (PEG 400, Mapric Produtos Farmacocosméticos Ltda, São Paulo, SP, Brazil) for using in biological preparations (Cintra-Francischinelli et al. 2008a).

Phytochemicals Antioxidant Capacity

Antioxidant activity was measured based on the reaction between 1,1-diphenyl-2-picrylhydrazyil (DPPH) and the oleanolic and ursolic acids solutions. Aliquots (250 µL) resulting from *ex vivo* experiments (see mouse phrenic nerve-diaphragm preparation) containing oleanolic and ursolic acids were added to 750 µL of ethanol solution (70% v/v) DPPH 0.1 mM. Absorbance at 515 nm was measured at 0, 15, 30, 45, and 60 minutes. As a control, 250 µL of 70% ethanol was added to 750 µL of DPPH solution (Brand-Willians et al. 1995). The ability of the acids in scavenging DPPH radical, expressed as percent inhibition, was calculated according to the mathematical equation:

$$\% \text{ of Inhibition} = (\text{Abs1-Abs2})/\text{Abs1} \times 100$$

where, Abs1 is the control absorbance and Abs2 is the sample absorbance.

In vitro Antimicrobial and Citotoxicological Profile of Oleanolic and Ursolic Acids

Oleanolic and ursolic acids (1 mg each) were firstly solubilized with 10 µL of dimethylsulfoxide (DMSO), and after that filled with NaCl 0.9%, to reach a final volume of 1 mL. To evaluate the absence of the antimicrobial activity, the solution with DMSO were prepared in NaCl 0.9%, without acids, filled to a final volume of 1 mL.

Minimal Inhibitory Concentration (MIC) analysis was performed using the microorganisms *Staphylococcus aureus*, *Escherichia coli*, and *Pseudomonas aeruginosa*. The microorganisms were grown in Erlenmeyer's with 50 mL of the TSB broth (tryptone soya broth) at 37°C for 24 hours (h). The suspension of each microorganism was diluted to the final concentration of 106 CFU/mL.

MIC was determined in 96-well microplates according to the procedure by Ataide et al. (2017) and Santos et al. (2018). The results were evaluated after 24 hours of incubation at 37°C, where 5 μL of each well was inoculated into petri plates with tryptic soy agar (TSA) culture medium. All the assays were made in triplicate.

In vitro Cholinesterase (ChE) Inhibition Assay

The ability of the facilitatory effect of oleanolic acid to inhibit ChE activity was assessed using a colorimetric assay (Labtest Diagnóstica S.A., Lagoa Santa, MG, Brazil) that was standardized with a human serum control of known ChE activity (Biocontrol N), as also carried out elsewhere (Werner et al. 2015). Biocontrol N (Bioclin Quibasa), a pool of normal human serum, was used as an internal quality control (20 μL), according to the manufacturer's recommendations. The ChE activity was determined spectrophotometrically (UV-M51 spectrophotometer, BEL Engineering) at 405 nm. The percentage of enzyme inhibition was calculated by comparing the enzymatic activity in the presence of 1.0 mg of facilitatory oleanolic acid/mL, with PEG 400 (20 μL) used as solubilizer, and with the 1.0 mg concentration of neostigmine (Sigma®), as a cholinesterase inhibitor. The assays were done in triplicate.

Ex vivo Pharmacological Experiments

Animals

Male white Swiss mice (25–30 g) were purchased from Anilab (Laboratory Animals, Paulinia, SP, Brazil). The animals were housed at 25°C ± 3°C (77°F ± 3°F) in a light/dark cycle of 12 hours, and had access to food and water *ad libitum*. This study was approved by the Animal Ethics Committee of Sorocaba University (protocol n° 093/2016), and the experiments were carried out according to the international guideline—ARRIVE (Animal Research: Reporting of *in vivo* Experiments (Kilkenny et al. 2010)).

Mouse phrenic nerve-diaphragm muscle preparation

The diaphragm and its phrenic nerve branch were obtained from mice anesthetized with Halothane (Cristália®) and sacrificed by exsanguination. Hemidiaphragms were mounted under a tension of 0.5 g in a 5 mL organ bath (Bülbring 1997) containing Tyrode solution, and aerated with 95% O_2 and 5% CO_2. Tyrode solution maintains the physiological conditions of the neuromuscular preparation at pH 7.0, and consists of (in mM): 137 NaCl, 2.7 KCl, 1.8 $CaCl_2$, 0.49 $MgCl_2$, 0.42 NaH_2PO_4, 11.9 $NaHCO_3$, and 11.1 Glucose. The preparation is indirectly stimulated through the phrenic nerve (ESF-15D double physiological stimulator), using supraximal stimuli and a frequency of 0.06 Hz with duration of 0.2 ms. Recording of muscle contraction is produced through the isometric cat. transducer 7003, coupled with a 2-Channel Recorder Gemini cat.7070, containing Basic Preamplifiers cat.7080 (Ugo Basile®). After recording under control conditions for 10 minutes during the stabilization of the preparation, the pharmacological protocols were performed.

The protocols for concentration-response curves of oleanolic and ursolic acids were—Tyrode control (n = 3); oleanolic acid at 200 μg/mL (n = 3), 300 μg/mL (n = 3), 400 μg/mL (n = 3); ursolic acid at 200 μg/mL (n = 4), 300 μg/mL (n = 4), and 400 μg/mL (n = 3).

The pharmacological protocols were—Tyrode control (n = 3), *B. jararacussu* venom (Bjssu, 40 µg/mL, n = 5), selected concentration of oleanolic acid for the neutralization assays (preincubation of Bjssu + oleanolic acid, 30 minutes, before addition to the bath, n = 3), and selected concentration of ursolic acid for the neutralization assays (preincubation of Bjssu + ursolic acid, 30 minutes, before addition to the bath, n = 4).

Histological analysis

Resulting muscles from *ex vivo* pharmacological assays were routinely processed for light microscopy analysis, according to Ferraz et al. (2014). The muscle tissue was fixed in 10% formaldehyde for 12 hours, and maintained in 70% alcohol. For the dehydration process, a growing alcoholic grade (70%, 85%, 96%, and absolute) was used, where the tissue remained in absolute alcohol for a minimum period of 24 hours. After this storage, the dehydration began without interruption. In the first wash, the piece was left immersed in absolute alcohol I for 50 minutes, and continuing with another 50 minutes in absolute alcohol II. The diaphragm muscles were placed in three xylol exchanges for 40 minutes per exchange. At the end, the tissues were cut into approximately three equal parts (beginning, middle, and end).

Muscles were submitted to inclusion in vials containing Paraplast Plus (Sigma-Aldrich®) kept in an oven at 56°C ± 3°C (132,8°F ± 3°F). Two exchanges of Paraplast Plus were performed, remaining for one hour in each bottle. The included muscle tissues were placed in silicone tray while drying and solidifying paraffin. The blocks were removed from the tray and fixed in a block of wood, which were taken to the microtome (Cryostat 300 Ancap®). It was thinned in sections of cuts of 30 µm until reaching the diaphragm muscle. At the time the muscle was reached, transversely sectioned, ultrathin 4 µm were placed in histological bath (Ancap®) at 39°C ± 3°C (102,2°F ± 3°F), and transferred to glass slides.

For the dewaxing step, the slides were organized into supports and baked at 100°C (212°F) for five minutes. The slides were then dipped into three wells containing sufficient xylene to cover them— remaining for five minutes in the first vat, and for 10 seconds in the other two vats. Sequentially, the pieces are hydrated by immersing the slides in vats containing ethyl alcohol in decreasing graduation (absolute, 96%, 85%, and 70%) for 10 seconds each, and ending with running water for 5 minutes. After that, the staining was made using Modified Harris Hematoxylin (Sigma-Aldrich®) (for one minute, running water for 5 minutes, and 70% alcohol for 10 seconds), and further, submitted to Eosin in alcoholic solution (Sigma-Aldrich®) (for 2 minutes and again under running water for 5 minutes), and finally covered with Entellan (Merck®).

Quantitative analysis was made selecting three random and virtual lines (Figure 7.2) by three examiners, and documented using the Zeiss AXIOSTAR Plus photomicroscope, at an increase of 400x (objective of 40x, where a bar of 1 cm = 40 µm).

In order to qualify the cell damage, a score (Figure 7.3) was adopted: (a) normal cells (N) presenting an integrated polygonal structure and peripheral nucleus, (b) edema (e), (c) the condensation of myofibrils (arrow), (d) delta lesions (similar to the Δ symbol coming from the Greek, arrows), (e) ghost cells or phantoms (g), represented by remains of cell membranes, and (f) myonecrosis (m).

Oxidative stress biomarkers

Samples (200 µL) from the organ bath containing neuromuscular preparations in Tyrode solution and submitted to pharmacological protocols were collected at times zero and 2 hours. In these solutions, reduced glutathione (GSH), glutathione peroxidase (GSH-Px), catalase (CAT), and thiobarbituric acid reactive substances (TBARS) were evaluated.

GSH was determined by quantification of total thiols (Ellman 1959). Tyrode solution (with each respective pharmacological protocol) (50 µL) was mixed with 900 µL of phosphate buffer, and it reacted with 50 µL of 5-5-dithio-bis-2-nitrobenzoic acid (DTNB) to form a yellow complex, which was read at 412 nm. GSH levels were expressed in µmol/mL solution.

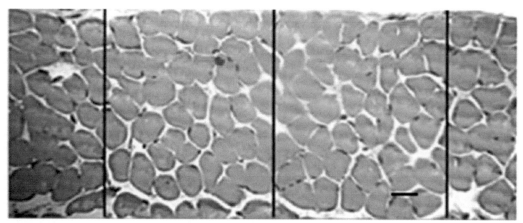

Figure 7.2: Cross-section of preparation in Tyrode control, stained with Hematoxylin and Eosin. Bar = 40 μm (Ferraz et al. 2014). Vertical lines indicate virtual areas for counting.

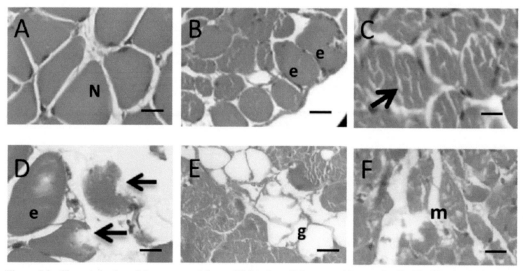

Figure 7.3: Characterization of the score used for qualifying the cell damage—(**a**) normal cells (N), (**b**) edema (e), (**c**) the condensation of myofibrils (arrow), (**d**) delta lesions (arrows), (**e**) ghost cells (g), and (**f**) myonecrosis (m). Bar = 40 μm (Ferraz et al. 2014).

To measure GSH-Px antioxidant enzyme activity, Tyrode solution was diluted in a solution containing GSH, glutathione reductase, NADPH, sodium azide, and 70 μL of H_2O_2. GSH-Px activity was monitored for two minutes at 340 nm (Paglia and Valentine 1967). GSH-Px activity was determined by decaying the absorbance of NADPH, which is proportional to the consumption of NADPH. Data was expressed as μmol NADPH/min.

Catalase method is based on the decomposition of H_2O_2 by the enzyme over three minutes, monitored at 240 nm. For this purpose, an aliquot of the Tyrode solution was diluted in potassium phosphate buffer, and 70 μL of H_2O_2 was added. A constant- κ assists in the expression of activity values (κ/min) (Aebi 1984).

Lipid peroxidation was measured by quantification of thiobarbituric acid reactive substances (TBARS method) (Ohkawa et al. 1979). An aliquot of 750 μL of H_3PO_4, 250 μL of thiobarbituric acid (TBA), and 50 μL of sodium dodecyl sulfate were mixed with 125 μL of Tyrode solution, which were taken in a 90°C bath for 45 minutes. Lipids-TBA product was monitored spectrophotometrically at 532 nm.

MIC results were evaluated through Basic Statistical Methods as mean ± standard deviation. All results from *ex vivo* pharmacological and histological assays were shown as mean ± SEM, and were statistically analyzed using *t*-Student's test. For oxidative stress biomarkers, results were shown as mean ± standard deviation, and Student's *t*-test (paired samples) were applied to compare initial (zero) and final (2 hours) times in the same group, and ANOVA One Way test for comparison among groups. The level of significance was 5% for all experiments.

Antioxidant Activity of Oleanolic and Ursolic Acids

The *in vitro* antioxidant activity test of ursolic and oleanolic acids was performed (Figure 7.4). Oleanolic acid showed an expressive antioxidant activity (mean of 52.1%) over one hour, while ursolic acid maintained basal levels (mean of 9.2%) throughout the experiment time, thus showing little or no antioxidant activity.

It is worth pointing out, according to Alam et al. (2013) "among *in vitro* free radical scavenging methods, DPPH method is furthermore rapid, simple (i.e., not involved with many steps and reagents), and inexpensive in comparison to other test models". Besides, regarding *ex vivo* antioxidant evaluation, this is the first study using the nutrient solution (Tyrode) as sample for bathing the phrenic-diaphragm nerve preparation, which justifies the importance of the study.

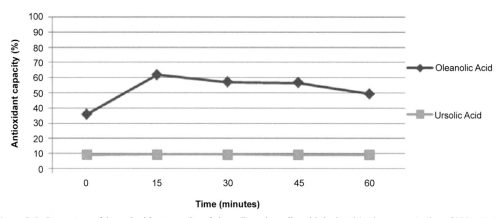

Figure 7.4: Percentage of the antioxidant capacity of oleanolic and ursolic acids isolated (at the concentration of 300 µg/mL), by the DPPH method.

Antimicrobial Activity of Oleanolic and Ursolic Acids

No activity was observed in DMSO solution, which demonstrated antimicrobial activity from ursolic acid. In this way, the ursolic acid showed antimicrobial activity only against *S. aureus*, Gram-positive bacteria, at 125 ug/mL.

Fontanay et al. (2008) were able to prove that oleanolic acid, and more particularly ursolic acid, showed moderate to good antibacterial activity, but limited to Gram-positive bacteria. Despite having a relatively similar chemical structure, ursolic acid, oleanolic acid, and betulinic acid harbored different antibacterial activities, most significant of which was that of ursolic acid.

Pereira (2015) used 16 compounds derived from ursolic acid, among them was oleanolic acid, which were also prepared using DMSO to obtain a final concentration of 1 mg/mL. With the studied concentration, the author obtained excellent results with the microdilution in 96 well plates. In the presence of ursolic and oleanolic acid, the MIC result was equivalent to 125 µg/mL for *E. coli* ATCC 25922. For *P. aeruginosa* ATCC 27853 in the presence of ursolic acid, the MIC value was 62.5 µg/mL, and with oleanolic acid, the MIC value was 125 µg/mL. And for *S. aureus* ATCC 29213, which in the presence of ursolic acid, presented 31.3 µg/mL, with oleanolic acid, the MIC value was 62.5 µg/mL.

Vieira (2003) analyzed the MIC of ursolic acid, which was dissolved only in DMSO, reaching a concentration of 2.56 mg/mL. It was observed that ursolic acid did not inhibit the growth of *S. aureus*, but there was inhibition of growth for *S. epidermidis* and *Proteus mirabilis* with MIC value of 32 µg/mL. The other microorganisms were tested, but there was no inhibition of the microorganisms, similar with our results.

Nascimento et al. (2014) observed the MIC value for *S. aureus* (ATCC 6538) of 32 µg/mL. Ursolic acid was also effective against *E. coli* (ATCC 25922), *K. pneumoniae*, and *S. flexneri*, with a MIC of 64 µg/mL in all three cases. For *P. aeruginosa* (ATCC 15442), the MIC value was 512 µg/mL. Kurek et al. (2012) also confirmed that oleanolic acid and ursolic acid have potential as a new class of antibacterial agents because they are active against many bacterial species, both Gram-positive and Gram-negative.

Comparing the results obtained during the research with the literature, we can conclude that the concentration of 500 µg/mL, diluted 1:10 in saline solution (NaCl 0.9%), was not enough to obtain a positive result in antimicrobial activity for Gram-negative microorganisms.

In vitro Cholinesterase (ChE) Inhibition Assay

Oleanolic acid (1 mg/mL) was evaluated in its ability to inhibit cholinesterase activity as neostigmine (1 mg/mL) does (Table 7.4), due to its facilitatory effect shown in *ex vivo* experiments (see next item).

It is known that one of the strategies available for treating Alzheimer's disease includes the increase in acetylcholine levels in brain regions, such as frontal cortex and hippocampus, and possibly preventing neuronal degeneration with antioxidants (Youdim and Buccafusco 2005). The progressive neurodegenerative disorder named Alzheimer's disease is related to genetic predisposition and characterized by the presence of neurofibrillary tangles, formation of extracellular deposition of β-amyloid peptide, oxidative stress, increased production of superoxide radicals, and reduced neurotransmitter levels (Mattson 2004, Verri et al. 2012).

Here, oleanolic acid showed no ability in inhibiting cholinesterase activity found as 2924 ± 33.5 U/L, when compared to neostigmine (87.1 ± 58.1 U/L), since the value is between the range of Bio Control N (2450 to 4000 U/L). PEG 400 used as solubilizing agent did not cause any effect on cholinesterase (2837 ± 197.7 U/L). The facilitatory effect elicited by oleanolic acid on mouse phrenic nerve-diaphragm preparation was not due to enzymatic inhibition, as shown after ChE evaluation, and the mechanism of action remains to be done in future studies.

Table 7.4: Cholinesterase determination (in triplicate).

Treatment	Time (min)				ChE (U/L)	Mean ± S.E.M
	A0	**A1**	**A2**	**A3**		
BioControl N (Range: 2450–4000 U/L)	1778	1742	1706	1669	3166	
	1761	1719	1684	1646	3341	3283.1 ± 100.6
	1753	1716	1676	1638	3341	
Polyethylene glycol 400	1708	1687	1653	1618	2614	
	1625	1594	1561	1525	2905	2837.6 ± 197.7
	1730	1699	1663	1627	2992	
Neostigmine 1 mg/mL	1815	1816	1813	1810	145.2	
	1813	1817	1815	1812	29.0	87.1 ± 58.1
	1843	1846	1843	1840	87.1	
Oleanolic acid 1 mg/mL	1795	1763	1730	1695	2905.4	
	1785	1754	1718	1683	2963.5	2924.7 ± 33.5
	1728	1696	1661	1628	2905.4	

S.E.M. = Standard error of mean.

Ex vivo Pharmacological Experiments

Continuing, the two triterpenes—oleanolic and ursolic acids—were exploited in their effects against the paralysis induced by *Bothrops jaracussu* venom, in an experimental nerve and muscle synapsis. Figure 7.5 shows the concentration-response curves of (a) oleanolic and (b) ursolic acids.

Oleanolic acid (300 μg/mL) showed a transitory facilitatory effect visualized by an augment of twitch tension amplitude, during the first 30 minutes. This response could elicit an anticholinesterasic effect, resulting in acetylcholine accumulation on synaptic cleft. Nowadays, acetylcholinesterase inhibitors have been revisited as a promise to treat Alzheimer's disease (AD) (Giménez-Liort et al. 2017), but the facilitatory effect of oleanolic acid is not due to anticholinesterasic action, as already commented above, and needs to be clarified in further studies.

Figure 7.6 shows experiments of venom + OA (Figure 7.6a) or venom + UA (Figure 7.6b) preincubated for 30 minutes before addition into the bath containing the biological preparation.

Note that oleanolic, but not ursolic, protected significantly against the blockade induced by snake venom, an effect which can be related to the facilitatory effect as seen with fractions of *Casearia*

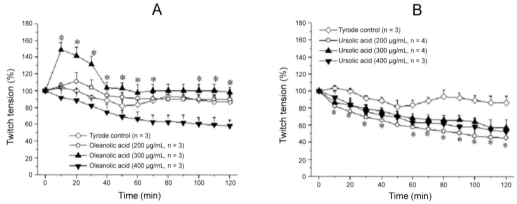

Figure 7.5: Mouse phrenic nerve-diaphragm preparation (indirect stimuli). Concentration-response of (a) oleanolic acid and (b) ursolic acid, at 120 minutes. The number of experiments (n) is shown in the legend of the figure. * $p < 0.05$ compared to Tyrode control (only shown for the selected concentration, 300 μg/mL for both phytochemicals).

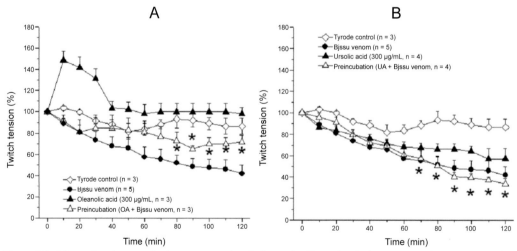

Figure 7.6: Mouse phrenic nerve-diaphragm preparation (indirect stimuli). Preincubation as a neutralizing model against the paralysis induced by *B. jararacussu* venom (Bjssu, 40 μg/mL) for 120 minutes. The concentration of 300 μg/mL of both, (a) oleanolic acid and (b) ursolic acid, respectively, was selected for preincubation assays. The number of experiments (n) is shown in the legend of figure. * $p < 0.05$ in a was compared to venom, wheras * in b was compared to Tyrode control.

sylvestris Sw. (Werner et al. 2015), a plant able to inhibit the paralysis induced by the same snake venom and experimental model (Cintra-Francischinelli et al. 2008b). It is known that these isomers differ in their protective effect depending on the adopted experimental model. For example, Senthil et al. (2007) found UA > OA activity against isoproterenol-induced myocardial ischemia in rats.

Histological analysis

Preparations resulting from pharmacological protocols were further analyzed histologically. Myofibril condensation and edema were the most common cell changes, in varying degrees, seen in all treatments, as shown in Table 7.5. Myonecrosis was a determinant type of cell damage to assign myotoxicity, showing the ability of oleanolic acid to avoid this effect of the venom. Ghost cells were practically absent in treatments involving oleanolic acid, but not involving ursolic acid when in mixture with the venom. The interaction (venom + ursolic acid) resulted in additive effect shown by intense edema and ghost cells.

Figure 7.7 shows the efficacy of oleanolic (#, $p < 0.05$ compared to the venom), but not ursolic acid to counteract against the myotoxic effects of *B. jararacussu* venom. The mixture of ursolic acid

Table 7.5: Score (results given in %) used as parameter to classify the damaged cells exposed to different treatments.

Treatments	Parameters					
	Normal	Myofibril condensation	Ghost cells	Edema	Delta lesion	Myonecrosis
Tyrode control	70.8	13.7	4.2	9.1	0	2.2
Venom (V)	23.2	31.6	5.4	25.3	1.1	13.4
Oleanolic acid (OA)	58.0	29.3	0.5	12.6	0	0.5
OA + V	45.7	29.6	1.0	18.3	0	4.0
Ursolic acid (UA)	41.1	30.1	2.9	0.1	0.2	7.7
UA + V	36.8	24.5	8.0	26.6	0.3	7.9

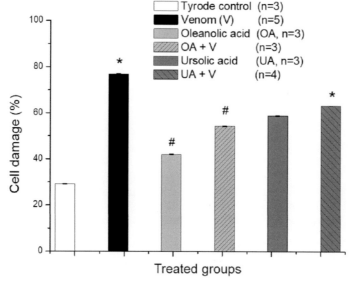

Figure 7.7: Neuromuscular preparations quantified by light microscopy. All type of injury was grouped to determine the efficacy of both phytochemicals, oleanolic acids (OA) and ursolic acids (UA), against the myotoxicity of *B. jararacussu* venom (V). Oleanolic acid statistically protected the preparation against the toxic effect of the venom (#, $p < 0.05$ compared to the venom). * $p < 0.05$ compared to the Tyrode control.

+ venom and protocols using venom alone were statistically significant (*, $p < 0.05$) when compared to Tyrode control.

Comparing the results from neurotoxic and myotoxic effects induced by snake venom, there was a positive correlation between these parameters, showing again the major ability of oleanolic rather than ursolic in protecting against myonecrosis-induced by *B. jararacussu* venom. Neurotoxicity *in vitro* and myotoxicity of this venom are intrinsically related effects, and there is a tendency to assume that *in vitro* neuroblocking is due to the presence of bothropstoxin-I (BthTX-I), a myotoxin with pre and post-synaptic action (Oshima-Franco et al. 2004, Correia-de-Sá et al. 2013).

Oxidative stress biomarkers

Redox biomarkers were evaluated only in the pharmacological protocols. Table 7.6 shows the GSH levels, an important endogenous antioxidant. It is observed that, comparing T0 to T1, the concentration of GSH increased significantly over the course of 120 minutes in all groups submitted to this test. The increase in GSH concentration can be attributed to its release from the phrenic nerve-diaphragm structure to the Tyrode solution. However, this increase occurred independently from the antioxidant acids action, and in the presence or absence of the venom.

When the groups within T0 are compared, there is no statistical difference. Likewise, no statistical difference was found when the groups within T1 were compared. This fact shows that *B. jararacussu* venom or ursolic/oleanolic acids had no effect on GSH levels. da Silva and colleagues corroborated our results; they also found no difference in GSH levels while evaluating the venom of *Crotalus durissus terrificus* in rat liver (da Silva et al. 2011).

The lipid peroxidation test evaluates thiobarbituric acid reactive substances (TBARS). These substances are products of lipid degradation, and data is presented in Table 7.7. A significant increase in the level of TBARS was observed in *B. jararacussu* group comparing T1 to T0. When the groups within T0 are compared, there is no statistical difference. However, a significant difference was

Table 7.6: Reduced Glutathione (GSH) levels, in mM, at time zero (T0) and after 120 minutes (T1) of venom action.

Groups	T0 (mean ± SD)	T1(mean ± SD)
TY (Tyrode)	0.40 ± 0.02	0.49 ± 0.04*
V (Venom *B.jssu*)	0.45 ± 0.03	0.53 ± 0.07*
Ursolic acid (UA)	0.41 ± 0.01	0.47 ± 0.02*
Oleanolic acid (OA)	0.42 ± 0.01	0.48 ± 0.01*
UA + V	0.43 ± 0.01	0.50 ± 0.04*
OA + V	0.39 ± 0.04	0.47 ± 0.04*

*: Statistically different from T0, using Student's t-test, $p < 0.05$.

Table 7.7: Thiobarbituric Acid Reactive Substances (TBARS) levels, in millimolar (mM), at time zero (T0) and after 120 minutes (T1) of venom action.

Groups	T0 (mean ± SD)	T1(mean ± SD)
TY (Tyrode)	3.53 ± 0.50	1.84 ± 0.74
V (Venom *B.jssu*)	3.38 ± 0.86	10.42 ± 0.09*[#]
Ursolic acid (UA)	4.76 ± 1.77	4.80 ± 0.58
Oleanolic acid (OA)	4.15 ± 1.40	3.72 ± 2.39
UA + V	5.05 ± 1.18	2.84 ± 0.21
OA + V	3.58 ± 0.89	3.01 ± 1.43

*: Statistically different from T0, using *t*-Student's tests ($p < 0.01$). [#]: Statistically different from all groups within T1, using ANOVA One-Way ($p < 0.05$).

found among venom group and all other groups within T1, which means that the venom affected the redox state and induced lipid peroxidation over time. On the other hand, TBARS levels decreased when ursolic and oleanolic acids were administered simultaneously to the venom, which is a very advantageous fact since the acids showed antioxidant protection on the lipid peroxidation induced by the venom.

In agreement, the increase in lipid peroxidation was also found by Asmari et al. (2015), in a study about the hepatotoxicity induced by *Echis pyramidum* venom in rats. The study was performed with another snake species, but the venom had the same neuroblocker effect of *B. jararacussu*, showing that the higher the venom concentration, the longer venom action in the tissue, and the greater the amount of fatty acids released and lipid peroxidation, increasing the amount of TBARS.

Regarding the enzymatic results of GPx and catalase, they were very low, even without reading in some samples. This fact leads us to believe that it is not possible to evaluate the activity of these biomarkers in the phrenic-diaphragm nerve model in Tyrode's solution. In this case, blood samples or tissues are more indicated. According to da Silva et al. (2011), catalase activity decreased when the concentration of *Crotalus durissus terrificus* venom increased, proving the oxidative effect of the venom. However, these authors used liver as a biological sample organ that contains great concentration of catalase.

Conclusions

Brazilian plants with antiophidic properties were contextualized in the world stage. Bioprospecting of plants leads naturally to phytochemicals, which can avoid certain effects of snake venoms. In this case-study, oleanolic acid (OA), but not ursolic acid (UA), minimized the blockade induced by the venom and also its myotoxicity. Antioxidant activity of OA and UA were observed only on lipid peroxidation, since TBARS levels increased in the group receiving only the venom, and decreased in groups receiving the acids + venom. Furthermore, the OA and UA could prevent the infections caused by bacteria from the skin flora and help in wound healing. These results can contribute not only to the knowledge on phytochemicals from Brazilian plants against *Bothrops* snake venom, but it may arouse interest against other snake venoms of the world.

References

Aebi, H. 1984. Catalase *in vitro*. Methods Enzymol. 105: 121–126.

Aguirre, M.C., Delporte, C., Backhouse, N., Erazo, S., Letelier, M.E., Cassels, B.K., Silva, X., Alegría, S., Negrete, R. 2006. Topical anti-inflammatory activity of 2alpha-hydroxy pentacyclic triterpene acids from the leaves of *Ugni molinae*. Bioorg. Med. Chem. 14(16): 5673–5677.

Aguirre-Crespo, F., Vergara-Galicia, J., Villalobos-Molina, R., Javier López-Guerrero, J., Navarrete-Vázquez, G., Estrada-Soto, S. 2006. Ursolic acid mediates the vasorelaxant activity of *Lepechinia caulescens* via. NO release in isolated rat thoracic aorta. Life Sci. 79(11): 1062–1068.

Alam, N., Bristi, N.J., Rafiquzzaman, M. 2013. Review on *in vivo* and *in vitro* methods evaluation of antioxidant activity. Saudi Pharm. J. 21: 143–152.

Allouche, Y., Jiménez, A., Uceda, M., Aguilera, M.P., Gaforio, J.J., Beltrán, G. 2009. Triterpenic content and chemometric analysis of virgin olive oils from forty olive cultivars. J. Agr. Food Chem. 57(9): 3604–3610.

Alvarado, H.L., Calpena, A.C., Garduño-Ramirez, M.L., Ortiz, R., Melguizo, C., Prados, J.C., Clares, B. 2018. Nanoemulsion strategy for ursolic and oleanic acids isolates from *Plumeria obtusa* improves antioxidant and cytotoxic activity in melanoma cells. Anti-Cancer Agent ME 18(6): 847–853.

Amico, V., Barresi, V., Condorelli, D., Spatafora, C., Tringali, C. 2006. Antiproliferative terpenoids from almond hulls (*Prunus dulcis*): identification and structure-activity relationships. J. Agr. Food Chem. 54(3): 810–814.

Amui, S.F., Puga, R.D., Soares, A.M., Giuliatti, S. 2011. Plant-antivenom: Database of anti-venom medicinal plants. Electron J Biotechn. 14(1): 1–9.

Andre, C.M., Legay, S., Deleruelle, A., Nieuwenhuizen, N., Punter, M., Brendolise, C., Cooney, J.M., Lateur, M., Hausman, J.F., Larondelle, Y., Laing, W.A. 2016. Multifunctional oxidosqualene cyclases and cytochrome P450 involved in the biosynthesis of apple fruit triterpenic acids. New Phytol. 211(4): 1279–1294.

Araújo, F.A., Santa-Lúcia, M., Cabral, R.F. 2003. Epidemiologia dos Acidentes por Animais Peçonhentos. pp. 6–12. *In*: Cardoso, J.L.C., Siqueira França, F.O., Wen, F.H., Sant'ana Malaque, C.M., Haddad, V.J. (eds.). Animais Peçonhentos no Brasil: Biologia, Clínica e Terapêutica dos Acidentes. São Paulo: Sarvier.

Asmari, A.K., Khan, H.A., Banah, F.A., Buraidi, A.A., Manthiri, R.A. 2015. Serum biomarkers for acute hepatotoxicity of *Echis pyramidum* snake venom in rats. Int. J. Clin. Exp. Med. 8(1): 1376–1380.

Ataide, J.A., de Carvalho, N.M., Rebelo, M.A., Chaud, M.V., Grotto, D., Gerenutti, M., Rai, M., Mazzola, P.G., Jozala, A.F. 2017. Bacterial nanocellulose loaded with bromelain: assessment of antimicrobial, antioxidant and physical-chemical properties. Sci. Rep. 7(1): 1–8.

Ayeleso, T.B., Matumba, M.G., Mukwevho, E. 2017. Oleanolic acid and its derivatives: biological activities and therapeutic potential in chronic diseases. Molecules. 22(11): p.ii: E1915. doi: 10.3390/molecules22111915.

Banihani, S., Swedan, S., Alguraan, Z. 2013. Pomegranate and type 2 diabetes. Food Nutr. Res. 33(5): 341–348.

Beaufay, C., Hérent, M.F., Quetin-Leclercq, J., Bero, J. 2017. *In vivo* anti-malarial activity and toxicity studies of triterpenic esters isolated form *Keetia leucantha* and crude extracts. Malaria J. 16(1): 406. doi 10.1186/s12936-017-2054-y.

Bernard, P., Scior, T., Didier, B., Hibert, M., Berthon J.Y. 2001. Ethnopharmacology and bioinformatic combination for leads discovery: application to phospholipase A2 inhibitors. Phytochemistry 58(6): 865–874. https://doi.org/10.1016/S0031-9422(01)00312-0.

Bero, J., Beaufay, C., Hannaert, V., Hérent, M.F., Michels, P.A., Quetin-Leclercq, J. 2013. Antitrypanosomal compounds from the essential oil and extracts of *Keetia leucantha* leaves with inhibitor activity on *Trypanosoma brucei* glyceraldehyde-3-phosphate dehydrogenase. Phytomedicine 20(3-4): 270–274.

Biondo, R., Soares, A.M., Bertoni, B.W., França, S.C., Pereira, A.M.S. 2004. Micropropagation by direct organogenesis of *Mandevilla illustris* (Vell) Woodson and effects of its aqueous extract on the enzymatic and toxic activities of *Crotalus durissus terrificus* snake venom. Plant Cell Rep. 22(8): 549–552.

Borges, C.C., Sadahiro, M., Dos-Santos, M.C. 1999. Aspectos epidemiológicos e clínicos dos acidentes ofídicos ocorridos nos municípios do Estado do Amazonas. Rev. Soc. Bras. Med. Trop. 32(6): 637–646.

Brand-Williams, W., Cuvelier, M.E., Berset, C. 1995. Use of a free radical method to evaluate antioxidant activity. LWT Food Sci. Technol. 28(1): 25–30.

Brandão, M.S., Pereira, S.S., Lima, D.F., Oliveira, J.P., Ferreira, E.L., Chaves, M.H., Almeida, F.R. 2013. Antinociceptive effect of *Lecythis pisonis* Camb. (Lecythidaceae) in models of acute pain in mice. J. Ethnopharmacol. 146(1): 180–186.

Breitbach, U.B., Niehues, M., Lopes, N.P., Faria, J.E., Brandão, M.G. 2013. Amazonian Brazilian medicinal plants described by C.F.P. von Martius in the 19th century. J. Ethnopharmacol. 147(1): 180–189.

Brendolise, C., Yauk, Y.K., Eberhard, E.D., Wang, M., Chagne, D., Andre, C., Greenwood, D.R., Beuning, L.L. 2011. An unusual plant triterpene synthase with predominant α-amyrin-producing activity identified by characterizing oxidosqualene cyclases from Malus × domestica. FEBS J. 278(14): 2485–2499.

Bülbring, E. 1997. Observation on the isolated phrenic nerve diaphragm preparation of the rat. Br. J. Pharmacol. 120(Suppl 1): 1–2.

Cáceres-Castillo, D., Mena-Rejón, G.J., Cedillo-Rivera, R., Quijano, L. 2008. 21beta-Hydroxy-oleanane-type triterpenes from *Hippocratea excelsa*. Phytochemistry 69(4): 1057–1064.

Camargo, T.M., Nazato, V.S., Silva, M.G., Cogo, J.C., Groppo, F.C., Oshima-Franco, Y. 2010. *Bothrops jararacussu* venom-induced neuromuscular blockade inhibited by *Casearia gossypiosperma* Briquet hydroalcoholic extract. J. Venom Anim. Toxins incl. Trop. Dis. 16(3): 432–441.

Cao, J., Peng, L.Q., Xu, J.J. 2016. Microcrystalline cellulose based matrix solid phase dispersion microextration for isomeric triterpenoid acids in loquat leaves by ultrahigh-performance liquid chromatography and quadrupole time-of-flight mass spectrometry. J. Chromatogr A. 1472: 16–26.

Cardoso, J.L.C., Franca, F.O.S., Wen, F.H., Málaque, A.S., Haddad, V.J. 2003. Animais peçonhentos no Brasil: biologia, clínica e terapêutica dos acidentes. São Paulo: Ed. Sarvier, 468p.

Carrasco, P.A., Mattoni, C.I., Leynaud, G.C., Scrocchi, G.J. 2012. Morphology, phylogeny and taxonomy of South American bothropoid pitvipers (Serpentes, Viperidae). Zool Scr. 41(2): 109–124.

Carvalho, B.M., Santos, J.D., Xavier, B.M., Almeida, J.R., Resende, L.M., Martins, W., Marcussi, S., Marangoni, S., Stábeli, R.G., Calderon, L.A., Soares, A.M., Da Silva, S.L., Marchi-Salvador, D.P. 2013. Snake venom PLA2s inhibitors isolated from Brazilian plants: synthetic and natural molecules. BioMed. Res. Int. 1–8. doi: 10.1155/2013/153045.

Catteau, L., Reichmann, N.T., Olson, J., Pinho, M.G., Nizet, V., Van Bambeke, F., Quetin-Leclercq, J. 2017. Synergy between ursolic and oleanolic acids from *Vitellaria paradoxa* leaf extract and β-lactams against methicillin-resistant *Staphylococcus aureus*: *in vitro* and *in vivo* activity and underlying mechanisms. Molecules. 22(12): pii: E2245. doi: 10.3390/molecules22122245.

Chippaux, J.P., Goyffon, M. 1998. Venoms, antivenoms and immunotherapy. Toxicon. 36(6): 823–846.

Cintra-Francischinelli, M., Silva, M.G., Andréo-Filho, N., Cintra, A.C.O., Leite, G.B., Cruz-Höfling, M.A., Rodrigues-Simioni, L., Oshima-Franco, Y. 2008a. Effects of commonly used solubilizing agents on a model nerve-muscle synapse. Lat. Am. J. Pharm. 27(5): 721–726.

Cintra-Francischinelli, M., Silva, M.G., Andréo-Filho, N., Gerenutti, M., Cintra, A.C., Giglio, J.R., Leite, G.B., Cruz-Höfling, M.A., Rodrigues-Simioni, L., Oshima-Franco, Y. 2008b. Antibothropic action of *Casearia Sylvestris* Sw. (Flacourtiaceae) extracts. Phytother Res. 22(6): 784–790.

Collaço, R.C.O., Cogo, J.C., Rodrigues-Simioni, L., Rocha, T. Oshima-Franco, Y., Randazzo-Moura, P. 2012. Protection by *Mikania laevigata* (guaco) extract against the toxicity of *Philodryas olfersii* snake venom. Toxicon. 60(4): 614–622.

Correia-de-Sá, P., Noronha-Matos, J.B., Timóteo, M.A., Ferreirinha, F., Marques, P., Soares, A.M., Carvalho, C., Cavalcante, W.L., Gallacci, M. 2013. Bothropstoxin-I reduces evoked acetylcholine release from rat motor nerve terminals: radiochemical and real-time video-microscopy studies. Toxicon. 61: 16–25.

Costa, H.N.R., Dos-Santos, M.C., Alcântara, A.F.C., Silva, M.C., França, R.C., Piló-Veloso, D. 2008. Constituintes químicos e atividade antiedematogênica de *Peltodon radicans* (Lamiaceae). Quím Nova. 31(4): 744–750.

Costa, T.S. 2009. Avaliação da atividade antiofídica do extrato vegetal de *Anacardium humile*: Isolamento e caracterização fitoquímica do ácido gálico com potencial antiofídico. Dissertação de Mestrado, do Programa de Pós-gradução em Toxicologia, USP, Ribeirão Preto.

Çulhaoğlu, B., Yapar, G., Dirmenci, T., Topçu, G. 2013. Bioactive constituents of *Salvia chrysophylla* Stapf. Nat. Prod. Res. 27(4-5): 438–447.

Cunha, W.R., de Matos, G.X., Souza, M.G., Tozatti, M.G., Andrade e Silva, M.L., Martins, C.H., da Silva, R., Da Silva Filho, A.A. 2010. Evaluation of the antibacterial activity of the methylene chloride extract of *Miconia ligustroides*, isolated triterpene acids, and ursolic acid derivatives. Pharm. Biol. 48(2): 166–169.

da Silva, J.G., da Silva Soley, B., Gris, V., do Rocio Andrade Pires, A., Caderia, S.M., Eler, G.J., Hermoso, A.P., Bracht, A., Dalsenter, P.R., Acco, A. 2011. Effects of the *Crotalus durissus terrificus* snake venom on hepatic metabolism and oxidative stress. J. Biochem. Mol. Toxicol. 25(3): 195–203.

da Silva, J.O., Fernandes, R.S., Ticli, F.K., Oliveira, C.Z., Mazzi, M.V., Franco, J.J., Giuliatti, S., Pereira, P.S., Soares, A.M., Sampaio, S.V. 2007. Triterpenoids saponins, new metalloproteinase snake venom inhibitors isolated from *Pentaclethra macroloba*. Toxicon. 50(2): 283–291.

Da Silva, A.J.M., Mattos, F.J.A., Silveira, E.R. 1997. 4'-dehydroxyeacabenegrin A-I from roots of *Harpalyce brasiliana*. Phytochemistry 46(6): 1059–1062.

Da Silva, T.P., Moura, V.M., De Souza, L.Y.A., Sousa, R.L., Mourão, R.H.V., Dos Santos, M.C. 2017. Espécies vegetais utilizadas no bloqueio da atividade hemorrágica induzida pelos venenos de serpentes do gênero *Bothrops* sp.: uma revisão da literatura. Sci. Amazon. 6(2): 36–57.

Daros, M.R., Matos, F.J., Parente, J.P. 1996. A New Triterpenoid Saponin, Bredemeyeroside B, from the Roots of *Bredemeyera floribunda*. Planta Med. 62(6): 523–527.

David, J.M., Souza, J.C., Silva Guedes, M.L., David, J.P. 2006. Phytochemical study of *Davilla rugosa*: flavonoids and terpenoids. Rev Bras Farmacogn. 16(1): 105–118.

De Paula, R.C., Sanchez, E.F., Costa, T.R., Martins, C.H.G., Pereira, P.S., Lourenço, M.V., Soares, A.M., Fuly, A.L. 2010. Antiophidican properties of plant extract against *Lachesis muta* venom. J. Venom. Anim. Toxins incl. Trop. Dis. 16(2): 311–323.

Dey, A., De, J.N. 2012. Phytopharmacology of antiophidian botanicals: A Review. Int. J. Pharmacol. 8(2): 62–79.

Dharmadasa, R.M., Akalanka, G.C., Muthukumarana, P.R.M., Wijesekara, R.G.S. 2015. Ethnopharmacological survey on medicinal plants used in snakebite treatments in Western and Sabaragamuwa provinces in Sri Lanka. J. Ethnopharmacol. 179: 110–127.

Domingues, R.M., Oliveira, E.L., Freire, C.S., Couto, R.M., Simões, P.C., Neto, C.P., Silvestre, A.J., Silva, C.M. 2012. Supercritical fluid extraction of *Eucalyptus globulus* bark-A promising approach for triterpenoid production. Int. J. Mol. Sci. 13(6): 7648–7662.

e Silva, M. de L., David, J.P., Silva, L.C., Santos, R.A., David, J.M., Lima, L.S., Reis, P.S., Fontana, R. 2012. Bioactive oleanane, lupane and ursane triterpene acid derivatives. Molecules. 17(10): 12197–12205.

Fan, H.W., Monteiro, W.M. 2018. History and perspectives on how to ensure antivenom accessibility in the most remote areas in Brazil. Toxicon. 151: 15–23.

Ellman, G.L. 1959. Tissue sulfhydryl groups. Arch. Biochem. Biophys. 82(1): 70–77.

Erbano, M., Ehrenfried, C.A., Stefanello, M.É., Dos Santos, E.P. 2012. Morphoanatomical and phytochemical studies of *Salvia lachnostachys* (Lamiaceae). Microsc. Res. Techniq. 75(12): 1737–1744.

Félix-Silva, J., Silva-Junior, A.A., Zucolotto, S.M., Fernandes-Pedrosa, M.F. 2017. Medicinal plants for the treatment of local tissue damage induced by snake venoms: an overview from traditional use to pharmacological evidence. Evid Based Complement Alternat. Med. 2017:1–52. doi: 10.1155/2017/5748256.

Ferraz, M.C., Celestino Parrilha, L.A., Duarte Moraes, M.S., Amaral Filho, J., Cogo, J.C., dos Santos, M.G., Franco, L.M., Groppo, F.C., Puebla, P., San Feliciano, A., Oshima-Franco, Y. 2012. Effect of lupane triterpenoids (*Dipteryx alata* Vogel) in the *in vitro* neuromuscular blockade and myotoxicity of two snake venoms. Curr. Org. Chem. 16(22): 2717–2723.

Ferraz, M.C., Yoshida, E.H., Tavares, R.V., Cogo, J.C., Cintra, A.C., Dal Belo, C.A., Franco, L.M., dos Santos, M.G., Resende, F.A., Varanda, E.A., Hyslop, S., Puebla, P., San Feliciano, A., Oshima-Franco, Y. 2014. An isoflavone from *Dipteryx alata* Vogel is active against the *in vitro* neuromuscular paralysis of *Bothrops jararacussu* snake venom and bothropstoxin-I, and prevents venom-induces myonecrosis. Molecules 19(5): 5790–5805.

Ferraz, M.C., de Oliveira, J.L., de Oliveira Junior, J.R., Cogo, J.C., dos Santos, M.G., Franco, L.M., Puebla, P., Ferraz, H.O., Ferraz, H.G., da Rocha, M.M., Hyslop, S., San Feliciano, A., Oshima-Franco, Y. 2015. The triterpenoid betulin protects against the neuromuscular effects of *Bothrops jararacussu* snake venom *in vivo*. Evid Based Complement Alternat. Med. 1–10. doi: 10.1155/2015/939523.

Fontanay, S., Grare, M., Mayer, J., Finance, C., Duval, R.E. 2008. Ursolic, oleanolic and betulinic acids: Antibacterial spectra and selectivity indexes. J. Ethnopharmacol. 120(2): 272–276. doi: 10.1016/j.jep.2008.09.001.

Fraga, C.N., Stehmann, J.R. 2010. Novidades taxonômicas para Dilleniaceae brasileiras. Rodriguésia. 61(1): S01–S06.

França, V.C., Vieira, K.V.M., Lima, E.O.L., Barbosa-Filho, J.M., Da-Cunha, E.V.L., Silva, M.S. 2005. Estudo fitoquímico das partes aéreas de *Aristolochia birostris* Ducht. (Aristolochiaceae). Rev. Bras. Farmacogn. 15(4): 326–330.

García-Risco, M.R., Vázquez, E., Sheldon, J., Steinmann, E., Riebesehl, N., Fornari, T., Reglero, G. 2015. Supercritical fluid extraction of heather (*Calluna vulgaris*) and evaluation of anti-hepatitis C virus activity of the extracts. Virus Res. 198: 9–14.

Gbaguidi, F., Accrombessi, G., Moudachirou, M., Quetin-Leclercq, J. 2005. HPLC quantification of two isomeric triterpenic acids isolated from *Mitracarpus scaber* and antimicrobial activity on *Dermatophilus congolensis*. J. Pharm. Biomed. Anal. 39(5): 990–995.

Gerardo, A., Aymard, C. 2007. Three new species of *Davilla* (Dilleniaceae) from Brazil. Novon. 17(3): 282–287.

Giménez, E., Juan, M.E., Calvo-Melià, S., Barbosa, J., Sanz-Nebot, V., Planas, J.M. 2015. Pentacyclic triterpene in *Olea europaea* L: a simultaneous determination by high-performance liquid chromatography coupled to mass spectrometry. J. Chromatogr. A. 1410: 68–75.

Giménez-Llort, L., Ratia, M., Pérez, B., Camps, P., Muñoz-Torrero, D., Badia, A., Clos, M.V. 2017. Behavioural effects of novel multitarget anticholinesterasic derivatives in Alzheimeer´s disease. Behav. Pharmacol. 28(2-3 spec. issue): 124–131.

Giovannini, P., Howes, M.J.R. 2017. Medicinal plants used to treat snakebite in Central America: Review and assessment of scientific evidence. J. Ethnopharmacol. 199: 240–256.

Gleńsk, M., Czapińska, E., Woźniak, M., Ceremuga, I., Włodarczyk, M., Terlecki, G., Ziółkowski, P., Seweryn, E. 2017. Triterpenoid acids as important antiproliferative constituents of European elderberry fruits. Nutr. Cancer. 69(4): 643–651.

Gordo, J., Máximo, P., Cabrita, E., Lourenço, A., Oliva, A., Almeida, J., Filipe, M., Cruz, P., Barcia, R., Santos, M., Cruz, H. 2012. *Thymus mastichina*: chemical constituents and their anti-cancer activity. Nat. Prod. Commun. 7(11): 1491–1494.

Gu, G., Barone, I., Gelsomino, L., Giordano, C., Bonofiglio, D., Statti, G., Menichini, F., Catalano, S., Andò, S. 2012. *Oldenlandia diffusa* extracts exert antiproliferative and apoptotic effects on human breast cancer cells through ERα/Sp1-mediated p53 activation. J. Cell Physiol. 227(10): 3363–3372.

Guimarães, A.C., Magalhães, A., Nakamura, M.J., Siani, A.C., Barja-Fidalgo, C., Sampaio, A.L. 2012. Flavonoids bearing an O-arabinofuranosyl-(1-->3)-rhamnoside moiety from *Cladocolea micrantha*: inhibitory effect on human melanoma cells. Nat. Prod. Commun. 7(10): 1311–1314.

Gutiérrez, J.M., Lomonte, B. 1989. Local tissue damage induced by *Bothrops* snake venoms. A review. Mem. Inst. Butantan. 51(4): 211–223.

Gutiérrez, J.M., Lomonte, B. 1995. Phospholipase A myotoxins from *Bothrops* snake venoms. Toxicon. 33(11): 1405–1421.

Gutiérrez, J.M., León, G., Rojas, G., Lomonte, B., Rucavado, A., Chavez, F. 1998. Neutralization of local tissue damage induced by *Bothrops asper* (Terciopelo) snake venom. Toxicon. 36(11): 1529–1538.

Gutiérrez, J.M. 2002. Comprendiendo los venenos de serpientes: 50 años de investigación en América Latina. Rev. Biol. Trop. 50(2): 377–394.

Gutiérrez, J.M., Lomonte, B., León, G., Rucavado, A., Chavez, F., Ângulo, Y. 2007. Trends in snakebite envenomation therapy: Scientific, tecnological and public heath considerations. Curr. Pharm. Des. 13(28): 2935–2950.

Gutiérrez-Rebolledo, G.A., Garduño-Siciliano, L., García-Rodríguez, R.V., Pérez-González, M.Z., Chávez, M.I., Bah, M., Siordia-Reyes, G.A., Chamorro-Cevallos, G.A., Jiménez-Arellanes, M.A. 2016. Anti-inflammatory and toxicological evaluation of *Moussonia deppeana* (Schldl. & Cham.) Hanst and Verbascoside as a main active metabolite. J. Ethnopharmacol. 187: 269–280.

Hattori, M., Kawakami, K., Akimoto, M., Takenaga, K., Suzumiya, J., Honma, Y. 2013. Antitumor effect of Japanese apricot extract (MK615) on human cancer cells *in vitro* and *in vivo* through a reactive oxygen species-dependent mechanism. Tumori. J. 99(2): 239–248.

Hegazy, A.K., Ezzat, S.M., Qasem, I.B., Ali-Shtayeh, M.S., Basalah, M.O., Ali, H.M., Hatamleh, A.A. 2015. Diversity of active constituents in *Cichorium endivia* and *Cynara cornigera* extracts. Acta Biol. Hung. 66(1): 103–118.

Houghton, P.J., Osibogun, I.M. 1993. Flowering plants used against snakebite. J. Ethnopharmacol. 39(1): 1–29.

Huang, J., Zhang, Y., Dong, L., Gao, Q., Yin, L., Quan, H., Chen, R., Fu, X., Lin, D. 2018. Ethnopharmacology, phytochemistry, and pharmacology of *Cornus officinalis* Sieb. et Zucc. J. Ethnopharmacol. 213: 280–301.

Indriunas, A.I., Aoyama, E.M. 2018. Systema Materiae Medicae Vegetabilis Brasiliensiesis de Martius: plantas empregadas para acidentes ofídicos. Ethnoscientia. 3: 1–7. doi 10.22276/ethnoscientia.v3i0.97.

Izidoro, L.F., Rodrigues, V.M., Rodrigues, R.S., Ferro, E.V., Hamaguchi, A., Giglio, J.R., Homsi-Brandeburgo, M.I. 2003. Neutralization of some hematological and hemostatic alterations induced by neuwiedase, a metalloproteinase isolated from *Bothrops neuwiedi pauloensis* snake venom, by the aqueous extract from *Casearia mariquitensis* (Flacourtiaceae). Biochimie. 85(7): 669–675.

Januário, A.H., Santos, S.L., Marcussi, S., Mazzi, M.V., Pietro, R.C., Sato, D.N., Ellena, J., Sampaio, S.V., França, S.C., Soares, A.M. 2004. Neoclerodane diterpenoid, a new metalloprotease snake venom inhibitor from *Baccharis trimera* (Asteraeae): Anti-proteolytic and anti-hemorrhagic properties. Chem. Biol. Interact. 150(3): 243–351.

Jiang, Q., Zou, S.Q. 2014. Determination of five triterpenic acids from Salvia chinensis of different parts by RP-HPLC-PDA. *Zhongguo zhongyao zazhi.* 39(22): 4379–4382.

Kadir, M.F., Karmoker, J.R., Alam, M.R., Jahan, S.R., Mahbub, S., Mia, M.M.K. 2015. Ethnopharmacological survey of medicinal plants used by traditional healers and indigenous people in Chittagong Hill Tracts, Bangladesh, for the treatment of snakebite. Evid Based Complement Alternat Med. 2015: 871675. doi: 10.1155/2015/871675.

Kalaycıoğlu, Z., Uzaşçı, S., Dirmenci, T., Erim, F.B. 2018. α-Glucosidase enzyme inhibitory effects and ursolic and oleanolic acid contents of fourteen Anatolian Salvia species. J. Pharm. Biomed. Anal. 155(5): 284–287.

Kalogeropoulos, N., Chiou, A., Pyriochou, V., Peristeraki, A., Karathanos, V.T. 2012. Bioactive phytochemicals in industrial tomatoes and their processing by products. LWT- Food Sci. Technol. 49(2): 213–216.

Katz, S.R., Newman, R.A., Lansky, E.P. 2017. *Punica granatum*: heuristic treatment for diabetes mellitus. J. Med. Food. 10(2): 213–217.

Kikuchi, T., Akazawa, H., Tabata, K., Manosroi, A., Manosroi, J., Suzuki, T. et al. 2011. 3-O-(E)-p-coumaroyl tormentic acid from *Eriobotrya japonica* leaves induces caspase-dependent apoptotic cell death in human leukemia cell line. Chem. Pharm. Bull. 59(3): 378–381.

Kilkenny, C., Browne, W.J., Cuthill, I.C., Emerson, M., Altman, D.G. 2010. Improving bioscience research reporting: the ARRIVE guidelines for reporting animal research. Plos. Biol. 8(6): e1000412.

Knudsen, C., Laustsen, A.H. 2018. Recent advances in next generation snakebite antivenoms. Trop. Med. Infect. Dis. 3(2): 42. doi:10.3390/tropicalmed3020000.

Kowalski, R. 2007. Studies of selected plant raw materials as alternative sources of triterpenes of oleanolic and ursolic acid types. J. Agr. Food Chem. 55(3): 656–662.

Kurek, A., Nadkowska, P., Pliszka, S., Wolska, K.I. 2012. Modulation of antibiotic resistance in bacterial pathogens by oleanolic acid and ursolic acid. Phytomedicine. 19(6): 515–519.

Lee, W.S., Im, K.R., Park, Y.D., Sung, N.D, Jeong, T.S. 2006. Human ACAT-1 and ACAT-2 inhibitory activities of pentacyclic triterpenes from the leaves of *Lycopus lucidus* TURCZ. Biol. Pharm. Bull. 29(2): 382–384.

Leite, J.P., Oliveira, A.B., Lombardi, J.A., Filho, J.D., Chiari, E. 2006. Trypanocidal activity of triterpenes from *Arrabidaea triplinervia* and derivatives. Biol. Pharm. Bull. 29(11): 2307–2309.

Li, J.Y., Xie, X.M., Li, Q.W., Zhang, Q., Chen, S.L., Wang, H.Q., Yu, W.X., Yang, M.2015. Dynamic change of four triterpenic acids contents in different organs of loquat (Eriobotrya japonica) and phenology. *Zhongguo zhongyao zazhi.* 40(5): 875–880.

Lin, Z., Wu, Z.F., Jiang, C.H., Zhang, Q.W., Ouyang, S., Che, C.T., Zhang, J., Yin, Z.Q. 2016. The chloroform extract of *Cyclocarya paliurus* attenuates high-fat diet induced non-alcoholic hepatic steatosis in Sprague Dawley rats. Phytomedicine 23(12): 1475–1483.

Liu, J. 1995. Pharmacology of oleanolic acid and ursolic acid. J. Ethnopharmacol. 49(2): 57–68.

Lorenz, P., Berger, M., Bertrams, J., Wende, K., Wenzel, K., Lindequist, U., Meyer, U., Stintzing, F.C. 2008. Natural wax constituents of a supercritical fluid CO(2) extract from quince (*Cydonia oblonga* Mill.) pomace. Anal. Bioanal. Chem. 391(2): 633–646.

Lu, H., Xi, C., Chen, J., Li, W. 2009. Determination of triterpenoid acids in leaves of Eriobotrya japonica collected at in different seasons. *Zhongguo zhongyao zazhi.* 34(18): 2353–2355.

Mahato, S.B., Garai, S. 1998. Triterpenoid saponins. Fortschr. Chem. Org. Naturst. 74: 1–196.

Maldonado, E., Galicia, L., Chávez, M.I., Hernández-Ortega, S. 2016. Neo-clerodane diterpenoids and other constituents of *Salvia filipes*. J. Nat. Prod. 79(10): 2667–2673.

Martins, D., Carrion, L.L., Ramos, D.F., Salomé, K.S., da Silva, P.E., Barison, A., Nunez, C.V. 2013. Triterpenes and the antimycobacterial activity of *Duroia macrophylla* Huber (Rubiaceae). BioMed. Res. Int. 2013: 1–7. doi: 10.1155/2013/605831.

Martz, W. 1992. Plants with a reputation against snakebite. Toxicon. 30(10): 1131–1142.

Marzouk, A.M. 2009. Hepatoprotective triterpenes from hairy root cultures of *Ocimum basilicum* L. Z. Naturf. C., 64(3-4): 201–209.

Mattson, M.P. 2004. Pathways towards and away from Alzheimer's disease. Nature 430(7000): 631–639.

Mebs D. 2002. Venomous and Poisonous Animals: A handbook for Biologists, Toxicologists and Toxinologists, Physicians and Pharmacists. Med. Pharm. Sci. Publishers, Suttgart, Germany.

Miao, J., Zhao, C., Li, X., Chen, X., Mao, X., Huang, H., Wang, T., Gao, W. 2016. Chemical composition and bioactivities of two common chaenomeles fruits in China: *Chaenomeles speciosa* and *Chaenomeles sinensis*. J. Food Sci. 81(8): H2049–H2058.

Michel, P., Owczarek, A., Matczak, M., Kosno, M., Szymański, P., Mikiciuk-Olasik, E., Kilanowicz, A., Wesołowski, W., Olszewska, M.A. 2017. Metabolite profiling of eastern teaberry (*Gaultheria procumbens* L.) lipophilic leaf extracts with hyaluronidase and lipoxygenase inhibitory activity. Molecules. 22(3): 412–428.

Molander, M., Nielsen, L., Søgaard, S., Staerk, D., Rønsted, N., Diallo, D., Kusamba, C., van Staden, J., Jäger, A.K. 2014. African plants for inhibition of necrotic enzymes from snake venom. Planta Med. 80(16): P2Y21.

Moreno-Anzúrez, N.E., Marquina, S., Alvarez, L., Zamilpa, A., Castillo-España, P., Perea-Arango, I., Torres, P.N., Herrera-Ruiz, M., Díaz García, E.R., García, J.T., Arellano-García, J. 2017. A Cytotoxic and anti-inflammatory campesterol derivative from genetically transformed hairy roots of *Lopezia racemosa* Cav. (Onagraceae). Molecules. 22(1):pii118. doi: 10.3390/molecules22010118.

Mors, W.B. 1991. Plants active against snakebite. p. 352–382. *In*: Wagner, H., Hikino, H., Farnsworth, N.R. (eds.). Economic and Medicinal Plant Research, v. 5, London: Academic Press.

Mors, W.B., Nascimento, M.C., Pereira, B.M.R., Pereira, N.A. 2000. Plant natural products active against snakebite the molecular. Phytochemistry. 55(6): 627–642.

Moura, V.M., Mourão, R.H. 2012. Aspectos do ofidismo no Brasil e plantas medicinais utilizadas como complemento à soroterapia. Sci. Amazon. 1(3): 17–26.

Moura, V.M., Freitas de Sousa, L.A., Dos-Santos, M.C., Almeida Raposo, J.D., Evangelista Lima, A., de Oliveira, R.B., da Silva, M.N., Veras Mourão, R.H. 2015. Plants used to treat snakebites in Santarém, western Pará, Brazil: an assessment of their effectiveness in inhibiting hemorrhagic activity induced by *Bothrops jararaca* venom. J. Ethnopharmacol. 161: 224–232.

Mulholland, D.A., Mohammed, A.M., Coombes, P.H., Haque, S., Pohjala, L.L., Tammela, P.S., Crouch, N.R. 2011. Triterpenoid acids and lactones from the leaves of *Fadogia tetraquetra* var. *tetraquetra* (Rubiaceae). Nat. Prod. Commun. 6(11): 1573–1576.

Nakagawa, M., Nakanishi, K., Darko, L.L., Vick, J.A. 1982. Structures of cabenegrins A-I and A-II, potent anti-snake venoms. Tetrahedron Lett. 23(38): 3855–3858.

Nascimento, P.G.G., Lemos, T.L., Bizerra, A.M., Arriaga, A.M., Ferreira, D.A., Santiago, G.M., Braz-Filho, R., Costa, J.G. 2014. Antibacterial and antioxidant activities of ursolic acid and derivatives. Molecules 19(1): 1317–1327.

Nishijima, C.M., Rodrigues, C.M., Silva, M.A., Lopes-Ferreira, M., Vilegas, W., Hiruma-Lima, C.A. 2009. Anti-hemorrhagic activity of four Brazilian vegetable species against *Bothrops jararaca* venom. Molecules 14(3): 1072–1080.

Nowak, R., Wójciak-Kosior, M., Sowa, I., Sokołowska-Krzaczek, A., Pietrzak, W., Szczodra, A., Kocjan, R. 2013. HPTLC-densitometry determination of triterpenic acids in *Origanum vulgare*, *Rosmarinus officinalis* and *Syzygium aromaticum*. Acta Pol. Pharm. 70(3): 413–418.

Ohkawa, H., Ohishi, N., Yagi, K. 1979. Assay for lipid peroxides in animal tissues by thiobarbituric acid reaction. Anal Biochem. 95(2): 351–358.

Olmo-García, L., Bajoub, A., Fernández-Gutiérrez, A., Carrasco-Pancorbo, A. 2016. Evaluating the potential of LC coupled to three alternative detection systems (ESI-IT, APCI-TOF and DAD) for the targeted determination of triterpenic acids and dialcohols in olive tissues. Talanta. 150: 355–366.

Oprean, C., Borcan, F., Pavel, I., Dema, A., Danciu, C., Soica, C., Dehelean, C., Nicu, A., Ardelean, A., Cristea, M., Ivan, A., Tatu, C., Bojin, F. 2016. *In vivo* biological evaluation of polyurethane nanostructures with ursolic and oleanolic acids on chemically-induced skin carcinogenesis. *In Vivo* 30(5): 633–638.

Oshima-Franco, Y., Leite, G.B., Belo, C.A., Hyslop, S., Prado-Franceschi, J., Cintra, A.C., Giglio, J.R., da Cruz-Höfling, M.A., Rodrigues-Simioni, L. 2004. The presynaptic activity of bothropstoxin-I, a myotoxin from *Bothrops jararacussu* snake venom. Basic Clin. Pharmacol. Toxicol. 95(4): 175–182.

Otero, R., Fonnegra, R., Jiménez, S.L. 2000a. Plantas utilizadas contra mordeduras de serpientes en Antioquia y Chocó, Colombia. Universidad de Antioquia. Medellín, 402 p.

Otero, R., Fonnegra, R., Jiménez, S.L., Nuñez, V., Evans, N., Alzate, S.P., García, M.E., Saldarriaga, M., Del Valle, G., Osorio, R.G., Díaz, A., Valderrama, R., Duque, A., Vélez, H.N. 2000b. Snakebites and ethnobotany in the northwest region of Colombia Part I: Traditional use of plants. J. Ethnopharmacol. 71(3): 493–504.

Otero, R., Nuñez, V., Jiménez, S.L., Fonnegra, R., Osorio, R.G., García, M.E., Díaz, A. 2000c. Snakebites and ethnobotany in the northwest region of Colombia Part II: Neutralization of lethal and enzymatic effects of *Bothrops atrox* venom. J. Ethnopharmacol. 71(3): 505–511.

Otero, R., Nuñez, V., Barona, J., Fonnegra, R., Jiménez, S.L., Osorio, R.G., Saldarriaga, M., Díaz, A. 2000d. Snakebites and ethnobotany in the northwest region of Colombia Part III: Neutralization of the haemorrhagic effect of *Bothrops atrox* venom. J. Ethnopharmacol. 73(1-2): 233–241.

Owuor, B.O., Kisangau, D.P. 2006. Kenyan medicinal plants used as antivenin: a comparison of plant usage. J. Ethnobiol. Ethnomed. 2(7): 8.

Paglia, D.E., Valentine, W.N. 1967. Study on the quantitative and qualitative caracterization of erythrocyte glutathione peroxide. Transl Res. 70(1): 158–169.

Passero, L.F., Bonfim-Melo, A., Corbett, C.E., Laurenti, M.D., Toyama, M.H., de Toyama, D.O., Romoff, P., Fávero, O.A., dos Grecco, S.S., Zalewsky, C.A., Lago, J.H. 2011. Anti-leishmanial effects of purified compounds from aerial parts of *Baccharis uncinella* C. DC (Asteraceae). Parasitol. Res. 108(3): 529–536.

Pereira, M.S. 2015. Preparação e avaliação da atividade antimicrobiana de triterpenóides semi-sintéticos. Dissertação de Mestrado em Química Farmacêutica Industrial, apresentada à Faculdade de Farmácia da Universidade de Coimbra, p. 16–18, 29–80.

Pereira, N.A., Jaccoud, R.J.S., Mors, W.B. 1996. Triaga brasilica: renewed interest in a seventeenth century panacea. Toxicon. 34(5): 511–516.

Pereira, J.R., Queiroz, R.F., Siqueira, E.A., Brasileiro-Vidal, A.C., Sant'ana, A.E.G., Silva, D.M., de Mello Affonso, P.R.A. 2017. Evaluation of cytogenotoxicity, antioxidant and hypoglycemiant activities of isolate compounds from *Mansoa hirsuta* D.C. (Bignoniaceae). An Acad. Bras. Cienc. 89(1): 317–331.

Peschel, S., Franke, R., Schreiber, L., Knoche, M. 2007. Composition of the cuticle of developing sweet cherry fruit. Phytochemistry 68(7): 1017–1025.

Pinho, F.O., Pereira, I.D. 2001. Ofidismo. Rev. Ass. Med. Bras. 47(1): 24–9.

Price, K.R., Johnson, L.T., Fenwick, G.R. 1987. The chemistry and biological significance of saponins in foods and feedingstuffs. Crit. Rev. Food Sci. Nutr. 26(1): 27–135.

Puangpraphant, S., De Mejia, E.G. 2009. Saponins in yerba mate tea (*Ilex paraguariensis* A. St.-Hil) and quercetin synergistically inhibit iNOS and COX-2 in lipopolysaccharide-induced macrophages through NFkappaB pathways. J. Agr. Food Chem. 57(19): 8873–8883.

Rodríguez, I.I.G. 2010. Avaliação da Atividade Antiofídica de *Aristolochia sprucei*: Isolamento e Caracterização Estrutural de Composto Bioativo. Dissertação de mestrado, Faculdade de Ciências Farmacêuticas, Universidade de São Paulo-USP, Ribeirão Preto-São Paulo.

Rungsimakan, S., Rowan, M.G. 2014. Terpenoids, flavonoids and caffeic acid derivatives from *Salvia viridis* L. cvar. Blue Jeans. Phytochemistry 108: 177–188.

Saaby, L., Nielsen, C.H. 2012. Triterpene acids from rose hip powder inhibit self-antigen- and LPS-induced cytokine production and CD4$^+$ T-cell proliferation in human mononuclear cell cultures. Phytother Res. 26(8): 1142–1147.

Saimaru, H., Orihara, Y., Tansakul, P., Kang, Y.H., Shibuya, M., Ebizuka, Y. 2007. Production of triterpene acids by cell suspension cultures of *Olea europaea*. Chem. Pharm. Bull. 55(5): 784–788.

Salah El Dine, R., Ma, Q., Kandil, Z.A., El-Halawany, A.M. 2014. Triterpenes as uncompetitive inhibitors of α-glucosidase from flowers of *Punica granatum* L. Nat. Prod. Res. 28(23): 2191–2194.

Santos, C.A., Dos Santos, G.R., Soeiro, V.S., Dos Santos, J.R., Rebelo, M.A., Chaud, M.V., Gerenutti, M., Grotto, D., Pandit, R., Rai, M., Jozala, A.F. 2018. Bacterial nanocellulose membranes combined with nisin: a strategy to prevent microbial growth. Cellulose 1(11): 1–9.

Santos, E.O., Lima, L.S., David, J.M., Martins, L.C., Guedes, M.L., David, J.P. 2011. Podophyllotoxin and other aryltetralin lignans from *Eriope latifolia* and *Eriope blanchetii*. Nat. Prod. Res. 25(15): 1450–1453.

Santos Rosa, C., García Gimenez, M.D., Saenz Rodriguez, M.T., De la Puerta Vazquez, R. 2007. Antihistaminic and antieicosanoid effects of oleanolic and ursolic acid fraction from *Helichrysum picardii*. Pharmazie. 62(6): 459–462.

Santhosh, M.S., Hemshekhar, M., Sunitha, K., Thushara, R.M., Jnaneshwari, S., Kemparaju, K., Girish, K.S. 2013. Snake venom induced local toxicities: plant secondary metabolites as an auxiliary therapy. Mini-Rev. Med. Chem. 13(1): 106–123.

Scalon Cunha, L.C., Andrade e Silva, M.L., Cardoso Furtado, N.A., Vinhólis, A.H., Martins, C.H., da Silva Filho, A.A., Cunha, W.R. 2007. Antibacterial activity of triterpene acids and semi-synthetic derivatives against oral pathogens. Z. Naturf C. 62(9-10): 668–672.

Senthil, S., Chandramohan, G., Pugalendi, K.V. 2007. Isomers (oleanolic and ursolic acids) differ in their protective effect against isoproterenol-induced myocardial ischemia in rats. Int. J. Cardiol. 119(1): 131–133.

Shi, H.H., Cai, X.P., Ju, J.M., Zhang, Z.H., Hua, J.L., Lv, H., Li, W.L. 2014. Preparation of solid dispersion of triterpenoid acids from *Eriobotrya japonica* leaf and study on their dissolution *in vitro*. Zhong Yao Cai. 37(8): 1467–1470.

Silva, L.L., Gomes, B.S., Sousa-Neto, B.P., Oliveira, J.P., Ferreira, E.L., Chaves, M.H., Oliveira, F.A. 2012. Effects of *Lecythis pisonis* Camb. (Lecythidaceae) in a mouse model of pruritus. J. Ethnopharmacol. 139(1): 90–97.

Simões, L.R., Maciel, G.M., Brandão, G.C., Kroon, E.G., Castilho, R.O., Oliveira, A.B. 2011. Antiviral activity of *Distictella elongata* (Vahl) Urb. (Bignoniaceae), a potentially useful source of anti-dengue drugs from the state of Minas Gerais, Brazil. Lett. Appl. Microbiol. 53(6): 602–607.

Soares, A.M., Januario, A.H., Lourenço, M.V., Pereira, M.A.S., Pereira, P.S. 2004. Neutralizing effects of Brazilian plants against snake venoms. Drugs Future, Barcelona 29(11): 1105–1117.

Soares, A.M., Ticli, F.K., Marcussi, S., Lourenço, M.V., Januário, A.H., Sampaio, S.V., Giglio, J.R., Lomonte, B., Pereira, P.S. 2005. Medicinal plants with inhibitory properties against snake venoms. Curr. Med. Chem. 12(22): 2625–2641.

Soares, D.B.S., Duarte, L.P., Cavalcanti, A.D., Silva, F.C., Braga, A.D., Lopes, M.T.P., Takahashi, J.A., Vieira-Filho, S.A. 2017. *Psychotria viridis*: chemical constituents from leaves and biological properties. An Acad. Bras. Ciênc. 89(2): 927–938.

Soares-Silva, J.O., de Oliveira, J.L., Cogo, J.C., Tavares, R.V.S., Oshima-Franco, Y. 2014. Pharmacological evaluation of hexane fraction of *Casearia gossypiosperma* Briquet: antivenom potentiality. J. Life Sci. 8(4): 306–315.

Srivastava, P., Sisodia, V., Chaturvedi, R. 2011. Effect of culture conditions on synthesis of triterpenoids in suspension cultures of *Lantana camara* L. Bioprocess Biosyst. Eng. 34(1): 75–80.

Stenholm, A., Göransson, U., Bohlin, L. 2013. Bioassay-guided supercritical fluid extraction of cyclooxygenase-2 inhibiting substances in *Plantago major* L. Phytochem. Anal. 24(2): 176–183.

Sunitha, K., Hemshekhar, M., Thushara, R.M., Santhosh, M.S., Sundaram, M.S., Kemparaju, K., Girish, K.S. 2015. Inflammation and oxidative stress in viper bite: an insight withi and beyond. Toxicon. 98: 89–97. doi: 10.1016/j.toxicon.2015.02.014.

Szakiel, A., Mroczek, A. 2007. Distribution of triterpene acids and their derivatives in organs of cowberry (*Vaccinium vitisidaea* L.) plant. Acta Biochim. Pol. 54(4): 733–740.

Szakiel, A., Pączkowski, C., Huttunen, S. 2012. Triterpenoid content of berries and leaves of bilberry *Vaccinium myrtillus* from Finland and Poland. J. Agr. Food Chem. 60(48): 11839–11849.

Szakiel, A., Niżyński, B., Pączkowski, C. 2013. Triterpenoid profile of flower and leaf cuticular waxes of heather *Calluna vulgaris*. Nat. Prod. Res. 27(15): 1404–1407.

Ticli, F.K., Hage, L.I.S., Cambraia, R.S., Pereira, O.S., Magro, A.J., Fontes, M.R.M., Stábeli, R.G., Giglio, J.R., França, S.C., Soares, A.M., Sampaio, S.V. 2005. Rosmarinic acid, a new snake venom phospholipase A2 inhibitor from *Cordia verbenacea* (Boraginaceae): antiserum action potentiation and molecular interaction. Toxicon. 46(3): 318–327.

Truong, N.B., Pham, C.V., Doan, H.T., Nguyen, H.V., Nguyen, C.M., Nguyen, H.T., Zhang, H.J,, Fong, H.H., Franzblau, S.G., Soejarto, D.D., Chau, M.V. 2011. Antituberculosis cycloartane triterpenoids from *Radermachera boniana*. J. Nat. Prod. 74(5): 1318–1322.

Upasani, S.V., Beldar, V.G., Pranjal, A.U., Upasani, M.S., Surana, S.J., Patil, D.S. 2017. Ethnomedicinal plants used for snakebites in India: An overview. Integr. Med. Res. 6(2): 114–130.

Ushanandini, S., Nagaraju, S., Nayaka, S.C., Kumar, K.H., Kemparaju, K., Giris, K.S. 2009. The anti-ophidian properties of *Anacardium occidentale* bark extract. Immunopharmacol. Immunotoxicol. 31(4): 607–615.

van Baren, C., Anao, I., Leo Di Lira, P., Debenedetti, S., Houghton, P., Croft, S., Martino, V. 2006. Triterpenic acids and flavonoids from *Satureja parvifolia*. Evaluation of their antiprotozoal activity. Z Naturforsch C 61(3-4): 189–192.

Vasconcelos, M.A., Royo, V.A., Ferreira, D.S., Crotti, A.E., Andrade e Silva, M.L., Carvalho, J.C., Bastos, J.K., Cunha, W.R. 2006. *In vivo* analgesic and anti-inflammatory activities of ursolic acid and oleanolic acid from *Miconia albicans* (Melastomataceae). Z. Naturforsch. C. 61(7-8): 477–482.

Venditti, A., Serafini, M., Nicoletti, M., Bianco, A. 2015. Terpenoids of *Linaria alpina* (L.) Mill. from Dolomites, Italy. Nat. Prod. Res. 29(21): 2041–2044.

Verri, M., Pastoris, O., Dossena, M., Aquilani, R., Guerriero, F., Cuzzoni, G., Venturini, L., Ricevuti, G., Bongiorno, A.I. 2012. Mitochondrial alterations, oxidative stress and neuroinflammation in Alzheimer disease. Int. J. Immunopath. Pharmacol. 25(2): 345–353.

Vieira, L.C. 2003. Obtenção de derivados semissintéticos triterpênicos do ácido ursólico visando a atividade biológica. Programa de Pós-Graduação de Ciências Farmacêuticas da Universidade Federal do Rio Grande do Sul.

Wang, B., Jiang, Z.H. 1992. Studies on oleanolic acid. Chin. Pharm. J. 27: 393–397.

Wenzig, E.M., Widowitz, U., Kunert, O., Chrubasik, S., Bucar, F., Knauder, E., Bauer, R. 2008. Phytochemical composition and *in vitro* pharmacological activity of two rose hip (*Rosa canina* L.) preparations. Phytomedicine 15(10): 826–835.

Werner, A.C., Ferraz, M.C., Yoshida, E.H., Tribuiani, N., Gautuz, J.A., Santana, M.N., Dezzotti BA, de Miranda, V.G., Foramiglio, A.L., Rostelato-Ferreira, S., Tavares, R.V., Hyslop, S., Oshima-Franco, Y. 2015. The facilitatory effect of *Casearia sylvestris* Sw. (guaçatonga) fractions on the contractile activity of mammalian and avian neuromuscular apparatus. Curr. Pharm. Biotechnol. 16(5): 468–481.

Woźniak, Ł., Skąpska, S., Marszałek, K. 2015. Ursolic Acid—A pentacyclic triterpenoid with a wide spectrum of pharmacological activities. Molecules 20(11): 20614–20641.

Yang, J., Hu, Y.J., Yu, B.Y., Qi, J. 2016. Integrating qualitative and quantitative characterization of Prunellae Spica by HPLC-QTOF/MS and HPLC-ELSD. Chinese J. Nat. Med. 14(5): 391–400.

Yang, P., Li, Y., Liu, X., Jiang, S. 2007. Determination of free isomeric oleanolic acid and ursolic acid in *Pterocephalus hookeri* by capillary zone electrophoresis. J. Pharm. Biomed. Anal. 43(4): 1331–1334.

Yang, B., Li, H., Zhao, Y.X., Li, M.L. 2004a. Changes in level of organic acids in fructus crataegi after processing. *Zhongguo zhongyao zazhi* 29(11): 1057–1060.

Yang, S.M., Liu, X.K., Qing, C., Wu, D.G., Zhu, D.Y. 2007b. Chemical constituents from the roots of Homonoia riparia. Yao Xue Xue Bao. 42(3): 292–296.

Yang, Y.C., Wei, M.C., Chiu, H.F., Huang, T.C. 2013. Development and validation of a modified ultrasound-assisted extraction method and a HPLC method for the quantitative determination of two triterpenic acids in Hedyotis diffusa. Nat. Prod. Commun. 8(12): 1683–1686.

Yang, J., Hu, Y.J., Yu, B.Y., Qi, J. 2016. Integrating qualitative and quantitative characterization of Prunellae Spica by HPLC-QTOF/MS and HPLC-ELSD. Chin. J. Nat. Med. 14(5): 391–400.

Yin, M.C., Chan, K.C. 2007. Nonenzymatic antioxidative and antiglycative effects of oleanolic acid and ursolic acid. J. Agr. Food Chem. 55(17): 7177–7181.

Yoshida, E.H., Ferraz, M.C., Tribuiani, N., Tavares, R.V.S., Cogo, J.C., Dos Santos, M.G., Franco, L.M., Dal-Belo, C.A., De Grandis, R.A., Resende, F.A., Varanda, E.A., Puebla, P., San-Feliciano, A., Groppo, F.C., Oshima-Franco, Y. 2015. Evaluation of the safety of three phenolic compounds from *Dipteryx alata* Vogel with Antiophidian Potential. Chinese Med. 6(1): 1–12.

Youdim, M.B.H., Buccafusco, J.J. 2005. Multi-functional drugs for various CNS targets in the treatment of neurodegenerative disorders. Trends Pharmacol. Sci. 26(1): 27–35.

Zalewski, C.A., Passero, L.F., Melo, A.S., Corbett, C.E., Laurenti, M.D., Toyama, M.H., Toyama. D.O., Romoff, P., Fávero, O.A., Lago, J.H. 2011. Evaluation of anti-inflammatory activity of derivatives from aerial parts of *Baccharis uncinella*. Pharm. Biol. 49(6): 602–607.

Zhang, Y., Zhang, Y., Sheng, Y., Zhao, D., Lv, S., Hu, Y., Tao, J. 2011. Herbaceous peony (*Paeonia lactiflora* Pall.) as an alternative source of oleanolic and ursolic acids. Int. J. Mol. Sci. 12(1): 655–667.

Zhang, Y., Xue, K., Zhao, E.Y., Li, Y., Yao, L., Yang, X., Xie, X. 2013a. Determination of oleanolic acid and ursolic acid in chinese medicinal plants using HPLC with PAH polymeric C18. Pharmacogn Mag. 9(Suppl. 1): S19–24.

Zhang, Y., Zhou, A., Xie, X.M. 2013b. Determination of triterpenoic acids in fruits of *Ziziphus jujuba* using HPLCMS with polymeric ODS column. *Zhongguo zhongyao zazhi* 38(6): 848–851.

Zheng, R.X., Xu, X.D., Tian, Z., Yang, J.S. 2009. Chemical constituents from the fruits of *Hippophae rhamnoides*. Nat. Prod. Res. 23(15): 1451–1456.

Zhou, C., Zhao, D., Sheng, Y., Liang, G., Tao, J. 2012. Molecular cloning and expression of squalene synthase and 2,3-oxidosqualene cyclase genes in persimmon (*Diospyros kaki* L.) fruits. Mol. Biol. Rep. 39(2): 1125–1132.

Zhu, Z.B., Yu, M.M., Chen, Y.H., Guo, Q.S., Zhang, L.X., Shi, H.Z., Liu, L. 2014. Effects of ammonium to nitrate ratio on growth, nitrogen metabolism, photosynthetic efficiency and bioactive phytochemical production of *Prunella vulgaris*. J. Pharm. Biol. 52(12): 1518–1525.

8

Latin American Endemic (Wild) Medicinal Plants with High Value

Ethnobotanical, Pharmacological, and Chemical Importance

Amner Muñoz-Acevedo,[1,]* María C. González,[1] Ricardo D.D.G. de Alburquerque,[2] Ninoska Flores,[3] Alberto Giménez-Turba,[3] Feliza Ramón-Farias,[4] Leticia M. Cano-Asseleih[5] and Elsa Rengifo[6]

Introduction

The traditional healer is a trained individual from rural (or urban) zone (where financial or cultural barriers persist that limit the access to healthcare systems) belonging to a specific community (indigenous/peasant/afro-descendant), and mainly uses plants (herb/shrub/tree) for the preparation and application of therapeutic substances in one or more primary healthcare activities. The term traditional healer includes the traditional midwives and birth attendants, and herbalists, but does not exclude the spiritual or faith healers (shamans), as all of them contribute in the provision of healthcare and family planning service. Therefore, *"traditional healers are an invaluable human resource"* (Raden and Werner 1985, Hoff 1992).

In most developing countries, the healthcare systems place the common people (least favored) in a dilemma because these countries could continue to provide the same type of healthcare that cannot be extended to those most in need; however, the indigenous/native population has a historical tradition in the use of medicinal plants as a treatment for different ailments/disorders/diseases (e.g., Ayurvedic, Unani, Kampo Siddha, or Chinese traditional medicines), which makes them an affordable/accessible resource as a palliative alternative to primary healthcare (Raden and Werner 1985, Hoff 1992, Rauf and Jehan 2015). At this moment, the traditional medicine has an imperative role to play in society; for more than two decades, the World Health Organization has encouraged the Programme on

[1] Departamento de Química y Biología, Universidad del Norte, Barranquilla, Colombia.
[2] Laboratório de Tecnologia em Productos Naturais, Universidade Federal Fluminense, Niterói, Brasil.
[3] Instituto de Investigaciones Fármaco Bioquímicas, Universidad Mayor de San Andrés, La Paz, Bolivia.
[4] Instituto de Ecología, Xalapa, México.
[5] Centro de Investigaciones Tropicales, Universidad Veracruzana, Xalapa, México.
[6] Instituto de Investigaciones de la Amazonía Peruana, Iquitos, Perú.
* Corresponding author: amnerm@uninorte.edu.co

Traditional and Complementary Medicine and their relevance to public health in 179 WHO member states (developed and developing countries), monitoring health trends by supporting countries in generating evidence-based policies and strategic plans; Bolivia, Brazil, Colombia, Mexico, and Peru are a part of this programme (WHO 2019).

Consequently, under the coverage of complementary medicine, herbal medicine is one of the most important forms of traditional medicine, and at the present time, almost 80% of humanity has been related to or used medicinal plants in the form of herbal drugs according to WHO; and it has stated that about 25% of modern medicines are developed from traditionally used plant sources, and the research on traditional medicinal plants leads to the discovery of 75% of the herbal drugs (WHO 2002, Alamgir 2018). These natural drugs contain chemical substances (active ingredients) responsible for their therapeutic effects (Rauf and Jehan 2015, Ezekwesili-Ofili and Okaka 2019). Despite the benefits that these plants can offer, most of them (about 67–90%) are found/acquired in the wild, as reported by Edwards (2004) and Vines (2004).

The high value of these plants could be attributed not only to the medicinal uses they have, but some can be used with other purposes (e.g., spices/seasoning/in culinary), and for a specific population (indigenous/peasant/afro-descendant), they can provide it an economic means for survival. Also, for different systems of traditional medicine, the same plants could be valuable. Eventually, these plants are a potential source of "lead" molecules, which are pharmacologically active, and used for the pharmaceutical industry and medicine.

Each co-author is representative of at least one Latin American country, and was invited to contribute to this chapter, selecting two/three wild medicinal plant of great importance for the indigenous/afro-descendant/peasant inhabitants of their countries (Bolivia, Brazil, Colombia and Venezuela, Mexico and Central America, and Peru), conforming the ethnobotanical uses (particular/specific) and therapeutical actions/properties (specific/unique), along with those chemical constituents isolated/identified (distinctive/unique structures) which are possibly responsible for the pharmacological effects. This chapter is divided into the scientometric analysis of the topic, so that the plants selected by each country (region) mentioned above are described in the text, together with exploitation/sustainability, and opportunity for drug development (patent analysis).

Analysis of Text Mining for Latin American Wild Medicinal Plants

In order to assess the science activity correlated with the ethnobotanical/pharmacological/chemical importance of wild medicinal plants in Latin America, an analysis of text mining was carried out on articles indexed in the Scopus database (Elsevier 2019) using the search equation: (*title-abs-key* (*"wild medicinal plants*"*) and title-abs-key (*"ethnobotany and pharmacology*"*), and title-abs-key (*Latin America*)) and doctype (ar) and pubyear > 1980. As a result, 717 records were found in the 1980–2019 timeline (Figure 8.1), with 2017 and 2018 being the most productive years, with 107 records and 86 records, respectively. In addition, an exponential increase (R^2: 0.857), as the dynamic behavior of this topic, was observed in Figure 8.1. On the other hand, the trend of publication in the 2011–2018 timeline (excluding 2017 and 2019) was relatively similar between these years according to the dispersion of the data (67 ± 10 articles/year).

Referring to the tendency of publications on this topic by countries, in the top three were Brazil (156 articles), United States (97 articles), and India (65 articles), which were highlighted for having the highest number of records. However, if only Latin American countries are considered (excluding Brazil), according to the number of published articles, in descending order were Argentina, Mexico, Colombia, Chile, and Bolivia each with 40, 34, 13, 10, and 6 articles, respectively. Another important classification was performed from the text mining pursuant to different areas of application/knowledge (Figure 8.1, pie chart); thus the main areas were agricultural and biological sciences (~ 23%), followed by pharmacology, toxicology, and pharmaceutics (~ 20%), and finally, medicine (~ 14%).

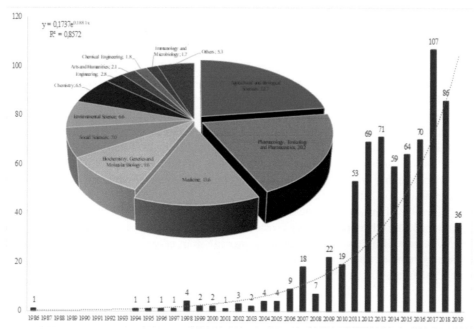

Figure 8.1: Distribution according to the number of articles published per year related to the topic of interest (1985–2019 timeline), along with areas of application/knowledge (pie chart). Calculations based on Scopus information (Elsevier 2019). Graph elaborated by authors.

The plants to be described below are native and very important in traditional medicine in at least one of the following countries: Bolivia, Brazil, Colombia/Venezuela, Mexico (or Central America), and Peru. However, it is important to mention that the same plants could be distributed in other Latin American countries.

Promising Plants from Bolivia

According to the expertise of the Bolivian co-authors (from Instituto de Investigaciones Fármaco Bioquímicas—UMSA) and the revised scientific literature, three Bolivian medicinal plants (*Galipea longiflora* K. Krause., *Hyptis brevipes* Poit., and *Tessaria integrifolia* Ruiz and Pav.—Figure 8.2) with high value for the indigenous populations were selected. Each of these plants will be briefly described, considering other scientific and common names, distribution, and botanical description.

***Galipea longiflora* K. Krause**—synonyms—*Angostura longiflora* (K. Krause) Kallunki. Common names—Yuruma huana epuna (Tacana language), evanta (Bolivia), matapira, mata-rabujo (Brazil). Tree or shrub (2–12 m), with pubescent young shoots, and alternate and trifoliate leaves; entire, and glabrous leaflets; narrow inflorescences. "Evanta" grows in the last spurs of the Andes Cordillera, between cities of La Paz and Beni, and it is distributed in Bolivia, Brazil, Peru, and Venezuela. The plant bark/leaves are used in traditional medicine by Tsimane, Mosetene, and Tacana communities (Bolivia) for the treatment of skin and digestive disorders, espundia (leishmaniasis), and as a fortifier for babies, children/adults (Fournet et al. 1989, Quenevo et al. 1999, Bourdy et al. 2000, Grandtner and Chevrette 2014).

***Hyptis brevipes* Poit.**—synonyms—*H. acuta* Benth., *H. lanceolata* Poir., *H. melanosticta* Griseb., *H. radiata* Kunth, *H. tweedii* Benth., *Lasiocorys poggeana* Baker, *Leucas globulifera* Hassk., *Mesosphaerum brevipes* (Poit.) Kuntze, *M. lanceolatum* (Poir.) Kuntze, *M. melanostictum* Kuntze, *M. tweedii* (Benth.) Kuntze, *Pycnanthemum subulatum* Blanco, *Thymus biserratus* Blanco. Common

(a) Taken by IIFB 2007.
Source: Bolivian co-authors.

(b) Taken by IIFB 2014.
Source: Bolivian co-authors.

(c) Taken by IIFB 2014.
Source: Bolivian co-authors.

Figure 8.2: Images of the three promising plants—(a) *G. longiflora*; (b) *H. brevipes*; (c) *T. Integrifolia*.

names—Id´ene eidhue, hierba bolita, coke, kikarugrag, lesser roundweed. Annual erect herb (up to 1 m), square stem, opposite and ovate/lanceolate leaves (both surfaces hairy), serrated margins, with glandular trichomes; racemose inflorescence, with axillar/terminal globose heads. *H. brevipes* is native to America, but it is widely distributed in another tropical region of the world (e.g., Asia); besides, the plant is listed in some traditional pharmacopeias to be used as a post-partum remedy, against asthma, malaria, cancer, intestinal parasites, and as mosquito repellents (Limachi et al. 2019, Gupta et al. 1996, Roosita et al. 2008, Quattrocchi 2012).

***Tessaria integrifolia* Ruiz and Pav.**—synonyms—*T. ambigua* DC.; *T. dentata* Ruiz and Pav.; *T. legitima* DC.; *T. mucronata* DC.; *Gynheteria dentata* (Ruiz and Pav) Spreng; *G. incana* Spreng; *G. salicifolia* Willd. ex Less.; *Conyza riparia* Kunt. Common names—Kkallakasa (quechua, Argentina) cahuara, mwirai, palo bonito, shita (Moseten and Tacana, Bolivia), pájaro bobo (Perú). Tree or shrub (3–5 m), leaves alternate, whole, bright green, with short petiole; terminal inflorescence with many orange-yellow flowers; green/dark-green globose fruits when ripe. This plant is used in the traditional medicine of Bolivia and Peru to treat digestive and respiratory disorders, infectious processes, against leishmaniasis, malaria, inflammation, stingray (Grandtner and Chevrette 2014, Muñoz et al. 2000, Arévalo-Lopéz et al. 2018, Vásquez-Ocmín et al. 2018).

Table 8.1 contains information about the pharmacological effects, the useful parts, and the ethnobotanical uses, together with the communities that exploit the plants. The reviewed literature revealed that 13 different quinoline alkaloids, substituted at the C-2 and C-4 positions, have been isolated and characterized (^{1}H-^{13}C-NMR, MS) as the active components in the extracts of different organs from Evanta (Fournet et al. 1989, 1993a). The novelty of four quinoline structures, named Chimanines A-D, found mainly in the leaves, together with the antiparasitic activities, is derived in a Franco-Bolivian patent on quinolines with leishmanicidal action (Fournet et al. 1996). The total alkaloid content of bark (TAB) was conformed mainly by low molecular weight quinolines. According to GC-MS analysis, 2-phenyl-quinoline (**1**; 67%) was the main alkaloid in all organs, followed by 4-methoxy-2-phenyl-quinoline (**2**; 11%), which together with 2-pentil-quinoline (**3**; 9%), 2-(3,4-methylendioxy-phenyl-ethyl)-quinoline (**4**), 4-methoxy-2-pentyl-quinoline (**5**), cusparine (**6**), and 2-propyl-quinoline (**7**) represented ca. 93 ± 3% of TAB composition. It is interesting to point out that *G. longiflora* contained compounds **1-3**, **5-7**, as main alkaloid constituents, with ratios changing with the age of the seedling, and that **1** has been detected at the dicotyledon stage of the plant (Quiroga-Selez et al. 2016).

The *in vitro* leishmanicidal (promastigote) and trypanocidal (epimastigote) activities were attributed to the alkaloid fractions, with reported IC_{90} values of 25 µg/mL [**Chimanines B** and **D**], 50 µg/mL [**7 and 2**], and 100 µg/mL [**1, 2** and others] (Fournet et al. 1993b, 1994a). *In vitro* studies indicated that TAB interfered with the activation of both mouse and human T cells, by reduction of interferon-gamma (INF-γ) production (Calla-Magarinos et al. 2009). Therefore, TAB had a direct leishmanicidal effect and due to the effect on INF-γ production, it might contribute

Table 8.1: Pharmacological effects, useful parts, ethnobotanical uses, and communities that exploit the selected Bolivian plants.

Plant (Family)	Communities	Ethnobotanical uses/plant part (preparation/application)		Pharmacological effects-in vitro/in vivo evaluation		References
Galipea longiflora (Rutaceae)	Chimane, Tacana, Mosetene, Tsimane	Diarrhea, diarrhea with blood, intestinal parasites, fortifier for babies, children and adults, espundia (leishmaniasis) Bark/powder (decoction/poultice)	Isolated compounds [Chimanine B and 2-Propylquinoline (7)]	Antiparasitic	*Trypanosoma cruzi*- IC$_{90}$ 25 µg/mL Balb/c mice infected with *T. cruzi* at 25 mg/kg	Fournet et al. 1994a, b, Nakayama et al. 2001, Martínez-Grueiro et al. 2005, Giménez et al. 2005, Gadisa et al. 2016
			Isolated compounds (2-substituted quinolines)	Nematocidal and trichomonacidal	*Caenorhabditis elegans*, *Heligmosomoides polygyrus*, *Trichomonas vaginalis*, 100 µM	
			Total alkaloids (bark)	Antiparasitic on *Leishmania* spp.	IC$_{50}$ (µg/mL) *L. amazonensis*: 12.2 ± 0.9, *L. braziliensis*: 12 ± 4, *L. aethiopica*: 7 ± 2, *L. lainsoni*: 10 ± 1	
Hyptis brevipes (Lamiaceae)	Tacana (Bolivia) Native regions of Panamá	Intestinal infections, post-partum remedy Leaves (decoction-heating/pounding)	MeOH extract (aerial part)	DNA intercalation	% DNA intercalation 20 ± 4	Gupta et al. 1996, Goun et al. 2003, Xu et al. 2013, Sakr et al. 2013, Sakr 2014, Suárez-Ortiz et al. 2017, Deng et al. 2009, Limachi et al. 2019
			CH$_2$Cl$_2$ extract/essential oil (aerial part)	Antimicrobial and antifungal	*S. aureus* 100 mg/mL, *Phytoptora parasitica* 100 mg/mL, *S. aureus*, *B. subtilis*, *P. aeruginosa*, *Fusarium graminearum*; MIC 3 µg/mL	
			CH$_2$Cl$_2$ extract 5% CH$_2$Cl$_2$ extract 4%	Insecticidal	Larval mortal. *Spodoptera littoralis*-100%, 77 ± 6%	
			EtOH extract (leaves)	Antiparasitic	IC$_{50}$ (µg/mL) *L. amazonensis*: 5.3 ± 0.4, *T. cruzi*: 15 ± 3	
			Essential oil	Free radical scavenging	DPPH· radical (µg/mL) SC$_{50}$: 2.0 ± 0.2	
			Isolated compound (**8-10**, brevipolides G,L,M)	Cytotoxic	IC$_{50}$ (µM) MCF-7: 4, HeLa: 3.3, KB: 2	
Tessaria integrifolia (Asteraceae)	Guarani, Moseten, Tacana (Bolivia) Ethnic groups of Amazon (Peru).	Stomach ache, diarrhea, liver injury, malaria, leishmaniasis, asthma, urinary tract infection, fever. Leaves, stem, aerial part (decoction/juices)	EtOH extract (leaves)	Antiparasitic	IC$_{50}$ (µg/mL) *L. donovani* axenic amastigoste: 0.51±0.09 intra-macrophage amastigote: 5±1 *L. braziliensis*: 32±15	Muñoz et al. 2000, Arévalo-Lopéz et al. 2018, Silva-Correa et al. 2018, Vásquez-Ocmín et al. 2018
					Inh. on *Leishmania* sp. *M. auratus*: 250 mg/kg, 95%	
			EtOH extract (leaves and roots)		Inh. on *P. falciparum* ClQR:10 µg/mL, 100% and 84%	
			EtOH extract (roots)		Inh. on *P. vincke*: 1000 mg/kg, 66%	

to the control of chronic inflammatory reaction that characterizes *Leishmania* infection pathology (Calla-Magariños 2012, Calla-Magariños et al. 2013). TAB and most of the pure components were tested on *in vivo* models, showing the absence of toxicity in mice; alkaloids **1, 3, 7**, and **Chimanine D**, provided promising results related to the reduction of the liver parasitic load (52–96%), depending on both the way of administration (oral or intralesional) and leishmania strains. The intralesional administration of raw alkaloid extract was the most efficient (96–99%) (Fournet et al. 1994b,1996). Other *in vivo* studies reported the gastro-protective and anti-nociceptive effects of TAB, and the action mechanism of **1** (Zanatta et al. 2009, Campos-Buzzi et al. 2010, Breviglieri et al. 2017). Interestingly, it has been found that quinolines substituted in position C-2, such as **1**, were not easily detected in plasma, suggesting these compounds can be sequestrated by blood components (Desrivot et al. 2007).

R_1

CHIMANINES A-D

	R_1	R_2		R_1	R_2		R_1	R_2
1	H	C_6H_5	**5**	H		**A**	OCH_3	$(CH_2)_2CH_3$
2	OCH_3	C_6H_5				**B**	H	$CH=CH-CH_3$
3	H	$(CH_2)_4CH_3$	**6**	OCH_3	$(CH_2)_4CH_3$	**C**	OCH_3	$CH=CH-CH_3$
4	H	$(CH_2)_2CH_3$	**7**	OCH_3		**D**	H	

Some studies of field validation with Evanta syrup on helminth parasites (*Ascaris* spp., *Strongyloides vermicularis*, *Trichuris trichura*, and *Uncinaria* spp.) in children from first to fifth primary grade (5–14 years old, with entero-parasite infestation diagnosed by copro-parasitology) at the Charcas II Community school (tropical zone Departamento de La Paz) showed promising results in their elimination/control. The efficacy was similar to those obtained with mebendazole and/or albendazole, but none of the treatments had an effect against *Hymenolepis nana*. While the efficacy on protozoan parasites was more complicated and less clear; despite that, Evanta was able to reduce populations of *Chilomastix mesnili, Endolimax nana*, and *Iodamoeba bütschlii*, but it had few effects against *Giardia lamblia, Entamoeba coli*, and *Blastocystis hominis* (IDH project 2010–2014).

Otherwise, some chemical studies on *H. brevipes* reported the brevipolides **A-O** as main constituents, whose framework is a 5,6-dihydro-α-pyrone. These kinds of compounds were isolated from plants from Indonesia (Deng et al. 2009), Mexico (Suárez-Ortiz et al. 2013, 2017), and Bolivia (Limachi et al. 2019); additionally, flavonoids, steroidal glycosides, and triterpenoids were isolated. The brevipolides G (**8**), L (**9**), M (**10**), and H (**11**) were active against MCF-7, HeLa, and KB cell lines.

Lastly, *Tessaria* species contain flavonoids, caffeoylquinic acid derivatives, and eudesmane-type sesquiterpenes as the main secondary metabolites (Peluso et al. 1995, Ono et al. 2000). The ethanol extract from *Tessaria integrifolia* leaves had leishmanicidal activity at a dose 250 mg/kg/day with significant inhibition of ulcerative lessons, when it was evaluated in an *in vivo* model on *Mesocricetus auratus* inoculated with *Leishmania* sp. strain isolated from Peruvian patients diagnosed with leishmaniasis (Silva-Correa et al. 2018). The main isolated compound was a eudesmane sesquiterpene (**12**), and the leishmanicidal activity could be attributed, at least in part, to the presence of this compound.

8/9 - E/Z

10 - Z; R: H
11 - E; R: CH₃

12

Plants with High Potential from Brazil

Three Brazilian medicinal plants (*Caesalpinia pyramidalis* Tul., *Eugenia punicifolia* Kunth., and *Jacaranda caroba* Vell.—Figure 8.3), with high value for the indigenous and/or afro-descendants communities were selected according to the scientific literature revised. Each plant will be briefly described below.

Caesalpinia pyramidalis **Tul.**—Synonyms—*C. gardneria* Benth., *Cenostigma pyramidale* E. Gagnon and G.P. Lewis. Common names—Catingueira, pau-de-rato, catinga-de-porco. The perennial plant developed as a tree (4–10 m) or a shrub (up to 2 m) depending on whether the environment is wet or semi-arid; leaves are biped, with leaflets and glandular dark brown/black hairs. Axillary-terminal yellow flowers, slightly hairy, presenting small glandular spots (back). Oblong-ellite fruit, minimal hairiness alva, and sparse yellow glandular trichomes. This species is native to Peru and Brazil, and particularly it is distributed in the caatinga biome (northeastern Brazilian states and part of Minas Gerais state). *C. pyramidalis* presents economic potential in these regions, due to its uses for reforestation, obtaining wood, and medicinal properties. Local populations and afro-descendant communities have used some parts (leaves, flowers, stem, and bark) to relieve certain disorders of respiratory and digestive systems. In addition, "catingueira" is the main vegetable species used by an afro-descendant community in the state of Bahia (Quilombolas from Raso da Catarina), and its use was cited by 100% of the respondents (Pereira et al. 2003, Maia 2004, Nishizawa et al. 2005, Sampaio et al. 2005, Agra et al. 2007a, Gomes and Bandeira 2012, Grandtner and Chevrette 2014, Braga 2015, Gagnon et al. 2016).

(a) Source: https://www.amigosjb.org.br/-wpcontent/uploads/2015/07/046Caesalpinia-pyramidalis.jpg.

(b) Taken by F. Souza; Source: http://faunaefloradorn.blogspot.com/2017/09/cereja-do-cerrado-eugenia-punicifolia.html.

(c) Source: http://www.abq.org.br/cb q/2017/trabalhos/7/1247924348.html.

Figure 8.3: Images of the three promising plants: (a) *C. pyramidalis*; (b) *E. punicifolia*; (c). *J. caroba*.

***Eugenia punicifolia* (Kunt) DC.**—Synonyms—*E. ovalifolia* Cambess., *E. pyramidalis* O. Berg., *Emurtia punicifiolia* (Kunt) Raf., *Myrtus punicifolia* Kunth., *Pseudomyrcianthes kochiana* (DC.) Kausel. Common names—pedra-ume-caá, pitanga-do-campo, cereja-do-cerrado, or murta-vermelha. Shrub (up to 1.5 m) with single leaves, bi-color, entire margin, primary vein and prominent secondary on both sides, abaxial/adaxial surfaces with hardening of the hair type; berry fruit when it is immature and has a persistent chalice with four lobes. "Pitanga-do-campo" has been found in Trinidad, Colombia, Venezuela, Peru, Bolivia, Guianas, Argentina, Paraguay, and Brazil. In the latter, the plant is widely/ mainly distributed in the northern region, and the natives and other communities have used it as medicine to alleviate some health problems, e.g., metabolic, respiratory, and liver disorders (Coelho de Souza et al. 2004, Cruz and Kaplan 2004, Grangeiro et al. 2006, Grandtner and Chevrette 2014, Morais et al. 2014, Sobral et al. 2015).

***Jacaranda caroba* (Vell.) DC.**—Synonyms—*J. clausseniana* Casar., *J. elegans* Mart. ex DC., *J. mendoncaei* Bureau and K. Schum., *J. oxyphylla* Cham., *Bignonia caroba* Vell. Common names— caroba, carobinha, caroba jacaranda, Brazilian caroba-tree, camboatá, and camboté. Shrub/tree (2–10 m) with long oblongolanceolate and coriaceous leaves, composite, bipinate with leaflets; purplish flowers, tubular and arranged in panicles; elliptical, dry and dehiscent fruits. "Caroba" is an endemic species of Brazil distributed in the Cerrado and Atlantic Forest biomes (states of Rio de Janeiro, São Paulo, Minas Gerais, Goiás, Bahia, and Federal District). In folk medicine, the plant is widely used as a treatment for syphilis, fungal infections, blood purification, skin wounds, and stomach ulcers, as tonic, diuretic, and astringent (Hiruma-Lima and Di Stasi 2002, Cesar et al. 2004, Botion et al. 2005, Fenner et al. 2006, Agra et al. 2007b, 2008, Gachet and Schühly 2009, Grandtner and Chevrette 2014, Lohmann 2015, Pereira 2018).

The ethnobotanical uses, the pharmacological effects (including *in vitro*/*in vivo* assessments), and the useful parts together with the communities that use the three plants are reported in Table 8.2. According to the continuous exploration of *C. pyramidalis*, based on bioprospecting and chemical analysis of extracts and derivatives, the species has contained phytosterols, phenolic acids, phenylpropanoids, lignans, flavonoids (e.g., caesalflavone (**13**, a new biflavonoid)), and tannins as the main secondary metabolites (Mendes et al. 2000, Bahia et al. 2005, Saraiva et al. 2012, Chaves et al. 2015, Oliveira et al. 2016). In addition, the anti-inflammatory (100 mg/kg v.o.), antinociceptive (30 mg/kg v.o.) (Santos et al. 2011, 2013, Santana et al. 2012,), antiulcerogenic (30 mg/kg; 29% protection) and gastroprotective properties of the ethanol extract of barks (Ribeiro et al. 2013), the antioxidant effect (IC_{50}: 43 ± 2 µg/mL) of the methanol extract from leaves by DPPH (1,1-diphenyl-2-picrylhydrazyl radical) method (Melo et al. 2010), antibacterial (Novais et al. 2003, Saraiva et al. 2012), and antifungal effectiveness from different extracts (Barbosa-Junior et al. 2015), anthelmintic activity of the aqueous leaf extract (on *H. contortus*: 2.5 mg/kg) (Borges-Santos et al. 2012), molluscicide and larvicidal actions of leaf and stem ethanol extracts on *B. glabrata* and *A. aegypti* (Luna et al. 2005), as well as the cytotoxic and antimutagenic activities of the aqueous extracts of barks (Silva et al. 2015) were established. Furthermore, lupeol (**14**), one of the main phytosterols of *C. pyramidalis* is related to the gastroprotective effect (30 mg/kg; 69% protection) (Lira et al. 2009), as well as anticancer, anti-inflammatory, and enzyme-inhibition activities, which in turn are also related to β-sitosterol (**15**) found in this plant (Gallo and Sarachine 2009, Saeidnia et al. 2014).

For the case of *E. punicifolia*, the volatile products of its secondary metabolism (essential oils, yields 0.2–0.8%) are one of the particular characteristics; that is, there are quantitative and chemical (regarding the major component) variabilities of the essential oils from plant leaves, depending on the collection site. Among the main constituents that have been identified are β-caryophyllene (10–34%, Amazonas, Pará), linalool (**16**) (44–61%, Pernambuco), and β-elemene (22%, restinga biome from Rio de Janeiro) (Maia et al. 1997, Oliveira et al. 2005, Pereira et al. 2010, Ramos et al. 2010). In addition, other non-volatile compounds (e.g., flavonoids), such as myricetin-3-O-rhamnoside (**17**), quercetin-3-O-galactoside, quercetin-3-O-xyloside, quercetin-3-O-rhamnoside, kaempferol-3-O-rhamnoside, along with phytol and gallic acid (**18**) were found (Sales et al. 2014). Chiefly, one of these compounds

Table 8.2: Ethnobotanical uses, pharmacological effects, useful parts, and communities that benefit from the selected Brazilian plants.

Plant (Family)	Communities	Ethnobotanical uses/plant part (preparation/application)	Pharmacological effects- *In vitro/in vivo* evaluation	References
Caesalpinia pyramidalis (Fabaceae)	Natives from Northeast Brazilian and Quilombolas	Treatment for diarrhea, dysentery and catarrhal infections, as diuretic, aphrodisiac. Stem, barks, leaves, flowers (decoction/maceration)	Antinociceptive and anti-inflammatory from EtOH extract of bark. Antiulcerogenic (EtOH extract of barks: 30 mg/kg). Free radical scavenging activity (MeOH extract of leaves, IC$_{50}$: 43 ± 2 µg/mL). Antibacterial (EtOAc extract from bark/leaves on *S. aureus*: φ inhibition: 10 mm). Antifungal (Leaves infusion against *Cryptococcus neoformans*). Anthelmintic (H$_2$O extract of leaves against *Haemonchus contortus*). Molluscicide (EtOH extract of stem: 100 ppm, 14% mortality of *Biomphalaria glabrata* eggs). Larvicidal (EtOH extract of stem: 500 ppm, 20% mortality of larvae from *Aedes aegypti*). Cytotoxic/Antimutagenic (H$_2$O extract of barks: 1 g/500 mL with IM: 8%)	Novais et al. 2003, Luna et al. 2005, Melo et al. 2010, Borges-Santos et al. 2012, Ribeiro et al. 2013, Barbosa Junior et al. 2015, Silva et al. 2015
Eugenia punicifolia (Myrtaceae)	Natives from different regions of Brazil	Treatment for diabetes, fever, cold and liver disorders. Fresh leaves, roots (infusion/decoction)	Cholinergic activity on nicotinic receptors (H$_2$O extract of leaves). Enzyme inhibition (α-amylase, α-glucosylase and xanthine-oxidase) and free radical scavenging activity (EtOH extract of leaves). Control of diabetes type 2 (Dried leaf powder). Antinociceptive (20–35% inhibition), anti-inflammatory (50–90%), and gastroprotective activities (88–99%) (EtOH extract of leaves: 125–500 mg/kg)	Grangeiro et al. 2006, Basting et al. 2014, Galeno et al. 2014, Sales et al. 2014
Jacaranda caroba (Bignoniaceae)	From Vale do Ribeira (São Paulo, Brazil) and natives from Mata Atlantica region.	Against syphilis, fungal skin infections, skin wounds, stomach ulcers, for blood purification, as tonic, diuretic, astringent. Leaves and barks (bath infusion, internal infusion, macerate spirit and decoction)	Enzyme inhibition (MAO-A: IC$_{50}$: 17 ± 3 µg/mL and 23 ± 2 µg/mL, AChE: EC$_{25}$: 6705 ± 48 µg/mL and 1006 ± 69 µg/mL, and BChE: EC$_{25}$: 1111 ± 103 µg/mL and 626 ± 55 µg/mL, respectively). (H$_2$O and H$_2$O/MeOH extracts of leaves, respectively). Antioxidant and anti-inflammatory (H$_2$O and H$_2$O/MeOH extracts of leaves). Anti-*Leishmania* (EtOH extract of leaves on *L. amazonensis*). Antimicrobial (EtOH extract of leaves on *Helicobacter pylori*). Anti-ulcerative (H$_2$O/EtOH extract and CH$_2$Cl$_2$ fraction from leaves: 70–100% reduction)	Bacchi et al. 1999, Ferreres et al. 2013, Ribeiro et al. 2014, Hernandes 2015

(**17**) showed a significant antioxidant capacity (IC_{50} 220 µg/mL by DPPH· method), inhibition (59%, 100 µg/mL) on xanthineoxidase, and insulin-like and insulin-sensitizing effects (from 0.08 µM) on adipocytes (Hayder et al. 2008, Manaharan et al. 2012).

As mentioned in the previous table, aqueous extract from *E. punicifolia* leaves showed cholinergic activity, reversing the nicotinic antagonism induced by gallamine or pancuronium in rats, more efficiently than neostigmine (5% of extract caused 89–94% inhibition) (Grangeiro et al. 2006). Furthermore, ethanol leaf extract showed different enzymatic inhibitions, as well as a remarkable antioxidant effect [IC_{50} (µg/mL): 10 ± 1, 28.8 ± 0.5, and 38 ± 3, by ABTS^{+}·, DPPH·, and O_2^{-}· scavenger methods, respectively], which may suggest the contribution of this plant to metabolic interferences, and among them, in the regulation of glycaemia, correlated to the popular medicinal use (Galeno et al. 2014). In fact, Sales et al. (2014) demonstrated the usefulness of *E. punicifolia* leaf powder (dry) as a regulator of type 2 diabetes (lower levels of glycosylated hemoglobin, basal insulin, thyroid stimulating hormone, C-reactive protein, and blood pressure) in an uncontrolled clinical trial during three months. Moreover, the ethanol extract showed antinociceptive, anti-inflammatory, and gastroprotective activities in rodents (Basting et al. 2014). Also, the anti-inflammatory power was observed by reducing the effects on muscle injury in mice treated with isolated pentacyclic triterpenes from plant leaves, decreasing levels of metalloproteases, and tumor necrosis, and NFκB transcription factors (Leite et al. 2014). Another interesting study related to metabolic parameters revealed that the leaf methanol extracts reduced liquid intake and glucose and urea levels in urine without altering the markers of liver function (Brunetti et al. 2006).

Regarding the chemical composition of *J. caroba*, the main components include derivatives of caffeic acids, flavonoids as quercetin, kaempferol, and isorhamnetin (**19**) (Ferreres et al. 2013), phytoquinoids (Martins et al. 2008), ursolic acid, oleanolic acid (**20**), 3-epichorosolic acid (**21**), and β-sitosterol (Valadares 2009, Pereira 2018). Based on the scientific evidence of the pharmacological effects of the aqueous/alcohol (methanol or ethanol) extracts from *J. caroba*, the inhibitory capacities (low to moderate) on enzymes could be mentioned, e.g., MAO-A, AChE, and BChE, as well as noticeable antioxidant and anti-inflammatory activities, specifically, inhibiting the action of superoxide-anion and releasing nitric oxide (ethanol extract—EC_{50} (µg/mL): 99 ± 4 and 113 ± 18; aqueous extract—EC_{50} (µg/mL): 31.5 ± 0.3 µg/mL and 165 ± 23, respectively) (Ferreres et al. 2013). The ethanol extract of leaves also showed important activities against *L. amazonensis* (IC_{50} 13 µg/mL) (Ribeiro et al. 2014) and against *H. pylori* (MIC 125 µg/mL) (Hernandes 2015). Moreover, in 1999, Bacchi et al. reported the *in vivo* antiulcerogenic activity of both the hydroalcohol extract and dichloromethane fraction from leaves, which were active on the reduction of ulcers induced by ethanol and HCl (70% and 100% reduction, respectively); thus, the last two science reports verified

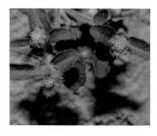

| 19 | 20 | 21 |

the effectiveness of the plant as a treatment for stomach disorders according to ethnomedicinal use. Probably, oleanolic acid (**20**) (main triterpene from *J. caroba*) could be responsible for the gastroprotective effect of the extract, as reported by Astudillo et al. (2002); **20** inhibited the appearance of gastric lesions induced by ethanol, aspirin, and pylorus ligature (50–200 mg/kg). Furthermore, **20** also had anti-inflammatory, antitumoral, and antibacterial properties, which could be seen as factors that promote gastric protection (Pollier and Goossens 2012).

High-potential Plants from Colombia and Venezuela

Based on the revised literature and the bibliometric analysis, three wild medicinal plants [*Acanthospermum australe* Kunth., *Jacaranda copaia* Aubl., and *Symphonia globulifera* L.—Figure 8.4] with high value for the indigenous, peasants, and/or afro-descendants communities from Colombia and Venezuela were selected. Each of the species is briefly described as follows.

Acanthospermum australe (**Loefl.**) **Kuntze.**—synonyms—*A. brasilum* Schrank, *A. hirsutum* DC., *A. xanthioides* (Kunth) DC., *Centrospermum xanthioides* Kunth, *Melampodium australe* Loefl., *Orcya adhaerens* Vells. Common names—cáncer de loma/yerba del cáncer, carrapicho/amor de negro/erva mijona/espinho de carneiro, tuyá/tapé, tapekue/tapequé/Paraguay-bur/Paraguay star-bur, sheepbur. Annual prostrate flowering shrub (10–60 cm), pilosa with pubescent stem with hairs and both sides of the leaves containing glands; the fruits in a star form are ellipsoid-fusiform-akenes, and they are covered with numerous uncinate hooks/spines. This species is native to South America from Colombia to Paraguay (at altitudes ranged 950–1800 m), except for Ecuador, Chile, and Argentina. The plant is used in traditional medicine by different communities (indigenous and/or peasants) as antitumor, ulcer, and antimalarial (García-Barriga 1992, Duke et al. 2009, Roth and Lindorf 2002, Quattrocchi 2012).

Jacaranda copaia **Aubl.**—synonyms—*J. amazonensis* Vattimo, *J. procera* (Willd.) R.Br., *J. paraensis* (Huber) Vattimo, *J. spectabilis* Mart. ex A. DC., *J. superba* Pittier, *Bignonia copaia* Aubl., *B. procera* Willd., *Kordelestris syphilitica* Arruda. Common names—aipay/cimarua(o), amchiponga/aspingo, arabisco, copaia, curnite(a)/chingalé, chapereke, guachipilin, huamans(z)amana, gualanchy, gualanday(i), jacaranda(á), palo de buba, pinguasí, vai-cuima-yek, wei-oima-yek. Perennial tree

(**a**) Taken by Germaine Parada; Source: http://www.tropicos.org/Image/100188419.

(**b**) Taken by A. Sanchún; Source: https://www.especiesrestauracion-uicn.org/data_especie_img.php?sp_name=Jacaranda%20copaia.

(**c**) Taken by Steven Paton; Source: https://biogeodb.stri.si.edu/bioinformatics-dfm/metas/view/10949.

Figure 8.4: Images of the three promising plants—(a) *A. australe*; (b) *J. copaia*; (c) *S. globulifera*.

(up to 45 m) with exestipulated, opposite/bipinnate leaves; paniculated inflorescences with bluish-purplish colors and dehiscent dried elliptical fruits. The plant is distributed in rainy forests (at altitudes 0–1000 m) from Central America to Bolivia/Brazil, including Suriname. In traditional medicine of the indigenous and/or peasant communities, different parts of the plants have been used to treat skin conditions (healing), and respiratory and digestive disorders (García-Barriga 1992, Duke et al. 2009, Quattrocchi 2012, Grandtner and Chevrette 2014, UEIA 2014, IUCN 2019).

***Symphonia globulifera* L.**—synonyms—*S. coccinea* (Aubl.) Oken, *S. gabonensis* (Vesque) Pierre, *S. microphylla* R.E. Schult., *S. utilissima* R.E. Schult., *Chrysopia microphylla* Hils. and Bojer ex Cambess., *Moronobea coccinea* Aubl., *M. globulifera* (L.f.) Schltdl., *M. grandiflora* Choisy. Common names—azufre, azufre caspi, barillo, bogum, brea amarilla, chuchuy, leche amarilla(o), machare, machasi, madroño, manie/miraña, pa(e)ramá(a)n, puenka, tomé, supute, yapí. Perennial tree (up to 30 m), with elliptical, opposite and simple leaves, no stipules; cymoso-subumbelated inflorescences, with obsolete peduncles and bracteoles; red-bright or pink flowers, chalices with quincuncial sepals; subglobose/ellipsoidal fruits, crowned by the style and its ramifications, yellowish to brown-pale when ripe; this species produces a thick yellow latex/resin that indigenous/peasant communities traditionally used it as medicine. Furthermore, other parts of the plant are also used for medicinal purposes. In Colombia, it is used to treat skin conditions (healing), and digestive disorder (Correa and Bernal 1993, García-Barriga 1992, Quattrocchi 2012, Grandtner and Chevrette 2014, UEIA 2014, IUCN 2019).

Table 8.3 registers the communities that use the three plants along with the useful parts, the ethnobotanical information, and the pharmacological effects (including *in vitro/in vivo* assessments). Based on the plants mentioned in Table 8.3, the extracts (ethanol/aqueous) from *A. australe* aerial parts presented *in vitro* antimicrobial, antiparasitic, antitumoral, and antiviral capacities together with enzymatic inhibition (on aldoreductase), with values from moderate to high (IC_{50}: 1×10^{-7}–5×10^{-6} M; EC_{50}: 6–70 µg/mL), with the antitumoral and antiviral properties standing out. Some molecules have been isolated which would be responsible for several bioactivities; thus, from *A. australe* (whole plant), an interesting sesquiterpene lactone of germacranolide type (acanthrostal **22**) with antineoplastic activity against L1210 cell (ED_{50} value: 5×10^{-6} M) was insulated, as reported by Matsunaga et al. (1996). The anticancer activities of this plant could be attributed to molecules of melampolide series (germacranolides); some of which were also isolated, e.g., acanthoaustralide derivatives (**23/24**) and acanthospermolide derivatives (**25–30**) (Bohlmann et al. 1981, Glasby 2005, Sánchez et al. 2009). Besides, Shimizu et al. (1987) isolated five flavonoids (type flavones) of *A. australe* aerial parts from Argentina; four of them were quercetin, rutin, hyperin, and trifolin. The most active flavonoid was 5,7,4´-trihydroxy-3,6-dimethoxyflavone (**31**), with IC_{50} value of 0.1 µM on rat lens aldose reductase and K_i value of 2×10^{-7} M (non-competitive inhibitor).

Meanwhile, "gualanday" had a distinctive chemical constituent called jacaranone (**32**, benzoquinone alkyl ester) insulated from methanol extract of trunk bark, along with ursolic acid (**33**). Both compounds were evaluated by *in vitro* Leishmaniasis test, and were effective (ED_{50}: 0.02 mM)

	R₁	R₂
23	H	H
24	CH₃CO	CH₃CO

	R₁	R₂
25	CH₃	COOH
26	CH₂OH	CHO

	R₁	R₂	R₃
27	H	COCH(CH₃)CH₂CH₃	H
28	COCH(CH₃)₂	H	H
29	COCH(CH₃)CH₂CH₃	H	H
30	COCH(CH₃)CH₂CH₃	COCH(CH₃)CH₂CH₃	H

Table 8.3: Useful parts, etnobotanical uses, pharmacological effects, and communities that use the selected Colombian/Venezuelan plants.

Plant (Family)	Communities	Ethnobotanical uses/plant part (preparation/application)	Pharmacological effects- *In vitro/in vivo* evaluation		References
Acanthospermum australe (Asteraceae)	Native (Venezuela; Brazil) or peasant (Colombia)	Against cancers/tumor, jaundice, vomiting, seizures, epilepsy, constipation, blenorrhea, malaria, snake bite, hepatobiliary disorders, microbial/viral infections, as tonic, diaphoretic, vermifuge, eupeptic, antianemic, antidiarrheal, antigonorrheic, and febrifuge. Whole plant [leaf, flower, or branch] (decoction/infusion-/maceration/poultice)	Essential oil (leaves)	Antimicrobial	*C. glabrata* - MIC 100 µg/mL	Shimizu et al. 1987, Carvalho et al. 1991, Matsunaga et al. 1996, Mirandola et al. 2002, Martins et al. 2011; Carvalho et al. 2014
			BuOH/EtOH:H_2O/H_2O/$CHCl_3$ extracts (aerial parts) Five flavonoids isolated	Aldosareductase inhibition (IC_{50})	Extracts: 2–4 µg Flavonoids: 0.1–9 µM K_i; 2×10^{-7} M (Flav. 1)	
			H_2O/EtOH, H_2O, BuOH, CH_3Cl extracts (roots and aerial parts)	Antitumoral	Ehrlich ascites tumor	
			Acanthrostal (**22**) isolated (hexane:EtOAc 2:1 fraction)		L1210 cell line IC_{50} 5×10^{-6} M	
			H_2O/EtOH extract (aerial parts)	Antiviral	Herpes and poliovirus EC_{50} 6–70 µg/mL	
				Antiparasitic	*P. falciparum* in rats	
Jacaranda copaia (Bignoniaceae)	Andoque, Wao-Shuar indigenous (amazon region); Vaupés river natives (Colombia); Los Jívaros (Perú); Alter do Chão (Pará, Brasil); Waimiri Atroari, Chácobo, El Tiriyó, Wayápi (French Guaina)	Against skin infections, colds, pneumonia, diarrhea, leishmaniasis, cancer, rheumatism, syphilis; as preventive, repellent, purgative. Leaves, bark, tubers, sap (infusion/decoction/fluid extract/dry powder/syrup/poultice/hot bath/fumes).	EtOH extract (young leaves)	Anticancer GI_{50}/LC_{50} (µg/mL)	36/223- PANC-1 74/228- HT-29 GI_{50}– 145 - PC-3; 15- MCF7; 32-4T1; 27-RAW-267	Sauvain et al. 1993, Villasmil et al. 2006, Gachet and Schühly 2009, Valadeau et al. 2009; Taylor et al. 2013; Roumy et al. 2015
			Jacaranone (**32**) and ursolic acid (**33**) isolated (CH_2Cl_2 fraction, trunk bark)	Antiparasitic	ED_{50} (mM) *L. amazonensis* **32**: 0.02 (promastigote) **33**: 0.02 (amastigote)	
			EtOH extract (leaves)		*L. amazonensis* - 16 ± 4 µg/mL *P. falciparum* ClQR- 8 ± 2 µg/mL	
			MeOH extract (leaves/branch bark)	Antimicrobial MIC (mg/mL)	5/0.6- *S. epidermidis* 0.2/1.2- *S. lugdinensis*	

Plant (family) / Source	Traditional use / Part (preparation)	Extract / Compounds	Activity	Results	References
Symphonia globulifera (Clusiaceae)	Emberá indigenous reserve; Orinoco river indigenous, Tukuna (Amazonas), Paracou (French Guiana)	MeOH extract (bark)	Antiviral	HSV (25 µg/mL)	Gustafson et al. 1992, Lopez et al. 2001, Ngouela et al. 2005, Ndjakou Lenta et al. 2007, Mkounga et al. 2009, Marti et al. 2010, Cottet et al. 2015, Fromentin et al. 2015, Sarquis et al. 2019
	Heart, lung, and stomach problems; anemia, malaria, as pain reliever		Cytotoxicity	IC_{50} (µg/mL)- L-6 cell line: 52 ± 6	
	Resin, leaves, and bark (decoction/poultice)		Antiparasitic	IC_{50} (µg/mL)- *P. falciparum* 4.1 ± 0.5- *P. falciparum* 11.5 ± 0.5- *T. brucei rhodesiense* 2.1 ± 0.8- *L. donovani*	
		Acylphloroglucinols/ benzophenones/xanthones/ biflavonoids (**34-59**) isolated from EtOAc/ MeOH extract (leaves/root bark/seed)		IC_{50} (µM)- *P. falciparum*- 1-4 *P. falciparum*- 2-7 *L. donovani*- 0.2-1 Guttiferone A (**34**) Xanthone V$_1$ (**50**)	
			Free radical scavenging	DPPH radical inhibition: **34-89%**	
			Antimicrobial MIC (µg/mL)	*S. aureus/B. subtili* **46, 48, 51, 52, 54-56** 0.6-14/1-12	
				E. feacalis/K. pneumoniae/E. coli **52**: 8/15/21 **55/56**: 0/0/4-8	
				300- seed extract on all strain	
			Anti-HIV	EC_{50} 1–10 µg/mL CH$_2$Cl$_2$/MeOH extract Guttiferones A-D	

31   32   33

on the promastigote and amastigote stages of *L. amazonensis*, correspondingly. However, in the *in vivo* test on *L. amazonensis* in mice, they had a weak anti-leishmanicidal activity. According to Table 8.3, all bioactivities registered of leaf/bark extracts (alcohol) from Jacaranda were promising-antitumoral [against MCF7 and 4T1 (mouse breast tumor)], antiparasitic (on *L. amazonensis* and *P. falciparum*), and antibacterial (against *Staphylococcus* spp.).

As a final point, the *S. globulifera* tree is a valuable species due to the remarkable chemistry it contains related to the acylphoroglucinols/xanthones/benzophenones type compounds (**34–59**), which would be responsible for promising biological actions, such as antimicrobial, antiviral, antiparasitic, cytotoxic, and anti-HIV.

	R_1	R_2
34	CH_2-CH=C(CH$_3$)$_2$	CH_2-CH=C(CH$_3$)$_2$
35	CH_2-CH(CH$_2$=CHCH$_3$)-CH$_2$-CH=C(CH$_3$)$_2$	CH_3

	R_1	R_2	R_3
36	H	CH_2-CH=C(CH$_3$)$_2$	CH_2-CH=C(CH$_3$)$_2$
37	OH	CH_2-CH=C(CH$_3$)$_2$	CH=CH-C(CH$_3$)=CH$_2$
38	OH	CH_2-CH=C(CH$_3$)$_2$	CH_2-CH=C(CH$_3$)-(CH$_2$)$_2$-CH=C(CH$_3$)$_2$
39	OH	CH_2-CH=C(CH$_3$)$_2$	CH=CH-C(OH)-(CH$_3$)$_2$
40	OH	CH=CH-C(OH)-(CH$_3$)$_2$	CH_2-CH=C(CH$_3$)$_2$
41/42	OH	CH_2CH(OH)-C(OH)-(CH$_3$)$_2$	CH_2-CH=C(CH$_3$)$_2$
43	OH	CH_2-CH=C(CH$_3$)$_2$	CH_2-CH=C(CH$_3$)$_2$

The main isolated secondary metabolites were polycyclic polyprenylated acylphloroglucinols (15 compounds from roots (barks)), followed by polyhydroxylated polyprenylated xanthones/benzophenones (21 compounds from roots (barks)/seed/leaves/heartwood), and biflavonoids (3 compounds from leaves). The higher demonstrated biological potential was against parasites (e.g., leishmaniasis and malaria) and microorganisms (*Escherichia coli*, *Staphylococcus aureus*, *Bacillus subtilis*, *Enterococcus feacalis*, and *Klebsiella pneumoniae*).

For instance, compounds **34,47,49,52** were highly active on *P. falciparum* W2 with IC$_{50}$ values of 1.3–3.9 µg/mL; while, symphonones A-I (**37–39**, **41–45**, **58**, **59**) along with molecules **35**, **36**, **40** were effective on *P. falciparum* FcB1, with IC$_{50}$ values of 2.1–10.1 µg/mL. Furthermore, on *L. donovani* strain, guttiferone A (**34**) and xanthone V1 (**50**) exhibited the highest effectiveness, with IC$_{50}$ values of 0.2 µg/mL and 1.4 µg/mL, respectively. In addition, globulixanthones A (**53**) and B (**57**) presented a good cytotoxic effect on KB cells with IC$_{50}$ values of 2.2 µg/mL and 1.8 µg/mL, individually.

A surprising biological potential was determined by Gustafson et al. (1992), when they isolated guttiferones A-D (e.g., **34**) from *S. globulifera* and established its anti-HIV activity. The four molecules

	R₁	R₂	R₃	R₄	R₅
46	OH	OH	H	H	H
47	H	OCH₃	OCH₃	OH	CH₂-CH=C(CH₃)₂

showed a good efficacy (similar level of activities) to inhibit the cytophatic effects against *in vitro* HIV infection in human lymphoblastoid CEM-SS cells, with EC$_{50}$ values of 1–10 μg/mL, which were lower than the cytotoxic concentration (> 50 μg/mL).

	R₁	R₂
49	OCH₃	CH₃
50	H	H

	R₁	R₂	R₃	R₄	R₅	R₆	R₇
51	H	CH₂-CH=C(CH₃)₂	OH	OCH₃	H	H	H
52	H	OCH₃	OCH₃	OH	CH₂-CH=C(CH₃)₂	OH	CH₂-CH=C(CH₃)₂
53	H	CH=CH-C(CH₃)=CH₂	OH	OCH₃	H	H	H
54	CH₂-CH=C(CH₃)₂	OH		OCH₃	OH	H	H

	R
55	H
56	OH

For closure, the tested microorganisms were highly susceptible to the globulixanthones C-F (**46, 48, 51, 54**) with MIC values of 4.5–14.0 μg/mL on *S. aureus*, and 1.2–12.5 μg/mL on *B. subtili*; as well as to the biflavonoids **55,56** with MIC values of 8.5 μg/mL on *S. aureus*, and 4.5–7.5 μg/mL on *E. coli*. Finally, globuliferin (**52**) was effective against *S. aureus*, *E. feacalis*, *K. pneumoniae*, and *E. coli* with MIC values of 0.6 μg/mL, 8.2 μg/mL, 15.2 μg/mL, and 21.2 μg/mL, respectively.

Promising Plants from Mexico and Central America

Based on reviewed literature and the bibliometric analysis, three wild medicinal plants (*Byrsonima bucidaefolia* Standl., *Croton draco* Schltdl. and Cham., and *Smilax aristolochiifolia* Mill.— Figure 8.5), with high value for the indigenous and/or peasants communities from Mexico and Central America were selected. Below, each plant will be described according to synonyms, common names, distribution, and botanical description.

***Byrsonima bucidaefolia* Standl.**—synonyms—none. Common names—sakpah, Saak´pah, sak bo' ob, nan che', craboo, nance agrio, nance blanco, nance silvestre, sak paj, zapote blanco, matasano, matasan, ajachel (K'aqchikel), ahache (Pocomchí). Small tree (5 m) with branches covered with soft and bushy fluff, rounded/broad leaves, reddish-yellowish flowers, and succulent globose yellow fruits. This species is native to Yucatan Peninsula (Mexico, Belize, and Guatemala), and it is distributed at altitudes of 5–400 m, mainly in savanna biome. The indigenous and/or peasants communities use the plant as food and against different skin and respiratory disorders and infections (Maya, Quintana Roo, and Yucatán) (Standley and Steyermark 1946, Argueta-Villamar et al. 1994, Arellano-Rodríguez et al. 2003, Castillo-Ávila et al. 2009, Polanco-Hernández et al. 2012, Tropicos 2019).

***Croton draco* Schltdl. and Cham.**—synonyms—*C. callistanthus* Croizat, *C. gossypiifolius* Vahl. *Croton panamensis* (Klotzsch) Müll. Arg., *Croton steyermarkianus* Croizat, *C. tacanensis* Lundell, *Croton triumfettoides* Croizat, *Cyclostigma denticulatum* Klotzsch, *C. draco* (Schltdl.) Klotzsch, *C. panamense* Klotzsch, *Oxydectes draco* (Schltdl.) Kuntze, *O. panamensis* (Klotzsch) Kuntze. Common names—targuá(a), sangre de draco/drago/perro, sangragrado, llora sangre, palo muela/de sangre, balsayu, cuate, sangreg(r)ado, escuahuitl (náhuatl), xix(z)t(l)e (huasteco), chucum (lacandón), peesnum-quina (qui-ui), pocsnum-quina (totonaca), negpinkuy (popoluca,). Shrub or tree (up to 18 m) with acute apex, irregularly dented and alternate/simple leaves, with stipules (the base of the leaf presents orange glands); racemose inflorescences, and white-creamy/yellow-green flowers; trilocular and subglobose fruits; the trichomes cover the entire plant. Sangregrado is native to Mexico and it is distributed in Central America (e.g., Guatemala, Nicaragua, Costa Rica to Panama) at altitudes ranged from 100–1700 m. The plant is used in ethnomedicine by indigenous and/or peasants communities, and once it is injured/cut, it emits a red exudate/latex that has medicinal properties against disorders of the digestive, skin, circulatory systems (Pennington and Sarukhán 1968, Castro et al. 1999, Murillo et al. 2001, González 2006, García and García 2008).

***Smilax aristolochiifolia* Mill.**—synonyms—*S. kerberi* F.W. Apt., *S. medica* Schltdl. and Cham., *S. milleri* Steud., *S. ornata* Lem. Common names—bigote de cozol/camalla/cosole, cuaumecapatli, mecapatli, cuculmeca, t'oknal ts'aah, es'co'ka, coco(l)meca, z(s)arz(s)aparrilla, gray/Mexican/ Veracruz sarsaparilla. Perennial climbing plant (up to 5 m) with prickly stems and tendrils, thin branches; leathery, extended, and alternate/ovate leaves; small and green dioecious flowers and red small globose fruits. The species is native to Mesoamerica region (Mexico, Belize, Guatemala, etc.) at altitudes of 100–800 m. The most used part of the plant is the root which is exploited as medicine and food by indigenous/peasant communities. The known therapeutic uses are against renal, digestive, reproductive, skin, respiratory and joint disorders, and cancer (Argueta-Villamar et al. 1994, Martínez et al. 2007, Duke et al. 2009, Ferrufino-Acosta and Greuter 2010, Tropicos 2019).

The pharmacological effects and ethnobotanical uses of the three selected plants along with useful parts are contained in Table 8.4. Based on the revised scientific literature, a few records on the active components of *B. bucidaefolia* were found; although, it is known that species of *Byrsonima* contain

(a) Taken by José L. Tapia, 2019; Source: Mexican co-authors.

(b) Taken by F. Ramón-Farías, 2019; Source: Mexican co-authors.

(c) Taken by Leticia M. Cano, 2018; Source: Mexican co-authors .

Figure 8.5: Images of the three promising plants—(a) *B. bucidaefolia*; (b) *C. draco*; (c) *S. aristolochiifolia*.

Tabla 8.4: Useful parts, pharmacological effects, ethnobotanical uses, and communities using the selected Mexican plants.

Plant (Family)	Communities	Ethnobotanical uses/plant part (preparation/ application)	Pharmacological effects- *in vitro/in vivo* evaluation		References
Byrsonima bucidaefolia (Malpighiaceae)	Mayas (Mexico, Belize, and Guatemala)	Fever, skin infections, dysentery Bark, leaves (infusion/ decoction)	MeOH extract (leaves) and isolated compounds	Free radical scavenging	DPPH· radical (μg/mL) EC_{50}: 0.9; **60,61**	Peraza-Sánchez et al. 2007, Castillo-Ávila et al. 2009, Polanco-Hernández et al. 2012
			MeOH extract (bark and leaves)	Antiparasitic	IC_{50} (μg/mL) extracts *L. mexicana* (promastigote) Bark: 36; Leaves: 60	
			EtOH extract (leaves, stem and bark)	Cytotoxic	*T. cruzi* (amastigote)- Leaf extract 100 Vero cell: 211 ± 4	
Croton draco (Euphorbiaceae)	Native of Mexico (Nahuatl, Totonacos, Popolucas, Zoques, Huastecos, Tzotzil, Lacandones, Tzeltal), Guatemala	Antimicrobial, wound healing, antitumoral, antihemorrhagic, anti-inflammatory, to reduce pain, against flu, cough, diarrhea, tuberculosis Latex, bark (maceration)	EtOAc/(Et)$_2$O/BuOH extracts (latex)	Hemolytic	IC_{50} (mg/mL) Latex/EtOAc: 0.43, (Et)$_2$O: 0.49, BuOH: 0.59 0.9 mM-%Inhibition: **62**, **63**, and cyclopeptides P$_1$/ P$_2$: 83,91-78/63	Castro et al. 1999, Tsacheva et al. 2004, Setzer et al. 2007, Cruz 2015, Alamillo 2017
			EtOH extract (roots)	Antiproliferative	0.001-1 μg/mL Inhibition > 50% on MDCK, Hep-2, MCF7, A549	
			EtOH:H$_2$O extract, hexane/EtOAc/H$_2$O fractions (latex, bark)	Antihemorrhagic	*Bothrops asper* venom applied on skin of Swis-Webster mice-%Inhibition 1 mg extract- Bark/latex: 100; H/EA/W fraction; Bark: 37/100/100; Latex: 0/0/100	
			Essential oil/ CH$_2$Cl$_2$:MeOH extract (bark) Individual constituents (essential oil)	Papain-like cysteine protease inhibition IC_{50} (μg/mL)	Cruzain inhibition Extract: 516 ± 9 Essential oil: 15.8 ± 0.1 **68**: 112 ± 14	
			Essential oil (bark)	Cytotoxic	100 μg/mL-%Inhibition HCT-15: 16, MCF-7: 37, MDA-MB-468: 9, SK-MEL-28: 10, SW620: 17, UACC-257: 16	

Table 8.4 contd. ...

...Table 8.4 contd.

Plant (Family)	Communities	Ethnobotanical uses/plant part (preparation/application)	Pharmacological effects-in vitro/in vivo evaluation		References
Smilax aristolochiifolia (Smilacaceae)	Mayas (Belize, Guatemala, and Mexico), nahuatl, tének y zoque-popolucas (Mexico)	Menstrual pain and other gynecological disorders, syphilis, kidney cleaning, depurative, diuretic, caugh, pneumonia, rheumatism, diabetes, antitumor, dermatosis, and dysentery Root (infusion)	$(CH_3)_2CO$ extract (root) **71**-riched fractions	Hypolipidemic, hypoglycemic, and hypotensive	% Reduction in mice **71**-fraction: 60% hyperlipidemic; 40% insulin resistance; 31% and 37% systolic and diastolic pressures	Sautour et al. 2005, 2006, Velasco-Lezama et al. 2009, Challinor et al. 2012, Botello et al. 2014, Pérez-Nájera et al. 2018
			Ethanol:H_2O extract (root) Chlorogenic acid/anastilbin-rich fractions	α-Amylase/α-glucosidase inhibition IC_{50} (µg/mL)	**α-amylase** Extract: 90 ± 4, **72**-F: 59 ± 1 α-glucosidase TE: 12.4 ± 0.3, **72**-F: 9 ± 2, **73**-F: 12.3 ± 0.9	
			H_2O extract (root)	Hematopoietic	Mice with aplastic anemia Dose 0.4 g/kg, erythrocytes, platelet, and bone marrow cell recovery (9 days)	
			Isolated saponins (aqueous root extract)	Antifungal MIC (µg/mL)	*C. albicans, C. glabrata, C. tropicalis* **74-76**: 12–50 **77-78**: 6–50	
				Cytotoxicity IC_{50} (µg/mL)	NF, HeLa, HT29, MCF7, MM96L, K562 **79-81**: 3.4-42	

saponins, flavonoids, tannins, and triterpenes as the main families of compounds, which have been related to antibacterial and cytotoxic activities. Castillo-Ávila et al. (2009) isolated and identified two compounds (methyl gallate (**60**) and methyl *m*-trigallate (**61**)), from the leaf extract when it was evaluated for its antioxidant power by DPPH·. However, methanol or ethanol extracts were effective againts the parasites *L. mexicana* or *T. cruzi*.

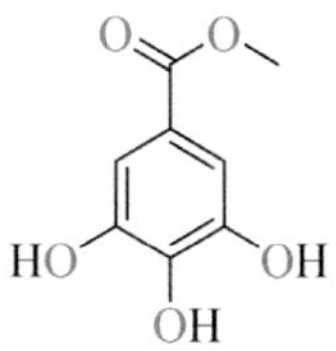

60 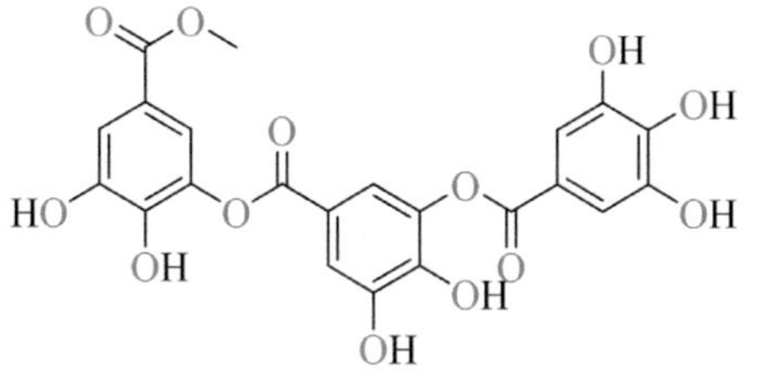   61 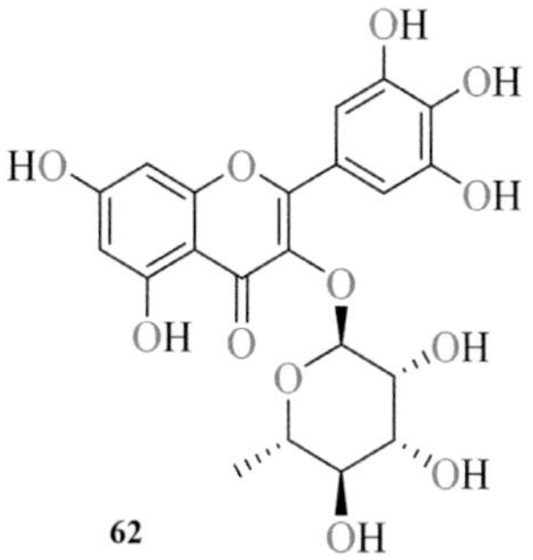   62

On the other hand, *C. draco* has a variety of chemical constituents in all parts of the plant from proanthocyanins (e.g., procyanidin B$_2$), anthocyanins, 1,4-naphthoquinones, sesquiterpene lactones, cardiotonic glycosides, saponines, terpenes to alkaloids (e.g., magnoflorine). Nevertheless, it is attributed that certain molecules are responsible for the biological properties, e.g., myricitrin (**62**), taspine (**63**), (epi-)catechin (**64/65**), and other flavonoids (Salatino et al. 2007). In accordance with Table 8.4, different authors described powerful antihemolytic and antihemorrhagic effects along with antiparasitic and antiproliferative/cytotoxic properties for both extracts/essential oils and isolated compounds. For the case of hemolytic effect of extracts, IC$_{50}$ values were 0.43–0.59 mg/mL and the isolated compounds with the highest activity values (%inhibition: 63–91%) at 0.9 mM were **63**, **62**, and cyclopeptides P$_1$/P$_2$ based on Tsacheva et al. (2004); while 1 mg of bark or latex extracts/fractions showed antihemorrhagic effect with inhibition values between 37–100%; flavonoids **64–67** were the isolated molecules from the ethyl acetate extract (most active) of plant bark (Castro et al. 1999). And, as a last point, Setzer et al. (2007) reported the astonishing inhibition of the cysteine protease cruzain (related to *T. cruzi*) by bark essential oil (IC$_{50}$ 15.8 ± 0.1 µg/mL), which was composed of β-caryophyllene (32%) and caryophyllene oxide (22%). Between some individual terpene constituents

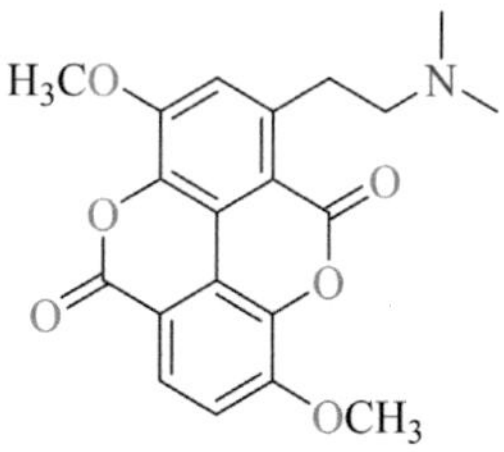

63 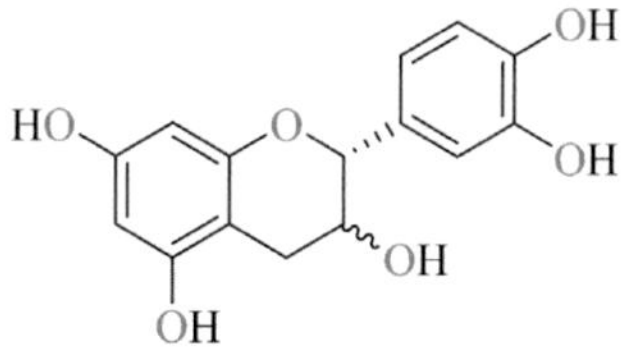   64/65 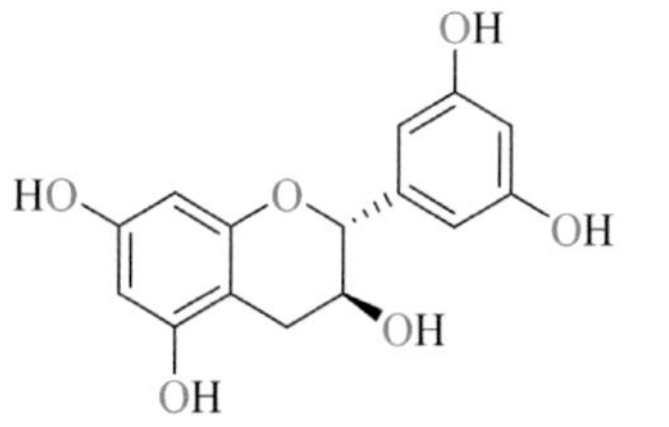   66

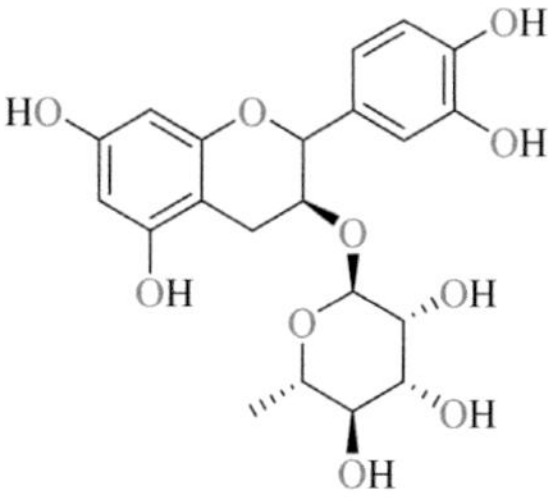

67 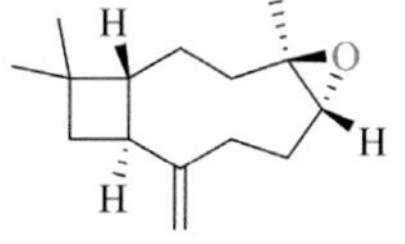   68 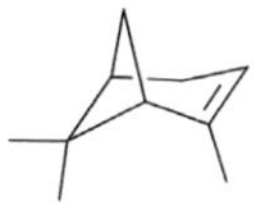   69    70

tested, three presented promising IC_{50} values (µg/mL), i.e., caryophyllene oxide (**68**) (112 ± 14), β-pinene (**69**) (155 ± 10), and α-pinene (**70**) (160 ± 9). The cytotoxicity evaluated for the essential oil on six cell lines showed low to moderate effects (100 µg/mL, %inhibition: 10–37%).

The extracts/fractions and isolated constituents from the endemic and wild medicinal plant *S. aristolochiifolia* root have evidenced particular bioactivities related to metabolic syndrome (e.g., antihyperlipidemy, antihypertensive, antidiabetes), anemia, antifungal, and cytotoxicity. In the first case, based on a report by Botello et al. (2014), the second fraction enriched with N-*trans*-feruloyl tyramine (**71**, 600 µg/kg) produced a decrease of 60% on trygliceride levels in mice; the acetone extract along with all N-*trans*-feruloyl tyramine-riched fractions decreased until 40% the insulin resistance parameter and among 31–37% systolic (78–98 mmHg) and diastolic (55–65 mmHg) blood pressures. Likewise, Pérez-Nájera et al. (2018) determined the inhibitions of α-amylase (IC_{50}: 59 ± 1–90 ± 4 µg/mL) and α-glucosidase (IC_{50}: 9 ± 2–12.3 ± 0.9 µg/mL) by the extract or enriched-fractions with chlorogenic acid (**72**) or anastilbin (**73**) as a measure of the potential for diabetes regulating. These components (**72** and **73**) were isolated and chemically characterized. The hematopoietic effectiveness of the aqueous extract of the root was reported by Velasco-Lezama et al. (2009); these authors demonstrated that the extract was able to increase the number of erythrocytes, platelets, and bone marrow cells on mice with aplastic anemia.

Lastly, an important group of constituents from *S. aristolochiifolia* root are saponins (steroidal compounds), which were insulated by Sautour et al (2005, 2006) and Challinor et al. (2012) and have presented antimicrobial and cytotoxic activities. The isolated smilagenins (spirostane-type saponins) (**74-78**) from "gray sarsaparilla" by Sautour et al. (2005, 2006) were tested against *C. albicans*, *C. glabrata*, and *C. tropicalis*; the MIC values determined on all strains were—**74/75**: 12.5–50 µg/mL, **76**: 25–50 µg/mL, **77**: 6.25–25 µg/mL, and **78**: 12.5–50 µg/mL. While, isolated sarsaparillosides (furostane-type saponins) (**79–81**) by Challinor et al. (2012) showed antineoplastic effects against six cell lines, i.e., for each cell line and three sarsaparillosides, the IC_{50} values were—NFF: 4.5–27 µg/mL, HeLa: 12–42 µg/mL, HT29: 4.8–14 µg/mL, MCF7: 3.4–24 µg/mL, MM96L: 3.8–23 µg/mL, and K562: 4.3–28 µg/mL.

Plants with High Potential from Peru

In agreement to the expertise of the Peruvian co-author (from Instituto de Investigaciones de la Amazonía Peruana—IIAP) and revised scientific literature, two Peruvian wild medicinal plants [*Ambrosia peruviana* Willd., and *Mansoa alliacea* Lam.—Figure 8.6] with high value for the Peruvian Amazon indigenous/peasant communities were selected. Some information on each of them can be found below.

***Ambrosia peruviana* Willd.**—synonyms—*Ambrosia cumanensis* Kunth; *Ambrosia elatior* L.; *Ambrosia orobanchifera* Meyen; *Ambrosia paniculata* Michx. var. *cumanensis* (Kunth) O.E. Schulz; *Ambrosia paniculata* var. *peruviana* (Willd.) O.E. Schulz. Common names—ajenjo, altamis(z)a, amargo, ambrosia silvestre, artemisa, cumana/peruvian ragweed, maki, mal(r)co(u), markhu/marquito, mashi paico. Annual herb or shrub plant (0.8–2 m), with alternate/ovate/lanceolate leaves (hairy both surface); spike-shaped inflorescences, hermaphrodite flowers, glabrous, yellowish; obovoid fruits, glandulous-hairy; the whole plant is fragrant. Although this species is native to Peru, it is distributed in Central and South America (at altitudes of 0–1500 m). The herb is used for treatment of respiratory, digestive, reproductive, and joint disorders, and against malaria, by the indigenous and/or peasant communities (Duke et al. 2009, Rengifo 2005, Mostacero et al. 2011, Quattrocchi 2012, Bussman and Sharon 2015, Tropicos 2019).

***Mansoa alliacea* Lam.**—synonyms—*Adenocalymma alliaceum* (Lam.) Miers, *A. pachypus* Bureau and K. Schum., *A. sagotii* Bureau and K. Schum., *Anemopaegma pachypus* K. Schum., *Bignonia alliacea* Lam., *Pachyptera alliacea* (Lam.) A.H. Gentry, *Pseudocalymma alliaceum* (Lam.) Sandwith, *P. sagotii* (Bureau and K. Schum.) Sandwith. Common names—ajo sacha/macho/silvestre, araruta, sacha árbol, boens, nia boens, bejuco de ajo, madre de Dios. Evergreen climbing shrub (up to 4 m), bifoliolate leaves, acute/obtuse apex; with trifid tendrils coming out of the stem; white-purplish flowers, with flared tubular corolla; racemose inflorescences; oblong linear capsular fruit; all vegetative parts are fragrant, smelling reminiscent of garlic/onion. The liana is native to Peru and it is distributed in Central and northern South America (at altitudes of 0–500 m). The Amazon indigenous/peasants

(a) Taken by Indiana Coronado; Source: http://www.tropicos.org/Image/100177393. **(b)** Taken by Rodolfo Vásquez; Source: http://www.tropicos.org/Image/100362051.

Figure 8.6: Images of the two promising plants—(a) *A. peruviana*; (b) *M. alliacea*.

traditionally use it as a treatment for skin, respiratory, digestive, and joint disorders, and as analgesic, antimalarial, etc. (Duke et al. 2009, Rengifo 2005, Mostacero et al. 2011, Quattrocchi 2012, Tropicos 2019).

The communities that use the two plants together with the ethnobotanical information, pharmacological effects (including *in vitro/in vivo* assessments), and useful parts are registered in Table 8.5. Referring to science literature consulted, the most abundant secondary metabolites isolated from *A. peruviana* leaves have been its essential oil. Based on Yánez et al. (2011), the EO that showed antibacterial potential was constituted by γ-curcumene (24%), *ar*-curcumene (14%), and bornyl acetate (10%), and the most susceptible strains were *Salmonella typhi* (MIC 350 μg/mL) and *S. aureus* (MIC 400 μg/mL). Furthermore, Mesa et al. (2017) determined the larvicidal effects on *A. aegypti* and antibacterial capacity against *B. cereus* and *B. subtilis* of extracts (hexane, dichloromethane, ethyl acetate, and ethanol) and EO from "altamisa". All extracts and EO (at 200 μg/mL) were effective on *A. aegypti* from the larvae to adult stages, with 100% of mortality, as well as on *B. subtilis* and *B. cereus*. At 500 μg, the inhibition halos (mm) were—hexane extract—10 ± 1–11 ± 1; dichloromethane extract—10 ± 2–15 ± 6; ethyl acetate extract—9 ± 1–11.5 ± 0.7; ethanol extract—10 ± 2 (it was not determined on *B. subtilis*); EO—9.5 ± 0.7–10 ± 2.

The most powerful extract was dichloromethane, which had an inhibition on *B. cereus* equivalent to the tetracycline antibiotic (positive control). The sesquiterpene lactones (pseudoguaianolides or ambrosanolide-class) that have been isolated from *A. peruviana* are interesting, e.g., damsin (**82**), confertin (**83**), cumanin (**84**), peruvin (**85**), peruvinin (**86**), ambrosin (**87**), psilostachyin C (**88**), and psilostachyin B (**89**). The first ambrosanolides isolated and structurally characterized were peruvin (**85**) and peruvinin (**86**) by Joseph-Nathan and Romo (1966), and Romo et al. (1967). In 1969, Herz et al. reported the isolation of three pseudoguaianolides: **87**, **82**, and **88** (Herz et al. 1969). Afterwards, **89** and a new aromadendrane-type sesquiterpene diol (*allo*-aromadendrane-4β,10α-diol (**90**)) were insulated and reported by Goldsby and Burke in 1987 (Goldsby and Burke 1987). Later, **83** was isolated/reported by Aponte et al. (2010), and in 2013, Sülsen et al. secluded/reported **84** from the aerial parts from the plant (Sülsen et al. 2013). Lastly, in 2016, Jiménez-Usuga et al. isolated three new sesquiterpene lactones (**91-93**) of methanol extract from aerial parts (Jiménez-Usuga et al. 2016).

Of all the ambrosanolides mentioned above, six (**82–85**, **88**, **93**) have been tested to determine their biological potential. Thus, damsin and confertin were active against the 14 cell lines evaluated (Aponte et al. 2010); nonetheless, the most significant GI_{50} values (μM) of **82** were 7.6 (on DU145 cells), 8.1 (on U937 line), and 10.3 (on DU145 cells). Meanwhile, **83** showed the highest GI_{50} values (3.6–9.2 μM) on 12 of the 14 lines tried (except MCF7 and HeLa). Other molecules, **84**, **85**, and **88** were active on BW5147 cell line (EC_{50}: 4.9–24.5 μg/mL) and normal T lymphocytes (CC_{50}: 35– > 50 μg/mL) (Martino et al. 2015); and **93** was highly cytotoxic against Jurkat (IC_{50}: 6 μM), U937 (IC_{50}: 8 μM), and HeLa (IC_{50}: 30 μM) lines (Jiménez-Usaga et al. 2016). Additionally, when the compounds **82** and **83** were tested on parasites *L. amazonensis* and *L. braziliensis*, both compounds showed promising activity on *L. amazonensis*, with IC_{50} values of 1.9 μM and 3.3 μM; while on

L. amazonensis, only **83** had a satisfactory IC_{50} value of 13.2 µM. Another valuable manuscript on bioproperties of a pseudoguaianolide (**84**) along with extract/fractions was reported by Sülsen et al. (2013). Firstly, these authors found that CH_2Cl_2:MeOH extract and its fractions (eight) from *A. peruviana* aerial parts inhibited the growth (%) of *T. cruzi* at different concentrations; that is, 1 µg/mL—extract: 9 ± 3%, F_{2-9}: 8/9–41 ± 3%; 10 µg/ mL—extract: 35 ± 1%, F_{2-9}: 25 ± 3–94.7 ± 0.5%; and 100 µg/mL—extract: 94 ± 1%, F_{2-9}: 40 ± 5–96.5 ± 0.2%. Once the most active fraction was established, then cumanin was isolated. This ambrosanolide was notably active against *T. cruzi* epimastigotes of two strains (RA and K98), with IC_{50} values of 12 µM and 4 µM, respectively; as well as on *L. amazonensis* and *L. braziliensis* epimastigotes, with IC_{50} values of 3 µM and 2 µM, correspondingly.

Finally, the last species under discussion is *M. alliacea*. In this way, Olivera-Condori et al. (2013) evaluated the antibacterial power and free-radical scavenging capacity of the "ajo sacha" EO from leaves; the authors found that EO presented low antibacterial power and anti-radical capacity. The EO was mainly constituted by allyl trisulfide (68%) and diallyl disulfide (19%). This chemical composition differed from that reported by Granados-Echegoyen et al. (2014), which identified to diallyl disulfide (50%), diallyl sulfide (12%), and di-2-propenyl trisulfide (10%). They determined the larvicidal effect (on *C. quinquefasciatus*) of EO/hydrolate/extracts (aqueous/MeOH/EtOH) from the shrub; the most active extracts were hydrolate ($LC_{50/90}$: 8–16 µg/mL, 24–72 hours) and MeOH/aqueous (42–46% mortality, at 10% concentration). Considering the determination of other potential bioproperties of the extracts (Table 8.5), the authors Freixa et al. (1998), Rana et al. (1999), Valadeau et al. (2009), Domínguez and Neves (2014), Towne et al. (2015), and Hamann et al. (2019) reported some of them. Thus, the antifungal effects on *T. mentagrophytes* and *M. gypseum* using CH_2Cl_2 and MeOH extracts at 5–10 mg/disk were measured by Freixa et al. (1998); both extracts were active on *M. gypseum* [φ inh.: 17–19 mm and 17–20 mm], while CH_2Cl_2 extract was only active on *T. mentagrophytes* [φ inh.: 21–23 mm]. Rana et al. (1999), for its part, determined the percentage inhibition on spore germination (as a measure of antifungal susceptibility) of eight fungal strains (*Alternaria alternata*, *A. brassicae*, *A. brassicicola*, *A. carthami*, *Colletotrichum capsici*, *Curvularia lunata*, *F. oxysporum*, and *F. udum*) of H_2O leaf extract; all of them were susceptible to the extract in a dose-dependent manner. The antinociceptive, immunostimulant, anti-inflammatory, and anticancer properties determined for the aqueous/alcohol extracts were promising, according to Hamann et al. (2019), Dominguez and Neves (2014), and Towne et al. (2015). The inhibitions were higher than 70% for the inflammation and nociception (based on allodynia prevention/reversion), and cell growth (T3-HA line). Lastly, Valadeau et al. (2009) and Ruiz et al. (2011) established the antiparasitic potential on *L. amazonensis* and *P. falciparum* (chloroquine resistance) of the ethanol extract from bark; these authors found IC_{50} values of 22 ± 9 µg/mL and 24 ± 10 µg/mL for each strain; and on ferriprotoporphyrin biocrystallization inhibition test (FBIT), a IC_{50} value of 2.0 µg/mL. At the end, from methanol extract of wood, two naphthoquinone-type molecules (**91** and **92**) were isolated, which showed a high cytotoxic potential (on V-79 line) with IC_{50} values of 5.6 µg/mL and 6 µg/mL, individually.

Exploitation/Sustainability and Opportunity of Drug Development (patents) from these Plants

As most of the medicinal plants used in traditional medicine are collected as wild, they are suffering inappropriated and unmeasured exploitation by unqualified/inexpert people, which is causing loss of biological availability (genetic diversity) and habitat destruction, and therefore, the non-sustainability of the natural resource. Additionally, certain populations related to medicinal plants (e.g., local groups, herbalists, and herbal traders) take some vital parts of plants (roots, stem, trunk) and/or whole plants, which would put the preservation of the species at risk (categorization as threatened), or in the worst case, lead to them becoming endangered. Part of the solution to those problems described above could be the domestication of plants and the implementation of cropping systems, which would

Table 8.5: Pharmacological effects, ethnobotanical uses, useful parts, and communities that use the selected Peruvian plants.

Plant (Family)	Communities	Ethnobotanical uses/ plant part (preparation/ application)	Pharmacological effects- *in vitro/in vivo* evaluation		References
Ambrosia peruviana (Asteraceae)	Shipibo-Conibo, Ashaninka, Amazon riverside mestizo Santa Clara-Loreto. San Antonio de Saniyacu-Loreto	Antimalarial, insecticide, intestinal parasitosis, vermifuge, emenagogue, amenorrhea, depurative, neuralgia, antidepressant, rheumatism. Leaves/aerial parts/flowers (infusión/decoction/bath/ poultice-macerate)	Antimicrobial	Essential oil (leaves)	φ inhibition-mm, MIC-μg/ mL $S.$ *aureus*: 8, 400; *E. faecalis*: 11, 500; *E. coli*: 7, 500; *S. typhi*: 8, 350	Aponte et al. 2010, Guauque et al. 2010, Yánez et al. 2011, Sülsen et al. 2013, Bussmann and Sharon 2015, Martino et al. 2015, Jimenez-Usuga et al. 2016, Mesa et al. 2017
				Hexane/CH$_2$Cl$_2$/EtOAc/ EtOH extracts, and essential oil (leaves)	φ inhibition-mm, 500 μg *B. cereus*: 10 ± 1–15 ± 6 *B. subtilis*: 9 ± 4–11 ± 1	
			Larvicidal		*A. aegypti*, %mortal., 200 μg/mL Extr. 5%, 24 hours; 100% 144 hours (larvae to adults)	
			Toxicity	EtOH extract (leaves)	LC$_{50}$ - *Artemia salina*, 64 μg/mL	
			Anthelmintic		100% mortaility- *Toxocara canis*, 50 μg/mL, 4 hours	
			Cytotoxity- antiproliferative		GI$_{50}$ (μM)- Active on 14 cell lines **82**: 8–23; **83**: 4–17	
			Antiparasitic	Extract/fractions/isolated ambrosanolides (leaves, aerial parts)	IC$_{50}$ (μM) *L. amazonensis*- **82**: 2; **83**: 3; *T. cruzi* - **82**: > 200; **83**: 13 %GI-*T. cruzi*; 1-100 μg/mL; ext.: 9 ± 3–94 ± 1; F$_{2-9}$; 25 ± 3–96.5 ± 0.2 IC$_{50}$ (μM), **84**: 12 (RA), 4 (K98); trypomast.- **84**: 180 (RA), 170 (K98); amastig.- **84**: 8. *L. amazonensis/L. braziliensis*- **84**: 3/1	

Plant (Family) / Ethnic groups & Traditional use	Extract	Activity	Results	References
Mansoa alliacea (Bignoniaceae) Shipibo, Conibo, Ashaninka, Lamas, Amuesha, Yanesha, Quechua, Lupuna, Tamshiyacu, Achuales, Ese´eja, Cuna (Panama), Wayapi (French Guiana), Gresol (Guyana), Tapajos (Brazil). Analgesic, antipyretic, antimalarial, repellent/fumigant, against arthritis, abdominal pain, epilepsy, and cephagia Aerial parts/leaves/bark/root/capitule/stems (infusión/decoction/tincture/poultice)	Essential oil (leaves)	Free radical scavenging	DPPH: Radical- CE$_{50}$: 234 mg/mL	Itokawa et al. 1992, Freixa et al. 1998, Rana et al. 1999, Valadeau et al. 2009, Ruiz et al. 2011, Olivera-Condori et al. 2013, Dominguez and Neves 2014, Granados-Echegoyen et al. 2015, Towne et al. 2015, Hamann et al. 2019
	Essential oil (leaves)	Antibacterial	φ and%inhibition 5 mg/mL *S. aureus*: 8 mm, 26% *B. subtilis*: 9 mm, 28%	
	CH$_2$Cl$_2$/MeOH extracts (leaves)	Antifungal	5–10 mg/disk, φ (mm) *Tricophyton mentagrophytes* and *Microsporum gypseum*: 0–23/17-20	
	H$_2$O extract (leaves)	Antifungal	% inhibition spore germ.; extract (1:2): 98-100% on all fungi tested.	
	H$_2$O extract (leaves)	Anticancer	T3-HA line; extract equal to 0.03–0.09 g plant/mL,%GI: 70–100	
	EtOH extract (bark)	Antiparasitic	IC$_{50}$ (μg/ mL)- *L. amazonensis*: 22 ± 9; *P. falciparum* ClQR: 24±10 *P. falciparum* ClQR: >10, FBIT: 2	
	H$_2$O:EtOH extract (leaves)	Antinociceptive/anti-inflammatory	CFA in mice Extract: 100/90% prevention/reversion (allodynia); reversed hyperalgesia (4 time); antiallodynic effect non-selective and δ-selective opioid rec. antagonists: 98% and 93%	
	Lyophilized H$_2$O extract (leaves)	Immunostimulant	% Activ. albino rats Holtzman-Doses 13.2/26.4 mg/kg, 75/76%	
	H$_2$O/MeOH/EtOH extracts/ Essential oil/ hydrolate (leaves)	Larvicidal (4th instar larvae)	LC$_{50}$/LC$_{90}$ (μg/mL) *Culex quinquefasciatus* EO: 267-147/494-312, 24–72 hours; hydrol: 10-8/16-12, 24-72 hours 10% extract,% mortal., RGI- MeOH: 32, 0.76; EtOH: 46, 0.7; H$_2$O: 42, 0.6	
	Isolated component (from MeOH extract of wood)	Cytotoxicity (V-79 cells)	IC$_{50}$ (μg/ mL) Extract: +; **91**: 5.6; **92**: 6	

satisfy the current and future demands for production of plants and/or herbal drugs, as well as to relieve the pressure of wild populations (Alamgir 2018). According to the query in the webpage of the IUCN red list of threatened species (IUCN 2019), the greater part (64%) of the plants included in this manuscript were not registered. The remaining part (36%) of the plants (*T. integrifolia*, *C. pyramidalis*, *J. caroba*, *S. globulifera*, and *C. draco*) were listed, although their status are of "least concern", because the current population trends are stable. Based on this information, it could be considered that the plants are not overexploited or at apparent risk.

The review of the current state of patents and/or marketing of phytotherapeutic products related to the plants under study generated the following results: Evanta herb (*G. longiflora*) recorded two patents related to (i) preparation method of chimanine A (CN102838538A) from China, and (ii) extraction of 2-substituted quinolines for the treatment of leishmaniasis (FR2682107A1) from France (expiration date: August 2019). In the case of *J. caroba*, it is marketed as an extract/mother tincture for homoeophatic medicine by SBL Pvt Ltd, Natural Heatlh Supply, Willmar Schwabe India Pvt Ltd, Washington Homeophatic Products, Rappen Apohteke and Herbal Foods; or as ground material by Folha e Raiz. The patent US20060165812A1 (method and topical formulation for treating headaches) includes *J. caroba* and *J. copaia* as active ingredients for such formulation. *J. copaia* is included as an active ingredient in the patent code JP2002316936A (antibacterial agent and anti-inflammatory agent). Although there is no exclusive patent on *J. copaia*, a patent (US4078145A, expired 1995) associated with jacaranone (**32**) from *J. caucana* (species native to Colombia and closely related to *J. copaia*) was found. The patent title was "phytoquinoid possessing anti-tumor activity". *C. draco* has been mentioned in three patents as an active ingredient—(i) code US20030099727A1, entitled "compositions and methods for reducing cytotoxicity and inhibiting angiogenesis", (ii) code US20020041906A1, entitled "compositions and methods for enhancing therapeutic effects", and (iii) code WO2006078848A1, titled "compositions containing botanical extracts rich in phlorizin and methods for using such compositions in blood glucose modification and to affect aging". A patent, entitled "smilagenin and its use" (code US7368137B2), mentions *S. aristolochiifolia* as a natural source for the isolation of this molecule. In addition, *A. peruviana* is marketed as a liquid herbal supplement (or tincture) called Marco (by Irae® herbal rejuvenation, Salvia Paradise, Dr. Clark) or solid (capsules) by Salvia Paradise, Herbis® and Čajový Dom. To conclude, Rainforest Pharmacy, Raintree Nutrition and Bioaurora market *M. alliacea* as a solid herbal supplement (capsules) called ajo sacha, ajos sacha, and Huanarpo macho, respectively, and Bio Deli Organico, as a liquid supplement called 20% ajo sacha. On *M. alliacea*, only one patent was found, entitled "tissue culture method of *Mansoa alliacea*" (code CN105724253A).

Conclusion

The species that were selected and described in this chapter are some examples of Latin American wild medicinal plants that have a strong rooting (are well-known) in the native and mestizo communities (of each country) that use them for different therapeutic purposes, which in most cases (or not) have been validated taking into account the results of the determined biological activities. Therefore, they would constitute a promising source to be candidates for phytotherapeutic products. In addition, these plants have a broad ethnomedicinal description in the different regions, varying their purposes of use, diversity of preparations, and chemical constituents. The 14 plants had some ethnobotanical uses (against intestinal parasites, leishmania, malaria, microbial infections, cancers/ tumors, snake bites, and/or rheumatism), preparations (decoction, maceration, and/or infusion), and wide pharmacological effects (anti-leishmanial/trypanosomial/plasmodial, anti-tumoral/cancer/ proliferative/cytotoxic, anti-HIV, anti-bacterial/fungal, antihemorrhagic/hemolytic, hematopoietic, larvicidal, anthelmintic, antinociceptive/anti-inflammatory) along with specific, isolated, and active chemical constituents (chimanines, brevipolides, melampolides, flavones/flavanones/biflavonoids, phytosterols, acylphloroglucinols, xanthones/benzophenones, alkaloids, smilagenins/sarsaparillosides, ambrosanolides, organosulfur compounds, naphthoquinones) in common. It is very important to note that due to the wild nature of these plants, it is advisable to domesticate the species and implement the sustainable farming systems for the best use of these important plant resources.

Acknowledgments

The authors thank the UMSA-SIDA strengthening program (75000553) through Biomolecules (Antiparasitic) and Tacana Bioprospection projects; Desparacitacion de niños en escuelas rurales I y II IDH 2010-2014 proyectos, and CIPTA, CIMTA.

References

Acanthospermum australe—photo 8.4a. Tropicos.org. Missouri Botanical Garden. http://www.tropicos.org/Image/100188419. Consulted on-line: May 2019.

Agra, M.F., França, P.F., Barbosa-Filho, J.M. 2007a. Synopsis of the plants known as medicinal and poisonous in Northeast of Brazil. Rev. Bras. Farmacogn. 17: 114–140.

Agra, M.F., Baracho, G.S., Basílio, I.J.D., Nurit, K., Coelho, V.P., Barbosa, D.A. 2007b. Sinopse da flora medicinal do Cariri Paraibano. Oecol. Bras. 11: 323–330.

Agra, M.F., Silva, K.N., Basílio, I.J.L.D., França, P.F., Barbosa-Filho, J.M. 2008. Survey of medicinal plants used in the region Northeast of Brazil. Rev. Bras. Farmacogn. 18: 472–508.

Alamillo, V.J. 2017. Evaluación del efecto antitumoral del extracto acuoso del látex de *Croton draco* var. *draco* Cham. and Schltdl y *Croton lechleri* Muell. Arg. sobre líneas celulares derivadas de cáncer. Tesis de maestría. Universidad Veracruzana. Veracruz, México. 69 p.

Alamgir, A.N.M. 2018. Therapeutic use of medicinal plants and their extracts: Vol II: Phytochemistry and bioactive compounds, progress in drug research. Springer International Publishing.

Ambrosia peruviana—photo 8.6a. Tropicos.org. Missouri Botanical Garden. http://www.tropicos.org/Image/100177393. Consulted on-line: May 2019.

Aponte, J.C., Yang, H., Vaisberg, A.J., Castillo, D., Málaga, E., Verástegui, M., Casson, L.K., Stivers, N., Bates, P.J., Rojas, R., Fernandez, I., Lewis, W.H., Sarasara, C., Sauvain, M., Gilman, R.H., Hammond, G.B. 2010. Cytotoxic and anti-infective sesquiterpenes present in *Plagiochila disticha* (Plagiochilaceae) and *Ambrosia peruviana* (Asteraceae). Planta Med. 76: 705–707.

Arellano-Rodríguez, J.A., Flores-Guido, J.S., Tun-Garrido, J., Cruz- Bojórquez, M.M. 2003. Nomenclatura, forma de vida, uso, manejo y distribución de las especies vegetales de la península de Yucatán. Etnoflora Yucatanense, Universidad Autónoma de Yucatán, Mérida.

Arévalo-Lopéz, D., Nina, N., Ticona, J.C., Limachi, I., Salamanca, E., Udaeta, E., Paredes, C., Espinoza, B., Serato, A., Garnica, D., Limachi, A., Coaquira, D., Salazar, S., Flores, N., Sterner, O., Giménez, A. 2018. Leishmanicidal and citotoxic activity from plants used in Tacana traditional medicine (Bolivia). J. Ethnopharmacol. 216: 120–133.

Argueta-Villamar, A., Cano, L.M., Rodarte, M.E., Gallardo, M.C. 1994. Atlas de las plantas de la medicina tradicional mexicana. Vol. II. Instituto Nacional Indigenista. Gobierno de México. 1413–1414.

Astudillo, L., Rodrigues, J.A., Schmeda-Hirschmann, G. 2002. Gastroprotective activity of oleanolic acid derivatives on experimentally induced gastric lesions in rats and mice. J. Pharm. Pharmacol. 54: 583–588.

Bacchi, E.M., Rios, J.P.C., Dias, T.G. 1999. Fraction responsible for *Jacaranda caroba* DC. antiulcer action. In: 2nd IUPAC International Conference on Biodiversity, UFMG, Belo Horizonte, MG. Programme and Abstracts, 173 p.

Bahia, M.V., Dos Santos, J.B., David, J.P., David, J.M. 2005. Biflavonoids and other phenolics from *Caesalpinia pyramidalis* (Fabaceae). J. Braz. Chem. Soc. 16: 1402–1405.

Barbosa Junior, A.M., Mélo, D.L.F.M., Almeida, F.T.C., Trindade, R.C. 2015. Estudo comparativo da susceptibilidade de isolados clínicos de *Cryptococcus neoformans* (Sanfelice, 1895) frente a alguns antifúngicos de uso hospitalar e extratos vegetais obtidos de plantas medicinais da região semiárida sergipana. Rev. Bras. Plant. Med. 17: 120–132.

Basting, R.T., Nishijima, C.M., Lopes, J.A., Santos, R.C., Lucena Périco, L., Laufer, S., Bauer, S., Costa, M.F., Santos, L.C., Rocha, L.R., Vilegas, W., Santos, A.R., Dos Santos, C., Hiruma-Lima, C.A. 2014. Antinociceptive, anti-inflammatory and gastroprotective effects of a hydroalcoholic extract from the leaves of *Eugenia punicifolia* (Kunth) DC. in rodents. J. Ethnopharmacol. 157: 257–267.

Bohlmann, F., Zdero, C., King, R.M., Robinson, H. 1981. Germacranolides, a guaianolide with a β-lactone ring and further constituents from *Grazielia* species. Phytochemistry 20: 1069–1075.

Borges-Santos, R.R., Santos, J.L.L., Farouk, Z., David, J.M., David, J.P., Lima, J.W.M. 2012. Biological effect of leaf aqueous extract of *Caesalpinia pyramidalis* in goats naturally infected with gastrointestinal nematodes. Evid.-Based. Compl. Alt. 1–6.

Botello, C.A., González-Cortazar, M., Herrera-Ruiz, M., Román-Ramos, R., Aguilar-Santamaría, L., Tortoriello, J., Jiménez-Ferrer, E. 2014. Hypoglycemic and hypotensive activity of a root extract of *Smilax aristolochiifolia*, standardized on N-*trans*-feruloyl-tyramine. Molecules 19: 11366–11384.

Botion, L.M., Ferreira, A.V.M., Côrtes, S.F., Lemos, V.S., Braga, F.C. 2005. Effects of the Brazilian phytopharmaceutical product Ierobina® on lipid metabolism and intestinal tonus. J. Ethnopharmacol. 102: 137–142.

Bourdy, G., DeWalt, S.J., Chávez de Michel, L.R., Roca, A., Deharo, E., Muñoz, V., Balderrama, L., Quenevo, C., Giménez, A. 2000. Medicinal plants uses of the Tacana, an Amazonian Bolivian ethnic group. J. Ethnopharmacol. 70: 87–109

Braga, R. 2015. Plantas do Nordeste, especialmente do Ceará. Tercer Ed. Mossoró: Série C.

Breviglieri, E., Mota da Silva, L., Boeing, T., Somensi, L.B., Cury, B.J., Gimenez, A., Cechinel Filho, V., de Andrade, S.F. 2017. Gastroprotective and anti-secretory mechanisms of 2-phenylquinoline, an alkaloid isolated from *Galipea longiflora*. Phytomedicine 25: 61–70.

Brunetti, I.L., Vendramini, R.C., Januario, A.H., Franca, S.C., Pepato, M.T. 2006. Effects and toxicity of *Eugenia punicifolia* extracts in streptozotocin-diabetic rats. Pharm. Biol. 44: 35–43.

Bussman, R.W., Sharon, D. 2015. Medicinal plants of the Andes and the Amazon. The magic and medicinal flora of northern Peru. 292 p.

Caesalpinia pyramidilis—photo 8.3a. Asociação de Amigos do Jardim Botânico—AAJA. 2019. https://www.amigosjb.org.br/wp-content/uploads/2015/07/046-Caesalpinia-pyramidalis.jpg. Consulted on-line: May 2019.

Calla-Magariños, J. 2012. Bioactive leishmanicidal alkaloids molecules from *Galipea longiflora* Krause with immunodulatory activity. Doctoral Thesis, Faculty of Science, The Wenner-Gren Institute, Stockholm University, Sweden.

Calla-Magarinos, J., Giménez, A., Troye-Blomberg, M., Fernández, C. 2009. An alkaloid extract of evanta, traditionally used as anti-leishmania agent in Bolivia, inhibits cellular proliferation and interferon-γ production in polyclonally activated cells. Scand. J. Immunol. 69: 251–258.

Calla-Magariños, J., Quispe, T., Giménez, A., Freysdottir, J., Troye-Blomberg, M., Fernández, C. 2013. Quinolinic alkaloids from *Galipea longiflora* Krause suppress production of proimflammatory cytokines *in vitro* and control inflammation *in vivo* upon *Leishmania* infection in mice. Scand. J. Immunol. 77: 30–38.

Campos-Buzzi, F., Fracasso, M., Clasen, B.K., Ticona, J.C., Giménez, A., Cechinel-Filho, V. 2010. Evaluation of antinociceptive effects of *Galipea longiflora* alkaloid extract and major alkaloid 2-fenilquinoline. Methods Find Exp. Clin. Pharmacol. 32: 707–711.

Carvalho, C.C., Turatti, I.C.C., Lopes, N.P., Salvador, M.J., do Nascimento, A.M. 2014. Chemical composition and antimicrobial activity of essential oil from Brazilian plants *Acanthospermum australe*, *Calea fruticosa* and *Mikania glauca*. Afr. J. Pharm. Pharmacol. 8: 392–398.

Carvalho, L.H., Krettli, A.U., Carvalho, L.H., Krettli, A.U. 1991. Antimalarial chemotherapy with natural products and chemically defined molecules. Mem. Inst. Oswaldo Cruz 86: 181–184.

Castillo-Ávila, G.M., García-Sosa, K., Peña-Rodríguez, L.M. 2009. Antioxidants from the leaf extract of *Byrsonima bucidaefolia.* Nat. Prod. Commun. 2009: 4–83.

Castro, O., Gutiérrez, J.M., Barrios, M., Castro, I., Romero, M., Umaña, E. 1999. Neutralización del efecto hemorrágico inducido por veneno de *Bothrops asper* (Serpentes: Viperidae) por extractos de plantas tropicales. Rev. Biol. Trop. 47: 605–616.

Cesar, A.T., Sollero, P.A., Pereira, C., Sollero, G. 2004. *Jacaranda caroba*, medicamento de Mure. Cult. Homeopatica 3: 6–7.

Challinor, V.L., Parsons, P.G., Chap, S., White, E.F., Blanchfield, J.T., Lehmann, R.P., De Voss, J.J. 2012. Steroidal saponins from the roots of *Smilax* sp.: structure and bioactivity. Steroids 77: 504–511.

Chaves, T.P., Silva, J.P.R., de Medeiros, F.D., da Costa Filho, J.H., Alencar, L.C.B., Santana, C.P., Felismino, D.C., Vieira, V.M., Dantas, A.C. 2015. Traditional use, phytochemistry and biological activities of *Poincianella pyramidalis* (Tul.) LP Queiroz. Afr. J. Biotech. 14: 52.

Coelho de Souza, G., Haas, A.P.S., Von Poser, G.L., Schapoval, E.E.S., Elisabetsky, E. 2004. Ethnopharmacological studies of antimicrobial remedies in the south of Brazil. J. Ethnopharmacol. 90: 135–143.

Correa, J.E., Bernal, H.Y. 1993. Especies vegetales promisorias de los países del Convenio Andrés Bello, Tomo IX. Ed. Secretaría Ejecutiva del Convenio Andrés Bello (SECAB), Bogotá, Colombia.

Cottet, K., Fromentin, Y., Kritsanida, M., Grougnet, R., Odonne, G., Duplais, C., Michel, S., Gaillard, M. 2015. Isolation of guttiferones from renewable parts of *Symphonia globulifera* by centrifugal partition chromatography. Planta Med. 81: 1604–1608.

Cruz, A.V.M., Kaplan, M.A.C. 2004. Uso medicinal de espécies das famílias Myrtaceae e Melastomataceae no Brasil. Floresta e Ambiente 11: 47–52.

Cruz, C.H.Y. 2015. Diseño de um esquema de atención farmacêutica de aplicación em medicina tradicional (caso *Croton draco*) Schltdl. and Cham. Tesis de licenciatura. Facultad de Farmacia. FES-Zaragoza. UNAM. 93 p.

Deng, Y., Balunas, M.J., Kim, J.A., Lantvit, D.D., Chin, Y.W., Chai, H., Sugiarso, S., Kardono, L.B.S., Fong, H.H.S., Pezzuto, J.M., Swanson, S.M., Carcache de Blanco, E.J., Kinghorn, A.D. 2009. Bioactive 5,6-dihydropyrone derivatives from *Hyptis brevipes*. J. Nat. Prod. 72: 1165–1169.

Desrivot, J., Herrenknecht, C., Ponchel, G., Garbi, N., Prina, E., Fournet, A., Bories C., Figadère, B., Hocquemiller, R., Loiseau, P.M. 2007. Antileishmanial 2-substituted quinolines: *in vitro* behaviour towards biological components. Biomed. Pharmacother. 61: 441–450.

Domínguez, C.F., Neves, J.L.A. 2014. Actividad inmunoestimulante del extracto acuoso liofilizado de las hojas de "*Mansoa alliacea* L. (ajo sacha)"; en ratas albinas Holtzmman, IMET—EsSalud, 2013. Tesis professional, Univ. Nac. Amazon. Perú.

Duke, J.A. 2009. Duke's Handbook of medicinal plants of Latin America. CRC Press, Boca Ratón. 962 p.

Edwards, D. 2004. No remedy in sight for herbal ransack. New Sci. 181: 10–11.

Elsevier. 2019. Scopus database. Avalaible: https://www.scopus.com/home.uri (consulted online: May–June 2019).

Eugenia punicifiolia—photo 8.3b. Fauna e Flora do RN. 2017. http://faunaefloradorn.blogspot.com/cereja-do-cerrado-eugenia-punicifolia.html. Consulted on-line: May 2019.

Ezekwesili-Ofili, J.O., Okaka, N.A. 2019. Herbal medicines in African traditional medicine. *In*: Builders, P.F. (ed.). Herbal Medicine. IntechOpen.

Fenner, R., Betti, A.H., Mentz, L.A., Rates, S.M.K. 2006. Plantas utilizadas na medicina popular brasileira com potencial atividade antifúngica. Rev. Bras. Ciên. Farm. 42: 369–394.

Ferreres, F., Grosso, C., Gil-Izquierdo, A., Valentão, P., Andrade, P.B. 2013. Phenolic compounds from *Jacaranda caroba* (Vell.) A. DC.: approaches to neurodegenerative disorders. Food Chem. Toxicol. 57: 91–98.

Ferrufino-Acosta, L., Greuter, W. 2010. Typification of the name *Smilax lanceolate* L. Taxon 59: 287–288.

Fournet, A., Barrios, A.A., Muñoz, V., Hocquemiller, R., Cavé, A., Bruneton, J. 1993b. 2-Substituted quinoline alkaloids as potential antileishmanial drugs. Antimicrob. Agents Chemother. 37: 859–863.

Fournet, A., Barrios, A.A., Muñoz, V., Hocquemiller, R., Roblot, F., Cavé, A. 1994a. Antiprotozoal activity of quinoline alkaloids isolated from *Galipea longiflora*, a Bolivian plant used as a treatment for cutaneous leishmanisis. Phytother. Res. 8: 174–178.

Fournet, A., Ferreira, M.E., Rojas de Arias, A., Torres de Ortiz, S., Fuentes, S., Nakayama, H., Schinini, A., Hocquemiller, R. 1996. *In vivo* efficacy of oral and intralesional administration of 2-Substituted Quinolines in experimental treatment of new world leishmanisis caused by *Lesihmania amazonensis*. Antimicrob. Agents Chemother. 40: 2447–2451.

Fournet, A., Gantier, J.C., Gautheret, A., Leysalles, L., Munos, L.H., Mayrargue, J., Moskowitz, H., Cavé, A. and Hocquemiller, R. 1994b. The activity of 2-substituted quinoline alkalids in BALB/c mice infected with *Leishmania donovani*. J. Antimicrob. Chemother. 33: 537–544.

Fournet, A., Hocquemiller, R., Roblot, F., Cavé, A., Richomme, P., Bruneton, J. 1993a. Les chimanines, nouvelles quinoleines substituees en 2, isolees d'une plante bolivienne atiparasitaire: *Galipea longiflora*. J. Nat. Prod. 56: 1547–1552.

Fournet, A., Vagneur, B., Vagner, B., Richomme, P., Bruneton, J. 1989. Aryl-2 et alkyl-2 quinoléines nouvelles isolees d'une Rutacée bolivienne: *Galipea longiflora*. Can. J. Chem. 67: 2116–2118.

Freixa, B., Vila, R., Vargas, L., Lozano, N., Adzet, T., Canigueral, S. 1998. Screening for antifungal activity of nineteen Latin American plants. Phytother. Res. 12: 427–430.

Fromentin, Y., Cottet, K., Kritsanida, M., Michel, S., Gaboriaud-Kolar, N., Lallemand, M.C. 2015. *Symphonia globulifera*, a widespread source of complex metabolites with potent biological activities. Planta Med. 81: 95–107.

Gachet, M.S., Schühly, W. 2009. Jacaranda-an ethnopharmacological and phytochemical review. J. Ethnopharmacol. 121: 14–27.

Gadisa, E., Salamanca, E., Aseffa, A., Ticona, J.C., Udaeta, E., Flores, N., Chuqui, R., Giménez, A. 2016. Estudios de suceptibilidad de cepas de *Leishmania aethiopica* frente a alcaloides de *Galipea longiflora* (Evanta). Rev. Con-Ciencia 4: 11–19.

Gagnon, E., Bruneau, A., Hughes, C.E., Queiroz, L.P., Lewis, G.P. 2016. A new generic system for the pantropical *Caesalpinia* group (Leguminosae). PhytoKeys. 71: 1–160.

Galeno, D.M.L., Carvalho, R.P., Boletí, A.P.A., Lima, A.S., Almeida, P.D.O., Pacheco, C.C., Souza, T.P., Lima, E.S. 2014. Extract from *Eugenia punicifolia* is an antioxidant and inhibits enzymes related to metabolic syndrome. Appl. Biochem. Biotechnol. 172: 311–324.

Gallo, M.B.C., Sarachine, M.J. 2009. Biological activities of lupeol. Int. J. Biomed. Pharm. Sci. 3: 46–66.

García-Barriga, H.G. 1992. Flora medicinal de Colombia: botánica médica. Vol II/Vol III, (Instituto de Ciencias Naturales, Universidad Nacional).

García, I.M., García, C.H.M. 2008. Etnobotánica del árbol "sangregado" *Croton draco* (Euphorbiaceae) en la Sierra de Santa Marta, México. Universciencia 15: 41–48.

Giménez, A., Avila, J.A., Ruiz, G., Paz, M., Udaeta, E., Ticona, J.C., Salamanca, E., Paredes, C., Rodriguez, N., Quints, K., Feraudy, C., Gutierrez, I., Chuqui, R., Quenevo, C., Dalence, M.F., Bascope, M. 2005. Chemical, biological and pharmacological studies on *Galipea Longiflora*, Krause. Rev. Bol. Quím. 22: 94–107.

Glasby, J.S. 2005. Dictionary of plants containing secondary metabolites. Taylor and Francis e-Library. 1644 p.

Goldsby, G., Burke, B.A. 1987. Sesquiterpene lactones and a sesquieterpene diol from jamaica Ambrosia peruviana. Phytochemistry 26: 1059–1063.

Gomes, T.B., Bandeira, F.P.S.F. 2012. Uso e diversidade de plantas medicinais em uma comunidade quilombola no Raso da Catarina, Bahia. Acta Bot. Bras. 26: 796–809.

González, J. 2006. Flora digital de la selva. Organización para estudios tropicales. Costa Rica. 1–35.

Goun, E., Cunningham, G., Chu, D., Nguyen, C., Miles, D. 2003. Antibacterial and antifungal activity of Indonesian ethnomedical plants. Fitoterapia 76: 592–596.

Granados-Echegoyen, C., Pérez-Pacheco, R., Soto-Hernández, M., Ruiz-Vega, J., Lagunez-Rivera, L., Alonso-Hernández, N., Gato-Armas, R. 2014. Inhibition of the growth and development of mosquito larvae of *Culex quinquefasciatus* (Diptera: Culicidae) treated with extract from leaves of *Pseudocalymma alliaceum* (Bignonaceae). Asian Pac. J. Trop. Med. 7: 594–601.

Granados-Echegoyen, C., Pérez-Pacheco, R., Alexander-Aguilera, A., Lagunez-Rivera, L., Alonso-Hernández, N., Chairez-Martinez, E.J. 2015. Effects of aqueous and ethanol extract of dried leaves of *Pseudocalymma alliaceum* (Bignonaceae) on haematological and biochemical parameters of wistar rats. Asian Pac. J. Reprod. 4: 129–134.

Grandtner, M.M., Chevrette, J. 2013. Dictionary of trees. Volume 2, South America. Nomenclature, taxonomy and ecology. Elsevier—Academic Press Publications. 1159 p.

Grangeiro, M.S., Calheiros-Lima, A.P., Martins, M.F., Arruda, L.F., Garcez-do-Carmo, L., Santos, W.C. 2006. Pharmacological effects of *Eugenia punicifolia* (Myrtaceae) in cholinergic nicotinic neurotransmission. J. Ethnopharmacol. 108: 26–30.

Guauque, M.P., Castaño, J.C., Gómez-Barrera, M. 2010. Detección de metabolitos secundarios en *Ambrosia peruviana* Willd. y determinación de la actividad antibacteriana y antihelmíntica. Rev. Infectio 14: 186–194.

Gupta, M.P., Monge, A., Karikas, G.A., Lopez de Cerain, A., Solis, P.N., de Leon, E., Trujillo, M., Suarez, O., Wilson, F., Montenegro, G., Noriega, Y., Santana, A.I., Correa, M., Sanchez, C. 1996. Screening of Panamanian medicinal plants for brine shrimp toxicity, crown gall tumor inhibition, cytotoxicity and DNA intercalation. Int. J. Pharmacogn. 34: 19–27.

Gustafson, K.R., Blunt, J.W., Munro, M.H.G., Fuller, R.W., Mckee, T.C., Cardellina, J.H., McMahon, J.B., Cragg, G.M., Boyd, M.R. 1992. The guttiferones, HIV-inhibitory benzophenones from *Symphonia globulifera*, *Garcinia livingstonei*, *Garcinia ovalifolia* and *Clusia rosea*. Tetrahedron 48: 10093–10102.

Hamann, F.R., Brusco, I., de Campos Severo, G., de Carvalho, L.M., Faccin, H., Gobo, L., Oliveira, S.M., Rubin, M.A. 2019. *Mansoa alliacea* extract presents antinociceptive effect in a chronic inflammatory pain model in mice through opioid mechanisms. Neurochem. Int. 122: 157–169.

Hayder, N., Bouhlel, I., Skandrani, I., Kadri, M., Steiman, R., Guiraud, P., Mariotte, A.M.G., Ghedira, K., Dijoux-Franca, M.G., Chekir-Ghedira, L. 2008. *In vitro* antioxidant and antigenotoxic potentials of myricetin-3-*o*-galactoside and myricetin-3-*o*-rhamnoside from *Myrtus communis*: modulation of expression of genes involved in cell defence system using cDNA microarray. Toxicol. *In Vitro* 22: 567–581.

Hernandes, L.S. 2015. Farmacologia e fitoquímica de extratos e formulações de *Jacaranda decurrens* Cham., *Jacaranda caroba* (Vell.) DC. e *Piper umbellatum* L. Doctoral thesis. Universidade de São Paulo.

Herz, W., Anderson, G., Gibaja, S., Raulais, D. 1969. Sesquiterpene lactones of some *Ambrosia* species. Phytochemistry 8: 877–881.

Hiruma-Lima, C.A., Di Stasi, L.C. 2002. Plantas medicinais na Amazônia e na Mata Atlântica. São Paulo: Editora UNESP. 323–330.

Hoff, W. 1992. Traditional healers and community health. World Health Forum 13: 182–187.

IDH project 2010-2014. IIFB-FCFB-UMSA. Desparasitación de niños en escuelas rurales I y II.

Itokawa, H., Matsumoto, K., Morita, H., Takeya, K. 1992. Cytotoxic naphthoquinones from *Mansoa alliacea*. Phytochemistry 31: 1061–1062.

IUCN. 2019. International Union for the Conservation of Nature. Species for restoration . https://www.iucn.org/es.

Jacaranda caroba—photo 8.3c. 57 Congresso Brasileiro de Química. 2017. http://www.abq.org.br/cbq/2017/trabalhos/7/12479-24348.html. Consulted on-line: May 2019.

Jacaranda copaia—photo 8.4b. UICN.org. Especies restauración. https://www.especiesrestauración-uicn.org/data_
    especie_img.php?sp_name=Jacaranda%20copaia. Consulted on-line: May 2019.
Jimenez-Usuga, N.S., Malafronte, N., Cotugno, R., De Leo, M., Osorio, E., De Tommasi, N. 2016. New sesquiterpene
    lactones from *Ambrosia cumanensis* Kunth. Fitoterapia 113: 170–174.
Joseph-Nathan, P., Romo, J. 1966. Isolation and structure of peruvin. Tetrahedron 22: 1723–1728.
Leite, P.E., Lima-Araújo, K.G., França, G.R., Lagrota-Candido, J., Santos, W.C., Quirico-Santos, T. 2014. Implant of
    polymer containing pentacyclic triterpenes from *Eugenia punicifolia* inhibits inflammation and activates skeletal
    muscle remodeling. Arch. Immunol. Ther. Exp. (Warsz) 62: 483–491.
Limachi, I., Condo, C., Palma, C., Nina N., Salamanca, E., Ticona, J.C., Udaeta, E., Flores, N., Serato, A., Marupa,
    N., Chao, B., Ibaguarid, G., Naye, C., Manner, S., Sterner, O., Giménez, A. 2019. Antiparasitic metabolites from
    Hyptis brevipes, a Tacana medicinal plant. Nat. Prod. Comm. 14: 1–4.
Lira, S.R.S., Rao, V.S., Carvalho, A.C.S., Guedes, M.M., Morais, T.C., Souza, A.L., Trevisan, M.T.S., Lima, A.F.,
    Chaves, M.H., Santos, F.A. 2009. Gastroprotective effect of lupeol on ethanol-induced gastric damage and the
    underlying mechanism. Inflammopharmacology 17: 221–228.
Lohmann, L.G. 2015. Bignoniaceae in lista de espécies da flora do Brasil. Jardim Botânico do Rio de Janeiro. Growing
    knowledge: an overview of seed plant diversity in Brazil. Rodriguésia 66: 1085–1113.
Lopez, A., Hudson, J.B., Towers, G.H.N. 2001. Antiviral and antimicrobial activities of Colombian medicinal plants.
    J. Ethnopharmacol. 77: 189–196.
Luna, J.S., Santos, A.F., Lima, M.R.F., Omena, M.C., Mendonça, F.A.C., Bieber, L.W., Santana, A.E.G. 2005. A study
    of the larvicidal and molluscicidal activities of some medicinal plants from northeast Brazil. J. Ethnopharmacol.
    97: 199–206.
Maia, G.N. 2004. Caatinga: árvores e arbustos e suas utilidades/Gerda Nickel Maia. 1st Ed. São Paulo, DandZ
    Computação Gráfica e Editora.
Maia, J.G.S., Zoghbi, M.G.B., Luz, A.I.R. 1997. Essential oil of *Eugenia punicifolia* (HBK) DC. J. Essent. Oil Res.
    9: 337–338.
Manaharan, T., Appleton, D., Cheng, H.M., Palanisamy, U.D. 2012. Flavonoids isolated from *Syzygium aqueum* leaf
    extract as potential antihyperglycaemic agents. Food Chem. 132: 1802–1807.
Mansoa alliacea—photo 8.6b. Tropicos.org. Missouri Botanical Garden. http://www.tropicos.org/Image/100362051.
    Consulted on-line: May 2019.
Marti, G., Eparvier, V., Moretti, C., Prado, S., Grellier, P., Hue, N., Thoison, O., Delpech, B., Guéritte, F., Litaudon,
    M. 2010. Antiplasmodial benzophenone derivatives from the root barks of *Symphonia globulifera* (Clusiaceae).
    Phytochemistry 71: 964–974.
Martins, L.R.R., Brenzan, M.A., Nakamura, C.V., Filho, B.P.D., Nakamura, T.U., Cortez, L.E.R., Cortez, D.A.G.,
    2011. *In vitro* antiviral activity from *Acanthospermum australe* on herpesvirus and poliovirus. Pharm. Biol.
    49: 26–31.
Martins, M.B.G., Castro, A.A., Cavalheiro, A.J. 2008. Caracterização anatômica e química de folhas de *Jacaranda
    puberula* (Bignoniaceae) presente na Mata Atlântica. Rev. Bras. Farmacol. 18: 600–607.
Martínez, M.A., Evangelista, V., Basurto, F., Mendoza, M., Cruz-Rivas, A. 2007. Flora útil de los cafetales en la
    Sierra Norte de Puebla, México. Rev. Mex. Biodivers. 78: 15–40.
Martínez-Grueiro, M., Giménez-Pardo, C., Gómez-Barrio, A., Franck, X., Fournet, A., Hocquemiller, R., Figadère, B.,
    Casado-Escribano, N. 2005. Nematocidal and trichomonacidal activities of 2-substituted quinolines. Farmaco
    60: 219–224.
Martino, R., Beer, M.F., Elso, O., Donadel, O., Sülsen, V., Anesini, C. 2015. Sesquiterpene lactones from *Ambrosia*
    spp. are active against a murine lymphoma cell line by inducing apoptosis and cell cycle arrest. Toxicol. *In Vitro*
    29: 1529–1536.
Matsunaga, K., Saitoh, M., Ohizumi, Y. 1996. Acanthostral, a novel antineoplastic *cis, cis, cis*-germacronalide from
    Acanthospermum australe. Tetrahedron Lett. 37: 1455–1456.
Melo, J.G., Araújo, T.A.S., Castro, V.T.N.A., Cabral, D.L.V., Rodrigues, M.V., Nascimento, S.C., Amorim, E.L.C.,
    Albuquerque, U.P. 2010. Antiproliferative activity, antioxidant capacity and tannin content in plants of semi-arid
    northeastern Brazil. Molecules 15: 8534–8542.
Mendes, C.C., Bahia, M.V., David, J.M., David, J.P. 2000. Constituents of *Caesalpinia pyramidalis.* Fitoterapia
    71: 205–207.
Mesa, A.M., Naranjo, J.P., Díez, A.F., Ocampo, O., Monsalve, Z.L. 2017. Actividad antibacterial y larvicida sobre
    Aedes aegypty L. de extractos de *Ambrosia peruviana* Willd (Altamisa). Rev. Cubana Plant. Med. 22: 1–13.
Mirandola, L., Justo, G.Z., Queiroz, M.L.S. 2002. Modulation by *Acanthospermum Australe* extracts of the tumor
    induced hematopoietic changes in mice. Immunopharmacol. Immunotoxicol. 24: 275–288.

Mkounga, P., Fomum, Z.T., Meyer, M., Bodo, B., Nkengfack, A.E. 2009. Globulixanthone F, a new polyoxygenated xanthone with an isoprenoid group and two antimicrobial biflavonoids from the stem bark of *Symphonia globulifera*. Nat. Prod. Commun. 4: 803–808.

Morais, L.M.F., Conceição, G.M., Nascimento, J.M. 2014. Família Myrtaceae: análise morfológica e distribuição geográfica de uma coleção botânica. Agrarian Academy. Goiânia, Centro Científico Conhecer. 346 p.

Mostacero, J., Castillo-Picon, F., Mejia, F.R., Gamarra, O.A., Charcape, J.M., Ramírez, R.A. 2011. Plantas medicinales del Perú. Taxonomia, ecogeografía, fenología y etnobotánica. Asamblea Nacional de Rectores. Instituto de estudios Universitarios. Trujillo.

Muñoz, V., Sauvain, M., Bourdy, G., Callapa, J., Rojas, I., Vargas, L., Tae, A., Deharo, E. 2000. The search for natural bioactive compounds through a multidiciplinary approach in Bolivia: antimalarial acitivy of some plants used by Mosetene Indians. J. Ethnopharmacol. 69: 139–155.

Murillo, R.M., Japukovic, J. Rivera, J., Castro, V.H. 2001. Diterpenes and other constituents from *Croton draco* (Euphorbiaceae). Rev. Biol. Trop. 49: 259–264.

Nakayama, H., Ferreira, M.E., Rojas de Arias, A., Vera de Bilbao, N., Torres, S., Schinini, A., Fournet, A. 2001. Experimental treatment of chronic *Trypanosoma cruzi* infection in mice with 2-*n*-propylquinoline. Phytother. Res. 15: 630–632.

Ndjakou Lenta, B., Vonthron-Sénécheau, C., Fongang Soh, R., Tantangmo, F., Ngouela, S., Kaiser, M., Tsamo, E., Anton, R., Weniger, B. 2007. *In vitro* antiprotozoal activities and cytotoxicity of some selected Cameroonian medicinal plants. J. Ethnopharmacol. 111: 8–12.

Ngouela, S., Ndjakou Lenta, B., Tchamo, D.N., Zelefack, F., Tsamo, E., Connolly, J.D. 2005. A prenylated xanthone with antimicrobial activity from the seeds of *Symphonia Globulifera*. Nat. Prod. Res. 19: 23–27.

Nishizawa, T., Tsuchiya, A., Pinto, M.M.V. 2005. Characteristics and utilization of tree species in the semiarid woodland of northeast Brazil. *In*: Nishizawa, T., Uitto, J.I. (eds.). The fragile tropics of Latin America: sustainable management of changing environments.

Novais, T.S., Costa, J.F.O., David, J.P.L., David, J.M., Queiroz, L.P., França, F., Giulietti, A.M., Soares, M.B.P., Santos, R.R. 2003. Atividade antibacteriana em alguns extratos de vegetais do semiárido brasileiro. Rev. Bras. Farmacogn. 13: 4–7.

Oliveira, J.C.S., David, J.M., David, J.P. 2016. Composição química das cascas das raízes e flores de *Poincianella pyramidalis* (Fabaceae). Quím. Nova 39: 189–193.

Oliveira, R.N., Dias, I.J.M., Câmara, C.A.G. 2005. Estudo comparativo do óleo essencial de *Eugenia punicifolia* (HBK) DC. de diferentes localidades de Pernambuco. Rev. Bras. Farmacog. 15: 39–43.

Olivera-Condori, M., Flores-Arizaca, J., Vasquez-Zavaleta, T., Ocsa-Borda, E. 2013. Propiedades fisicoquímicas y bioactivas *in vitro* del aceite esencial de *Mansoa alliacea* (Lam.) A. Gentry. El Ceprosimad 2: 96–102.

Ono, M., Masuoka, C., Odake, Y., Ito, Y., Nohara, T. 2000. Eudesmane derivatives from *Tessaria integrifolia*. Phytochemistry 53: 479–484.

Patent CN102838538A. Preparation method of chimanine A.

Patent CN105724253A. Tissue culture method of Mansoa alliacea.

Patent FR2682107A1. 2-Substituted quinolines for the treatment of leishmaniasis.

Patent JP2002316936A. Antibacterial agent and anti-inflammatory agent.

Patent US20020041906A1. Compositions and methods for enhancing therapeutic effects.

Patent US20030099727A1. Compositions and methods for reducing cytotoxicity and inhibiting angiogenesis.

Patent US20060165812A1. Method and topical formulation for treating headaches.

Patent US4078145A. Phytoquinoid possessing anti-tumor activity.

Patent US7368137B2. Smilagenin and its use.

Patent WO2006078848A1. Compositions containing botanical extracts rich in phlorizin and methods for using such compositions in blood glucose modification and to affect aging.

Peluso, G., De Feo, V., De Simone, F., Bresciano, E., Vuotto, M.L. 1995. Studies on the inhibitory effects of caffeoylquinic acids on monocyte migration and superoside ion production. J. Nat. Prod. 58: 639–646.

Pennington T.D., Sarukhán, J. 1968. Árboles tropicales de México. INIF, FAO. México, D.F.

Peraza-Sánchez, S.R., Cen-Pacheco, F., Noh-Chimal, A., May-Pat, F., Simá-Polanco, P., Dumonteil, E., García-Miss, M.R., Mut-Martín, M. 2007. Leishmanicidal evaluation of extracts from native plants of the Yucatan peninsula. Fitoterapia 78: 315–318.

Pereira, N.R. 2018. Isolamento de ácidos terpênicos de *Jacaranda caroba*, síntese e avaliação da atividade antimicrobiana de derivados do ácido ursólico. Master thesis. Programa de Pós-Graduação em Química, Universidade Federal dos Vales do Jequitinhonha e Mucuri, Diamantina. 98 p.

Pereira, S.C., Gamarra-Rojas, C.F.L., Gamarra-Rojas, G., Lima, M., Gallindo, F.A.T. 2003. Plantas úteis do Nordeste do Brasil. Recife: Centro Nordestino de Informações sobre Plantas—CNIP. Associação Plantas do Nordeste. 140 p.

Pereira, R.A., Zoghbi, M.G.B., Bastos, M.N.C. 2010. Essential oils of twelve species of Myrtaceae growing wild in the Sandbank of the Resex Maracanã, State of Pará, Brazil. J. Essent. Oil Bear. Plant. 13: 440–450.

Pérez-Nájera, V.C., Gutiérrez-Uribe, J.A., Antunes-Ricardo, M., Hidalgo-Figueroa, S., Del Toro-Sánchez, C.L., Salazar-Olivo, L.A., Lugo-Cervantes, E. 2018. *Smilax aristolochiifolia* root extract and its compounds chlorogenic acid and astilbin inhibit the activity of α-amylase and α-glucosidase enzymes. Evid.-Based. Compl. Alt. 2018: 1–13.

Pollier, J., Goossens, A. 2012. Oleanolic acid. Phytochemistry 77: 10–15.

Polanco-Hernández, G., Escalante·Erosa, F., García·Sosa, K. Acosta, K., Chan·Bacab, M., Sagua, H., Gonzalez, J., Osario, L., Peña-Rodríguez, L.M. 2012. *In vitro* and *in vivo* trypanocidal activity of native plants from the Yucatan península. Parasitol. Res. 110: 31–35.

Quattrocchi, U. 2012. CRC World dictionary of medicinal and poisonous plants: common names, scientific names, eponyms, synonyms, and etymology (5 volume set). CRC Press, Boca Ratón. 4038 p.

Quenevo, C., Bourdy, G., Giménez, A. 1999. "Tacana: ecuanasha aquí, ecuanasha id'rene cuana, me schanapaque" (Tacana: conozcan nuestros árboles, nuestras hierbas). UMSA -CIPTA-IRD-FONAMA-EIA, La Paz, Bolivia. 499 p.

Quiroga-Selez, G., Parra-Lizarazu, C., Salamanca-Capusiri, E., Flores-Quisbert, E., Giménez-Turba, A. 2016. Chemical study of seedlings of *Galipea longiflora* (Evanta), a bio-guided approach. Rev. Bol. Quim. 33: 134–141.

Raden, S.H.R., Werner, R. 1985. The role of traditional healers in the provision of health care and family planning services: Malay traditional and indigenous medicine. Malays. J. Reprod. Health 3: S82–S89.

Ramos, M.F.S., Monteiro, S.S., Silva, V.P., Nakamura, M.J., Siani, A.C. 2010. Essential oils from Myrtaceae species of the Brazilian Southeastern maritime forest (Restinga). J. Essent. Oil Res. 22: 109–113.

Rana, B.K., Taneja, V., Singh, U.P. 1999. Antifungal activity of an aqueous extract of leaves of garlic creeper (*Adenocaymma alliaceum* Miers.). Pharm. Biol. 37: 13–16.

Rauf, A., Jehan, N. 2015. Chapter: The folkloric use of medicinal plants in public health care. SM Group Open Access eBooks, 160 Greentree Drive, Suite 101, Dover, DE 19904, USA. 1–13.

Rengifo, E. 2005. Análisis de experiencias y registros, 2000–2004.

Ribeiro, A.R.S., Diniz, P.B., Estevam, C.S., Pinheiro, M.S., Albuquerque-Júnior, R.L., Thomazzi, S.M. 2013. Gastroprotective activity of the ethanol extract from the inner bark of *Caesalpinia pyramidalis* in rats. J. Ethnopharmacol. 147: 383–388.

Ribeiro, T.G., Chávez-Fumagalli, M.A., Valadares, D.G., Franca, J.R., Lage, P.S., Duarte, M.C., Andrade, P.H., Martins, V.T., Costa, L.E., Arruda, A.L., Faraco, A.A., Coelho, E.A., Castilho, R.O. 2014. Antileishmanial activity and cytotoxicity of Brazilian plants. Exp. Parasitol. 143: 60–68.

Romo, J., Joseph-Nathan, P., Romo de Vivar, A., Álvarez, C. 1967. The structure of peruvinin—a pseudoguaianolide isolated from *Ambrosia peruviana* Willd. Tetrahedron 23: 529–534.

Roosita, K., Kusharto, C.M., Sekiyama, M., Fachrurozi, Y., Ohtsuka, R. 2008. Medicinal plants used by villagers of a Sundanese community in West Java, Indonesia. J. Ethnopharmacol. 115: 72–81.

Roth, I., Lindorf, H. 2002. South American medicinal plants. Botany, remedial properties and general uses. Springer-Verlag Berlin Heidelberg. 498 p.

Roumy, V., Gutierrez-Choquevilca, A.L., Lopez Mesia, J.P., Ruiz, L., Ruiz Macedo, J.C., Abedini, A., Landoulsi, A., Samaillie, J., Hennebelle, T., Rivière, C., Neut, C. 2015. *In vitro* Antimicrobial activity of traditional plant used in Mestizo Shamanism from the Peruvian Amazon in case of infectious diseases. Pharmacogn. Mag. 11: S625–S633.

Ruiz, L., Ruiz, L., Maco, M., Cobos, M., Gutierrez, A.L., Roumy, V. 2011. Plants used by native Amazonian groups from the Nanay River (Peru) for the treatment of malaria. J. Ethnopharmacol. 133: 917–921.

Saeidnia, S., Manayi, A., Gohari, A.R., Abdollahi, M. 2014. The story of beta-sitosterol—a review. Eur. J. Med. Plant. 4: 590–609.

Sakr, H.H. 2014. Effect of *Hyptis brevipes* dichloromethane extract on feeding and histological structure of the midgut and malpighian tubules of *Spodoptera littoralis* Larvae (Lepidoptera: Noctuidae). Sci. J. Fac. Sci. XXVI: 1–19.

Sakr, H.H., Roshdy, S.H., El-Seedi, H.R. 2013. *Hyptis brevipes* (Lamiaceae) extracts strongly inhibit the growth and development of *Spodoptera littoralis* (Boisd.) larvae (Lepidoptera: Noctuidae). J. Appl. Pharm. Sci. 3: 83–88.

Salatino, A., Faria, M.L., Negri, G. 2007. Traditional uses, chemistry and pharmacology of *Croton* species (Euphorbiaceae). J. Braz. Chem. Soc. 18: 11–33.

Sales, D.S., Carmona, F., Azevedo, B.C., Taleb-Contini, S.H., Bartolomeu, A.C.D., Honorato, F.B., Martinez, E.Z., Pereira, A.M.S. 2014. *Eugenia punicifolia* (Kunth) DC. as an adjuvant treatment for type-2 diabetes mellitus: a non-controlled, pilot study. Phytother. Res. 28: 1816–1821.

Sampaio, E.V.B., Pareyn, F.G.C., Figucrôa, J.M., Santos-Júnior, A.G. 2005. Espécies da flora nordestina de importância econômica potencial. Recife, Associação Plantas do Nordeste, 331 p.

Sánchez, M., Kramer, F., Bargardi, S., Palermo, J.A. 2009. Melampolides from Argentinean *Acanthospermum australe*. Phytochem. Lett. 2: 93–95.

Santana, D.G., Santos, C.A., Santos, A.D.C., Nogueira, P.C.L., Thomazzi, S.M., Estevam, C.S., Antoniolli, A.R., Camargo, E.A. 2012. Beneficial effects of the ethanol extract of *Caesalpinia pyramidalis* on the inflammatory response and abdominal hyperalgesia in rats with acute pancreatitis. J. Ethnopharmacol. 142: 445–455.

Santos, C.A., Passos, A.M., Andrade, F.C., Camargo, E.A., Estevam, C.S., Santos, M.R., Thomazzi, S.M. 2011. Antinociceptive and anti-inflammatory effects of *Caesalpinia pyramidalis* in rodents. Rev. Bras. Farmacogn. 21: 1077–1083.

Santos, C.A., Santos, D.S., Santana, D.G., Thomazzi, S.M. 2013. Evaluation of mechanisms involved in the antinociception of the ethanol extract from the inner bark of *Caesalpinia pyramidalis* in mice. J. Ethnopharmacol. 148: 205–209.

Saraiva, A.M., Saraiva, M.G., Gonçalves, A.M., Sena Filho, J.G., Xavier, H.S., Pisciottano, M.N.C. 2012. Avaliação da atividade antimicrobiana e perfil fitoquímico de *Caesalpinia pyramidalis* Tull. (Fabaceae). BioFar- Rev. Biol. Farm. 7: 52–60.

Sarquis, R.S.F.R., Sarquis, I.R., Sarquis, I.R., Fernandes, C.P., da Silva, G.A., Silva, R.B.L., Jardim, M.A.G., Sánchez-Ortíz, B.L., Carvalho, J.S.T. 2019. The use of medicinal plants in the riverside community of the Mazagão River in the Brazilian Amazon, Amapá, Brazil: ethnobotanical and ethnopharmacological studies. Evid.-Based Compl. Alt. 2019: 6087509.

Sautour, M., Miyamoto, T., Lacaille-Dubois, M.A. 2005. Steroidal saponins from and their antifungal activity. J. Nat. Prod. 68: 1489–1493.

Sautour, M., Miyamoto, T., Lacaille-Dubois, M.A. 2006. Bioactive steroidal saponins from *Smilax medica*. Planta Med. 72: 667–670.

Sauvain, M., Dedet, J.P., Kunesch, N., Poisson, J., Gantier, J.C., Gayral, P., Kunesch, G. 1993. *In vitro* and *in vivo* leishmanicidal activities of natural and synthetic quinoids. Phytother. Res. 7: 167–171.

Setzer, W., Stokes, S., Bansal, A., Haber, W., Caffrey, C., Hansell, E., McKerrow, J. 2007. Chemical composition and cruzain inhibitory activity of *Croton draco* bark essential oil from Monteverde, Costa Rica. Nat. Prod. Commun. 2: 685–689.

Shimizu, M., Horie, S., Arisawa, M., Hayashi, T., Suzuki, S., Yoshizaki, M., Kawasaki, M., Terashima, S., Tsuji, H., Wada, S., Ueno, H., Morita, N., Berganza, L.H., Ferro, E., Basualdo, I. 1987. Chemical and pharmaceutical studies on medicinal plants in Paraguay. I. Isolation and identification of lens aldose reductase inhibitor from "Tapecué" *Acanthospermum australe*. Chem. Pharm. Bull. 35: 1234–1237.

Silva, F.D.B., Sales, M.A.G., Sá, O.R.M., de Deus, M.S.M., Sousa, J.M.C., Peron, A.P., Ferreira, P.M.P. 2015. Potencial citotóxico, genotóxico e citoprotetor de extratos aquosos de *Caesalpinia pyramidalis* Tul, *Caesalpinia ferrea* Mart., e *Caesalpinia pulcherrima* Sw. Rev. Bras. Biociências. 13: 101–109.

Silva-Correa, C.R., Cruzado-Rasco, J.L., Gonzalez-Blas, M.V., Garcia-Armas, J.M., Ruiz-Reyes, S.G., Villarreal-La Torre, V.E., Gamarra-Sanchez, C.D. 2018. Identification and structural determination of a sesquiterpene of *Tessaria integrifolia* Ruiz and Pav. Leaves and evaluation of its leishmanicidal activity. Rev. Peru. Med. Exp. Salud Pública 35: 221–227.

Sobral, M., Proença, C., Souza, M., Mazine, F., Lucas, E. 2015. *Myrtaceae* in Lista de Espécies da Flora do Brasil. Jardim Botânico do Rio de Janeiro. Avaliable in: http://floradobrasil.jbrj.gov.br/jabot/floradobrasil/FB10515>. Access: july 2019.

Standley, P.C., Steyermark, J.A. 1946. Malpighiaceae. *In*: Standley, P.C., Steyermark, J.A. (eds.). Flora of Guatemala. Part V. Fieldiana, Bot. 24: 468–500.

Suárez-Ortiz, G.A., Cerda-García-Rojas, C.M., Fragoso-Serrano, M., Pereda-Miranda, R. 2017. Complementarity of DFT calculations, NMR anisotropy, and ECD for the configurational analysis of Brevipolides K-O from *Hyptis brevipes*. J. Nat. Prod. 80: 181–189.

Suárez-Ortiz, G.A., Cerda-García-Rojas, C.M., Hernández-Rojas, A., Pereda-Miranda, R. 2013. Absolute configuration and conformational analysis of brevipolides, bioactive 5,6-dihydro-α-pyrones from *Hyptis brevipes*. J. Nat. Prod. 76: 72–78.

Sülsen, V.P., Cazorla, S.I., Frank, F.M., Laurella, L.C., Muschietti, L.V., Catalán, C.A., Martino, V.S., Malchiodi, E.L. 2013. Natural terpenoids from *Ambrosia* species are active *in vitro* and *in vivo* against human pathogenic trypanosomatids. PLoS Negl. Trop. Dis. 7: e2494.

Symphonia globulifera—photo 8.4c. Neotropical plant portal detailed collection record information. https://serv.biokic.asu.edu/imglib/neotrop/misc/201406/10949_1403068451_web.jpg. Consulted on-line: May 2019.

Taylor, P., Arsenak, M., Abad, M.J., Fernández, Á., Milano, B., Gonto, R., Ruiz, M.C., Fraile, S., Taylor, S., Estrada, O., Michelangeli, F. 2013. Screening of Venezuelan medicinal plant extracts for cytostatic and cytotoxic activity against tumor cell lines. Phytother. Res. 27: 530–539.

Towne, C., Dudt, J.F., Ray, D.B. 2015. Effect of *Mansoa alliacea* (Bignonaceae) leaf extract on embryonic and tumorigenic mouse cell lines. J. Med. Plant Res. 9: 799–805.

Tropicos org. 2019. Missouri Botanical Garden. June-august 2019. http://www.tropicos.org.

Tsacheva I., Rostan, J., Iossifova, T., Vogler, B., Odjakova, M., Navas, H., Kostova, I., Kojouharova, M., Kraus, W. 2004. Complement inhibiting properties of dragon's blood from *Croton draco*. Z. Naturforsch 59c: 528–532.

UEIA. 2014. Catálogo Virtual de flora del Valle de Aburrá. https://catalogofloravalleaburra.eia.edu.co. Consulted on-line: june-august 2019.

Valadares, Y.M. 2009. *Remijia ferruginea* D.C, *Jacaranda caroba* D.C e *Solanum paniculatum L*: fitoquímica, atividades biológicas e síntese de derivados dos ácido ursólico e oleanólico. Master thesis. Universidade Federal de Minas Gerais. 323 p.

Valadeau, C., Pabon, A., Deharo, E., Albán-Castillo, J., Estevez, Y., Lores, F.A., Rojas, R., Gamboa, D., Sauvain, M., Castillo, D., Bourdy, G. 2009. Medicinal plants from the Yanesha (Peru): Evaluation of the leishmanicidal and antimalarial activity of selected extracts. J. Ethnopharmacol. 123: 413–422.

Vásquez-Ocmín, P., Cojean, S., Rengifo, E., Suyyagh-Albouz, S., Amasifuen-Guerra, C.A., Pomel, S., Cabanillas B., Mejía, K., Loiseau, P.M., Figadère, B., Maciuk, A. 2018. Antiprotozoal activity of medicinal plants used by Iquitos-Nauta road communities in Loreto (Peru). J. Ethnopharmacol. 210: 372–385.

Velasco-Lezama, R., Muñoz Torres, A., Tapia Aguilar, R., Flores Sáenz, J.L., Fregoso Padilla, M., Vega Avila, E., Barrera Escorcia, E. 2009. Hematopoietic activity of *Smilax aristolochiaefolia* (Zarzaparrilla) in mice with aplastic anemia. Proc. West. Pharmacol. Soc. 52: 83–87.

Villasmil, J., Abad, M.J., Arsenak, M, Fernandez, A., Ruiz, M.C., Williams, B., Michelangeli, F., Herrera, F., Peter, T. 2006. Cytotoxic and antitumor activities of Venezuelan plant extracts *in vitro* and *in vivo*. Pharmacologyonline 3: 808–816.

Vines, G. 2004. Herbal harvests with a future: towards sustainable sources for medicinal plants. Plant life International. 1–10. http://www.plantlife.org.uk.

World Heatlh Organization. 2019. WHO global report on traditional and complementary medicine. Geneva: World Health Organization. 228 p.

World Heatlh Organization. 2002. WHO general guidelines for methodologies on research and evaluation of traditional medicine. Geneva: World Health Organization.

Xu, D.H., Huang, Y.S., Jiang, D.Q., Yuan, K. 2013. The essential oils chemical compositions and antimicrobial, antioxidant activities and toxicity of three *Hyptis* species. Pharm. Biol. 51: 1125–1130.

Yánez, C.A., Rios, N., Mora, F., Rojas, L., Diaz, T., Velasco, J., Rios, N., Melendez, P. 2011. Composición química y actividad antibacteriana del aceite esencial de *Ambrosia peruviana* Willd. de los llanos venezolanos. Rev. Peru. Biol. 18: 149–151.

Zanatta, F., Gandolfi, R.B., Lemos, M., Ticona, J.C., Gimenez, A., Clasen, B.K., Cechinel-Filho, V., de Andrade, S.F. 2009. Gastroprotective activity of alkaloid extract and 2-phenylquinoline obtained from the bark of *Galipea longiflora* Krause (Rutaceae). Chem. Biol. Interact. 180: 312–317.

9

Phytochemicals from Wild Medicinal and Aromatic Plants of Argentina

María Paula Zunino,[1] *Andrés Ponce,*[2] *Alejandra Omarini*[3]
and *Julio Alberto Zygadlo*[1],*

Introduction

In the rural and indigenous communities of Argentina, medicinal plants play a very important role in the care of people's health (Goleniowski et al. 2006, Martinez and Barboza 2010, Martinez and Escobar 2017, Suarez 2018, Lujan and Martinez 2019). This strong relationship between people and medicinal plants could be explained because some rural or aboriginal communities are isolated not only as a result of the territorial extension, but also because of economic factors and the absence of modern medicine. Moreover, in these communities the traditional medicine is better accepted from a spiritual and cultural point of view. The number of medicinal plants collected and used by the first inhabitants of the current Argentine territory was increased with the contribution received by the European migrants related to their pharmacological or medicinal properties (Kujawska et al. 2017, Bermejo et al. 2019, Lujan and Martinez 2019). Currently, the market related to the aromatic and medicinal plants has grown in urban areas as a result of increasing interest to use natural products for healthcare (Bach et al. 2014, Lujan and Martinez 2019). In Argentina, the use of traditional medicine involves more than 60% of the population, and many of the medicinal plants are sold in pharmacies, street markets, or natural products shops (Bach et al. 2014, Lujan and Martinez 2019) (Figure 9.1). In recent years, the advancement of the agricultural practices over natural areas has led to a significant loss of the medicinal plants' diversity, and this fact negatively impacts the continuous supply of the urban markets.

A systematic literature survey was carried out in scientific databases, PubMed, Scopus, and Google Scholar about the research carried out on the folkloric practice versus evidence-based medicine, using medicinal plants. Thus, this chapter provides molecular, phytochemical, genetic, and pharmacological evidence to support the traditional indigenous phytotherapy practiced in Argentina.

---

[1] Universidad Nacional de Córdoba, Facultad de Ciencias Exactas, Físicas y Naturales. Instituto Multidisciplinario de Biología Vegetal (CONICET-IMBIV). Avenida Vélez Sarsfield 1611. Córdoba. Argentina; paula.zunino.254@unc.edu.ar

[2] Universidad Nacional de Córdoba, Cátedra de Fisiología Humana. Facultad de Medicina. Universidad Nacional de Córdoba. E-mail: andresaponce@daad-alumni.de

[3] Instituto de Ciencias de la Tierra y Ambientales de La Pampa (CONICET-UNLPam). Mendoza 109 (L6302EPA) Santa Rosa, La Pampa; aomarini@yahoo.com.ar

* Corresponding author: jzygadlo@unc.edu.ar

Figure 9.1: Photograph of a shop selling aromatic and medicinal plants.

In this chapter, we have not provided information about aromatic plants that produce essential oils responsible for bioactivity. This topic was previously reviewed by our group (Bigliani et al. 2012, Zunino et al. 2012, Zygadlo et al. 2017).

Wild Medicinal and Aromatic Plants of Argentina

Each indigenous community culture in Argentina makes its own selection of medicinal plants. This indicates that the same plant can be used as medicinal by one community, but it is not accepted by others. For example, the selection of medicinal plants made by the native inhabitants of the north-west of Argentina follows an analogy between the plant to be used and the disease. However, in the Mapuches community, in Patagonia, the sense of smell and taste played a very important role in the selection of medicinal plants. Cosmological concepts exist between medicinal plants and the Comechingones community (Goleniowski et al. 2006). In the Guarani culture, the plants operate as a vehicle for healing to occur (Kujawska et al. 2017). The province of Neuquén (Argentina), where the influence of the Mapuche community was strong, led the integration of the traditional with modern medicine, which concluded with the creation of an intercultural hospital called Ranguiñ Kien.

The main phytochemical compounds studied from medicinal plants were essential oils, flavonoids, phenolic acids, and phenols. In some groups, such as *Solanaceae* or *Apocynaceae*, the emphasis was placed on nitrogenated compounds, such as alkaloids and its derivatives. The most popular medicinal plants used in traditional medicine by aboriginal communities, to treat the signs or symptoms of ailments related to microbial infection, are *Asteraceae, Fabaceae, Polygonaceae, Solanaceae,* and *Euphorbiaceae*. However, 85% of the extracts did not show activity against *Candida* spp., or dermatophytes (Cordisco et al. 2019).

Phytochemicals from Wild Medicinal and Aromatic Plants of Argentina

The objective of this chapter was to validate the medicinal use of plants traditionally used in Argentina. To accomplish this objective, we only reviewed those plants that have been characterized (chemical, pharmacological, molecular biology, or bioactivity test (antimicrobial or antiparasitic)). Several studies revealed the information about the traditional medicinal plants native from Argentina and their use for medicinal purposes (Goleniowski et al. 2006, Barboza et al. 2009, Martinez and Barboza 2010, Martinez 2015, Martinez and Escobar 2017, Suarez 2018, Lujan and Martinez 2019). The scientific names of the medicinal plants were revised carefully using different websites, such as www.plantlist.org, www.tropicos.org, www.gbif.org, www.darwin.edu.ar (Darwinian Institute), and www.floraargentina.edu.ar (Flora Argentina).

The number of medicinal plants described by different authors related to the ethnobotanical used by native communities from Argentina exceeds 600 species (Goleniowski et al. 2006, Barboza et al. 2009, Martinez and Barboza 2010, Martinez 2015, Martinez and Escobar 2017, Suarez 2018, Lujan and Martinez 2019). However, it is interesting to highlight that only a few studies have been conducted to know the active ingredients present in the plant extracts or their pharmacological properties (Alvarez 2019), leaving to date only the empirical information. The phytochemical composition revealed the presence of flavonoids (quercetin, kaempferol, rutin), coumarin (umbelliferone), organic acids (quinic, chlorogenic, caffeic acids), amines (ephedrine, pseudoephedrine), alkaloids (aspidospermine, quebrachamine, crotosparine), and anthraquinones, among other compounds. However, *Solanaceae* and *Apocynaceae* showed a big diversity of alkaloids than other medicinal plants, whereas in *Ephedraceae*, the amines were the most abundant metabolite.

Acanthaceae

Ruellia ciliatiflora Hook. *R. hygrohila* Mart., *R. simplex* Wright and *Justicia comata* (L.) Lam. are species within the *Acanthaceae* family with medicinal properties, but only *J. pectoralis* Jacq. was chemically studied; several secondary metabolites were founded, such as umbelliferone, betain, flavonoids, saponins, and amino acids (Barboza et al. 2009).

Justicia pectoralis Jacq.

J. pectoralis is popularly used for stomach upset, anxiolytics, expectorant, soothing, hypotensive, leg pain, cough, or aphrodisiac problems (Leal et al. 2017). Phytochemical studies revealed the presence of coumarins, flavonoids, steroids, triterpenoids, and alkaloids. Ellagic acid and coumarin contributed to the anti-inflammatory activity (Nunes et al. 2018). The toxicology studies of hydroalcoholic extracts showed that the lethal dose (LD_{50}) in rats was 3.0 g/kg, while the oral administration of 2.0 g/kg was not lethal. The Global Harmonized System of Classification and Labeling of Chemicals classified the *J. pectoralis* extract as nontoxic (Leal et al. 2017). Nunes et al. (2018) showed lower toxicity and cytotoxicity of *J. pectoralis* extracts. Manchishi (2017) evaluated the anticonvulsant effects of this plant. Results also showed potentiation of GABAergic activity as the mode of action. Acetone/water extracts showed bacteriostatic activity against *Acinetobacter baumannii* (MIC 500 µg/mL) and *Klebsiella pneumoniae* (MIC 500 µg/mL).

Aquifoliaceae

Ilex argentina Lillo, *I. brevicuspis* Reissek, *I. dumosa* var. *guaranina* Loes, *I. theazans* Mart. ex Reissek and *I. paraguariensis*, are the most important medicinal species of this family. However, biological and pharmacological studies were carried out only with *I. paraguariensis*. Phytochemical studies revealed the presence of trimethylxanthine, dimethylxanthine, quercetin, kaempferol, rutin, chlorogenic and caffeic acids, matesaponin, matesaponin 2, 3, 4 and 5, guaicin B, nudicaucin, dicaffeoylquinic acids, and glycosides of ursolic acids (Alvarez 2019).

Ilex paraguariensis St. Hill

It is a medicinal plant commonly known as "Yerba Mate" (YM), which is consumed as a tea in Argentina, Uruguay, Paraguay, and Brazil. The tea modulates redox homeostasis of the immune and central nervous systems (Cittadini et al. 2015). Methylxanthines are heterocyclic compounds known as purine alkaloids (caffeine, theobromine, and theophylline). The contents of theobromine and caffeine in ethanolic extracts of leaves were 895.79 mg/kg and 5100 mg/kg, whereas in the aqueous extracts, the values were 837 and 4800 mg/kg of leaves, respectively (Mateos et al. 2018, Meinhart et al. 2019). Besides the alkaloids, the YM infusions are a rich source of phenolic acids and flavonoids, such as chlorogenic acid and rutin (Correa et al. 2017, Gan et al. 2018, Mateos et al. 2018). These compounds contributed to a weight loss with additional hypocholesterolemic effects (Balsan et al. 2019), and to avoid cardiovascular risk factors (Cardozo Junior and Morand 2016). It was demonstrated that the oral intake of YM exerted palliative effects on the neurological paraneoplastic syndrome, and it can contribute to neuroprotection (Cittadini et al. 2018, 2019). The quinic acid derivatives represented the most important fraction, with more than 70 percent. The most abundant compounds were 5-caffeoylquinic, 4-caffeoyl quinic acid, 1-caffeoylquinic acid, 1.3- and 1,4-dicaffeoylquinic acids, whereas hydroxycinnamic acid derivatives were present in lower quantities. The flavonoids quercetin rutinoside and isorhamnetin rutinoside were identified in the YM tea (Cittadini et al. 2018). Chlorogenic acid and quercetin from dietary YM were available and bioactive in brain, and showed a significant reduction of interleukin-6, thereby attenuating neuroinflammation and damage (Cittadini et al. 2019, 2018, de Lima et al. 2019). The extraction temperature of the water produced changes in the composition of phenolic acids and flavonoids (Gerke et al. 2018). Therefore, the different temperatures used for preparation and consumption of the mate infusions (hot or room temperature) have a specific composition (Correa et al. 2017). Moreover, the phenolic composition also changes with the YM varieties (Mateos et al. 2018). The post *in vitro* digestion, in the large and small intestines and in the liver, reduced about 20 to 33% of the total phenolic compounds (Cardozo-Junior et al. 2016, Correa et al. 2017). In the biological fluids, mainly sulfated conjugates of ferulic and caffeic acids were identified (Gómez-Juaristi et al. 2018). The colonic microbiota is mainly involved in the bioavailability of phenolic compounds (Cardozo-Junior et al. 2016, Gómez-Juaristi et al. 2018). The 13% of ingested phenolic compounds were excreted after 24 hours by urine (Gómez-Juaristi et al. 2018). Beverages prepared with YM showed antioxidant capacity by increasing serum levels of paraoxonase-1, which is associated with the antioxidant functions of high-density lipoprotein and weight loss. The YM infusions showed a strong beneficial effect in the treatment of obesity (Kim et al. 2015) by reducing lipogenesis, improvement in glucose tolerance (Oh et al. 2016), and anorexigenic effects in the short term (Rocha et al. 2018, Sahebkar-Khorasani et al. 2019). The YM infusion (1 g/kg body weight/day) adjusted antioxidant enzyme activities and decreased lipid peroxidation in overfeeding Wistar rats (Conceição et al. 2017). Furthermore, the consumption of YM infusions prevented atherosclerotic diseases (Balsan et al. 2019). The YM aqueous extract reduced the hyperglycaemia, a clinical condition caused by a carbohydrate metabolic disorder. This result was explained by the reestablishment of the redox status and the reduction of glycosylated proteins levels in the blood (de Lima et al. 2018, 2019). de Lima et al. (2019) found that YM extract reduced peripheral neuropathy. The major constituents of the flavor of YM were 2,6-dimethyl-1,7-octadien-3-ol, linalool, and α-terpineol (Polidoro et al. 2016). In addition, a polysaccharide was identified as a rhamnogalacturonan I with anti-inflammatory and antimicrobial properties (Kungel et al. 2018). The 4, 5-dicaffeoylquinic acid was the most active molecule against DU-145 prostate cancer cell, with an IC_{50} = 5 μM (Lodise et al. 2019). When the patients drink YM infusion, there is decreasing bone resorption because the YM inhibiting the osteoclastogenesis reduces oxidative stress by decreasing the receptor activator of nuclear factor kappa-B ligant and increasing osteoprotegerin (Pereira et al.

2017a). However, da Veiga et al. (2018) described a neutral effect on the bone metabolism without changes in the serum levels of total calcium of the subjects drinking YM infusions. The antimicrobial activity of the YM aqueous extract is related to the high content of phenolic acids and flavonoids. Rempe et al. (2017) suggested that the mechanism of action of the YM as an antimicrobial impacted the carbon metabolism and not the cell membrane.

Anacardiaceae

The secondary products identified in the medicinal species of this family were—α-amyrin, catechin, quercetin, rutin, kaempferol, quercetin-3-O-galactoside, shikimic acid, acylated quercetin glycosides, chamaejasmin, and isomasticadienonic acid (Romero et al. 2016, Alvarez 2019). Species of this family are used in traditional medicine, such as *Schinopsis lorentzii* (Griseb.) Engl., *Schinus bumelioides* I. M. Johnst, *S. johnstonii* F.A. Barkley, *S. longifolius* (Lindl.) Speg., *S. meyeri* F.A. Barkley, and *S. odonellii* F.A. Barkley, with a lack of biological or pharmaceutical studies.

Schinus sp.

Many illnesses in Peru are treated with aqueous or alcoholic extracts of leaves or fruit of *S. molle* L. The extracts showed low toxicity (LC_{50}= > 10000 mg/mL (Bussmann et al. 2011). This medicinal plant is used in folk medicine for the treatment of depression. It was demonstrated that the treatment with ethanolic extract showed antidepressant-like effects through serotonergic, dopaminergic, and noradrenergic systems, and it was attributed to the presence of high content of rutin (Machado et al. 2008). The administration of 0.3–3.0 mg/kg of rutin showed antidepressant effect in mice (Rabiei and Rabiei 2017), and also displayed important cytotoxic effects on human leukemic monocyte lymphoma cells line (EC_{50} from 9.5 mg/kg) (Calzada et al. 2018). The extracts of *S. areira* L. and *S. fasciculata* (Griseb) I.M. Johnst. showed antifungal properties against dermatophytes with MIC values of 0.5 to 1.0 mg/mL (Svetaz et al. 2010). The cholesterol esterase plays an important role in the hypercholesterolemia of obese people, and it was inhibited by aqueous or methanolic extracts of leaves or fruit of *S. molle*. The inhibition percentage of pancreatic cholesterol esterase was in the range of 8.98 to 15.47% (Asmaa and Ream 2016). Several phytochemicals with antibacterial activities, such as terebinthene and pinicolic acid were isolated from bark resin of *S. molle* (Malca-Garcia et al. 2017), whereas lupeol and phenolic lipids showed antifungal activity (Aristimuño Ficoseco et al. 2014). Brazilian traditional medicine used *S. lentiscifolia* Marchand as antiseptic medicament. The aqueous extract of *S. weinmannifolia* leaves exhibited antibacterial activities (MIC from 125 to 250 µg/mL), whereas the hexane extract was very effective against *Candida* species (Gehrke et al. 2013). The main phytochemical isolated from *S. weinmanniifolia* was moronic acid (3-oxoolean-18-en-28-oic acid), a diterpene with antimicrobial activity (Gehrke et al. 2013). Terán Baptista et al. (2018) founded an IC_{50} of 0.9 mg/mL against several bacterial strains using ethyl acetate extract. The main bioactive compounds identified in the ethyl acetate extract of *S. fasciculata* were kaempferol, quercetin, and agathisflavone. The agathisflavone was also founded in *S. polygamus* (Cav.) Cabrera. Dumitru et al. (2019) found that this flavone ameliorated memory and decreased anxiety by regulation of AChE activity and by controlling the oxidative stress. The methanolic extract of *S. polygamus* showed higher antipyretic, anti-inflammatory, and analgesic activities. Quercetin was isolated from methanolic extract, and this flavonol could be contributing to its pharmacological effects (Erazo et al. 2006). The presence of β-sitosterol in the hexane extract was associated with the analgesic effects (Erazo et al. 2006). Previous reports revealed that β-sitosterol decreased myeloperoxidase activity, inhibited neutrophil migration, and increased pain tolerance (Erazo et al. 2006). The chagas disease represents a major health issue in Latin America, and is caused by *Trypanosoma cruzi*, a protozoan parasite. It was reported that *S. molle* represents a valuable source of compounds that can be used for the treatment of trypanosomiasis, and methanolic extract showed 100% of growth inhibition at a dose of 150 µg/mL with an IC_{50} = 16.31 µg/mL (Molina-Garza et al. 2014).

Apocynaceae

Phytochemicals compounds isolated from species of this family were caudatin glycoside, araujiain hI, hII, hIII, bufadienolides, flavonoids, kaurene, aspidospermine, quebrachamine, normacusine B, coronaridine, aspidosamine, aspidospermine, quebrachamine, vobasine, hydroxyindolenineine, alkahimineine, alkaline alkaloimine (Alvarez 2019). Medicinal species without biological or pharmacological evaluations were *Amblyopetalum coccineum* (Griseb.) Malme, *Araujia angustifolia* (Hook. and Am.) Decne., *Asclepias flava* Lillo, *A. mellodora* A. St.-Hil., *Forsteronia glabrescens* Müll. Arg., *Funastrum flavum* (Decne.) Malme, *F. gracile* (Decne.) Schltdl., *Macrosiphonia petraea* (a. S. –Hil) K. Schum. and *Mandevilla laxa* (Ruiz and Pav.) Woodson.

Aspidosperma quebracho blanco Schltdl.

Aspidosperma quebracho-blanco is used in folk medicine for respiratory diseases. The indole alkaloids are the main phytochemical founded in the bark, and they exhibited antimalarial activity against *Plasmodium falciparum* (Bourdy et al. 2004) and analgesic activity (Benoit et al. 1973). The intake of *A. quebracho-blanco* extracts resulted in a wide variation of the redox state in the different encephalic regions (Canalis et al. 2014). For centuries in Latin America, the male impotence was treated in folk medicine with the bark of *Aspidosperma*; Sperling et al. (2002) showed that the effect may be caused by its yohimbine content.

Asteraceae

Asteraceae represents the botanical family with the largest number of medicinal species, and constitutes more than 39% of the total medicinal flora of Argentina. The principal genus reported were *Baccharis*, *Senecio*, and *Eupatorium* (Barboza et al. 2009). However, most of the phytochemical studies were related to the essential oil composition of *Senecio* (Zhao et al. 2015) and *Eupatorium* species (Zygadlo et al. 2017). The principal phytochemicals identified in different medicinal species of the *Asteraceae* family were 5-7-4'-trihydroxy-3-6-dimethoxyflavone, apigenin, fhamnazin, chlorogenic acid, caffeic acid, nevadensin, quercetin, 3-0-methyl quercetin, isochlorogenic acid, dicaffeoyl quinic acids, glucuronic acids, 3-methoxy galangin, 3,7-dimethoxy-5,8-dihydroxyflavone, alkamides, ageconyflavones, chalcones, jaceidin-7-methylether, chrysoeriol, bartemidiolide, articulinudies, articulin acetate, and eupafolin (Barboza et al. 2009, Romero et al. 2016, Alvarez 2019). Although, a large number of ethnobotanical studies showed that *Asteraceae* represents a botanical family with a large number of species used by traditional medicine. Only a few species have been investigated in order to identify their chemical composition, or have had their biological or pharmacological effects evaluated, for example, the species of the genus *Gamochaeta* or *Heterosperma* (Barboza et al. 2009).

Achyrocline sp.

Ethanolic extracts of the leaves or flowers of *A. satureioides* (Lam.) DC. showed lysis effects on *T. cruzi* (Rojas de Arias et al. 1995), and *A. flaccida* (Weinm.) DC. showed antileishmanial activity against *L. braziliensis* (Rocha et al. 2005). However, contradictory results related to the toxicity and mutagenic effects were reported (Rojas de Arias et al. 1995, Bussmann et al. 2011). The extracts of *A. satureioides* or *A. tomentosa* Rusby showed inhibitory activity on AChE. The organic fraction of *A. tomentosa* inhibited 85% of AChE, while the aqueous extract only inhibited 21.4% of AChE (Carpinella et al. 2010). The apigenin, rhamnazin, 7-hydroxy-3, 5, 40-trimethoxyflavanone, quercetin-3-methyl ether, galangin-3-methyl ether, isokaempferide, quercetine, and 3, 5-dihydroxy-6, 7, 8-trimethoxy flavone were extracted from *A. alata*. The presence of these flavonoids explains the potential to treat respiratory or digestive diseases. Many flavonoids and caffeic acid derivatives also showed antiproliferative activity on human hepatocellular carcinoma cell line Hep3B ($IC_{50} = 16.6\,\mu g/mL$)

(Carraz et al. 2015). Murine T-Lymphoma cells were inhibited by aqueous extract of *A. flaccida* (IC$_{50}$ = 30.2 µg/mL) (Fernández et al. 2002). Diabetics, alcoholics, smokers, or older persons represent the sector of the population that is susceptible to have chronic wounds. The extracts of *A. alata* (Kunth) DC., *A. flaccida*, and *A. satureioides* exhibited reepithelization and collagen remodelling of wounds. The mechanism of action showed a reduction in the inflammatory response in combination with an induction in the proliferation of keratinocytes (Cazander et al. 2012, Pereira et al. 2017, Carvalho et al. 2018). The high quantity of chlorogenic acid and quercetin, with antioxidant and anti-inflammatory properties, in the extracts of *A. alata* and *A. satureioides*, can explain the fast close of the wound (Pereira et al. 2017b). The extract of *A. alata* with a large concentration of gnaphaliin, helipyrone, obtusifolin, and lepidissipyrone showed to be effective against *S. mutans*, a caries-producing bacteria (Demarque et al. 2015). The results of Santin et al. (2010) related to the anti-ulcer effect showed that the hydroalcoholic extract of the inflorescences *A. satureioides* (500 mg/kg of extract) displayed curative ratio of 86.2%, although it was not related to the anti-secretor mechanisms. The effect was associated with an increase of mucus production (Santin et al. 2010), and with the presence of luteolin, quercetin, and 3-O-methyl-quercetin in the extracts (Santin et al. 2014).

Baccharis sp.

The diterpene ent-3α, 19-disuccinyloxy-kaur-16-ene, oleanolic acid, and the flavones irsimaritin and cirsiliol were isolated from *B. rufescens* Spreng (Simirgiotis et al. 2003). *B. trimera* (Less.) DC. (Figure 9.2) inhibited reactive oxygen species production through the PKC signalling pathway and inhibition subunit p47[phox] phosphorylation of nicotinamide adenine dinucleotide phosphate oxidase (de Araujo et al. 2017). Furan neo-clerodane diterpenes obtained from *B. flabellata* Hook. and Arn. showed superoxide radical scavenging activities—the monomer was more effective toward ROS, and the dimers were an excellent RNS scavenger (Funes et al. 2018). Hydroalcoholic extracts of *B. burchellii* Baker and *B. crispa* Spreng. showed antioxidant activity and antiradical capacity (Oliveira et al. 2014). Cold aqueous extract of *B. articulata* (Lam.) Pers. induced the death of human peripheral blood mononuclear cells by apoptosis, increased the frequency of micro-nuclei in the bone marrow, and exerted low mutagenic effects. The phytochemicals presented in the cold aqueous extract were luteolin, acacetin, chlorogenic acid, and tannins (Cariddi et al. 2012), with antioxidant activities (Borgo et al. 2010). When obese rats were treated with methanolic extract of *B. trimera*, inhibition of the pancreatic lipasa, the enzyme that hydrolyzed triglycerides and α- or β-glucosidases was observed (Cercato et al. 2015, Rabelo et al. 2018). Moreover, *B. trimera* extracts also exhibited anti-adipogenic effects (de Souza Marinho Do Nascimento et al. 2017). Echinocystic acid (saponin),

Figure 9.2: Photograph of *Baccharis trimera* plant in the paleontological park of Ischigualasto (San Juan, Argentina).

rutin, apigenin, quercetin, luteolin, and eupafolin hispidulin were the main compounds presented in aqueous or alcoholic extracts of *B. trimera* (Cercato et al. 2015). According to a previous report, the supplementation of diet with luteolin, apigenin, and quercetin reduced the body weight (Cercato et al. 2015). On the other hand, genotoxic effects were described when a dose of 42 mg/kg of aqueous extract of *B. trimera* was administered. The administration of 8.4 mg/kg of *B. trimera* extract showed toxic effects on kidney and liver cells (Cercato et al. 2015). The 4, 10-aromadendranediol isolated from *B. gaudichaudiana* DC. showed benefits to treat brain diseases, because it induced neurite outgrowth activity in neurons via activation of the ERK signalling pathway (Chang et al. 2017). The antimicrobial activity of *B. trimera* against *S. aureus* and methicillin-resistant *S. aureus* (MIC = 6.56 mg/mL) was explained by the high concentration of flavonoids (da Silva et al. 2018). Two diterpenes isolated from *B. grisebachii* Hieron. showed activity toward dermatophytes with MIC value of 12.5 µg/mL (Feresin et al. 2003). *Baccharis articulata*, *B. crispa*, *B. phyteumoides* (Less.) DC., and *B. trimera* were tested as antifungal against dermatophytes; the active compounds identified were the flavonoids aforementioned and genkwanin, ent-clerodanes bacrispin, bacchotricuneatin A, and hawtriwaic acid. The bacchotricuneatin A and bacrispine showed synergistic antifungal effects (Rodriguez et al. 2013). Organic and aqueous extracts of *B. gaudichaudiana* and *B. spicata* (Lam.) Baill. were active against PV-2 and VSV virus. Apigenin was identified as the main compound in the organic extract of *B. gaudichaudiana* with a strong antiviral effect (Visintini Jaime et al. 2013). The antinociceptive activity of *B. flabellata* extract was reported by Funes et al. (2018); the analgesic activity is mainly due to the presence of ent-clerodane and its dimer. The hydroethanolic extract of *B. trimera* exhibited lower toxicity than lapachol (do Nascimento Kaut et al. 2018), and it was able to decrease glycemia and increased the insulin after 7 days of treatment. This hypoglycaemic effect could be associated with the presence of flavonoids and chlorogenic acids (do Nascimento Kaut et al. 2018, Rabelo et al. 2018). The supplementation of aqueous extract of *B. trimera* leaves (100 mg/kg of animal) showed 70% liver regeneration after hepatectomy (Lima et al. 2017). The oxidative stress in brain and liver could be prevented by infusions of *B. trimera* (100 mg/kg) (Sabir et al. 2017). The hydroalcoholic extract of *B. trimera* reduced the lesion area induced by acute and chronic ethanol consumption (Rabelo et al. 2018), but it did not protect against gastric wall mucus depletion. The better anti-ulcerogenic activity was observed with 30 mg/kg (Livero et al. 2016). *B. grisebachii* extract exhibited the highest gastroprotective effect (750 mg/kg) with 93% damage inhibition, using the ethanol-induced gastric damage in a standard rat model (Gómez et al. 2019).

Senecio sp.

The parasitic infection of *Haemonchus contortus*, an endoparasitic nematode, was related to the death of sheep. On the other hand, endoparasitic diseases result in economic losses linked to use of anthelmintic drugs. An aqueous extract of *Senecio brasiliensis* (Spreng.) Less. displayed ovicidal activity against *H. contortus*, but its larvicidal property was less pronounced (Soares et al. 2019). Thus, the egg hatching inhibition was associated with high concentration of integerrimine (a macrolide alkaloid) from leaves extract of *S. brasiliensis*. (Soares et al. 2019). Among liver diseases, the hepatic sinusoidal obstructive syndrome is mainly linked in rural areas with ingestion of pyrrolidine alkaloid, present in *S. brasiliensis*. The treatment of menopause by drinking *Senecio* tea daily is another cause of the disease (Barcelos et al. 2019). These poisoning products of the alkaloids are frequent with the consumption of *Senecio* as a medicinal plant, although in the ethnobotany studies, the use of this plant in different communities was mentioned (Bolzan et al. 2007).

Ephedreacea

Boff et al. (2008) reported the absence of ephedrine and pseudoephedrine in *Ephedra triandra*, whereas N-methylephedrine, ephedrine, 6-hydroxykynurenic acid, and pseudo-ephedrine were isolated from

E. americana Humb. and Bonpl. ex Willd., *E. breana* Phil., *E. chilensis* C. Presl., *E. ochreata* Miers and *E. tweediana* Fisch. and C.A. Mey.

Ephedra triandra Tul

In Argentina, *E. triandra* is called "tramontane" or "pico de loro", and its aerial parts are used in alcoholic preparation as an anti-inflammatory, with veterinary use. Moreover, it is used as a decoction to reduce uric acid in the blood or to treat antirheumatic, stomatic, and antidiarrheal effects in humans. In modern medicine, *Ephedra* sp. is used in the treatment of the respiratory tract and bronchospasm diseases; its medical properties are connected with the presence of alkaloids.

Euphorbiaceae

Most phytochemical studies conducted in the medicinal species of this family are focused on the analysis of essential oils (Zygadlo et al. 2017). However, the following compounds were reported-crotosparine and isoquinoline alkaloid in *Croton bonplandianus* Baill., quercetin and tannic acid in *Euphorbia collina* var. *andina* (Phil.) Subils and *E. serpens* var. *microphylla* Müll. Arg., carpinusin, astragalin, chlorogenic acid and glucogallin in *E. helioscopia* L. and *Sapium haematospermum* Müll. Arg., quercetin, gallic, syringic, caffeic acids, kaempferol, isorhamnetin, coumarin, and scopoletin in *Sebastiania brasiliensis* Spreng. and *S. commersoniana* (Baill.) L. B. Sm. and Downs. The 36% of the medicinal species of this family do not have biological evaluations or phytochemical analysis (Alvarez 2019).

Acalypha communis Müll. Arg.

Its popular name is "albahaquilla del campo". In Córdoba's hills (Argentina), healing properties are attributed to this species, the leaves are used in a decoction to treat wounds, sores, and ulcers (Martínez 2015). *A. communis* was evaluated for its antioxidant activity, which was related to the total phenol content (Aguirre and Borneo 2013). Three cycloartane-type triterpenes isolated from aerial parts of the plant showed moderated antimicrobial activity against Gram-positive and Gram-negative bacteria. They exhibited a MIC of 8, 32, and 8 µg/mL, respectively, against vancomycin-resistant enterococci, and the 16-α-hydroxymollic was active against methicillin-resistant staphylococci (Gutierrez-Lugo et al. 2002). Seebaluck et al. (2015) reported anti-candidiasis activity of the methanolic extract of *A. communis*.

Croton sp.

The popular name "Cachamia" corresponds to *C. argentinus* Müll. Arg., which is an endemic species. In Córdoba province (Argentina), its aerial parts are used in decoction and infusion to treat digestive and liver disorders (Martínez 2015). Borneo et al. (2009) determined that the water extract presented a high antioxidant power (554.6 µmol of Fe(II)/g). The extract had 50.5 mg gallic acid/g of phenolics compounds, and it could be used as a natural substitute for artificial antioxidants currently used in food processing. The methanol extract of *C. hieronymi* Griseb. showed strong activity against lung carcinoma cells A-549 (IC_{50} = 0.25 mg/mL), mouse lymphoma (IC_{50} = 1.0 mg/mL), and human colon carcinoma (IC_{50} = 2.5 mg/mL). Catalán et al. (2003) described new compounds, the acetophenone derivative xanthoxylin, and the peptide derivatives aurentiamide acetate and N-benzoylphenylalanyl-N-benzoylphenylalaninate.

Euphorbia sp.

Euphorbia hirta var. *ophthalmica* (Pers.) Allem and Irgang have 47 patents in the last five years about medicinal uses. In the folk medicine of Córdoba province, Argentina, its latex is used to treat skin conditions, such as warts and mycosis (Martinez 2015). Scientific studies reported that E. hirta L. possesses antibacterial, antiasthmatic, sedative, antidiarrheal, antispasmodic, anti-inflammatory, antifungal, anti-allergic, diuretic, antioxidant, antitumor, antidiabetic, anxiolytic, sedative, antiplasmodial, and galactogenic properties, and anti-snake venom activity (Kausar et al. 2016, Ndjonka et al. 2018). The ethanolic extracts presented high inhibitory activity against *C. albicans* and *S. aureus* (MIC of 12.5 mg/mL and 25.0 mg/mL, respectively) (Gupta et al. 2018). Silver nanoparticles containing ethanolic extract showed anticancer activity against neuroblastoma and breast cancer cells (Selvam et al. 2019), whereas the nanoparticles containing flavonoid and tannin were toxic against *P. aeruginosa* and *B. subtilis*. All the nanoparticles were able to disturb the cell membrane, released internal proteins, and were more effective against Gram-negative than Gram-positive bacteria (Raji et al. 2019a). The ethanolic extract showed the presence of 9, 12, 15-octadecatrien-1-ol, pentadecylic acid, ethyl linoleate, 1, 2, 3-trihydroxy benzene, γ-tocopherol, 5-hydroxymethyl-2-furancarboxaldehyde, myristic acid, 7, 10-octadecadienoic acid methyl ester, phytol, ethyl palmitate, and squalene. The extract showed anti-inflammatory and anxiolytic effects on neonatal asthmatic rats with inflammation (Xia et al. 2018). Sharma et al. (2018) showed that ethanolic extracts induced cardiorenal protection, which may be associated with its antihyperglycemic, antidyslipidemia, and antioxidant potential. Promprom and Chatan (2018) showed that ethanolic extracts presented weak estrogenic effects in ovariectomized rats, and it could be useful for health benefits during menopause. Abdelkhalek et al. (2018) reported that ethyl acetate extract of *E. hirta* was effective against *S. aureus* (MRSA) (MIC 25 mg/mL). The anti MRSA compounds were identified as hydroquinone and O-coumaric acid. Moreover, this extract showed anti-onchocerca activity (Ndjonka et al. 2018). The saponin fraction (from alcoholic extract) had antibacterial activity against *P. aeruginosa*, and the flavonoid and tannin fractions (from aqueous extract) showed antioxidant properties (Raji et al. 2019b). On the other hand, from ethyl acetate and methanol extracts, four pure compounds were isolated—quercitrin, luteolin, quercetin, and caffeic acid. The results showed that the ethyl acetate extract and quercetin possess strong protective effects with cell viability of 81% and 82% at the dose of 0.1 mg/mL. This plant could be used in the treatment of oxidative stress, which could be related to neurodegenerative diseases (Bach et al. 2018). Methanol extract presented antioxidant and anti-psoriasis activity. The maximum cell death (88.37%) was observed at 0.781 µg/mL concentration, and the IC_{50} was 12.20 µg/mL (Jeba et al. 2018). Methanol extract induced apoptotic cell death, which suggests that *E. hirta* could be used as an apoptosis-inducing anticancer agent for breast cancer treatment and malignancy cell line MCF-7 (Kwan et al. 2016, Behera et al. 2016). *E. portulacoides* L. has 25 patents about skin treatments and cancer. Only a few researches about phytochemistry, and the presence of diterpenes and phloroacetophenones were described. Other compounds, such as lathyrane, abietane, kaurane, and ingenane were described (Bittner et al. 2001). The root infusion of *E. serpens* Kunth is used as diuretic by the people of Córdoba hills (Argentina), however in Yemen Republic, the leaves are used as infusions for female infertility (Al-Fatimi 2019). In Brazil, this plant is used in popular medicine to treat kidney stones, bladder affections, inflammation in the kidneys, and as a diuretic. The chemical composition of leaves and latex indicated the presence of phenols, flavonoids, cyanidines, tannins, and saponins (Aita et al. 2009).

Phyllanthus niruri L.

A review about the use and phytochemistry of this species was conducted by Zunino et al. (2003), Tewari et al. (2017), and Kaur et al. (2017). The clinical evidence demonstrating the beneficial properties of *P. niruri* as immunomodulatory for the treatment of various infectious diseases were reviewed by Tjandrawinata et al. (2017). The anti-hyperglycemic and antioxidant potential effect was demonstrated by Kumar et al. (2019) for streptozotocin-induced diabetic rats. In northeastern Brazil, this species is used in infusion as abortive, or to treat urinary calculus, diabetes, loss of appetite, and cholecystitis (Magalhães et al. 2019). A 35 kDa herbal antioxidant protein molecule (PNP) was isolated and purified from this plant as therapeutic agents. The two active ingredients, phyllanthin and corilagin were isolated and characterized (Bhattacharyya et al. 2017). This PNP could confer protection against indomethacin (non-steroidal anti-inflammatory drugs) mediated hepatic oxidative impairments (Bhattacharyya et al. 2017). Moreover, the plant has long been used as a hepatoprotection and for treatment of hepatitis B. Li et al. (2017) showed that the ethanol fraction inhibited the growth of HBV-infected HepG2/C3A cells, and its active compound ellagic acid exerted a cytotoxic effect against those cells, but did not affect HBV replication. Baiguera et al. (2018) investigated the efficacy and safety of 12-month treatment with aqueous extract of *P. niruri* (250 mg, 10% lignans) in subjects with chronic hepatitis B virus infection. This study does not support the use of *P. niruri* for the treatment of chronic hepatitis B. Besides, two lignans (hypophyllanthin and phyllanthin) were responsible for the *in vitro* anticancer activity shown against human lung cancer cell line A549, hepatic cancer cell line SMMC-7721, and gastric cancer cell line MGC-803. On the other hand, the methanolic extract showed selective cytotoxicity against MCF-7 breast cancer cells (Wan Omar and Zain 2018). The phytochemicals identified in ethanolic extract of *P. niruri*, including hypophyllanthin, catechin, epicatechin, rutin, quercetin and chlorogenic, caffeic, malic, and gallic acids showed a positive correlation with antioxidant and α-glucosidase inhibitory activities (Mediani et al. 2017). The tannin corilagin was a major component extracted from *P. niruri*, and it inhibited the growth of ovarian cancer cells via the TGF-β/AKT/ERK signalling pathways. It was demonstrated that corilagin enhanced the sensitivity of ovarian cancer cells to chemotherapy (Jia et al. 2017). Klein-Júnior et al. (2017) showed that this tannin reduced the lesion area of ethanol-induced gastric ulcers in mice by 88 percent. The use of *P. niruri* as infusion intake was safe and did not cause significant adverse effects on serum metabolic parameters in humans. The consumption of *P. niruri* contributed to the elimination of urinary calculi (Pucci et al. 2018). The property of promoting protection against ulcers is attributed to the regeneration of the mucosal layer and substantial prevention of the formation of hemorrhage and edema (Mostofa et al. 2017).

Fabaceae

Apigenin, tannin, quercetin, cynaroside, rhamnetin, rutin, quercitrin, diosmetin, kaempferol, protocatechuic acid, saponins, anthraquinones, and metabolites with nitrogen as a characteristic structure, such as pyrrolizidine, isoquinoline, quinolizidine, and cyanogenic glycosides were founded in Fabaceae species (Alvarez 2019). More than 100 medicinal plants were reported that belong to this botanical family, however there is a lack of information about t the pharmacological action or chemical composition of 65% of the species (Barboza et al. 2009).

Bauhinia forficata Link

This plant is used in Argentina and Chile for diabetes diseases. It is known as "pezuña de vaca" (Figure 9.3). Fuentes Mardones and Alarcon Enos (2010) suggested that the leaves extract increases endothelium-dependent relaxation of aortic rings of ALX-rats, and this effect may be due to its antioxidant activity. Also, this extract significantly reduced fasting blood sugar. Butanol extract performed a stimulatory effect of glucose uptake in isolated gastric glands of normal and alloxan-diabetic rabbits (Fuentes and Alarcón 2006).

Caesalpinia gilliesii (Hook.) D. Dietr

Its popular name is "*Lagaña de Perro*" or "*Bird of Paradise*". The compounds described for this plant were diterpenoids, isovouacaperol, sitosterol, flavonoids, gallic acid, tannins, resin, benzoic acid, and homoisoflavonoids (Kheiri Manjili et al. 2012). Moreover, cardiac glycosides, cyanogenic glycosides, saponins, and coumarins were identified in extracts studied by Osman et al. (2013), with different ratios. On the other hand, Osman et al. (2016) described a new 12, 13, 16-trihydroxy-14(Z)-octadecenoic

Figure 9.3: Photograph of flowers, branches, and leaves of *Bahuinia forficata*, popular name "pezuña de vaca or "cow hoof", in the natural reserve of the city of Cordoba.

acid, and Emam et al. (2019) identified a new polyoxygenated flavonol. The proteins derived from the seeds showed that it was possible to upregulate and to counteract the inflammatory process, and to minimize the damage of the liver (Rizk et al. 2016). The polysaccharide galactomannan isolated from seed aqueous extract improved inflammatory and apoptotic markers (Abdel-Megeed et al. 2019).

Erythrina crista-galli L.

It is the national tree and flower of the Argentine Republic. Methanolic extract of the stems showed the presence of phytosterols, stigmast-4-en-3-one, stigmast-4, 22-dien-3-one, 6β-hydroxystigmast-4-en-3-one, lup-20 (29)-en-3-one, oleanonic acid, oleanolic acid, olean-12-en-3β,28-diol, and olean-12-en-3β,22β,24-triol (Lee and Huang 2004). Erythrinan alkaloids from the bark exhibited a range of pharmacological properties. Three phenolic compounds have also been isolated—phaseollidin, sandwicensin, and lonchocarpol A. They showed antimalaria and antioxidant activities (Tjahjandarie et al. 2014). The alkaloid erythraline showed an effect on inflammatory diseases (rheumatism and hepatitis) through inhibition of TAK1 (Etoh et al. 2013). The compounds apigenin-7-O-rhamnosyl-6-C-glucoside, a flavonoid glycoside, and luteolin-6-C-glucoside were described in aqueous leaf extracts (Ashmawy et al. 2016).

Prosopis alba Griseb.

This tree is known in Argentina as "*algarrobo blanco*". The fruits and bark are used in folk medicine as diuretic. *Prosopis* is considered a multipurpose tree and shrub by FAO. The exudate gum showed a higher concentration of phenolics, flavonoids, and tannins compared to the Arabic gum, which was positively correlated with antioxidant properties (Vasile et al. 2019). The edible ripe pods contain mainly quercetin O-glycosides and apigenin-based C-glycosides (Pérez et al. 2014).

Hypericaceae

The flavonoids isoquercetrin, kaempferol, quercertin, hyperine, quercitrin, amentoflavone, and guaijaverine were founded in the medicinal species of this family (Alvarez 2019).

Hypericum connatum Lam.

The rutin and apigenin were the main components presented in the ethanolic extract, while caffeic acid, (–)-epicatechin, and p-coumaric acid were abundant in the ethyl acetate extract. These extracts exerted antibacterial activity, whereas the ethanolic extract showed antiquorum-sensing *Chromobacterium violaceum* activity, and this effect was attributable to the presence of rutin and apigenin (Fratianni et al. 2013). The compound luteoforol inhibited the cytopathic effect and reduced the viral titer of HSV-1 DNA viral strains KOS and VR733 (ATCC) (Fritz et al. 2007). On the other hand, *H. connatum* was used as an antidepressant, and the extracts could be used as a natural source of antidepressant medication for pregnant women. Da Conceição et al. (2014) showed that methanolic and hexane extracts can interfere with trophoblast differentiation and Ca^{2+} influx. The effects were concentration-dependent. This study suggested that attention must be paid to the potential toxic effects of this plant. It has been shown that tincture of this plant (ethanol 70% extract) exhibited an important antispasmodic effect mainly due to non-competitive antagonism of the agonist and of Ca^{2+} influx to smooth muscle (Matera et al. 2016).

Solanaceae

The literature on phytochemical studies is numerous; the most important chemical groups identified were pyridyl-pyrrolidine, steroidal, pyrrolidine, tropane alkaloids, glycoalkaloids, coumarins, lignans, sapogenins, steroidal saponins, spirosten-δ-lactone saponin, and withanolides (Barboza et al. 2009, Romero et al. 2016, Alvarez 2019)

Capsicum annuum L.

This species is widely studied for its extensive pharmacological actions, for example, antibacterial (Alinia-Ahandani 2018), antifungal, anticancer, antioxidant, antiprotozoal, hypocholesterolaemic/ hypolipidemic (Arumugam et al. 2008), immunomodulatory, antimutagenic, and pesticidal. Recent studies showed that carotenoid extract exhibited good anti-inflammatory activity (Boiko et al. 2017). The capsaicin (*trans*-8-methyl-*N*-vanillyl-6- nonenamide) is the major pungent ingredient of the fruits and it was effective against pain in rheumatoid arthritis, post-herpectic neuralgia, diabetic neuropathy, and anticancer (Kundu and Surh 2009, Zheng et al. 2016).

Solanum sisymbriifolium Lam.

It is known as "*Espina colorada*", and it is used in traditional medicine in South America for antihypertensive and diuretics purposes. Other pharmacological actions included soothing, colics, liver disease, jaundice, cirrhosis, gallstones, rheumatism, anti-inflammatory, or as diuretic. Recent studies have partially validated the antihypertensive effects of the extracts, making the species a promising natural source for future developments (Simões et al. 2016). The nuatigenin-3-O-β-chacotriose was identified as the main hypotensive compound (Ibarrola et al. 2011). Another compound isolated from this plant was solasodine, which showed anticonvulsant and sedative properties (Chauhan et al. 2011). However, the toxicity of the unripe fruits, together with the popular belief that the plant possesses contraceptive properties, suggested that "*Espina colorada*" should be used with precaution.

Table 9.1 describes the bioactivity of different plant extracts and their popular uses, while Table 9.2 shows the antimicrobial and antiparasitic activities.

Table 9.1: Bioactivity of different plant extracts with evidence based-medicine and folkloric uses (*).

Plant Scientific Name	Family	Ethanolic extract	Aqueous extract *(infusion or decoction)	Methanolic extract	Other extracts	References
Alternanthera pungens Kunth	Amaranthaceae	100 mg/kg extract/body weight reduced 80% blood glucose concentration[1]	* Antidiarrheal, carminative propierties [2]			[1] Olugbemiga et al. 2016 [2] Zunino et al. 2003
Gomphrena celosioides Mart.		Inhibition HepG2 cell IC$_{50}$ = µg/mL[1] Dose 100 mg/kg reduce mean arterial pressure. GC extract acted as diuretic.[2]	Inhibition HepG2 cell IC$_{50}$ = 250 µg / mL [1] Doses between 250/750 mg/kg of body weight decrease hepatotoxic effect of carbon tetrachloride [3] * carminative, anticonceptive [4]			[1] Promraksa et al. 2019 [2] de Paula Vasconcelos et al. 2018, 2017 [3] Sangare et al. 2014 [4] Zunino et al. 2003
Amaranthus hybridus L.		200 and 400 mg/kg caused reduction in blood glucose levels[3]	Anticarcinogenic effect [1] IC$_{50}$ values (µg /mL) activity inhibitory, angiotensin-1= 53.4; OH radical scavenging act. = 58.0; DPPH radical scavenging act. 36.3 [4] * diuretic [5]	Antioxidant effect, IC$_{50}$ = 28 µg/mL. Ehrlich's ascites carcinoma cells were inhibited with 25 µg of extract.[2]		[1] Adewale and Olorunju 2013 [2] Al-Mamun et al. 2016 [3] Balasubramanian et al. 2017 [4] Oboh et al. 2016 [5] Martínez 2015
Lithraea molleoides (Vell.) Engl	Anacardiaceae	250 g extract/100 mL showed antiulcerogenic effect[2]	20g extract/100mL showed antiulcerogenic effect[2] * fruits, digestive [4]		Hidro alcoholic extract: Dose 1 g/kg anti-ulcerogenic activity[1] IC$_{50}$ = 50 ug/mL against human hepatocellular carcinoma cell line, HepG2[3]	[1] Araujo et al. 2006 [2] Garro et al. 2015 [3] Ruffa et al. 2002 [4] Martínez 2015
Mulinum crassifolium Phil.	Apiaceae		Infusion, dose 100mg/kg, Lesion reduction 74%. Gastroprotective effect *antidiabetes, bronchial and intestinal disorders			Areche et al. 2019
Aristolochia argentina Griseb.	Aristolochiaceae		62.5 mg/kg antidiarrheal activity[1] *anti-inflammatory, to relieve hemorrhoids [2]			[1] Paredes et al. 2016 [2] Martínez 2015

Table 9.1 contd. ...

...Table 9.1 contd.

Plant Scientific Name	Family	Ethanolic extract	Aqueous extract *(infusion or decoction)	Methanolic extract	Other extracts	References
Artemisia copa Phil	Asteraceae		Antispasmodic activity on gastrointestinal system[1] Vasorelaxing and hypotensive effects[2] Sedative principles, potential anxiolytic and anticonvulsant activities[3] *reduce blood pressure, antirheumatic, antispasmodic[4]			[1]Gorzalczany et al. 2013a [2]Gorzalczany et al. 2013b [3]Miño et al. 2010 [4]Zunino et al. 2003
Tecoma ipe Mart. ex K. Schum. *T. stans* (L.) Juss. ex Kunth	Bignonaceae	Nephroprotective capacity[3]	Anticancer, anti-inflammatory properties[1] Lipase inhibition[2] *barks, leaves: astringent, antiseptic, diuretic, antidiabetic; flowers: expectorant[4]			[1]Niwa et al. 2013 [2]Mopuri and Islam 2017 [3]Chandra Mo et al. 2016 [4]Zunino et al. 2003
Ephedra triandra Tul.	Ephedreaceae	Anti-inflammatory (aerial parts)[1]	*Reduce uric acid in blood, antirheumatic[2]			[1]Martinez and Luján 2011 [2]Martinez 2015
Adesmia boronioides Hook.	Fabaceae		Antioxidant activity (aereal part), antiproliferative activity on human cancer cell lines[1] *rheumatic pains, hair loss, colds, digestive disorders, aphrodisiac[2]	Inhibit 5-lipoxygenase (proinflammatory pathways)[2]		[1]Gastaldi et al. 2018 [2]González et al. 2003
Geoffroea decorticans (Gill ex Hook & Arn.) Burkart			Antinociceptive action (fruits extract)[3] *bark: respiratory conditions[4]		Methanol:water 70:30 fruit extract: Antioxidant activity and inhibited pro-inflammatory enzymes[1,2]	[1]Costamagna et al. 2016 [2]Jiménez-Aspee et al. 2017 [3]Reynoso et al. 2013 [4]Martínez 2015
Petunia nyctaginiflora Juss.	Solanaceae	Anti-virus action against Herpes simplex virus-1 (HSV-1)				Padma et al. 1998
Urtica circularis (Hicken) Soraru	Urticaceae	Anti-inflammatory activity[1] antinociceptive effect[3]	*Astringent, diuretic, antirheumatic and anti-inflammatory[2]		hydro-ethanolic (20:80) of aerial parts, sedative activity[2]	[1]Marrassini et al. 2011 [2]Anzoise et al. 2013 [3]Gorzalczany et al. 2011

Table 9.2: Antimicrobial and antiparasitic properties of different medicinal plant extracts.

Plant Scientific Name	Family	Ethanolic extract	Aqueous extract	Methanolic extract	Other extract	References
L. molleoides (Vell.) Engl.	Anacardiaceae		*H. pylori* HP105 and HP109 MIC = 16 ug/mL			Ibañez et al. 2017
Muinum spinosum (Cav.) Pers	Apiaceae	Doses 100 µg/mL. 97.9% growth inhibition *T. cruzi.* (Aerial parts) 54.0% growth inhibition *T. cruzi.* (Roots)[2]			Slime *S. aureus* MIC/MBC = 0.5/1 mg/mL[1] *S. aureus* methicillin resistant MIC/MBC = 0.5/1 mg/mL[1]	[1]Echenique et al. 2014 [2]Sülsen et al. 2006
Araujia brachystephana (Griseb.) Fontella & Goyder (syn. *Morrenia brachystephana* Griseb.)	Apocynaceae			Antifungal activity (dermatophytes) MIC > 1		Muschietti et al. 2005
Peschiera australis (Mull. Arg.) Miers					IC$_{50}$ = 12 µg/mL against *Leishmania amazonensis*	de Oliveira et al. 2017
Vallesia glabra (Cav.) Link.		MIC = 64 mg/mL *E. coli* [3] MIC = 16 mg/mL *S aureus* [3]	MIC = 32 mg/mL *E. coli* [3]		MIC against dermatophytes 0.5 to 1 mg/mL[1] IC$_{50}$ = 3.75 µg/mL against *P. falciparum*[2]	[1]Svetaz et al. 2010 [2]Bourdy et al. 2004 [3]Bussmann et al. 2010
Aristolochia argentina Griseb.	Aristolochiaceae	MIC = 0.03 mg/spot *F. verticillioides*[1]	*H. pylori* HP105, MIC = 8 ug/mL[4]	MIC = 500 ug/mL Dermatophytes[3]	IC$_{50}$ = 22.3µg/mL against *L. amazonensis*[2]	[1]Carpinella et al. 2010 [2]Cortés et al. 2006 [3]Cordisco et al. 2018 [4]Ibañez et al. 2017
Aspilia silphioides (Hook. & Arn.) Benth. & Hook. f.	Asteraceae				93.4% growth inhibition epimastigotes T. cruzi (10 ug/mL)	Selener et al. 2019
Ambrosia tenuifolia Spreng.					[1]Anti-*T cruzi* activity [2]IC$_{50}$ hispidulin = 46.7 µM against *T. cruzi,* and 6.0 µM against *L. mexicana*	[1]Sülsen et al. 2016 [2]Sülsen et al. 2007

Table 9.2 contd. ....

...Table 9.2 contd.

Plant Scientific Name	Family	Ethanolic extract	Aqueous extract	Methanolic extract	Other extract	References
Artemisia copa Phil.		antifungal activity MIC = 80 to 100 µg/mL	antifungal activity MIC = 80 to 100 µg/mL			Ortiz et al. 2019
Gaillardia megapotamica (Spreng.) Baker.					98% growth inhibition epimastigotes T. cruzi (10 ug/mL)	Selener et al. 2019
Tessaria absinthioides (Hook & Arn.) ex DC.		66% anti-biofilm activity against *Bacillus* spp. MIC = 250 ug/mL				Romero et al. 2016
Acacia caven Molina	Fabaceae	*Staphylococcus aureus(MIC = 4)*	*Staphylococcus aureus ATCC (MIC=2)*			Martinez et al. 2014
Geoffroea decorticans (Gill ex Hook & Arn.) Burkart				Bark extract: *Staphylococcus aureus* ATCC 8095 (MIC = 0.125); *Enterococcus faecium* (MIC = 0.31); *Pseudomas aeroginosa* (MIC = 0.5); *Salmonella typhimurium* (MIC = 0.31); *Klebsiella pneumonia* (MIC > 1); *Escherichia coli* (MIC = 0.8)		Salvat et al. 2004

Key: MIC, Minimum Inhibitory Concentration (mg/mL)

Conclusion

Based on all the species cited here, only 34% had scientific studies to support their bioactivities or pharmacological actions, and half of them matched well between popular use and the experimentally proven studies. The 32% of the species were not scientifically evaluated for their medicinal properties. Most of the studies only used one species to evaluate different bioactivities. In general, the extracts with anti-inflammatory effects also showed antioxidant activities; the main compounds presented in the extracts were phenols and flavonoids. Recent studies are focused on the use of plant extracts against different cancer types. It is important to note that only 6% of the species were evaluated for their toxic effects.

Acknowledgments

We want to thank Universidad Nacional de Córdoba and Consejo Nacional de Investigaciones Científicas y Técnicas for financial support.

References

Abdel-Megeed, R.M., Hamed, A.R., Matloub, A.A., Kadry, M.O., Abdel-Hamid, A.H.Z. 2019. Regulation of apoptotic and inflammatory signaling pathways in hepatocellular carcinoma via *Caesalpinia gilliesii* galactomannan. Mol. Cell Biochem. 451: 173–184.

Abdelkhalek, E.S., El-Hela, A.A., El-Kasaby, A.H., Sidkey, N.M., Desouky, E.M., Abdelhaleem, H.H. 2018. Antibacterial activity of Polygonum plebejum and *Euphorbia hirta* against Staphylococcus aureus (MRSA). J. Pure Appl. Microbiol. 12: 2205–2216.

Adewale, A., Olorunju, A. 2013. Modulatory of effect of fresh *Amaranthus caudatus* and *Amaranthus hybridus* aqueous leaf extracts on detoxify enzymes and micronuclei formation after exposure to sodium arsenite. Pharmacognosy Res. 5: 300–305.

Aguirre, A., Borneo, R. 2013. Antioxidant capacity of medicinal plants. pp. 527–535. *In*: Ross Watson, R., Preedy, V.R. (eds.). Bioactive Food as Dietary Interventions for Liver and Gastrointestinal Disease (eds). Elsevier, San Diego: Academic Press.

Aita, A.M., Matsuura, H.N., Machado, C.A., Ritter, M.R. 2009. Espécies medicinais comercializadas como "quebra-pedras" em Porto Alegre, Rio Grande do Sul, Brasil. Rev. Bras Farmacogn. 19: 471–477.

Alvarez. 2019. Pharmacological properties of native plants from Argentina. Springer pp. 1–242. DOI: 10.1007/978-3-030-20198-2.

Al-Fatimi, M. 2019. Ethnobotanical survey of medicinal plants in central Abyan governorate, Yemen. J. Ethnopharmacol. 241: 111973. DOI:10.1016/j.jep.2019.111973.

Al-Mamun, M.A., Husna, J., Khatun, M., Hasan, R., Kamruzzaman, M., Hoque, K.M.F., Reza, M.A., Ferdousi, Z. 2016. Assessment of antioxidant, anticancer and antimicrobial activity of two vegetable species of *Amaranthus* in Bangladesh. BMC Complement. Altern. Med. 16, 157. DOI:10.1186/s12906-016-1130-0.

Alinia-Ahandani, E. 2018. Medicinal plants with disinfectant effects. J. Pharm. Sci. Res. 10, I.

Anzoise, M.L., Marrassini, C., Ferraro, G., Gorzalczany, S. 2013. Hydroalcoholic extract of *Urtica circularis*: A neuropharmacological profile. Pharm. Biol. 51: 1236–1242.

Araujo, C.E.P., Bela, R.T., Bueno, L.J.F., Rodrigues, R.F.O., Shimizu, M.T. 2006. Anti-ulcerogenic activity of the aerial parts of *Lithraea molleoides*. Fitoterapia. 77: 406–407.

Areche, C., Fernandez-Burgos, R., De Terrones, T.C., Simirgiotis, M., García-Beltrán, O., Borquez, J., Sepulveda, B. 2019. *Mulinum crassifolium* Phil; Two new mulinanes, gastroprotective activity and metabolomic analysis by uhplc-orbitrap mass spectrometry. Molecules. 24: 1–12. DOI:0.3390/molecules24091673_rfseq1.

Aristimuño Ficoseco, M.E., Vattuone, M.A., Audenaert, K., Catalán, C.A.N., Sampietro, D.A. 2014. Antifungal and antimycotoxigenic metabolites in Anacardiaceae species from northwest Argentina: isolation, identification and potential for control of *Fusarium* species. J. Appl. Microbiol. 116: 1262–1273.

Arumugam, M., Vijayan, P., Raghu, C., Ashok, G., Dhanaraj, S.A., Kumarappan, C.T. 2008. Anti-adipogenic activity of *Capsicum annum* (Solanaceae) in 3T3 L1. J. Complement. Integr. Med. 5. DOI:10.220/1553-3840.1151.

Ashmawy, N.S., Ashour, M.L., Wink, M., El-Shazly, M., Chang, F.R., Swilam, N., Abdel-Naim, A.B., Ayoub, N. 2016. Polyphenols from *Erythrina crista-galli*: Structures, molecular docking and phytoestrogenic activity. Molecules 21: 1–14. DOI:10.3390/molecules21060726.

Asmaa, B.H., Ream, N. 2016. *In vitro* screening of the pancreatic cholesterol esterase inhibitory activity of some medicinal plants grown in Syria. Int. J. Pharmacogn. Phytochem. Res. 8: 1432–1436.

Bach, H.G., Wagner, M.L., Ricco, R.A., Fortunato, R.H. 2014. Sale of medicinal herbs in pharmacies and herbal stores in Hulingham district, Buenos Aires, Argentina. Rev. Bras Farma. 24: 258–264.

Bach, L.T., Dung, L.T., Tuan, N.T., Phuong, N.T., Kestemont, P., Quetin-Leclercq, J., Hue, B.T.B. 2018. Antioxidant activity against hydrogen peroxide-induced cytotoxicity of *Euphorbia hirta* L. AIP Conf. Proc. 2049. DOI:10.1063/1.5082519.

Baiguera, C., Boschetti, A., Raffetti, E., Zanini, B., Puoti, M., Donato, F. 2018. *Phyllanthus niruri* versus placebo for chronic hepatitis B virus infection: A randomized controlled trial. Complement Med. Res. 25: 376–382.

Barboza, G.E., Cantero, J.J., Nuñez, C., Pacciaroni, A., Ariza Espinar, L. 2009. Medicinal plants. A general review and a phytochemical and ethnopharmacological screening of the native argentine flora. Kurtziana 34: 7–365.

Barcelos, T., Dall´Oglio, M., de Araujo, A., Cerski, T., da Silva, M. 2019. Sinusoidal obstruction syndrome secondary the intake of *Senecio brasiliensi*s: A case report. A. Hepat. 1. 1. DOI: 10.1016/j.aohep.2019.08.009.

Balasubramanian, T., Karthikeyan, M., Muhammed Anees, K.P., Kadeeja, C.P., Jaseela, K. 2017. Antidiabetic and Antioxidant Potentials of *Amaranthus hybridus* in Streptozotocin-Induced Diabetic Rats. J. Diet. Suppl. 14: 395–410.

Balsan, G., Pellanda, L.C., Sausen, G., Galarraga, T., Zaffari, D., Pontin, B., Portal, V.L. 2019. Effect of yerba mate and green tea on paraoxonase and leptin levels in patients affected by overweight or obesity and dyslipidemia: a randomized clinical trial. Nutr. J. 18:5. DOI:10.1186/s12937-018-0426-y.

Behera, B., Dash, J., Pradhan, D., Tripathy, G., Pradhan, R. 2016. Apoptosis and necrosis of human breast cancer cells by an aqueous extract of *Euphorbia hirta* leaves. J. Young Pharm. 8: 186–193.

Bermejo, J.E., Delucchi, G., Charra, G., Pochettino, M.L., Hurrell, J.A. 2019. "Cardos" of two worlds: transfer and re-signification of the uses of thistles from the iberian peninsula to Argentina. Ethnob. Conserv. 8: 5–10. DOI: 10.1545/ec2019-03-8.05-1-22.

Benoit, P.S., Angry, G., Lyon, R.L., Fong, H.H.S., Farnsworth, N.R. 1973. Biological and pytochemical evaluation of plants XIII: Preliminary estimation of analgesic activity of Rhazinilam, a novel alkaloid isolated from *Aspidosperma quebracho-bianco* leaves. J. Pharm. Sci. 62:1889. DOI:10.1002/jps.2600621138.

Bhattacharyya, S., Banerjee, S., Guha, C., Ghosh, S., Sil, P.C. 2017. A 35 kDa *Phyllanthus niruri* protein suppresses indomethacin mediated hepatic impairments: Its role in Hsp70, HO-1, JNKs and Ca2+ dependent inflammatory pathways. Food Chem. Toxicol. 102: 76–92.

Bigliani, M.C., Grondona, E., Ponce, A.A. 2012. Inflamación y aceites esenciales. pp. 181–192. *In*: J.A. Zygadlo (ed.). Aceites esenciales. Química, ecología, comercio, producción y salud. Editorial Universitas. Córdoba, Argentina.

Bittner, M., Alarcón, J., Aqueveque, P., Becerra, J., Hernández, V., Hoeneisen, M., Silva, M. 2001. Estudio quimico de especies de la familia Euphorbiaceae en Chile. Boletín la Soc Chil Química. 46: 419–431.

Boff, B.D.S., Sebben, V.C., Paliosa, P.K., Azambuja, I., Singer, R.B., Limberger, R.P. 2008. Investigação da presença de efedrinas em *Ephedra tweediana* Fisch andamp; C.A. Meyer e em *E. triandra* Tul. (Ephedraceae) coletadas em Porto Alegre/RS. Rev. Bras Farmacogn. 18: 394–401.

Boiko, Y.A., Kravchenko, I.A., Shandra, A.A., Boiko, I.A. 2017. Extraction, identification and anti-inflammatory activity of carotenoids out of *Capsicum anuum* L. J. Herb. Med. Pharmacol. 6: 10–15.

Bolzan, A., Silva, C., Francescato, L., Murari, A., Silva, G., Heldwein, G. 2007. Species de *Senecio* na medicina popular da America Latina e toxicidade relacionada a sua utilizacao. Lat. Ame. J. Pharm. 26: 619–625.

Borgo, J., Xavier, C.A.G., Moura, D.J., Richter, M.F., Suyenaga, E.S. 2010. Influência dos processos de secagem sobre o teor de flavonoides e na atividade antioxidante dos extratos de *Baccharis articulata* (Lam.) Pers., Asteraceae. Rev. Bras Farmacogn. 20: 12–17.

Borneo, R., León, A.E., Aguirre, A., Ribotta, P., Cantero, J.J. 2009. Antioxidant capacity of medicinal plants from the Province of Córdoba (Argentina) and their *in vitro* testing in a model food system. Food Chem. 112: 664–670.

Bourdy, G., Oporto, P., Gimenez, A., Deharo, E. 2004. A search for natural bioactive compounds in Bolivia through a multidisciplinary approach: Part VI. Evaluation of the antimalarial activity of plants used by Isoceño-Guaraní Indians. J. Ethnopharmacol. 93: 269–277.

Bussmann, R.W., Malca-García, G., Glenn, A., Sharon, D., Chait, G., Díaz, D., Pourmand, K., Jonat, B., Somogy, S., Guardado, G., Aguirre, C., Chan, R., Meyer, K., Kuhlman, A., Townesmith, A., Effio-Carbajal, J., Frías-Fernandez, F., Benito, M. 2010. Minimum inhibitory concentrations of medicinal plants used in Northern Peru as antibacterial remedies. J. Ethnopharmacol. 132: 101–108.

Bussmann, R.W., Malca, G., Glenn, A., Sharon, D., Nilsen, B., Parris, B., Dubose, D., Ruiz, D., Saleda, J., Martinez, M., Carillo, L., Walker, K., Kuhlman, A., Townesmith, A. 2011. Toxicity of medicinal plants used in traditional medicine in Northern Peru. J. Ethnopharmacol. 137: 121–140.

Calzada, F., Solares-Pascasio, J., Valdes, M., Garcia-Hernandez, N., Velázquez, C., Ordoñez-Razo, R., Barbosa, E. 2018. Antilymphoma potential of the ethanol extract and rutin obtained of the leaves from *Schinus molle* linn. Pharmacognosy Res. 10: 119–123.

Canalis, A.M., Cittadini, M.C., Albrecht, C., Soria, E.A. 2014. *In vivo* redox effects of *Aspidosperma quebracho-blanco* schltdl., Lantana grisebachii stuck and ilex paraguariensis A. St.-Hil. on blood, thymus and spleen of mice. Indian J. Exp. Biol. 52: 882–889.

Cardozo Junior, E.L., Morand, C. 2016. Interest of mate (*Ilex paraguariensis* A. St.-Hil.) as a new natural functional food to preserve human cardiovascular health—A review. J. Funct. Foods 21: 440–454.

Cariddi, L., Escobar, F., Sabini, C., Torres, C., Reinoso, E., Cristofolini, A., Comini, L., Núñez Montoya, S., Sabini, L. 2012. Apoptosis and mutagenicity induction by a characterized aqueous extract of *Baccharis articulata* (Lam.) Pers. (Asteraceae) on normal cells. Food Chem. Toxicol. 50: 155–161.

Carpinella, M.C., Andrione, D.G., Ruiz, G., Palacios, S.M. 2009. Screening for acetylcholinesterase inhibitory activity in plant extracts from Argentina. Phyther Res. 24: 259–263.

Carpinella, M.C., Ruiz, G., Palacios, S.M. 2010. Screening of native plants of central Argentina for antifungal activity. Allelopath J. 25: 423–432.

Carraz, M., Lavergne, C., Jullian, V., Wright, M., Gairin, J.E., Gonzales De La Cruz, M., Bourdy, G. 2015. Antiproliferative activity and phenotypic modification induced by selected Peruvian medicinal plants on human hepatocellular carcinoma Hep3B cells. J. Ethnopharmacol. 166: 185–199.

Carvalho, A.R., Diniz, R.M., Suarez, M.A.M., Figueiredo, C.S.S.S., Zagmignan, A., Grisotto, M.A.G., Fernandes, E.S., da Silva, L.C.N. 2018. Use of some asteraceae plants for the treatment of wounds: From ethnopharmacological studies to scientific evidences. Front Pharmacol. 9: 1–11.

Catalán, C.A.N., de Heluani, C.S., Kotowicz, C., Gedris, T.E., Herz, W. 2003. A linear sesterterpene, two squalene derivatives and two peptide derivatives from *Croton hieronymi*. Phytochemistry 64: 625–629.

Cazander, G., Jukema, G.N., Nibbering, P.H. 2012. Complement ativation and inhibition in wound healing. Clin. Dev. Immunol. 2012: 1–14.

Cercato, L.M., White, P.A.S., Nampo, F.K., Santos, M.R.V., Camargo, E.A. 2015. A systematic review of medicinal plants used for weight loss in Brazil: Is there potential for obesity treatment? J. Ethnopharmacol. 176: 286–296.

Chandra Mo, S., Anand, T., Mini Priya, R. 2016. Protective Effect of *Tecoma stans* Flowers on Gentamicin-Induced Nephrotoxocity in Rats. Asian J. Biochem. 11: 59–67.

Chang, S., Ruan, W.C., Xu, Y.Z., Wang, Y.J., Pang, J., Zhang, L.Y., Liao, H., Pang, T. 2017. The natural product 4,10-aromadendranediol induces neuritogenesis in neuronal cells *in vitro* through activation of the ERK pathway. Acta Pharmacol Sin. 38: 29–40.

Chauhan, K., Sheth, N., Ranpariya, V., Parmar, S. 2011. Anticonvulsant activity of solasodine isolated from *Solanum sisymbriifolium* fruits in rodents. Pharm. Biol. 49: 194–199.

Cittadini, M.C., Canalis, A.M., Albrecht, C., Soria, E.A. 2015. Effects of oral phytoextract intake on phenolic concentration and redox homeostasis in murine encephalic regions. Nutr. Neurosci. 18: 316–322.

Cittadini, M.C., García-Estévez, I., Escribano-Bailón, M.T., Rivas-Gonzalo, J.C., Valentich, M.A., Repossi, G., Soria, E.A. 2018. Modulation of fatty acids and interleukin-6 in glioma cells by South American tea extracts and their phenolic compounds. Nutr. Cancer. 70: 267–277.

Cittadini, M.C., Albrecht, C., Miranda, A.R., Mazzuduli, G.M., Soria, E.A., Repossi, G. 2019. Neuroprotective effect of *Ilex paraguariensis* intake on brain myelin of lung adenocarcinoma-bearing male Balb/c mice. Nutr. Cancer. 71: 629–633.

Conceição, E.P.S., Kaezer, A.R., Peixoto-Silva, N., Felzenszwalb, I., De Oliveira, E., Moura, E.G., Lisboa, P.C. 2017. Effects of *Ilex paraguariensis* (yerba mate) on the hypothalamic signalling of insulin and leptin and liver dysfunction in adult rats overfed during lactation. J. Dev. Orig. Health Dis. 8: 123–132.

Cordisco, E., Sortino, M., Svetaz, L. 2019. Antifungal activity of traditional medicinal plants from Argentina: Effect of their combiantion with antifungal drugs. Curr. Trad. Med. 5: 75–95.

Correa, V.G., Gonçalves, G.A., de Sá-Nakanishi, A.B., Ferreira, I.C.F.R., Barros, L., Dias, M.I., Koehnlein, E.A., de Souza, C.G.M., Bracht, A., Peralta, R.M. 2017. Effects of *in vitro* digestion and *in vitro* colonic fermentation on stability and functional properties of yerba mate (*Ilex paraguariensis* A. St. Hil.) beverages. Food Chem. 237: 453–460.

Cortés, M.J., Armstrong, V., Barrero, A.F., Bandoni, A.E., Priestap, H.A., Fournet, A., Prina, E. 2006. Configuration and Leishmanicidal Activity of (−)-Argentilactone Epoxides. Nat. Prod. Res. 20: 1008–1014.

Costamagna, M.S., Zampini, I.C., Alberto, M.R., Cuello, S., Torres, S., Pérez, J., Quispe, C., Schmeda-Hirschmann, G., Isla, M.I. 2016. Polyphenols rich fraction from *Geoffroea decorticans* fruits flour affects key enzymes involved in metabolic syndrome, oxidative stress and inflammatory process. Food Chem. 190: 392–402.

da Conceição, A.O., von Poser, G.L., Barbeau, B., Lafond, J. 2014. *Hypericum caprifoliatum* and *Hypericum connatum* affect human trophoblast-like cells differentiation and Ca2+ influx. Asian Pac. J. Trop. Biomed. 4: 367–373.

da Silva, A.R.H., Lopes, L.Q.S., Cassanego, G.B., de Jesus, P.R., Figueredo, K.C., Santos, R.C.V., Lopes, G.H.H., de Freitas Bauermann, L. 2018. Acute toxicity and antimicrobial activity of leaf tincture *Baccharis trimera* (Less). Biomed. J. 41: 194–201.

da Veiga, D.T.A., Bringhenti, R., Bolignon, A.A., Tatsh, E., Moresco, R.N., Comim, F.V., Premaor, M.O. 2018. The yerba mate intake has a neutral effect on bone: A case–control study in postmenopausal women. Phyther Res. 32: 58–64.

de Araújo, G.R., Rabelo, A.C.S., Meira, J.S., Rossoni-Júnior, J.V., Castro-Borges, W. de, Guerra-Sá, R., Batista, M.A., Silveira-Lemos, D. da, Souza, G.H.B. de, Brandão, G.C., Chaves, M.M., Costa, D.C. 2017. *Baccharis trimera* inhibits reactive oxygen species production through PKC and down-regulation p47 phox phosphorylation of NADPH oxidase in SK Hep-1 cells. Exp. Biol. Med. 242: 333–343.

de Lima, M.E., Colpo, A.Z.C., Rosa, H., Salgueiro, A.C.F., da Silva, M.P., Noronha, D.S., Santamaría, A., Folmer, V. 2018. *Ilex paraguariensis* extracts reduce blood glucose, peripheral neuropathy and oxidative damage in male mice exposed to streptozotocin. J. Funct. Foods. 44: 9–16.

de Lima, M.E., Ceolin Colpo, A.Z., Maya-López, M., Rangel-López, E., Becerril-Chávez, H., Galván-Arzate, S., Villeda-Hernández, J., Sánchez-Chapul, L., Túnez, I., Folmer, V., Santamaría, A. 2019. Comparing the effects of chlorogenic acid and *Ilex paraguariensis* extracts on different markers of brain alterations in rats subjected to chronic restraint stress. Neurotox. Res. 35: 373–386.

de Oliveira, L.F.G., Pereira, B.A.S., Gilbert, B., Corrêa, A.L., Rocha, L., Alves, C.R. 2017. Natural products and phytotherapy: an innovative perspective in leishmaniasis treatment. Phytochem. Rev. 16: 219–233.

de Paula Vasconcelos, P.C., Spessotto, D.R., Marinho, J.V., Salvador, M.J., Junior, A.G., Kassuya, C.A.L. 2017. Mechanisms underlying the diuretic effect of *Gomphrena celosioides* Mart. (Amaranthaceae). J. Ethnopharmacol. 202: 85–91.

de Paula Vasconcelos, P.C., Tirloni, C.A.S., Palozi, R.A.C., Leitão, M.M., Carneiro, M.T.S., Schaedler, M.I., Silva, A.O., Souza, R.I.C., Salvador, M.J., Junior, A.G., Kassuya, C.A.L. 2018. Diuretic herb *Gomphrena celosioides* Mart. (Amaranthaceae) promotes sustained arterial pressure reduction and protection from cardiac remodeling on rats with renovascular hypertension. J. Ethnopharmacol. 224: 126–133.

de Souza Marinho Do Nascimento, D., Oliveira, R.M., Camara, R.B.G., Gomes, D.L., Monte, J.F.S., Costa, M.S.S.P., Fernandes, J.M., Langassner, S.M.Z., Rocha, H.A.O. 2017. *Baccharis trimera* (Less.) DC exhibits an anti-adipogenic effect by inhibiting the expression of proteins involved in adipocyte differentiation. Molecules 22: 1–16.

Demarque, D.P., Fitts, S.M.F., Boaretto, A.G., Da Silva, J.C.L., Vieira, M.C., Franco, V.N.P., Teixeira, C.B., Toffoli-Kadri, M.C., Carollo, C.A. 2015. Optimization and technological development strategies of an antimicrobial extract from *Achyrocline alata* assisted by statistical design. PLoS One 10. DOI:10.1371/journal.pone.0118574.

do Nascimento Kaut, N.N., Rabelo, A.C.S., Araujo, G.R., Taylor, J.G., Silva, M.E., Pedrosa, M.L., Chaves, M.M., Rossoni Junior, J.V., Costa, D.C. 2018. *Baccharis trimera* (Carqueja) improves metabolic and redox status in an experimental model of type 1 diabetes. Evidence-Based Complement Altern. Med. 2018: 1–12. DOI:10.1155/2018/6532637.

Dumitru, G., El-Nashar, H.A.S., Mostafa, N.M., Eldahshan, O.A., Boiangiu, R.S., Todirascu-Ciornea, E., Hritcu, L., Singab, A.N.B. 2019. Agathisflavone isolated from *Schinus polygamus* (Cav.) Cabrera leaves prevents scopolamine-induced memory impairment and brain oxidative stress in zebrafish (Danio rerio). Phytomedicine. 58:152889. DOI:10.1016/j.phymed.2019.152889.

Echenique, D., Chiaramello, A., Rossomando, P., Mattana, C., Alcaráz, L., Tonn, C., Laciar, A., Satorres, S. 2014. Antibacterial Activity of *Mulinum spinosum* Extracts against Slime-Producing *Staphylococcus aureus* and Methicillin-Resistant *Staphylococcus aureus* Isolated from Nasal Carriers Sci. World J. 2014: 1–6. DOI:10.1155/2014/342143.

Emam, M., El Raey, M.A., El-Haddad, A.E., El Awdan, S.A., Rabie, A.G.M., El-Ansari, M.A., Sobeh, M., Osman, S.M., Wink, M. 2019. A new polyoxygenated flavonol gossypetin-3-o-β-d-robinobioside from *Caesalpinia gilliesii* (Hook.) D. Dietr. and *in vivo* hepatoprotective, anti-inflammatory, and anti-ulcer activities of the leaf methanol extract. Molecules 24(1): 138. DOI:10.3390/molecules24010138.

Erazo, S., Delporte, C., Negrete, R., García, R., Zaldívar, M., Iturra, G., Caballero, E., López, J.L., Backhouse, N. 2006. Constituents and biological activities of *Schinus polygamus*. J. Ethnopharmacol. 107: 395–400.

Etoh, T., Kim, Y.P., Ohsaki, A., Komiyama, K., Hayashi, M. 2013. Inhibitory effect of erythraline on toll-like receptor signaling pathway in RAW264.7 cells. Biol. Pharm. Bull. 36: 1363–1369.

Feresin, G.E., Tapia, A., Gimenez, A., Gutierrez Ravelo, A., Zacchino, S., Sortino, M., Schmeda-Hirschmann, G. 2003. Constituents of the Argentinian medicinal plant *Baccharis grisebachii* and their antimicrobial activity. J. Ethnopharmacol. 89: 73–80.

Fernández, T., Cerdá Zolezzi, P., Risco, E., Martino, V., López, P., Clavin, M., Hnatyszyn, O., Canigueral, S., Hajos, S., Ferraro, G., Alvarez, E. 2002. Immunomodulating properties of Argentine plants with ethnomedicinal use. Phytomedicine 9: 546–552.

Fratianni, F., Nazzaro, F., Marandino, A., Fusco, M.D.R., Coppola, R., De Feo, V., De Martino, L. 2013. Biochemical Composition, Antimicrobial Activities, and Anti–Quorum-Sensing Activities of Ethanol and Ethyl Acetate Extracts from *Hypericum connatum* Lam. (Guttiferae). J. Med. Food. 16: 454–459.

Fritz, D., Venturi, C.R., Cargnin, S., Schripsema, J., Roehe, P.M., Montanha, J.A., von Poser, G.L. 2007. Herpes virus inhibitory substances from *Hypericum connatum* Lam., a plant used in southern Brazil to treat oral lesions. J. Ethnopharmacol. 113: 517–520.

Fuentes Mardones, O., Alarcón Enos, J. 2010. *Bauhinia candicans* improves the endothelium-dependent relaxation in aortic rings of alloxan-diabetic rats. Bol. Latinoam. y del Caribe Plantas Med. y Aromat. 9: 485–490.

Fuentes, O., Alarcón, J. 2006. *Bauhinia candicans* stimulation of glucose uptake in isolated gastric glands of normal and diabetic rabbits. Fitoterapia. 77: 271–275.

Funes, M., Garro, M.F., Tosso, R.D., Maria, A.O., Saad, J.R., Enriz, R.D. 2018. Antinociceptive effect of neo-clerodane diterpenes obtained from *Baccharis flabellata*. Fitoterapia. 130: 94–99.

Gan, R.Y., Zhang, D., Wang, M., Corke, H. 2018. Health benefits of bioactive compounds from the genus *Ilex*, a source of traditional caffeinated beverages. Nutrients 10(11):1682. DOI:10.3390/nu10111682.

Garro, M.F., Salinas Ibáñez, A.G., Vega, A.E., Arismendi Sosa, A.C., Pelzer, L., Saad, J.R., Maria, A.O. 2015. Gastroprotective effects and antimicrobial activity of *Lithraea molleoides* and isolated compounds against *Helicobacter pylori*. J. Ethnopharmacol. 176: 469–474.

Gastaldi, B., Marino, G., Assef, Y., Silva Sofrás, F.M., Catalán, C.A.N., González, S.B. 2018. Nutraceutical Properties of Herbal Infusions from Six Native Plants of Argentine Patagonia. Plant Foods Hum. Nutr. 73: 180–188.

Gehrke, I.T.S., Neto, A.T., Pedroso, M., Mostardeiro, C.P., Da Cruz, I.B.M., Silva, U.F., Ilha, V., Dalcol, I.I., Morel, A.F. 2013. Antimicrobial activity of *Schinus lentiscifolius* (Anacardiaceae). J. Ethnopharmacol. 148: 486–491.

Gerke, I.B.B., Hamerski, F., de Paula Scheer, A., da Silva, V.R. 2018. Solid–liquid extraction of bioactive compounds from yerba mate (*Ilex paraguariensis*) leaves: Experimental study, kinetics and modeling. J. Food Process. Eng. 41: 1–10.

Goleniowski, M.E., Bongiovanni, G.A., Palacio, L., Nuñez, C.O., Cantero, J.J. 2006. Medicinal plants from the "Sierra de Comechingones", Argentina. J. Ethnopharm. 107: 324–341.

Gómez-Juaristi, M., Martínez-López, S., Sarria, B., Bravo, L., Mateos, R. 2018. Absorption and metabolism of yerba mate phenolic compounds in humans. Food Chem. 240: 1028–1038.

Gómez, J., Simirgiotis, M.J., Lima, B., Paredes, J.D., Villegas Gabutti, C.M., Gamarra-Luques, C., Bórquez, J., Luna, L., Wendel, G.H., Maria, A.O., Feresin, G.E., Tapia, A. 2019. Antioxidant, gastroprotective, cytotoxic activities and uhplc pda-q orbitrap mass spectrometry identification of metabolites in baccharis grisebachii decoction. Molecules 24. DOI:10.3390/molecules24061085.

González, S.B., Houghton, P.J., Hoult, J.R.S. 2003. The activity against leukocyte eicosanoid generation of essential oil and polar fractions of *Adesmia boronioides* Hook.f. Phyter Res. 17: 290–293.

Gorzalczany, S., Marrassini, C., Miño, J., Acevedo, C., Ferraro, G. 2011. Antinociceptive activity of ethanolic extract and isolated compounds of *Urtica circularis*. J. Ethnopharmacol. 134: 733–738.

Gorzalczany, S., Moscatelli, V., Acevedo, C., Ferraro, G. 2013a. Spasmolytic activity of *Artemisia copa* aqueous extract and isolated compounds. Nat. Prod. Res. 27: 1007–1011.

Gorzalczany, S., Moscatelli, V., Ferraro, G. 2013b. *Artemisia copa* aqueous extract as vasorelaxant and hypotensive agent. J. Ethnopharmacol. 148: 56–61.

Gupta, D., Kumar, M., Gupta, V. 2018. An *in vitro* investigation of antimicrobial efficacy of *Euphorbia hirta* and *Murraya koenigii* against selected pathogenic microorganisms. Asian J. Pharm. Clin. Res. 11: 359. DOI:10.22159/ajpcr.2018.v11i5.24578.

Gutierrez-Lugo, M.-T., Singh, M.P., Maiese, W.M., Timmermann, B.N. 2002. New antimicrobial cycloartane triterpenes from *Acalypha communis*. J. Nat. Prod. 65: 872–875.

Ibañez, S., Sosa, A., Ferramola, F., Paredes, J., Wendel, G., Maria, A., Vega, E. 2017. Inhibition of *Helicobacter pylori* and its associated urease by two regional plants of San Luis Argentina. Int. J. Curr. Microbiol. App. Sci. 6: 2019–2106. DOI: 10.20546/ijcmas.2017.609.258.

Ibarrola, D.A., Hellión-Ibarrola, M.C., Montalbetti, Y., Heinichen, O., Campuzano, M.A., Kennedy, M.L., Alvarenga, N., Ferro, E.A., Dölz-Vargas, J.H., Momose, Y. 2011. Antihypertensive effect of nuatigenin-3-O-β-chacotriose from *Solanum sisymbriifolium* Lam. (Solanaccac) (ñuatî pytâ) in experimentally hypertensive (ARH+DOCA) rats under chronic administration. Phytomedicine. 18: 634–640.

Jeba, C.R., Ilakiya, A., Deepika, R., Sujatha, M., Sivaraji, C. 2018. Antipsoriasis, antioxidant, and antimicrobial activities of aerial parts of *Euphorbia hirta*. Asian J. Pharm. Clin. Res. 11: 513–517.

Jia, L., Zho, J., Zhao, H., Yanjin, H., Lv, M., Zhao, N., Zheng, Z., Lu, Y., Ming, Y., Yu, Y. 2017. Corilagin sensitizes epithelial ovarian cancer to chemotherapy by inhibiting Snail-glycolysis pathways. Oncol. Rep. 38: 2464–2470.

Jiménez-Aspee, F., Theoduloz, C., Soriano, M. del P.C., Ugalde-Arbizu, M., Alberto, M.R., Zampini, I.C., Isla, M.I., Simirigiotis, M.J., Schmeda-Hirschmann, G. 2017. The native fruit *Geoffroea decorticans* from arid Northern Chile: Phenolic composition, antioxidant activities and *in vitro* inhibition of pro-inflammatory and metabolic syndrome-associated enzymes. Molecules. 22: 1–18.

Kaur, N., Kaur, B., Sirhindi, G. 2017. Phytochemistry and Pharmacology of *Phyllanthus niruri* L.: A Review. Phyther Res. 31: 980–1004.

Kausar, J., Muthumani, D., Hedina, A., S, S., Anand, V. 2016. Review of the phytochemical and pharmacological activities of *Euphorbia hirta* Linn. Pharmacogn. J. 8: 310–313.

Kheiri Manjili, H., Jafari, H., Ramazani, A., Davoudi, N. 2012. Anti-leishmanial and toxicity activities of some selected Iranian medicinal plants. Parasitol. Res. 111: 2115–2121.

Kim, S.Y., Oh, M.R., Kim, M.G., Chae, H.J., Chae, S.W. 2015. Anti-obesity effects of Yerba Mate (*Ilex Paraguariensis*): A randomized, double-blind, placebo-controlled clinical trial. BMC Complement Altern. Med. 15: 1–8.

Klein-Júnior, L., da Silva, L., Boeing, T., Somensi, L., Beber, A., Rocha, J., Henriques, A., Andrade, S., Cechinel-Filho, V. 2016. The protective potential of *Phyllanthus niruri* and corilagin on gastric lesions induced in rodents by different harmful agents. Planta Med. 83: 30–39.

Kumar, A., Kumar Rana, A., Singh, Amit, Singh, Alok. 2019. Effect of methanolic extract of *Phyllanthus niruri* on leptin level in animal model of diabetes Mellitus. Biomed. Pharmacol. J. 11: 57–63.

Kundu, J.K., Surh, Y.-J. 2009. Capsaicin—A Hot Spice in the Chemoprevention of Cancer. pp. 311–339. *In*: Molecular Targets and Therapeutic Uses of Spices. World Scientific. DOI:0.1142/9789812837912_0012.

Kungel, P.T.A.N., Correa, V.G., Corrêa, R.C.G., Peralta, R.A., Soković, M., Calhelha, R.C., Bracht, A., Ferreira, I.C.F.R., Peralta, R.M. 2018. Antioxidant and antimicrobial activities of a purified polysaccharide from yerba mate (*Ilex paraguariensis*). Int. J. Biol. Macromol. 114: 1161–1167.

Kujawska, M., Hilgert, N.I., Keller, H.A., Gil, G. 2017. Medicinal plant diversity and inter-cultural interactions between indigenous guarani, criollos and polish migrants in the subtropics of Argentina. PLoS ONE 12: 1–21. DOI: 10.1371/journal.pone.0169373.

Kwan, Y.P., Saito, T., Ibrahim, D., Al-Hassan, F.M.S., Ein Oon, C., Chen, Y., Jothy, S.L., Kanwar, J.R., Sasidharan, S. 2016. Evaluation of the cytotoxicity, cell-cycle arrest, and apoptotic induction by *Euphorbia hirta* in MCF-7 breast cancer cells. Pharm. Biol. 54: 1223–1236.

Leal, L.K.A.M., Silva, A.H., Viana, G.S. de B. 2017. *Justicia pectoralis*, a coumarin medicinal plant have potential for the development of antiasthmatic drugs? Brazilian J. Pharmacogn. 27: 794–802.

Lee, S.-Y., Huang, K.-F. 2004. Constituents of stems of *Erythrina crista-galli* Linn. Chinese Pharm J. 56: 159–162.

Li, Y., Li, X., Wang, J.K., Kuang, Y., Qi, M.X. 2017. Anti-hepatitis B viral activity of *Phyllanthus niruri* L. (Phyllanthaceae) in HepG2/C3A and SK-HEP-1 cells. Trop. J. Pharm. Res. 16: 1873–1879.

Lima, S.O., Figueiredo, M.B.G. de A., de Santana, V.R., Santana, D.P.A., Nogueira, M. de S., Porto, E.S., de Andrade, R.L.B., Santos, J.M., de Albuquerque Junior, R.L.C., Cardoso, J.C. 2017. Effect of aqueous extract of the leaves of *Baccharis trimera* on the proliferation of hepatocytes after partial hepatectomy in rats. Acta Cir. Bras. 32: 263–269.

Lívero, F.A. do. R., Martins, G.G., Queiroz Telles, J.E., Beltrame, O.C., Petris Biscaia, S.M., Cavicchiolo Franco, C.R., Oude Elferink, R.P.J., Acco, A. 2016. Hydroethanolic extract of *Baccharis trimera* ameliorates alcoholic fatty liver disease in mice. Chem. Biol. Interact. 260: 22–32.

Lodise, O., Patil, K., Karshenboym, I., Prombo, S., Chukwueke, C., Pai, S.B. 2019. Inhibition of prostate cancer cells by 4,5-dicaffeoylquinic acid through cell cycle arrest. Prostate Cancer. 2019. DOI:10.1155/2019/4520645.

Lujan, C., Martinez, G. 2019. Etnobotánica médica urbana y periurbana de la ciudad de Córdoba (Argentina). Blacpma. 18. 155–196.

Machado, D.G., Bettio, L.E.B., Cunha, M.P., Santos, A.R.S., Pizzolatti, M.G., Brighente, I.M.C., Rodrigues, A.L.S. 2008. Antidepressant-like effect of rutin isolated from the ethanolic extract from *Schinus molle* L. in mice: Evidence for the involvement of the serotonergic and noradrenergic systems. Eur. J. Pharmacol. 587: 163–168.

Magalhães, K. do N., Guarniz, W.A.S., Sá, K.M., Freire, A.B., Monteiro, M.P., Nojosa, R.T., Bieski, I.G.C., Custódio, J.B., Balogun, S.O., Bandeira, M.A.M. 2019. Medicinal plants of the Caatinga, northeastern Brazil: Ethnopharmacopeia (1980–1990) of the late professor Francisco José de Abreu Matos. J. Ethnopharmacol. 237: 314–353.

Malca-García, G.R., Hennig, L., Ganoza-Yupanqui, M.L., Piña-Iturbe, A., Bussmannd, R.W. 2017. Constituents from the bark resin of *Schinus molle*. Brazilian J. Pharmacogn. 27: 67–69.

Manchishi, S.M. 2017. Recent advances in antiepileptic herbal medicine. Curr. Neuropharmacol. 16: 79–83.

Marrassini, C., Davicino, R., Acevedo, C., Anesini, C., Gorzalczany, S., Ferraro, G. 2011. Vicenin-2, a potential anti-inflammatory constituent of *Urtica circularis*. J. Nat. Prod. 74: 1503–1507.

Martinez, G.J., Barboza, G.E. 2010. Natural pharmacopoeia used in traditional Toba medicine for the treatment of parasitosis and skin disorders (Central Chaco, Argentina). J. Ethnopharmacology 132: 86–100.

Martínez, G.J., Luján, M.C. 2011. Medicinal plants used for traditional veterinary in the Sierras de Córdoba (Argentina): An ethnobotanical comparison with human medicinal uses. J. Ethnobiol. Ethnomed. 7: 23. DOI:0.1186/1746-4269-7-23.

Martínez, G.J. 2015. Las Plantas en la Medicina Tradicional de las Sierras de Córdoba., 2015th ed. Editorial Detodoslosmares, Córdoba.

Martinez, G.J., Escobar, D.J. 2017. Plantas de interés veterinario en la cultura campesina de la Sierra de Ancasti (Catamarca, Argentina). Blacpma 16: 329–346.

Martinez, M.A., Mattana, C.M., Satorres, S.E., Sosa, A., Fusco, M.R., Laciar, A.L., Alcaraz, L.E. 2014. Screening phytochemical and antibacterial activity of three San Luis native species belonging at the fabaceae family. Pharmacologyonline 3: 1–6.

Mateos, R., Baeza, G., Sarriá, B., Bravo, L. 2018. Improved LC-MSn characterization of hydroxycinnamic acid derivatives and flavonols in different commercial mate (*Ilex paraguariensis*) brands. Quantification of polyphenols, methylxanthines, and antioxidant activity. Food Chem. 241: 232–241.

Matera, S., Bruno, F., Bayley, M., Pérez, V., Querini, C., Ragone, M.I., Consolini, A.E. 2016. Intestinal antispasmodic effects of three argentinian plants: *Hypericum connatum*, *Berberis ruscifolia* and *Cecropia pachystachya*: Mechanisms of action and comparison with the effects of *Brugmansia arborea*. Pharmacologyonline. 2016: 91–99.

Mediani, A., Abas, F., Maulidiani, M., Khatib, A., Tan, C.P., Safinar Ismail, I., Shaari, K., Ismail, A. 2017. Characterization of metabolite profile in *Phyllanthus niruri* and correlation with bioactivity elucidated by nuclear magnetic resonance based metabolomics. Molecules 22. DOI:10.3390/molecules22060902.

Meinhart, A.D., Damin, F.M., Caldeirão, L., Godoy, H.T. 2019. Methylxanthines in 100 Brazilian herbs and infusions: Determination and consumption. Emirates J. Food Agric. 31: 125–133.

Miño, J.H., Moscatelli, V., Acevedo, C., Ferraro, G. 2010. Psychopharmacological effects of *Artemisia copa* aqueous extract in mice. Pharm. Biol. 48: 1392–1396.

Molina-Garza, Z.J., Bazaldúa-Rodríguez, A.F., Quintanilla-Licea, R., Galaviz-Silva, L. 2014. Anti-Trypanosoma cruzi activity of 10 medicinal plants used in northeast Mexico. Acta Trop. 136: 14–18.

Mopuri, R., Islam, M.S. 2017. Medicinal plants and phytochemicals with anti-obesogenic potentials: A review. Biomed. Pharmacother. 89: 1442–1452.

Mostofa, R., Ahmed, S., Begum, M.M., Sohanur Rahman, M., Begum, T., Ahmed, S.U., Tuhin, R.H., Das, M., Hossain, A., Sharma, M., Begum, R. 2017. Evaluation of anti-inflammatory and gastric anti-ulcer activity of *Phyllanthus niruri* L. (Euphorbiaceae) leaves in experimental rats. BMC Complement Altern Med. 17: 267. DOI:10.1186/s12906-017-1771-7.

Muschietti, L., Derita, M., Sülsen, V., de Dios Muñoz, J., Ferraro, G., Zacchino, S., Martino, V. 2005. *In vitro* antifungal assay of traditional Argentine medicinal plants. J. Ethnopharmacol. 102: 233–238.

Ndjonka, D., Djafsia, B., Liebau, E. 2018. Review on medicinal plants and natural compounds as anti-Onchocerca agents. Parasitol. Res. 117: 2697–2713.

Niwa, Y., Matsuura, H., Murakami, M., Sato, J., Hirai, K., Sumi, H. 2013. Evidence That Naturopathic Therapy Including *Cordyceps sinensis* Prolongs Survival of Patients With Hepatocellular Carcinoma. Integr. Cancer Ther. 12: 50–68.

Nunes, T.R.D.S., Cordeiro, M.F., Beserra, F.G., Souza, M.L. De, Silva, W.A.V. Da, Ferreira, M.R.A., Soares, L.A.L., Costa-Junior, S.D., Cavalcanti, I.M.F., Pitta, M.G.D.R., Pitta, I.D.R., Rêgo, M.J.B.D.M. 2018. Organic extract of *Justicia pectoralis* Jacq. leaf inhibits interferon- γ secretion and has bacteriostatic activity against *Acinetobacter baumannii* and *Klebsiella pneumoniae*. Evidence-based Complement Altern. Med. 2018. DOI: 10.1155/2018/5762368.

Oboh, G., Akinyemi, A.J., Adeleye, B., Oyeleye, S.I., Ogunsuyi, O.B., Ademosun, A.O., Ademiluyi, A.O., Boligon, A.A. 2016. Polyphenolic compositions and *in vitro* angiotensin-I-converting enzyme inhibitory properties of common green leafy vegetables: A comparative study. Food Sci. Biotechnol. 25: 1243–1249.

Oh, K.E., Shin, H., Jeon, Y.H., Jo, Y.H., Lee, M.K., Lee, K.S., Park, B., Lee, K.Y. 2016. Optimization of pancreatic lipase inhibitory and antioxidant activities of *Ilex paraguariensis* by using response surface methodology. Arch. Pharm. Res. 39: 946–952.

Oliveira, S., Souza, G.A., Eckert, C.R., Silva, T.A., Edmar Silva Sobra, E.S., Fávero, O.P., Ferreira, M.J.P., Romoff, P., Baader, W. 2014. Evaluation of antiradical assays used in determining the antioxidant capacity of pure Artigo. Quim. Nov. 37: 497–503.

Olugbemiga, O.S., Grace, O.D., Adeola, T.A., Ibibia, E.I.T., Akhere, O.J., Adeiza, O.D., Oluchi, A.Y., Obiora, N.C., Stephen, A.O. 2016. Antidiabetic and antidyslipidemic effect of ethanolic extract of *Alternanathera pungens* on alloxan-induced diabetic rats. Asian. J. Biochem. 11: 82–89.

Ortiz, S., Lecsö-Bornet, M., Michel, S., Grougnet, R., Boutefnouchet, S. 2019. Chemical composition and biological activity of essential oils from *Artemisia copa* Phil. var. copa (Asteraceae) and *Aloysia deserticola* (Phil.) Lu-Irving and O'Leary (Verbenaceae), used in the Chilean Atacama's Taira Community (Antofagasta, Chile). J. Essent. Oil Res. 00: 1–7. DOI:10.1080/10412905.2019.1572549.

Osman, S., Alazzouni, A., Khalek, S.M., Koheil, M., El-Haddad, A. 2013. Phytoconstituents and biological activities of the aerial parts of *Caesalpinia gilliesii* (Hook) Family Caesalpinacae growing in Egypt. Life Sci. J. 10: 2418–2430.

Osman, S., El-Haddad, A., El-Raey, M., Abd El-Khalik, S., Koheil, M., Wink, M. 2016. A new octadecenoic acid derivative from *Caesalpinia gilliesii* flowers with potent hepatoprotective activity. Pharmacogn. Mag. 12: 332. DOI:10.4103/0973-1296.185752.

Padma, P., Pramod, N.P., Thyagarajan, S.P., Khosa, R.L. 1998. Effect of the extract of *Annona muricata* and *Petunia nyctaginiflora* on *Herpes simplex* virus. J. Ethnopharmacol. 61: 81–83.

Paredes, J.D., Sosa, Á., Fusco, M., Teves, M.R., Wendel, G.H., Pelzer, L.E. 2016. Antidiarrhoeal activity of *argentina* Gris. (Aristolochiaceae) in rodents. J. Appl. Pharm. Sci. 6: 146–152.

Pereira, A.A.F., Tirapeli, K.G., Chaves-Neto, A.H., da Silva Brasilino, M., da Rocha, C.Q., Belló-Klein, A., Llesuy, S.F., Dornelles, R.C.M., Nakamune, A.C. de M.S. 2017a. *Ilex paraguariensis* supplementation may be an effective nutritional approach to modulate oxidative stress during perimenopause. Exp. Gerontol. 90: 14–18.

Pereira, L.X., Silva, H.K.C., Longatti, T.R., Silva, P.P., Di Lorenzo Oliveira, C., de Freitas Carneiro Proietti, A.B., Thomé, R.G., Vieira, M. do C., Carollo, C.A., Demarque, D.P., de Siqueira, J.M., dos Santos, H.B., Parreira, G.G., de Azambuja Ribeiro, R.I.M. 2017b. *Achyrocline alata* potentiates repair of skin full thickness excision in mice. J. Tissue Viability. 26: 289–299.

Pérez, M.J., Cuello, A.S., Zampini, I.C., Ordoñez, R.M., Alberto, M.R., Quispe, C., Schmeda-Hirschmann, G., Isla, M.I. 2014. Polyphenolic compounds and anthocyanin content of *Prosopis nigra* and *Prosopis alba* pods flour and their antioxidant and anti-inflammatory capacities. Food Res. Int. 64: 762–771.

Polidoro, A. dos, S., Scapin, E., Malmann, M., do Carmo, J.U., Machado, M.E., Caramão, E.B., Jacques, R.A. 2016. Characterization of volatile fractions in green mate and mate leaves (*Ilex paraguariensis* A. St. Hil.) by comprehensive two-dimensional gas chromatography coupled to time-of-flight mass spectrometry (GC × GC/ TOFMS). Microchem. J. 128: 118–127.

Promprom, W., Chatan, W. 2018. Estrogenic effects of *Euphorbia hirta* l. Extract in ovariectomized rats. Pharmacogn. J. 10: 435–438.

Promraksa, B., Phetcharaburanin, J., Namwat, N., Techasen, A., Boonsiri, P., Loilome, W. 2019. Evaluation of anticancer potential of Thai medicinal herb extracts against cholangiocarcinoma cell lines. PLoS One 14: e0216721. DOI:10.1371/journal.pone.0216721.

Pucci, N.D., Marchini, G.S., Mazzucchi, E., Reis, S.T., Srougi, M., Evazian, D., Nahas, W.C. 2018. Effect of *Phyllanthus niruri* on metabolic parameters of patients with kidney stone: A perspective for disease prevention. Int. Braz. J. Urol. 44: 758–764.

Rabelo, A.C.S., de Pádua Lúcio, K., Araújo, C.M., de Araújo, G.R., de Amorim Miranda, P.H., Carneiro, A.C.A., de Castro Ribeiro, É.M., de Melo Silva, B., de Lima, W.G., Costa, D.C. 2018. *Baccharis trimera* protects against ethanol induced hepatotoxicity *in vitro* and *in vivo*. J. Ethnopharmacol. 215: 1–13.

Rabiei, Z., Rabiei, S. 2017. A review on antidepressant effect of medicinal plants. Bangladesh J. Pharmacol. 12: 1–11.

Raji, P., Samrot, A.V., Keerthana, D., Karishma, S. 2019a. Antibacterial activity of alkaloids, flavonoids, saponins and tannins mediated green synthesised silver nanoparticles against *Pseudomonas aeruginosa* and *Bacillus subtilis*. J. Clust. Sci. 30: 881–895.

Raji, P., Samrot, A.V., Bennet Rohan, D., Divya Kumar, M., Geetika, R., Kripu Sharma, V., Keerthana, D. 2019b. Extraction, characterization and *in vitro* bioactivity evaluation of alkaloids, flavonoids, saponins and tannins of *Cassia alata, Thespesia populnea, Euphorbia hirta* and *Wrightia tinctoria*. Rasayan J. Chem. 12: 123–137. DOI:10.31788/RJC.2019.1214054.

Rempe, C.S., Lenaghan, S.C., Burris, K.P., Stewart, C.N. 2017. Metabolomic analysis of the mechanism of action of yerba mate aqueous extract on *Salmonella enterica* serovar Typhimurium. Metabolomics 13: 1–13.

Reynoso, M.A., Vera, N., Aristimuño, M.E., Daud, A., Sánchez Riera, A. 2013. Antinociceptive activity of fruits extracts and arrope of *Geoffroea decorticans* (chañar). J. Ethnopharmacol. 145: 355–362.

Rizk, M.Z., Ally, H.F., Abo-Elmatty, D.M., Mohammed, M.M.D., Ibrahim, N., Younis, E.A. 2016. Hepatoprotective effect of *Caesalpinia gilliesii* and *Cajanus cajan* proteins against acetoaminophen overdose-induced hepatic damage. Toxicol. Ind. Health. 32: 877–907.

Rocha, D.S., Casagrande, L., Model, J.F.A., dos Santos, J.T., Hoefel, A.L., Kucharski, L.C. 2018. Effect of yerba mate (*Ilex paraguariensis*) extract on the metabolism of diabetic rats. Biomed. Pharmacother. 105: 370–376.

Rocha, L.G., Almeida, J.R.G.S., Macêdo, R.O., Barbosa-Filho, J.M. 2005. A review of natural products with antileishmanial activity. Phytomedicine 12: 514–535.

Rodriguez, M.V., Sortino, M.A., Ivancovich, J.J., Pellegrino, J.M., Favier, L.S., Raimondi, M.P., Gattuso, M.A., Zacchino, S.A. 2013. Detection of synergistic combinations of *Baccharis* extracts with Terbinafine against *Trichophyton rubrum* with high throughput screening synergy assay (HTSS) followed by 3D graphs. Behavior of some of their components. Phytomedicine 20: 1230–1239.

Rojas de Arias, A., Ferro, E., Inchausti, A., Ascurra, M., Acosta, N., Rodriguez, E., Fournet, A. 1995. Mutagenicity, insecticidal and trypanocidal activity of some Paraguayan Asteraceae. J. Ethnopharmacol. 45: 35–41.

Romero, C.M., Vivacqua, C.G., Abdulhamid, M.B., Baigori, D., Slanis, C., Allori, G., Tereschuk, L. 2016. Biofilm inhibition activity of traditional medicinal plants from northwestern Argentina against native pathogen and environmental microorganisms. Rev. Soc. Bras Med. Trop. 49: 703–712. DOI: 10.1590./0037-8682-0452-2016.

Ruffa, M., Ferraro, G., Wagner, M., Calcagno, M., Campos, R., Cavallaro, L. 2002. Cytotoxic effect of Argentine medicinal plant extracts on human hepatocellular carcinoma cell line. J. Ethnopharmacol. 79: 335–339.

Sabir, S.M., Athayde, M.L., Boligon, A.A., Rocha, J.B.T. 2017. Antioxidant activities and phenolic profile of *Baccharis trimera*, a commonly used medicinal plant from Brazil. South African. J. Bot. 113: 318–323.

Sahebkar-Khorasani, M., Jarahi, L., Cramer, H., Safarian, M., Naghedi-Baghdar, H., Salari, R., Behravanrad, P., Azizi, H. 2019. Herbal medicines for suppressing appetite: A systematic review of randomized clinical trials. Complement Ther. Med. 44: 242–252.

Salvat, A., Antonacci, L., Fortunato, R.H., Suarez, E.Y., Godoy, H.M. 2004. Antimicrobial activity in methanolic extracts of several plant species from northern Argentina. Phytomedicine 11: 230–234.

Sangare, M.M., Sina, H., Bayala, B., Baba-Moussa, L.S., Ategbo, J.M., Senou, M., Dramane, K.L. 2014. Évaluation de la dose efficace de l'extrait aqueux de *Gomphrena celosioides* face à une hépatopathie induite par le tétrachlorure de carbone. Phytothérapie 12: 393–398.

Santin, J.R., Lemos, M., Júnior, L.C.K., Niero, R., de Andrade, S.F. 2010. Antiulcer effects of *Achyrocline satureoides* (Lam.) DC (Asteraceae) (Marcela), a folk medicine plant, in different experimental models. J. Ethnopharmacol. 130: 334–339.

Santin, J.R., Lemos, M., Klein-júnior, L.C., Oliveira, D., Silveira, A.C., Petreanu, M., Zermiani, T., Bruz, A., Niero, R., de Andrade, S.F. 2014. Gastro protective and anti-helicobacter pylori effects of a flavonoid rich fraction obtained from *Achyrocline satureoides* (LAM) D.C. Int. J. Pharm. Sci. 6: 417–422.

Seebaluck, R., Gurib-Fakim, A., Mahomoodally, F. 2015. Medicinal plants from the genus *Acalypha* (Euphorbiaceae)–A review of their ethnopharmacology and phytochemistry. J. Ethnopharmacol. 159: 137–157.

Selener, M.G., Elso, O., Grosso, C., Borgo, J., Clavin, M., Malchiodi, E.L., Cazorla, S.I., Redko, F., Sülsen, V.P. 2019. Anti-*Trypanosoma cruzi* activity of extracts from argentinean *Asteraceae* species. Iranian J. Pharm. Res. 18: 1–7. DOI: 10.22037/ijpr.2019.14491.12430.

Selvam, P., Vijayakumar, T., Wadhwani, A., Muthulakshmi, L. 2019. Bioreduction of silver nanoparticles from aerial parts of *Euphorbia hirta* L. (EH-ET) and its potent anticancer activities against neuroblastoma cell lines. Indian J. Biochem. Biophys. 56: 132–136.

Sharma, G., Ashhar, U., Aeri, V., Katare, D.P. 2018. Effect of ethanolic extract of *Euphorbia hirta* on chronic diabetes mellitus and associated cardiorenal damage in rats. Int. J. Green Pharmacy 12(3): 191–199.

Simirgiotis, M.J., García, M., Sosa, M.E., Giordano, O.S., Tonn, C.E. 2003. An ent-kaurene derivative from aerial parts of baccharis rufescens. An des la Asoc Quim Argentina. 91: 109–116.

Simões, L.O., Conceição-Filho, G., Ribeiro, T.S., Jesus, A.M., Fregoneze, J.B., Silva, A.Q.G., Petreanu, M., Cechinel-Filho, V., Niero, R., Niero, H., Tamanaha, M.S., Silva, D.F. 2016. Evidences of antihypertensive potential of extract from S*olanum capsicoides* All. in spontaneously hypertensive rats. Phytomedicine 23(5): 498–508.

Soares, M., Domingues, R., Gaspar, B., dos Santos, A., Canuto, M., Minho, P., Vieira B. 2019. *In vitro* ovicidal effect of a Senecio brasiliensis extract and its fractions on Haemonchus contortus. BMC Vet. Res. 15. 99. DOI: 10.1186/s12917-019-1843-7.

Sperling, H., Lorenz, A., Krege, S., Arndt, R., Michel, M.C. 2002. An extract from the bark of *Aspidosperma quebracho blanco* binds to human penile α-adrenoceptors. J. Urol. 168: 160–163.

Suárez, M.E. 2018. Medicines in the forest: ethnobotany of wild medicinal plants in the pharmacopeia of the Wichi people of Salta province (Argentina). J. Ethnopharmacology, https://doi.org/10.1016/j.jep.2018.10.026.

Sülsen, V., Güida, C., Coussio, J., Paveto, C., Muschietti, L., Martino, V. 2006. *In vitro* evaluation of trypanocidal activity in plants used in Argentine traditional medicine. Parasitol. Res. 98: 370–374.

Sülsen, V.P., Muschietti, L.V., Martino, V.S., Malchiodi, E.L., Anesini, C.A., Coussio, J.D., Redko, F.C., Frank, F.M., Cazorla, S.I. 2007. Trypanocidal and Leishmanicidal Activities of Flavonoids from Argentine Medicinal Plants. Am. J. Trop. Med. Hyg. 77: 654–659.

Sülsen, V.P., Puente, V., Papademetrio, D., Batlle, A., Martino, V.S., Frank, F.M., Lombardo, M.E. 2016. Mode of Action of the Sesquiterpene Lactones Psilostachyin and Psilostachyin C on *Trypanosoma cruz*i. PLoS One. 11: 0150526. DOI:10.1371/journal.pone.0150526.

Svetaz, L., Zuljan, F., Derita, M., Petenatti, E., Tamayo, G., Cáceres, A., Cechinel Filho, V., Giménez, A., Pinzón, R., Zacchino, S.A., Gupta, M. 2010. Value of the ethnomedical information for the discovery of plants with antifungal properties. A survey among seven Latin American countries. J. Ethnopharmacol. 127: 137–158.

Terán Baptista, Z.P., Gómez, A. de los A., Kritsanida, M., Grougnet, R., Mandova, T., Aredes Fernandez, P.A., Sampietro, D.A. 2018. Antibacterial activity of native plants from Northwest Argentina against phytopathogenic bacteria. Nat. Prod. Res. 0: 1–4. DOI:10.1080/14786419.2018.1525716.

Tewari, D., Mocan, A., Parvanov, E.D., Sah, A.N., Nabavi, S.M., Huminiecki, L., Ma, Z.F., Lee, Y.Y., Horbańczuk, J.O., Atanasov, A.G. 2017. Ethnopharmacological approaches for therapy of jaundice: Part II. Highly used plant species from Acanthaceae, Euphorbiaceae, Asteraceae, Combretaceae, and Fabaceae families. Front Pharmacol. 8: 1–14.

Tjahjandarie, T.S., Pudjiastuti, P., Saputri, R.D., Tanjung, M. 2014. Antimalarial and antioxidant activity of phenolic compounds isolated from *Erythrina crista-galli* L. J. Chem. Pharm. Res. 6: 786–790.

Tjandrawinata, R., Susanto, L., Nofiarny, D. 2017. The use of phyllanthus niruri L. as an immunomodulator for the treatment of infectious diseases in clinical settings. Asian Pacific J. Trop. Dis. 7: 132–140.

Vasile, F.E., Romero, A.M., Judis, M.A., Mattalloni, M., Virgolini, M.B., Mazzobre, M.F. 2019. Phenolics composition, antioxidant properties and toxicological assessment of *Prosopis alba* exudate gum. Food Chem. 285: 369–379.

Visintini Jaime, M.F., Redko, F., Muschietti, L.V., Campos, R.H., Martino, V.S., Cavallaro, L.V. 2013. *In vitro* antiviral activity of plant extracts from Asteraceae medicinal plants. Virol. J. 10: 1–10.

Wan Omar, W.A., Mohd Zain, S.N.D. 2018. Therapeutic index of methanolic extracts of three Malaysian *Phyllanthus* species on MCF-7 and MCF-10A cell lines. Pharmacogn. J. 10: s30–s32. DOI:10.5530/pj.2018.6s.5.

Xia, M., Liu, L., Qiu, R., Li, M., Huang, W., Ren, G., Zhang, J. 2018. Anti-inflammatory and anxiolytic activities of *Euphorbia hirta* extract in neonatal asthmatic rats. AMB Express 8. 179. DOI:10.1186/s13568-018-0707-z.

Zhao, G., Cao, Z., Zhang, W., Zhao, H. 2015. The sesquiterpenoids and their chemotaxonomic implications in *Senecio* L. (*Asteraceae*). Biochem. Syst. Ecol. 59. 340–347. DOI: 10.1016/j.bse.2015.02.001.

Zheng, J., Zhou, Y., Li, Y., Xu, D.P., Li, S., Li, H. Bin. 2016. Spices for prevention and treatment of cancers. Nutrients 8(8): 495. DOI: 10.3390/nu8080495.

Zunino, M.P., López, M.L., Zygadlo, J. 2003. Medicinal plants of argentina. pharmacological properties and phytochemistry. pp. 209–245. *In*: Imperato, F. (ed.). Advances in Phytochemisty. Research Singpost Trivandrum Editorial, Kerala, India.

Zunino, M.P., Bregonzio, C., Baiardi, G. 2012. Efectos de los aceites esenciales naturales sobre el sistema nervioso central. pp. 165–179. *In*: J.A. Zygadlo (ed.). Aceites Esenciales. Química, Ecología, Comercio, Producción y Salud. Editorial Universitas, Córdoba.

Zygadlo, J.A., Zunino, M.P., Pizzolitto, R.P., Merlo, C., Omarini, A., Dambolena, J.S. 2017. Antibacterial and anti-biofilm activities of essential oils and their components including modes of action. *In*: Rai, M., Zacchino, S., Derita, M. (eds.). Essential Oils and Nanotechnology for Treatment of Microbial Diseases. CRC Press/ Taylor and Francis. ISBN 978-1138630727.

10

The Zigzag Trail of Symbiosis among Chepang, Bat, and Butter Tree

An Analysis on Conservation Threat in Nepal

Tirth Raj Ghimire,[1,]* *Roshan Babu Adhikari*[2] *and Ganga Ram Regmi*[2]

Introduction

The Oxford Advanced Learner's Dictionary has defined "indigenous" (*adj*) as belonging to a particular place rather than coming to it from somewhere else and "ethnic" (*adj*) as connected with or pertaining to a nation, race, or tribe that shares a cultural tradition (Hornby 2015). Although these terms are difficult to define unequivocally, depending on countries, "indigenous peoples" are also referred to as "indigenous ethnic minorities", "aboriginals", "hill tribes", "minority nationalities", "scheduled tribes", or "tribal groups" (World Bank 2013). It has been estimated that 370 million indigenous peoples constitute more than 5,000 different indigenous cultures, and more than 4,000 languages in more than 90 countries (United Nations 2009). Notably, their identities and cultures are exclusively connected to the lands and the available natural resources, which play a crucial role in their sustainable development (World Bank 2013). Therefore, their future is closely connected to the solutions to the crises in biodiversity and climate change.

While the roles of indigenous people in participatory biodiversity conservation around the globe have already been an exciting topic, they have been recently prioritized in Nepal. Equal and inclusive participation might be difficult, probably due to the presence of multireligious, multiethnic, multiracial, multilingual, and multicultural people within this small country. The country possesses 125 castes/ ethnic groups, 123 language speakers, and ten religious groups (CBS 2012). There are officially 59 groups (37% of the population) that are recognized as indigenous communities (*Adivasi Janajati* in Nepali) (CBS 2012). The indigenous groups include Chepang and others, such as Tamang, Kumal, Sunuwar, Majhi, Danuwar, Thami/Thangmi, Darai, Bhote, Baramu/Bramhu, Pahari, Kusunda, Raji, Raute, Hayu, Magar, Chyantal, Rai, Sherpa, Bhujel/Gharti, Yakha, Thakali, Limbu, Lepcha, Bhote, Byansi, Jirel, Hyalmo, Walung, Gurung, and Dura living in the hill and mountains in Nepal (Bennett et al. 2008). Chepang possesses direct and close contact with natural resources, biodiversity, and

[1] Animal Research Laboratory, Faculty of Science, Nepal Academy of Science and Technology, Khumaltar, Lalitpur, Nepal.
[2] Third Pole Conservancy, GPO Box 26288, Kathmandu, Nepal.
* Corresponding author: tirth.ghimire@nast.gov.np.

nature. For example, they are exclusively dependent on the fauna, such as bats, monkeys, dogs, birds, and others, and flora, such as butter tree (BT) (*Diploknema butyracea*), fiddlehead ferns, *Dioscorea deltoidea, D. alata, Urtica dioica,* and others in many areas within the country. However, in this chapter, we aim to explain the trends of symbiotic pathways among Chepang, bats (*Eonycteris spelaea* and others), and BT.

Chepang, their Life, and Environment

The term Chepang (*Che*: dog, *Pang*: bow; Chepang language) is derived by their lifestyle, because they live by hunting animals with the help of a dog and the bow. Chepang was first mentioned as *"Amid the dense forests of the central region of Nepal, to the westward of the great valley, dwell, in scanty numbers and nearly in a state of nature, ... they pay no taxes, acknowledge no allegiance, but, living entirely upon wild fruits and the produce of the chase, ... They have bows and arrows, of which the iron arrow-heads are procured from their neighbors, but almost no other implement of civilization, and it is in the very skillful snaring of the beasts of the field and the fowls of the air that all their little intelligence is manifested"* by Brian Houghton Hodgson, a British resident in Colonial India (Hodgson 1848). Although it is believed that Chepang were the first settlers in the Mahabharat ranges, they may be living since time immemorial (Gurung 1990). Chepang is a highly marginalized group in Nepal. The majority of them lived a semi-nomadic life, marked by hunting and gathering, fishing, slash and burn cultivation, and spent most of their lives in the forest and caves (Gurung 1990, Rijal 2011). They reside in the northern part of Chitwan, the western part of Makwanpur, the southern part of Dhading, and the southern part of Gorkha districts (Thapa 2013), as well as in the upper hills of Lamjung and Tanahu districts of central Nepal (Sharma 2011). Their population is 68,399, which represents 0.26% of the total population of the country (CBS 2012). In the year 1977, late King Birendra visited Chepang settlements in Makawanpur and provided them with the surname "*Praja*" (King's Subjects), and He officially launched "Praja Development Program" to empower them (Maharjan et al. 2010, Acharya 2015). Since then, Praja had been a matter of dignity for these groups, although they prefer to use Chepang instead of Praja to maintain their ethnic identity and dignity (Maharjan et al. 2010, Acharya 2015).

Although Chepang is one of the oldest groups of the country, it started agricultural practices only 80 to 120 years ago (Gurung 1990, Thapa 2013). Its agriculture is still the primitive type, and prefers along the slopes and nearby forest areas (Gurung 1990, Maharjan et al. 2010, Sharma 2011, Thapa 2013), and is a significant source for their livelihood (Gurung 1990, Nakarmi 1995, Aryal 2013). However, due to arid and stony landscapes, unproductive land with no irrigation at all, their annual harvest is lower, and it lasts just for a few months (e.g., six months) (Rijal 2011). It has also been reported that the harvest covers only 60% of the families for this duration, although only one percent of them have cereal food surplus (Aryal 2013). That is why Chepang has been facing severe starvation and chronic food deficiency for many years (Aryal 2013, Thapa 2013). To cope with the dearth of nutrition, Chepang depends on the forest and its products (Maharjan et al. 2010, Rijal 2011). They use forest products, such as fiddlehead ferns, *Dioscorea deltoidea, D. alata, Asparagus racemosus,* and *Urtica dioica* for personal use, for barter with rice or grains, and for business (Maharjan et al. 2010, Rijal 2011, Thapa 2013; Acharya et al. 2017, Lamichhane 2017). Many of them are experts in fishing and catching honeycombs, wasps, hornets, bats, and wild birds, and fulfill their necessities of food as well as cash.

While Chepang are said to be traditional, their trends have changed over a few years. For example, males temporarily migrate to other neighboring cities and follow labor and driving. Many households rear pigs, goats, cows, oxen, and chicken as a part of their additional source of income. They sell honey, black gram, beans, mustard, ginger, cabbage, tomato, cucumber, and BT fruits. They also prepare and trade alcoholic products derived from BT fruits, millet, and rice for an alternative income, although all their income gets spent on buying food.

Bats

Bats (*Raniwain* for *Pteropus*; *Rowin* for *Cynopterus*; *Syawin* for *Myotis*; and *Dhankacha* for *Rhinolophus; Win* for all bats: Chepang; *Chamero*: Nepali) are highly diversified mammals that fall in the Order Chiroptera. The Order contains 227 genera and 1,411 species of bats, and is quantitatively the second largest after the Order Rodentia (Jones and MacLarnon 2001, American Society of Mammalogists 2019). They are traditionally classified into megachiroptera and microchiroptera (Mickleburgh et al. 2009, Fenton and Simmons 2015). The formerly known megabats include mostly fruit-eating and non-echolocating bats, whereas the latter microbats include mostly insectivorous and echolocating bats. However, molecular evidences show that they are to be classified into the suborders Yinpterochiroptera and Yangochiroptera (Tsagkogeorga et al. 2013). Yinpterochiroptera includes species of megabats, and five of the microbat families (Tsagkogeorga et al. 2013). In contrast, Yangochiroptera consists of the megabats, and most of the microbat families (Tsagkogeorga et al. 2013). By feeding habit, they may be frugivory, insectivory, nectarivory, carnivory, or sanguivory. In Nepal, there are 53 valid species with the probable occurrence of 7 species of bats (SMCRF 2010). Our field survey recorded few dominant species, such as *Cynopterus sphinix*, *Eonycteris spelaea*, *Nyctalus noctula*, *Pipistrellus javanicus*, *P. coromandra*, *P. tenuis*, *Rhinolophus pusillus*, *Rhinolophus macrotis*, *R. pearsonni*, and *Rousettus leschenaultii* in the Chepang-dwelling regions in the study areas, although other species may be present.

Butter Trees

The butter tree (BT) or Nepal Butter Fruit (*Chiuri*: Nepali; *Madhupushpa*: Sanskrit; *Yoshi, Yelsi, or Waksi*: Chepang). It is a deciduous tree that belongs to the family Sapotaceae. Its scientific name is synonymously called *Aesandra butyracea*, *Bassia butyracea*, *Diploknema butyracea*, *Illipe butyracea*, and *Madhuca butyracea*. There are ten species of the Genus *Diploknema*, although two species (*D. butyracea* and *D. butyraceoides*) are predominant in Nepal, Bhutan, and India (Shu 1996). It is naturally present in subtropical and warm temperate areas throughout the Himalayan landscapes from Nepal to Sikkim, Darjeeling, and Arunanchal Pradesh in India, and Bhutan. It is distributed in many parts, principally in the sub-Himalayan tracts on open hillsides (150–1620 m asl) (Devkota et al. 2012) in almost 50 districts of Nepal (Joshi 2010). Out of 5,859 square kilometers estimated potential areas for BTs in Nepal, only 1,934 square kilometer areas are covered with an estimation of 1,08,13,713 trees (Joshi 2010). Notably, the current study areas and their periphery possess 117 sq km BT potential forest area, with an estimated 39 sq km BT area comprising about 216,167 BTs (Joshi 2010). In Chitwan and Makwanpur districts, these trees have been naturally found near the inhabitants of Chepang. Interestingly, it has been quantitatively estimated that average fruit yield per tree is 67.33 kg, butter yield per 100 kg is 39.35 kg, oil yield per weight is 44.5%, nectar production per tree is 13 liters, and nectar secretion per flower per day is 27.9 mg (Joshi 2010). Another study [De la Court 1995 cited in (Paudel and Wiersum 2002)] reported that a mature BT can produce 13 to 40 kg seeds that can generate 4–12 kg of butter per annum. In Nepal, the overall production of seeds is 1,825 tonnes, which can generate 640 tonnes of butter per annum, indicating its potential role in the livelihood of Chepang and other associated inhabitants.

Symbiosis among Chepang, Butter Tree, and Bat

(i) Existing Strengths

Since ancient times, the relationship among Chepang, BTs, and bats have been an indispensable part along with various landscapes of the country. The relation can be expressed in terms of a triad involving the close-association among these three members of the ecosystem (Figure 10.1). This triad forms triangular regulations involving caring and being cared for, and benefiting and being benefitted by one another. These three members in an area determine the symbiotic relationship. Their associations have been shared or represented by the triangular structure in the figure.

Interestingly, BT has been planted, loved, and cared for like their own children by Chepang for many years, and has been culturally and traditionally associated with them. They worship BTs as the milk-producing cattle. There was a popular tradition of providing at least two butter trees as dowry by the parents to their daughter at her marriage, and this tradition is still in practice. The Chepang possessing large numbers of butter trees are regarded as affluent and superior in their community. They are so connected with these trees that they never chop off the plants, although they use the naturally dead plants as firewood. It is remarkable that if any new BT is observed in the forest area, the Chepang who first visited and saw it cleans the surrounding and claims it. In this situation, nobody can use that plant and its products without his/her permission.

Ethnologic researches suggest that BTs have been associated with the livelihood of Chepang. In the study area, there are four types of BTs depending on the differences in the seasonal fall off of leaves, flowering, and fruiting. For example, their leaves fall down from September to November, the flowering season continues from December to February, and fruiting occurs from June to September. Thus, when the leaves of most of the plants are fallen, especially from February to May, those of BTs remain green, as well as create heavenly landscapes by blooming (Figure 10.2). The blooming allures many bats, bees, and birds that visit the flowers for nectars and juices, which consequently results in an eco-friendly environment. It is impressive that the income collecting their seeds, processing them to butter, and selling them would be four times as much as that obtained by the grain cultivation [De la Court 1995 cited in (Paudel and Wiersum 2002)]. Its seeds are given to the priests who perform

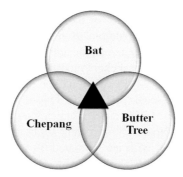

Figure 10.1: Symbiotic Triad of Chepang-Bat-Butter Tree.

Figure 10.2: BT flower blooming.

religious activities in Chepangs' houses. Butter extracted from the seeds is of a high quality, possessing an agreeable smell and flavor with 46% olein and 54% palmitin (FAO 1982). Field observation and interviews taken by us found that Chepang are efficient in generating butter or seed oil and oilcake by using a seed-squeezing indigenous equipment (Figure 10.3). BT seed butter is used as an ingredient for cooking vegetables, as fuel for lighting lamps, and as oil for hair beauty. Besides, it is used for treating boils, cuts, cracked feet, fungal infections, headaches, pimples, and rheumatism. Importantly, the seeds are one of the best parasiticides fed to the goats in a controlled dose-dependent manner. The oilcake is used as fertilizer and pesticide in the crop fields, especially during rice seedlings, as repellants to the leech infestation to the animals and humans, and poison fish. Several studies support the information obtained from our surveys (Khanka et al. 2009, Koirala et al. 2009, Joshi 2010).

Similarly, its fruit is an important source of a dietary supplement and is the only food for some Chepang households during its ripening season. The fruits are also used to prepare alcohol after distillation, especially to use during their festivals. The flowers are used to treat cough, relieve constipation, heart pain, burns, and earache, and increase sexual potentialities. The juice of the corolla is boiled, resulting in the sweet liquid that is a crucial source of alternative sugar. The leaves are used to serve food, to prepare leafy bowl (*tapari:* Nepali language), to make bread and steamed-rice, to feed domestic animals, such as goats and cows, and to make bedding materials. Its bark juice is used to treat asthma, boils, diabetes, helminthic infestations, hemorrhage, indigestion, leprosy, and tonsillitis. These applications have also been listed in the earlier literatures (Khanka et al. 2009, Koirala et al. 2009, Joshi 2010).

In addition to cultural, traditional, and social values, BT is an economically significant plant for the Chepang. For example, they sell BT fruits (USD 0.7/kg) and exchange oilcakes with rice in equal quantity. In recent years, beekeeping and honey production have been developing into a professional system by outsiders, and they bring their hives in the BT areas for foraging activities (Figure 10.4). It is because honey obtained from the collected nectar of BTs has been in high demand in the city areas. In this context, Chepang is allowing the beekeepers to forage their bees in the BT forests seasonally; in turn, they charge about USD 4–6 per hive. Besides, few local people have sustained traditional honey production (rock honeybee), and sell honey (USD 7–9/kg), which has also supported the financial status of the Chepang. Income generation via honeybee has been explained earlier (Joshi 2010).

Not only the BT, but also bats are connected to the traditional, social, and economic contexts of the livelihood of the Chepang. Chepang has traditionally believed that bat meat can be used for nutrition, for the treatment of asthma, gastrointestinal illnesses, joint pain, renal diseases, and tuberculosis. Few traditional healers consume its meat to prevent various diseases and to get energy. Our field observation recorded that very few youths were found to drink its raw blood, believing the fact that it could increase the sexual potentialities. They also sell the captured bat (USD 0.5–1.0/bat) to the local restaurant that can sell the cooked meat to the restaurant visitors or outsiders (USD 1–1.5/cooked bat), indicating its medicinal and economic values for Chepang.

It has been observed that the bat-consuming habit of the Chepang is unusual. They steam bats in boiling water for about 10–15 minutes. They pinch it via a sharp wooden stick. Then, they dry the bat in the fire for a few minutes and hang them from the ceiling of their houses for further drying and future use (Figure 10.5). They can consume the dried bats till 3–4 months. They fry the dried bats in oil. It is interesting that they also consume visceral masses, including intestinal materials. Few people are also found to consume raw meat, fresh blood, smoked meat, and powdered meat directly, or by turning into pickles in daily dishes.

Since time immemorial, there has been a long relation among these three groups—Chepang, bats, and BTs. BT produces fleshy fruits and provides humans and animals, including bats, nutrition and several other chemicals required for them. In turn, different bats help in the dispersal of this plant in several ways. Firstly, nectarivores initiate pollination, especially the cross-one that leads to better development and production of fruits and seeds. Compared to the pollination performed by insects and birds, bats deposit a large amount and variety of pollen genotypes on the plant stigmas

Figure 10.3: A locally installed seed-squeezing equipment to release butter and oilcake.

Figure 10.4: Bee hives established for foraging activities near BTs.

Figure 10.5: Roast of frugivorous/nectarivorous bats.

and disperse the pollen to a considerable distance (Fleming et al. 2009). Bats are sophisticated, faster, and reliable pollinators because they have a larger body size, higher energy requirements, and can carry larger pollen loads compared to any other pollinators (Fleming et al. 2009). Thus, bats play an essential role in maintaining genetic continuity as well as promote outcrossing in plant populations. In these contexts, BT flowers may control the visits by bats, and in turn, bats effectively manage the productivity of the BT flowers, seeds, and fruits. The BT flower is composed of the calyx with 4–5 sepals, corolla with a campanulate tube surrounded by 8–12 petaloid lobes, which are longer than tubes (FAO 1982). It has 30–45 stamens with glabrous filaments and 10-locular ovary surrounded by a linear style (FAO 1982). Bats have been shown to be attracted by large white (or brown) flowers with a musty or no smell. Besides, for nocturnal feeding, bats are equipped with visual sensitivity for sensing either reflection or absorption of the wavelengths by flowers at dusk (Giraudoux 2007). Both echolocation and olfaction are critical to locate food in nectarivores and frugivores (von Helversen and von Helversen 1999), suggesting the coevolution of both bats and their host trees.

Secondly, it is interesting to know that frugivore bats do not eat the fruits hanging on the parent plant, but pick the fruit and carry them to near or far distances, hang on to another plant, and consume them. They suck up all the inner fleshy materials and let the seeds and peel drop down to the ground, ensuring the continuous propagation of the BT. *Eonycteris spelaea*, a cave-dweller and long distance-traveler for foraging activities, seems to be a critical mammal in pollinating and dispersing seeds of the BT in the current study area (SMCRF 2010, Acharya et al. 2015, Sharma et al. 2018). Long-distance pollination by bats is crucial in the conservation of the species. It is because human disturbance fragments plant populations and increases geographic isolation. In the absence of long-distance pollination, plants within habitat fragments experience self-fertilization compared to those in continuous forests (Fleming et al. 2009). Although frugivores such as *Eonycteris spelaea*, *Cynopterus sphinx*, and *Rousettus leschanaultii* take part in the pollination and seed dispersal of many plants, such as banana, jackfruit, litchi, mango, papaya, palm, sacred fig, East Indian shade tree, *Eucalyptus*, Indian lilac, they may assist in the livelihood of the Chepang alternatively.

While the association of frugivores, nectarivores, and BT are found in ample literature, how insectivores play a role in Chepangs' livelihood is lacking. For a few years, Nepalese chiroptologists have argued that insectivore bats could be used as an important biological control method. Scientists believe insectivore bats consume mosquitoes and many insects (Schalk and Brigham 1995, Fenton and Simmons 2015), which would be critical in controlling outbreaks of vector-borne diseases, such as malaria, Japanese encephalitis, dengue, and others. Although we are researching the role of bats in zoonotic transmission of the parasites (Adhikari et al. 2018, 2019), it is not included in this chapter.

(ii) The Zigzag Trail of Symbiosis

The field and interview surveys found that when negative pressure on bats (hunting, natural calamities, diseases) was increased, bat populations, BT productivity, and their economic status decreased (Figure 10.6). In contrast, when the pressure on bats was reduced, bat populations, BT productivity, and their economic status were subsequently increased. The ups and downs patterns can be observed to be seasonal—for example, during bat visits in winter and summers, BT productivity is increased if negative pressure for bats is decreased. However, when this pressure is increased, BT productivity is reduced. There are mainly four consequences on the survival and productivity of bats and BTs—via the hunting pressure, overlap niche, and other factors in bats and BT productivity.

The lines are expressed on the basis of views of the local indigenous Chepang people. The slightly tilted lines with respect to time represent that the trends of bat populations, BT productivity, and economic status have been decreasing for many years.

Hunting Pressure on Bats

While bats are one of the members in symbiotic triads, they have been suffering from the hunting pressure by the Chepang as well as other local people in the study area. For example, for many years, extreme hunting pressure has been experienced by *Rousettus leschenaultii*, which is listed as

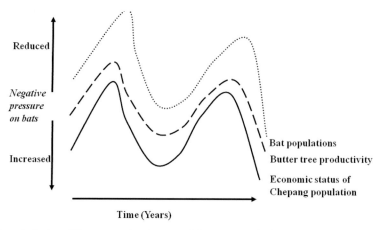

Figure 10.6: A hypothetical trend line representing the ups and downs of symbiosis among the Chepang, bats, and BTs with respect to time.

Least Concern by IUCN (Bates and Helgen 2008) and as Near Threatened by the National Red List (Jnawali et al. 2011). Another species *Eonycteris spelaea* listed as Least Concern by IUCN (Francis et al. 2008) and as Data Deficient by the National Red List (Jnawali et al. 2011). In these contexts, one or more of the socio-economic factors, such as lack of awareness, lack of sufficient foods, existing ethnic values of bat meat, unemployment, recreational experiences, entertainment, grouping among adolescents and youths, increasing bat meat demands at the nearby restaurants and city, and poverty might be critical in determining the initiation of bat hunting in the area. To capture bats, although boys of age groups 10–30 years are engaged, older men also actively take part. They usually hunt bats during foraging on BTs. During the flowering and ripening periods, when nectarivores and frugivores visit flowers or ripening fruits, hunters put the fine and strong net (*Bhuwa*: Nepali) near the BTs at night (8–12 pm) (Figure 10.7). They keep nails in their mouths and produce bat-calling sound. The young boys also use unique rubber toys to imitate bat sounds. Bats are then attracted to the sound. When they approach the area, they get trapped in the net. The hunters immediately tie up their wings, put them inside their bags, squeeze their hearts, or break their necks. Each household may, thus, capture hundreds of bats during nights. The scenarios are pathetic when people catch the cave-rooster insectivore bats. The situation of hunting pressure becomes worse if hunters catch the pregnant female bats during predominant seasons and consume them to fulfill their hunger and taste (Figure 10.8). Thus, bat hunting pressure has resulted in the reduction of the number of bats visiting

Figure 10.7: Net installed at the BT to capture frugivorous bats.

Figure 10.8: A captured pregnant bat.

the area. Old and experienced people were aware of this, and they enumerated that the trend of bat visits had been lowering compared to previous years.

Hunting Pressure of Bats on BT Productivity

Increasing hunting pressure on bats has resulted in the reduction of the productivity of fruits of BT each year compared to the past years. There is a negative trend in BT production, due to decreasing fruit production per tree (reduced by 1/3rd to 1/5th over the last ten years) (Paudel and Wiersum 2002). It has been reported that due to a reduction in fruit production, the amount of seeds sold per household in the southern parts of the study area fell from about 40 kg in 1990 to 250 kg in 1995, and 70 kg in 2000 (Paudel and Wiersum 2002). Some farmers hypothesized that catching of birds and bats reduced pollination and decreased productivity (Paudel and Wiersum 2002). Local people believe that although flowering is best in each season, this does not synchronize with the fruit production, indicating reducing pollination via bats.

Overlapped Niche and BT Productivity

In addition to hunting, another factor is the competition between bat and honeybee for the flower niche. Chepang is earning cash by allowing the beekeepers to forage the bees of outsiders. As a result, both nectarivorous bats and bees have to share similar food sources. Although hybrid bees have increased honey production, the productivity of BT fruits has decreased, indicating that the bees are not the best pollinators for these plants. Also, bees obtain the optimum nectar from the single BT flower at a time and return to the hive. In contrast, sucking a single flower is not enough for the bat due to large body size. Therefore, the bats revisit another flower of either the nearby or distant BT flowers, which enhances the probability of pollination. In our study, a few Chepang suspected that if the number of bat visits declines with the current frequency, the BTs will suffer a lot with zero pollination and zero production of fruits as well as seeds soon.

Other Pressures on Bats and Consequences

Recently, due to the decentralization and local development of indigenous populations, several factors, such as the construction of roads, deforestation, and destruction of many bat caves are inducing pressures to bats and BT productivity. Human disturbance accompanied by deforestation and shaking the bat-roosting trees may also add additional problems. It has been predicted that flying foxes usually roost in a huge colony outside the protected areas and may undergo mass mortality because of either habitat clearance or human-induced activities (SMCRF 2010). A few local people believe that the aging of BTs, slow regeneration, and lack of reforestation might be the cause of declining BT productivity. However, we have observed that local people in Siddi area have recently started plantation of BTs, indicating the Chepangs' awareness of the importance of BT conservation.

The area has experienced many forest fires in the dry season, leading to a reduction in the numbers of bats. Landslides during heavy rains destroyed many old caves, and ignorance of concerned authorities to restore them led to nothing in conservations. Besides, human encroachment and frequent visits to the caves, and either local or foreign tourism, have led bats to migrate to new places for the search of survival factors. Storing dry straw of rice, maize, and wheat inside the cave also hinder the entry and exit of the bat and may enhance stress. Cave-dwelling bats, especially those that are pregnant, are extremely sensitive to microclimatic conditions such as temperature and humidity. The situation is harsh for those species which have meager fertility rates; for example, many species produce one pup per birth and one birth in a year (SMCRF 2010).

Besides, climate change may also have negative and positive effects on the survival of bats. For example, it can affect the flowering characteristics of plants, and subsequently may break migratory behavior, foraging strategy, and reproductive patterns of bats (Frick et al. 2019). On the other hand, it may induce the mobility of these fauna, leading to increased adaptation for the survival in the diverse habitats (Frick et al. 2019). Interestingly, these dichotomous views can be applied for the relation between BTs and bats in changing climates.

Conservation Threat in Nepal: Current Gaps and Future Directions

Chepang are rich in their indigenous knowledge of medicinal values of bats and BT. Thus, searching for an evidence-based indigenous culture on the conservation of bats and BTs and its roles in the health of these members of the society will be the best protocols for future research. We have formulated the following specific topics for the analysis:

- How does climate change affect the livelihood of the Chepang, bats, and BTs?
- How are Chepang, bats, and BTs distributed in the landscape, and what are the most critical landscape co-variates for their distribution?
- How does taxonomic status of bats and their preferences affect the exploitation of BTs?
- How could the consumption of bats affect the pathology, physiology, and immunology of the body?
- What is the role of bats in the control and prevention of human diseases?
- What are the biochemical, physiological, and immunological effects of the butter, oilcake, fruits, flowers, and barks *in vitro* and *in vivo*?
- Can One Health concept be used for the welfare of Chepang, bats, and BTs at the landscape level?
- What are the possible threats to BTs and community-based conservation of bats?

A Case Study in Chitwan, Nepal

The current study was conducted in Shaktikhor and Siddi areas of Eastern Chitwan in central Nepal (Figure 10.9). These areas are almost 152 kilometers away from Kathmandu, the capital city of Nepal. The area is rich in vegetation, such as *Acacia catechu*, *Artemisia* spp., *Bambusa* spp., *Bombax ceiba*, *Diploknema butyracea*, *Mangifera* spp., *Prunus* spp., *Shorea robusta*, *Utrica dioica*, and

Figure 10.9: Landscape of the study area.

U. parviflora. Although the study area has been reported to harbor a total of 13 species of birds, eight species of mammals, and six species of reptiles by the latest survey (IDML 2019), we only focused on bats because of our purpose of the study. To find out the associations among bats, BT, and Chepang, we conducted a two-year field observation, data collection via interviews, and questionnaires to the focus groups. The groups included the persons of age 14 to 94 years. The interviews were taken in the Nepali language and individually to ensure the independence of data, and lasted for 20–30 minutes each. The interview data have been expressed in Annex-1 of this chapter.

Conclusions

Conservation is solely determined by the psychology and necessities of the concerned people. After a two-year survey, we found that rather than awareness, necessities have played significant roles in the hunting of bats. The hand-to-mouth problem is stronger than the knowledge of these indigenous people. We observed that people are fully aware of the bat-human interaction and its biodiversity values. Although school awareness programs, radio-awareness programs, hoarding board displays, and training have been conducted around the country, their roles during BT flowering and producing seasons have not fully worked. These initiatives are just imposing them to initiate conservation. Conservation of small mammals such as bats should be felt necessary in daily lifestyle rather than being imposed by the government and others. Thus, we think there is a gap of something that could stop their preference to kill bats. The reduction in poverty should be targeted to address the existing gap. In this context, if the State focuses on the production of BT seeds, and subsequently butter and oilcake via the household, and indigenous technology training, such as installation and processing of seed pressure or squeezer and others, as well as subsidies in the form of grants and fellowship, it could enhance their knowledge on mass production of seeds and sustainable income generation. That could act as an inducer of their feelings on the necessity of bats for the improved rates of the output of BT pollination, fruits, and seeds. Besides, BT nectar is an important food source for commercial beekeeping. Therefore, the beekeeping training to the Chepang community and local women groups would be beneficial, and open alternative livelihood options for them. Such development of alternative income generation source fosters the Chepang community in the conservation of both BTs and bat species. In addition, the State should think about establishing a breeding center of bats, although these mammals are common and least concerned. The breeding center will help increase their populations as well as rare species. By creating an indigenous focus group, the breeding center can be managed, run, and regulated.

Bats, BT, and the local environment are the properties of Chepang. We should respect, protect, nurture, and manage their traditional, cultural, religious, and economic values by initiating the best approach of sustainable development programs. Only by developing all the members of the triad will we be able to conserve them and their rights to live in a natural habitat.

Acknowledgments

We are grateful to Dr. Pushpa Raj Acharya, Central Campus of Science and Technology (CCST), Faculty of Science and Technology, Mid-Western University, Nepal for his discussion about bat identification and distribution, and to the local people who voluntarily took part in our study and gave lots of information regarding the conservation of bats, BTs, and Chepang.

References

Acharya, A.K., Joshi, B.K., Gauchan, D., Poudel, Y.P. 2017. Chapter V, Associated biodiversity for food and agriculture in Nepal. pp. 60–77. *In*: Joshi, B.K., Acharya, A.K., Gauchan, D., Chaudhary, P. (eds.). The state of Nepal's biodiversity for food and agriculture. Ministry of Agricultural Development (MoAD), Kathmandu, Nepal.

Acharya, P.R. (ed.). 2015. Chepang, Chiuri ra Chamera. 1 ed. Kathmandu, Nepal: Friends of Nature, welt hunger hilfe.

Acharya, P.R., Racey, P.A., McNeil, D., Sotthibandhu, S., Bumrungsri, S. 2015. Timing of cave emergence and return in the dawn bat (*Eonycteris spelaea*, Chiroptera: Pteropodidae) in Southern Thailand. Mammal Study 40(1): 47–52.

Adhikari, R.B., Maharjan, M., Ghimire, T.R. 2018. Presence of gastrointestinal parasites in the stool of bats in Shaktikhor area, Chitwan, Nepal. 23rd International Conference of International Academy of Physical Sciences (CONIAPS XXIII) on Advances in Physical Sciences to Achieve Sustainable Development Goals November 16–18, 2018; Kathmandu, Nepal.

Adhikari, R.B., Maharjan, M., Ghimire, T.R. 2019. Assessment of gastrointestinal parasites in bat in Shaktikhor area, Central Southern Nepal. National Young Scientists Conference (NYSC)-2019 (April 23–24, 2019) organized by Ministry of Industry, Tourism, Forests and Environment, State, no 3, Makwanpur, Nepal; The National Trust for Nature Conservation (NTNC), Khumaltar, Lalitpur, Nepal.

American Society of Mammalogists. 2019. Taxon Summary Statistics. [accessed November 2, 2019]. https://mammaldiversity.org/summary.

Aryal, B. 2013. State of Food (in) Security in Chepang Community: A Case of Dahakhani VDC, Chitwan. Econ. Lit. XI: 60–66.

Bates, P., Helgen, K. 2008. Rousettus leschenaultii. The IUCN Red List of Threatened Species 2008: e.T19756A9011055. https://dx.doi.org/10.2305/IUCN.UK.2008.RLTS.T19756A9011055.en. Downloaded on 01 April 2020.

Bennett, L., Dahal, D.R., Govindasamy, P. 2008. Caste, ethnic and regional identity in Nepal: Further analysis of the 2006 Nepal demographic and health survey Macro International Inc., Calverton, Maryland, USA.

CBS. 2012. National Population and Housing Census 2011: National Report. Kathmandu, Nepal.

Devkota, H.P., Watanabe, T., Malla, K.J., Nishiba, Y., Yahara, S. 2012. Studies on Medicinal Plant Resources of the Himalayas: GC-MS Analysis of Seed Fat of Chyuri (*Diploknema butyracea*) from Nepal. Pharmacog. J. 4(27): 42–44.

FAO. 1982. Fruit-bearing Forest Trees: Technical Notes. Food and Agriculture Organization of the United Nations. (FAO Forestry Paper 34).

Fenton, M.B., Simmons, N.B. 2015. Bats: A World of Science and Mystery. University of Chicago Press.

Fleming, T.H., Geiselman, C., Kress, W.J. 2009. The evolution of bat pollination: a phylogenetic perspective. Ann. Bot. 104(6): 1017–1043.

Francis, C., Rosell-Ambal, G., Tabaranza, B., Carino, P., Helgen, K., Molur, S., Srinivasulu, C. 2008. *Eonycteris spelaea*. The IUCN Red List of Threatened Species 2008: e.T7787A12850087. https://dx.doi.org/10.2305/IUCN.UK.2008.RLTS.T7787A12850087.en. Downloaded on 01 April 2020.

Frick, W.F., Kingston, T., Flanders, J. 2019. A review of the major threats and challenges to global bat conservation. Ann. NY Acad. Sci.

Giraudoux, J. 2007. Chapter 7, Flowers. pp. 159–197. *In*: Lee, D. (eds.). Nature's Palette : The Science of Plant Color. The University of Chicago Press, Ltd., London.

Gurung, G.M. 1990. Economic modernization in a Chepang village in Nepal. Occ. Pap. in Soc. and Anthr. 2: 32–39.

Hodgson, B.H. 1848. On the Chepang and Kusandas tribes of Nepal. J. of the As. Soc. of Beng. 17(2): 650–658.

Hornby, A.S. 2015. Oxford Advanced Learner's Dictionary of Current English. pp. 1–1820. *In*: Deuter, M., Turnbull, J., Bradbery, J. (eds.). 9 ed. Oxford, UK: Oxford University Press.

IDML. 2019. Draft Report: Environmental Impact Assessment of Shaktikhor Industrial District, Chitwan. Submitted to Government of Nepal, Ministry of Forests and Environment, Singh Durbar, Kathmandu, Nepal: Industrial District Management Limited (IDML), Balaju, Kathmandu, Nepal.

Jnawali, S.R., Baral, H.S., Lee, S., Acharya, K.P., Upadhyay, G.P., Pandey, M., Shrestha, R., Joshi, D., Laminchhane, B.R., Griffiths, J., Khatiwada, A.P., Subedi, N., Amin, R. (compilers) 2011. The Status of Nepal Mammals. The National Red List Series, Department of National Parks and Wildlife Conservation, Kathmandu, Nepal.

Jones, K.E., MacLarnon, A. 2001. Bat life histories: testing models of mammalian life-history evolution. Evol. Ecol. Res. 3(4): 487–505.

Joshi, S.R. 2010. Resource Analysis of Chyuri (*Aesandra butyracea*) in Nepal Micro-Enterprise Development Programme (MEDEP-NEP 08/006), UNDP/Ministry of Industry, Government of Nepal.

Khanka, M.S., Tewari, L., Kumar, S., Singh, L., Nailwal, T.K. 2009. Extraction of high quality DNA from *Diploknema butyracea*. Food Chem. 1: 33–35.

Koirala, P.N., Mishra, S., Chaudhary, S., Barme, C. 2009. Assessment of Chiuri (*Diploknema Butyracea*) for its commercialization in Rolpa District. District Forest Office, Rolpa.

Lamichhane, D. 2017. Chapter IV, Forest biodiversity for food and agriculture in Nepal. pp. 49–59. *In*: Joshi, B.K., Acharya, A.K., Gauchan, D., Chaudhary, P. (eds.). The State of Nepal's Biodiversity for Food and Agriculture. Ministry of Agricultural Development (MoAD), Kathmandu, Nepal.

Maharjan, K.L., Piya, L., Joshi, N.P. 2010. Annual subsistence cycle of the Chepangs in mid-hills of Nepal: An integration of farming and gathering. Him. J. of Sociol & Anthr. IV: 105–133.

Mickleburgh, S., Waylen, K., Racey, P. 2009. Bats as bushmeat: a global review. Oryx. 43(2): 217–234.

Nakarmi, A. 1995. Socioeconomic Status of Prajas of Dhading District. Kirtipur, Nepal: Tribhuvan University.

Paudel, S., Wiersum, K.F. 2002. Tenure arrangements and management intensity of Butter tree (*Diploknema butyracea*) in Makawanpur district, Nepal. The Int. For. Rev. 4(3): 223–230.

Rijal, A. 2011. Surviving on Knowledge: Ethnobotany of Chepang community from midhills of Nepal. Ethnobot. Res. & Appl. 9: 181–215.

Schalk, G., Brigham, R.M. 1995. Prey selection by insectivorous bats: are essential fatty acids important? Can. J. of Zool. 73(10): 1855–1859.

Sharma, B., Baniya, S., Subedi, A., Gyawali, K., Panthee, S., Ghimire, P., Bist, B.S., Budha, M. 2018. First record of dawn bat *Eonycteris spelaea* (Dobson, 1871) (Mammalia: Chiroptera: Pteropodidae) from western Nepal. J. of Bat. Res. & Cons. 11: 1.

Sharma, D.P. 2011. Understanding the Chepangs and Shifting Cultivation: A Case Study from Rural Village of Central Nepal. Dhaul. J. of Sociol and Anthr. 5: 247–260.

Shu, Z.L. 1996. *Diploknema* Pierre, Arch. Néerl Sci. Exact. Nat. 19: 104. 1884. Flora of China. 15: 208–209.

SMCRF. 2010. Bats of Nepal: A field guide. Nepal: Small Mammals Conservation and Research Foundation (SMCRF).

Thapa, R.B. 2013. Field Research Report on Food and Nutrition Security of the Forest Dependent Households from the Forests of Nepal. Renaissance Society Nepal (RSN), Bhaktapur, Nepal.

Tsagkogeorga, G., Parker, J., Stupka, E., Cotton, J.A., Rossiter, S.J. 2013. Phylogenomic analyses elucidate the evolutionary relationships of bats. Curr. Biol. 23(22): 2262–2267.

United Nations. 2009. State of the world's indigenous peoples. Department of Economic and Social Affairs, Division for Social Policy and Development, Secretariat of the Permanent Forum on Indigenous Issues, The United Nations, New York.

von Helversen, D., von Helversen, O. 1999. Acoustic guide in bat-pollinated flower. Nat. 398(6730): 759. World Bank. 2013. Operational Manual: OP 4.10—Indigenous Peoples. [accessed September 18, 2019]. https://policies.worldbank.org/sites/ppf3/PPFDocuments/090224b0822f89d5.pdf.

Annex 1

Interview and Self-description of the Zigzag Trail of Symbiosis among the Chepang, Bats, and Butter Trees

Male; 32; Farmer

My family has been living in this place for 100 years. We usually see them during the evening and nights. However, the numbers have been decreasing year-by-year due to deforestation, construction activities, and landslides. Bee keepers from outside visit the BT forest and give about USD 0.5 per hive for keeping the hive for foraging in BTs. Thus, the plantation of BT has been increased, and its numbers are increasing. BT is useful for bats because it is the main food. Conserving bats is helpful for Chepang because it is a source of food, seed dispersal, and means of collecting seeds of BT fruits. Conserving bats is useful for nature because it helps in insect control and pollination. Conserving BT is also helpful for Chepang because it is the main source of income. The cost of two BT plants is equal to one newly delivered buffalo. Conserving BT is also helpful for nature because it provides shelter to birds, food for birds, oxygen, and food for rocky (wild) bees. We can conserve bats and BT by increasing awareness and plantations.

Male; 90; Traditional Healer

My family has been living in this place for 90 years. The numbers of bat visits have been decreasing year-by-year due to population growth and extreme hunting, encroachment in the forest and caves. For example, during the construction of the road from Kaule to Hugdi, several caves were destroyed. We have not planted, but reared or taken care of BTs because they provide shelter to birds, food for birds, oxygen, and food for rock (wild) bees. Beekeepers from outsides visit the BT forest and give about USD 0.5 per hive for keeping the hive for foraging in BTs. Thus, plantation and numbers of BT have been increased. Conserving bats is helpful as it is a source of seed dispersal. Conserving bats is useful for nature because it helps in insect control and pollination. Conserving BT is also helpful for Chepang because it is the primary source of income. The cost of ten BT plants is equal to one newly delivered buffalo. We can conserve bats and BT by increasing awareness and plantations.

Male; 45; Farmer

My family has been living in this place for 100 years. We usually see the bats during the flowering and ripening season of BTs in the evening and nights. However, the number of bats visiting these areas has decreased in recent years. Bats might have migrated to other better places in search of food and water sources. Many bats were killed due to habitat destruction during the construction of roads and hunting. Oilcakes are used to treat infection caused by rice bug and rice borer. Butter products such as nectar and fruits are the main and nutricious foods for bats. BTs provide more oxygen than other plants. It also serves as fodder for domestic animals such as goats and cows. Conservation of bats is important because bats are beautiful and unique creatures. Bats are taken as the Chepang's traditional flying bird. Bats are also the foods for *Panthera pardus* found in this area. BTs are homes of birds, monkeys, and others. However, the introduction of hybrid breeds of large body-sized bees in the butter forests had led to a reduction in the production of butter fruits and seeds. BTs have been cut down in few local areas for the construction of roads. Afforestation and public awareness programs can play a better role in the conservation of both bats and BTs.

Male; 43; Farmer

My family has been living in this place for 60 years. Bats are usually seen in this area, but the number is on a decreasing trend in the past few years. The reason might be extensive hunting during the seasons. Local people used to hunt hundreds of bats daily during peak season; they used to even hunt pregnant bats. The present BTs in the forest were planted by our forefathers, and were successfully handed over to us, and we owned most of the BTs in the forest. The major benefit of BTs includes butter and fruits.

Further, oilcakes can also be used as manure in the fields. BTs provide the nectar and fruits for bats, bees, birds, and others. Bats are one of our traditional food, and they help in pollination and insect control by consuming them. BTs call bats and serve as shelter and food for animals. Similarly, bats are also part of nature and naturally help in pollination. Thus, their conservation is a must. An awareness program may be effective for it.

Male; 35; Laborer

My family has been living in this place for 100 years. We usually see the small bats that sometimes come to our houses. However, the fruit-eating bats are seen only during the flowering and ripening period of BTs. The number of bats visiting the area has decreased in the past few years significantly. Kids usually kill the small bats in caves for fun and consume them. The locals extensively hunt the larger fruit-eating bats during the flowering and ripening periods of BTs. BTs are the sources of fodder for domestic animals and the sources of butter, fruits, and medicines. BTs are liked by bats as the nectar and fruits are their favorite food. Bats help in seed dispersal, natural pollination, and control the insects. Afforestation and awareness programs can play a role in the conservation of both bats and BTs.

Male; 14; School Dropout (Bat Hunter)

I started hunting bats when I was 10. I enjoy hunting at night with friends. I like the sweet taste of bats. I also sell bats and buy food for the family. I have heard from my grandfather and his friends that they used to hunt more than 50–80 bats per day, but in the past few years, it is deficient, for example, 6–8 bats per day. We have lots of BTs in the forest, but still, only a few bats visit the areas. Extensive hunting is the primary cause of their decreased population. BTs are relevant, especially because they call bats and are the foods of birds, bees, monkeys, and domestic animals. BTs protect us from the extreme heat of the sun during summer days. Conserving bats can be useful for Chepang as it is a popular food, and source of income. It also helps us by controlling the mosquito populations. Conservation of BTs helps protect the shelter and meals for animals, such as dogs, monkeys, bees, and birds. Afforestation and awareness programs can be useful for the conservation of both BTs and bats.

Female; 28; Farmer and Household Worker

My family has been living in this area for 50 years. I usually see the bats during evening time while roaming around, but I have heard that the number of bats visiting our areas has decreased, from the people who typically go for hunting at night. Increased population growth, deforestation, natural calamities such as landslides in bat habitat, and hunting may be the reason for the decreased bat population. BTs are highly essential for us as their seeds are pressed to extract oil. The oilcakes are of high economic value. We exchange oilcakes with rice in the nearby market. BTs are also the right food for bats, bees, birds, monkeys, and many others. Conservation of bats doesn't affect Chepang; however, their presence makes an environmental balance by controlling the insect population. The introduction of a good market for butter products and generation of public awareness can be an effective way for the conservation of both BTs and bats.

Male; 47; Agriculture and Goat keeper

My family has been living in this area for 200 years. I usually see bats flying around. However, their numbers have drastically changed, as few bats are seen these days. I used to hunt 40–50 bats per night a few years back regularly, but now, although I rarely go hunting, it is challenging to kill 8–10 bats. Extensive hunting, destruction of habitats as a result of landslide, and construction of roads are the major causes of the decline of bats. Most of the BTs in the forest are natural, and bats disperse the seeds and help in the growth of the plants to new areas. The nectar and the ripe BT fruits are the most favorite foods of bats. The tree provides us butter, oilcakes, fodders, and firewood. Bats are an essential part of nature and help to maintain environmental balance by seed dispersal, pollination, and controlling insect population. BTs also serve as sources of food and shelters for bees, birds, monkeys, and dogs. Decreasing the hunting activities and increasing afforestation can be useful for the conservation of both BTs and bats.

Specific Plants and Ailments

11

Role of Wild Plants in Curing and Healing the Skin Diseases

Mudassar Mehmood[1],* and *Rao Zahid Abbas*[1]

Introduction

Man and nature are the two sides of the same coin. The relationship between man and nature works on both levels, the spiritual and the physical. It enlightens the human soul on one hand and nourishes the human body on the other. Wordsworth, the poet of nature, expresses the strength of this relation as "Away from nature, man is poor creature". Having the mother-like tenderness, nature celebrates both colors of man's life, as all the shades of happiness and the pangs of sorrow. As nature is very kind, so it is this power that can only offer the most effective and the least harmful treatment to all the physical diseases of human beings. The association among mankind and restorative plants exists legitimately from the beginning stages of the universe. As nature is the mother of creation, so recognition with the use of restorative plants is an eventual outcome of the various significant lots of man's fight against diseases that urges man to look for active compounds in roots, seeds, and aerial parts of plants.

Present-day science has attested to their dynamic movement and gives it a status in current pharmacotherapy by showing the extent of drugs from plants. Man's treatment of remedial plants was totally established in his regular procedure because there was not satisfactory information either concerning the clarifications behind the infirmities or concerning which plant and how it could be utilized as a fix. The reasons behind the utilization of helpful plants for treatment of explicit diseases were being found, and the use of therapeutic plants to gradually enhance. The unique and complex structures in the extracts of wild plants show their action. In the 16th century, iatrochemistry asserted the plants as a wellspring of treatment and prophylaxis. The utilization of normal medicine has ended up being topical again as a result of the reducing reasonability of fabricated medicines, and the growing contraindications of their usage.

Ethnomedicinal plants have excelled the synthetic medicines due to their fewer side effects, rapid action, and low price. In old times, the magic and superstition overwhelmed the ethnomedicinal practice. Today, the scientific tests have proven the remarkable curative power of many traditionally used herbs. Nowadays, the dangerous and costly drugs are replaced by the safe alternative medication in the form of ethnomedicinal plants. Ethnomedicine has imparted a significant contribution to the

[1] Department of Parasitology, University of Agriculture, Faisalabad
* Corresponding Author: mudassar2711@gmail.com

world of medicine. Ethnobotany enjoys the features of the wide scope application and understanding of primitive societies and plant utilization. Skin is the most sensitive organ and covers all the body of the human. In all animals and humans, skin serves as the first line of defense, and combats the infection when it tries to enter the body through it. The skin contains numerous specific cells and structures. It is secluded into three rule layers, as epidermis, dermis, and hypodermis. Each layer has a substitute assignment to do in keeping up the skin prosperity. The aim of this chapter is to analyze the treatment of skin diseases with the help of wild plant extracts, their inhibitory concentrations, active ingredients, and mode of action. So this discussion has confirmed the role of wild plants and their secondary metabolites as therapeutic agents. This herbal treatment saves us from any drug resistance and the side effects of drugs, which is the part and parcel of allopathic treatment.

Herbal Drugs for Skin Diseases

Extracts of medicinal wild plants prove their significant potential as compared to antimicrobial drugs used against skin diseases. The most engaging nature of common medication is their few symptoms and a better understanding of resilience. Thus, a few plants have been examined for the treatment of skin sicknesses, extending from tingling to skin disease. There are numerous kinds of wild plants that are highly utilized for treatment of skin maladies. Some wild plants are talked about as follows.

Bauhinia variegata L.

Vernacular names: Kachnar (Hindi); Devakanchanamu (Telugu); Arisinaaaatige (Kannada); Shemmandarai (Tamil) (Figure 11.1.).

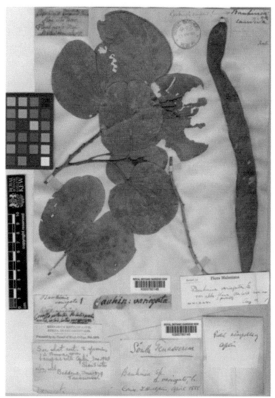

Figure 11.1: Scientific name: *Bauhinia variegata* L. Specimen: K000780146. Image credit to: "© copyright of the Board of Trustees of the Royal Botanic Gardens, Kew."

General characteristics

It is a medium deciduous tree, having a bark with longitudinal splits. Leaves are straightforward, alternate, bilobed, and more extensive than long (Vaz 1979). Flowers are variously colored in lateral sessile or short peduncle corymbs. Fruit is long, hard, flat, dehiscent, and glabrous. Seeds are flat and numerous (Desai and kapoor 2010).

Medicinal uses

Bark concentrates are remotely regulated for treating skin maladies and dermis *abscess.* In the dermis papillary tumor model, huge anticipation, with the deferred appearance, and decrease in the combined number of papillomas was observed in the DMBA + *B. varigata* + croton oil regarded bunch when contrasted with the DMBA + croton oil gathering (Das et al. 2004).

Different skin diseases, wound healing, leprosy, and stomatitis have been traditionally cured in India by *B. variegata* (Rajkapoor et al. 2006). The calming capability of the leaves, bark, and underlying foundations of this tree are generally utilized *in vitro* models, and various distributed reports have declared its status (Singh et al. 2019). The isolation of a bioactive triterpenesaponin from the leaves and a flavonol glycoside from the roots has also been confirmed by authentic reporting (Zaka et al. 2006).

Curcuma longa L.

Vernacular names: Haldi (Hindi); Haridra (Telugu); Arishina (Kannada); Manjal (Tamil) (Figure 11.2).

Figure 11.2: Scientific name: *Curcuma longa* L. Specimen: K0001096893. Image credit to: "© copyright of the Board of Trustees of the Royal Botanic Gardens, Kew."

General characteristics

Curcuma longa is a flowering plant having light colored yellow flowers, about 5 cm long. Its plants are 1 m tall. Rhizomes have many branches having an orange or bright yellow color. The initial structure of the rhizome is ovate-oblong, or pear-shaped. It also has an aromatic characteristic. Peduncles are white and green in color (Kirtikar et al. 1993).

Medicinal uses

The juice of the new rhizome is applied to late injuries, wounds, and parasite nibbles (Mortellini et al. 2006). When it is blended with ginger oil, it avoids dermal infections. A glue of turmeric alongside the mash of *Azadirachta indica* leaves is utilized in the treatment of ringworm disease tingling, dermatitis, and other parasitic infections of the skin. For the treatment of unending dermatitis and tingles, a balm that is made of *C. longa*, Cannabis leaves, onion, and gushing mustard oil gives prompt and enormous alleviation (Apisariyakul et al. 1995).

Cichorium intybus L.

Vernacular names: Kasni (Hindi); Kasini (Telugu); Chikory (Kannada); Kasinikkerai (Tamil).

General characteristics

Cichorium intybus is a ragged perennial herb with blue or lavender blooms. Its stature is 1 to 4 feet. Blooms exist as single on almost leafless branches, and furthermore in bunches in leaf axils. Its plants have the characteristics of bitter and milky sap. Its root resembles a tail of a cow and is beefy, having caramel shading from outside and white shading from the inside (Barcaccia et al 2016).

Medicinal uses

The dose of aqueous extracts of *C. intybus* (CIAE) inhibited mast cell-immediate allergic reactions. These extracts dose-dependently prohibited the anaphylactic reactions induced by compound 48/80 in mice. At the dose rate of 1000 mg/kg it is also remarkably reduced local anaphylactic reactions that are initiated by anti-dinitrophenyl IgE. The watery concentrate of C. *intybus* denies shaft cell-intervened brisk sort of unfavorable susceptible responses *in vivo* and *in vitro*. It is found that *C. intybus* prohibits prostaglandin E (2) and cyclooxygenase (2), and besides diminishing immunotoxicity incited by ethanol, have moderating properties (Jippo et al. 2000). A cosmetic composition is additionally created that anticipates maturing of the skin in which the active fixing is an extract of the elevated pieces of *C. intybus*. Its effectiveness consists of its capacity to preclude radical reactions, in particular by the chelation of iron.

Dodonaea viscosa (L.) Jacq.

Vernacular names: Sanatta (Hindi); Pullena (Telugu); Bandaru (Kannada); Virali (Tamil).

General characteristics

Dodonaea viscose is a bush that consists of a single stem short tree, which is up to 7 m high. It has a black bark, of fluctuating harshness, dainty, and shedding in extended slim strips (ICRAF 1992). Florets have shining blue or often ultra-white or light pink color. Leaves are simple and alternate, excreting the gummy exudate that adds apparent sheen to the leaves; the petiole is extremely short, up to 2.5 mm long (Turnbull 1986).

Medicinal uses

In the event of both extracted and chiseled injury model in rodents, the injury healing movement could be animated by the utilization of ethanolic concentrate of dried leaves. In the extraction model, a quicker pace of wound withdrawal and epithelization could be seen in 10% concentrate treated injuries. In wound models, such as smashing quality of dermis, neoplasm, and lesion withdrawal are the outstanding reactions achieved by the utilization of ethanolic suspension. It also delivered a critical reaction against hostile the healing properties of dexamethasone (Prassana et al. 2007). 30% to 60% of edema and aggravation could be diminished by the compelling utilization of *D. viscosa*. A portion of 1,000 mg/kg of half ethanolic concentrate demonstrated greatest (56.67%) mitigating impacts with carrageenan-incited edema in the rear paw of rodents and that were equal to 100 mg/ml of phenyl butazone (66%) given Intraperitoneal route. A critical mitigating action inside the carrageenan-initiated rodent paw edema method was also expanded by the water removed in the portion of 100 mg/kg (Rani et al. 2009).

Various sorts of parasites, such as Aspergillusniger, Aspergillusflavus, Paecilomycesvarioti, and Trichophytonrubrum causing skin illnesses have been treated with the concentrates of leaves and shoot of *D. viscosa*. All the rough concentrates were demonstrated as a critical enemy of fungal impact against these tried organisms. Various kinds of concentrates, for example, chloroform, ethanol, methanol, ethyl acetate, and fluid concentrates are utilized as a viable enemy of fungal operator. Among this enemy of fungal specialists, chloroform has noteworthy hindrance action against growths.

Azadirachta indica A. Juss.

Vernacular names: Nim (Hindi); Veppachetta (Telugu); Turakabevu (Kannada); Vembu (Tamil) (Figure 11.3).

Figure 11.3: Scientific name: *Azadirachta indica* A. Juss. Specimen: K000657065. Image credit to: "© copyright of the Board of Trustees of the Royal Botanic Gardens, Kew."

General characteristics

It is a medium evergreen tree. Blooms are small and white with shading, and nectar-scented. The fruit is a drupe, single-seeded, beefy, which at that point turns yellow on aging. Leaves are densely present around the end of the branches. These are about 20 to 40 cm in length and are light green in color. Seeds are hard and ellipsoid (Muñoz et al. 2007).

Medicinal uses

Seed oil is utilized to treat scabies and infection. It is also applied to the head to advance hair development. Delicate leaves are bitten to control allergies (Bhowmik et al 2010). For the treatment of wounds, leaf glue is topically connected to them. Dandruff and loss of hair are also treated by its leaf paste (Lodha 2019). A paste of the leaf mixed with turmeric powder is externally applied to treat skin infections, smallpox, and chickenpox (Pai et al. 2004). Flowers of *Azadirachta indica* boiled in gingelly oil are applied on the head against dandruff once a day after taking bath till recovery (Niharika et al. 2010). Gum of *Azadirachta indica* is mostly successful against skin infections, such as ringworms, scabies, wounds, and ulcers. A glue arranged with neem and turmeric was observed to be successful in the treatment of scabies. With no unfavorable impacts, the glue was found to fix scabies within 3 to 15 days (Charles and Charles 1992). The biological action of neem is usually finished with its unrefined concentrates just as its various parts from leaf, bark, root, seed, and oil (Anyaehie 2009). The chloroform concentrates of stem bark chiefly demonstrated calming activities. This concentrate is compelling against carrageenan–incited paw edema in rodent and mouse-ear irritation (Mahabub et al. 2009). Fiery stomatitis in kids is relieved by its bark separately (Reardon 2016).

Ficus carica L.

Vernacular names: Anjir (Hindi); Anjooramu (Telugu); Anjura (Kannada); Simaiyatti (Tamil); (Figure 11.4).

General characteristics

It is a huge bush that grows up to 7 to 10 meters tall, with smooth white bark that droops as the tree grows. Its flowers are mini in structure and unnoticeable. The structure of the leaf blade consists of the broken upper surface and smooth underside. Its fragrant foliage is 12–25 centimeters in length and profoundly lobed with 3 or 5 flaps (Papadopoulou et al. 2002).

Medicinal uses

In the treatment of warts, the fig tree is traditionally used in few rustic areas of Iran. Low recurrence rate, patient compliance, ease of use, no detail of any reactions, and short-duration therapy are the useful effects observed in the patients with warts due to this therapy of *F. carica* (Bohlooli et al. 2007). The removal of lump action of fig tree latex is probably going to be the consequence of proteolytic movement of the latex compounds.

Murraya koenigii (L.) Spreng

Vernacular names: Kathnim (Hindi); karipaku (Telugu); Gandhabevu (Kannada); karivempu (Tamil).

General characteristics

It is a gigantic aromatic shrub or small tree. Leaves are imparipinnate, fragrant, and gland-dotted, leaflets, 9–12 in number, ovate or lanceolate pubescent below (Mhaskar et al. 2000). Fragrant white

Figure 11.4: Scientific name: *Ficus carica* L. Specimen: K001050164. Image credit to: "© copyright of the Board of Trustees of the Royal Botanic Gardens, Kew."

flowers bloom unevenly throughout the year. Usually, it is cultivated for its scented leaves. Their fruits are edible, but the seeds are inedible. Fruits are purple to black and have two seeds (Handral et al. 2012).

Medicinal uses

Its leaves boiled in gingelly oil along with Lawsoniainermis leaves is applied (as a hair tonic) on the head to prevent hair loss (Kumar et al. 1999). The paste of its leaf is used for the treatment of bruises, discoloration, and wounds. Wound healing activity was checked by male albino rats, and screened by ethanolic extracts of leaves of *Murraya koenigii*. It is found from the injury recovery model that the three groups taken for the injury relief activity show a decrease in wound area each day. Due to the treatment with *Murraya koenigii*, the entry point model demonstrated an expansion in the elasticity of the injury which is 12-day old. Thus, the leaves of *Murraya koenigii* have a significant healing capacity of wounds (Anand et al. 2011).

The most important point of the present-day study is the investigation of the drug which reduces inflammation and the substance that inhibits oxidation activities of *Murraya koenigii* leaves. Alkaloids have a wide range of pharmacological properties, including the anti-inflammatory activity. Sub planter injection of carrageenan (Khan et al. 1996) produced hind paw edema in rats. Pet ether extract (PMK) of *M. koenigii* leaves and alkaloids (AMK) isolated from PMK at doses of 100 and 300 mg/kg/day, was given for 11 days to observe the percentage of inhibition of paw edema which was comparable to aspirin, used as a reference drug. PMK and AMK produced a significant inhibition of paw edema. PMK and AMK treatment significantly reversed the carrageenan induced and also decreased Superoxide dismutase (SOD), Catalase (CAT), glutathione (GSH) levels in paw as compared to Carageenan treated rats. Leaves of *Murraya koenigii* were dried in the shade and powdered mechanically. Powdered leaves were defatted with the petroleum ether. The filtrate was concentrated to get the pet ether extract of *M. koenigii* (PMK). The extract was further subjected to isolation of alkaloids according to the method of Cordell GA (Reddy et al. 2012).

The detailing of cream with essential oil of leaf of *M. koenigii* has a sun insurance factor. It was postulated that sun pigment factor for curry leaf oil cream has minimum sun protection ability against sunlight and erythema as compared to *M. koenigii* leaf oil cream.. The natural skin pigmentation can be kept up by the utilization of this cream. It can also be utilized as the expansion in an arrangement of something to upgrade the movement (Handral et al. 2012).

Melia azedarach L.

Vernacular names: Bakain (Hindi); TurkaVepa (Telugu); Bevu (Kannada); MalaiVembu (Tamil).

General characteristics

The grown-up tree has an adjusted crown, and regularly measures 7–12 meters tall, and in exceptional conditions *M. azedarach* can achieve a stature of 45 meters. Young trees are easy prey to uncertain weather, whereas the old trees can fight. The leaflets are dull green above and lighter green underneath, with serrated edges. The blossoms are small and fragrant, with five pale purple or lilac petals, developing in groups. The seeds can sustain their vitality up to two years (Rubae 2009).

Medicinal uses

M. azedarach possesses significant wound healing potential in alloxan-induced diabetic rats. The result shows that its methanol leaf extract has a strong activity of wound healing in alloxan-induced diabetic rats. Some basic mechanisms are responsible for the delay of the wound healing process in diabetic Mellitus. Such mechanisms increase blood sugar, weaken local immune systems, lower cell defenses, and increase the chance of microbial infections. It has been shown in this study that the topical application of its leaf extract is the powerful wound healer in diabetic rats, and its effect was comparable in certain aspects to standard povidone-iodine. The enhancement of wound healing activity in diabetic rats may be due to the antimicrobial activity of *M. azedarach* (Vijaya et al. 2012).

The extracts of *M. azedarach* flower were prepared and used for the treatment of bacterial skin diseases in children (Rahman et al. 1991). For the cream preparation, methanolic extracts of flowers were used. Neomycin was used for the activity comparison of skin drug and the prepared cream. The diameter of the infected area before and after two weeks of treatments was measured. The results showed that in several cases the prepared cream was significantly more powerful, and the flower extract was a potent cure for rabbits suffering from a skin infection that was produced by Staphylococcus aureus (Saleem et al. 2008).

Plumbago zeylanica L.

Vernacular names: Chitrak (Hindi); Agnimaata (Telugu);Chitramulika (Kannada); Kodivaeli (Tamil) (Figure 11.5).

General characteristics

It is a widely spread evergreen bush that ranges around 6 feet in nature. Dull green leaves are 6 inches in length. The stems bear the lax habit and often a more climbing habit (Manu et al. 2012). They are 1 to 2 m long. They are quickly developing plants.

Figure 11.5: Scientific name: *Plumbago zeylanica* L. Specimen: K001134413. Image credit to: "© copyright of the Board of Trustees of the Royal Botanic Gardens, Kew."

Medicinal uses

Caustic, vesicant, and aphrodisiac are the characteristics of *P. zeylanica* leaves. They are used in the treatments of diseases, such as scabies, swelling, and infectious skin. Its paste is effective in the cure of painful rheumatic areas and itchy skin problems. The plant is crushed and the prepared paste is topically administrated over the affected area. It is investigated that a medicinal plant plumbagin (5-hydroxy-2-methyl-1,4-napthoquinone) is separated from the roots of the *P. zeylanica* showing that, in mice, the topical application of plumbagin prevented UV-induced development of squamous cell carcinomas (Sand et al. 2012). The antiviral exercises of the 80% methanolic extractions of *Plumbago zeylanica* were tried against Coxsackie Virus B3 Coxsackie Virus B3 (CVB3), influenza A virus and herpes simplex virus type1 Kupka (HSV-1) utilizing cytopathic impact (CPE) inhibitory measures in HeLa, MDCK, and GMK cells separately. The plaque decreasing measures were utilized as a corroborative analysis of their antiviral movement.

It is accounted for that plumbagin, a naphthoquinone disconnected from Plumbago species, demonstrated its inhibitory movement against amastigotes of Leishmaniadonovani and L. amazonesis. The base of *P. zeylanica* is harsh and furthermore valuable in the treatment of scabies (Sharma and Kaushik 2014).

Abutilon indicum (L.)Sweet

Vernacular names: kanghi (Hindi); Adavibenda (Telugu); Srimudrigida (Kannada); Thuththi (Tamil).

General characteristics

It is a small bush. Leaves are straightforward, alternate, shaggy, have a serrated edge, and pinnacle intense. Fruit is round in shape, having 11 to 20 carpels that change into brown color in dry form (Rahuman et al. 2008). Its seeds are kidney-like in shape. Flowering happens in September-April. Blooms are yellow, axillary, and the stamens various and monodelphous.

Medicinal uses

To treat the ringworm infection, a paste made from fresh leaves with water is applied externally on the skin thrice a day (Abdul et al. 2010). Leaf paste is also applied over the spot of a snake bite not scorpion sting (Shrikanth et al. 2014).

Portulaca oleracea L.

Vernacular names: Lunia (Hindi); Boddupavilikoora (Telugu); Dudagorai (Kannada); Paruppukeerai (Tamil) (Figure 11.6).

Figure 11.6: Scientific name: *Portulaca oleracea* L. Specimen: K000313628. Image credit to: "© copyright of the Board of Trustees of the Royal Botanic Gardens, Kew."

General characteristics

It is a summer annual plant that has a thick series of branches at the base. The leaves of *P. oleraceae* can alternate. Seeds are brownish dark and kidney-shaped in form. Blossoms have five customary parts. They are yellow in shading. Blossoms initially show up pre-summer and proceed into mid-fall (Amirul et al. 2014). The morning sun rays open the loosely hanging petals for a small time period.

Medicinal uses

During scorching heat, this herb protects the body from rashes and skin inflammations. It also has natural cooling and soothing effects. A viable mixture of the leaves is utilized in the treatment of burns and skin emissions, such as bubbles and carbuncles. Its fluid concentrates go about as antibacterial and antifungal specialists during the topical application onto the skin (El-Sayed et al. 2019). Its herb is utilized as a poultice in the treatment of bug stings, irritations, skin injuries, ulcers, tingling skin, dermatitis, and abscesses.

The essential injury healing capacity of *P. oleracea* was tried on Musmusculus JVI-1. The extraction wound surface was topically covered with the new homogenized unadulterated aerial pieces of *Portulaca oleracea* by utilizing its single and numerous dosages. The impact of *P. oleracea* on wound healing was evaluated by observing the injury withdrawal and rigidity measurements. It is inferred that the injury healing capacity of *P. oleracea* was animated by diminishing the surface territory of the injury and expanding the rigidity. A solitary portion of 50 mg of *P. oleracea* brought about the biggest compression of the wound. A similar consequence of wound withdrawal was also achieved by utilizing its two dosages of 25 mg (Rashed et al. 2003).

Atopic dermatitis is an endless fiery skin sickness. An examination confirmed that *P. oleracea* concentrate was filling in as a viable specialist against LPS-treated Raw 264.7 cells and keratinocytes and the skin of NC/Nga mice with atopic dermatitis, just as smooth mice with pruritus. The levels of NO, PGE2, and pro-inflammatory cytokines were measured in the media after the treatment of different concentrations for *P. oleracea* extract in LPS-treated Raw 264.7 cells and keratinocytes. H&E staining and toluidine blue staining were used for the skin tissue identification of all NC/Nga mice. The compound 48/80 having an antipruritic effect was treated by the number of scratching behaviors of the hairless SKH-1 mice. It is concluded that there is a remarkable reduction in the productions of NO and PGE2 due to LPS- and IFN-gamma-treated Raw 264.7 cells as compared to non-treated Raw 264.7 cells. So in the case of atopic dermatitis in NC/Nga mice, there is a great reduction in the thicknesses of epidermis and dermis that is revealed by H&E and toluidine blue staining while treating with *P. oleracea* extract (XueYuan et al. 2010).

Rubia cordifolia L.

Vernacular names: Majith (Hindi); Chiranji (Telugu); Chitravalli (Kannada); Manjitti (Tamil) (Figure 11.7).

General characteristics

Rubia cordifolia is a climbing herb that develops to 1.5 m in height. Its evergreen leaves are 5–10 cm long and 2–3 cm expansive and are heart-like in shape (Santhan 2014). In structure, the stem is long, irregular, and woody at the end. The blooms are small, with five greenish-yellow or pale yellow petals, in thick racemes.

Medicinal uses

Rubia cordifolia is utilized for the treatment of malignant growths, ulcers, and swellings. It also functions as a disinfectant specialist for wounds (Karodi et al. 2009). There are numerous employments

Figure 11.7: Scientific name: *Rubia cordifolia* L. Specimen: K001123313. Image credit to: "© copyright of the Board of Trustees of the Royal Botanic Gardens, Kew."

of *R. cordifolia* in present-day pharmacology. A gel formulation having anthraquinone rich fraction of *R. cordifolia* showed anti-acne ability against Propiobacterium acne, *Staphylococcus epidermidis* and *Malassezia furfur* when it is compared with standard Clindamycin gel. Various portions of *R. cordifolia* roots concentrates were demonstrated enemies of malignancy exercises *in vitro* and in animal models. The development prohibitory action on chosen malignancy cell lines, just as on normal human mammary epithelial cells, was shown by its unrefined watery concentrates (Shoemaker et al. 2005). The quinones and RC-18 demonstrated astounding enemy of malignancy movement against L1210, L5178Y, P388 leukemia, B16 melanoma (Adwanker and Chitnis 1982), S-180 and the cyclic hexapeptides against leukemia. The blockage of protein union was shown by restricting the hexapeptides to eukaryotic 80S ribosomes, bringing about preclusion of aminoacyl-tRNA authoritative and peptidyl-tRNA translocation. An enemy of tumor movement was the cyclic hexapepetide disengaged from dried roots (Itokawa et al. 1984). Human nasopharynx carcinoma, P388 lymphocytic leukemia, and MM2 mammary carcinoma cells were influenced by the alkyl ether and ester subordinates of RA-V.

The proximity of rubimallin in *R. cordifolia* root concentrate has been going about as a calming specialist. It is investigated that the rats with carrageenan paw edema treated with aqueous extract of *R. cordifolia* root, in a dose dependent manner, show significant anti-inflammatory ability by comparing with standard drug phenylbutazone (Antarkar et al. 1983). The lipoxygenase protein pathway, which invigorates the generation of various fiery arbiters, for example, leukotrienes that are engaged with numerous provocative issue, and the creation of cumene-hydroperoxides, is also restrained by this watery concentrate (Tripathi et al. 1995).

   R. cordifolia is mainstream all over the world for its medicinal uses in wound healing. The injury healing proficiency on the extraction twisted model in mice was examined by its alcoholic concentrate and the hydrogel. A solitary portion of alcoholic concentrate was connected to the outside of the extraction wound. Its impact on wound healing was evaluated by watching the injury zone and histopathology. In the treatment of the mice, the various impacts were created by this gel in type of wound conclusion, the decline in surface region of the wound, and wound contracting capacity, tissue recovery at the injury site, and histopathological attributes. *R. cordifolia* is also used as an anti-aging agent. It is mainly used in the treatment of photoaging that results in a form of skin wrinkles. For the treatment of photoagaing, the "Anti-Wrinkle cream" is formulated with *R. cordifolia* and other ingredients that have antioxidant, anti-inflammatory, and UVR protective properties. Beta-sitosterol and daucosterol (Qiao et al. 1990), gallic acid, rubimallin, hydroxyanthraquinones, tannins aliz (Cai et al. 2004) are the main parts of *Rubiacordifolia*. Rubiadin, isolated from *R. cordifolia*, has a great power of antioxidant property that keeps lipid peroxidation from happening in a dose-dependent manner (Tripathi et al. 1997).

Sesbania sesban (L.) Merr.

Vernacular names: Jainti (Hindi); Samintha (Telugu); Arisina (Kannada); Chittagathi (Tamil) (Figure 11.8).

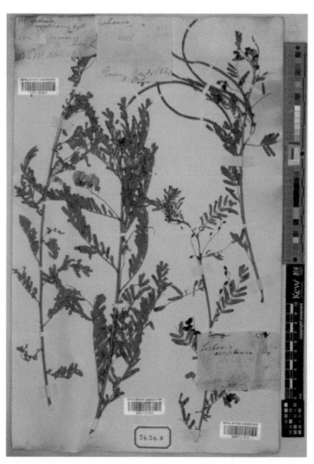

Figure 11.8: Scientific name: *Sesbania sesban* (L.) Merr. Specimen: K001121605. Image credit to: "© copyright of the Board of Trustees of the Royal Botanic Gardens, Kew."

General characteristics

It is a delicate, lush, and quickly developing bush. Its root system is penetrative and the length of the stem is probably 12 cm in diameter (Manjunath and Habte 1991). It has fruits that are a bit turned and have length up to 30 cm. Leaves are paripinnate compound, leaflets, direct to elliptical, glabrous. It has yellow flowers with brown lining.

Medicinal uses

A high quantity of saponin content is present in the leaves of *S. sesban* (Dande et al. 2010). The leaves of *S. sesban* are a laxative, demulcent, maturant, and helpful for the treatment of all pains and inflammations. A legitimate examination was proposed to assess the topical calming action of the rough saponins removal via carrageenan actuated rodent paw edema strategy by setting up the gel definition. There is a significant anti-inflammatory activity shown by the group treated with crude saponins extract of 2% w/w gel formulation, as compared to control gathering, and the outcomes were practically identical to the action that appeared by the reference drug (Kendra 2000).

Santalum album L.

Vernacular names: Chanda (Hindi); Bhadrasri (Telugu); Agarugandha (Kannada); Anukkam (Tamil) (Figure 11.9).

Figure 11.9: Scientific name: *Santalum album* L. Specimen: K000880539. Image credit to: "© copyright of the Board of Trustees of the Royal Botanic Gardens, Kew."

General characteristics

Sandalwood is an evergreen tree, and its bark surface is black with irregular cracks. Leaves are simple, opposite 12 to 18 mm long. Flowers are reddish-purple in color (Rakesh et al. 2010). α-santalol that is isolated from Sandalwood has useful therapeutic abilities as anti-inflammatory, anti-oxidant, anti-viral and anti-bacterial.

Medicinal uses

Sandalwood oil and their derivatives are used in preparing medicaments for the prevention and treatment of viral-induced tumors in humans. This oil and its components are also used to cure genital warts and HPV of the genital tract, and also help in the prevention of skin cancer. Its capacity to initiate cell-cycle capture and apoptosis in diseased cells is its most detailed anticancer component of activity. In India, sandalwood helps to cure the eruptive skin diseases. A study is done to investigate, the chemopreventive effects of sandalwood oil on 7,12-dimethylbenz(a)anthracene-(DMBA)-initiated and 12-O-tetradecanoyl phorbol-13-acetate(TPA)-promoted skin papillomas, and TPA-induced ornithine decarboxylase (ODC) activity in CD1 mice. There was a significant decrease in papilloma incidence by 67%, stops its multiplication by 96%, and TPA-induced ODC activity is lowered by 70% by this oil. Thus, it is proved that sandalwood is a very effective chemopreventive agent used for skin diseases, mainly cancer (Dwivedi and Abu-Ghazaleh 1997).

Sida acuta Burm.f.

Vernacular names: Bariara (Hindi); Muttavapulagamu (Telugu); Bheemanakaddi (Kannada); Arivalmanaipoondu (Tamil).

General characteristics

S. acuta is a bush having a place with *Malvaceae* family. Structure wise, its petals are of light yellow color, having a length of 6 to 8 mm. Seeds are trigonous and 2 mm in length. In the subtropical areas, the plant is broadly spread in shrubs, in ranches, and around homes (Mann et al. 2003).

Medicinal uses

In India, it is known as Pillavaltichedi, and traditionally it is used for the treatment of skin diseases and ulcer (Ignacimuthu et al. 2006). For killing dandruff and for the strengthening of the hair, the paste made of its leaves mixed with coconut oil is applied regularly on the head. It is helpful for curing wounds, cancer, and different inflammatory skin diseases. A study showed the antimicrobial effect of the ethanolic and aqueous extracts of *S. acuta*. The extracts of *S. acuta* consist of saponins; tannins, cardiac glycosides, alkaloids, and anthraquinones that were revealed by photochemical analysis. Test isolates from human skin infections were Vacillussubtilis, *Escherischia coli*, *Aspergillus niger*, and *Aspergillus fumigatus*. The zone of inhibition shows the different potential of ethanolic and aqueous extracts against these skin infections (Ekpo and Etim 2009).

Sapindus emarginatus Vahl.

Vernacular names: Reetha (Hindi); Kukudu-kayalu (Telugu); Kookatakayi (Kannada); Ponnankottai (Tamil).

General characteristics

The *Sapindus emarginatus* tree is 10 m high with a dim, dark-colored bark. Leaves are paripinnate, alternate, forceful, tomentose, and swollen at the base (Arora et al. 2012). Its seeds are yellow and brown in color. Pollination of the flowers is done by insects. Blooms are polygamous, greenish-white.

Medicinal uses

Squashed products of this tree are utilized for the removal of dandruff (Harsha et al. 2002). Natural products are also utilized for the treatment of face patches. Exocarp of organic products are kept in water and attached on these patches for the treatment (Upadhye et al. 1986). An examination demonstrated the analgesic and mitigating action of the methanolic concentrate of pericarps of *S. emarginatus.* The nearness of saponins, terpenoids, tannins, flavonoids, glycosides, and sugars was inspected by the phytochemical screening of the pericarps (Gogte 2000). By utilizing formalin test and swirl's hot plate method, the central analgesic action of the concentrate was looked at. It was considered that the concentrate was utilized for calming movement in carrageenan-instigated rear paw edema in rodents, and the value of the paw was estimated plethysmometrically. The investigation has executed the portion of (200 and 400 mg/kg, p.o) of this concentrate. For the centrally acting analgesic action, pentazocin (10 mg/kg, i.p.) is utilized as standard medication. Furthermore, for peripheral acting analgesics and calming movement, indomethacin (10 mg/kg, i.p.) is utilized as standard medication. The methanolic concentrate of *S. emarginatus* was fundamentally utilized for the decrease of carrageenan-incited paw edema in rodents and analgesic action demonstrated by an increment in the response time by swirl's hot plate method. This methanolic concentrate demonstrated astounding analgesic and mitigating impact similar to the standard medications (Chah et al. 2006).

Thymus vulgaris L.

Vernacular names: Jangliajwain (Hindi); Maruvam (Telugu); Balukambi (Kannada); Omam (Tamil); (Figure 11.10).

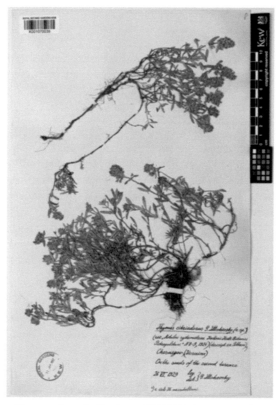

Figure 11.10: Scientific name: *Thymus vulgaris* L. Specimen: K001070039. Image credit to: "© copyright of the Board of Trustees of the Royal Botanic Gardens, Kew."

General characteristics

T. vulgaris is a blooming plant in the mint family Lamiaceae. Its leaf has a length of 4 to 12 mm and width of 3 mm. It has aromatic odor. It is developed in a large portion of the European nations. It is a small perennial bush with a semi-evergreen groundcover (Christopher 2008). The stems become woody with age and its leaves are pretty much nothing.

Medicinal uses

It is a curing agent of bacterial skin infections that leads to pain, tenderness, edema, and reddening of the skin. It is also affected in the treatment of anti-fungal infections, but has no beneficial effects on cellulitis (Renu 2011). The oil of *T. vulgaris* is a combination of monoterpenes. The natural terpenoidthymol and its phenol chemical compound carvacrol (Nickavar et al. 2005, Amiri 2012) are the main constituents of this oil. It is a medicinal drug that has many beneficial effects, such as antioxidative, anti-tissue, antimicrobial, and antibacterial. There were some additional acids, such as terpenoids, flavonoids, glycosides, and synthetic resin found in *Thymus* spp.

The aqueous extracts from species of the *Lamiaceae* family were examined for their antiviral activity against Herpes simplex virus (HSV). It showed inhibitory activity against Herpes simplex virus type 1 (HSV-1) and type 2 (HSV-2). The acyclovir-resistant strain of HSV-1 was tested *in vitro* on RC-37 cells in a plaque reduction assay (Nolkemper et al. 2006).

Tephrosia purpurea (L.) Pers.

Vernacular names: Dhamasia (Hindi); Vempali (Telugu); Empali (Kannada); Kolingi (Tamil) (Figure 11.11).

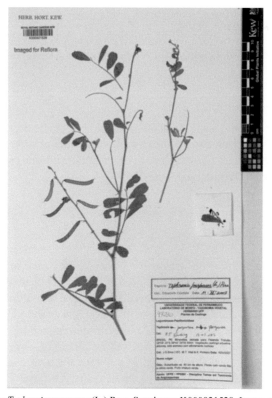

Figure 11.11: Scientific name: *Tephrosia purpurea* (L.) Pers. Specimen: K000921528. Image credit to: "© copyright of the Board of Trustees of the Royal Botanic Gardens, Kew."

General characteristics

It is a perennial herb or subshrub. Flowers are 7 mm long, found in different colors. They have compound leaves which are 5 to 15 cm in length. The upper surface of leaflets is smooth and lower surface is silky and tapered (Santhan 2014). Seeds are ellipsoid and dark brown in color.

Medicinal uses

An outstanding herb, *T. purpurea* is utilized in the treatment of malignant growth and ulcer. An investigation was led in rodents by utilizing three sorts of wound models, for example, entry point wound, extraction wound, and dead space wound that was treated with ethanolic concentrate of *T. purpurea* as a basic balm. A standard medication Fluticasone propionate treatment was contrasted, and the after-effects of this salve (ethanolic concentrate of *T. purpurea*), regarding wound withdrawal, elasticity, histopathological and biochemical parameters was studied. There is a wonderful increment in fibroblast cells, collagen strands, and veins arrangement demonstrated by histopathological study (Svobodova et al. 2003).

The screening of *T. purpurea* areial parts was traditionally used for curing burn wounds was the topic of present investigation was the topic of the present investigation. Basic salve base B.P was used for the planning of Flavonoid division treatment for topical application. Partial thickness and full thickness burn wound models were successfully restored by the silver sulphadiazine treatment and straightforward salve base B.P. The treatment of partial-thickness burn was trailed by wound compression and rigidity, while protein, hydroxyproline, hostile to oxidant compounds were determined by full-thickness model. In case of flavonoid rich fraction, when it is compared with control group, the wound contraction and tensile strength of skin tissue were observed significantly greater (Sinha et al. 1982).

Vernonia scorpioides (Lam.) Pers.

Vernacular names: Sahadevi (Hindi); Sahadevi (Telugu); Menasina kase (Kannada); Naichottepoonde (Tamil).

General characteristics

V. cinerea is a small shrub growing worldwide in the temperate climate. Its height is 0.5 feet to 3 feet. Its fruit is kernel-like in shape and has 1.5 mm length with white tuft-like appendage. Petioles are small and curved. Its leaves have a hairy surface beneath and the flowers are purple or pink-colored, blooming in the rainy season (Keeley and Jones 1979).

Medicinal uses

Skin issues, including mending of unending injuries, for example, ulcers of the lower appendages and diabetic injuries are treated with *Vernonia scorpioides*. It is investigated that treatment having 20% of the ethanol concentrate of the leaves of *V. scorpioides* were utilized every day for the recuperating procedure of extraction wounds in the skin of mice, compared with the control. About a 4 mm skin wound region was extracted on anesthetized mice, and after a treatment of 3, 7, and 14 days, the injuries were surgically evacuated and histologically analyzed. The level of putrefaction zone, mononuclear incendiary cells, fibroblasts, and veins determined injury healing action. In the intense period of healing, the sores were broadened and the rot region was escalated by treating with *V. scorpioides* compared with the control group. At any rate, the treatment did not deny either the enrollment and incitement of incendiary cells or the healing procedure. The expanded territory of necrotic tissue, ordering and exudates framed in the treated gatherings are the unsafe consequences of the quick utilization of this concentrate on extracted tissue. In any case, the arrangement of granulation tissue was not restricted by this concentrate (Dalazen et al. 2005).

An examination was conducted on the topical mitigating impact of the ethanolic concentrate, of *Vernonia scorpioides* (EEVS) on intense and incessant cutaneous irritation models in mice. The topical anti-inflammatory activity of EEVS was checked out against acute models (12-O-tetradecanoylphorbol acetate (TPA)- and arachidonic acid (AA)- induced mouse ear oedema) and chronic models (multiple applications of croton oil). A portion related disallowance of edema in both the TPA-and AA-initiated intense models (DI50 = 0.24 and 0.68 mg/ear with the restraint of $80 \pm 5\%$ and $65 \pm 5\%$, separately, for 1 mg/ear) was coming about by the impact of the ethanolic concentrate of V. scorpioides (EEVS) (Laryssa et al. 2011). In addition, the topical utilization of EEVS diminished the TPA-initiated increment in myeloperoxidase action (MPO) in the ear. In the interminable model, all parameters evaluated: edema development ($31 \pm 2\%$), epidermal hyperproliferation (histology), and MPO ($25 \pm 10\%$) were decreased by the EEVS. In this way, the EEVS is working reasonably in exceptional and perpetual ignitable systems, and its working is additionally strikingly affected by the prevention of neutrophil migration into energized tissue, similar to epidermal hyper-proliferation (Young et al. 1989).

Waltheria indica L.

Vernacular names: Ratti (Hindi); Nallebenda (Telugu); Gulaganji (Kannada); Shembudu (Tamil) (Figure 11.12).

Figure 11.12: Scientific name: *Waltheria indica* L. Specimen: K000447628. Image credit to: "© copyright of the Board of Trustees of the Royal Botanic Gardens, Kew."

General characteristics

An erect pubescent perennial herb or undershrub. Leaves are simple and margin serrated. Flowers are small, yellow, sessile in dense axillary clusters (Verdoorn 1981). The seed is oval-shaped, 2.5 mm in length, 1.5 mm in width, and black in color. Fruit capsule is one-seeded and enclosed within hairy calyx (Irvine 1961).

Medicinal uses

The powder of this plant is used for drying and healing of wounds. Pain due to inflammation is controlled by using root decoction. An investigation demonstrated that rough concentrates and disengaged mixes diminished the pain-relieving, calming, antibacterial, and antifungal exercises (Ahmadiani et al. 1998). Phytochemical examinations demonstrated that the unrefined concentrates and confined mixes comprise of the components of cyclopeptide alkaloids, flavonoids, quercetin, kaempferol, tannins, sterols, terpenes, saponins, anthraquinones (Zongo et al. 2013).

Ziziphus oenoplia (L.) Mill

Vernacular names: Makai (Hindi); Paragi (Telugu); Barige (Kannada); Suraimullu (Tamil) (Figure 11.13).

Figure 11.13: Scientific name: *Ziziphus oenoplia* (L.) Mill. Specimen: K001038516. Image credit to: "© copyright of the Board of Trustees of the Royal Botanic Gardens, Kew."

General characteristics

A large, dense, straggling thorny pubescent shrub. Branchlets are densely tomentose and thorns are in pairs. The leaves are simple, alternate, and 3–6 cm long. Its fruit has a smooth surface and is 8–10 mm in length. Its seeds are uncovered with endocarp (Hosne et al. 2008). The flowers are small, greenish, and short in length.

Medicinal uses

The aqueous and alcoholic extracts of fruits of *Z. oenoplia*, affected the wound healing activity and also reported that activity was established in alcoholic, followed by aqueous extracts when related to control. As a result, it is stated that alcoholics followed by aqueous extracts seemed to control the powerful wound healing activity. The results were compared to the farm cetin sulphate cream as a reference standard drug. The essence of present examination is to peep through the antiulcer activity of *Z. oenoplia* (Jena et al. 2012). One of the natural remedies for ulcers is powdered root of *Z. oenoplia* that was extracted with alcohol.

Conclusion

Succinctly, due to the decreasing efficacy of many synthetic drugs due to the development of drug-resistant strains and increasing contraindications of their usage on the skin due to awful side effects, people needed an alternative for their skin problems. Extracts from plants proved to be an alternative, as nature is the best medical caretaker to fix all human and animal illnesses. Medicines from plants have a traditional foundation that demonstrates the way that over 80% of individuals are reliant on traditional healthcare, especially for skin-related issues. Herbal items are not too expensive, as compared to allopathic medications, and offer advantages to all its consumers. Herbals serve mankind by offering dynamic fixings and viable treatment for skin sicknesses, extending from rashes to loathsome skin malignancy. These valuable gems of human comfort exist in the backwoods and the materialistic exercises, for example, deforestation, living space pulverization, and urbanization add toxic substances to these life-sparing medications, offered for free. The conscious use of these plants and present-day researches are the two stages to enliven the possibilities of herbal medications for treatment of skin ailments and all the other physical sicknesses.

References

Abdul, M.M., Sarker, A.A., Saiful, I.M., Muniruddin, A. 2010. Cytotoxic and antimicrobial activity of the crude extracts of Abutilon indicum. Int. J. Pharmacol. Phytochem. Res. 2: 1–4.

Adwankar, M.K., Chitnis, M.P. 1982. *In vivo* Anti-cancer activity of RC-18: A Plant Isolate from Rubia cordifolia, Linn, against a spectrum of experimental tumour models. Chemotherapy 28: 291–293.

Ahmadiani, A., Fereidoni, M., Semnanian, S., Kamalinejad, M., Serami, S. 1998. Antinociceptive and anti-inflammatory effects of Sambucus ebulus rhizome extract in rats. J. Ethnopharmacol. 61: 229–235.

Amiri, H. 2012. Essential oils composition and antioxidant properties of three Thymus species. Evid.- Based Complement Altern. Med. 2012: 728065.

Amirul, A., Abdul, S.J., Mohd, R.Y., Azizah, A.H., Hakim, A. 2014. Morpho-physiological and mineral nutrient characterization of 45 collected Purslane (*Portulaca oleracea* L.) accessions. Bragantia 73: 426–437.

Anand, T., Kalaiselvan, A., Gokulakrishnan, K. 2011. Wound healing activity of Murraya koenigii in Male Albino Rats. IJCR. 3: 425–427.

Antarkar, D.S., Tasneem, C., Narendra, B. 1983. Anti-inflammatory activity of Rubia cordifolia Linn in rats. Indian J. Pharmacol. 15: 185–188 .

Anyaehie, U.B. 2009. Medicinal properties of fractionated acetone/water neem Azadirachta indica leaf extract from Nigeria: a review. Niger J. Physiol Sci. 24: 157–159.

Apisariyakul, A., Vanittanakom, N., Buddhasukh, D. 1995. Antifungal activity of turmeric oil extracted from Curcuma longa (Zingiberaceae) J. Ethnopharmacol. 49: 163–169.

Ara, H., Hassan, M., Khanam, M. 2008. Taxonomic study of the genus *Ziziphus* Mill. (Rhamnaceae) of Bangladesh. BANGL. J. Plant. Taxon. 15: 47–61.

Arora, B., Bhadauria, P., Tripathi, D., Sharma, A. 2012. Sapindus emarginatus: Phytochemistry and various biological activities. I. G. J. P. S. 2: 250–257.

Barcaccia, G., Ghedina, A., Lucchin, M. 2016. Current advances in genomics and breeding of leaf chicory (*Chichorium intybus* L.). Agriculture 6: 1–24.

Barrett, D.R., Fox, J.E.D. 1997. Santalum album: kernel composition, morphological and nutrient characteristics of preparasitic seedlings under various nutrient regimes. Ann. Bot. 79: 59–66.

Bhowmik, D., Yadav, C.J., Tripathi, K.K., Kumar, K.P.S. 2010. Herbal Remedies of *Azadirachta indica* and its medicinal application. J. Chem. Pharm. Res. 2: 62–72.

Bohlooli, S., Mohebipoor, A., Mohammadi, S., Kouhnavard, M., Pashapoor, S. 2007. Comparative study of fig tree efficacy in the treatment of common warts (*Verruca vulgaris*) vs. cryotherapy. Int. J. Dermatol. 46: 524–26.

Brototi, B., Kaplay, R.D. 2011. Azadirachta indica(Neem): it's Economic utility and chances for commercial planned plantation in Nanded District. Int. J. Pharma. 1: 100–104.

Brown, J.P. 1980. A review of the genetic effects of naturally occurring flavonoids, anthraquinones and related compounds. Mutat Res. 75: 243–277.

Cai, Y., Sun, M., Xing, J., Corke, H. 2004. Antioxidant Phenolic constituents in roots of Rheum *Offiicinale* and Rubia Cordifolia: Structure-radical scavenging activity relationships. J. Agric. Food Chem. 52: 7884–7890.

Chah, K.F., Eze, C.A., Emuelos, C.E., Esimone, C.O. 2006. Antibacterial and wound healing properties of methanolic extracts of some Nigerian medicinal plants. J. Ethnopharmacol. 104: 164–167.

Charles, V., Charles, S.X. 1992. The use and efficacy of *Azadirachta indica* ADR (Neem) and Curcuma longa (Turmeric) in scabies. A pilot study. Trop. Geogr. Med. 44: 178–181.

Christopher Brickell. 2008. RHS A-Z Encyclopedia of Garden Plants. Dorling Kindersley, United Kingdom.

Dalazen, P., Molon, A., Biavatti, M.W, Kreuger, M.R.O. 2005. Effects of the topical application of the extract of vernonia scorpioides on excisional wounds in mice. REV. BRAS. FARMACOGN 15: 82–87.

Dande, P.R., Talekar. V.S., Chakraborthy, G.S. 2010. Evaluation of crude saponins extract from leaves of Sesbania sesban (L.) Merr. for topical anti-inflammatory activity. Int. J. Res. Pharm. Sci. 1: 296–299.

Danin, A., Baker, I., Baker, H.G. 1978. Cytogeography and taxonomy of the Portulaca oleracea l. polyploidy complex. ISR J Plant Sci. 27: 177–211.

Das, R.K., Ghosh, S., Sengupta, A., Das, S., Bhattacharya, S. 2004. Inhibition of DMBA/croton oil-induced two-stage mouse skin carcinogenesis by diphenylmethyl selenocyanate. Eur. J. Cancer Prev. 13: 411–417.

Desai, Y.G., Kapoor, M.K. 2010. Impact of building construction dust on leaf morphology and flowering in *Bauhinia purpurea* Linn. Adva. Plant. Sci. 23: 569–572.

Dwivedi, C., Abu-Ghazaleh, A. 1997. Chemopreventive effects of sandalwood oil on skin papillomas in mice. Eur. J. Cancer. Prev. 6: 399–401.

Ekpo, M.A., Etim, P.C. 2009. Antimicrobial activity of ethanolic and aqueous extracts of Sida acuta on microorganisms from skin infections. Corpus ID: 55329505.

El-Sayed, M., Awad, S., Ibrahim, A. 2019. Impact of Purslane (*Portulaca oleracea* L.) extract as antioxidant and antimicrobial agent on overall quality and shelf life of greek-style youghurt. EJFS. 47: 51–64.

Gogte, V.M. 2000. Ayurvedic Pharmacology and Therapeutic Uses of Medicinal Plants (Dravyagunavigyan), First ed. Bharatiya Vidya Bhavan (SPARC), Mumbai Publications, pp. 421–422.

Gupta, A.K., Chauhan, J.S. 1984. Constituents from the stem of *Bauhinia variegata.* Natl. Acad. Sci. Lett. 7: 15–16.

Gupta, M., Mazumder, U.K., Manikandan, L., Haldar, P.K., Bhattacharya, S., Kandar, C.C. 2003. Antibacterial Activity of Vernonia Cinerea. Fitoterapia. 74: 148–150.

Handral, H., Pandith, A., Shruthi, S.D. 2012. A review on Murraya koenigii: Multipotential medicinal plant. AJPCR. 5: 5–14.

Handral, H.K., Pandith, A., Shruthi, S.D. 2012. A review on Murraya koenigii multipotential, medicinal plant. Asian J. Pharm. Clin Res. 5: 5–14.

Harbone, J.B., Baxter, H. 1993. Phytochemical Dictionary. A Hand Book of Bioactive compound from plants. Taylor and Fransic, Washington, D.C., U.S.A. pp. 237.

Harsha, V.H., Hebbar, S.S., Hedge, G.R., Shripathi, V. 2002. Ethnomedical knowledge of plants used by kunabi tribe of Karnataka in India. FITOTERAPIA. 73: 281–287.

ICRAF. 1992. A selection of useful trees and shrubs for Kenya. International Centre for Research in Agroforesty. Nairobi.

Ignacimuthu, S., Ayyanar, M., Sankara-Sivaramann, K. 2006. Ethnobotanical investigations among tribes in Madurai District of Tamil Nadu (India). J. Ethnobiol. Ethnomed. 2: 25.

Irvine, F.R. 1961. Woody Plants of Ghana. Oxford University Press, London, pp.185.

Itokawa, H., Takeya, K., Mori, N., Hamanaka, T., Sonobe, T., Mihara, K. 1984. Isolation and antitumour activity of cyclic hexapeptides isolated from Rubiae Radix. Chem. Pharm. Bull. 32: 284–290.

Jena, B., Ratha, B., Kar, S. 2012. Wound healing potential of ziziphus xylopyrus wild (Rhamnaceae) stem bark ethanol extract using *in vitro* and *in vivo* model. JDDT. 2: 41–46.

Jippo, T., Nomura, S., Kitamura, Y. 2000. Cichorium intybus mast cell-mediated immediate-type allergic reactions. Int. J. Orient. Med. 1: 82–88.

Karodi, R., Jadhav, M., Rub, R., Bafna, A. 2009. Evaluation of the wound healing activity of a crude extract of Rubia cordifolia L. (Indian madder) in mice. I.J.A.R. 2: 12–18.

Karou, D., Dicko, M.H., Simpore, J., Traore, A.S. 2005. Antioxidant and antibacterial activities of polyphenols from ethnomedicinal plants of Burkina Faso. Afr. J. Biotechnol. 4: 823–828.

Keeley, S.C., Jones, S.B. 1979. Distribution of pollen types in vernonia. Syst. Bot. 4: 195–202.

Kendra, M. 2000. 'Clinical assessment of application of Sesbania sesban (Jayanti) leaves in vicharchika', Proceedings of International congress on "Ayurveda-2000", Chennai, pp.245.

Khan, B.A., Abraham, A., Leelamma, S. 1996. Biochemical response in rats to the addition of curry leaf (Murraya koenigii) and mustard seeds (Brassica juncea) to the diet. Plant Food Hum. Nutr. 49: 295–299.

Kirtikar, K.R., Basu, B.D., Blatter, E., Caius, J.F., Mhaskar, K.S. 1993. Indian Medicinal Plants. 2nd Ed. Vol II. Lalit Mohan Basu, Allahabad, India, 1182.

Kumar, V.S., Sharma, A., Tiwari, R., Kumar, S. 1999. Murraya koenigii: A review. J. Med. Aromat Plant Sci. 21: 1139–1144.

Lodha, G. 2019. Formulation and Evaluation of polyherbal shampoo to promote hair growth and provide antidandruff action. JDDT. 9: 296–300.

Mahabub, U.Z., Ahmed, M., Akter, R.A., Aziz, M., Ahmed, M. 2009. Studies on Anti inflammatory, antinociceptive and antipyretic activities of ethanol extract of *Azadirachta indica* leaves. Bangladesh J. Sci. Ind. Res. 44: 199–206.

Manjunath, A., Habte, M. 1991. Root morphological characteristics of host species having distinct mycorrhizal dependency. Can. J Bot. 69: 671–676.

Mann, A., Gbate, M., Umar, A.N. 2003. Sida acuta subspecie *acuta*. Medicinal and economic palnt of Nupeland, Jube Evans Books and Publication, pp. 241.

Manu, P., Ankita, L., Swati, R., Anju, R. 2012. *Plumbago zeylanica* L.: A mini review. I.J.P.A. 3: 399–405.

Mhaskar, K.S., Blatter, E., Caius, J.F. 2000. Kirtikar and Basu's Illustrated Indian Medicinal Plants. Indian J. Med. Sci. 1: 86–96.

Mortellini, R., Foresti, R., Bassi, R., Green, C.J. 2006. Curcumin, an antioxidant and anti-inflammatory agent, induces heme oxygenase-1 and protects endothelial cells against oxidative stress. Free Radic. Biol. Med. 28: 1303–1312.

Muñoz, V.S., Alberto, A.I.L., Luis, M.R.S., Davila, H.V., Jesus, B.F. 2007. Neem tree morphology and oil content. issues in new crops and new uses. pp. 126–128. *In*: Janick, J., Whipkey, A. (eds.). A.SHS Press, Alexandria, VA.

Nickavar, B., Mojab, F., Dolat-Abadi, R. 2005. Analysis of the essential oils of two *Thymus* species from Iran. Food Chemistry 90: 609–611.

Niharika, A., Aquicio, J.M., Anand, A. 2010. Antifungal properties of neem (*Azadirachta indica*) leaves extract to treat hair dandruff. I.S.R.J. 2: 244–252.

Nolkemper, S., Reichling, J., Stintzing, F., Carle, R., Schnitzler, P. 2006. Antiviral effect of aqueous extracts from species of the Lamiaceae family against Herpes simplex virus type 1 and type 2 *in vitro*. Planta. Med. 72: 1378–1382.

Obiefuna, I., Young, R. 2005. Concurrent administration of aqueous Azadirachta indica (Neem) leaf extract with DOCA salt prevents the development of hypertension and accompaning electrocardiogram changes in the rat. Phytother. Res. 19: 792–795.

Pai, M.R., Acharya, L.D., Udupa, N. 2004. Evaluation of antiplaque activity of Azadirachta indica leaf extract gel—a 6-week clinical study, J. Ethnopharmacol. 90: 99–103.

Papadopoulou, K., Ehaliotis, C., Tourna, M., Kastanis, P., Karydis, I., Zervakis, G. 2002. Genetic relatedness among dioecious *Ficus carica* L. cultivars by random amplified polymorphic DNA analysis, and evaluation of agronomic and morphological characters. GENETICA. 114: 183–194.

Prassana, V., Habbu, Hanumanthachary Joshi, Patil, B.S. 2007. Potential Wound Healers from Plant Origin. Pharmacognosy Reviews 1(2): 271–281.

Qiao, Y.F., Wang, S.X., Wu, L.J., Li, X., Zhu, T.R. 1990. Studies on antibacterial constituents from the roots of Rubia Cordifolia L. Yao XueXue Bao. 6: 51–57.

Rahman, Ahmad, S., Qureshi, S., Ranman, A.U., Badar, Y. 1991. Toxicological studies of Melia azedarach (flowers and berries) Pak. J. Pharma. Sci. 4: 153–158.

Rahuman, A.A., Gopalakrishnam, G., Venkatcsan, P., Geetha, K. 2008. Isolation and identification of mosquito larvicidal compound from Abutilon indicum (Linn.) Sweet. Parasitol. Res. 102: 981–988.

Rajkapoor, B., Jayakar, B., Murgeshand, N.D. 2006. Akthisekaran Chemoprevention and cytotoxic effect of Bauhinia variegatea against N-nitrosodiethylamine induced liver tumors and human cancer cell lines. J. Ethnopharmacol. 104: 407–409.

Rakesh, K.S., Upma., Ashok, k., Sahil, A. 2010. Santalum Album Linn: A review on morphology phytochemistry and pharmacological aspects. Int. J. Pharm. Tech. Res. 2: 914–919.

Rani, M.S., Pippalla, R.S., Mohan, K. 2009. *Dodonaea Viscosa* Linn.—an overview. Asian J. Pharm. 1: 97–112.

Rasheed, A.N., Afifi, F.U., Disi, A.M. 2003. Simple evaluation of the wound healing activity of a crude extract of *Portulaca oleracea* L. (growing in Jordan) in Mus musculus JVI1. J. Ethnopharmacol. 88: 131–136.

Rauh, L.K., Horinouchi, C.D.S, Loddi, A.M.V., Pietrovski, E.F., Neris, R., Guimaraes, F.S.F., Buchi, D.F., Biavatti, M.W., Otuki, M.F., Cabrini, D.A. 2011. Effectiveness of Vernonia scorpioides ethanolic extract against skin inflammatory processes. J. Ethnopharmacol. 138: 390–397.

Reardon, S. 2016. Veppachetta. Its solutions and usage for the developing world. Nature. 509: 56–149.

Reddy, V.P., Sahana, N., Uroj, A. 2012. Antioxidant activity of Aegle marmelos and Psidium guajava leaves. Int. J. Med. Arom. Plants 2: 155–160.

Renu, S. 2011. Treatment of skin diseases through medicinal plants in different regions of the world. Int. J. Compr. Pharm. 4: 1–4.

Ross, I.A. 2001. Medicinal plants of the world: Chemical constituents, Traditional and modern medicinal uses, Totowa, New Jersy 2: 81–85.

Ross, J.A., Kasum, C.M. 2002. Dietry Flavoniods: bioavailability, metabolic effects, and safety. Annu. Rev. Nutr. 22: 19–34.

Rubae, A.Y. 2009. The potential uses of melia *Azedarach* L. as pesticidal and medicinal plant, review. American-Eurasian J. Sus. Agri. 3: 185–194.

Saini, M.L., Saini, R., Roy, S., Kumar, K. 2008. Comparative pharmacognostical and antimicrobial studies of Acacia species (Mimosaceae). J. Med. Plant Res. 2: 378–386.

Saleem, R., Rani, R., Ahmed, M., Sadaf, F., Ahmad, S.I., Zafar, N., Khan, S.S., Siddiqui, B.S., Lubna, F., Ansari, S.A., Khan, Faizi, S. 2008. Effect of a cream containing Melia azedarach flowers on skin diseases in children. Phytomed. 15: 231–236.

Sand, J.M., Bin Hafeez, B., Jamal, M.S., Witkowsky, O., Siebers, E.M., Fistcher, J., Verma, A.K. 2012. Plumbagin(5-hdroxy 2-methyl-1, 4-naphthoquinone), isolated from Plumbago zeylanica, inhibits ultra-violet radiation induced development of squamous cell carcinomas. Carcinogenisis. 33: 184–190.

Santhan, P. 2014. Leaf structural characteristics of important medicinal plants. IJRAP. 5: 673–679.

Schmutterer, H. 1996. The neem tree: source of unique natural products for integrated pest management, medicine, industry and other purposes. J. Am. Chem. Soc. 188: 3999–4000.

Selvakumar, V., Anbudurai, P.R., Balakumar, T. 2001. *In vitro* propagation of the medicinal plant *Plumbago zeylanica* L. through nodal explants. *In Vitro* Cell. Dev. Biol. Plant. 37: 280–284.

Sharma, N., Kaushik, P. 2014. Medicinal, biological and pharmacological aspects of Plumbago zeylanica (Linn.). J. Pharmacogn. Phytochem. 3: 117–120.

Shoemaker, M., Hamilton, B., Dairkee, S.H., Cohen, I., Campbell, M.J. 2005. *In vitro* anticancer activity of Twelve Chinese Medicinal herbs. Phytother. Res. 19: 649–651.

Shrikanth, V.M., Janardhan, B., More, S.S., Muddapur, U.M., Mirajkar, K.K. 2014. *In vitro* anti snake venom potential of Ablution indicum Linn leaf extracts against Echis carinatus. J. Pharmacogn Phytochem. 3: 111–117.

Shrivastava, S., Bera, T., Roy, A., Singh, G., Ramachandrarao, P., Dash, D. 2007. Characterisation of Sandalwood Tree (Santalum album). Nanotechnology 18: 1–9.

Shukla, A., Garg, A., Mourya, P., Jain, C.P. 2016. Zizyphus oenoplia Mill: A review on Pharmacological aspects. A.P.G. 1: 8–12.

Singh, N. Singh, A., Pabla, D. 2019. A Review on medicinal uses of Bauhinia variegata Linn. Pharma. Tutor. 7: 12–17.

Sinha, B., Natu, A.A., Nanavati, D.D. 1982. Prenylated flavonoids from Tephrosia purpurea seeds. Phytochemistry 21: 1568–1570.

Svobodova, A., Psotova, J., Walterova, D. 2003. Natural phenolics in the prevention of Uv-induces skin damage: A review. Biomed 47: 137–145.

Tripathi, Y.B., Sharma, M., Manickam, M. Rubiadin, 1997. A new antioxidant from Rubia cordifolia. Indian J. Biochem. Biophys. 34: 302–306.

Tripathi, Y.B., Sharma, M., Shukla, S., Tripathi, P., Thyagaraju, K., Reddanna, P. 1995. 1Rubia cordifolia inhibits potato lipoxygenase. Indian J. Exp. Biol. 33: 109–112.

Turnbull, J.W. 1986 (Editor). Multipurpose Australian trees and shrubs; lesser-known species for fuel wood and agro forestry. ACIAR. pp.316 .

Upadhye, A., Kumbhojkar, M.S., Vartak, V.D. 1986. Observations on wild plants used in folk medicine in the rural areas of the Kolhapur district. Anc. Sci. Life 6: 119–121.

VAZ, A.M.S.F. 1979. Considerações sobre a taxonomia do gênero Bauhinia L. sect. Tylotaea Vogel (Leguminosae-Caesalpinioideae) do Brasil. Rodriguesia 31: 127–234.

Verdoon, I.C. 1981. The genus Waltheria in southern Africa. Bothalia 13: 275–276 .

Vidya, V., Srinivasan, Sengottuvelu. 2012. Wound healing potential of Melia azedarach L. leaves in alloxan induced diabetic rats. GJRMI. 1: 265–271.

Vijaya, V.T., Srinivasan, D., Sengottuvelu, S. 2012. Wound healing potential of *Melia azedarach* L. leaves in alloxan induced diabetic rats. G.J.R.M.I. 1: 265–271.

XueYuan, Y., LiMin, X., AiMin, Z., Hai, W., Yan, W., Fei, H., HongFu, X., Rui, Y., Dan, J., Xiao, M., FeiFei, W. 2010. The adjuvant treatment of atopic dermatitis withmedical skin preparation containing extracts from Portulaca oleracea and avocado. Am. J. Clin. Dermatol. 39: 460–462.

Young, L.M. De, Kheifets, J.B., Ballaron, S.J., Young, J.M. 1989. Edema and cell infiltration in the phorbol ester-treated mouse ear are temporally separate and can be differentially modulated by pharmacologic agents. Agents Actions 26: 335–341.

Zaka, S.H., Saleem, M., Shakir, S., Khan. 2006. A. Fatty acid composition of Bauhinia variegataand Bauhniamalabarica seed oils-comparsion of their physico-chemicalproperties. Fette Seifen Anstrichm. 85: 169–170.

Zongo, F., Ribuot, C., Boumendjel, A., Guissou, I. 2013. Botany, traditional uses, phytochemistry and pharmacology of Waltheria indica L. (syn. Waltheria americana): A review 148: 14–26.

12

Choerospondias axillaris (Hog plum)
Multiple Health Benefits

Sajan L. Shyaula

Introduction

In recent years, there has been growing interest in studying phytochemicals in fruits and vegetables due to their health-promoting properties. Epidemiological studies have indicated that consumption of fruits and vegetables is negatively associated with some age-related and non-communicable diseases in humans. The arrays of secondary metabolites that are restricted in distribution within the particular plant species are responsible for its potent protective properties. Glucosinolates from Brassicaceae, cucurbitacins from Cucurbitaceae, capsaicinoids from capsicum, thiosulfides from allium, and spirostanol glycosides from Dioscoreaceae are some representative classes of secondary metabolites having significant biological activities (João 2012). The inclusion of diversity of fruits and vegetables in the food habit can thus contribute significantly to improve health by providing these arrays of secondary metabolites. Many secondary metabolites isolated from fruits and vegetables are relatively nontoxic and can be consumed as dietary supplements too. In addition, the extracts prepared from these fruits and vegetables are typically compositionally complex materials and can exhibit biological activities by different mechanisms than those of isolated drug molecules. It is thus necessary to include varieties of fruits and vegetables in daily consumption as they can provide more effective ways to struggle against diseases than conventional drugs because of their multi-targeting attributes, low cost, low toxicity, and wide availability.

Choerospondias axillaris (Roxb.) B.L. Burtt and A.W. Hill is an important plant with multiple health benefits, but remains to be explored more scientifically. To date, *C. axillaris* fruit is considered as an underutilized fruit and is included in the diet of a limited ethnic population. It has been used in various traditional medicinal systems either as a major constituent or as a part of constituents of herbal formulations for treatment of cardiac and other problems. *C. axillaris* is known as a hog plum, lapsi, chanchin, modoki, jujube, and many others that are specific to host country and ethnicity. Lapsi is a large, deciduous, edible native fruit tree of the family *Anacardiaceae*, and grows between 900 m and 2,000 m asl in the Himalayan range (Jackson 1994, Paudel et al. 2002a). Lapsi trees are reported to be native to Nepal. Its distribution is not restricted to the Himalayas, and they are also found in Thailand (Jackson 1994), Vietnam (Nguyen et al. 1996), India, and China (Hau et al. 1997, Zhou et al. 1997, Feng et al. 1999). The tree is largely known for its delicious fruits, timber, and medicinal

Faculty of Science, Nepal Academy of Science and Technology, Khumaltar, Lalitpur, Nepal; shyaulasajan@gmail.com

values in Nepal, China, Vietnam, and Mongolia. Seed stones are used as the fuel in brick kilns. It has a growing popularity and economic importance due to its nutritive value and medicinal effects, but comprehensive information on the chemical composition and bioactivity of its fruits is still lacking.

Considering the multiple health benefits of *Choerospondias axillaris*, botanical characteristics, commercial products, domesticating techniques, phytochemicals, and pharmacological properties have been described in this chapter. With the aim of utilization of wild plant resources, the literature has been reviewed, and we hope that the knowledge offered in this chapter serves as an updated comprehensive database contributing to the development of plant-derived foods from *C. axillaris* with multiple health benefits.

Botany

This deciduous lapsi tree can grow up to 20 meters tall. The outer bark of lapsi is dark grey or red-brown and the inner bark is red. The bark is cracked, and peeling in vertical flakes. Branchlets are observed to be red-brown to gray-brown. Juvenile lapsi trunks are green, lenticellate, and smooth. The midribs of juvenile leaves displayed red-brown to red-orange coloration. Lapsi leaves as described by Shu are petiolulate, imparipinnate compound with opposite leaflets (Shu 2008). The leaf can be ovate, to ovate-lanceolate, or oblong-ovate. Lapsi leaves are age dimorphic, with the young leaves with scattered teeth, and mature leaves without teeth. The leaf petiole is inflated at the base. The bases of the leaflets are rounded to cuneate with an acuminate leaf apex.

Gardner stated that lapsi flowers are 0.4–0.5 cm and dark red (Gardner et al. 2000). The trees producing pistillate flowers are called pothilapsi (female tree), and others producing staminate flowers are called bhalelapsi (male trees). Pistillate flowers have empty anthers and staminate flowers lack gynoecium. The pollens are transported by insects, honey bees, and wind. The male flowers are found in branched clusters at the end of twigs and upper leaf axils. Bisexual flowers are found in leaf axils in groups of 2–3. The calyx is less than 2 mm and has 5 lobes. The female flowers are dark red-purple, smooth on the outer surface, and glandular-hairy on the inside. Lapsi flowers bloom from February to March and to a lesser degree in April and May. The duration of flowering is about two weeks long. The majority of the growing season coincides with the monsoon season. Lapsi trees begin to lose their leaves beginning mid to late November, with the majority of leaf drop occurring early to mid-December. Dormancy lasts until February, when bud break occurs. Flowers develop soon after bud break and continue for about two weeks. Flowering only occurs after 7 to 10 years of growth (Seber 2016).

Fruit Characteristics

Lapsi fruit ranges in mass from 8 to 18 g. The fruit produced by female lapsi trees is drupes. Lapsi fruit can be ellipsoidal, obovate-ellipsoidal, or spherical. The fruit is green until maturity, at which point it turns yellow. The flesh is light yellow in color, fibrous, and acidic, having a specific aromatic flavor. The endocarp of lapsi fruit contains five seeds enclosed in a woody mesocarp capsule. Each seed is isolated from the others by woody septa. Inside the endocarp box, the seeds are fused to the mesocarp at their bases. The superior portions of the seeds are unattached and free inside the cavities of the stone (Hill 1937). During the germination of lapsi seeds, longitudinal slits at the apices of the woody mesocarp separate due to cell expansion of the hypocotyls. The separated seed cavity becomes a pore for the embryo to grow out from it. Lapsi is a climacteric fruit. It can be harvested unripe or ripe. The fruit quality is extremely variable. The genetic diversity of the trees, elevation, light availability, and water availability are factors affecting fruit quality. Many types of frugivores consume lapsi fruits (Brodie et al. 2009, Lai et al. 2014). Farmers have categorized lapsi into different types according to their indigenous indicators that are based on fruit size being small/large, time of maturity being early and late, taste being sweet and sour, and pulp content being high and low. The fruit had a pulp content range of 23–45%, with an average of 37.6 percent. The peel content ranged from 18–33%, with an average of 22.8 percent. The stone weight ranged from 18% to 33% of the total weight of the fruit. The average weight of the stone was 27% of the total mass. The seed weight comprised 20–38% of the fruit weight, with an average of 22.8 percent.

Horticulture

Lapsi grows best in full sun and saturated soil. The lapsi tree has a low tolerance of shade and frost, and is moderately tolerant of low fertility and drought (Tyystjarvi 1981). Lapsi trees used to be grown on slopes between terraces, mixed with other fruit trees and fodder trees. Many lapsi trees are grown on the margins of unirrigated farms. This is a common practice because harvesting lapsi is perceived to be damaging to crops on the farm. Over a century ago, farmers started protecting lapsi trees on their farmlands, mostly to make use of their timber. These days the trees are used mainly for their fruit; they are only occasionally cut for timber and never the fruiting trees. As such the use of lapsi trees in agroforestry systems has made a gradual shift from agrosilviculture to agrohorticulture. Lapsi is propagated primarily by seed. Lapsi can be vegetatively propagated by chip budding, grafting, propagation of hardwood and softwood cuttings, and by tissue culture.

Expansion of lapsi cultivation for quality production is limited due to associated risk of non bearingness, as only female trees produce fruits. Local people are consulting forest officers looking for methods to identify female trees and to select the best seedlings, but the forest offices have limited experience with domestication techniques. Therefore, there is a need for research regarding the identification of the fruiting tree and its management. The selection of female plants is a vital part of domestication, and farmers have developed their own techniques for identifying male and female trees. Their assumptions in this regard are: (i) female plants sprout earlier than male plants under the same conditions at the beginning of the growing season; (ii) only female plants release milky latex when leaves are pricked; (iii) wood from female plants does not blast while burning whereas that from males does so loudly; and (iv) wood from male trees splits easily. Though these assumptions were found to be valid, they are yet to be studied further (Paudel 2003a, Paudel et al. 2003b, Paudel and Parajuli 1999).

Cultural Importance

The tree has social, cultural, ecological, and economic value in Nepal. The existence of lapsi as the prehistoric vegetation types of Sleshmantak Ban (meaning lapsi forest) around Pashupatinath in Kathmandu has been quoted in Swathani Brata katha, an anthology of mythical stories. Lapsi is said to be important to many Nepali people and people of the Hindu faith. Lapsi fruits are used in Hindu rituals, Newari feasts, festivals, and celebrations. In Hinduism, the fruits are used as offerings to the gods and goddesses, and are grown on many religious sites in the Kathmandu valley (Chhetri and Gauchan 2007).

Lapsi may have some cultural significance to Newari people. A traditional feast (Sukubhoye) for Newari people contains lapsi soup, which is believed to aid in digestion and "purify the elements" (Bajracharya 2015). Due to the superstition that the presence of lapsi tree makes a site prone to lightning strikes during thunderstorms, people have tended not to plant the tree around their homesteads. However, lapsi trees have been protected as fruit trees.

Commercial Products

Lapsi is recognized as having great potential as a cash-generating commodity. Lapsi is still a relatively unknown commodity outside Nepal and China. Observations of lapsi production resulted in an estimated average yield per a tree of 200 kg. The average weight of the fruit at maturity was estimated at 13 g. Lapsi is increasing in value and popularity, and varieties of processed products are available in the market. Further research should aim to increase efficiency of processing techniques based on local knowledge and skills. Processing facilities turn the raw fruit into products, such as mada, candy, pickle, powder, pastilles, vinegar, and drugs (Gautam 1997, 2004).

Mada

Mada is a collective name for dried Lapsi mat prepared from the pulp and peel of Lapsi fruits. The fruits are boiled in water for two hours in a drum, and salt is added. Seeds are separated from the boiled fruits manually. The pulp and peel are placed on a polythene sheet and laid in a pit in the ground, where it can be stored for a longer time. Whenever it is convenient,the pulp is taken out, spread on wooden planks and dried in the sun for two days. The mada is used for the sour topping in vegetable pickles.

Candy or Titaura

Only fully ripened fruits are selected. They are washed thoroughly with water and boiled in a minimum amount of water for 20 minutes. The boiled fruits are then emptied into a bamboo basket for filtration. The fruits are peeled and the seeds are separated. A quantity of sugar equal to the quantity of pulp is mixed in, and the mixture is stored in a polythene container. A tray with thick lining on all sides is taken, and the surface is greased with cooking oil. The stored mixture is then spread over the greased surface of the tray and solar dried for 4 to 19 days, depending upon the intensity of sunlight. The degree of dryness can be checked by testing the stickiness of the surface. Once dry, the mixture is diced and packaged. It is one of the most popular dry foods in supermarkets in Nepal.

Lapsi Pickle

Fruits are washed and boiled in water. The skins are split from fruit and strained for a few minutes. The appropriate amount of mustard oil is taken and brought to a simmer. Usually fenugreek seeds are fried in oil. Black cumin, fennel, asafetida, turmeric, salt, and other additives are added into it. Lapsi is then added to the mixture and stirred thoroughly. Chilly powder and sugar are added. The thick paste pickle is allowed to cool overnight, then packed into jars.

Powdered Skins

The producers of lapsi candy manually separate the skin and seeds. The dried skin is made to a powder by grinding, which is widely used as the sour toppings.

Skin Vinegar

Using *C. axillaris* skin-like material, skin vinegar was made by submerged fermentation with the inoculation of bifidobacteria. The skin vinegar was bright, glossy, smooth, and uniform with a delectable flavor of sour and sweet, and without any peculiar smell. The optimal fermentation conditions were as follows: fermentation temperature of 45°C, inoculum of 5%, fermentation time of 72 hours,and the total acid being 3.15 g/100 ml (Dingjian et al. 2009).

Pastilles

A pastille is a type of candy made of a thick liquid that has been solidified and is meant to be consumed by light chewing and allowing it to dissolve in the mouth. In China, the fruits of *C. axillaris* are used for production of pastilles. The texture properties (hardness, chewiness, gumminess, stickiness, and springiness) of *C. axillaris* pastilles gradually increased with the decrease of moisture content (Wu 2012).

Taste and Flavor

The fruits are sour in taste. The content of total organic acids is in significant amounts, upto 8.13% (Liu and Chen 2000). The fresh pulps of fruits are light yellow in color, but on storage it slowly turns into yellow-brown. The browning index and 5-hydroxymethylfurfural content kept positive correlation with temperature and storage period. Reducing ascorbic acid and total polyphenol content was negatively associated with temperature and storage period. Under the conditions of low temperature, reducing ascorbic acid oxidation was the major browning reaction. When the system temperature rose to 30°C, Maillard reaction was the chief factor that caused browning (Yu et al. 2013).

Nutritional Value

The nutrient content of lapsi fruits can vary according to the genetic variation and the environmental conditions that affect growth and development. There are wide variations in the analytical results carried out by the different groups of researchers. A comparison of the analytical results is presented in a tabulated form (Table 12.1).

From the above comparative results, lapsi is rich in potassium, calcium, and magnesium content. The results also show that the fruit of lapsi contains less fat and a higher amount of protein. *C. axillaris* pulp contains indispensible amminoacids and high vitamin C 88.7 mg/100 g which is a high nutrition material for health protection foods.

Table 12.1: Comparison of nutritional composition in fruits of *C. axillaris*.

	Gupta et al. 2019 Pandey et al. 2018	Chen et al. 2001	Zhai et al. 2015	Bhutia et al. 2011	Rai et al. 2005	Paudel et al. 2002b
Moisture	78.7 %	83%			84%	83%
Total sugar		3%				3.4%
Crude fat	2.10%	0.6%		0.05%	0.9%	
Crude protein	6.31%	2.7%		4.11%	2.2%	
Total soluble solid		3.0%		9.99%	6.37%	6.22%
Ash	2.89%					
Total titratable acid		1.7%				3.53%
Total crude fiber	1.55%	1.9%				
Total acid						6.8%
Ca	12.37 µg/L	0.7%	269.88 µg/g		202 mg/100 g	57 mg/100 g
Mg	40.16 µg/L	0.2%	195.30 µg/g			34.6 mg/100 g
K	280.12 µg/L	1.6%	2,970.11 µg/g		639 mg/100g	355 mg/100 g
Mo	2.63 µg/L					
Na	6.96 µg/L	130.6 µg/g			5 mg/100g	2.05 mg/100 g
Zn	0.11 µg/L	28.3 µg/g	1.51 µg/g			0.08 mg/100 g
Fe	26.47 µg/L	0.2%	27.37 µg/g			0.1 mg/100 g
Cu	9.06 µg/L	30.9 µg/g	2.45 µg/g			0.07 mg/100 g
Mn	11.57 µg/L		2.45 µg/g			
P		0.1%	133.94 µg/g			

During the analysis of the elements for the bioavailability of minerals for human body in *C.axillaris* flesh, peels, aqueous extractives, and gastric digesta, determined by the inductively coupled plasma atomic emission spectrometry (ICP-AES), there is difference of distribution and release of mineral elements between peels and flesh of *C. axillaris*. The content of minerals are lower in flesh than that of peel, and the release rates of minerals in the flesh were found to be higher than those in the peels (Zhai et al. 2015). The contents of microelements in the extracts are also determined by atomic absorption spectrophotometry (Zhang and Meiyan 2003).

Phytochemical Composition

The uses of different parts of *Choerospondias* in traditional medication system have resulted in considerable chemical analysis of the plant and their active principles (Lian et al. 1984, Chong et al. 2008, Tang and Eisenbrand 2013, Wang et al. 2014, Li et al. 2008, Shi et al. 2007). The phytochemical investigation of *C. axillaris* has resulted in different classes of compounds, which include organic acids, phenolics, flavonoids, proanthocyanins, ellagic acids, aromatics, polysaccaharides, alkenyled bridge ring ketones, sterols, proteases and fatty acids, as shown in Table 12.2.

Organic Acids

The sweet and sour taste of *C. axillaris* is mainly due to the significant amount of organic acid present in it. The content of total organic acid amounts up to 8.13%. Citric acid and L-malic acid (Figure 12.1) are two main organic acids of *C. axillaris*, the content of which accounted for 26.36% and 22.95% of total organic acids, respectively. These organic acids have cardioprotective effects (Tang et al. 2013, Xiaogeng and Yousheng 2000). Besides citric and malic acids, the fruits are rich in vitamin C (Shah 1978). The vitamin C contents are significantly different in the ultrasonic treatment and the non-ultrasound treatment when measured by the xylene-dichloro-indophenol colorimetry. The average content of vitamin C with the ultrasound treatment group and no ultrasound treatment group are 51.626 mg/mL and 33.113 mg/mL, respectively (Jun et al. 2012). Other organic acids and their derivatives reported in *C. axillaris* are quininic acid, succinic acid, monoglyceride citrate, methylcitrate, comenic acid, triethyl citrate, 2-isopropyl malic acid, and jasmonic acid (Figure 12.1).

Phenolics

The total phenolic content of *C. axillaris* was reported to be higher than that of other fruits (Li et al. 2016a). The quantitative assay demonstrated that peel contained a significantly higher amount of phenolics than flesh. The phenolic compound comprises of simple phenol to a different class of compounds and its condensed polymer forms. These phenolic constituents were considered to be the major contributor to the reported medicinal use. The most abundant phenolic acid in peel was ellagic acid followed by gallic acid, while the most abundant phenolic acid in flesh was gallic acid followed by protocatechuic acid. The variety of phenolic compounds in flesh was a little more than that in the peel, though the total phenolic content (TPC) and total flavonoid content (TFC) of flesh were much lower than those of the peel. The simple phenolics (Figure 12.2) found in *C. axillaris* are protocatechuic acid, vanillic acid, syringaldehyde, *p*-hydroxybenzoic acid, protocratechualdehyde, and salicylic acid. The galloylglucosidic constituents have been isolated from stem barks of *C. axillaris* by chromatographic technique (Li et al. 2014a), and further supported by HPLC-Q-TOF-MS/MS-based analysis (Yang et al. 2016). The reported gallic acid derivatives (Figure 12.3) are gallic acid, ethyl gallate, gallic acid ethyl ether, l-O-galloyl-β-D-glucose, 1,6-di-O-galloyl-β-D-glucose, 1,4-di-O-galloyl-β-D-glucose, 1,4,6-tri-O-galloyl-β-D-glucose, and 1,3,4,6-tetra-O-galloyl-β-D-glucose.

In a study, oral ration of *Choerospondias* fruit with Sandalwood extract results in earlier plasma peaks of protocatechuic acid and gallic acid, lower clearance rate, and longer half-life, indicating

Table 12.2: Different classes of compounds isolated from *C. axillaris*.

Plant parts	Class of compounds	Name of compounds	References
	Organic acids	Citric acid L-Malic acid	Tang et al. 2013
		Vitamin C	Jun et al. 2012
		Quininic acid Succinic acid Monoglyceride citrate Methylcitrate Comenic acid Triethyl citrate 2-Isopropyl malic acid Chebulic acid Jasmonic acid	Yang et al. 2016
	Phenolic	Protocatechuic acid	Li et al. 2016a
		Vanillic acid	Zhang et al. 2013 Deng et al. 2006
		Syringaldehyde	Shen et al. 2009
		p-Hydroxybenzoic acid Protocratechualdehyde Salicylic acid	Yang et al. 2016
	Gallic acid derivatives	Gallic acid Ethyl gallate l-O-galloyl-β-D-glucose 1, 6-di-O-galloyl-β-D-glucose 1, 4-di-O- galloyl-β-D-glucose 1,4,6-tri-O-galloyl-β-D-glucose 1, 3, 4, 6-tetra-O-galloyl-β-D-glucose	Li et al. 2014a
	Flavonoids and proanthocyanins	(+)-Catechin	Li et al. 2016a Li et al. 2016c
		Epicatechin	
		Epicatechin gallate	
		Dimer of catechin	
		Trimer of epicatechin	
		Dimer of catechin and epicatechin gallate	
		(+)-Catechin-7-O-β-D-glucopyranoside (+)-Catechin-4'-O-β-D-glucopyranoside	Li et al. 2009a
		(+)-Catechin (6'-8) (+)-catechin	Li and Cui 2014
		Dehydrodicatechin A	
		Dihydroquercetin-7-*O*-β-D-glucopyranoside	
		Narigenin-4'-*O*-(6''-*O*-galloyl-β-D-glucopyranoside)	
		Pinocembrin-7-*O*-β-D-glucopyranoside	
		Naringenin-4'-*O*-β-D-glucopyranoside	
		Dihydrokaempferol-7-*O*-β-D-glucopyranoside	
		Pinocembrin Naringenin Choerospondin	Lü et al. 1983
		Gambiriin A3	Li and Cui 2014

Table 12.2 contd. ...

...Table 12.2 contd.

Plant parts	Class of compounds	Name of compounds	References
	Flavonoids and proanthocyanins	Gambiriin A1	
		Kaempferol	Zhang et al. 2013
		Kaempf ero1-5-O -arabinoside	Khabir et al. 1987
		Kaempferol-7-O-glucopyranoside	Gao et al. 2017
		Chrysin	Li et al. 2005 Lü et al. 1983
		Quercetin Quercetin-3-O-glucoside Quercetin-3-O-arabinoside Quercetin-3-O-rhamnoside	Yang et al. 2016
		Quercetin-3- *O* - (arabinoside - glucoside	
		Rutinum	Li et al. 2009b
		Quercetin-7-*O*-β-D-glucopyranoside	Li and Cui 2014
		Dihydroquercetin	Lian et al. 2003
		Hyperin	
		Lueolin-3'-O-beta-D-glucopyranoside	Li et al. 2009b
	Ellagic acid	3-Methylellagic acid Methyl ellagic acid glucoside Dimethoxy ellagic acid glycoside 3,3' Di- O- methylellagic acid Ellagic acid	Yang et al. 2016 Shen et al. 2009
	Sterols	β-Sitosterol	Zhang et al. 2013
		Daucosterol	Li et al. 2009b
		Ursolic acid	Yang et al. 2016
		Ergosterol	Gao et al. 2017
	Aromatics	Tetradecyl-*E*-ferulate	Li et al. 2005
		Dibutyl phthalate	
	Fatty acids	Hexadecanoic acid Correctitude forty-two alkyl acid	Li et al. 2009b
		Stearic acid Triacontanoic acid Octacosanol	Lian et al. 2003 Zhu et al. 2003
	Miscellaneous	Deoxyuridine	Fan et al. 2005
		Jatrorrhizine	Wang et al. 1983
		5-Hydroxymethylfurfural 6-Hydroxy indole lactic acid Scopoletin	Yang et al. 2016
		Proteases	Upadhyay et al. 2013a Upadhyay et al. 2013b Karki et al. 2009
		Polysaccharides	Wang et al. 2009
Stem barks	Alkenyled bridged-ring ketones	Choerosponins A and B	Li et al. 2017

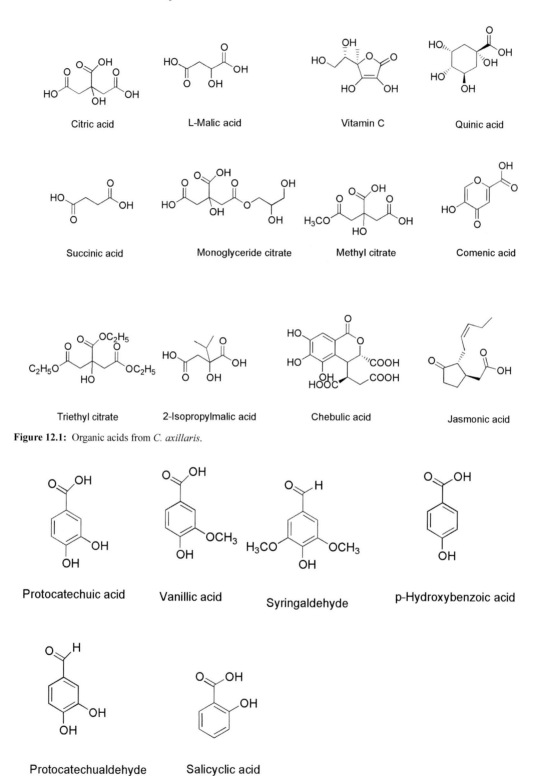

Citric acid   L-Malic acid   Vitamin C   Quinic acid

Succinic acid   Monoglyceride citrate   Methyl citrate   Comenic acid

Triethyl citrate   2-Isopropylmalic acid   Chebulic acid   Jasmonic acid

Figure 12.1: Organic acids from *C. axillaris*.

Protocatechuic acid   Vanillic acid   Syringaldehyde   p-Hydroxybenzoic acid

Protocatechualdehyde   Salicyclic acid

Figure 12.2: Phenolic class of compounds from *C. axillaris*.

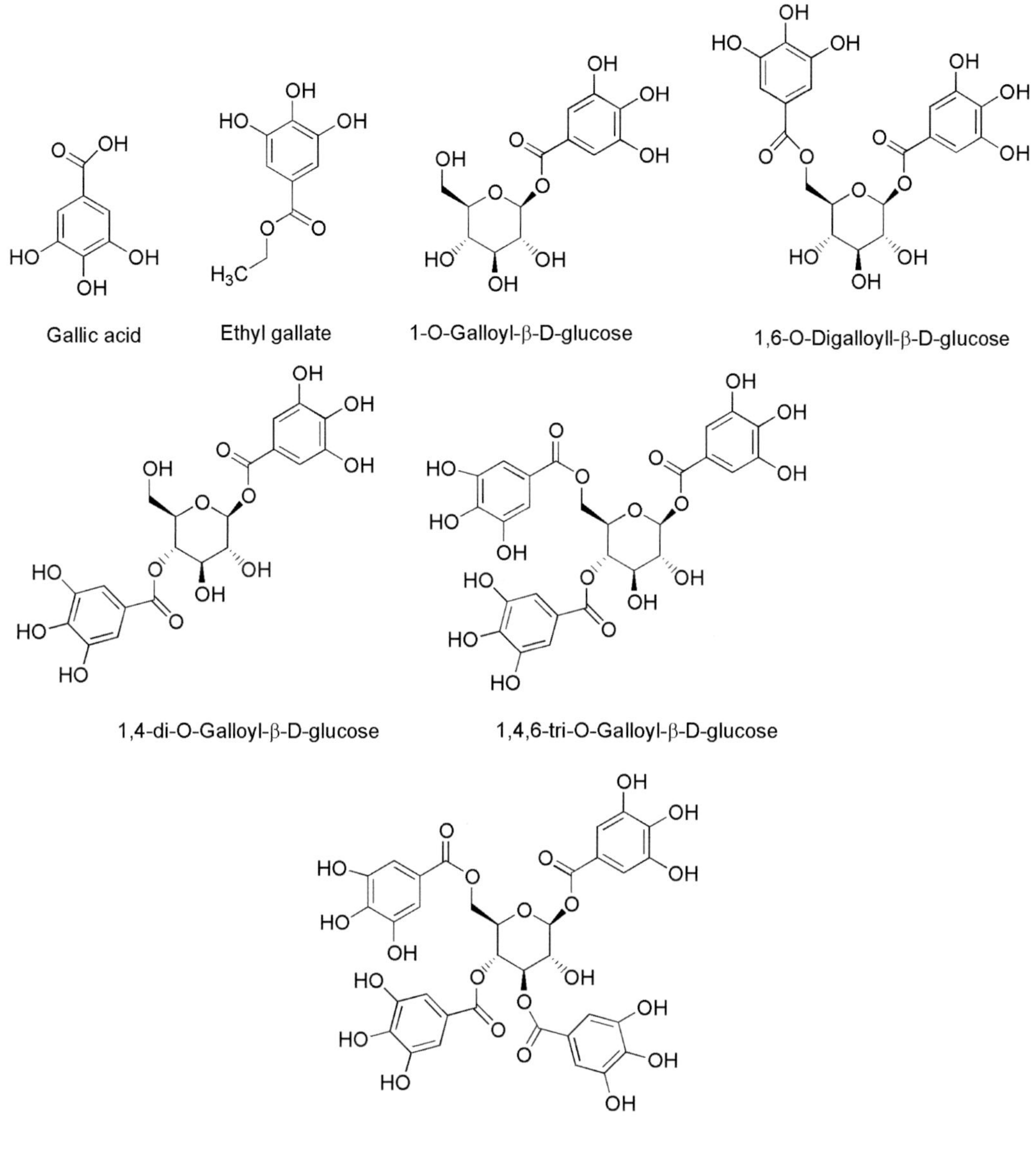

Gallic acid   Ethyl gallate   1-O-Galloyl-β-D-glucose   1,6-O-Digalloyll-β-D-glucose

1,4-di-O-Galloyl-β-D-glucose   1,4,6-tri-O-Galloyl-β-D-glucose

1,3,4,6-O-tetragalloyl-β-D-glucoside

Figure 12.3: Gallic acid derivatives from *C. axillaris*.

Sandalwood can promote the absorption of phenolic compounds in *Choerospondias* fruit (Liang et al. 2010).

Flavonoids and Proanthocyanins

The content of proanthocyanins (PAs) in *C. axillaris* peels is relatively high, with a yield of 17.8 percent. The main phenolic components within the peels are proanthocyanins, which were composed of (+)-catechin, (−)-epicatechin, (−)-epicatechin gallate, and (−)-epigallocatechin and their galloylderivatives. Proanthocyanins, also known as the condensed tannins, are oligomers and polymers of flavan-3-ol that are bound together with B-type and A-type linkages. The mean degree of polymerization (mDP) of the total PAs from *C. axillaris* peels was 4.61 (Li et al. 2016b). Compositional analysis indicated that the proanthocyanins had extension units mainly consisting of epicatechin gallate or epicatechin, and terminal units mainly consisting of catechin (Li et al. 2015a). Most of the *C. axillaris* fruits are peeled when processed as the raw material for the food industry, which leads to a large amount of fruit peel being generated as a by-product. Peels have an astringent taste due to their high PAs content. There is a direct relationship between the bioactivities of PAs and structural factors, such as monomer compositions, the linkage types of the interflavan bonds, degree of polymerization, and galloylation. Numerous studies have reported that the degree of polymerization and galloylation of PAs influence their bioactivities, although these results are often contradictory. MALDI-TOF-MS and HPLC-MS analysis of different fractions of *C. axillaris* have shown that degree of polymerization upto 14 units and with various gallyol derivatives have been detected (Li et al. 2018). The total flavonoids were always assumed to be the effective constituents for its medicinal use. Quercetin, kaempferol, naringenin, and its derivatives are also present in a significant amount as the flavonoid constituents. Different methods have been applied for extracting flavonoids from *C. axillaris* (Xu et al. 2013). Active constituents in *C. axillaris* were extracted using cellulase, with the aim of exploring a more efficient method for extraction of polyphenols from *C. axillaris*. Enzymatic hydrolysis with cellulase can break down β-D-glucose bonds in plants more gently, damage cell walls, decompose plant tissues, and accelerate the release of active constituents, thereby improving the extraction yield (Sun et al. 2015b). The isolated flavonoids and proanthocyanin (Figures 12.4a–d) compounds are presented in Table 12.2. The *in vitro* stability, bioaccessibility, and biological activity of PAs from *C. axillaris* peels were investigated following passage through a simulated gastrointestinal tract. Simulated gastric digestion caused little change in the total phenolic content (TPC) and mean degree of polymerization (mDP) of the extracts. Polyphenols from the *C. axillaris* peels are relatively stable during passage through a simulated gastrointestinal tract and maintain their biological activities. After the simulated intestinal digestion, the TPC and mDP of both extracts decreased compared to the non-digested initial extracts, which was attributed to polyphenol-protein interactions (Li et al. 2015b).

Others

Similarly, the ellagic acid and its derivatives (Figure 12.5) have also been reported, such as ellagic acid, 3,3'-di O-methylellagic acid, methylellagic acid, methyl ellagic acid glucoside, dimethoxy ellagic acid glycoside. Sterols (Figure 12.6) reported from *C. axillaris* are β-sitosterol, daucosterol, ursolic acid, and ergosterol. The aromatic derivatives (Figure 12.7) isolated are tetradecyl-*E*-ferulate and dibutyl phthalate (Li et al. 2005). Fatty acid derivatives (Figure 12.8) isolated are stearic acid, octacosanol, and triacotanoic acid, etc. 5-Hydroxymethylfurfural, deoxyuridine, jatrorrhizine, scopoletin, and 6-hydroxyindol-3-lactic acid (Figure 12.9) have also been isolated by a different group of researchers. Activity-guided isolation of antitumor compounds from the $CHCl_3$ extract partitioned from the 95% ethanol extract led to isolation of two new cytotoxic alkenyled bridged-ring ketones, namely choerosponins A and B (Figure 12.10). These compounds possess rare dioxatricyclo skeleton (Li et al. 2017).

Enzymes and Polysaccharides

The protease was extracted from the root of *C. axillaris* with 0.1 M phosphate buffer of pH 7 and then precipitated successively with TCA and ammonium sulfate (Upadhyay et al. 2013a, b Karki et al. 2009). The water-soluble polysaccharide was extracted by water immersion, ethanol precipitation method preliminarily purified, and precipitated by adding calcium solution under hydrothermal synthesis to form polysaccharide-Ca(II) (Wang et al. 2009).

a

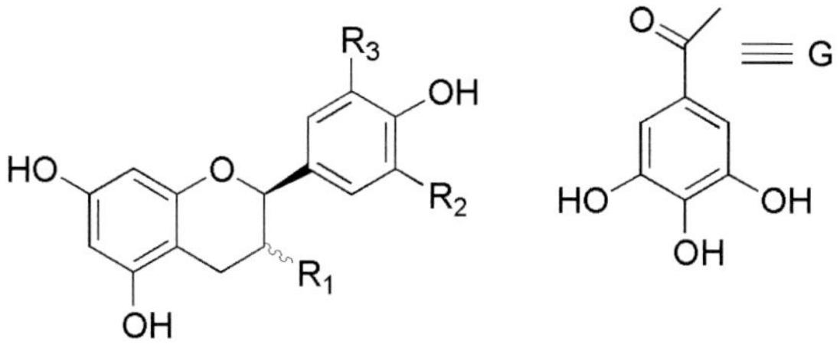

	R1	R2	R3
Catechin	""OH	H	OH
Epicatechin	—OH	H	OH
Epicatechin gallate	—OG	H	OH

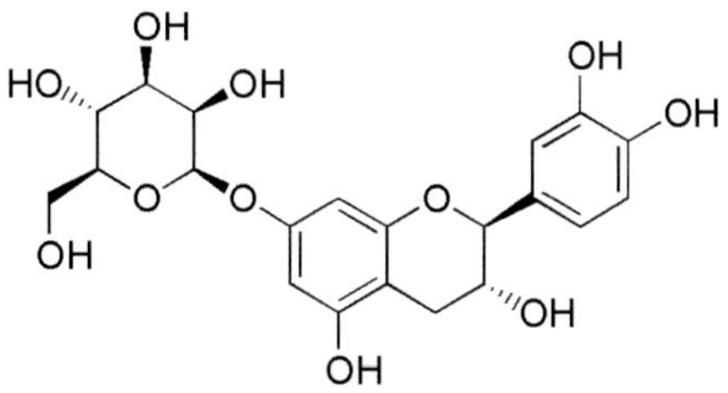

(+) Catechin-7-O-β-D-glucoside

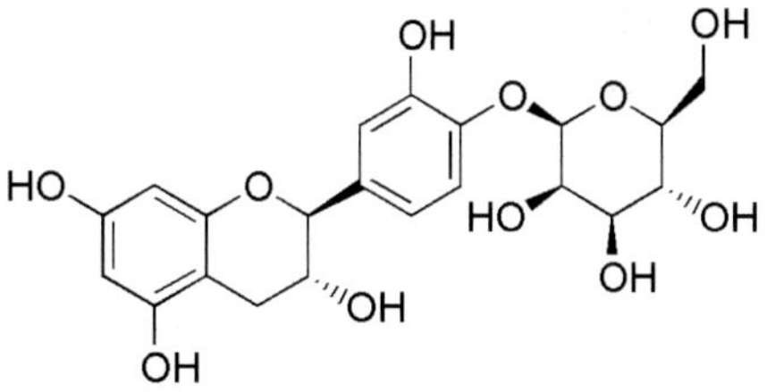

(+) Catechin-4'-O-β-D-glucoside

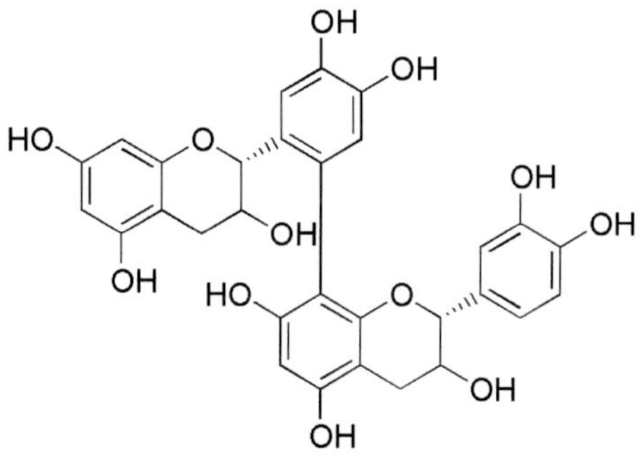

(+)-Catechin (6'-8) (+)-catechin

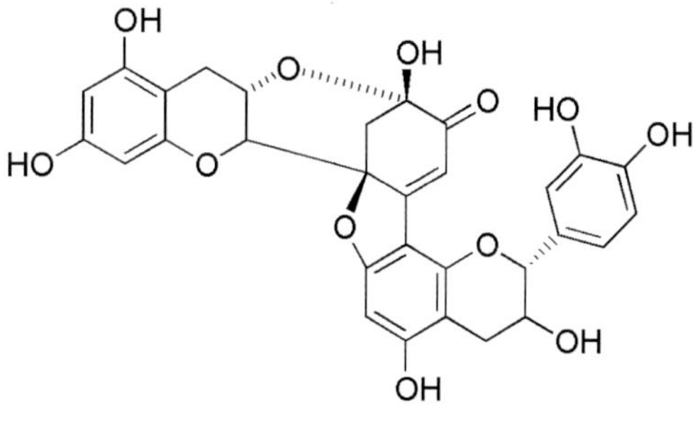

Dehydrodicatechin A

b

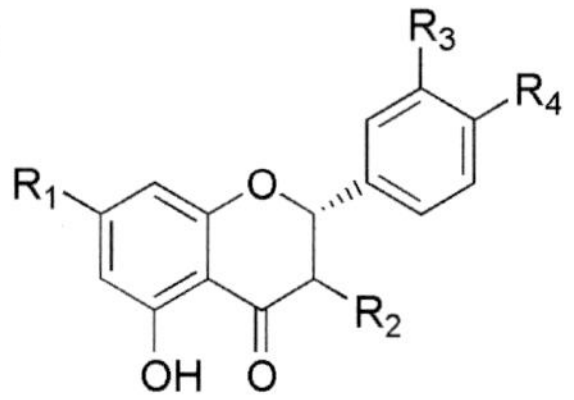

R1	R2	R3	R4
OGlc	H	H	H
OH	H	H	OGlc
OGlc	OH	H	OH
OGlc	Oh	OH	OH

Pinocembrin-7-O-β-D-glucopyranoside
Naringenin-4'-O-β-D-glucopyranoside
Dihydrokaempferol-7-O-β-D-glucopyranoside
Dihydroquercetin- 7-O-β-D-glucopyranoside

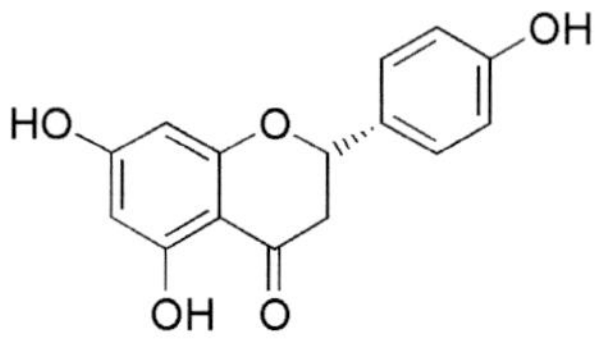

Naringenin

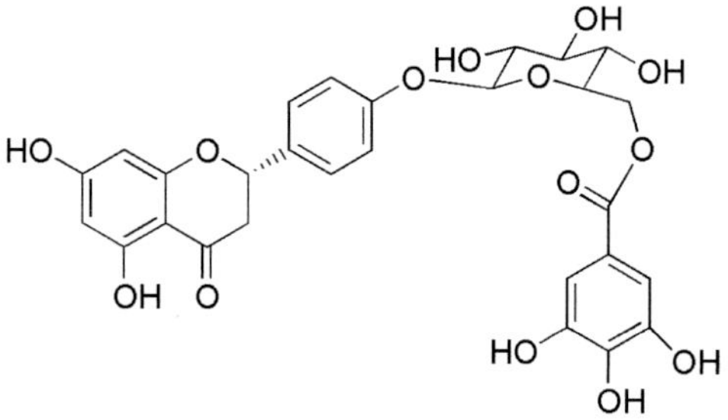

Narigenin-4'-O-(6"-O-galloyl-β-D-glucopyranoside).

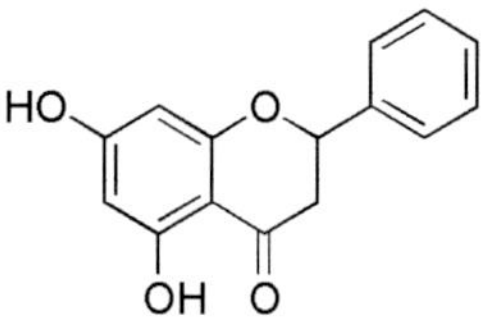

Pinocembrin

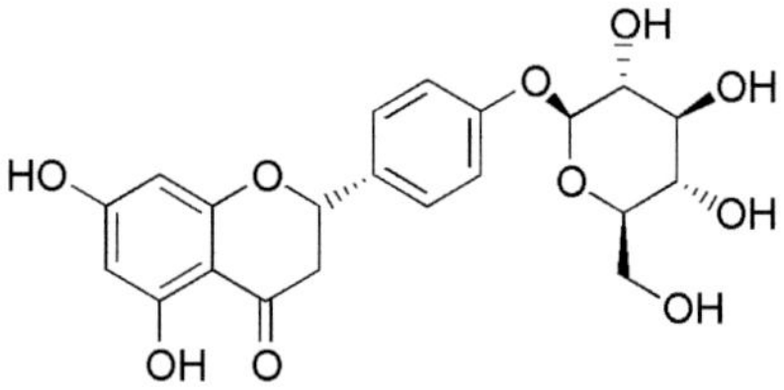

Choerospondin

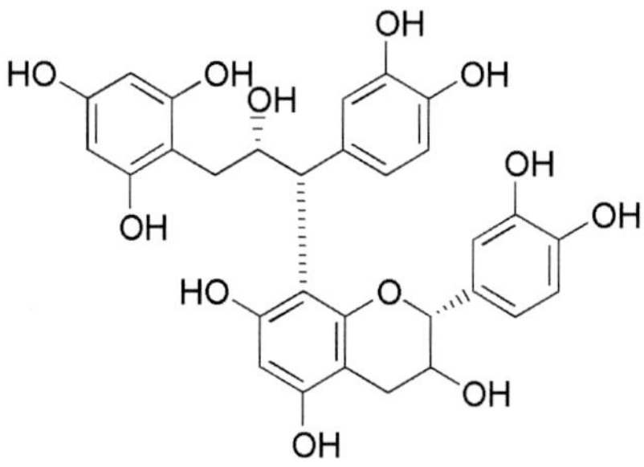

Gambiriin A1

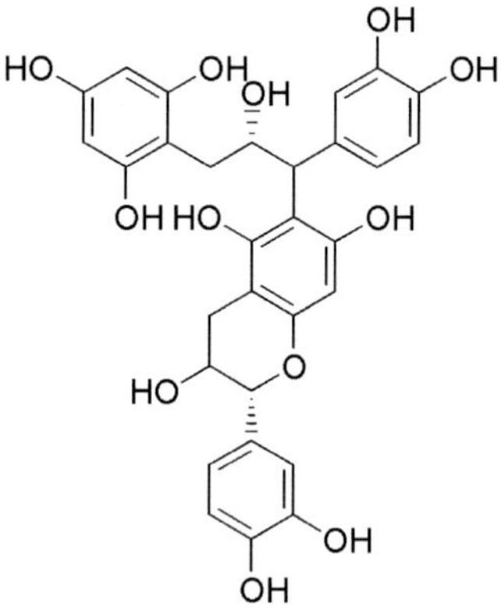

Gambiriin A3

c

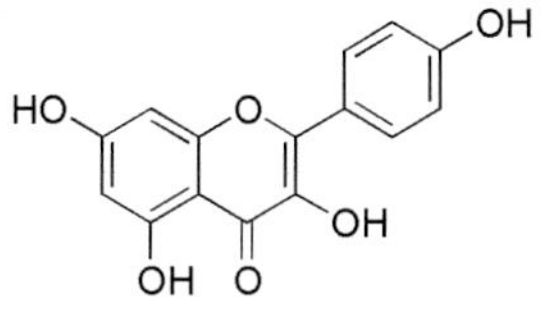

Kaempferol

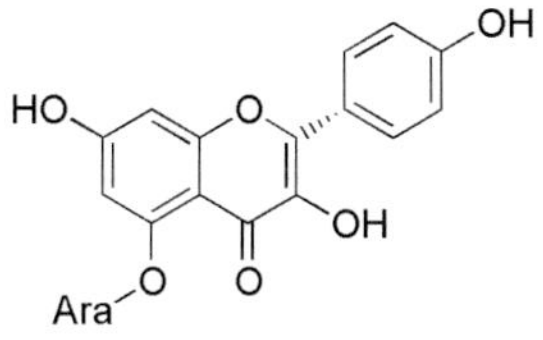

Kaempferol-1-O-5-arabinoside

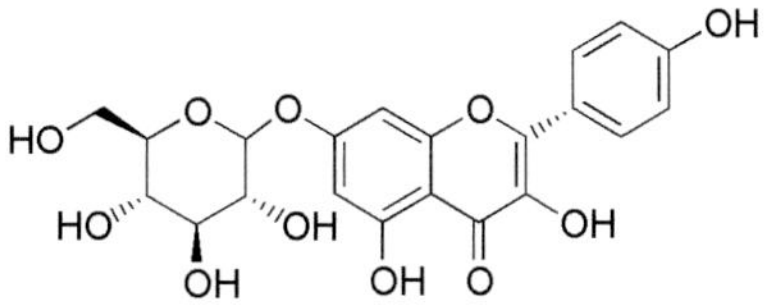

Kaempferol-7-O-glucoside

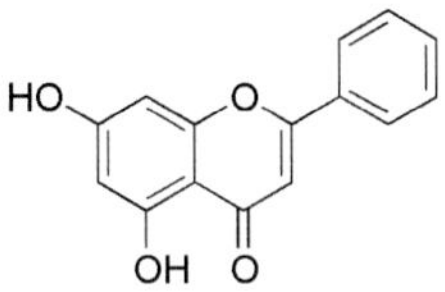

Chrysin

d

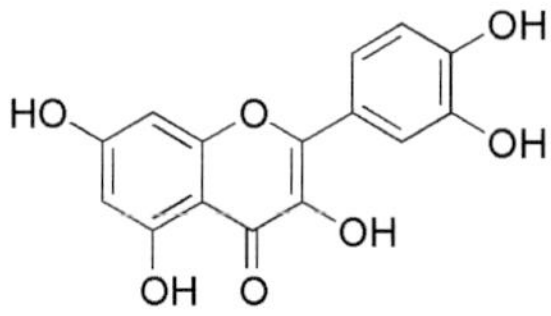

Quercetin

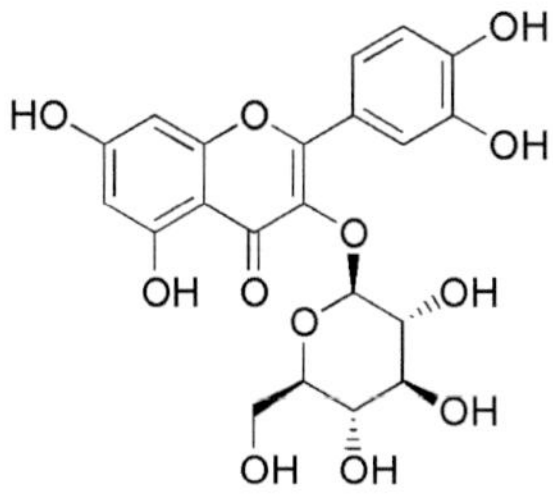

Quercetin-3-O-glucoside

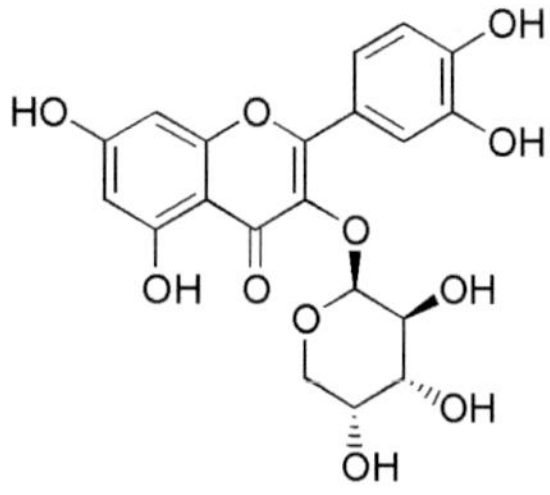

Quercetin-3-O-arabinoside

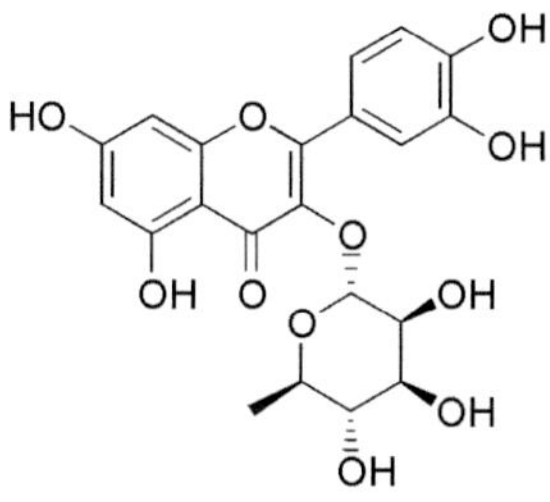

Quercetin-3-O-rhamnoside

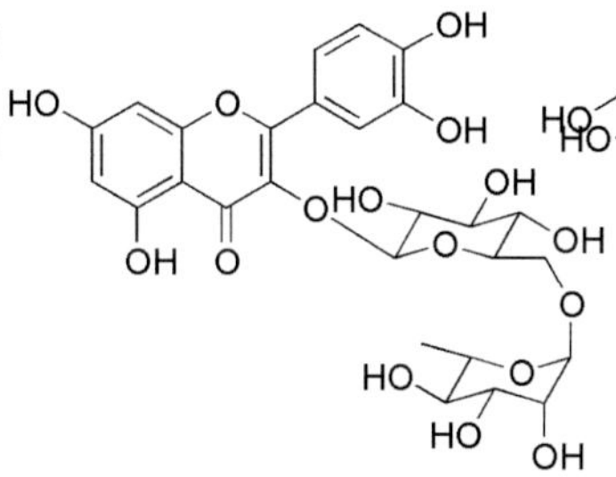

Rutinum

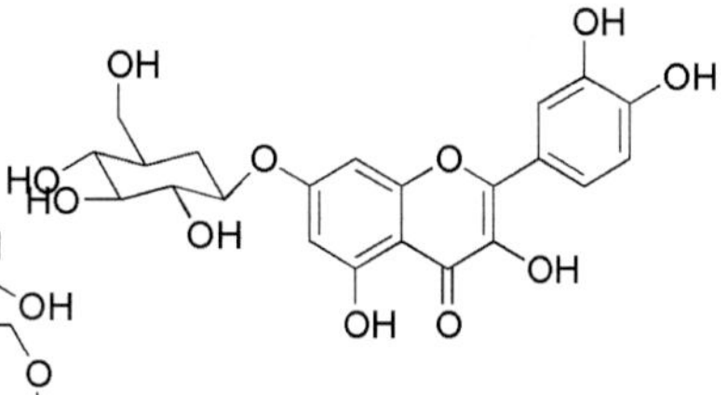

Quercetin-7-O-β-D-glucopyranoside

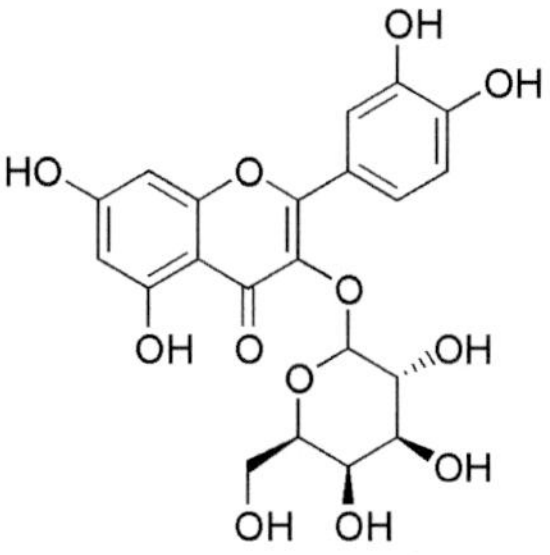

Hyperin

Figure 12.4a-d: Flavonoids and proanthocyanins from *C. axillaris*.

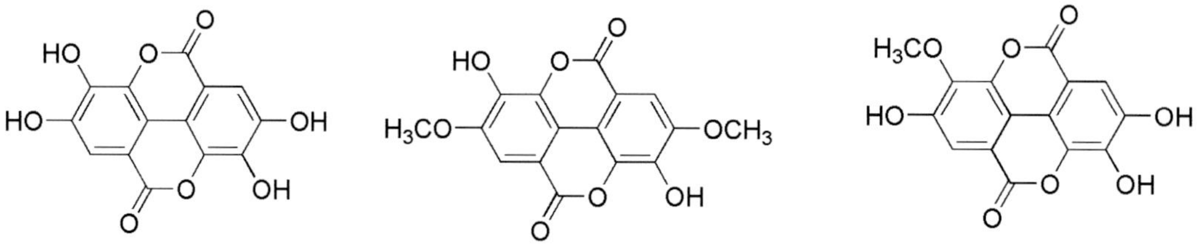

Ellagic acid 3,3'-Di-O-methylellagic acid 3-Methyl ellagic acid

Figure 12.5: Ellagic acid derivatives of *C. axillaris*.

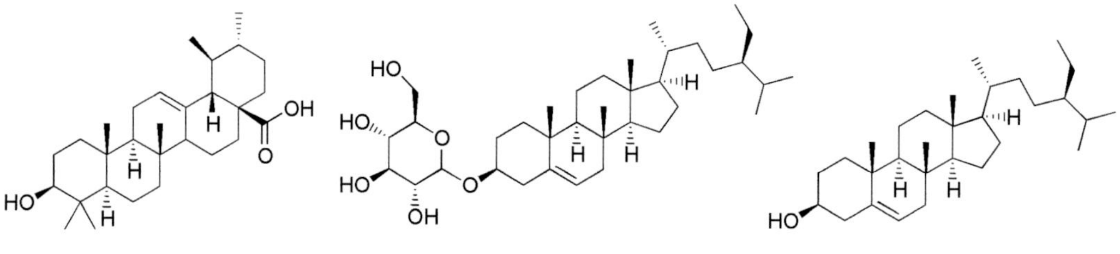

Ursolic acid Daucosterol β-sitosterol

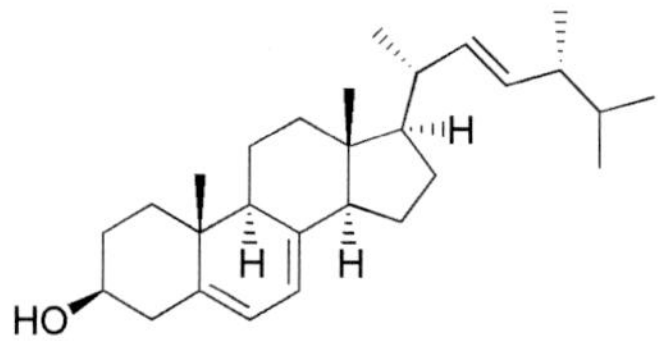

Ergosterol

Figure 12.6: Sterols from *C. axillaris*.

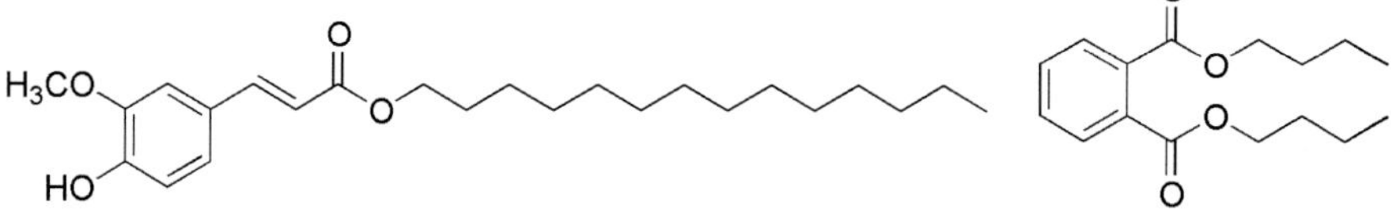

Tetrdecyl-*E*-ferulate Dibutyl phthalate

Figure 12.7: Aromatics from *C. axillaris*.

Traditional Uses

Choerospondias has been used as traditional Chinese medicine (TCM) to treat cardiovascular diseases for a long time (China Pharmacopoeia Committee 2010). The Guan-Xin-Shu-Tong capsule (GXSTC) is well-known in traditional Chinese medicine, and is used for the treatment of coronary heart disease and angina pectoris, which mainly consists of *C. axillaris*. Angina pectoris usually happens because one or more of the heart's arteries is narrowed or blocked, also called ischemia (Gao et al. 2017). Similarly, Guang-Zao-Qi-Wei-Wan (Meng 2012), San-Wei-Guang-Zao capsule (Liga et al. 2002), Guangzaofufang (Li 2009) and GuangZao and RouDouKou (Lu et al. 2018), etc. are used for angina pectoris and coronary heart diseases.

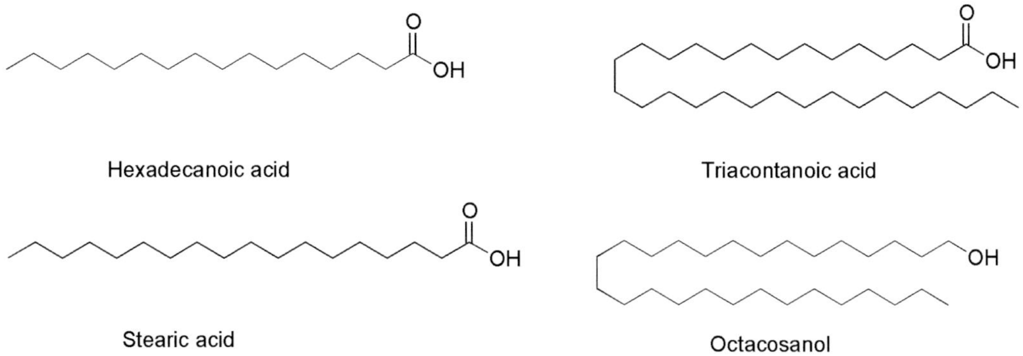

Hexadecanoic acid

Triacontanoic acid

Stearic acid

Octacosanol

Figure 12.8: Fatty acids from *C. axillaris*.

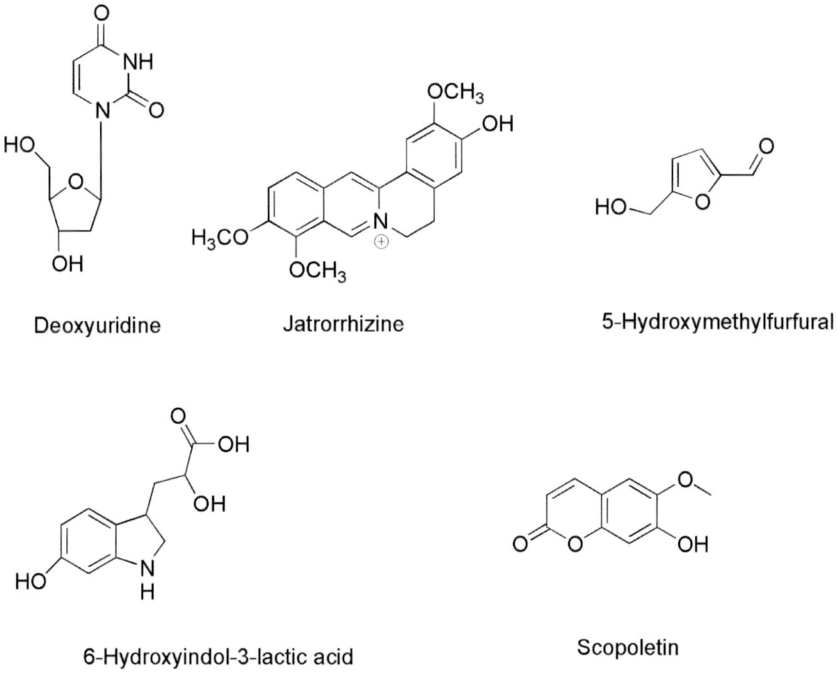

Deoxyuridine

Jatrorrhizine

5-Hydroxymethylfurfural

6-Hydroxyindol-3-lactic acid

Scopoletin

Figure 12.9: Miscellaneous class of compounds from *C. axillaris*.

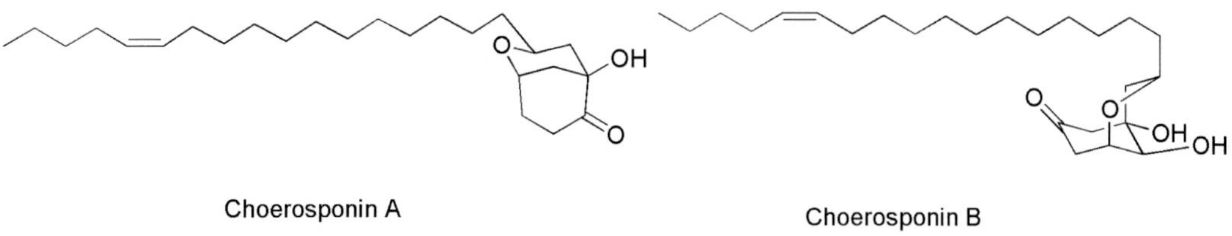

Choerosponin A

Choerosponin B

Figure 12.10: New alkenyled bridged-ring ketones from *C. axillaris*.

Several properties, such as treatment of myocardial ischemia, calming nerves, ameliorating blood circulation, and improving microcirculation, have been reported for *C. axillaris*. *C. axillaris* is one of the commonly used medicinal materials in Mongolian medicine with the effects of improving *Qi* and blood circulation, nourishing heart, and tranquilization, and has the functions of treating the stagnation of *Qi* and blood stasis, the obstruction of *Qi* in the chest with pain, the shortness of breath, feeling uneasy, and so on. According to preliminary statistics, 101 kinds of medicines for the oral administration containing *C. axillaris* were recorded based on Inner Mongolia Medicine Standard (Liu et al. 2013, Shi et al. 1985, Wang and Yang 2004). The medicinal importance of *C. axillaris* has been mentioned by Labh and Shakya (2016d). The bark has medicinal value for treating secondary burns. The mean healing time was significantly shorter for patients treated with *C. axillaris* compared to patients treated with saline gauze, and the number of wound infections was significantly lower in the *C. axillaris* group (Quang 1994, Nguyen et al. 1996).

Pharmacology

Various reports provide evidence for cardiovascular protective effects of *C. axillaris*, which have been demonstrated using *in vitro* and *in vivo* assays (Tang et al. 2009). These cardiovascular protective effects have been shown using such parameters as anti-arrhythmatic, hypoxic tolerance, myocardial ischemia protection, antioxidative and immune function (Table 12.3). Different other activities are also coupled with cardiovascular protective property, which is described briefly in the following headings.

Cardioprotective Properties

Myocardial infarction (MI) remains a major cause of morbidity and mortality worldwide. Preventing or even reversing myocardial fibrosis has been a key goal in the prevention and treatment of severe cardiovascular events. Nuclear factor-kappa B (NF-κB) plays a vital role in a variety of physiological and pathological processes, and has been shown to be activated after MI. Activation of NF-κB induces the activation of the genetic program that leads to the transcription of chemokines, cytokines, and matrix metalloproteinases (MMPs), and further promotes inflammatory and fibrotic response that participate in the progression of ventricular remodeling. Total flavonoid of *Choerospondias* (TFC) significantly improved cardiac dysfunction, the heart coefficient, and myocardial fibrosis in MI rats. TFC also decreased the levels of tumor necrosis factor (TNF-α) and interleukin 6 (IL-6), but increased interleukin 10 (IL-10) content. Moreover, treatment with TFC protected the heart from chronic MI injury by decreasing the expressions of MMP-2, 9, transforming growth factor (TGF-β1), and phosphor IKBα (p-IKBα). The results suggested that TFC attenuated cardiac dysfunction and myocardial interstitial fibrosis by modulating the nuclear factor-kappa B (NF-κB) signaling pathway. TFC could be a promising candidate for therapies against myocardial fibrosis and progression of initiate myocardial injury and dysfunction (Sun et al. 2015a).

Acute myocardial infarction (AMI) is the sharp decline in the coronary artery or the interruption of blood supply to a part of the heart, resulting in heart cells to die. The resulting ischemia (restriction in blood supply) and ensuing oxygen shortage induce myocardium infarction or heart damage, and can induce a cascade of events that will paradoxically produce additional myocardial cell dysfunction and cell apoptosis. Inhibiting cardiomyocyte apoptosis and oxidative stress could serve as the basis for the potential development of drugs for ischemic heart diseases. TFC could protect the heart from ischemia/reperfusion (I/R) injury by increasing the levels of catalase, glutathione peroxidase, and superoxide dismutase in hearthomogenate and decreasing that of malondialdehyde level. These beneficial effects were associated with the decrease in TUNEL-positive nuclear staining, Bax and caspase-3 levels, and the increase in Bcl-2 expression and decreased activation of p38 mitogen-activated protein kinase (MAPK) and Jun N-terminal kinase. TFC improved ischemia/reperfusion-induced myocardium impairment via antioxidative and anti-apoptotic activities, and these beneficial effects were intervened by MAPK signaling pathway (Li et al. 2014b, Zhou et al. 1994).

Total flavonoids are considered to be the main active constituents responsible for the pharmacological actions of *C. axillaris*. The content of total flavonoids in *C. axillaris* is comparatively low, whereas the contents of total organic acids are in significant amounts, up to 8.13 percent. Thus, protective effects of two organic acids, citric acid, and L-malic acid, which are the main components of *C. axillaris*, were investigated on myocardial ischemia/reperfusion injury and the underlying mechanisms. The *in vivo* results showed that citric acid and L-malic acid have protective effects on myocardial ischemia/reperfusion injury by anti-inflammatory, antiplatelet aggregation, and direct cardiomyocyte protective effects. *In vitro* experiments revealed that both citric acid and L-malic acid significantly reduced LDH release, decreased apoptotic rate, downregulated the expression of cleaved caspase-3, and upregulated the expression of phosphorylated Akt in primary neonatal rat cardiomyocytes subjected to hypoxia/reoxygenation injury. These results suggest that both citric acid and L-malic acid have protective effects on myocardial ischemia/reperfusion injury (Tang et al. 2013).

In an experiment, the myocardial ischemia was induced by isoproterenol in the rat. The prophylactically intragastrical administration of TFC at the dose of 200 mg/kg body weight effectively suppressed the variation of J points in electrocardiogram and inhibited the upregulated serum level of creatine kinase, creatine kinase-MB, and lactate dehydrogenase in myocardial ischemia, revealing its cardioprotective effect. Part of its cardioprotective mechanism may relate to the induction of TGF-β1 to competitively inhibit NF-κB signaling pathway. TGF-β1, a multifunctional polypeptide, is believed to influence cardiac development and function. Preventive treatment with TFC significantly increased TGF-β1, TβRI, and TβRII mRNA. Prophylactically exogenous administration with the *C. axillaris* component may serve as a novel therapeutic strategy for ischemic cardiovascular diseases (Ao et al. 2007, Li 1985, Dai et al. 1992).

TFC can protect the heart and liver by maintaining the integrity of the structure and function of cytomembrane, scavenging free radicals, enhancing activity of antioxidation enzyme, inhibiting lipid peroxidation. By producing free radicals from adriamycin, it injures the rat's heart and liver *in vivo*. The activity of antioxidant enzymes superoxide dismutase (SOD), phospholipid hydroperoxide glutathione peroxidase (GSH-Px) in rat cardiac muscle and liver is lower in adriamycin groups than the control group, and content of malonaldehyde (MDA) is higher. After added TFC in the system, activity of SOD, GSH-Px goes up, and content of MDA falls (Bagenna et al. 2002).

In another experiment, adriamycin-induced rat cardiac peroxidation, activity of lactate dehydrogenase (LDH), aspartate aminotransferase (AST), creatine kinase (CK) in serum, and content of MDA in cardiac muscle decreased, activities of SOD and GSH-Px in cardiac muscle increased, and activities of LDH and content of MDA in cultured fluid decreased. These results showed that TFC has a cardiac muscle protecting effect (Zhang et al. 2001).

TFC may improve heart function in hypoxic conditions. The heart rate and the amplitude of cardiac contraction and electrocardiogram were measured by polygraph in the isolated perfused rat heart of Langendorff. It decreases the heart rate and the amplitude of cardiac contraction caused by perfusing with a hypoxic solution (Yang et al. 2000).

Antiarrhythmic

Lapsi fruit is commonly used for the treatment of cardiovascular diseases in Vietnam, Mongolia, and China. The maintenance of balanced ion channels in cardiac myocytes is essential for normal cardiac functions. If the balance among ion channels is disturbed under pathological conditions, consequently induced arrhythmia develops. Drugs that acted to restore normal balance in ion channels produced an effective antiarrhythmic effect. The arrhythmogenic effects of aconitine include various ventricular rhythm disorders. Total flavones derived from *C. axillaris* folium produced antiarrhythmic effects using a rat model of aconitine-induced arrhythmia. With respect to hemodynamics, high-dose TFC were effective in reducing heart rate (HR) without associated changes in blood pressure (BP) in all groups. TFC decreased left ventricular systolic pressure and maximal velocity rate of ventricular pressure with no marked effect on left ventricular end-diastolic pressure. It is worth noting that TFC produced actions equivalent to those of verapamil, a standard therapeutic drug used currently (Qiu et al. 2016). The alcoholic extract has inhibitive effects in arrhythmia induced by aconitine, ouabain, and myocardial ischemia (Zhang et al. 2013).

Table 12.3: Pharmacological activities of *C. axillaris.*

Plant parts/ constituents	Activities	Assays	Results	References
	Carcioprotective properties	Blood samples were collected to determine tumor necrosis factor-α (TNF-α) and interleukin 6, 10 (IL-6, IL-10) levels. Expressions of matrix metalloproteinases-2, 9, phosphor-IKBα (p-IKBα) and transforming growth factor-β1 (TGF-β1) were assayed by Western blot	Block of NF-κB signaling pathway, resulting in the inhibition of MMPs levels, and p-IKBα and TGF-β1 express ions	Sun et al. 2015a
Total flavonoid		AMI rat model was used to assess the cardioprotective effects of TFC on hemodynamics and histopathological changes, and focused on the correlate to oxidative damage and cell apoptosis	Increase the levels of catalase, glutathione peroxidase and superoxide dismutase in heart homogenate, and decrease that of malondialdehyde level. Decrease in TUNEL-positive nuclear staining, Bax and caspase-3 levels, and the increase in Bcl-2 expression decreased activation of p38 mitogen-activated protein kinase (MAPK) and Jun N-terminal kinase	Li et al 2014b
Malic and citric acid		*In vivo* and *in vitro* experiments rat model of myocardial ischemia/reperfusion injury	Significantly reduced myocardial infarct size, serum levels of TNF-α, and inhibited ADP-induced platelet aggregation Significantly reduced LDH release, decreased apoptotic rate, downregulated the expression of cleaved caspase-3, and upregulated the expression of phosphorylated Akt in primary neonatal rat cardiomyocytes subjected to hypoxia/reoxygenation injury	Tang et al. 2013
TFC		Effect on myocardial ischemia induced by isoproterenol in rat	Cardioprotective effect during ischemia injury by inhibiting the variation of J joint in ECG and in preventing the increase in serum CK, CK-MB and LDH level.	Ao et al. 2007
		Reverse transcriptase-polymerase chain reaction (RT-PCR) methods for expression of transforming growth factor β1	Significantly increased TGF-β1, TβRIand TβRII mRNA levels,	Bagenna et al. 2002
TFC		Adriamycin-induced peroxidation model on rat	Activity of SOD, GSH-PX goes up, and content of MDA falls gradually	
		Adriamycin-induced peroxidation model on rat	Activity of LDH, AST, CK in serum and content of MDA in cardiac muscle decreased; activities of SOD and GSH-Px in cardiac muscle increased; activities of LDH and content of MDA in cultured fluid decreased	Zhang et al. 2001
TFC		Measured by polygraph in the isolated perfused rat heart of Langendorff	Decrease of the heart rate and the amplitude of cardiac contraction caused by perfusing with a hypoxic solution	Yang et al. 2000

Total flavonoid	Antiarrhythmic actions		
	Antiarrhythmic effects using a rat model of aconitine-induced arrhythmia	Inhibition of ventricular contraction without altering ventricular diastolic function	Qiu et al. 2016
		Inhibitive effects in arrhythmia induced by aconitine, ouabain and myocardial ischemia	Zhang et al. 2013
	Resisted the occurrence of arrhythmias induced by aconitine, ouabain and ligation of coronary artery	Altered the time of ventricular ectopic beats, ventricular tachycardia, ventricular fibrillation and heart arrest	Yang et al. 2008
	Langendorff perfuse applied on the isolated cardiac function	Slowed the heart rate and don't affect the myocardial contraction	Wang et al. 2005b
		Resisted the occurrence of arrhythmias induced by aconitine	Wang et al. 2005a
		Counteracted significantly the arrhythmia caused by perfusing with an anoxic solution	Xu et al. 2001
		Counteracted the atrial fibrillation induced by I.V. $CaCl_2$-ACh in mice	Li et al. 1984
		Arrhythmia induced by I.V. aconitine in anesthetized rats was markedly retarded	
		Elevated the doses of I.V. ouabain to induce VP, VT, VF, and HS in anesthetized guinea pigs	
		Adrenaline induced arrhythmia in conscious rabbits was reduced	
		Arrhythmia induced by I.V. $BaCl_2$ 13 mg/kg in rats was immediately recovered to a normal sinus rhythm	
	L-type Ca^{2+} current and transient outward K^+ were recorded by pach-clamp whole cell recording technique	L-type Ca^{2+} current had not markedly changed transient outward K^+ was markedly inhibited	Yang et al 2004
	Intracellular free Ca^{2+} was measured by calcium fluorescent probe Fluo-3/AM and laser confocal microscope	Intracellular free Ca^{2+} of resting phase and systole phase was decreased	

Table 12.3 contd. ...

...Table 12.3 contd.

Plant parts/ constituents	Activities	Assays	Results	References
Proanthocyanin	Anticancer activity	Caco-2 cell viability test	Induced morphological changes of Caco-2 cells in a dose-dependent manner	Li et al. 2018
		Inhibition of HepG2 and Caco-2 cancer cell proliferation	Reduced in number and appeared to be less dense. Exhibited the cell shrinkage, aggregation and partial detachment	Li et al. 2016a
Total flavonoid		Inhibit angiotensin II-induced proliferation of cardiac fibroblasts	Proliferation of TFC-treated fibroblasts was significantly less

Inhibitory effects were partly blocked by pretreatment with NG-nitro-L-arginine methyl ester (L-NAME) and 1H-[1,2,4]-oxadiazole-[4,3-a]-quinoxalin-1-one (ODQ) | Lei et al. 2015 |
			Inhibited collagen synthesis induced by angiotensin II in cardiac fibroblasts	Yang et al. 2012
Gambiriin A3 and gambiriin A1		Proliferation-inhibiting effect on K562 were evaluated by the MTT method	Inflated cell membranes and cell content leakage	Li and Cui. 2014 Li et al. 2009a
Pinocembrin, Naringenin		Inhibited the proliferation of human cancer HCT-15 and HeLa cells	Inhibited the cell cycle of tsFT210 cells at the G_2/M phase	Li et al. 2005
Proanthocyanidins		Antiangiogenic effects using HUVECs *in vitro*, and zebrafish embryo angiogenesis model *in vivo*	Attenuated the phosphorylation of Akt, ERK, and p38MAPK dose-dependently in endothelial cells from human umbilical veins	Li et al. 2016c
Choerosponins A and B		Flow cytometry and SRB methods were employed against tsFT210, HCT-15, HeLa, A2780 and MCF-7 cell lines	Showed strong cytotoxicity	Li et al. 2017
		In vitro differentiation of human umbilical cord blood stem cells	Promote *in vitro* proliferation of hUCBSCs	Sa et al. 2010

Flavan-3-ol monomers, procyanidins	Antioxidant activity	DPPH scavenging activity phosphomolybdenum assay FRAP assay	Antioxidant activity of peel was significantly higher than that of flesh	Li et al. 2016b
		DPPH scavenging activity	Ethanolic extract scavenged more than aqueous extract	Labh et al. 2015
Proanthocyanins		DPPH, ABTS radical scavenging activity, ferric-reducing antioxidant power, and phosphomolybdate assay Cellular antioxidant activity PBS wash protocol/no PBS wash protocol	Positive correlation existed between activity and the total phenolics contents	Li et al. 2016b
			In cellular-based method, activity increased as their molecular weight decreased	
		DPPH assay	Significant antioxidant activity	Chalise et al. 2010
		Tested *in vitro* by using pyrogallol autoxidation method, potassium ferricyanide reduction method and Fenton method	Crude extracts had strong capacity to scavenge both ·OH and O^{-2}, while the ion exchange chromatography extracts had strong scavenging capability of ·OH, but weak capacity to scavenge O^{-2}.	Di et al. 2010
		Scavenging effects on active oxygen species	Scavenge active oxygen efficiently	Wang et al. 2008b
		In vivo D-galactose induced mouse aging model	Inhibited D-galactose induced oxidative damage	Wang et al. 2008a
		Scavenging effects on the superoxide anions	High antioxidant effect	
		RBC's oxidate injury	TFFC inhibited RBC'S self oxidation and cultivated oxidation, protected hemoglobin, inhibited the formation of LPO and green pigments	Wu et al. 2002

Table 12.3 contd. ...

...Table 12.3 contd.

Plant parts/constituents	Activities	Assays	Results	References
	Immunological Effects	Carbon clearance method, cutaneous delayed hypersensitivity reaction method, serum hemolysin method, and index of immune organs	Enhance the phagocytic function of mononuclear macrophage and the cutaneous delayed hypersensitivity reaction of mice, and increase the content of hemolysin antibody and the thymus index in mice	Liu et al. 2013 Hou et al. 1998
		Immunity and survival of juvenile tilapia (Oreochromis niloticus	Significant improvement offeed conversion ratio, protein digestibility and energy retention in tilapia. Decreased AST and ALT activity in both liver and muscle activity	Labh et al. 2017
		Respiratory burst activity	Elevated phagocytic	
		Survival, growth and protein profile of common carp *Cyprinus carpio* fingerlings	Increased in weight gain, SGR, total protein	Labh and Shakya 2016a
		Survival, growth and hepatic enzyme activities in *Cyprinus carpio* fingerlings	Significant decreasing trend in SGOT, SGPT and ALP	Labh and Shakya 2016b
		Effect of higher levels of dietary vitamin C on growth and protein levels in the brain and liver of common carp	Higher weight gain and specific growth rate Concentrations of vitamin C was found higher in liver as compared to brain.	Labh and Shakya 2016c
		Immunologic function and sports endurance of mice	Significantly differences in antihypoxic effect, weight-carrying swimming time, biochemical index of blood serum and immunologic index for experimental mouse	Deng and Ji 2002
Total flavonoid		Cells of apoptosis were observed with microscope for counting the proportion of apoptosis Adenine deaminase defficiency activity was detected with spectrophotometer	Inhibit dexamethason induced thymocyte apoptosis TFC also could facilitate the restoration of ADA activation	Li et al. 1998
			TFC strengthened functions of cellular immunity and humoral immunity in normal mice as well as in immunodepressed mice induced by cyclophosphamide	Wang et al. 1991

gambiriin A1 (+)-catechin (6'-8) (+)-catechin	Antibacterial	Agar diffusion test	Inhibit Gram-positive *S. aureus* and *B. subtilis* bacteria and Gram-negative *E. coli* bacteria	Li et al. 2016a
		Paper disc method	Inhibited the growth of *Staphylococcus aureus* ATCC6538	Li and Cui 2014
			Inhibition effects on gram positive bacteria than Gram-negative bacteria.	Xu et al. 2013
Proanthocyanins	Antidiabetic activity	Inhibitory effects on α-amylase and α-glucosidase	Inhibition of α-amylase and α-glucosidase by PAs occurred in a dose-dependent manner	Li et al 2015a
Flavonoids	Anti-hypoxia activities	MTT assay	Cell viabilities of ECV304 cells and PC12 cells increased	Li and Cui. 2014 Li et al. 2009a
Galloyal derivatives		MTT Assay	Protective effects on anoxia-induced injury in cultured ECV304 and PC12 cells	Li et al. 2014a
	Protection of injured neuron	Detected by MTT method	Improve the activity of injured neuron decrease the extraction of lactate dehydrogenase (LDH) released by injured neuron	Guo et al. 2007
	Treatment of renal calculus	Determination of concentration of calcium and oxalate in mice fed with 1% ethylene alcohol and 1α (OH)VitD3	Inhibit renal calculus	Yang et al. 2010
	Antiviral effect	Anti-Coxsackievirus B_3 effects by Cell-based viral myocardial model	Significantly inhibit CVB_3 viral reproduction, decrease release of LDH and CK-MB and suppress secretion of TNF-α from cardiomyocytes infected by CVB_3	Liu et al. 2007
		Anti-herpes simplex virus activities by plaque reduction assay using Vero cells	Exhibited appreciable inhibitory activities against HSV-1	Jo et al. 2005

Aconitine was used to induce arrhythmia in the rats, ouabain was used to induce arrhythmia in guinea pigs, and arrhythmia was induced by ligation of coronary artery in the rats. *C. axillaris* extract significantly prolonged the lasting time of ventricular ectopic beats, ventricular tachycardia, ventricular fibrillation, and heart arrest induced by using aconitine in the rats. Similarly, the extract significantly prolonged the lasting time of arrhythmias induced by using ouabain in guinea pigs. The extract greatly reduced ectopic beats, delayed the beginning time of arrhythmias, shortened the persisting time of ventricular tachycardia, and decreased the frequency of ventricular fibrillation in the rats of ligation of coronary artery (Yang et al. 2008).

Langendorff perfuse is a predominant *in vitro* technique used for examination of cardiac contractile strength and heart rate. Langendorff perfuse was applied in the experiment to observe the influence of the three flavone ingredients in *C. axillaris* on the isolated cardiac function, and the flavone ingredient slowed the heart rate and didn't affect the myocardial contraction. The flavones resisted the occurrence of arrhythmias induced by aconitine by concentration-dependent ways (Wang et al. 2005b).

TFC significantly counteracted the arrhythmia caused by perfusing with an anoxic solution. TFC markedly prolonged arrhythmia appearance time, and evidently decreased the frequency of arrhythmia and cardiac arrest. TFC also markedly increased ventricular fibrillation threshold, and showed a good dose-effect and time-effect relation (Xu et al. 2001).

Total flavones have anti-arrhythmic properties. Arrhythmia induced by $CaCl_2$-ACh (acetylcholine), ovabain, adrenaline, $BaCl_2$ and aconitine was markedly retarded. The action of intravenous (I.V.) TF against arrhythmia was more remarkable than that of diphenylhydantoin, propranolol, and lidocaine (Li et al. 1984). The actions of anti-arrhythmia and anti-myocardia-ischemia are concerned with the L-type Ca 2^+ current, transient outward K^+ and intracellular free Ca of ventricular myocytes in rats. TFC could markedly prolongate the action potential phase because of the reduction of transient outward K^+, but TFC has no effect on L-type Ca 2^+ current. TFC also could markedly decrease intracellular free Ca 2^+ of resting phase and systole phase in myocardia cells (Yang et al. 2004).

Anticancer Activity

C. axillaris peel is a potential source of natural chemopreventive agents for the treatment of cancer. When proanthocyanin (PAs) fractions isolated from *C. axillaris* fruit peels with a different mean degree of polymerization (mDP) were investigated for antiproliferative effects on Caco-2 cells, the results indicated that proanthocyanidin fractions induced dose and time-dependent reductions of Caco-2 cell viability. There was a positive correlation between the degree of polymerization and galloylation of the PAs and their antiproliferative activity *in vitro*. The observed reduction in Caco-2 cell viability was due to apoptosis via the activation of caspase-9, caspase-3, caspase-8, and the elevation of intracellular ROS generation (Li et al. 2018). Both peel and flesh polyphenolic extracts from *C. axillaris* fruit possessed antiproliferative properties on cancer cells, with the effects of peel being superior to those of flesh. Phenolic extracts inhibited the growth of HepG2 and Caco-2 cells in a dose- and time-dependent manner (Li et al. 2016a).

The proliferation of cardiac fibroblasts and the accumulation of excessive amounts of proteins in the extracellular matrix are the basic pathologic processes of myocardial fibrosis. Furthermore, Angiotensin (Ang II) activates a series of signaling molecules to induce cardiac fibrosis. Total flavonoid of *C. axillaris* inhibited angiotensin II-induced proliferation of cardiac fibroblasts via a mechanism that probably involves activation of the NO-cyclic guanosine monophosphate signaling pathway (Lei et al. 2015, Bao et al. 2014). Collagen types I and III are the major fibrillar collagens that comprise approximately 80% and 10% of the extracellular matrix. TFC inhibited collagen synthesis in cardiac fibroblasts in a dose-dependent manner, and the inhibitory effects were blocked by pretreatment with NG-nitro-L-arginine methyl ester (L-NAME) and 1H-[1,2,4]-oxadiazole-[4,3-a]-quinoxalin-1-one (ODQ). The inhibitory effect might associate with the activation of the NO/cGMP signaling pathway (Yang et al. 2012). The proliferation-inhibiting effect on human myeloid leukemia (K562) cells of compounds was detected by an MTT assay. Compounds gambiriin A3 and gambiriin A1 had apparent cytotoxicity on K562 cells. The morphology of the cells treated with compounds showed inflated cell membranes and cell content leakage, and thus had apparent cytotoxicity on K562 cells (Li and Cui 2014, Li et al. 2014a).

The production of new capillaries (angiogenesis) is critical for tumor growth and metastasis. Angiogenesis involves a number of highly coordinated processes, such as endothelial cell proliferation, migration, tubule formation, and remodeling. The coordinated activation of numerous signaling pathways is necessary during angiogenesis, such as the PI3K/Akt and MAPK/ERK/p38MAPK signaling pathways.

The antiangiogenic effect of PAs extract was demonstrated by the inhibition of migration of human umbilical vein endothelial cells (HUVECs) and tube formation. The origin of inhibition was attributed to a reduction in ROS production and to inhibition of the activation of MAPK/ERK/ p38MAPK and PI3K/Akt signaling pathways. The antiangiogenic effect of PAs was also seen in an *in vivo* zebrafish model (Li et al. 2016c).

Human umbilical cord blood stem cells have been widely used in the study of spinal cord injury. In Mongalian medicine, *C. axillaris* fruit extract could promote *in vitro* proliferation of human umbilical cord blood stem cells (Sa et al. 2010).

Flow cytometry and sulforhodamine B (SRB) methods were employed to evaluate the antitumor activity of the compounds. In SRB assay, the compounds inhibited the proliferation of human cancer HCT-15 and HeLa cells. Flow cytometric analysis indicated that compounds pinocembrin and naringenin slightly inhibited the cell cycle of tsFT210 cells at the G2/M phase at higher concentrations, while dibutyl phthalate showed strong cytotoxicity at a higher concentration, but at a lower concentration, inhibited the cell cycle at the G0/G1 phase (Li et al. 2005).

Antioxidant

The oxidative damage caused by reactive oxygen species (ROS), such as the superoxide radical and hydroxyl radicals, on lipids, proteins, and nucleic acids may trigger various diseases, including cardiovascular disease. Epidemiological studies have shown that the administration of antioxidants may decrease the probability of cardiovascular diseases. *C. axillaris* fruit extracts have excellent antioxidant activity.

PAs fractions from *C. axillaris* peels were obtained by solvent extraction and further fractionated on size-exclusion gel column chromatography. The total phenolics contents of the five collected fractions were determined, and their antioxidant activities were evaluated by both chemical-based methods (DPPH, ABTS, FRAP, PM) and a cellular antioxidant assay. Each fraction exhibited potent antioxidant activity in a dose-dependent manner as determined by chemical-based methods. A significant positive correlation existed between the total phenolics contents of all fractions and reducing powder. However, the antioxidant activity of the PAs determined by the cellular-based method increased as their molecular weight decreased, which was not consistent with the results observed in the chemical-based methods (Li et al. 2016b). When the antioxidant activity was evaluated by DPPH radical scavenging, phosphomolybdenum assay, and FRAP assay, the peel had higher radical-scavenging activity than the flesh. UPLC/ESI-QTOF-MS analysis revealed that flavan-3-ol monomer and oligomeric procyanidins were the most abundant compounds in peel and flesh. The peel has more flavan-3-ol monomers and procyanidins than the flesh. Thus, the antioxidant potential was dependent on the contents of flavan-3-nol monomers and procyanidins, and there is a dose-dependent relation between antioxidant effect and the contents of flavan-3-ol monomers and procyanidins (Li et al. 2016b). During another antioxidant study, the ethanolic extract has shown greater DPPH antioxidant activity than aqueous extract (Labh et al. 2015). The antioxidant activities of each extract from Mongolian medicine *C. axillaris* fruit were tested *in vitro* by using pyrogallol autoxidation method, potassium ferricyanide reduction method, and Fenton method. The crude extracts and ion exchange chromatography extracts exhibited antioxidant activity by different mechanisms (Di et al. 2010). Among the 15 edible wild fruits of Nepal studied by Chalise, the antioxidant activities have shown a direct correlation with the total phenolic content (TPC). Antioxidants and polyphenols rich fruits are used as a part of their culture and *C. axillaries* also showed significant antioxidant activity (Chalise et al. 2010). Scavenging effects of *Choerospondias* on active oxygen species was studied by Wang et al., and they found that *Choerospondias* could scavenge active oxygen efficiently (Wang et al. 2008b).

Rodents injected with D-galactose display symptoms that resemble accelerated aging. The long-term administration of galactose induced changes in these redox-related biomarkers in mice, including decrease in SOD, GSH-Px activities, and GSH levels, as well as increase of the MDA level. *C. axillaries* could increase the activity of SOD and decrease the level of MDA. Thus, the intragastric administration of the extract inhibited D-galactose induced oxidative damage. For its *in vitro* scavenging effects on the superoxide anions, DPPH, H_2O_2, OH⁻, the reducing power, and Fe^{2+-} chelating ability, as well as the inhibition of lipid peroxidation was also evaluated, and showed a high antioxidant effect (Wang et al. 2008a).

During RBC's oxidate injury study, TFC inhibited RBC's self oxidation and cultivated oxidation, protected hemoglobin, and inhibited the formation of lipid peroxidation (LPO) and green pigments (Wu et al. 2002).

Immunological Effects

In the development of aquaculture, the main concern is the control of infectious diseases and maintenance of the health of cultured fish. Immune-protection by dietary manipulation has emerged as an important area of research. The feeding trial with different concentration of lapsi fruit extract (LFE) in fingerlings of *O. niloticus,* resulted in improvement in the growth, heamato-immunological responses, and protected the animals against *Aeromonashydrophila* infection at 0.2% LFE, while a higher dose of LFE incorporation led to stress and immunosuppression (Labh et al. 2017). When experiments were conducted to study the effects of lapsi on some hematological parameters of common carp *Cyprinuscarpio* fingerlings, a minimum amount 0.4 g kg⁻1 of lapsi fruit extracts in fish feeds elicited a higher increase in hematological parameters of common carp (Labh and Shakya 2016a). In an experiment, carp were fed with basal diet containing 40% protein supplemented with ethanol extract of lapsi fruit at 0, 0.1, 0.2, 0.4, 0.8, and 1.6 g kg⁻¹ at the rate of 3% of their body weight twice daily for 70 days. The minimum amount of 0.4 g lapsi fruit extracts per kg is sufficient to be added to a diet for good serum enzyme levels and growth performances of common carp (Labh and Shakya 2016b). To examine the effect of higher levels of dietary vitamin C on growth and protein levels in the brain and liver of common carp, *Cyprinuscarpio* was supplemented in the diets through lapsi fruits. Fish fed with a diet supplemented with lapsi fruits showed higher weight gain and specific growth rate (Labh and Shakya 2016c).

The study of the mononuclear phagocyte system, thymus, and serum hemolysin gives information about the immunity system. The mononuclear phagocyte system has phagocytosis and bactericidal action and antitumor effects. The thymus is the central immune organ for the differentiation and maturation of T lymphocytes. Serum hemolysin is a sensitive marker to reflect and test the humoral immune function. TFC may influence cellular immunity and humoralimmunity by enhancing the phagocytic function of mononuclear macrophage. TFC may enhance cellular immunity and increase the thymus weight. TFC increased the content of serum hemolysin in normal mice and the antibody titer induced by contact again antigen, indicating that TFC could enhance humoral immunity, relating to IgM and IgG. Thus, TFC could improve the celiac macrophage activity and specific immunity of mice (Liu et al. 2013). *C. axillaries* could enhance cellular, humoral immune function, and sports endurance of mice (Deng and Ji 2002).

Dexamethason induces immunodeficiency in the patients who take it for a long time due to thymus atrophy, and as a result, thymocyte apoptosis as well as adenine deaminase deficiency activity is lowered. TFC can promote the immune responses of the body, which provides powerful evidence for the treatment of immunodeficiency. TFC could inhibit dexamethason induced thymocyte apoptosis and promote proliferation differentiation of the thymocytes in different periods. TFC could also facilitate the restoration of adenine deaminase activation of the thymocytes in the thymus atrophy mice (Li et al. 1998).

TFC markedly strengthened functions of cellular immunity and humoral immunity in normal mice, as well as in immune depressed mice induced by cyclophosphamide. TFC caused a significant increase of the weights of spleen and thymus and the production of serum hemolysin in normal and immune depressed mice. TFC elevated normal titer of antibody induced by secondary antigen stimulation, and increased alfa-naphthyl acetate esterase ANAE (+) cell percentage of lymphocyte and phagocytic activity of macrophages of the abdominal cavity in normal and immune depressed mice (Wang et al. 1991).

Antibacterial

In the agar diffusion test, peel phenols (PP) exhibited a higher antibacterial potential than flesh phenol (FP). PP showed a significantly higher antimicrobial effect against Gram-positive *S. aureus* and *B. subtilis* bacteria, and somewhat weaker against Gram-negative *E. coli* bacteria in a dose-dependent manner. In the case of *S. typhimurium* and *L. monocytogenes*, no inhibition was obtained (Li et al. 2016a). In another experiment done by disc method, gambiriin A1(+)-catechin (6'-8) (+)-catechin isolated from *C. axillaries* inhibited the growth of *Staphylococcus aureus* ATCC6538 (Li and Cui 2014, Li et al. 2014a).

When bacteriostatic activities of the extracts on foodborne pathogen were investigated, the extracts had better inhibition effects on Gram-positive bacteria than Gram-negative bacteria (Xu et al. 2013).

Inhibitory Effects on α-amylase and α-glucosidase

Bioactive proanthocyanins were isolated from the peel of *C. axillaris* fruit, which is a waste product of the food processing industry. Inhibition of enzymes capable of digesting carbohydrates within the human gastrointestinal tract (GIT), such as α-amylase and α-glucosidase, may be an effective therapeutic tool for the prevention or treatment of type 2 diabetes. The inhibition of α-amylase and α-glucosidase by PAs occurred in a dose-dependent manner. Furthermore, there was a close relationship between the ability of the PAs to inhibit α-amylase and α-glucosidase, and their degree of polymerization and galloylation (Li et al. 2015a).

Anti-hypoxia activities

Hypoxia is a deficiency in the amount of oxygen reaching the tissues. Anti-hypoxia activity was tested by the MTT method. Cell viabilities of ECV304 cells and PC12 cells, two cell lines treated with isolated compounds, notably increased, which suggested that the compounds exhibited good anti-hypoxia activities (Li and Cui 2014, Li et al. 2009a). In another similar type of experiment, glycol derivatives showed protective effects on anoxia-induced injury in cultured ECY304 and PC12 cells (Li et al 2014a).

Protection of Injured Neuron

Experiment on rat cultured cortex neuron injury induced by Serum-Free DMEM and effects of astrocytes culture stimulated showed that *C. axillaris* significantly improves the activity of injured neurons detected by MTT method, and decreases the extraction of lactatedehydrogenase (LDH) released by injured neuron (Guo et al. 2007).

Treatment of Renal Calculus

Mice were fed with 1% ethylene alcohol and 1α(OH)VitD3 to cause renal calculus. The concentrations of calcium and oxalate acid in urine and kidney were determined in mice to evaluate the effects of *C. axillaris*. *C. axillaris* demonstrated a decrease in the concentration of oxalic acid and calcium in urine and kidney (Yang et al. 2010).

Anti-Coxsackievirus B_3 effects

The cell-based viral myocardial model was established using Coxsackievirus B (CVB_3) to infect cardiomyocytes. The protection effects of TFC on virus-infected Hela cells and cardiomyocytes were detected by MTT assays. TFC could protect Hela cells and cardiomyocytes from CVB_3. TFC at high and middle dosages can significantly inhibit viral reproduction, decrease the release of LDH and CK-MB, and suppress secretion of TNF-α from cardiomyocytes infected by CVB_3 (Liu et al. 2007).

Anti-herpes Simplex Virus Activities

Mice were infected cutaneously with an anti-herpes simplex virus (HSV-1), and the extracts were orally administrated three times daily. When extracts were screened for HSV-1 activity determined by using a plaque reduction assay using Vero cells, methanol extracts of *C. axillaries* showed therapeutic effects (Jo et al. 2005).

Conclusion

C. axillaris is a potential agroforestry tree species for domestication for a human nutrient supplementation and income generation. There is an incredible market opportunity for processed lapsi products, and further research should be carried out to increase the efficiency of processing and domesticating techniques based on local knowledge and skills. The multiple health benefits of *C axillaris* should be enough to tap its potential benefits. Scientific research demonstrated by usage *in vitro* and *in vivo* provides evidence for cardioprotective effects and other health-promoting effects, and further research should be done to validate its claimed medicinal properties. Comprehensive information on the chemical compositions and bioactivities of this plant are lacking, and in this situation, a compilation of cumulative research efforts is attempted for knowledge sharing on *C. axillaris*.

References

Ao, J., Feng, H., Xia, F. 2007. Transforming growth factor and nuclear factor Kappa B mediated prophylactic cardioprotection by total flavonoids of *fructusChoroespondiatis* in myocardial ischemia. Cardiovasc. Drugs Ther. 21: 235–241.

Bagenna, Jilinbaiyi, Y.X. Bai, Li, B.S. 2000. Study on extracting technology of total flavone of Mongolian medicine *FructusChoroespondiatis* with orthogonal design. Chin. Tradit. Patent. Med. 22(4): 253–255.

Bagenna, W.Q. Tai, Z.M. Wang, L. Xu, Li, B.S. 2002. Protective function of total flavonoids of *FructusChoerospondiatis* to adriamycin-induced rat heart and liver peroxidation. J. Inner Mongolia Univ. Nat. (Natural Sciences, Quarterly) 17(3): 267–269.

Bajracharya, S. 2015. Lapsi: Nepal's best kept secret ECS Nepal 159 http://ecs.com.np/features/lapsi-nepals-best-kept-secret. Accessed on Sep. 16, 2019.

Bao, J., Jin, M., Yang, Y., Gao, X., Shu, L. Xing, H. 2014. Effects of total flavones of *Choerospondias axillaris* on mRNA and protein expression of type I and type III collagen in rat cardiac fibroblasts. Acta Pharma. Sin. 49: 136–141.

Bhutia, K., Suresh, C., Amar, R., Subba, P. 2011. Nutritional composition of some minor fruits of the Sikkim Himalayas. Proceedings of the International Symposium on Minor Fruits and Medicinal Plants for Health and Ecological Security (ISMF & MP), West Bengal, India. 22(19): 344–346.

Brodie, F., Helmy, O., Brockelman, Y., Maron, L. 2009. Functional differences within a guild of tropical mammalian frugivores. Ecology 90(3): 688–698.

Chalise, J.P., Acharya, K., Gurung, N., Bhusal, R.P., Gurung, R., Basnet, N.S., Basnet, P. 2010. Antioxidant activity and polyphenol content in edible wild fruits from Nepal. Int. J. Food Sci. Nutr. 61(4): 425–32.

Chen, J., Deng, X., Bai, Z., Yang, Q., Chen, G., Liu, Y., Liu, Z. 2001. Fruit characteristics and *Muntiacusmuntijak vaginalis* (Muntjac) visits to individual plants of *Choerospondias axillaris*. Biotropica 33(4): 718.

Chhetri, R., Gauchan, D. 2007. Traditional knowledge on fruit pulp processing of lapsi in Kavrepalanchowk district of Nepal. I JT K6(1): 46–49.

China Pharmacopoeia Committee. 2010. Pharmacopoeia of the People's Republic of China, China Medical Science Press: Beijing, China, Volume I, p. 41.

Chong, X.T., Chen, Z.L., Yao, Q.Q. 2008. Review on chemical constituents and pharmcological activities of *Choerospondias*Burtt et Hill. J. Baotou Med. Coll. Qilu Pharma. Affairs 1.

Dai, H.Y., Li, Q.A., Chen, L.F., Deng, H.W. 1992. Protective effect of extract from *Choerospondias axillaris* fruit on myocardial ischemia of rats. Chin. Tradit. Herb. Drugs 23: 641–643.

Deng, K., Zhang, Y., Wang, P., Yang, Z. 2006. Simultaneous determination of gallic acid and protocatechuic acid in *Choerospondias axillaries* by RP-HPLC. Se Pu 24(6): 652–3.

Deng, G.L., Ji, L.J. 2002. Effect of extract from *Choerospondias axillaries* on immunologic function and sports endurance of mice. Spor. Sci. 23(5): 53–54.

Di, J.J., Zuo, G.L., Wang, Y., Zhang, S.J., Ding, H.M., Wei, Y.C. 2010. Antioxidant activity of extracts from Mongolian medicine axillary *Choerospondias* fruit *in vitro*. J. Anhui Agri. Sci. 34: 19325–19330.

Dingjian, C., Shan, W., Hui, L., Linchun, M. 2009. Fermentation technology of *Choerospondias axillaris* skin vinegar. Chin. Brew. 9.

Fan, H.Y., Song, Y.T., Sai, Y. 2005. Isolation and structure identification of a deoxyuridine from *Choerospondias axillaris* (Roxb.) Burtt et Hill. Chin. J. Nat. Med. 2: 83–85.

Feng, J.G., Xu, Y.T., Chen, Y.T. 1999. Growth performance of eight native broad leaved species on hill country in southwestern Zhejiang. For. Res. Beijing 12(4): 438–441.

Gao, X., Mu, J., Li, Q., Guan, S., Liu, R., Du, Y., Zhang, H., Bi, K. 2017. Comprehensive identification of Guan-Xin-Shu-Tong capsule *via* a mass defect and fragment filtering approach by high resolution mass spectrometry: *in vitro* and *in vivo* study. Molecules 22: 1007.

Gardner, S., Sidisunthorn, P., Anusarnsunthorn, V. 2000. A field guide to forest trees of Northern Thailand. Bangkok. Kobfai Pub, Project.

Gautam, K. 1997. The sweet and sour tale of lapsi domesticating and commercializing *Choerospondias axillaris*. Agrofor. Today 9(3): 13–16.

Gautam, K. 2004. Lapsi (*Choerospondias axillaris*) emerging as a commercial non-timber forest product in the hills of Nepal. Forest Products, Livelihoods, and Conservation: Case Studies of Non-timber Forest Product Systems. Bogor Barat, Indonesia: CIFOR, 2004.

Guo, H., Yao, W.B., Wang, H., Gao, X.D., Zhu, Y., Wang, L.Q. 2007. Effect of *Choerospondias axillaries* fruit on rat cultured cortex neuron injury induced by Serum-Free DMEM. Chin. J. Biochem. Pharma. 2(28): 87.

Gupta, S., Jhala, V.S., Singh, C., Chauhan, N., Bhatt, S.S. 2019. Nutritional importance of underutilized fruits: '*Spondiasaxillaris*and *Eriolobusindica*' of Uttarakhand hills. J. Pharmacogn. Phytochem. 8(1): 335–339.

Hau, C.H., Parrotta, J.A., Turnbull, J.W. 1997. Tree seed predation on degraded hillsides in Hong Kong. For. Ecol. Manag. 99(1/2): 215–221.

Hill, W. 1937. The method of germination of seeds enclosed in a stony endocarp. Ann. Bot.1(2): 239–56.

Hou, H.Y., Qing, R., Wang, Y.Z. 1998. Studies on effects of total flavones of *Choerospondias axillaris* on immunological function in mice body liquid. Chin. J. Ethnomed. Ethnopharm. 4(4): 38–39.

Jackson, J.K. 1994. Manual of afforestation in Nepal. Forest Research and Survey Centre, Kathmandu.

Jo, M., Nakamura, N., Kurokawa, M., Komatsu, K., Shiraki, K., Hattori, M. 2005. Anti-herpes simplex virus activities of traditional Chinese medicines, used in Yunnan and Tibetan provinces of China. Med. Pharma. Soc.WAKAN-YAKU 22: 321–328.

João, S.D. 2012. Major classes of phytonutriceuticals in vegetables and health benefits: a review. J. Nutr. Ther. 1: 31–62.

Jun, D.J., Jun, Z.S., Mai, D.H., Chun, W.Y., Yan, X. 2012. Determination of the Vitamin C in *Fructus Choerospondiatis*. J. Inner Mongolia Univ. Nat. (Natural Sciences) 4: 434–436.

Karki, S., Shakya, R., Agrawal, V.P. 2009. A novel class of protease from *Choroespondias axillaris* (Lapsi) leaves. IJLSCI 3: 1–5.

Khabir, M., Khatoon, F., Ansari, W.H. 1987. Kaempferol-5-O-arabinoside a new flavonol glycoside from the leaves of *Choerospondias axillaris*. Indian J. Chem. B 26(1): 85.

Labh, S.N., Shakya, S.R., Kayasta, B.L. 2015. Extract of medicinal lapsi *Choerospondias axillaris* (Roxb.) exhibit antioxidant activities during *in vitro* studies. J. Pharmacogn. Phytochem. 4(3): 194–197.

Labh, S.N., Shakya, S.R. 2016a. Effects of dietary lapsi (*Choerospondias axillaris*Roxb.) on survival, growth and protein profile of common carp (*Cyprinuscarpio* L.) fingerlings. Int. J. Zool. Stud. 1(5): 36–41.

Labh, S.N., Shakya, S.R. 2016b. Effects of lapsi*Choerospondias axillaris* (Roxb.) on survival, growth and hepatic enzyme activities in *Cyprinuscarpio* fingerlings. Int. J. Appl. Res. 2(9): 01–07.

Labh, S.N., Shakya, S.R. 2016c. Fruits of lapsi*Choerospondias axillaris* enhances ascorbic acid level in brain and liver of common carp (*Cyprinuscarpio* L.) during intensive aquaculture. Int. J. Chem. Stud. 4(4): 199–205.

Labh, S.N., Shakya, S.R. 2016d. Medicinal importance of *Choerospondias axillaris* (Roxb.) Burtt& Hill fruits in Nepal. Trop. Plant Res. 3(2): 463.

Labh, S.N., Shakya, S.R., Gupta, S.K., Kumar, N., Kayastha, B.L. 2017. Effects of lapsi fruits (*Choerospondias axillaris* Roxburgh, 1832) on immunity and survival of juvenile tilapia (*Oreochromisniloticus* Linnaeus, 1758) infected with *Aeromonas hydrophila.* Int. J. Fish.Aquat. Stud. 5(2): 571–577.

Lai, X., Guo, C., Xiao, Z. 2014. Trait-mediated seed predation, dispersal and survival among frugivore-dispersed plants in a fragmented subtropical forest, Southwest China. Integr. Zool. 9(3): 246–254.

Lei, J., Yang, Y., Zhu, D., Gao, X., Wang, X., Xing, H., Bao, J., Shu, L. 2015. Anti-proliferative effect of the extract of Guangzao (Fructus Choerospondiatis) on cultured at cardiac fibroblasts. J. Tradit. Chin. Med. 35(6): 685–689.

Li, Z.X. 1985. Effect of the compound of *Choerospondias axillaris* on the tolerance to hypoxia and protection from acute myocardial ischemia in animals. Zhongyao Tong bao 10(3): 42–44.

Li, B., Tian, G., Lin, N. 1998. Effects of total flavones of *Choerospondias axillaris* on thymocyte apoptosis and ADA activation of mice. Zhonghua Weishengwuxue He MianyixueZazhi 18(5): 386–391.

Li, C.W., Cui, C.B., Cai, B., Han, B., Dou, D.Q., Chen, Y.J. 2005. Aromatic chemical constituents of *Choerospondias axillaris* and their *in vitro* antitumor activity. Chin. J. Med. Chem. 15: 138–147.

Li, C.W., Cui, C.B., Cai, B., Yao, Z.W. 2008. The research progress of *Choerospondias axillaris*. Pharma. J. Chin. PLA 24: 231–234.

Li, C.W., Cui, C.B., Cai, B., Han, B., Li, M.M., Fan, M. 2009a. Flavanoidal constituents of *Choerospondias axillaris* and their *in vitro* antitumor and anti-hypoxia activities. Chin. J. Med. Chem. 19(64): 48–51.

Li, C.W., Cui, C.B. 2014. One new and nine known flavonoids from *Choerospondias axillaris* and their *in vitro* antitumor, anti-hypoxia and antibacterial activities. Molecules 19: 21363–21377.

Li, C.W., Cui, C.B., Cai, B., Han, B., Li M.M., Fan, M.J. 2014a. Galloylglucosidic constituents of *Choerospondias axillaries* and their *in vitro* anti-tumor, anti-hypoxia and anti-bacteria activities. J. Int. Pharm. Res. 41(4): 449–455.

Li, C., He, J., Gao, Y., Xing, Y., Hou, J., Tian, J. 2014b. Preventive effect of total flavones of *Choerospondias axillaris* on ischemia/reperfusion-induced myocardial infarction-related MAPK signaling pathway. Cardiovas. Toxicol. 14(2): 145-52.

Li, C.W., Han, B., Cai, B., Cui, C.B. 2017. Choerosponins A and B, two new cytotoxic bridged-ring ketones and the determination of their absolute configurations. Molecules 22: 531.

Li, Q., Chen, J., Li, T., Liu, C., Zhai, Y., McClements, D.J., Liu, J. 2015a. Separation and characterization of polyphenolics from underutilized byproducts of fruit production (*Choerospondias axillaris* peels): inhibitory activity of proanthocyanidins against glycolysis enzymes. Food Funct. 6(12): 3693–701.

Li, Q., Chen, J., Li, T., Liu, C., Wang, X., Dai, T., JulianMcClements, D., Liu, J. 2015b. Impact of *in vitro* simulated digestion on the potential health benefits of proanthocyanidinsfrom *Choerospondias axillaris*peels. Food Res. Int. 78: 378–387.

Li, Q, Chen, J., Li, T., Liu, C., Liu, W., Liu, J. 2016a. Comparison of bioactivities and phenolic composition of *Choerospondias axillaris* peels and fleshes. J. Sci. Food Agric. 96(7): 2462–71.

Li, Q., Wang, X., Chen, J., Liu, C., Li, T., McClements, D.J., Dai, T., Liu, J. 2016b. Antioxidant activity of proanthocyanidins-rich fractions from *Choerospondias axillaris* peels using a combination of chemical-based methods and cellular-based assay. Food Chem. 1(208): 309–17.

Li, Q., Wang, X., Dai, T., Liu, C., Li, T., McClements, D.J., Chen, J., Liu, J. 2016c. Proanthocyanidins, isolated from *Choerospondias axillaris* fruit peels, exhibit potent antioxidant activities *in vitro* and a novel anti-angiogenic property *in vitro* and *in vivo*. J. Agric. Food Chem. 64(18): 3546–56.

Li, Q., Liu, C., Li, T., Mc Clements, D.J., Fu, Y., Liu, J. 2018. Comparison of phytochemical profiles and antiproliferative activities of different proanthocyanidins fractions from *Choerospondias axillaris* fruit peels. Food Res. Int. 113: 298–308.

Li, S.H., Wu, X.J., Zheng, Y., Jiang, C.L. 2009b. Studies on the chemical constituents from the bark of *Choerospondias axillaris.* Zhong Yao Cai 32(10): 1542–4.

Li, Z.M. 2009. The clinical research of Guangzaofufang capsule on Angina. Chin. J. Integ. Med. Cardio/cerebrovascular Dis. 12: 1387–1388.

Li, Z.X., Tian, F.J., Wu, X.Y., Zhang, Y.P., Tian, L., Shi, S. 1984. Anti-arrhythmic action of total flavones of *Choerospondias axillaris*fructus. Acta Pharma. Sin. 5(4): 251–254.

Lian, Z., Zhang, C., Li, C., Zhou, Y. 1984. Studies on chemical constituents of *Choerospondias axillaris*. Zhongguo Yao Li Xue Bao 5(4): 251–4.

Lian, Z., Zhang, C., Li, C., Zhou, Y. 2003. Studies on chemical constituents of *Choerospondias axillaris*. Zhong Yao Cai 26(1): 23–4.

Liang, Y.J., Sang, B., Shen, X.J., Lan, W., Wang, S.X., Zheng, X.H. 2010. Effects of Sandalwood on pharmacokinetics of gallic acid and protocatechuic acid in *Choerospondiatis* fruit AJSMMU 8.

Liga, Qihe, Eerdun, Hasi, Mei, H. 2002. Clinical observation of San-Wei-Guang-Zao capsule on coronary heart disease. J. Med. Pharm. Chin. Minor. 8: 17–18.

Liu, X.G., Chen, Y.S. 2000. Component analysis of *Fructus Chorospondiatis.* Chin. Wild Plant Res. 19(3): 35–40.

Liu, X.L., Liang, B., Zhang, Y., Liu, X.L. 2007. Anti-Coxsackievirus B_3 effects of total flavones of *Fructus Choerospondiatis in vitro.* Chin. J. Hospital Pharm. 12.

Liu, X.Y., Yang, Y.M., Han, F., Zhang, H.N., Du, J.X., Hao, X.M. 2013. Immunological effects of total flavones from leaves of *Choerospondias axillaris* on mice. Chin. Herb Med. 5(2): 121–124.

Lu, J., Hu, Y., Wang, L., Wang, Y., Na, S., Wang, J., Shun, Y., Wang, X., Xue, P., Zhao, P., Su, L. 2018. Understanding the multitarget pharmacological mechanism of the traditional mongolian common herb pair GuangZao-RouDouKou acting on coronary heart disease based on a bioinformatics approach. Evid. Based Complement. Alternat. Med. Article ID 7956503.

Lü, Y.Z., Wang, Y.L., Lou, Z.X., Zu, J.Y., Liang, H.Q., Zhou, Z.L. 1983. The isolation and structural determination of naringenin and choerospondin from the bark of *Choerospondias axillaris*. Yao XueXueBao 18(3): 199–202.

Meng, Q.C. 2012. The effect of Guang-Zao-Qi-Wei-Wan on stable angina pectoris. J. Med. Pharm. Chin. Minor. 5: 1–2.

Ming, N.C., Hai, L.Z., You, Z., Yian, W.S. 1996. Studies on nutritive components of *Choerospondias axillaris* fruit and its opening utilization. Quart. For. By-product Spec. Chin. 3.

Nguyen, D.D., Nguyen, N.H., Nguyen, T.T., Phan, T.S., Nguyen, V.D., Grabe, M., Johansson, R., Lindgren, G., Stjernstrom, N.E., Soderberg, T.A. 1996. The use of a water extract from the bark of *Choerospondias axillaris* in the treatment of second degree burns. Scand. J. Plast. Recons. 30(2): 139–144.

Pandey, Y., Upadhyay, S., Bhatt, S.S., Muddarsu, V.R., Debbarma, N. 2018. Screening of nutritional composition and phytochemical content of underutilized fruits 'Spondias axillaris and Eriolobus indica' of Sikkim Himalayas. Pharm. Innovat. J. 7(4): 1146–1150.

Paudel, K.C., Parajuli, D.P. 1999. Domestication and commercialization of lapsi tree; a potential income source through agroforestry in midil hill of Nepal. Sci. World 116–120.

Paudel, K.C., Pieber, K., Klumpp, R., Laimer, M. 2003b. Evaluation of Lapsi tree (*Choerospondias axillaris*, Roxb.) for fruit production in Nepal. Die Bodenkultur 54(1): 3–9.

Paudel, K., Pieber, K., Klumpp, R., Laimer, M. 2002a. Collection and evaluation of germplasm of lapsi (*Choerospondias axillaris* (Roxb.) B.L. Burtt and A.W. Hill), an indigenous fruit tree of Nepal. Plant Gen. Res. Newslett. 130: 36–46.

Paudel, K., Eder, R., Paar, E., Pieber, K. 2002b. Chemical composition of Lapsi (*Choerospondias axillaris*) fruit from Nepal. Mitteilungen Klosterneuburg Rebe und Wein, Obstbau und Fruchteverwertung 52: 45–53.

Paudel, K. 2003a. Domesticating Lapsi, Choerospondias axillarisRoxb (B. L. Burtt& A. W. Hill) for fruit production in the middle mountain agroforestry system in Nepal. Himal. J. Sci. 1(1): 55–58.

Qiu, M., Dong, Y.H., Han, F., Qin, J.M., Zhang, H.N., Du, J.X., Hao, X.M., Yang, Y.M. 2016. Influence of total flavonoids derived from *Choerospondias axillaris* folium on aconitine-induced antiarrhythmic action and hemodynamics in wistar rats. J. Toxicol. Environ. Heal. A79(19): 878–83.

Quang, C.M. 1994. Phytochemical and pharmacological evaluation of *Choerospondia axillaris*: A vietanamese medicinal plant used to treat burns. Licentiate thesis 17, Monograph, Uppsala University, Sweden.

Rai, K., Sharma, R., Tamang, J. 2005. Food value of common edible wild plants of Sikkim. J. Hill Res. 18(2): 99–103.

Sa, R., Bulin, B., Chen, J.M. 2010. Effects of three kinds of Mongalian medicine axillary *Choerospondias* fruit extracts on differentiation of human umbilical cord blood stem cells *in vitro*. J. Clin. Rehab. Tis. Engin. Res. 14(6): 1078–1081.

Seber, C.W.C. 2016. An analysis of *Choerospondias axillaris* "lapsi" regarding production, product development and its effect on rural lapsi farmers in Nepal. M Sc Thesis submitted to Jordan College of Agricultural Sciences and Technology, California State University.

Shah, D. J.1978. Ascorbic acid (Vitamin C) content of Lapsi-pulp and peel at different stage of maturation. Research Bulletin 2035 B. S. Food Research Section, HMGN, Dept. of Food and Agricultural Marketing Services, Kathmandu.

Shen, X.J., Ge, R.L., Wang, J.H. 2009. Chemical constituents from *Choerospondias axillaris* (Roxb.) Burtt et Hill. J. Henan Univ. (Medical Science) 28: 195–199.

Shi, R.J., Dai, Y., Fang, M.F., Zhao, X., Zheng, J.B., Zheng, X.H. 2007. HPLC-ESI-MS analysis of the water soluble extracts of *Fructus Choerospondiatis*. J. Chin. Med. Mater. 30(3): 294–297.

Shi, S., Li., Z.X., Tian, F.J., Bai, Y.F., Tian, L., Yang, Y.M. 1985. Effect of flavanoid from *Choerospondias axillaries* fruit on left ventricle function and hemodynamics of anaesthesia dog. Inner Mongolia Pharma. J. 2: 14–15.

Shu, N.S.Z. 2008. Choerospondias B. L. Burtt& A. W. Hill, Ann. Bot. (London), ser. 2, 1: 254. 1937. Flor. Chin. 11: 341–342.

Sun, B., Xia, Q., Gao, Z. 2015a. Total flavones of *Choerospondias axillaris* attenuate cardiac dysfunction and myocardial interstitial fibrosis by modulating NF-κB signaling pathway. Cardiovasc. Toxicol. 15(3): 283–9.

Sun, Z., Zhang, L., Fang, Y., Huang, H., Wang, M. 2015b. Study on enzymatic extraction of polyphenol constituents in *Fructus choerospondiatis*. Biomed. Res. 26 (2): 321–327.

Tang, L., Li, G.Y., Yang, B.Y., Kuang, H.X. 2009. Studies on the chemical constituents of *Choerospondiatis Fructus*. Chin. Tradit. Herb Drugs 40(4): 541–543.

Tang, W., Eisenbrand, G. 2013a. *Choerospondias axillaris.* Chinese drugs of plant origin, chemistry, pharmacology and use in traditional and modern medicine. p. 307–309.

Tang, X., Liu, J., Dong, W., Li, P., Li, L., Lin, C., Zheng, Y., Hou, J., Li, D. 2013b. The cardioprotective effects of Citric acid and L-Malic acid on myocardial ischemia/reperfusion injury. Evid. Based Complement. Alternat. Med. Article ID 820695.

Tyystjarvi, P.K. 1981. Silvicultural aspects of community forestry development in the hills of Nepal. Community Forestry Development Project, Kathmandu.

Upadhyay, S.K., Magar, R.T., Thapa, C.J. 2013a. Biochemical characterization of protease isolated from different parts of *Choerospondias axillaris* (Lapsi). Biochem. Anal. Biochem. 2: 135.

Upadhyay, S.K., Magar, R.T., Thapa, C.J. 2013b. Characterization of protease isolated from root of *Choreospondias axillaries* (Lapsi). J. Plant Sci. 1(3): 39–42.

Wang, X.Q., Wang, L.W., Zhao, Y. 2014. Progressinchemical constituents and pharmacological activity of *Choerospondias axillaris* fruit. Food Sci. 35(13): 281–285.

Wang, F.H., Yang, Y.M., Xu, J.H., Qin, J.M., Ying, K., Zhang, C.Z., Song, Y.T., Yu, T.F. 2005a. Comparing the actions of the three flavone ingredients in *Choerospondias axillaris* on arrhythmias induced by aconitine. ZhongguoZhong Yao ZaZhi 30(14): 1096–8.

Wang, F.Y.Y., Xu, J., Qin, J., Ying, K., Zhang, C., Song, Y. 2005b. Influence of the three flavone ingredients in *Choerospondias axillaris* on the isolated cardiac function. J. Med. Pharm. Chin. Minor. 05: 27–29.

Wang, H., Gao, X.D., Zhou, G.C., Cai, L., Yao, W.B. 2008a. *In vitro* and *in vivo* antioxidant activity of aqueous extract from *Choerospondias axillaris* fruit. Food Chem. 106(3): 888–895.

Wang, Y.C., Zhao, B.Y., Tan, C.J., Wang, J.J. 2008b. Scavenging effects of *Choerospondias* on active oxygens and its antioxidation *in vitro*. Stud. Trace Elem. Heal. 5.

Wang, Y.G., A, L.T., Xu, X.T. 2009. The extraction of *Choerospondias axillaris* polysaccharide and the synthesis of *Choerospondias* polysaccharide-Ca(II). Chin. J. Spect. Lab. 5.

Wang, Y., Lü, Y., Lou, Z., Zu, J., Liang, H., Zhou, Z. 1983. Isolation and identification of tangerine and jatrorrhizine from bark of *Choerospondias axillaris*. ActaPhysicaSin. 18: 199–202.

Wang, Y., Ren, J., Wang, R. 1991. Effects of flavones of *Choerospondias axillaris* fructus on immunological function in mice. Chin. Pharma. Bull. 3: 214–217.

Wang, F.H., Yang, Y.M. 2004. Development of study on Mongolia medicine *Choerospondiatis Fructus* and its flavonoids. J. Baotou Med. Coll. 20(3): 258–260.

Wu, R.N., Li, D.L., Na, Y.H. 2002. Inhibiting effects of total flavonoids in *Fructus Choerospondiatis* on RBC's oxidate injury. Li Shi Zhen Med. Mater. Medica Res. 13: 653–654.

Wu, S.S. 2012. Effects of moisture content on the texture and color of *Choerospondias axillaris* pastilles. Food Machin. 6(66): 12–15.

Xiaogeng, X., Yousheng, C. 2000. Analysis of constituents in *Choerospondias axillaris* fruits. Chin. Wild Plant Res. 3.

Xu, J.L., Li, Q., Wang, Z.J., Liu, C.M., Liu, J.Y. 2013. Study on extraction and bacteriostatic activities of flavonoids from *Choerospondias axillaris* Peel. Sci. Tech. Food Indust. 11.

Xu, J., Yang, Y., Qin, J.M., Xu, C., Li, Z. 2001. Antiarrhythmic action of total flavones of *Choerospondias axillaries* fructus in rat heart *in vitro*. J. Med. Pharm. Chin. Minor. 2.

Yang, K., Zeng, C.H., Li, W.Z., Li, Z.X. 2010. Effect of *Choerospondias axillaris* on the urinary calculus in mice. Chin. Tradit. Patent Med. 5: 719–722.

Yang, L.M., Yang, L.J., Jia, P., Lan, W., Zhang, Y.J., Wang, S.X., Zhang, P., Zheng, X.H. 2016. HPLC-Q-TOF-MS/MS-based analysis of chemical constituents in *Choerospondiatis fructus*. J. Second Milit. Med. Univ. 37(2): 159–166.

Yang, Y.M., Qin, J.M., Xu, J.H., Zhou, E.F., Feng, G.Q., Bai, Y.F., Song, Y.T., Li, Z.X. 2004. Effects of total flavones of *Choerospondias axillaris* fructus on L-type Ca^{2+} current, transient outward K^+ and intracellular free Ca^{2+} of ventricular myocytes in rats. Chin. Pharma. Bull. 20(7): 784–788.

Yang, Y., Xu, J., Qin, J. 2000. Effects of total flavones from fruit of *Choerospondias axillaris* of Mongolian drugs on function of isolated heart in rat. J. Baotou Med. Coll. 16(4): 259–62.

Yang, Y., Ying, K., Wang, F., Guan, H., Qin, J.M., Xu, J., Zhang, C., Bai, Y., Song, Y. 2008. Comparing the antiarrhythmia action of the extracts from *Choerospondias axillaris* (Roxb.) Burtt et Hill. J. Baotou Med. Coll. 01: 4–7.

Yang, Y., Gao, X., Wang, X., Su, L., Xing, H. 2012. Total flavonoids of fructus *Choerospondias* inhibits collagen synthesis of cultured rat Cardiac fibroblasts induced by angiotensin II: correlated withno/cgmp signaling pathway. Eur. J. Pharma. Sci. 47(1): 75–83.

Yi, Y., Yang, H., Zhao, Y., Bai, Z. 2010. Extracting flavonoids from *Choerospondias axillaris* by percolation. ZhongguoZhong Yao ZaZhi. 35(14): 1806–8.

Yu, W., Liang, R.H., Li, T., Liu, J.Y., Liu, C.M., Liu, W. 2013. Research on non-enzymatic browning reaction of *Choerospondias axillaris* fruit cake during storage. Sci. Tech. Food Indust. 34: 319–326.

Zhai, Y.X., Chen, J., Li, T., Liu, J.Y., Wang, X.Y., Cheng, C., Liu, C.M. 2015. Determination of mineral elements in *Choerospondias axillaris* and its extractives by ICP-AES.Guang Pu Xue Yu Guang Pu Fen Xi 35(4): 1052–5.

Zhang, H., Yang, Y., Wu, G., Zhang, H., Gao, B. 2013. Separation of antiarrhythmic constituents of *Choerospondias axillaris* (Roxb.) Burtt et Hill. J. Baotou Med. Coll. 29: 1–4.

Zhang, X., Bao, B., Wang, Z., Na, R., Sun, F., Li, B. 2001. Protective effect of total flavonoids of *fructus choerospondiatis* on adriamycin-induced rat cardiac peroxidation. Zhong Yao Cai 24(3): 185–7.

Zhang, Y., Meiyan, L. 2003. Determination of microelements in different kinds of extracts of total flavones of *Choerospondias axillaris* fructus. J. Baotou Med. Coll. 4.

Zhou X., Zhan, Y., Jiang, Y., Hu, J. 1994. Myocardial protective effect of Guangzao-potassium cardiac arrest solution in dogs. Bull. Hunan Med. 19: 398–400.

Zhou, D.M., Zhu, G.Q., Wu, S.Y., Wu, Y.F., Hu, H.J., Ye, C.W. 1997. Preliminary report on tree species selection for the cultivation of *Lentinusedodes*. J. Zhejiang For. Sci. Tech. 17(1): 18–23.

Zhu, L., Chengzhong, Z., Chong, L., Yawei, Z. 2003. Studies on chemical constituents of *Choerospondias axillaris*. J. Chin. Med. Mater. 1.

13

Artemisia Species
Medicinal Values with Potential Therapeutic Uses

Suroowan Shanoo, Jugreet B. Sharmeen and *Mahomoodally M. Fawzi**

Introduction

Artemisia is one of the largest heterogeneous genus of the plant species and an important member of the Asteraceae (Koul et al. 2018). It occurs well in temperate regions of the globe, such as in Asia, Africa, Australia, China, Europe, India, Iran, Japan, North America (Canada, Mexico, and the United States), and Turkey with around 500 known species (Koul et al. 2018, Watson et al. 2002, Liu et al. 2009). The origin of the name *Artemisia* can be traced back from the ancient Greek word "Artemis" which means goddess, and "absinthium" which relates to unenjoyable or without sweetness. Generally, *Artemisia* is known as "Worm wood", as it is traditionally employed to treat intestinal worms. Other common names of the genus include "Mug word", "Sagebrush", or "Tarragon" (Obistioiu et al. 2014, Tajadod et al. 2012).

Most *Artemisia* species occur as annual, herbaceous, ornamental, aromatic, biannual, medicinal, perennial plants or shrubs (Abad et al. 2012). Their colors range from blue-green, dark green, as well as silver green, with a bitter taste and pungent smell, given the biosynthesis of terpenoids and sesquiterpene lactones as part of the plant's metabolism. In general, their morphological characteristics include an alternate leaf with a small capitula and tubular florets, obovoid achenes with an absent pappus, or in some species a small scarious ring occurs (Heywood et al. 1977, Mucciarelli and Maffei 2002).

Traditionally, several species of the Artemisia are cultivated as crops, which are subsequently prepared as tonics, teas, and medicinal potions (Koul et al. 2018). In this advent, the most medicinally employed and scientifically investigated species around the world include *Artemisia absinthium, A. annua, A. afra, A. arborescens, A. capillaris, A. arboratum, A. asiatica, A. douglasiana, A. dracunculus, A. indica, A. japonica, A. judaica, A. tripartite, A. verlotiorum, A. vestita,* and *A. vulgaris* (Bora and Sharma 2011).

Indeed, these species are rich in a panoply of secondary metabolites, such as acetylenes, caffeoylquinic acids, coumarins, flavonoids, sterols, and terpenoids, among others. *Artemisia* species are also profuse in volatile phyto-constituents, which constitute a fountain of treasured pharmacologically active compounds. The most common essential oils emanating from the genus

---

Department of Health Sciences, Faculty of Science, University of Mauritius, Réduit, Mauritius.
* Corresponding author: f.mahomoodally@uom.ac.mu

include cadinene, phellandrene, pinene, thuiyl alcohol, and thujone (Koul et al. 2018, Watson et al. 2002, Liu et al. 2009, Bora and Sharma 2011).

The biological activities of these essential oils are diverse, ranging from being acaricidal, anti: arthritis, cancer, convulsant, diabetic, fertility, fungal, herpes virus, hyperlipidemic, hypertensive, malaria, migraine, oxidant, parasitic, pyretic, rheumatic, spasmodic, tumor, viral, hepato, and neuro protective. They are also abortifacient, analgesic, choleretic, urine stimulant, and an antidote against insect poisoning (Rajeshkumar and Hosagoudar 2012, Mojarrab et al. 2016, Taherkhani 2014, Tariku et al. 2010).

Given the profuse number of members in the *Artemisia* genus and the panoply of pharmacologically active secondary metabolites they biosynthesize, it is worth highlighting in a single documentation the ethno-pharmacological, phytochemical, and therapeutic potential of the most affluent species.

Essential Oils and *Artemisia* Species

Essential oils (EOs) are colorless liquids composed mostly of volatile and aromatic compounds present naturally in all parts of the plants, including bark, flowers, peel, seeds, stem, and the whole plant (Sánchez-González et al. 2011). EOs are secondary metabolites which are important as a part of plants' defence mechanism, thus having various medicinal properties in addition to antimicrobial properties (Tajkarimi et al. 2010). Furthermore, as potent free radical scavengers, they play an imperative role in the prevention of diseases arising from cellular damage due to free radicals, such as cancer, brain dysfunction, decline in immune system, and cardiovascular diseases (Aruoma 1998, Kamatou and Viljoen 2010). Their significance in aromatherapy has also been highlighted as therapeutic agents (Ali et al. 2015). In addition, the antiviral, antidiabetic, anti-inflammatory, antispasmodic, as well as hepatoprotective and antiallergic properties of EOs have been revealed by several studies (Özbek et al. 2003, Pérez et al. 2011, Mitoshi et al. 2014, Gavanji et al. 2015, Al-Hajj et al. 2016, Morales-López et al. 2017, Guo et al. 2018, Heghes et al. 2019).

The distinctive odor of EOs relies on the plants' origin, organ, and species. Moreover, these volatile oils have a high refractive index and optimal rotation, due to the presence of several asymmetrical compounds. In addition, EOs commonly have a relative density lower than that of water, although there are some exceptions. EOs are usually acknowledged as being hydrophobic, however they are mostly soluble in alcohols, fats, and organic solvents. Moreover, they are sensitive to oxidation forming resinous products via polymerization (Li et al. 2014). About 3,000 EOs have been produced from at least 2,000 species of plants, out of which 300 are important commercially (Djilani and Dicko 2012), for instance in industries such as cosmetics, perfumery, food, and agriculture (Burt 2004).

Broadly, EOs constituents can be classified into two distinct chemical classes, namely terpenes and phenylpropanoids. Terpenes can be divided into two main groups: (1) terpenes containing a hydrocarbon structure (the mono-, sesqui-, and di-terpenes) and (2) their oxygenated derivatives (alcohols, acids, aldehydes, ketones, esters, lactones, oxides, and phenols) (Moghaddam and Mehdizadeh 2017). Additionally, there are three biosynthetic pathways from which the main components of EOs are derived, notably (1) the mevalonate pathway leading to sesquiterpenes, (2) the methylerythritol pathway leading to monoterpenes and diterpenes, and finally (3) the shikimic acid pathway leading to phenylpropenes (Baser and Buchbauer 2010). Although EOs are very complex mixtures containing about 20–60 chemical compounds at quite different concentrations, they are characterized by 2–3 major components at relatively high concentrations (20–70%), compared to other components present in trace amounts. In general, these major components determine the biological properties of the EO (Abad et al. 2012).

EOs can vary greatly in their composition and yield, depending on several internal (type of soil, plant maturity, genetics) and external factors (geographical origin, climate, seasonal variation, method of extraction, etc.) (Marotti et al. 1994, Hussain et al. 2008, Anwar et al. 2009). Besides, the quality of EOs strongly depends on all these factors that may interfere and also limit their yield (Zuzarte and Salgueiro 2015). Like EOs from other plants, the composition and yield of EOs from *Artemisia* species are also subjected to variation, as evidenced by several studies. Thus, this section will aim to provide a brief overview of factors influencing the yield and composition of *Artemisia* EOs.

For instance, Padalia et al. (2014) showed in their study that EOs obtained from the aerial parts of *A. nilagirica* var. *septentrionalis* differed significantly in their composition during different seasons (autumn, spring, summer, rainy, and winter). Although the *Artemisia* ketone was the major constituent present in the EOs irrespective of the season, it was found in the highest amount in winter (61.2%), while its content varied from 60.7–38.3% in the other seasons. The other major components, namely germacra-4,5,10-trien-1-α-ol, germacrene D, Artemisia alcohol, β-caryophyllene, and chrysanthenone also varied as a result of seasonal variations (1.9–4.9%, 3.1–6.8%, 1.4–3.6%, 1.9–6.8%, and 1.5–7.7%, respectively). Besides, changes in the yield of EOs in different seasons were also noted, whereby the highest EOs yield was obtained during the rainy season (0.70%), followed by summer (0.68%), while the lowest yield was in winter and autumn (0.45%), followed by spring (0.58%) season.

The influence of different extraction techniques, such as steam- and hydro-distillation, including extraction by organic solvent and headspace technique on flowers and leaves of the wild grown *A. annua* EO was also investigated (Vidic et al. 2018). Even though EOs extracted by steam- and hydro-distillation demonstrated significant resemblance in their chemical composition, they fluctuated in quantity, particularly in the content of one of the major constituents, camphor, which was 24.0% and 16.9% in steam-distilled and hydro-distilled EO, respectively. Furthermore, only EOs obtained by distillation yielded oxygenated sesquiterpene compounds, of which caryophyllene oxide was the most abundant (8.2% in EO obtained by extraction with petroleum ether as a solvent, followed by steam distillation). In addition, all samples of EOs contained oxygenated monoterpenes (70.6% for headspace of plant material EO, and 42.6% for steam-distilled fraction of petroleum ether extract). Additionally, steam-distilled EO yielded the highest number of 47 total identified components, while only 11 components were identified in EO obtained by headspace of petroleum ether extract. Also, while *Artemisia* ketone was the major compound of steam- and hydro-distilled EO as well as headspace sample of plant material (30.2%, 28.3%, and 46.4%), camphene (25.6%) was the dominant compound headspace of petroleum ether extract (Vidic et al. 2018).

In the study of Badoni and colleagues (2009), the effect of altitudinal variation (500 m, 1,200 m, 2,000 m) on *A. nilagirica* EOs was inspected. Interestingly, EOs isolated from plants collected from the lowest altitude (500 m) contained α-thujone (36.94%) as the principal constituent, whereas only a small amount of α-thujone was identified in the plants collected from the other two altitudes. Furthermore, the highest concentration of 4-nitrobenzoic acid-4-methoxyphenyl ester (22.12%) was afforded by EO of plants collected at an altitude of 1,200 m, compared to those at 500 m and 2,000 m altitudes, which yielded only 1.77 and 3.59%, respectively. Likewise, *l*-linalool (32.47%) was the leading constituent of plants at 2,000 m altitude, but was detected in very low proportions in others.

Similarly, Behtari et al. (2012) demonstrated that growth stages (flowering and vegetative), altitudes (1,100 m, 1,200 m, 1,280 m, and 1,380 m), and their interactions positively affected the EOs content (ml g^{-2}) of *A. herba-alba*. For example, the maximum mean values of EOs content (0.8 and 0.92 ml g^{-2}) were gained at 1,280 m altitude during vegetative and flowering stages, respectively. Additionally, *cis*-pinocarveol and *Artemisia* ketone were identified as the main components of the EOs. However, the highest content of main components was obtained during the flowering stage. The comparative level for several constituents were increased, decreased, or disappeared within the EOs of plants at different growth stages.

Rana et al. (2013) also reported considerable variation in *A. annua* EOs at various growth stages, notably at vegetative, pre-bloom, bloom, and post-bloom phases. The percentage (%) yield ranged from 0.14–0.64% (w/w), with the highest yield at the bloom stage and the lowest at the vegetative stage. Oxygenated monoterpenes (39.0–57.0%) were found to be the principal EO fractions, followed by sesquiterpene hydrocarbons (11.8–26.2%), and monoterpene hydrocarbons (4.2–15.1%). Besides, although all EOs samples contained the same major constituents, they did show variation in quantity-camphor (28.6–31.7%), 1,8-cineole (2.1–20.8%), germacrene D (3.8–12.0%), β-caryophyllene

(2.8–6.9%), trans-β-farnesene (0.7–4.5%), α-pinene (0.5–2.4%), p-cymene (0.8–2.3%), and terpinen-4-ol (0.9–2.1%).

The effect of two different types of soil on the EOs composition and content of *Artemisia sieberi* grown in central Iran (Hossein Abad, site A and Golchegan, site B) was also evaluated by Bidgoli et al. (2013). The percentage of potassium (K), nitrogen (N), phosphorus (P), and organic carbon (OC) in the soils, as well as other soil factors, such as pH, percent of soil particles (clay, sand, and silt), electrical conductivity (EC), and exchangeable sodium percentage (ESP) were determined. Air-dried aerial parts of *A. sieberi* were subjected by hydro-distillation for EOs extraction, followed by GC/GC-MS analysis. The highest percentage of EOs yield was obtained at site B (0.79% w/w), while only 0.32% (w/w) was yielded at site A. Furthermore, soil analysis revealed that the soil at site A contained higher levels of K, N, P, and OC than site B. Similarly, greater EC, pH, and EPS was observed in the soil at site A. For instance, the soil at site A had an EPS and EC of 18.07 and 8.45, whilst they were 12.91 and 7.43 in the soil at site B, respectively. Besides, while 23 compounds were identified in EOs from plants collected at site A, only 14 were detected in EOs extracted from plants in site B. Variation in the major constituents of the EOs from plants at site A (trans-Methyl isoeugenol, 32.60%; trans-caryophyllene, 9.62%; myrcene, 6.92%; allo-ocimene, 6.37% and α-pinene, 6.05%) and site B (β-Bisabolene, 33.53%; α-pinene, 32.20%; trans-isodillapiole, 9.73% and myrcene, 8.98%) was also noted. Hence, from the comparative analysis of EOs compounds and soil characteristics, it can be deduced that the physical (texture and structure) and the chemical (EC, ESP, and pH) properties of the soil can significantly affect the quantity and quality of the plants' EOs.

Moreover, the drying process at different temperatures was seen to have an effect on EOs amount and composition of *A. annua* harvested at full blooming stage (Khangholil and Rezaeinodehi 2008). The aerial parts were subjected to complete drying by placing the plants in the shade (room temperature) and in the oven at temperatures 35, 45, 55, and 65°C, followed by extraction using hydro-distillation in a Clevenger apparatus and GC/MS analysis. The results demonstrated that higher drying temperatures caused the EO yield to reduce, notably 1.12% at room temperature, 0.88% (35°C), 0.55% (45°C) to 0.50% (55°C), and 0.37% (65°C). The monoterpenes content was also seen to decrease gradually while sesquiterpenes content increased. Besides, at 35°C, 45°C, and 55°C, *Artemisia ketone* (14.4–17.2%), 1,8-cineole (9.9–11.4%), and camphor (9.3–9.8%) were the three most dominant components. At room temperature, the EOs afforded the highest amount of artemisia ketone (21.6%) and 1,8-cineole (14.7%), but contained pinocarvone (8.8%) rather than camphor as the third abundant compound. On the contrary, β-caryophyllene (12.5%), germacrene D (9.0%), and trans pinocarveol (7.8%) were the major compounds present at 65°C.

Younsi et al. (2018) also investigated the relationship between the chemotypic and genetic diversity of natural populations of *Artemisia herba-alba* growing wild in Tunisia. For this purpose, 80 individuals collected from eight populations growing wild in different geographic areas were included to evaluate the intraspecific variability of EOs composition, genetic diversity, and population structure of *A. herba-alba*. The EOs composition was observed to differ considerably between populations. Moreover, EOs chemical profiles were categorized into four chemotypes, namely, camphor, α-thujone/trans-sabinyl acetate, trans-sabinyl acetate, and α-thujone/camphor/β-thujone. Despite significant relationship between a set of climatic data and the quantity of some EOs compounds, the global chemical deviation among populations was not linked to their geographic and bioclimatic appurtenances. Besides, a high level of genetic diversity within populations was detected. The level of genetic diversity also varied across populations and chemotypes. While populations from the α-thujone/trans-sabinyl acetate chemotype displayed the highest genetic diversity, populations from α-thujone/camphor/β-thujone chemotype showed significant genetic variation. Also, an important genetic differentiation was noted among populations as well as chemotypes. The combined analysis demonstrated a significant link between the molecular and chemical markers. The PCA, conducted on percentages of major oil components and the frequencies of polymorphic RAPD and ISSR bands, enabled the division of populations in relation to their chemotypic classification. Table 13.1 shows common essential oil components derived from *Artemisia* species.

Table 13.1: Summary of common *Artemisia* species and their corresponding essential oils.

Artemisia Species bearing essential oil	Country	Part(s) used	Extraction procedure used	Main components	Biological activities reported	References
A. sieberi Besser	Iran		Hydrodistillation	1, 8 cineole (45.88%), 4-terpineol (3.89%), camphor (3.40%), chrysanthenone (3.00%), α-terpineol (2.97%), methyleugenol (6.44%) and eugenol (2.75%)	Antimicrobial	Sardashti et al. (2015)
A. santolina Schrenk	Iran			1, 8-cineole (21.07%), comphor (13.13%), chrysanthenone (6.98 %), trans-methyl cinnamate (5.56%), lyratyl acetate (5.20%), 4-terpineol (4.39%), and borneol (3.75%)		Sardashti et al. (2015)
A. dracunculus L.	Iran	Aerial parts	Steam distillation	Hinokitiol (17.47%), estragole (17.28%), pulegone (10.23%), limonene (7.57%), methyl eugenol (7.46%) and bornyl acetate (7.12%)	Antispasmodic	Jalilzadeh-Amin et al. (2012)
	Romania	Leaves		Sabinene (42.38%), isoelemicin (12.91%), methyl eugenol (9.09%), elemicin (7.95%), and betaocimene (6.46%)	Antioxidant	Fildan et al. (2019)
	Turkey	Aerial parts	Hydrodistillation	1,8-cineole (35.88%), camphor (32.28%), camphene (9.13%), borneol (7.07%), thymene (3.31%), terpinen-4-ol (3.26%), γ-terpinene (1.32%), α-terpineol (1.29%), caryophyllene oxide (1.28%), and β-pinene (1.10%)	Antibacterial	Kumlay et al. (2015)
	Iran			Trans-anethole (21.1%), α-trans-ocimene (20.6%), limonene (12.4%), α-pinene (5.1%), allo ocimene (4.8%), methyl eugenol (2.2%), β-pinene (0.8%), α-terpinolene (0.5%), bornyl acetate (0.5%), and bicyclogermacrene (0.5%)	Anti-convulsant	Sayyah et al. (2004)
	Canada	Aerial parts		Methyl eugenol (35.8%), Terpinolene (19.1%), and methyl chavicol (16.2%)	Antimicrobial	Lopes-Lutz et al. (2008)
A. abrotanum L.	India	Leaves and Flowering tops	Hydrodistillation	1, 8-cineole, davanone, and nerolidol	Anti-convulsant	Dhanabal et al. (2007)
A. vulgaris L.	Brazil	Leaves	Hydrodistillation	Caryophyllene (37.45%), germacrene-D (16.17%), and humulene (13.66%)	Antimicrobial	Malik et al. (2019)
	America	Leaves		Germacrene D (25%), caryophyllene (20%), alpha-zingiberene (15%), and borneol (11%)	Anticancer	Williams et al. (2012), Saleh et al. (2014)
		Buds		1,8-cineole (32%), camphor (16%), borneol (9%), and caryophyllene (5%)		

Species	Country	Plant part	Extraction method	Composition	Activity	Reference
A. gmelinii Weber ex Stechm	China	Aerial parts	Hydrodistillation	Cyclobutaneethanol, endo-borneol, germacrene D, eucalyptol, selin-6-en-4α-ol, bisabolone oxide A, caryophyllene, and terpinen-4-ol.	Antidiabetic	Xu et al. (2019)
A. absinthium L.	India	Leaves	Hydrodistillation	Chrysanthenyl acetate (49.15%) and L-β-pinene (39.62%)	Antioxidant	Wani et al. (2014)
	Canada			*Trans*-Sabinyl acetate (26.4%), myrcene (10.8%), trans-Thujone (10.1%)	Antimicrobial	Lopes-Lutz et al. (2008)
A. judaica L.	Jordan	Aerial parts	Hydrodistillation	Piperitone (30.4%), camphor (16.1%), and ethyl cinnamate (11.0%)	Antifungal and anti-inflammatory	Abu-Darwish et al. (2016)
A. herba-alba Asso.	Morocco	Aerial parts	Hydrodistillation	Verbenol (21.83%), bisabolone oxide (17.55%), farnesene epoxide (17.08%), and β-thujone (6.14%)	Anti-proliferative	Tilaoui et al. (2011)
A. iwayomogi Kitam	Korea	Aerial parts	Hydrodistillation	Camphor (19.31%), 1,8-cineole (19.25%), borneol (18.96%), camphene (4.64%), and β-caryophyllene (3.46%)	Antibacterial	Yu et al. (2003)
		Aerial parts				
A. annua L.	China	Flowers (post-flowering stage)	Hydrodistillation	Camphor (16.62%), caryophyllene (16.27%), β-caryophyllene oxide (15.84%), β-farnesene (9.05%), and (-)-spathulenol (7.21%)	Anti-acetylcholinesterase	Yu et al. (2011)
A. scoparia Waldst. & Kit.	India	Residues	Hydrodistillation	Citronellal (15.2%), acenaphthene (11.08%), b-citronellol (11.02%), caryophyllene oxide (10.03%), b-caryophyllene (9.37%), and eugenol (6.03%)	Antioxidant	Singh et al. (2009)
A. Montana (Nakai) Pamp.	Japan	Leaves	Hydrodistillation	1,8-cineole, camphor, borneol, α-piperitone, and caryophyllene oxide	Sedative	Kunihiro et al. (2017)
A. biennis Willd.	Canada	Aerial parts	Hydrodistillation	(E)-beta-Farnesene (40.0%), (Z)-beta-Ocimene (34.7%), (Z)-en-yn-Dicycloether (10.0%)	Antimicrobial	Lopes-Lutz et al. (2008)
A. cana Pursh		Aerial parts		Camphor (37.3%), 1,8-cineole (21.5%)		Lopes-Lutz et al. (2008)
A. frigida Willd.		Aerial parts		1,8-cineole (25.1%), camphor (20.6%)		Lopes-Lutz et al. (2008)
A. longifolia Nutt.		Aerial parts		1,8-cineole (27.6%), camphor (18.5%)		Lopes-Lutz et al. (2008)
A. ludoviciana Nutt.		Aerial parts		1,8-cineole (22.0%), camphor (15.9%), davanone (11.5%)		

Table 13.1 contd. ...

...Table 13.1 contd.

Artemisia Species bearing essential oil	Country	Part(s) used	Extraction procedure used	Main components	Biological activities reported	References
A. arborescens L.	Italy	Aerial parts	Distillation	Camphor (35.73%), β-thujone (23.97%), and chamazulene (7.6%)	Antiviral	Sinico et al. (2005)
A. campestris L.	Tunisia	Leaves and Stems	?? Hydrodistillation	β-pinene (36.4%), 2-undecanone (14.7%), limonene (10.57%), and benzene (3.6%)	Anthelmintic	Abidi et al. (2018)
Artemisia argyi Lévl. et Vant	China	Leaves	Hydrodistillation	Neointermedeol (9.652%), caryophyllene oxide (8.713%), α-Thujone (7.989%), β-Caryophyllene (7.495%), and borneol (6.482%)	Antimicrobial	Guan et al. (2019)
		Leaves	Subcritical Extraction	β-Caryophyllene (20.022%), α-Thujone (11.312%), borneol (8.273%),(+)-2-Bornanone (7.253%) neointermedeol (1.16%), and caryophyllene oxide (0.133%)		
		Leaves	Simultaneous distillation-extraction	Caryophyllene oxide (21.553%), neointermedeol (16.779%), borneol (16.356%), α-Thujene (14.551%), α-Thujone (14.551%), and β-caryophyllene (13.687%), (+)-2-Bornanone (10.022%)		

Other Artemisia Species: Traditional uses and Pharmacological Activities

Artemisia abrotanum L. (Southernwood)

This species is characterized by a cylindrical, erect, and green color stem. The leaves are petiolate, pubescent on the underside, and glabrous on the upper side. Being a perennial undershrub, the plant grows up to 1 meter high, while the bark is smooth and brown. The leaves are alternate with a long footstalk, while the flowers are greenish in color, and the seeds are naked and solitary (Quattrocchi 2012). It has in general a lemon-like odor. Among the lower leaves, around 2–3 are pinnatipartite, while the upper leaves are non-auriculate and pinnatipartite (French Pharmacopoeia 2008).

It is also commonly known as European sage, lady's love, oldman wormwood, and is grown in France, India, Italy, Spain, Saudi Arabia, and the United States. In ancient Greek and Roman systems of traditional medicine, *Artemisia abrotanum* L. was employed to relieve respiratory complications to enhance the clearing of the respiratory tract and improve breathing, and as a spasmolytic. The leaves are useful for gastrointestinal problems and aid digestion, menstrual flow as a febrifuge, antispasmodic, and serve as an anthelmintic (Thomson 1826).

Various secondary metabolites have been isolated and identified from the plant, with the most bountiful being 1,8-cineole, davanone, germacrene D, piperitone, and silphiperfol-5-en-3-ol A. Other constituents present in minor proportions are—1.8-cineole, 2E-hexanal, α-thujene, α-pinene, α-phellandrene, α-terpinene, β-pinene, dehydro-1.8-cineole, p-cymenene, γ-terpinene, z-myroxide, benzene acetaldehyde, camphene, camphor, cis-p-menth-2-en-1-ol, cis-piperitol, cis-sabenene hydrate, cryptone, linalool, menthone, myrcene, nonanal, o-cymene, piperitone, sabina ketone, sabinene, terpinolene, trans-sabinene hydrate, σ-terpineol, borneol, terpinen-4-ol, trans-piperitol, trans-carveol, terpnen-4-ol acetate (Kowalski et al. 2007).

Pharmacological investigations of the plant have revealed spasmolytic and anthropod repellant activity. Isolated flavonols from the plant have been demonstrated to normalize trachea and smooth muscle contraction induced by carbacholine in guinea pigs. The toluene extract is effective against ticks and the fever mosquito *Aedes aegypti*, with the coumarins and the thujyl alcohol being the most prominent insect repellants (Bergendorff and Sterner 1995, Tunón et al. 2006).

When tested clinically as a nasal spray among 12 patients suffering from allergic conjunctivitis, rhinitis, and other bronchial symptoms, it relieved the patients to the same efficacy as other antihistamine or cromoglicate preparations the patients had employed previously (Remberg et al. 2004).

Laboratory investigations of this species have demonstrated that its ethanolic extracts bear antibacterial and antifungal activities. The formulations from this species are employed as astringent, febrifuge, antiseptic, stimulant, among other uses (Abad et al. 2012, Suresh et al. 2010). Secondary metabolites derived from this species, including cineole, borneol, and *p*-cymene protect against bites from *Aedes aegypti* (Mohamed et al. 2010). Medicinal preparations from the essential oil of *A. abrotanum* are useful against allergic rhinitis and other respiratory complications (Koul et al. 2018).

Artemisia herba-alba Asso (White Wormwood)

Artemisia herba-alba Asso is a perennial shrub known as "white herb" in Latin, given its white and woolly stems and leaves, grows between 20–40 cm in height (Samy and Francis 1999). It occurs mostly in Northern Africa, South Western Europe, and Western Asia (Ali et al. 2019). The leaves of the species are extremely aromatic and covered with glandular hairs that reflect sunlight. The shoots bear leaves which are grey, petiolate, ovate to orbicular. Flowering stems occur mostly in winter and are much smaller (Samy and Francis 1999).

Traditionally known as the desert wormwood in English, or Armoise Blanche in French, and even as Shih in Arabic culture, this plant is traditionally employed by many cultures to manage and/or treat cold, diabetes, hypertension, respiratory disorders, including bronchitis, cough, and infectious diseases of the skin, such as syphilis and scabies (Moufid and Eddouks 2012, Mighri et al. 2010). The aqueous extracts possess both antioxidant and antimicrobial activities (Gurib-Fakim and Mahomoodally 2013). The herbal tea is known to be analgesic, antibacterial, and antispasmodic. In Algeria, it is employed as a fodder plant for livestock (Bora and Sharma 2011).

The essential oil from *A. herba-alba* is rich in oxygen, containing monoterpenes 1,8-cineole (20.1%), α-thujone (22.9%), β-thujone (25.1%), and camphor (10.5%), and was active against the fungal strains *Epidermophyton floccosum* and *Trichophyton rubrum*. It also inhibited the formation of germ tube in *Candida albicans*. Interestingly, the oil has the potential to inhibit nitric oxide production induced by lipolysaccharides without cytotoxicity up to a concentration of 1.25 µL/ml in macrophages and 0.32 µL/ml in microglia, respectively (Abu-Darwish et al. 2015).

Artemisia absinthium L. (Wormwood)

This species is known by various names in different parts of the globe. The most common names include wormwood, grand wormwood, absinthe, absinthium, and absinthe wormwood. It grows in Canada, Eurasia, Northern Africa, and the United States. The stems grow up to 1.2 meters high, and are branched and silvery green. The leaves are arranged in spirals, being greenish-grey above and white below. The basal leaves are up to 2.5 cm long. The flowers are clustered, pale yellow and tubular, with the flowering season being in early autumn or summer (Goud and Swamy 2015).

This species has a long-standing use in Turkish traditional medicine against fever, sepsis, stomachache, and as a diuretic (Joshi 2013). On the other hand, in traditional Chinese medicine, it is known to relieve gastric pain and enhance cardiac and cognitive functions (Tajehmiri et al. 2014). Based on its desirable aromatic nature, it has been employed in various alcoholic drinks, such as the spirit absinthe, foods, and soft drinks, with as close as 206 flavoring agents derived from the plant. The leaves are rich in caffeic, ferulic, and gallic acid, as well as myricetin, which confers them a strong antioxidant potential (Altunkaya et al. 2018).

The essential oil from *A. absinthium* has demonstrated good antibacterial activity, being useful against *Saccharomyces cerevisiae* and *Candida albicans* (Seddiek et al. 2011). When subjected to an antioxidant assay, the methanolic extract of the dried plant demonstrated a positive effect as an antioxidant. In sheep intestines nematodes, the aqueous and the ethanolic extract exhibited significant anthelmintic activity when compared to the conventional anthelmintic drug albendazole (Hristova et al. 2013, Lee et al. 2013). Several other biological activities are reported from the ethanolic and aqueous extracts of this species, with the most common being hepatoprotective, antimicrobial, and antiparasitic (Altunkaya et al. 2018, Lee et al. 2013).

Clinical studies conducted on the plant have demonstrated that it is a good antiparasitic agent and reduces the number of *Toxocara cati* or *Toxocara canis* eggs in rat feces (Tariq et al. 2009). It is also lethal against *Trichinella spiralis*, and hence offers protection against roundworm infection, as shown in a study conducted on rats. It protects the liver and restores the potential of its enzymes, such as catalase, glutathione, and superoxide dismutase. It may also be of benefit in Crohn's disease and against yeast infection (Omer et al. 2007, Juteau et al. 2003).

Artemisia afra Jacq ex Wild (African Wormwood)

This species grows in clumps with ridged woody stems of height 0.5 to 2 meters. The leaves are dark green in color and are fern-shaped. Below, the leaves are light green and covered with white bristles. The flowers are butter-colored, and are between 3–5 millimeter in diameter, and the flowering season is in late summer. When the plant is bruised, it diffuses a mixture of a sweet and pungent smell (Watt and Breyer-Brandwijk 1962, Van Wyk et al. 1997).

It is one of the most used and oldest medicinal plants in South Africa (Koul et al. 2018). It is a perennial woody shrub growing up to 2 meters high (Van Wyk et al. 1997). Despite its widespread use, limited research has been conducted on this species. For example, in the year 2008, only 42 publications and two patents were available for this plant (Van Wyk 2008). A wide array of traditional uses are associated with this plant, ranging from minor ailments, such as cold, cough, dyspepsia, headaches, to chronic ailments, including diabetes, diseases of the bladder and kidney, and malaria. Modern uses of the species overlap with the ancient ones, and it is still employed to treat colds, coughs, and diabetes, and also respiratory disorders (Koul et al. 2018). In different parts of Africa, it is known by different names based on its ethnobotanical uses and the language spoken in that region. The aqueous leaf extracts exhibit antimicrobial potential (Muleya et al. 2014).

Various traditional formulations are prepared from this plant in Africa. For example, its syrup relieves bronchial symptoms, while an infusion of the roots is claimed to be effective against diabetes; the lotion is employed for the management of hemorrhoids and ear problems, while the fresh tips inserted between the nose and teeth soothe cold and flu, as well as toothache (Erasto et al. 2005, Mahop and Mayet 2007).

Artemisia afra is rich in secondary metabolites, as demonstrated by metabolites retrieved by extraction techniques, such as hydro-distillation and microwave-assisted and ultrasound techniques. The main monoterpenoids detected in the plant include artemisia acetate, alcohol and ketone, ascaridole, azulene, borneol, bornyl acetate, camphene, camphor, cis-carveol, caryophylla-2(12),6(13)-dien-5-one, cis-chrysanthenol, chrysanthenone, cis-chrysanthenyl acetate, 1,8-cineole, cumin alcohol, cumin alcohol, dehydro carvyl acetate, dehydro-1,8-cineole, dehydrosabinaketone, limonene, linalool, myrcene, myrtenal, myrtenol. It is also a reservoir of sesquiterpenes, such as davanone, calamenene, cubebol, germacrene, germacrene D, globulol, intermedeol, intermediol, t-muurolol, spathulenol, and other metabolites, such as artemisal, berbenone, cuminaldehyde, p-cymene, octadecanol, among others (Liu et al. 2009).

Thujone present in the volatile oil of *Artemisia* is a toxic constituent, and prolonged use or when employed in large doses can give rise to unpleasant adverse effects, most of which include convulsions, fatty degeneration of the liver, restlessness, vertigo, and vomiting. Thujone is a mixture of α and β isomers, with the first one being the most toxic with an LD_{50} of 87.5 mg/kg when administered subcutaneously in mice, compared to 442.2 mg/kg for the second isomer (Oyedeji et al. 2009). Thujone has a very low solubility in water, and despite this fact, it is better to limit the use of *A. annua* to not more than two weeks, and it should not be employed among pregnant women (McGaw et al. 2000).

Artemisia annua L. (Sweet Wormwood, Sweet Annie, Annual Wormwood, qinghao, huang hua hao)

With a brownish or violet-brown erect stem, the plant is hairless, growing between 30 to 100 centimeter (cm) tall. When cultivated, it has been recorded that this plant can grow up to 200 cm in height. The leaves are divided by profound cuts into 2 or 3 small leaflets, and range between 3 to 5 cm in length, and have an intense aromatic smell (Anonymous 2015).

The plant grows well in China, Korea, Japan, and Vietnam, among other countries and is famous for its use in traditional medicine. In Chinese traditional medicine, *A. annua* has been employed to treat fevers and chills where malaria has been the root cause for more than 2,000 years (Abad et al. 2010). It is widely cultivated in Africa and has been naturalized in Europe, South America, and the United States. It biosynthesizes the secondary metabolite Artemisinin, which is a key molecule in the global fight against malaria. Indeed, in 2015, the Chinese scientist Youyou Tu was awarded the Nobel prize in medicine for discovering artemisinin and its efficacy against malaria (Dlugónska 2015).

The spread of malaria is still a major challenge, given an increase in resistance of Plasmodium falciparum to conventional drug agents, such as amodiaquine, chloroquine, and sulfadoxine–pyrimethamine (SP). In addition, the development of the multi-drug resistant Plasmodium falciparum in South East Asia and South America has complicated the treatment of the disease. Hence, combination artemisinin-based therapies are a favorable approach to fight off the Plasmodium parasite, and is also advocated by the World Health Organization (WHO 2006).

Compared to conventional drug agents employed, artemisinin has almost no side–effects, with dihydroartemisinic acid being its precursor (Tian et al. 2017). Artemisin has been an invaluable contribution, and has enabled significant relief among malaria sufferers following the development of resistance among malaria parasites against quinines (Bhakuni et al. 2001, Sen et al. 2007). Besides the generalized pharmacological activities of the Artemisia genus, A. annua is also a potent anti-cancer and anti-leishmaniasis agent (Bhakuni 2001, Sen et al. 2007, Crespo-Ortiz and Wei 2011). In an attempt to prevent the development of resistance against artemisinin, its combination with other antimalarial drugs can be envisaged, following results retrieved from validated clinical trials to prevent any occurrences of possible hepatotoxicity (Efferth 2017, Steketee and Eisele 2017).

As the yield of artemisin within the dried aerial parts of the plant is generally low, a few techniques are employed to boost the yield. For example, crop improvement and microbial synthesis of the compound is a viable approach to meet the demands of pharmaceutical companies (Alejos-Gonzalez et al. 2011). Breeding the plant and mapping its genetic map is another route that can be employed in this effort (Graham et al. 2010). In addition, techniques which enable the production of seeds of good quality can also be explored (Graham et al. 2010, Wetzstein et al. 2014).

Dried leaves of *A. annua* have demonstrated antiplasmodial activity and have been found effective in malaria patients unresponsive to artemisinin combination therapy and intravenous artesunate (Desrosiers and Weathers 2016). When the essential oil was combined with the dried leaves of *A. annua*, an enhanced solubility and availability of artemisinin was obtained. These techniques and the use of artemisinin in general is more cost-effective than conventional drug regimens, and is hence more affordable to patients residing in low income countries (Aderibigbe 2017).

Artemisia capillaris Thunb. (Capillary wormwood, Yerba Lenna Yesca)

Artemisia capillaris Thunb. can grow up to 1 meter high, with the stems becoming woody mostly at the base. It is an important component of Chinese and Korean medicine, where it is employed to cure fever, hepatitis, inflammation, jaundice, and malaria, among other ailments. The plant is rich in secondary metabolites, such as apigenin, capillarisin, coumaric acid, and hesperidin, all of which possess important anticancer and antimicrobial activities. The tablet derived from the plant is an effective treatment for hepatitis B infection since it prevents the replication of the virus (Tajehmiri et al. 2014).

Techniques such as GC-MS and TLC have enabled the discovery of other pharmacologically valuable metabolites from the plant. These include achillin, 1-borneol, camphor, and coumarin, all with an interesting anti318 carcinogenic activity. On the other hand, the use of the abovementioned techniques has also permitted the discovery of five other antibacterial compounds, which are α-pinene, β-pinene, β-caryophyllene, capillin, and piperitone. Analysis of its essential oil has shown the presence of germacrene D, which has significant fumigant activity (Koul et al. 2018).

Artemisia dracunculus L. (Estragon)

It is a perennial plant in the sunflower family. It grows between 120–150 cm in height, and possesses slender branches. Its leaves are glossy green, lanceolate with dimensions—2–8 cm long and 2–10 mm broad, and has an entire margin. The flowers occur in small capitula, which are around 2–4 mm in diameter, where each capitulum can house up to 40 yellow or greenish yellow florets (Stuckey and McGee 2002).

Artemisia dracunculus has been employed in traditional Arabic, Asian, and Russian medicine for diverse ill-health conditions. This herb is famous for its use in allergic rashes, as antiepileptic, carminative, anticoagulant, antihyperlipidemic, dermatitis and other skin irritations and fevers, laxative, and antispasmodic properties, vermifuge, and against wounds. The plant's extract is a good candidate as a potential coronary heart disease risk reducer. Two isolated compounds from the plant, namely estragole and methyleugenol demonstrated hyperglycemic activity in rats *in vivo* (Koul et al. 2018).

Following GC-MS (gas chromatography-mass spectrometry) analysis of the plant's essential oil, it was found that it is copious in (E)-β-ocimene (3.1%), limonene (3.1%) methyleugenol (1.8%), (Z)-anethole (81.0%), and (Z)-β-ocimene (6.5%). When the essential oils of *A. dracunculus, A. absinthium, A. santonicum, A. spicigera,* and *A. indica* were investigated for their antimicrobial potential, the weakest activity was recorded for *A. dracunculus* essential oil (Kordali et al. 2005).

Altogether 32 components have been identified following GC-MS of the aerial parts in the full flowering stage found in the Himalayan region. The essential oils is most abundant in davanone and other volatile phytoconstituents present in small amounts, namely 1,8-cineol, β-thujone, cis chrysanthenyl acetate, davanone oil, estragole, herniarin, sabinyl acetate, and terpineol have been identified from the plant—all of which are known to be antifungals. The alcoholic extract is rich in selin-11-en-ol and 1,8-cineole, which are both known to be antibacterial and antifungals. The star component artemisolide is an inhibitor of the NF-KB. Another compound from the plant eupatilin extracted from different artemisia species has potential anticancer activity (Haider et al. 2014).

Artemisia verlotiorum Lamotte (Chinese mugwort, Chinese wormwood, Mugwort, Verlot's mugwort)

In general, the leaves of the *Artemisia* species are described as alternate, capitula small, usually racemouse, paniculate or capitate, inflorescence, rarely solitary; involucral bracts in few rows, receptacle flat to hemispherical, without scales and sometimes hirsute; florets all tubular, achenes obovoid, pappus absent or sometimes a small scarious ring (Heywood et al. 1977, Bora and Sharma 2011). Strongly rhizomatous, *Artemisia verlotiorum* grows up to 30–60 cm in height, while the branches are sparsely distributed. In addition, it has a potent wormwood scent (Bittencourt De Souza et al. 2010).

The plant grows well in almost all parts of the northern hemisphere, and is native to eastern Asia and found mostly in the southwest of China, where it originated. It is also widely available in South Central Europe (Geissman 1970). It grows naturally in various countries as weed, for instance in Australia, Mauritius, and Rodrigues (Gurib-Fakim 1996).

In Italy, mainly in Tuscany, the plant infusion is administered as a remedy against hypertension (Martinotti et al. 1997). In Mauritius, the plant decoction is employed against fever, psoriasis, and influenza (Gurib-Fakim 1996). On the other hand, in Brazil, it is considered as a highly beneficial plant medically, since it can treat any disease related to the circulatory, digestive, genito-urinary, and respiratory disorders (Bittencourt De Souza et al. 2010).

The primary investigation of the *A. verlotiorum* demonstrated that it is rich in three crystalline sesquiterpenoid lactones, namely artemorin, anhydrovertolorin, and vertolorin, respectively (Geissman 1970). It also biosynthesizes a plethora of volatile constituents—α-phellandrene, β-thujone, cadinene, 1 camphene, cineole, fenchone, and thujyl alcohol. In addition, it contains the following fatty acids-palmitic and valeric acid. Investigation of the volatile constituents of *A. verlotiorum* of French origin have identified the following constituents present in majority—α-thujone, 1,8-cineole, and β-thujone. Other populations, from Italy, have characterized the constituents, including caryophyllene oxide, borneol, camphor, and 1,8-cineole as the main constituents of the whole plant oil, among others. In Mauritius, Gurib-Fakim has identified germacrene D and myrcene occurring as the major constituents from the Mauritian samples (Gurib-Fakim 1996).

In vitro and *in vivo* investigations have demonstrated that *A. verlotiorum* exerts anti-hypertensive, mycotic, and viral activities, respectively (Calderone et al. 1999, 1998, Macchioni et al. 1999).

In addition, *A. verlotiorum* has demonstrated possible muscarinic stimulation action *in vitro*. It is hypothesized that it contains a metabolite which serves as a remarkable agonist of muscarinic receptors, and results in vasorelaxant and negative inotropic actions that can evoke the antihypertensive response. Nonetheless, since this metabolite is unknown and no antihypertensive drugs act on muscarinic receptors, efforts to establish a plausible link between these are in progress (Martinotti et al. 1997). When the aqueous infusion of the plant was assayed in normotensive rats *in vivo*, a strong and transient reduction of the mean arterial pressure was recorded. The marked hypotensive activity was inhibited by atropine. It is believed that *A. verlotiorum* extract mediates a strong vasodilatory activity by inducing the release of nitric oxide after activation of the nitric oxide-guanosine 3'–5'-cyclic monophosphate (cGMP) pathway, following muscarinic receptor agonism (Calderone et al. 1999).

An *in vitro* investigation was conducted of the aqueous and the methanolic extracts of *A. verlotiorum* against *Saprolegnia ferax*. A minimum inhibitory concentration of 1% was recorded from the aqueous extract, while the methanolic extract was more active with an MIC of 0.25% (Macchioni et al. 1999).

Artemisia verlotiorum extract exhibited strong antiviral activity against the feline immunodeficiency virus model, which is a significant model of the human immunodeficiency virus type 1, which causes AIDS infection in humans. The aqueous lyophilized extract of *A. verlotiorum* inhibited the syncytia, viral reverse transcriptase activity, and the viral capsid protein P24 expression significantly. The cytotoxicity assay resulted in a negative finding. Hence, it may be concluded that the plant metabolizes phytochemicals which have a potent antiviral activity, but they must also be isolated, and the mechanisms through which they act must be identified (Calderone et al. 1998).

Conclusion

The *Artemisia* family is undeniably a large family, incorporating a plethora of medicinally valuable plant species. Based on their long term traditional use by the world's population, its members can be considered as being generally safe for consumption. Nonetheless, the presence of thujone, a toxic metabolite, in certain species warrants that a safe and standard dose in humans is established for each species. Results from diverse laboratory and clinical investigations demonstrate that species from this genus hold promise as antimicrobial, antimalarial, and anticancer agents. Undeniably, the vast traditional and pharmacological actions associated with their members make them valuable candidates for fueling up the drug discovery pipeline.

References

Abad, M.J., Bedoya, L.M., Apaza, L., Bermejo, P. 2012. The *Artemisia* L. genus: a review of bioactive essential oils. Molecules 17(3): 2542–2566.

Abidi, A., Sebai, E., Dhibi, M., Alimi, D., Rekik, M., B'chir, F., Maizels, R.M., Akkari, H. 2018. Chemical analyses and anthelmintic effects of *Artemisia campestris* essential oil. Vet. 263: 59–65.

Abu-Darwish, M.S., Cabral, C., Gonçalves, M.J., Cavaleiro, C., Cruz, M.T., Efferth, T., Salgueiro, L. 2015. *Artemisia herba-alba* essential oil from Buseirah (South Jordan): Chemical characterization and assessment of safe antifungal and anti-inflammatory doses. J. Ethnopharmcol. 174: 153–160.

Abu-Darwish, M.S., Cabral, C., Gonçalves, M.J., Cavaleiro, C., Cruz, M.T., Zulfiqar, A., Khan, I.A., Efferth, T., Salgueiro, L. 2016. Chemical composition and biological activities of *Artemisia judaica* essential oil from southern desert of Jordan. J. Ethnopharmcol. 191: 161–168.

Aderibigbe, B. 2017. Design of drug delivery systems containing artemisinin and its derivatives. Molecules 22(2): 323.

Alejos-Gonzalez, F., Qu, G., Zhou, L.L., Saravitz, C.H., Shurtleff, J.L., Xie, D.Y. 2011. Characterization of development and artemisinin biosynthesis in self-pollinated *Artemisia annua* plants. Planta 234(4): 685–697.

Al-Hajj, N.Q.M., Sharif, H.R., Aboshora, W., Wang, H. 2016. *In vitro* and *in vivo* evaluation of antidiabetic activity of leaf essential oil of *Pulicaria inuloides*-Asteraceae.

Ali, B., Al-Wabel, N.A., Shams, S., Ahamad, A., Khan, S.A., Anwar, F. 2015. Essential oils used in aromatherapy: A systemic review. Asian Pacific Journal of Tropical BiomedicineAsian. Pac. J. Trop. Biomed. 5(8): 601–611.

Ali, R.F., Alaila, A.K., Aldaaiek, G.A. 2019. The potential of benefiting variation between the same species of *Artemisia Herba-alba* from different location in Northeast of Libya. J. Appl. Life Sci. Int. 1–6.

Altunkaya, A., Yıldırım, B., Ekici, K., Terzioğlu, Ö. 2018. Determining essential oil composition, antibacterial and antioxidant activity of water wormwood extracts. GIDA. 39(1): 17–24.

Anonymous. "*Artemisia annua* (sweet wormwood)". Royal Botanic Gardens. Archived from the original on October 6, 2015.

Anwar, F., Hussain, A.I., Sherazi, S.T.H., Bhanger, M.I. 2009. Changes in composition and antioxidant and antimicrobial activities of essential oil of fennel (*Foeniculum vulgare* Mill.) fruit at different stages of maturity. J. Herbs Spices Med. Plants 15(2): 187–202.

Aruoma, O.I. 1998. Free radicals, oxidative stress, and antioxidants in human health and disease. JAOCS. 75(2): 199–212.

Badoni, R., Semwal, D.K., Rawat, U. 2009. Altitudinal variation in the volatile constituents of *Artemisia nilagirica*. Int. J. Essent. Oil Therapeutics 3: 66–68.

Baser, K., Buchbauer, G. 2010. Handbook of Essential Oils: Science, Technology, and Applications. CRC Press, Boca Raton, FL.

Behtari, B., Gholami, F., Khalid, K.A., Tilaki, G.D., Bahari, R. 2012. Effect of growth stages and altitude on *Artemisia herba-alba* Asso essential oil growing in Iran. J. Essent. Oil Bear. Pl. 15(2): 307–313.

Bergendorff, O., Sterner, O. 1995. Spasmolytic flavonols from *Artemisia abrotanum*. Planta Med. 61(04): 370–371.

Bhakuni, R.S., Jain, D.C., Sharma, R.P., Kumar, S. 2001. Secondary metabolites of *Artemisia annua* and their biological activity. Curr. Sci. 35–48.

Bidgoli, R.D., Pessarakli, M., Heshmati, G.A., Barani, H., Saeedfar, M. 2013. Bioactive and fragrant constituents of *Artemisia sieberi* Besser grown on two different soil types in Central Iran. Commun. Soil Sci. Plan. 44(18): 2713–2719.

Bittencourt De Souza, L.F., Laughinghouse IV, H.D., Pastori, T., Tedesco, M., Kuhn, A.W., Canto-Dorow, T.S.D., Tedesco, S.B. 2010. Genotoxic potential of aqueous extracts of Artemisia verlotorum on the cell cycle of *Allium cepa*. Int. J. Environ. Stud. 67(6): 871–877.

Bora, K.S., Sharma, A. 2011. The genus *Artemisia*: a comprehensive review. Pharm. Biol. 49(1): 101–109.

Burt, S. 2004. Essential oils: their antibacterial properties and potential application in foods—A review. Int. J. Food Microbiol. 94: 223–253.

Calderone, V., Nicoletti, E., Bandecchi, P., Pistello, M., Mazzetti, P., Martinotti, E., Morelli, I. 1998. *In vitro* antiviral effects of an aqueous extract of *Artemisia verlotorum* Lamotte (Asteraceae). Phytother. Res. 12(8): 595–597.

Calderone, V., Martinotti, E., Baragatti, B., Cristina Breschi, M., Morelli, I. 1999. Vascular effects of aqueous crude extracts of *Artemisia verlotorum* Lamotte (Compositae): *in vivo* and *in vitro* pharmacological studies in rats. Phytother. Res. 13(8): 645–648.

Crespo-Ortiz, M.P., Wei, M.Q. 2011. Antitumor activity of artemisinin and its derivatives: from a well-known antimalarial agent to a potential anticancer drug. BioMed Res. Int., 2012.

Desrosiers, M.R., Weathers, P.J. 2016. Effect of leaf digestion and artemisinin solubility for use in oral consumption of dried *Artemisia annua* leaves to treat malaria. J. Ethnopharmcol. 190: 313–318.

Dhanabal, S.P., Paramakrishnan, N., Manimaran, S., Suresh, B. 2007. Anticonvulsant potential of essential oil of *Artemisia abrotanum*. Curr. Trends. Biotechnol. Pharm. 1(1): 112–116.

Djilani, A., Dicko, A. 2012. The therapeutic benefits of essential oils. *In*: Nutrition, Well-Being and Health, 7: 155–179.

Dlugónska, H. 2015. The Nobel Prize 2015 in physiology or medicine for highly effective antiparasitic drugs. Ann. Parasitol. 61(4).

Efferth, T. 2017. From ancient herb to modern drug: *Artemisia annua* and artemisinin for cancer therapy. *In*: Seminars in Cancer Biology, 46: 65–83. Academic Press.

Erasto, P., Adebola, P.O., Grierson, D.S., Afolayan, A.J. 2005. An ethnobotanical study of plants used for the treatment of diabetes in the Eastern Cape Province, South Africa. Afr. J. Biotechnol. 4(12).

Fildan, A.P., Pet, I., Stoin, D., Bujanca, G., Lukinich-Gruia, A.T., Jianu, C., Jianu, A.M., Radulescu, M., Tofolean, D.E. (2019). *Artemisia dracunculus* Essential Oil Chemical composition and antioxidant properties. Chim-Bucharest 70(1): 59-62.

French Pharmacopoeia 2008. Southernwood For Homoeopathic Preparations Abrotanum For Homoeopathic Preparations. The General Chapters and General Monographs of the European Pharmacopoeia and Preamble of the French Pharmacopoeia apply.

Gavanji, S., Sayedipour, S.S., Larki, B., Bakhtari, A. 2015. Antiviral activity of some plant oils against herpes simplex virus type 1 in Vero cell culture. J. Acute Med. 5(3): 62–68.

Geissman, T.A. 1970. Sesquiterpene lactones of Artemisia—*A. verlotorum* and *A. vulgaris*. Phytochemistry 9(11): 2377–2381.

Goud, B.J., Swamy, B.C. 2015. A review on history, controversy, traditional use, ethnobotany, phytochemistry and pharmacology of Artemisia absinthium Linn. Int. J. Adv. Res. Eng. Appl. Sci. 4(5): 77–107.

Graham, I.A., Besser, K., Blumer, S., Branigan, C.A., Czechowski, T., Elias, L., Guterman, I., Harvey, D., Isaac, P.G., Khan, A.M., Larson, T.R. 2010. The genetic map of *Artemisia annua* L. identifies loci affecting yield of the antimalarial drug artemisinin. Science 327(5963): 328–331.

Guan, X., Ge, D., Li, S., Huang, K., Liu, J., Li, F. 2019. Chemical Composition and Antimicrobial Activities of *Artemisia argyi* Lévl. et Vant Essential Oils Extracted by Simultaneous Distillation-Extraction, Subcritical Extraction and Hydrodistillation. Molecules 24(3): 483.

Guo, R.H., Park, J.U., Jo, S.J., Yang, J.Y., Lee, S.S., Park, M.J., Kim, Y.R. 2018. Anti-allergic inflammatory effects of the essential oil from fruits of *Zanthoxylum coreanum* Nakai. Front. Pharmacol. 9: 1441.

Gurib-Fakim, A. 1996. Volatile constituents of the leaf oil of *Artemisia verlotiorum* Lamotte and *Ambrosia tenuifolia* Sprengel (syn.: *Artemisia psilostachya* auct. non L.). J. Essent. Oil. Res. 8(5): 559–561.

Gurib-Fakim, A., Mahomoodally, M.F. 2013. African Flora as potential sources of medicinal plants: towards the chemotherapy of major parasitic and other infectious diseases: a review. JJBS 147(624): 1–8.

Haider, S.Z., Mohan, M., Andola, H.C. 2014. Constituents of *Artemisia indica* Willd. from Uttarakhand Himalaya: A source of davanone. Pharmacogn. Res. 6(3): 257.

Heghes, S.C., Vostinaru, O., Rus, L.M., Mogosan, C., Iuga, C.A., Filip, L. 2019. Antispasmodic effect of essential oils and their constituents: A review. Molecules 24(9): 1675.

Heywood, V.H., Harborne, J.B., Turner, B.L. 1977. *In*: Biology and Chemistry of the Compositae. Academic Press.

Hristova, L., Damyanova, E., Doichinova, Z., Kapchina-Toteva, V. 2013. Effect of 6-benzylaminopurine on micropropagation of *Artemisia chamaemelifolia* Vill. (Asteraceae). Bulg. J. Agric. Sci. 19(2): 57–60.

Hussain, A.I., Anwar, F., Sherazi, S.T.H., Przybylski, R. 2008. Chemical composition, antioxidant and antimicrobial activities of basil (*Ocimum basilicum*) essential oils depends on seasonal variations. Food Chem. 108(3): 986–995.

Jalilzadeh-Amin, G., Maham, M., Dalir-Naghadeh, B., Kheiri, F. 2012. *In-vitro* effects of *Artemisia dracunculus* essential oil on ruminal and abomasal smooth muscle in sheep. Comp. Clin. Path. 21(5): 673–680.

Joshi, R.K. 2013. Antimicrobial activity of volatile oil of *Artemisia capillaris* growing wild in Uttrakhand Himalaya. J. Pharmacogn. Phytochem. 1(6).

Juteau, F., Jerkovic, I., Masotti, V., Milos, M., Mastelic, J., Bessiere, J.M., Viano, J. 2003. Composition and antimicrobial activity of the essential oil of *Artemisia absinthium* from Croatia and France. Planta Med. 69(02): 158–161.

Kamatou, G.P., Viljoen, A.M. 2010. A review of the application and pharmacological properties of α-Bisabolol and α-Bisabolol-rich oils. JAOCS 87(1): 1–7.

Khangholil, S., Rezaeinodehi, A. 2008. Effect of drying temperature on essential oil content and composition of sweet wormwood (*Artemisia annua*) growing wild in Iran. PJBS. 11(6): 934–937.

Kordali, S., Kotan, R., Mavi, A., Cakir, A., Ala, A., Yildirim, A. 2005. Determination of the chemical composition and antioxidant activity of the essential oil of *Artemisia dracunculus* and of the antifungal and antibacterial activities of Turkish *Artemisia absinthium*, *A. dracunculus*, *Artemisia santonicum*, and *Artemisia spicigera* essential oils. J. Agric. Food. Chem. 53(24): 9452–9458.

Koul, B., Taak, P., Kumar, A., Khatri, T., Sanyal, I. 2018. The artemisia genus: a review on traditional uses, phytochemical constituents, pharmacological properties and germplasm conservation. J. Glycom. Lipidom. 7: 1–7.

Kowalski, R., Wawrzykowski, J., Zawislak, G. 2007. Analysis of essential oils and extracts from *Artemisia abrotanum* L. and *Artemisia dracunculus* L. Herba Pol. 53(3).

Kumlay, A.M., Yildirim, B.A., Ekici, K., Ercisli, S. 2015. Screening biological activity of essential oils from *Artemisia dracunculus* L. Oxid. Commun. 38(3): 1320–1328.

Kunihiro, K., Myoda, T., Tajima, N., Gotoh, K., Kaneshima, T., Someya, T., Toeda, K., Fujimori, T., Nishizawa, M. (2017). Volatile components of the essential oil of *Artemisia montana* and their sedative effects. J. Oleo Sci. 66(8): 843–849.

Sánchez-González, L., Vargas, M., González-Martínez, C., Chiralt, A., Cháfer, M. 2011. Use of essential oils in bioactive edible coatings: a review. Food Eng. Rev. 3: 1–16.

Lee, Y.J., Thiruvengadam, M., Chung, I.M., Nagella, P. 2013. Polyphenol composition and antioxidant activity from the vegetable plant 'Artemisia absinthium' L. AJCS. 7(12): 1921.

Li, Y., Fabiano-Tixier, A.-S., Chemat, F. 2014. Essential Oils: From Conventional to Green Extraction. Springer, New York, NY.

Liu, N.Q., Van der Kooy, F., Verpoorte, R. 2009. *Artemisia afra*: a potential flagship for African medicinal plants? South African Journal of BotanyS. Afr. J. Bot. 75(2): 185–195.

Lopes-Lutz, D., Alviano, D.S., Alviano, C.S., Kolodziejczyk, P.P. 2008. Screening of chemical composition, antimicrobial and antioxidant activities of Artemisia essential oils. Phytochemistry 69(8): 1732–1738.

Macchioni, F., Perrucci, S., Flamini, G., Cioni, P.L., Morelli, I. 1999. Antimycotic activity against Saprolegnia ferax of extracts of *Artemisia verlotorum* and *Santolina etrusca*. Phytother. Res. 13(3): 242–244.

Mahop, T.M., Mayet, M. 2007. Enroute to biopiracy? Ethnobotanical research on antidiabetic medicinal plants in the Eastern Cape Province, South Africa. Afr. J. Biotechnol. 6(25): 2945–2952.

Malik, S., de Mesquita, L.S.S., Silva, C.R., de Mesquita, J.W.C., de Sá Rocha, E., Bose, J., Abiri, R., de Maria Silva Figueiredo, P., Costa-Júnior, L.M. 2019. Chemical Profile and Biological Activities of Essential Oil from *Artemisia vulgaris* L. Cultivated in Brazil. Pharmaceuticals 12(2): 49.

Marotti, M., Piccaglia, R., Giovanelli, E., Deans, S.G., Eaglesham, E. 1994. Effects of variety and ontogenic stage on the essential oil composition and biological activity of fennel (*Foeniculum vulgare* Mill.). J. Essent. Oil Res. 6(1): 57–62.

Martinotti, E., Calderone, V., Breschi, M.C., Bandini, P., Cioni, P.L. 1997. Pharmacological action of aqueous crude extracts of *Artemisia verlotorum* Lamotte (Compositae). PTR. 11(8): 612–614.

McGaw, L.J., Jäger, A.K., Van Staden, J. 2000. Antibacterial, anthelmintic and anti-amoebic activity in South African medicinal plants. J. Ethnopharmacol. 72(1-2): 247–263.

Mighri, H., Hajlaoui, H., Akrout, A., Najjaa, H., Neffati, M. 2010. Antimicrobial and antioxidant activities of *Artemisia herba-alba* essential oil cultivated in Tunisian arid zone. C. R. Chim. 13(3): 380–386.

Mitoshi, M., Kuriyama, I., Nakayama, H., Miyazato, H., Sugimoto, K., Kobayashi, Y., Jippo, T., Kuramochi, K., Yoshida, H., Mizushina, Y. 2014. Suppression of allergic and inflammatory responses by essential oils derived from herbal plants and citrus fruits. Int. J. Mol. Med. 33(6): 1643–1651.

Moghaddam, M., Mehdizadeh, L. 2017. Chemistry of Essential Oils and Factors Influencing Their Constituents. pp. 379–419. *In*: Soft Chemistry and Food Fermentation. Academic Press.

Mohamed, A.E.H.H., El-Sayed, M., Hegazy, M.E., Helaly, S.E., Esmail, A.M., Mohamed, N.S. 2010. Chemical constituents and biological activities of Artemisia herba-alba. Rec. Nat. Prod. 4(1).

Mojarrab, M.A.H.D.I., Emami, S.A., Gheibi, S., Taleb, A.M., Heshmati Afshar, F. 2016. Evaluation of anti-malarial activity of *Artemisia turcomanica* and *A. kopetdaghensis* by cell-free β-hematin formation assay. RJP. 3(4): 59–65.

Morales-López, J., Centeno-Álvarez, M., Nieto-Camacho, A., López, M.G., Pérez-Hernández, E., Pérez-Hernández, N., Fernández-Martínez, E. 2017. Evaluation of antioxidant and hepatoprotective effects of white cabbage essential oil. Pharm. Biol. 55(1): 233–241.

Moufid, A., Eddouks, M. 2012. *Artemisial herbal allbal*: A Popular Plant with Potential Medicinal Properties. PJBS, 15(24): 1152–1159.

Mucciarelli, M., Maffei, M. 2002. Introduction to the genus. 1–50 in CW Wright, *Artemisia*. Medicinal and aromatic plants-industrial profiles. V. 18.

Muleya, E., Ahmed, A.S., Sipamla, A.M., Mtunzi, F.M., Mutatu, W. 2014. Evaluation of anti-microbial, anti-inflammatory and anti-oxidative properties *Artemisia afra, Gunnera perpensa* and *Eucomis autumnalis*. NFS 4(6): 1.

Obistioiu, D., Cristina, R.T., Schmerold, I., Chizzola, R., Stolze, K., Nichita, I., Chiurciu, V. 2014. Chemical characterization by GC-MS and *in-vitro* activity against *Candida albicans* of volatile fractions prepared from *Artemisia dracunculus, Artemisia abrotanum, Artemisia absinthium* and *Artemisia vulgaris*. Chem. Cent. J. 8(1): 6.

Omer, B., Krebs, S., Omer, H., Noor, T.O. 2007. Steroid-sparing effect of wormwood (*Artemisia absinthium*) in Crohn's disease: a double-blind placebo-controlled study. Phytomedicine 14(2-3): 87–95.

Oyedeji, A.O., Afolayan, A.J., Hutchings, A. 2009. Compositional variation of the essential oils of Artemisia afra Jacq. from three provinces in South Africa-a case study of its safety. Nat. Prod. Commun. 4(6): 1934578X0900400622.

Özbek, H., Uğraş, S., Dülger, H., Bayram, I., Tuncer, I., Öztürk, G., Öztürk, A. 2003. Hepatoprotective effect of *Foeniculum vulgare* essential oil. Fitoterapia. 74(3): 317–319.

Padalia, R.C., Verma, R.S., Chauhan, A., Chanotiya, C.S. 2014. Seasonal Variation in Essential oil Composition of *Artemisia nilagirica* var. septentrionalis from Foot Hills of Western Himalaya. Rec. Nat. Prod. 8(3): 281.

Pérez, G.S., Zavala, S.M., Arias, G.L., Ramos, L.M. 2011. Anti-inflammatory activity of some essential oils. J. Essent. Oil. Res. 23(5): 38–44.

Quattrocchi, U. 2012. CRC world dictionary of medicinal and poisonous plants: common names, scientific names, eponyms, synonyms, and etymology (5 Volume Set). CRC press.

Rajeshkumar, P.P., Hosagoudar, V.B. 2012. Mycorrhizal fungi of *Artemisia japonica*. Bull. Basic Appl. Plant Biol. 2(1): 7–10.

Rana, V.S., Abirami, K., Blázquez, M.A., Maiti, S. 2013. Essential oil composition of *Artemisia annua* L. at different growth stages. JOSAC. 22(2): 181–187.

Remberg, P., Björk, L., Hedner, T., Sterner, O. 2004. Characteristics, clinical effect profile and tolerability of a nasal spray preparation of *Artemisia abrotanum* L. for allergic rhinitis. Phytomedicine 11(1): 36–42.

Saleh, A.M., Aljada, A., Rizvi, S.A., Nasr, A., Alaskar, A.S., Williams, J.D. 2014. *In vitro* cytotoxicity of *Artemisia vulgaris* L. essential oil is mediated by a mitochondria-dependent apoptosis in HL-60 leukemic cell line. BMC Complement. Altern. Med. 14(1): 226.

Samy, Z., Francis, G. 1999. A walk in Sinai (PDF). Egypt. J. Nat. Hist., 1, [ISSN1110-6867] 13.

Sardashti, A., Bazerafshan, I., Ganjali, A. 2015. Photochemical composition and antimicrobial of essential oils from two *Artemisia* species for their application in drinking-water. IJMRR. 3(5): 417–427.

Sayyah, M., Nadjafnia, L., Kamalinejad, M. 2004. Anticonvulsant activity and chemical composition of *Artemisia dracunculus* L. essential oil. J. Ethnopharmcol. 94(2-3): 283–287.

Seddiek, S.A., Ali, M.M., Khater, H.F., El-Shorbagy, M.M. 2011. Anthelmintic activity of the white wormwood, *Artemisia herbaalba* against *Heterakis gallinarum* infecting turkey poults. J. Med. Plants Res. 18;5(16): 3946–57.

Sen, R., Bandyopadhyay, S., Dutta, A., Mandal, G., Ganguly, S., Saha, P., Chatterjee, M. 2007. Artemisinin triggers induction of cell-cycle arrest and apoptosis in Leishmania donovani promastigotes. J. Med. Microbiol. 56(9): 1213–1218.

Singh, H.P., Mittal, S., Kaur, S., Batish, D.R., Kohli, R.K. 2009. Chemical composition and antioxidant activity of essential oil from residues of *Artemisia scoparia*. Food Chem. 114(2): 642–645.

Sinico, C., De Logu, A., Lai, F., Valenti, D., Manconi, M., Loy, G., Bonsignore, L., Fadda, A.M. 2005. Liposomal incorporation of *Artemisia arborescens* L. essential oil and *in vitro* antiviral activity. Eur. J. Pharm. Biopharm. 59(1): 161–168.

Steketee, R.W., Eisele, T.P. 2017. Watching the availability and use of rapid diagnostic tests (RDTs) and artemisinin-based combination therapy (ACT). Malaria J. 16(1): 165.

Stuckey, M., McGee, R.M.N. 2002. McGee & Stuckey's Bountiful Container: Create Container Gardens of Vegetables, Herbs, Fruits, and Edible Flowers. Workman Publishing.

Suresh, J., Vasavi, V.A., Rajan, D., Ihsanullah, M., Khan, M.N. 2010. Antimicrobial Activity of *Artemisia abrotanum* and Artemisia pallens. IJPPE. 3: 18–21.

Taherkhani, M. 2014. *In vitro* cytotoxic activity of the essential oil extracted from *Artemisia absinthium*. IJT. 8(26): 1152–1156.

Tajadod, G., Mazooji, A., Salimpour, F., Samadi, N., Taheri, P. 2012. The essential oil composition of *Artemisia vulgaris* L. in Iran. Ann. Biol. Res. 3(1): 385–389.

Tajehmiri, A., Issapour, F., Moslem, M.N., Lakeh, M.T., Kolavani, M.H. 2014. *In-vitro* antimicrobial activity of *Artemisia annua* leaf extracts against pathogenic bacteria. Adv. Stud. Biol. 6(3): 93–97.

Tajkarimi, M.M., Ibrahim, S.A., Cliver, D.O. 2010. Antimicrobial herb and spice compounds in food. Food Control, 21: 1199–1218.

Tariku, Y., Hymete, A., Hailu, A., Rohloff, J. 2010. Essential-oil composition, antileishmanial, and toxicity study of *Artemisia abyssinica* and *Satureja punctata* ssp. *punctata* from Ethiopia. Chem. Biodivers. 7(4): 1009–1018.

Tariq, K.A., Chishti, M.Z., Ahmad, F., Shawl, A.S. 2009. Anthelmintic activity of extracts of *Artemisia absinthium* against ovine nematodes. Vet. Parasitol. 160(1-2): 83–88.

Thomson, A.T. 1826. The London Dispensatory. Longman, Rees, Orme, Brown, and Green.

Tian, N., Tang, Y., Tian, D., Liu, Z., Liu, S. 2017. Determination of dihydroartemisinic acid in *Artemisia annua* L. by gas chromatography with flame ionization detection. Biomed. Chromatogr. 31(3): 3824.

Tilaoui, M., Mouse, H.A., Jaafari, A., Aboufatima, R., Chait, A., Zyad, A. 2011. Chemical composition and antiproliferative activity of essential oil from aerial parts of a medicinal herb *Artemisia herba-alba*. Rev. Bras. Farmacogn. 21(4): 781–785.

Tunón, H., Thorsell, W., Mikiver, A., Malander, I. 2006. Arthropod repellency, especially tick (*Ixodes ricinus*), exerted by extract from *Artemisia abrotanum* and essential oil from flowers of *Dianthus caryophyllum*, Fitoterapia 77(4): 257–261.

Van Wyk, B.E. 2008. A broad review of commercially important southern African medicinal plants. J. Ethnopharmcol. y 119(3): 342–355.

Van Wyk, B.E., Oudtshoorn, B.V., Gericke, N. 1997. Medicinal Plants of South Africa. Briza.

Vidic, D., Čopra-Janićijević, A., Miloš, M., Maksimović, M. 2018. Effects of Different Methods of Isolation on Volatile Composition of *Artemisia annua* L. Int. J. Anal. Chem.

Wani, H., Shah, S.A., Banday, J.A. 2014. Chemical composition and antioxidant activity of the leaf essential oil of *Artemisia absinthium* growing wild in Kashmir, India. Aust. J. Pharm. 3(2): 90–94.

Watson, L.E., Bates, P.L., Evans, T.M., Unwin, M.M., Estes, J.R. 2002. Molecular phylogeny of subtribe Artemisiinae (Asteraceae), including Artemisia and its allied and segregate genera. BMC Evol. Biol. 2(1): 17.

Watt, J.M., Breyer-Brandwijk, M.G. 1962. The medicinal and poisonous plants of southern and eastern Africa being an account of their medicinal and other uses, chemical composition, pharmacological effects and toxicology in man and animal. The Medicinal and Poisonous Plants of Southern and Eastern Africa being an Account of their Medicinal and other Uses, Chemical Composition, Pharmacological Effects and Toxicology in Man and Animal., (Edn 2).

Wetzstein, H.Y., Porter, J.A., Janick, J., Ferreira, J.F. 2014. Flower morphology and floral sequence in *Artemisia annua* (Asteraceae). AJB. 101(5): 875–885.

Williams, J.D., Saleh, A.M., Acharya, D.N. 2012. Composition of the essential oil of wild growing *Artemisia vulgaris* from Erie, Pennsylvania. Nat. Prod. Commun. 7(5): 1934578X1200700524.

World Health Organization. 2006. WHO monograph on good agricultural and collection practices (GACP) for *Artemisia annua* L.

Xu, Q., Zhang, L., Yu, S., Xia, G., Zhu, J., Zang, H. 2019. Chemical composition and biological activities of an essential oil from the aerial parts of *Artemisia Gmelinii* weber ex Stechm. Nat. Prod. Res. 1–4.

Younsi, F., Rahali, N., Mehdi, S., Boussaid, M., Messaoud, C. 2018. Relationship between chemotypic and genetic diversity of natural populations of *Artemisia herba-alba* Asso growing wild in Tunisia. Phytochemistry 148: 48–56.

Yu, H.H., Kim, Y.H., Kil, B.S., Kim, K.J., Jeong, S.I., You, Y.O. 2003. Chemical composition and antibacterial activity of essential oil of *Artemisia iwayomogi*. Planta Med. 69(12): 1159–1162.

Yu, Z., Wang, B., Yang, F., Sun, Q., Yang, Z., Zhu, L. 2011. Chemical Compositionand Anti-acetyl cholinesterase Activity of Flower Essential Oils of *Artemisia annua* at Different Flowering Stage. IJPR. 10(2): 265.

Zuzarte, M., Salgueiro, L. 2015. Essential oils chemistry. *In*: Bioactive essential oils and cancer (19–61). Springer, Cham.

14

The Potential Use of Mandacaru (*Cereus* spp.) Bioactive Compounds

Maria Gabrielly de Alcântara Oliveira, Giovanna Morghanna Barbosa do Nascimento and *Gleice Ribeiro Orasmo**

Introduction

Mandacaru (*Cereus* spp.) is a characteristic columnar cactus of the Caatinga biome (Rizzini 1992), but throughout Brazil, this genus is very expressive in the flora constitution of several states. The Cactaceae family is composed of 100 genus and 1,500 species, distributed almost exclusively in the dry regions of the Americas (Barthlott and Hunt 1993). The family is divided into three subfamilies-Opuntioideae, Pereskioideae, and Cactoideae. The latter contains the genus *Cereus*, which comprises upright and succulent stem plants, and was first described by Hermann in 1698 and later by Miller in 1754, and includes 900 species published (Davet 2005).

Mandacaru plants have similar morphology, but are termed as distinct species. Among the best-known species of *Cereus* are *C. adeemani*, *C. bicolor*, *C. comarapanus*, *C. friccie*, *C. jamacaru*, *C. hildmannianus*, *C. peruvianus*, *C. repandus*, *C. trigonodendron*, and *C. vargasianus* (Davet 2005). The species *Cereus jamacaru* De Candolle is found in the Brazilian northeast region, having been classified by Taylor and Zappi (2004) as a predominantly caatinga species, although it also occurs in other types of environments. Also noteworthy is the species *Cereus peruvianus* Miller, found in southern Brazil, as well as *C. repandus* Miller; and in the southeast region mandacaru is classified as *C. hildmannianus* K. Schum, but this species also occurs in Santa Catarina (southern region), where it is known as "tuna" (Colonetti 2012).

According to Britton and Rose (1963), plants of the genus *Cereus* can be trees or shrubs with erect stems, being described as consisting of columnar type stem with number and arrangement of variable longitudinal ribs, where the axillary buds containing thorns are inserted, known as areolas. It is also noteworthy that the mandacaru stem is richly mucilaginous (Colonetti 2012). This mucilage plays a very important role in plant physiology, ensuring low transpiration for adaptation in arid climates, as cacti normally grow under stress conditions (Alvarez et al. 1992), and this mucilage is commercially exploited in the manufacture of cosmetics (Figure 14.1).

These plants are considered important because they are widely used as ornamental and forage, mainly in northeastern Brazil, but also because they have a number of characteristics that are of

Department of Biology, Natural Sciences Center, Federal University of Piauí, Teresina, Piauí, Brazil.
* Corresponding author: gleice@ufpi.edu.br

Figure 14.1: Mandacaru plant. Source: author's collection.

economic, commercial, industrial, and medicinal interest. Studies aiming to search for natural products (metabolites) from mandacaru showed that *C. peruvianus* plants produce amine alkaloids (Vries et al. 1971, Oliveira and Machado 2003), esters with potential for application as a waterproofing barrier (Rezanka and Dembitsky 1998), and a viscous gum with various industrial applications (Alvarez et al. 1992, 1995, Nozaki et al. 1993, Barros and Nozaki 2002). The species *Cereus jamacaru* presents a great diversity of biological compounds, such as alkaloids, steroids, triterpenes, glycosides, oils, and waxes that are used by the pharmaceutical industry (Davet 2005).

Mandacaru also has great medicinal importance, and is widely used in the traditional medicine of northeastern Brazil. In other countries, such as Mexico, plants of the genus *Cereus* are widely used in folk medicine. According to Hollis and Scheinvar (1995), cactus is used by healers in Mexico as analgesics, antibiotics, diuretics, for the treatment of bowel problems, coughs, and cardiac and nervous disorders, to cure some types of ulcers, and also to control diabetes and cholesterol.

Mandacaru plants in northeastern Brazil, species *C. jamacaru* De Candolle, remain in edaphoclimatic conditions characterized by high temperatures, irregular rainfall, and low natural soil fertility. According to Cavalcanti and Resende (2007), mandacaru presents good development in areas degraded soils, and may repopulate areas where traditional crops are no longer possible, thus being important in the sustainability and conservation of the Caatinga biome's biodiversity.

Use of Mandacaru Plants and their Importance in the Semiarid

In the northeastern region of Brazil, Cactaceae are of great economic importance in forage activity (Davet 2005), especially during the dry season, and are used to feed cattle, goats, and sheep (Rocha and Agra 2002, Santos 2007). *C. jamacaru* is the main species used as animal feed. The highlight of these plants in the semiarid regions is their efficiency in remaining succulent during the drought, ensuring their continuous use as fodder due to a large amount of mucilage (Colonetti 2012) (Figure 14.2).

Mandacaru is harvested manually by removing the thorns using a machete, and preserving the main stem. The material is ground and offered to the animals, pure or mixed with forage sorghum

Figure 14.2: Use of mandacaru as forage. Source: www.calilanoticias.com/2012/05/mandacaru-vem-sendo-a-ultima-alternativa-para-alimentacao-do-gado.

silage, and/or aggregated with other foods. Silva et al. (2010a, b) showed higher values of weight gain in cattle and heifers fed with mandacaru and sorghum silage. Milk yield in goats also increased with a silk flower hay diet associated with mandacaru (Silva et al. 2011). These data show the nutritional value of mandacaru, and in fact, Cavalcanti and Resende (2007) concluded that mandacaru biomass is an important component of drought feeding.

Andrade et al. (2006) add that mandacaru plants, besides being used for animal feeding, have other economic highlights in northeastern Brazil, such as human food, ornamentation of squares and gardens, filling saddles and pillows, painting of houses, and as a source of wood for the manufacture of doors, windows, slats, and rafters. In fact, the species *C. jamacaru* can reach 10 m in height and its core is used as a raw material for the manufacture of doors (Scheinvar 1985). Silva (2015) also reports 12 categories of use of mandacaru in the semiarid, highlighting its use as medicinal, forage, and food.

In the jewelry design sector, a significant amount of products has been developed using the mandacaru thorn, in which local designers and artisans associate conventional materials (silver and crystals) and regional products (stone and leather) with the mandacaru thorn, producing jewelry with regional cultural identity, and differentiating their products from others (Lopes 2016).

Mandacaru Fruits and Genetic Improvement

Mandacaru fruits are exported to Asian countries for candy manufacturing (Santos 2007), and fruits produced by breeding in Israel are exported to Europe for fresh consumption, and sold as exotic fruit at a high cost (Mizrahi 2014). *C. peruvianus* has been domesticated since the 1990s in the Israeli region (Nerd et al. 1993, Weiss et al. 1993, Mizrahi and Nerd 1999), where it is characterized as a fruit crop, and commercially grown on a small scale.

Due to the high commercial value of its fruits, a breeding program has been implemented in Israel to obtain larger, tastier fruits that remain intact (without early cracking) until their full ripeness. In this program, the species *C. peruvianus* was pointed to as being limited by the low genetic variability found in clones that have been cultivated for this purpose (Gutman et al. 2001).

The predominant form of propagation of mandacaru plants in the northern and northeastern regions of Brazil should also contribute to the reduction of genetic variability, which is an important factor for the formation of strategic reserve banks and for improvement proposals. The development of faster-growing mandacaru varieties with reduced thorn numbers and size should lead to a reduction in production costs to provide food to animals and humans, as well as providing raw material of industrial interest. Thus, broadening the genetic basis of the species of *Cereus* may even provide an increase in the diversity of substances of interest, which these plants already exhibit (Figure 14.3).

Figure 14.3: Mandacaru fruit. Source: http://jardimdesuculentas.net76.net/apostila/07.html.

Mandacaru Mucilage: Important in Cosmetics Manufacturing

The production of mucilage is the main characteristic of Cactaceae, being composed of complex polysaccharides, with varied composition, depending on the species (Sáenz et al. 2004). Mucilage polysaccharides swell in the presence of water, taking gummy consistency, with adhesive and thickening properties, and may be considered as a potential hydrocolloid (Colonetti 2012). Many of these polysaccharides found in Cactaceae have been used to modify the rheological properties of some products. In traditional medicine, they are widely used for treating skin and epithelial wounds and also for mucosal irritation (Cai et al. 2008).

Plant polysaccharides are generally an interesting source of additives for various industries, especially the food and pharmaceutical industries. In the food industry, it is used in the preparation of jams and jellies, and in the pharmaceutical industry to give stability to emulsions and ointments, as well as in cosmetology for the production of creams based on the mucilaginous content of these plants (Hou et al. 2002).

As reported by Lopes (2016), since mid-2013 the French industry L'Occitane® has been commercially exploiting mandacaru extract for the development of products in cosmetology, such as the production of moisturizers, soaps, scrubs, hand creams, and deodorants, according to the author, and the company has provided the population of the semiarid with income generation (Figure 14.4).

Habit Cosméticos® has developed products based on mandacaru mucilage, and according to the company, the shampoo ensures deep hydration in damaged hair, which besides moisturizing, forms a film preventing water loss, stimulating the natural form of hydration, reports Bittes Cosméticos (2019). The moisturizing cream "Flor de Mandacaru", developed by the company Pedaços de Aromas Brasil®, is made with natural ingredients, free of artificial colors and parabens and, according to the manufacturer, is considered a vegan product, being free of testing in animals and ingredients of animal origin (Pedaços de Aromas Brasil 2019).

Figure 14.4: Moisturizing cream based on mandacaru mucilage, L'Occitane company. Source: https://br.loccitaneaubresil.com/product/creme-hidratante-desodorante-corporal-mandacaru-2.html.

The Potential Use of Compounds Found in Mandacaru

Among the most frequently found substances in Cactaceae are phenylethylamine alkaloids, such as hordenine, mescaline, and lofophorine. Phenylethylamine alkaloids have been found in *Echinocereus merkeri* (Agurell et al. 1969). Mandacaru plants of the species *Cereus jamacaru* have a great diversity of biological compounds—alkaloids, steroids, triterpenes, glycosides, oils, and waxes that are used by the pharmaceutical industry (Davet 2005).

In *Cereus jamacaru* was also identified the presence of tyramine alkaloid, known for its sympathomimetic activity and probably responsible for cardiotonic activity (Brhun and Lindgren 1976). This alkaloid is also common in other cactus species (Scheinvar 1985).

Phytochemical studies in mandacaru, in the species *C. jamacaru*, detected the amines- tyramine, hordein, and N-methylthiramine (Davet 2005). Brhun and Lindgren (1976) identified tyramine (or 2-p-hydroxyphenylethylamine) on fresh stems of *Cereus jamacaru*. According to Burret et al. (1982), the species *C. jamacaru* has no flavones, with predominance of kampferol and methyl-3-flavonols. Studies with the crude ethanolic extract of *C. jamacaru* stem detected β-sitosterol and tyramine (Davet et al. 2009). In the ethanolic and aqueous extracts of *Cereus jamacaru* cladodes, the alkaloids—tyramine, n-methyl tyramine and hordein, and amino acid tyrosine were also detected, and only in the ethanolic extract, the irritating compounds—anthracnone, hydroquinone, phenol, and geranyl acetone were detected (Medeiros 2011).

The metabolic profile of *Cereus peruvianus* mandacaru callus culture using mass spectrometry confirmed the presence of tyramine alkaloids in this species (Ferrarezi et al. 2015). The use of plant cell culture for the production of substances of interest has greatly contributed to advances in various areas of plant physiology and biochemistry. Different strategies using *in vitro* culture systems have been studied with the objective of increasing the production of secondary metabolites. Thus, an increase in alkaloid production was achieved in callus culture medium in *Cereus peruvianus* (Machado et al. 2006, Oliveira and Machado 2003, Rocha et al. 2005).

Flavonoids are a class of secondary metabolites commonly found in plants. Araújo et al. (2008) evaluated mandacaru, among other medicinal plants of Caatinga, for the number of tannins and flavonoids. The study showed a low tannin index for *C. jamacaru*, considering the average for the other evaluated plants, as well as the flavonoid indices, were considered below the average obtained by the evaluated plants. In order to establish the presence and distribution of flavonoids as a factor for chemotaxonomy, Burret et al. (1982) concluded that the species *C. jamacaru* would fall into the group Cereoideae, a group characterized by not having flavones.

Similar to the work of Araújo et al. (2008), Burret et al. (1982) detected a little amount of flavonoids in *C. jamacaru*. However, studies using High-Performance Liquid Chromatography (HPLC) revealed the presence of flavonoids in mandacaru samples, which showed relevant bands of these compounds in *C. jamacaru* (Nascimento and Orasmo 2017, Oliveira and Orasmo 2018) (Figure 14.5).

In order to chemically evaluate the fruits of the species *Cereus fernambucensis*, known as "manacaru" in southeastern Brazil, for possible use as a functional food, Souza (2013) identified the substances 3-O-rubinosidio and isoramnetina-3-O-raminosidium from the aqueous and methanolic extract by High-Performance Liquid Chromatography (HPLC). The study also revealed that the antioxidant capacity was high, ranging from 60 to over 90%, depending on the fraction and/or extract evaluated. The hydrolyzable tannins in the extract were not detected and the vitamin C content of the "manacaru" fruit is more concentrated in the peel than in the pulp, presenting a significant concentration compared to other Cactaceae and other commonly used fruits, since the sugars were concentrated significantly in the fruit pulp.

Mayworm and Salatino (1996) reported that the oils of the seeds of the *Cereus jamacaru* are rich in unsaturated fatty acids, mainly oleic acid and linoleic acid. In addition, saturated oils, such as palmitic and stearic oils were found. In order to improve the potential use of mandacaru polysaccharides in the industry, Alvarez et al. (1995) evaluated the pectic content quantitatively and qualitatively.

Figure 14.5: Ethanolic extract from mandacaru samples for HPLC analysis. Source: author's collection.

The authors concluded that the waxy pecto-cellulosic cuticle of columnar cactus cladodes *Cereus peruvianus* is a source of α-D-polygalacturonic acid or pectic acid. Cellulose nanowhishers were obtained from mandacaru spines, providing a new renewable source of reinforcement with potential nanocomposite applications (Nepomuceno et al. 2017).

Industrial Use of Mandacaru

Mandacaru has been used as a source of primary metabolism products as well as secondary metabolites for the pharmaceutical, food, and chemical industries. The stems of *C. peruvianus* plants produce wax esters with potential application as an impermeable barrier (Dembitsky and Rezanka 1996, Rezanka and Dembitsky 1998), and a viscous gum with various industrial applications (Alvarez et al. 1992, 1995). From the gum produced, arabinogalactan was isolated, which inhibited the formation of gastric lesions in ethanol-treated mice, suggesting its potential use in phytotherapeutic processes (Tanaka et al. 2010).

The studies by Nozaki et al. (1993) and Barros and Nozaki (2002) showed that the complex heteropolysaccharides constituting the stem of *C. peruvianus* can replace the application of synthetic polyelectrolytes used in industrial wastewater treatment processes. In a similar study, Zara et al. (2012) reports that mandacaru has also been used as a low-cost natural polymer for turbidity removal and as an aid in coagulation and flocculation in water treatment, making it clear by balancing its pH and total alkalinity. The author concludes that mandacaru polymers are efficient and are an alternative for water treatment, especially in the Brazilian semiarid region, where mandacaru is abundant.

According to Almeida et al. (2006), the fruit of mandacaru (*C. jamacaru*) has great potential for industrial use, as it has relatively high levels of total soluble solids and total sugars, and important constituents in biotechnological processes, such as alcoholic fermentation. Aiming at the physicochemical and chromatographic characterization of mandacaru fermented beverage, Almeida et al. (2011) found that fermented mandacaru had qualities comparable to other fermented fruit, such as cashews, oranges, and "cajá" (*Spondias mombin*) produced by other researchers. The authors reported that the production of fermented mandacaru is a way to obtain products with higher added value, generate profit, contribute to the development of the Northeast region, and enable the use of mandacaru fruit in agroindustry.

Phytochemical study using extracts from the peel and pulp of the fruits, as well as seeds of *C. jamacaru*, showed that the fruit has potential antioxidant compounds, and can be used both in natural consumption, in the manufacture of juices from pulp, as well as industry for the extraction of antioxidant compounds, which can be used as food additives (Brito 2015).

Use of Mandacaru in Folk Medicine

Mandacaru also has great medicinal importance, being widely used in the folk medicine of northeastern Brazil. Tourinho (2000) described the Cactaceae as constituting of therapeutic properties and cite the mandacaru, of the species *Cereus jamacaru* as of great efficiency to treat renal diseases. Agra (1996) also reported the effectiveness of *C. jamacaru* in the treatment of kidney problems, especially kidney stones, as well as its use as a syrup for the treatment of coughs, bronchitis, and ulcers. Andrade et al. (2006) also report that the association of mandacaru with legumes *Senna uniflora* and *Senna obtusifolia* in the form of tea is efficient to alleviate intestinal problems.

In folk medicine, in the northeastern semiarid, the roots and stems of the *C. jamacaru* cactus are used as diuretics, and the stem is used to reduce blood pressure as it has emenagogue properties. Whole plant syrup is used to combat scurvy and treat respiratory tract disorders, such as coughs, bronchitis, and ulcers (Scheinvar 1985).

According to Lucena et al. (2012), as for the use of *C. jamacaru* in folk medicine, users report that the most used parts of the plant are the pulp of the stem and the root, which are macerated, used as sauces, sitz bath, to relieve abnormalities in the intimate region, and by decoction, which consists in extracting the active ingredients from the plant through cooking. Such processes are used for various purposes, such as clearing veins, wounds in the womb, gastritis, inflammation, kidney problems, ulcer, discharge or "infection of the woman" (Figure 14.6).

Popular reports and studies of ethnopharmacology and ethnobotany have shown numerous uses of aqueous extracts of mandacaru. Its stem and root, used as infusions or decocos, are touted as diuretics and improve heart and kidney disease. Stem barks are scraped and macerated, diluted with water, and also used for kidney disorders and cholesterol control (Albuquerque et al. 2007a).

According to Guedes et al. (2009), popular culture also utilizes the infusion of mandacaru stem, the control of albuminuria, diabetes, the treatment of vesicular problems, and the alleviation of respiratory problems, such as cough and bronchitis. Gonzáles and Villarreal (2007) report that the needy population of northeastern Brazil makes use of flowers of *C. jamacaru*, infused or in nature, for the treatment of worms, boils, abscesses, and the mitigation of fevers.

Cereus jamacaru hydroethanolic extract showed evident tumor inhibition on mouse-induced tumors (Sarcoma 180), however, the authors consider further pharmacological studies are necessary to evaluate the potentiality of this species as antitumor (Souza et al. 2001). Messias et al. (2010) found that *C. jamacaru* methanolic extract showed no toxic reactions on most of the hematological and biochemical parameters studied in pregnant adult Wistar rats. However, increased serum levels appear to have liver overload, so the authors suggest further investigation.

Figure 14.6: Flowers of mandacaru are used in folk medicine. Source: https://www.flickr.com/photos/egbertoaraujo /6006758911.

Medeiros et al. (2019) reported a new phytochemical characterization of *C. jamacaru*, indicating its use as a herbal medicine in the treatment of obesity. The study showed that mandacaru extract reduced food intake and body weight gain in rats. However, the extract showed significant intrinsic genotoxic potential. The treatment also altered the expression of the enzymes ABCB1 and CYP2D4, suggesting to contribute to the pharmacokinetic effects of *C. jamacaru* extract.

Via Farma® produces a dried extract of mandacaru or koubo (*Cereus* spp.), marketed as an appetite suppressant due to the presence of the tiramide alkaloid, according to the manufacturer's information; The site also reports that *Cereus* cactus extract is diuretic because it has betalain and indicaxanthin, which eliminate fluids and toxins besides acting in the reduction of the cholesterol, because it contains omega 6 and 9, and has antioxidant activity because it is rich in vitamin C (Via Farma 2019).

Andrade et al. (2006) concluded that *C. jamacaru* in natural stem has been shown to have anti-inflammatory and contraceptive properties. The authors also reported that the action of tyramine, found on the stem and roots, has healing and antifungal action on rodent skin. Likewise, Santana (2016) showed anti-inflammatory activity from *C. jamacaru* extract, showing an effect on the reduction of acute inflammatory processes. In addition, the results show that mandacaru has contraceptive properties.

Cereus peruvianus constitutes 1.7% of the herbal medicine Sanativo® (SAN) of traditional use in the northeast region of Brazil, and the mandacaru is responsible for the asepsis of the affected regions. SAN is indicated for the treatment of wounds, burns, sore throats, and injured epithelial tissues, and is also composed of 20% "angico" (*Piptadenia colubrina*), which has hemostatic and healing action, 20% "aroeira" (*Schinus terebinthifolius*), anti-inflammatory and antibacterial, 1.7% of "camapu" (*Physalis angulata*), with balsamic and analgesic activity. In order to evaluate the healing activity and the possible toxic effects of oral administration of herbal medicine, Lima (2006) concluded that Sanativo® has a significant healing property and that oral treatment produces low toxicity in Wistar rats.

An extensive review by Silva (2015) recorded several cactus species, classifying them by number of users and categories of use. The species *C. jamacaru* stood out, being the most cited by the members of needy communities approached, for its use as medicinal, as well as the diversity of use and the number of citations. According to the author, communities in northeastern Brazil cited the use of mandacaru in the treatment of genitourinary, digestive, and respiratory disorders (Andrade et al. 2006, Agra et al. 2007a, 2008, Albuquerque et al. 2007a, Almeida et al. 2010, Lucena et al. 2014), as well as in the treatment of renal diseases (Albuquerque et al. 2007b, Marinho et al. 2011, Cordeiro and Felix 2014), of stomach ulcer and indication as diuretic (Agra et al. 2007a, 2008), and inflammation in general (Lucena et al. 2014). The species is still cited in wound healing and inflammation of the urethra (Andrade et al. 2006, Albuquerque et al. 2007a, Lucena et al. 2014), in the treatment of rheumatism (Albuquerque et al. 2007b, Marinho et al. 2011) and enteritis (Marinho et al. 2011), liver problems (Agra et al. 2007b, Alves and Nascimento 2010), and care after snake bite (Cordeiro and Felix 2014).

Lucena et al. (2015) listed the use categories of various cactus species cited by residents of the rural Santa Rita community in the municipality of Congo, Paraíba, northeastern Brazil. The species *C. jamacaru* was considered the most versatile, falling into the 11 categories recorded. For medicinal use, informants cited its use for influenza and cough, allergy, back problems, diabetes, rheumatism, kidney and worm problems, bronchitis, skin problem and tuberculosis, appendix and gallbladder, prostate disease, colic, and menstrual problems.

In order to investigate the anticholinesterase action of plant species with the potential to inhibit enzymes responsible for the emergence of degenerative diseases, such as Alzheimer's disease, Queiroz et al. (2011) tested in vitro ethanolic extracts of *Croton urucurana*, *Heteropterys aphrodisiac*, *Chenopodium ambrosioides*, and *Cereus jamacaru*. *C. urucurana* extract provided superior inhibitory action than *Solanum tuberosum* extract, used as a positive control, and *C. jamacaru* cactus exerted discrete anti-acetylcholinesterase activity. However, the authors report that further studies are needed for better and further information.

According to Davet (2005), Cactaceae are rich in steroids, which may be related to the antimicrobial activity presented by mandacaru, and mandacaru also presents tyramine, which has bactericidal action. So the prospects for obtaining natural antibiotics from the Brazilian cactus seem to be good, concludes Lopes (2016).

Conclusion

From north to south of Brazil, mandacaru plants, classified as different species, are of great economic, industrial, pharmacological, and medicinal importance. However, further studies are needed to detect new bioactive compounds and their mechanisms of action for their use. Its occurrence and uses in the northeast of Brazil are noteworthy, where the needy population of the semiarid makes use of mandacaru in various activities, such as the whole plant or its parts, as well as in folk medicine, albeit empirically. Thus, the proper management of mandacaru plants is important for the conservation of the genetic diversity of the species, as this also implies greater diversity of biomolecules and active compounds for use. Therefore, it is worth providing cultivation programmed for the use of the whole plant, for the consumption of fruits, or as a source of raw material for industry.

References

Agra, M.F. 1996. Plantas da medicina popular dos Cariris Velhos, Paraíba, Brasil: espécies mais comuns. Ed. União.

Agra, M.F., Freitas, P.F. e Barbosa-Filho, J.M. 2007a. Synopsis of the plants known as medicinal and poisonous in Northeast of Brazil. Rev. Bras. Farmacogn. 17(1): 114–140.

Agra, M.F., Baracho, G.S., Nurit, K., Basilio, I.J.L.D., Coelho, V.P.M. 2007b. Medicinal and poisonous diversity of the flora of "Cariri Paraibano", Brazil. J. Ethnopharmacol. 111: 383–395.

Agra, M.F., Silva, K.N., Basílio, I.J.L.D., Freitas, P.F., Barbosa-Filho, J.M. 2008. Survey of medicinal plants used in the region Northeast of Brazil. Rev. Bras. Farmacogn. 18(3): 472–508.

Agurell, S., Lundström, J., Masoud. 1969. Cactaceae alkaloids VII: alkaloids of *Echinocereus merkeri*. J. Pharm. Sci. 58(11): 1413–1414.

Albuquerque, U.P., Monteiro, J.M., Ramos, M.A. e Amorim, E.L.C. 2007a. Medicinal and magic plants from a public market in Northeastern Brazil. J. Ethnopharmacol. 110: 76–91.

Albuquerque, U.P., Medeiros, P.M., Almeida, A.L.S., Monteiro, J.M.M., Neto, E.M.F.L., Melo, J.G., Santos, J.P. 2007b. Medicinal plants of the caatinga (semiarid) vegetation of NE Brazil: A quantitative approach. J. Ethnopharmacol. 114: 325–354.

Almeida, C.F.C.B.R., Ramos, M.A., Amorim, E.L.C. e Albuquerque, U.P. 2010. A comparison of knowledge about medicinal plants for three rural communities in the semi-arid region of the northeast of Brazil. J. Ethnopharmacol. 127: 674–684.

Almeida, M.M., Silva, F.L.H., Conrado, L. de S., Freire, R.M.M., Valença, A.R. 2006. Caracterização físico-química de frutos do mandacaru. Rev. Bras. Prod. Agroind. 8(1): 35–42.

Almeida, M.M., Silva, F.L.H., Conrado, L. de S., Mota, J.C., Freire, R.M.M. 2011. Estudo cinético e caracterização da bebida fermentada do *Cereus jamacaru* P. DC. Revista Verde. 6(2): 176–183.

Alvarez, M., Costa, S.C., Huber, A., Baron, M., Fontana, J.D. 1995. The cuticle of the cactus *Cereus peruvianus* as a source of a homo-α-D-galacturonan. Appl. Biochem. Biotechnol. 51(1): 367–377.

Alvarez, M., Costa, S.C., Utumi, H., Huber, A., Beck, R., Fontana, J.D. 1992. The anionic glycan from the cactus *Cereus peruvianus*—structural features and potential uses. Appl. Biochem. Biotechnol. 34: 283–295.

Alves, J.A., Nascimento, S.S. 2010. Levantamento fitogeográfico das plantas medicinais nativas do Cariri paraibano. Rev. Geogr. Acad. 4(2): 73–85.

Andrade, C.T.S., Marques, J.G.W. e Zappi, D.C. 2006. Utilização medicinal de cactáceas por sertanejos baianos. Rev. Bras. Plantas Med. 8(3): 36–42.

Araújo, T.A. de S., Alencar, N.L., Amorim, E.L. de, Albuquerque, U.P. de. 2008. A new approach to study medicinal plants with tannins and flavonoids contents from the local knowledge. J. Ethnopharmacol. 120(1): 72–80.

Barros, M.J., Nozaki, J. 2002. Pollutants abatement from effluents of paper and pulp industries by flocculation/coagulation and photochemical degradation. Quim. Nova 25: 736–740.

Barthlott, W., Hunt, D.R. 1993. Cactaceae. pp. 161–197. *In*: Kubiztki, K., Rohwer, J.G., Bittrich, V. (eds.). The families and genera of vascular plants, v. 2. Springer-Verlag, Berlin, Germany.

Brhun, J., Lindgren, J. 1976. Cactaceae Alkaloids XXIII: Alkaloids of Pachycereus pectinaboriginum and *Cereus jamacaru*. *Lloydia*, 39(2-3): 175–177.

Brito, M.S. de. 2015. Avaliação do potencial antioxidante e citotóxico das partes do fruto de Cereus jamacaru DC, Cactaceae. Dissert. Programa de Pós-Graduação em Biotecnologia. Universidade Federal do Ceará. Sobral, CE.

Bittes Cosméticos. 2019. Avaliable: https://www.bittescosmeticos.com.br. Accessed: 28/10/2019.

Britton, N.L., Rose, J.N. 1963. The Cactaceae: descriptions and illustrations of plants of the cactus family. Dover Publications.

Burret, F., Lebreton, P., Voirin, B. 1982. Les aglycones flavoniques de Cactees: distribution, signification. Lloydia. 45(6): 687–693.

Cai, W., Gu, X., Tang, J. 2008. Extraction, purification, and characterization of the polysaccharides from *Opuntia milpa alta*. Carbohydr. Polym. 71(3): 403–410.

Cavalcanti, N.B., Resende, G.M. 2007. Efeito de diferentes substratos no desenvolvimento de mandacaru (*Cereus jamacaru* P. DC.), facheiro (*Pilosocereus pachycladus* Ritter), xiquexique (*Pilosocereus gounelli* (A. Webwr ex K. Schum.) Bly. ex Rowl.) e coroa-de-frade (*Melocactus bahiensis* Britton & Rose), R.C. 20: 28–35.

Colonetti. Vivian Caroline. 2012. Caracterização da mucilagem do fruto e cladódio de Cereus hildmaniannus K. Schum. Disser. (Pós-Graduação em Engenharia Química). Florianópolis: Universidade Federal de Santa Catarina. 83p.

Cordeiro, J.M.P. e Félix, L.P. 2014. Conhecimento botânico medicinal sobre espécies vegetais nativas da caatinga e plantas espontâneas no agreste da Paraíba. Brasil. Rev. Bras. Plantas Med. 16(3, supl. I): 685–692.

Davet, A. 2005. Estudo fitoquímico e biológico do cacto Cereus jamacaru De Candolle, Cactaceae. Disser. (Mestrado em Ciências Farmacêuticas). Curitiba: Universidade Federal do Paraná, 121p.

Davet, A., Carvalho, J.L.S., Dadalt, R.C., Virtuoso, S., Dias, J.F.G.D., Miguel, M.D., Miguel, O.G. 2009. *Cereus jamacaru*: a non buffered LC quantification method to nitrogen compounds. Chromatogr. 69: 245–247.

Dembitsky, M.V., Rezanka, T. 1996. Molecular species of wax esters in *Cereus peruvianus*. Phytochenistry, 42: 1075–1080.

Ferrarezi, A.A., Porto, C., Pilau, E.J., Mangolin, C.A., Oliveira, A.J.B., Gonçalves, R.A.C. 2015. Perfil metabólico da cultura de calos de Cereus peruvianus usando espectrometria de massas. *In*: XXIV EAIC Encontro Anual de Iniciação Científica. UEM. Maringá.

Gonzáles, O.H., Villarreal, O.B. 2007. Crassulacean acid metabolism photosynthesis in columnar cactus seedlings during ontogeny: the effect of light on nocturnal acidity accumulation and chlorophyll fluorescence. Am. J. Bot. 94(8): 1344–1351.

Guedes, R.S., Alves, E.U., Gonçalves, E.P., Bruno, R.D.L.A., Braga-Júnior, J.M., Medeiros, M.S.D. 2009. Germinação de sementes de *Cereus jamacaru* D.C. em diferentes substratos e temperaturas. Acta Sci. Biol. Sci. 31(2): 159–164.

Gutman, F., Bar-Zvi, D., Nerd, A., Mizrahi, Y. 2001. Molecular typing of *Cereus peruvianus* clones and their genetic relationship with other *Cereus peruvianus* species evaluated by RAPD analysis. J. Hortic. Sci. Biotechnol. 76: 709–713.

Hollis, H., Scheinvar, L. El interesante mundo de las Cactáceas. 1995. Mexico: Consejo Nacional de Ciencia y Tecnologia y Fondo de Cultura Economica, p. 92–96.

Hou, W.C., Hsu, F.L., Lee, M.H. 2002. Yam (*Dioscorea batatas*) tuber mucilage exhibited antioxidant activities *in vitro*. Planta Med. 68(12): 1072–1076.

Lima, C.R. de. 2006. Atividade cicatrizante e avaliação toxicológica pré-clínica do fitoterápico Sanativo®. Disser. (Pós-Graduação em Ciências Farmacêuticas). Recife: Universidade Federal de Pernambuco. 77p.

Lopes, J.de A. 2016. O Mandacaru e sua utilização como material expressivo e alternativo renovável no design e na arte. Disser. (Pós-Graduação em Artes Visuais). Salvador: Universidade Federal da Bahia. 105p.

Lucena, C.M., Carvalho, T.K.N., Marín, E.A., Nunes, E.N., Oliveira, R.S., Melo, J.G., Casas, A. e Lucena, R.F.P. 2014. Potencial medicinal de cactáceas en la región semiárida del Nordeste de Brasil. Gaia Sci. Ed. Esp.: Populações Tradicionais. 36–50.

Lucena, C.M., Carvalho,T.K.N., Ribeiro, J.E.S., Quirino, Z.G.M., Casas, A., Lucena, R.F.P. 2015. Conhecimento botânico tradicional sobre Cactáceas no Semiárido do Brasil. Gaia Sci. Ed. Esp.: Cactaceae. 9(2): 77–90.

Lucena, C.M., Costa, G.M., Sousa, R.F., Carvalho,T.K.N., Marreiros, N.A., Alves, C.A.B., Pereira, D.D., Lucena, R.F.P. 2012. Conhecimento local sobre Cactáceas em comunidades rurais na mesorregião do sertão da Paraíba (Nordeste, Brasil). Biotemas 25(3): 281–291.

Machado, F.A.P.S.A., Capelasso, M., Oliveira, A.J.B., Gonçalves, R.A.C., Zamuner, M.L.M., Mangolin, C.A., Machado, M.F.P.S. 2006. Alkaloid production and Isozymes Expression from Cell suspension culture of *Cereus peruvianus* Mill. (Cactaceae). J. Plant Sci. 1(4): 324–331.

Marinho, M.G.V., Silva, C.C., Andrade, L.H.C. 2011. Levantamento etnobotânico de plantas medicinais em área de caatinga no município de São José de Espinharas, Paraíba, Brasil. Rev. Bras. Plantas Med. 13(2): 170–182.

Mayworm, M., Salatino, A. 1996. Teores de óleo e composição de ácidos graxos de sementes de *Cereus jamacaru* DC (Cactaceae), *Zizyphus joazeiro* Mart. (Rhamnaceae) e *Anadenanthera colubrina* (Benth) *Brenan* var cebil (Griseb.) Von Altschul. (Mimosaceae). Sitientibus. 15: 201–209.

Medeiros, I.U. de, Medeiros, R.A. de, Bortolin, R.H., Queiroz, F.M.de, Silbiger, V.N., Pflugmacher, S., Schwarz, A. 2019. Genotoxicity and pharmacokinetic characterization of *Cereus jamacaru* ethanolic extract in rats. Biosci. Rep. 39 BSR20180672.

Medeiros, I.U. de. 2011. Identificação dos princípios ativos presentes no extrato etanólico de C. jamacaru e avaliação em ratos dos possíveis efeitos tóxicos e/ou comportamentais da exposição prolongada. Disser. (Pós-graduação em Ciências Farmacêuticas). Natal: Universidade Federal do Rio Grande do Norte. 122p.

Messias, J.B., Caraciolo, M.C.M., Oliveira, I.M.de, Montarroyos, U.R., Bastos, I.V.G.A., Guerra, M. de O., Souza, I.A. 2010. Avaliação dos parâmetros hematológicos e bioquímicos de ratas no segundo terço da gestação submetidas à ação de extrato metanólico de *Cereus jamacaru* D.C., Cactaceae. Rev. Bras. Farmacogn. 20: 478–483.

Mizrahi, Y. 2014. *Cereus Peruvianus* (Koubo) new Cactus Fruit for the world. Rev. Bras. Frutic. Jaboticabal, SP, 36(1): 068–078.

Mizrahi, Y., Nerd, A. 1999. Climbing and columnar cacti: new arid land fruit crops. pp. 358–366. *In:* Janick, J. (ed.). Perspectives on new crops and new uses. Alexandria: ASHS Press.

Nascimento, G.M.B. do, Orasmo, G.R. 2017. Perfil genético e metabólico em populações de mandacaru usando marcadores microssatélites e CLAE. *In*: XXVI SIC-Seminário de Iniciação Científica UFPI, Teresina-PI. Anais... Teresina: UFPI, 08–10 nov.

Nepomuceno, N.C., Santos, A.S.F., Oliveira, J.E., Glenn, G.M., Medeiros, E. S. 2017. Extraction and characterization of cellulose nanowhiskers from mandacaru (*Cereus jamacaru* DC.) spines. Cellulose. 24(1): 119–129.

Nerd, A., Raveh, E., Mizrahi, Y. 1993. Adaptation of five columnar cactus species to various conditions in the Negev desert of Israel. Econ. Bot. 43: 31–41.

Nozaki, J., Messerschmidt, I., Rodrigues, D.G. 1993. Tannery waters cleaning with natural polyeletrolytes: chemical speciation studies of chromium. Arq. Biol. Tecnol. 36: 761–770.

Oliveira, A.J.B., Machado, M.F.P.S. 2003. Alkaloid production by callous tissue cultures of *Cereus peruvianus* (Cactaceae). Appl. Biochem. Biotechnol. 104: 149–155.

Oliveira, M.G.de A., Orasmo, G.R. 2018. Detecção de compostos químicos de interesse e uso medicinal popular do mandacaru. *In*: XXVII SIC—Seminário de Iniciação Científica UFPI, Teresina-PI. Anais. Teresina: UFPI, 06–11 nov.

Pedaços de Aromas Brasil. 2019. Avaliable: http://www.pedacosdearomasbrasil.com.br Accessed: 28/10/2019.

Queiroz, F. M., Nascimento, M. A., Schawarz, A. 2011. Estudo preliminar *in vitro* da atividade antiacetilcolinesterásica de extratos etanólicos de plantas: possíveis alternativas no tratamento da Doença de Alzheimer. Rev. Biol. Farm. 6: 96–106.

Rezanka, T., Dembitsky, V.M. 1998. Very-long-chain alkyl esters in *Cereus peruvianus* wax. Phytochemistry, 42: 1145–1148.

Rizzini, C.T. 1992. Cactáceas: Os segredos da sobrevivência. *Ciên. Hoje* (Ed. Especial). Rio de Janeiro, p. 62–72.

Rocha, E.A., Agra, M.F. 2002. Flora do Pico do Jabre, Paraíba, Brasil: Cactaceae Juss. Acta Bot. Bras. 16: 15–21.

Rocha, L.K. da, Oliveira, A.J.B. de, Mangolin, C.A., Machado, M.F.P.S. 2005. Effect of different culture medium components on production of alkaloid in callus tissues of *Cereus peruvianus* (Cactaceae). Acta Sci. Biol. Sci. Maringá 27(1): 37–41.

Sáenz, C., Sepúlveda, E., Matsuhiro, B. 2004. *Opuntia* spp. mucilage's: a functional component with industrial perspectives. J. Arid Environm. 57(3): 275–290.

Santana, A.F. (2016). Composição química e atividade antioxidante das frações clorofórmica e hidrometanólica do extrato da raiz de *Cereus jamacaru* DC. (Cactaceae). 114 f. Dissertação (Mestrado Acadêmico em Recursos Genéticos Vegetais)—Universidade Estadual de Feira de Santana, Feira de Santana, Bahia.

Santos, P.R.G. 2007. Aspectos florísticos e vegetacionais de um inselbergue no semiárido de Pernambuco, Brasil. Recife: Universidade Federal de Pernambuco. Monografia (Bacharelado em Ciências Biológicas). 60p.

Scheinvar, L. 1985. Cactáceas. Flora Ilustrada Catarinense, Itajaí.

Silva, J.G.M., Lima, G.F.C., Aguiar, E.M., Melo, A.A.S., Rego, M.M.T. 2010b. Native cacti associated with sabiá and flor de seda shrub hays in male lamb feeding. Rev. Caat. 23: 123–129.

Silva, J.G.M., Lima, G.F.C., Paz, L.G., Aguiar, E.M., Melo, A.A.S., Rego, M.M.T. 2010a. Utilization of native cacti associated with sorghum silage on cattle feeding. Rev. Eletron. Cien. Centauro. (1): 01–09.

Silva, J.G.M., Melo, A.A.S., Rego, M.M.T., Lima, G.F.C., Aguiar, E.M. 2011. Native cacti associated with sabiá and flor de seda shrub hays in dairy goats' feeding. Rev. Caat. 24: 158–164.

Silva, V.A. 2015. Diversidade de uso das Cactáceas no Nordeste do Brasil: Uma Revisão. Gaia Sci. Ed. Esp.: Cactaceae. 9(2): 137–154.

Souza, I.A., Lima, M.C.A., Melo, U.B.C., Higino, J.S. 2001. Antitumour properties of *Cereus jamacaru* on an experimental model of câncer. *In*: Fundamental & Clinical Pharmacology. 3rd Meeting of the Federation of the European Pharmacological Societies, v. 15. Lyon, France.

Souza, R.D. de. 2013. Estudo dos Pigmentos polares do Extrato das cascas dos Frutos de Manacaru (*Cereus fernambucensis – Cactaceae*). Disser. Universidade Estadual do Norte Fluminense Darcy Ribeiro. Campos dos Goytacazes. Rio de Janeiro, RJ.

Tanaka, L.Y.A., Oliviera, A.J.B., Gonçalves, J.E., Cipriani, T.R., Souza, L.M., Marques, M.C.A., Werner, M.F.P., Baggio, C.H., Gorin, P.A.J., Sassaki, G.L., Iacomoni, M. 2010. An arabinogalactan with anti-ulcer protective effects isolated from *Cereus peruvianus*. Carbohydr. Polym. 82: 714–721.

Taylor, N.P., Zappi, D.C. 2004. Cacti of Eastern Brazil. Richmond: The Royal Bot. Gard. 460p.

Tourinho, M.J. 2000. Abordagem etnofarmacológica das plantas medicinais diuréticas no povoado de Capim Grosso, Município do Canindé de São Francisco, Sergipe, Curituba, Aracaju-SE 3(1): 34–47.

Via Farma. 2019. Avaliable: <http://viafarmanet.com.br>. Accessed: 28 out 2019.

Vries, J.X., Moyna, P., Diaz, V. 1971. Alkaloides cactos Uruguay. Rev. Latinoam. Quím. 3: 21–23.

Weiss, J., Nerd, A., Mizrahi, Y. 1993. Development of *Cereus peruvianus* (apple cactus) as a new crop for Negev desert of Israel. pp. 471–486. *In*: Janick, J., Simon, J.E. (eds.). In New Crops. New York: Wiley.

Zara, R.F., Thomazini, M.H., Lenz, G.F. 2012. Estudo da eficiência de polímero natural extraído do cacto mandacaru (*Cereus jamacaru*) como auxiliar nos processos de coagulação e floculação no tratamento de água. *Revista de Estudos Ambientais—REA*, 4(2esp): 75–83 (online).

15

Subfamily Bombacoideae:

Traditional Uses, Secondary Metabolites, Biological Activities, and Mechanistic Interpretation of the Anti-Inflammatory Activity of its species

Mariam I. Gamal El-Din, Fadia S. Youssef, Mohamed L. Ashour,*
Omayma A. Eldahshan and *Abdel Nasser B. Singab**

Introduction

Natural products from plants have been used for centuries as rich sources for curing various ailments. They are regarded as the backbone of traditional medicinal systems used throughout the whole world since antiquity (Thabet et al. 2018b). Traditional medicinal practices have formed the basis of most of the early medicines, followed by subsequent clinical, pharmacological, and chemical studies. Probably the most famous and well known example to date would be the synthesis of the anti-inflammatory agent, acetylsalicylic acid (aspirin), derived from the natural product salicin, isolated from the bark of the willow tree *Salix alba* L. The investigation of *Papaver somniferum* L. (opium poppy) resulted in the isolation of several alkaloids, including morphine, a commercially important drug, first reported in 1803 (Rishton 2008).

Recently, there is a focus on alternative therapies and the consumption of natural products for remedial purposes, especially those derived from plants. More than 13,000 plants have been studied during the last five years. Moreover, there has been a rapid elevation in the discovery of molecular targets that may be applied to the discovery of novel tools for the prevention and treatment of various human diseases (Rates 2001). Besides, natural products continue to provide unique structural diversity in comparison to standard combinatorial chemistry, which presents opportunities for discovering novel low molecular weight lead compounds. A huge number of natural product-derived compounds in various stages of clinical development highlighted the existing viability and significance of the use of natural products as sources of new drug candidates (Veeresham 2012).

The past 15 years have evidenced an explosion in the number of phylogenetic and taxonomic studies of the order Malvales. According to a recent Kubitzki system (2003) and the APG III system (APG, 2009), the previously recognized family Bombacaceae is recently being treated as subfamily

Department of Pharmacognosy, Faculty of Pharmacy, Ain-Shams University, 11566, Cairo, Egypt.
* Corresponding authors: dean@pharma.asu.edu.eg; ashour@pharma.asu.edu.eg

Bombacoideae under family Malvaceae (Bayer and Kubitzki 2003). Bombacoideae (alternatively Bombacaceae) is unchanged, except for the exclusion of the Asian tribe Durioneae in the recent Durionaceae.

Bombacoideae comprises around 30 genera of tropical trees mostly present in tropical regions of the world, especially tropical America. The largest genera include Bombax, Ceiba, and Adansonia (Baum and Oginuma 1994). Bombacoideae members are mostly present as large trees, with the tallest *Ceiba pentandra*, reaching a height of 70 meters (m). Trees are often spiny, buttressed, or have swollen trunks due to storage of water in the parenchyma. Leaves are often spiral, alternate, with undistinguished stipules, and pulvinate petioles. The leaves lamina are usually compound palmate with entire margin except for *Ochroma* species that are simple lobed. The existence of large showy actinomorphic flowers is a characteristic feature of Bombacoideae, with their short cylindrical epicalyx and pentamerous petals (Robyns 1964).

Phytochemical investigations of Bombacoideae resulted in the identification of almost 160 secondary metabolites belonging to various chemical classes, including flavonoids, lignans, procyanidinds, naphthoquinones, sesquiterpenes, and triterpenes. Different promising biological properties have been investigated, including antiproliferative, immunomodulatory, antiviral, antimicrobial, and anti-inflammatory activities. The aim of this chapter is to shed light on the recently isolated constituents, as well as the pharmacological activities of different genera of this medicinally valuable Bombacoideae. Data was collected from different databases, comprising Web of Knowledge (http://www.webofknowledge.com), PubMed (http://www.ncbi.nlm.nih.gov/pubmed/), and Scifinder (https://scifinder.cas.org/scifinder/login) until August 2019.

Ethnopharmacological and Traditional uses of Bombacoideae

Bombacoideae is the subfamily of many economically important plants. Many of its members are of considerable ecological and economic importance.

Adansonia digitata L. (Baobab) is also termed the chemist tree as well as the small pharmacy, and is famous for its countless medicinal and non-medicinal uses. Different plant parts have been widely used in folk medicine; the pulp was traditionally used for treating dysentery and fevers. Pulp extract was administered as eye-drops to alleviate measles. Various herbal remedies are composed mainly of leaves. Meanwhile, its dried powdered roots can be formulated as a mash, and is administered as a tonic to patients suffering from malaria. Moreover, its bark is found to be effective in treating sores owing to its semi-fluid gum (De Caluwé et al. 2010, Gebauer et al. 2002, Kaboré et al. 2011, Kamatou et al. 2011, Wickens 1980).

Bombax ceiba L. plant is an economical plant source of cotton and timber in addition to its use in the manufacture of matchsticks. According to Ayurveda, Bombax pods have stimulant, haemostatic, and astringent effects. The bark is mucilaginous and was given as infusion for demulcent and emetic effects. A paste of flowers and leaves is employed externally for the relief of skin troubles. The roots are used as a tonic in syphilis. Meanwhile, its bark gum is used in menorrhagia and leucorrhoea. Besides, the stem bark is used as a tonic in boils, acne, and pimples, whereas the seeds are applied on the skin to alleviate smallpox and chickenpox (Hossain et al. 2011a, Singh and Panda 2005, Williamson 2002).

Ceiba pentandra L. Gaertn., the tropical tree, gains its popularity in the industry of lifejackets and cushions owing to the water-resistant and elastic floss of their capsules that are known as kapok. Besides, it is used in folk medicine to relieve fever, parasitic infections, diarrhea, as well as gonorrhea in Central America, Asia, Oceania, and Africa. Water infusion of *C. pentandra* bark is traditionally used for the management of asthma in the Samoan Islands (Cox 1993, Dunstan et al. 1997, Noreen et al. 1998).

Chiranthodendron pentadactylon Larreat., is well known as the monkey's, the Devil's, or Mexican hand tree for its distinct red flowers, resembling human hands. The flowers were used for the treatment of heart diseases and the management of lower abdominal pain. Besides, they were used to lower edema and serum cholesterol levels. In Mexican traditional medicine, it was used to relieve different ailments, including headache, eye pain, diarrhea, dysentery, epilepsy, and cancer (Etkin 2000, Velázquez et al. 2009).

Pachira *P. aquatica* and *P. glabra* are commercially sold under the name money trees as they are supposed to bring good luck and money into the owner's home (Cheng et al. 2017). Medicinally, the skin of the immature green fruit of *P. aquatica* is sometimes used in the treatment of hepatitis. The seeds are used as an anesthetic. A cold water infusion of the crushed leaves is used to treat burning sensation in the skin. The bark is sometimes used medicinally to treat stomach complaints and headaches, and herbal tea of the boiled bark is used as a blood tonic (Barwick and Schokman 2004).

Secondary Metabolites Isolated from Subfamily Bombacoideae

(i) Flavonoids

Flavonoids constitute the major class of secondary metabolites that are predominant in Bombacoideae (Tables 15.1–15.5). Forty seven flavonoid compounds were isolated from different organs of *Adansonia, Bombax, Bombacopsis, Ceiba, Chorisia, Pachira, and Ochroma*. Flavones and flavonols are highly represented along with their hydroxy and methoxy derivatives, amounting up to 30% of the total previously isolated phytoconstituents of the subfamily (Tables 15.1 and 15.2). Polyhydroxy and methoxy derivatives occur mostly as flavones and flavonols isolated from different organs.

Only four flavonols have been reported, including the benzopyran dimer, shamimicin flavonol **(41)** that has been isolated from *Bombax ceiba* stem bark (Saleem et al. 2003). Flavanones, flavanonols, and isoflavones are not highly represented in Bombacoideae, although a few compounds have been reported, represented by hesperidin **(43)**, vavain **(45)**, and its glycosides **(46, 47)** (Ngounou et al. 2000, Noreen et al. 1998, Ueda et al. 2002, Qi et al. 1996).

(ii) Xanthones

Xanthones represent another class of complex phenolic compounds that is closely related to flavonoids, both in structure and chromatographic behavior. Despite the rareness of xanthone C-glycosides in the Malvaceae (Negi et al. 2013), four xanthone-*C*-glucosides, namely, 2-C-β-D-glucopyranosyl-1,3,6,7-tetrahydroxy-xanthone (Mangiferin) **(48)**, 2-*C*-β-D-glucopyranosyl-1,6,7-trihydroxy-3-*O*-(*p*-hydroxybenzoyl)-xanthone (49), 4-*C*-β-D-glucopyranosyl-1,6,8-trihydroxy-3,7-di-*O*-(*p*-hydroxybenzoyl)-xanthone **(50)**, and 4-*C*-β-D-glucopyranosyl-1,3,6, 8-tetrahydroxy-7-*O*-(*p*-hydroxybenzoyl)-xanthone **(51)** were isolated from *Bombax ceiba* leaves (Zoghbi et al. 2003) (Table 15.6). Moreover, the acetoxy xanthone, xanthone-3-acetoxy-1-hydroxy-6-methoxy-8-*O*-β-D-glucopyranosyl-(1→3)-α-L-rhamnoside **(52)**, was isolated from *Bombax ceiba* flowers (Sati et al. 2011).

(iii) Procyanidins

Procyanidins (condensed tannins) have gained a great attention in the last few years owing to their strong observed radical scavenging, antioxidant, anti-HIV, and antiviral activities (Shahat 2006). Four oligomeric procyanidins were isolated from the methanol extract of *Adansonia digitata* fruits, which are epicatechin-(4-β-8)-epicatechin **(53)**, epicatechin-(4-β-6)-epicatechin **(54)**, epicatechin-

Table 15.1: Chemical structures of the isolated flavones and their distribution in subfamily Bombacoideae.

Compound name	R_1	R_2	R_3	R_4	R_5	R_6	Plant name	Parts used	References
Apigenin (1)	OH	H	OH	H	H	OH	*Bombax ceiba*	Flowers	(El-Hagrassi et al. 2011)
Apigenin-7-*O*-α-L-rhamnoside (2)	OH	H	*O*-α-Rha.	H	H	OH	*Chorisia crispiflora*	Leaves	(Ashmawy et al. 2012)
Apigenin 7-*O*-β-D-glucoside (Cosmetin) (3)	OH	H	*O*-β- Glc.	H	H	OH	*Bombax ceiba*	Flowers	(El-Hagrassi et al. 2011)
							Chorisia crispiflora	Leaves	(Ashmawy et al. 2012)
Apigenin 7-*O*-neohesperidoside (Rhoifolin) (4)	OH	H	*O*-neo-hesperidoside	H	H	OH	*Chorisia insignis* C. crispiflora C. speciosa C. pubiflora	Leaves	(Ashmawy et al. 2012, Coussio 1964)
							C. chodatii C. speciosa	Flowers	(Refaat et al. 2015a)

Table 15.1 contd. ...

...Table 15.1 contd.

Compound name	Structure						Plant name	Parts used	References
Apigenin 7-O-β-D-rutinoside (5)	OH	H	O-β-rutinoside	H	H	OH	*Chorisia insignis*	Leaves	(El-Alfy et al. 2010)
Vitexin (6)	OH	H	OH	C-β-Glc.	H	OH	*Bombax ceiba*	Flowers	(Joshi et al. 2013)
							Ochroma pyramidale	Leaves	(Vázquez et al. 2001)
Isovitexin (Saponaretin) Apigenin-6-C-β-D-glucoside (7)	OH	C-β-Glc.	OH	H	H	OH	*Bombax ceiba*	Flowers	(El-Hagrassi et al. 2011, Joshi et al. 2013)
Apigenin-6-C-β-D-glucosyl-7-O-β-D-glucoside (Saponarin) (8)	OH	C-β-Glc.	O-β-Glc.	H	H	OH	*Bombax ceiba*	Flowers	(El-Hagrassi et al. 2011)
Vicenin 2 (9)	OH	C-β-Glc.	OH	C-β-Glc.	H	OH	*Bombax ceiba*	Flowers	(El-Hagrassi et al. 2011, Joshi et al. 2013)
Apigenin-4'-methylether-7-O-β-rutinoside (Linarin) (10)	OH	H	O-β-rutinoside	H	H	OMe	*Bombax ceiba*	Flowers	(El-Hagrassi et al. 2011)
Luteolin 7-O-β-D-glucoside (11)	OH	H	O-β-Glc.	H	OH	OH	*Chorisia chodatii*	Flowers	(Refaat et al. 2015a)
Luteolin 7-O-β-D-rutinoside (12)	OH	H	O-β-rutinoside.	H	OH	OH	*Chorisia insignis*	Leaves	(El-Alfy et al. 2010)
5,7-Dimethoxyflavone (13)	OMe	H	OMe	H	H	H	*Bombax anceps*	Roots	(Sichaem et al. 2010)
5-Hydroxy-7,4'-dimethoxyflavone (14)	OH	H	OMe	H	H	OMe	*Bombax anceps*	Roots	(Sichaem et al. 2010)
Xanthomicrol (4',5-dihydroxy-6,7,8-trimethoxyflavone) (15)	OH	OMe	OMe	OMe	H	OH	*Bombax ceiba*	Flowers	(El-Hagrassi et al. 2011)

Table 15.2: Chemical structures of the isolated flavonols and their distribution in subfamily Bombacoideae.

Compound name	Structure	R_1	R_2	R_3	R_4	R_5	R_6	R_7	R_8	R_9	Plant name	Parts used	References
Kaempferol (16)	OH	OH	OH	H	OH	H	H	H	OH	H	*Bombax ceiba*	Flowers	(Jain and Verma 2012)
Kaempferol-3-*O*-β-D-glucoside (Astragalin) (17)	*O*-β-Glc.	*O*-β-Glc.	OH	H	OH	H	H	H	OH	H	*Chorisia chodatii* *Chiranthodendron pentadactylon*	Flowers Flowers	(Refaat et al. 2015a) (Velázquez et al. 2009)
Kaempferol-3-*O*-β-D-rutinoside (18)	*O*-β-rutinoside	*O*-β-rutinoside	OH	H	OH	H	H	H	OH	H	*Chorisia crispiflora* *Bombax ceiba*	Leaves Flowers	(Ashmawy et al. 2012) (Joshi et al. 2013)
Kaempferol-3-*O*-β-D-glucuronoside (19)	*O*-β-Glucurono.	*O*-β-Glucurono.	OH	H	OH	H	H	H	OH	H	*Bombax ceiba*	Flowers	(Joshi et al. 2013)
Kaempferol-3-*O*-β-D-(6''-E-p-coumaroyl)-glucopyranoside (Tiliroside) (20)	*O*-β-(6''-p-coumaroyl)-Glc.	*O*-β-(6''-p-coumaroyl)-Glc.	OH	H	OH	H	H	H	OH	H	*Chorisia chodatii* *Chiranthodendron pentadactylon*	Flowers Flowers	(Refaat et al. 2015a) (Velázquez et al. 2009)
Kaempferol-3-*O*-β-D-(6''-acetyl)-glucoside (21)	*O*-β-(6''-acetyl)-Glc.	*O*-β-(6''-acetyl)-Glc.	OH	H	OH	H	H	H	OH	H	*Chorisia chodatii*	Flowers	(Refaat et al. 2015a)
Sexangularetin-3-*O*-sophoroside (22)	2(β-glucosyl)-*O*-β-Glc.	2(β-glucosyl)-*O*-β-Glc.	OH	H	OH	OMe	H	H	OH	H	*Bombax ceiba*	Flowers	(Joshi et al. 2013)
Quercetin (23)	OH	OH	OH	H	OH	H	H	OH	OH	H	*Bombax ceiba*	Flowers	(Joshi et al. 2013)
Quercetin-3-*O*-β-D-glucuronoside (24)	*O*-β-glucuronide	*O*-β-glucuronide	OH	H	OH	H	H	OH	OH	H	*Bombax ceiba*	Flowers	(Joshi et al. 2013)

Table 15.2 contd. ....

...Table 15.2 contd.

Compound name	Structure							Plant name	Parts used	References
Quercetin-3-O-β-D-gluco side (Isoquercitrin) (25)	O-β-Glc.	OH	H	OH	H	OH	OH	Adansonia digitata Bombax ceiba Chiranthodendron pentadactylon	Fruits Flowers Flowers	(Shahat 2006) (Joshi et al. 2013) (Velázquez et al. 2009)
Quercetin-3-O-β-D-glucuronoside (26)	O-β-glucuronide	OH	H	OH	H	OH	OH	Bombax ceiba	Flowers	(Joshi et al. 2013)
Quercetin-3-O-β-D-galacturonoside (27)	O-β-galacturonide	OH	H	OH	H	OH	OH	Bombax ceiba	Flowers	(Said et al. 2011)
Quercetin-3-O-β-D-rutinoside (Rutin) (28)	O-β-rutinoside	OH	H	OH	H	OH	OH	Bombax ceiba Chorisia insignis	Flowers Leaves	(Joshi et al. 2013) (El-Alfy et al. 2010)
Santin-7-methyl ether (29)	OMe	OH	OMe	OMe	H	H	OMe	Pachira aquatica	Stems	(Cheng et al. 2017)
3,5,7-Trimethoxyflavone (30)	OMe	OMe	H	OMe	H	H	H	Bombax anceps	Roots	(Sichaem et al. 2010)
5-Hydroxy-3,7,3',4'-tetramethoxyflavone (31)	OMe	OH	H	OMe	H	OMe	OMe	Bombacopsis glabra Bombax ancepskae Pachira aquatica	Stem barks Root barks Roots Stems	(Paula et al. 2002) (Sichaem et al. 2010) (Cheng et al. 2017)
5-Hydroxy-3,7,4'-trimethoxy-flavone (Retusin) (32)	OMe	OH	H	OMe	H	H	OMe	Pachira aquatica	Stems	(Cheng et al. 2017)
5-Hydroxy-3,6,7,4'-tetramethoxyflavone (33)	OMe	OH	OMe	OMe	H	H	OMe	Bombacopsis glabra	Stems Root barks	(Paula et al. 2006a)
5,4'-Dihydroxy-3,6,7,8-tetramethoxyflavone (34)	OMe	OH	OMe	OMe	OMe	H	OH	Pachira aquatica	Stems	(Cheng et al. 2017)
5-Hydroxy-3,6,7,8,4'-pentamethoxyflavone (5-Hydroxyauranetin) (35)	OMe	OH	OMe	OMe	OMe	H	OMe	Bombacopsis glabra Pachira aquatica	Stem barks Stems	(Paula et al. 2002) (Cheng et al. 2017)
5,4'-Dihydroxy-3,7-dimethoxyflavone (36)	OMe	OH	H	OMe	H	H	OH	Pachira aquatica	Stems	(Cheng et al. 2017)
3,5,6,7,8,3',4'-Hepta-methoxyflavone (37)	OMe	OMe	OMe	OMe	OMe	OMe	OMe	Pachira aquatica	Stems	(Cheng et al. 2017)
Shamimin (38)	OH	OH	C-β-Glc.	OH	H	OH	OH	Bombax ceiba	Leaves	(Shahat et al. 2003)

Table 15.3: Chemical structures of the isolated flavanols and their distribution in subfamily Bombacoideae.

Compound name	Structure		Plant name	Parts used	References

	R_1	R_2			
(+)-Catechin (**39**)	β-OH	H	*Ceiba pentandra*	Stem barks	(Noreen et al. 1998)
			Chiranthodendron pentadactylon	Flowers	(Velázquez et al. 2009)
			Ochroma pyramidale	Leaves	(Vázquez et al. 2001)
(-)+Epicatechin (**40**)	α-OH	H	*Adansonia digitata*	Fruits	(Shahat 2006)
			Chiranthodendron pentadactylon	Flowers	(Velázquez et al. 2009)
			Ochroma pyramidale	Leaves	(Vázquez et al. 2001)
5,7,3',4'-Tetrahydroxy-6-methoxyflavan-3-O-β-D-glucopyranosyl-α-D-xyloside (**41**)	O-β-Glc.-α-Xyl.	OMe	*Bombax ceiba*	Roots	(Chauhan et al. 1980)
Shamimicin (**42**) (1''',1''''''-bis-2-(3,4-dihydroxy-phenyl)-3,4-dihydro-3,7-dihydroxy-5-O-xylopyranosyloxy-2H-1-benzopyran)			*Bombax ceiba*	Stem barks	(Saleem et al. 2003)

Table 15.4: Chemical structures of the isolated flavanones and flavanonols and their distribution in subfamily Bombacoideae.

Compound name	Structure						Plant name	Parts used	References
	R_1	R_2	R_3	R_4	R_5				
Hesperidin (**43**)	H	OH	*O*-rutinoside	OH	OMe		*Bombax ceiba*	Roots	(Qi et al. 1996)
3,7-Dihydroxy flavan-4-one-5-*O*-β-D-galactopyranosyl (1→4)-β-D-glucoside (**44**)	OH	4-*O*-(β–galactosyl) β-Glc.	OH	H	H		*Adansonia digitata*	Roots	(Chauhan et al. 1984)

Table 15.5: Chemical structures of the isolated isoflavones and their distribution in subfamily Bombacoideae.

Compound name	Structure		Plant name	Parts used	References
Vavain (**45**) (5,3'-dihydroxy-7,4',5'-trimethoxyisoflavone)	R: OH		*Ceiba pentandra*	Stem barks	(Ueda et al. 2002), (Ngounou et al. 2000)
Vavain 3'-*O*-β-D-glucoside (5-hydroxy-7,4',5'-trimethoxyisoflavone-3'-*O*-β-D-glucoside) (**46**)	R: *O*-β-Glc.		*Ceiba pentandra*	Stem barks	(Ueda et al. 2002), (Ngounou et al. 2000)
5-Hydroxy-7,4',5'-trimethoxyisoflavone 3'-*O*-α-L-arabinofuranosyl(1→6)-β-D-glucoside (**47**)	R: 6-*O*-(α-arabinosyl)-β-Glc.		*Ceiba pentandra*	Stem barks	(Ueda et al. 2002)

Table 15.6: Chemical structures of the isolated xanthones and their distribution in subfamily bombacoideae.

Compound name	R_1	R_2	R_3	R_4	R_5	R_6	Plant name	Parts used	References
2-C-β-D-Glucopyranosyl-1,3,6, 7-tetrahydroxy-xanthone (Mangiferin) (**48**)	C-β-Glc.	OH	H	OH	OH	H	*Bombax ceiba*	Leaves	(Versiani 2004)
2-C-β-D-Glucopyranosyl-1,6,7-trihydroxy-3-O-(p-hydroxybenzoyl)-xanthone (**49**)	C-β-Glc.	O-p-hyroxy-benzoyl	H	OH	OH	H	*Bombax ceiba*	Leaves	(Versiani 2004)
4-C- β-D-Glucopyranosyl-1,6,8-trihydroxy-3,7-di-O-(p-hydroxybenzoyl)-xanthone (**50**)	H	O-p-hyroxy-benzoyl	C-β-Glc.	OH	O-p-hyroxy-benzoyl	OH	*Bombax ceiba*	Leaves	(Versiani 2004)
4-C-β-D-Glucopyranosyl-1,3,6, 8-tetrahydroxy-7-O-(p-hydroxybenzoyl)-xanthone (**51**)	H	OH	C-β-Glc.	OH	O-p-hyroxy-benzoyl	OH	*Bombax ceiba*	Leaves	(Versiani 2004)
Xanthone 3-acetoxy-1-hydroxy-6-methoxy-8-O-β-D-glucopyranosyl-(1→3)-α-L-rhamnoside (**52**)	H	OAc	H	OMe	H	3-O-(α-Rham.)-β-Glc.	*Bombax ceiba*	Flowers	(Sati et al. 2011)

(2β-O-7,4β-8)-epicatechin **(55)**, and epicatechin-(4-β-8)-epicatechin-(4-β-8)-epicatechin **(56)**. It is noteworthy to mention that they mainly consist of 2–3 interlinked 2–3 flavan-3-ol units (Table 15.7).

(iv) Simple Phenolics and Phenylpropanoids

Simple phenolic acids and phenolic esters have been reported from *Bombax ceiba* and *Chorisia crispiflora*, including protocatechuic acid **(57)**, gallic acid **(58)**, ethyl gallate **(59)**, 1-galloyl-β-glucose **(60)**, *trans*-3-(*p*-coumaroyl) quinic acid **(61)**, and neochlorogenic acid **(62)** (Ashmawy et al. 2012, Dhar and Munjal 1976, Wu et al. 2008).

Other miscellaneous phenolic compounds, including Shamiminol **(63)**, Bombalin **(64)**, and (2R, 3R, 4R, 5S)-5-(6-(2, 3-dimethylbutyl)-7-hydroxy-2-(4-hydroxyphenyl)-2H-chromen-5-yloxy)-6-methyl-tetrahydro-2H-pyran-2,3,4-triol **(65)**, were isolated from *B. ceiba* leaves and flowers (Table 15.8) (Faizi and Ali 1999, Khan et al. 2012, Wu et al. 2008).

(v) Lignans

Lignans, a subgroup of non-flavonoid polyphenols, are widely distributed in *Bombax* and *Ochroma*. Six lignan compounds, including (+)-pinoresinol **(66)**, matairesinol **(67)**, 5,6-dihydro-xymatairesinol **(68)**, bombasin **(69)**, bombasin-4-O-β-glucoside **(70)**, and dihydro-dehydro-diconiferylalcohol- 4-O-β-D-glucoside **(71)** were isolated from the flowers and roots of *B. ceiba* (Wu et al. 2008, Wang et al. 2013). Besides, another ten lignans compounds were isolated from *Ochroma lagopus* heartwood ethanol extract, including boehmenan A-D **(72-75)**, carolignan A-F **(76-81)**, and secoisolariciresinoyl diferulate **(82)** (Paula et al. 1995) (Table 15.9).

(vi) Coumarins

Coumarins are not widely represented in subfamily Bombacoideae, except for scopoletin **(83)** and cleomiscosin **(84)**, which were reported from the heartwood of *Ochroma lagopus* and *Pachira aquatica* stem (Cheng et al. 2017, Paula et al. 1996) (Table 15.10).

(vii) Naphthoquinones

Naphthoquinone, a class of compounds biosynthesized via shikimic acid pathway, is not widely represented in the subfamily Bommbacoideae. Only six compounds were isolated restricted to three genera *Bombax*, *Ceiba*, and *Pachira*. Bombaxquinone B (2-*O*-Methyl-isohemigossypolone) **(86)** was the most representative naphthoquinone isolated from roots, heartwood of *Bombax ceiba*, roots of *Bombax anceps*, as well as the root barks of *Ceiba pentandra* and *Pachira aquatica*. Isohemigossypolone **(85)** and other methyl derivatives, including 11-nor-2-*O*-methylisohemigossypolone **(87)**, 2,7-Dimethoxy-8-formyl-5-isopropyl-3-methyl-1,4-naphthoquinone **(88)**, and bombamalone D **(87)** were isolated from *B. ceiba* (Table 15.11).

(viii) Sesquiterpenes

Cadinane sesquiterpenes constitute the second most representative constituents of this subfamily after flavonoids. Hemigossypol **(90)**, isohemigossypol **(93)** and their methyl ether derivatives, hemigossypol-6-methylether **(91)**, hemigossypol-1,6,7-trimethy) ether **(92)**, isohemigossypol-1-methyl ether **(94)**, isohemigossypol-2-methylether **(95)**, isohemigossypol-1,2-methylether **(96)**, 2-acetyl-isohemigossypol-1-methylester **(97)**, 7-hydroxycadalene **(98)**, and Lacinilene C **(99)** have been isolated from *Bombax ceiba*, *B. anceps* (Sankaram et al. 1981, Seshadri et al. 1973) and *Ceiba pentandra* roots (Sichaem et al. 2010). In addition, sesquiterpene lactones are widely distributed in

Table 15.7: Chemical structures of the isolated procyanidins and their distribution in subfamily Bombacoideae.

Compound name	Structure	Plant name	Parts used	References
Epicatechin-(4-β8)-epicatechin (**53**)		*Adansonia digitata*	Fruits	(Shahat 2006)
Epicatechin-(4-β6)-Epicatechin (**54**)		*Adansonia digitata*	Fruits	(Shahat 2006)

Epicatechin-(2-β-O-7,4 β -8)-epicatechin
(**55**)

Adansonia digitata   Fruits   (Shahat 2006)

Epicatechin-(4-β-8)-epicatechin -(4-β-8)-
epicatechin (**56**)

Adansonia digitata   Fruits   (Shahat 2006)

Table 15.8: Chemical structures of miscellaneous phenolic compounds and their distribution in subfamily Bombacoideae.

Compound name	Structure	Plant name	Parts used	References

	R₁	R₂			
Protocatechuic acid (57)	H	H	*Chorisia crispiflora*	Leaves	(Ashmawy et al. 2012)
Gallic acid (58)	OH	H	*Bombax ceiba*	Seeds	(Dhar and Munjal 1976)
Ethyl gallate (59)	OH	-C₂H₅	*Bombax ceiba*	Seeds	(Dhar and Munjal 1976)
1-Galloyl-β-glucose (60)	OH	-Glc.	*Bombax ceiba*	Seeds	(Dhar and Munjal 1976)

trans-3-(*p*-coumaroyl) Quinic acid (61) — *Bombax ceiba* — Flowers — (Wu et al. 2008)

Neochlorogenic acid (62) — *Bombax ceiba* — Flowers — (Wu et al. 2008)

Shamiminol
(3,4,5-trimethoxy-phenol 1-*O*-β-D-
xyloside-(1→2)-β-D-glucoside) (**63**)

Bombax ceiba   Leaves   (Faizi and Ali 1999)

Bombalin
(6-*O*-4'-(hydroxycinnamoyl)-3-methyl-
D-gulono-γ-lactone) (**64**)

Bombax ceiba   Flowers   (Wu et al. 2008)

(2R, 3R, 4R, 5S)-5-(6-(2,
3-dimethylbutyl)-7-Hydroxy-2-(4-
hydroxyphenyl)
-2H-chromen-5-yloxy)-6-methyl-
tetrahydro-2H-pyran-2,3,4-triol (**65**)

Bombax ceiba   Flowers   (Khan et al. 2012)

Table 15.9: Chemical structures of isolated lignans and their distribution in subfamily Bombacoideae.

Compound name	Structure	Plant name	Parts used	References
(+)-Pinoresinol (**66**)		*Bombax ceiba*	Roots	(Wang et al. 2013)
Matairesinol (**67**)	R:H	*Bombax ceiba*	Roots	(Wang et al. 2013)
5,6-Dihydro-xymatairesinol (**68**)	R: OH	*Bombax ceiba*	Roots	(Wang et al. 2013)

	R₁	R₂			
Bombasin (**69**)	H	-COMe	*Bombax ceiba*	Flowers	(Wu et al. 2008)
Bombasin-4-*O*-β-glucoside (**70**)	Glc.	-COMe	*Bombax ceiba*	Flowers	(Wu et al. 2008)
Dihydro-dehydro-diconiferylalcohol- 4-*O*-β-D-glucoside (**71**)	Glc.	-(CH₂)₃OH	*Bombax ceiba*	Flowers	(Wu et al. 2008)

Boehmenan A (**72**)	R: H; $\Delta^{7,8}$(E), $\Delta^{7'',8''}$(E)		*Ochroma lagopus*	Heartwood	(Paula et al. 1995)
Boehmenan B (**73**)	R: H; ; $\Delta^{7,8}$(E), $\Delta^{7'8''}$(Z)		*Ochroma lagopus*	Heartwood	(Paula et al. 1995)
Boehmenan C (**74**)	R: H; ; $\Delta^{7,8}$(Z), $\Delta^{7'8''}$(E)		*Ochroma lagopus*	Heartwood	(Paula et al. 1995)
Boehmenan D (**75**)	R: OMe; ; $\Delta^{7,8}$(E), $\Delta^{7'8'}$(E)		*Ochroma lagopus*	Heartwood	(Paula et al. 1995)

Table 15.9 contd. ...

...Table 15.9 contd.

Compound name	Structure	Plant name	Parts used	References
Carolignan A (**76**)	R: H; 7´ α-OH; $\Delta^{7,8}$ (E), $\Delta^{7'8''}$ (E)	*Ochroma lagopus*	Heartwood	(Paula et al. 1995)
Carolignan B (**77**)	R: H; 7´ β-OH; $\Delta^{7,8}$ (E), $\Delta^{7''8'''}$ (E)	*Ochroma lagopus*	Heartwood	(Paula et al. 1995)
Carolignan C (**78**)	R: H; 7´ β-OH; $\Delta^{7,8}$ (E), $\Delta^{7'8''}$ (Z)	*Ochroma lagopus*	Heartwood	(Paula et al. 1995)
Carolignan D (**79**)	R: H; 7´ β-OH; $\Delta^{7,8}$ (Z), $\Delta^{7'8''}$ (E)	*Ochroma lagopus*	Heartwood	(Paula et al. 1995)
Carolignan E (**80**)	R: OMe; 7´ α-OH; $\Delta^{7,8}$ (E), $\Delta^{7'8''}$ (E)	*Ochroma lagopus*	Heartwood	(Paula et al. 1995)
Carolignan F (**81**)	R:OMe; 7´ β-OH; $\Delta^{7,8}$ (E), $\Delta^{7'8''}$ (E)	*Ochroma lagopus*	Heartwood	(Paula et al. 1995)
Secoisolariciresinoyl diferulate (**82**)		*Ochroma lagopus*	Heartwood	(Paula et al. 1995)

Table 15.10: Chemical structures of isolated coumarins and their distribution in subfamily Bombacoideae.

Compound name	Structure	Plant name	Parts used	References
Scopoletin (83)		*Ochroma lagopus*	Heartwood	(Paula et al. 1996)
		Pachira aquatica	Stems	(Cheng et al. 2017)
Cleomiscosin (84)		*Ochroma lagopus*	Heartwood	(Paula et al. 1996)

subfamily Bombacoideae, where 16 compounds **(100-115)** have been isolated from *B. ceiba*, *Pachira aquatica* (Zhang et al. 2008, Zhang et al. 2007) and *C. pentandra* (Puckhaber and Stipanovic 2001, Rao et al. 1993) (Table 15.12).

(ix) Triterpenes

Different triterpenes with a C27 and C30 skeleton have been isolated from different Bombacoideae. Ursolic acid **(116)**, α-amyrin **(117)**, and β-amyrin palmitate **(123)** were isolated from *Adansonia digitata* pulp (Al-Qarawi et al. 2003). Lupeol **(118)**, lupenone **(120)**, 2α, 3β-dihydroxylup-20(29)-ene **(121)**, (24R)-9,19-cyclolanost-25-ene-3β,24-diol **(126)**, (24S)-9,19-cyclolanost-25-ene-3β,24-diol **(127)**, and 9,19-cyclolanost-23-ene-3β,25-diol **(128)** were reported from *Bombacopsis glabra* stem bark and *Pachira aquatica* stem (Cheng et al. 2017, Paula et al. 2002).

The asymmetric bis-norsesquiterpenoid, aquatidial **(130)**, was isolated from the outer bark of *P. aquatica* root together with lupeol **(118)** (Paula et al. 2006b). Oleanolic acid **(125)**, gossypol **(129)**, as well as lupeol **(118)**, lupeol aceate **(119)**, α-amyrin **(117)**, and β-amyrin **(122)** were isolated from *Bombax ceiba* root, stem bark, and flowers (Jain and Verma 2012, Qi et al. 1996, Seshadri et al. 1973). β-Amyrone **(124)** and lupeol **(118)** were isolated from *Chorisia crispiflora* flowers (Hassan 2009). Lupeol **(118)** was also isolated from *Bombax anceps* roots, *Cavanillesia hylogeiton*, and *Ochroma pyramidale* leaves, representing the major prevalent triterpene in Bombacoideae (Sichaem et al. 2010, Bravo et al. 2002, Vázquez et al. 2001) (Table 15.13).

(x) Sterols

The sterol content of seed lipid unsaponifiable matter of six *Adansonia* s (*Adansonia digitata*, *A. fony*, *A. za*, *A. madagascariensis*, *A. suarezensis*, *A. grandidiera*) was investigated. In all s, β-sitosterol (24-ethyl-cholesterol) **(133)** was the major component, followed by campesterol (24-Methyl-cholesterol) **(132)**, and isofucosterol (24-ethylidene-cholesterol) **(136)**. Cholesterol **(131)**, stigmasterol (24-ethyl-5,22-cholestadien-3-β-ol) **(137)**, Δ^7-avenasterol (24-Ethylidene-7-cholesten-3-β-ol) **(139)**, and Δ^7-stigmasterol (24-ethyl-7-cholesten-3-β -ol) **(140)** were found at much lower concentrations at all investigated *Adansonia* (Bianchini et al. 1982). β-Sitosterol **(133)** was also isolated from *Bombax ceiba*, *Ceiba pentandra*, *Chorisia chodatii*, *C. crispiflora*, *Ochroma pyramidale*,

Table 15.11: Chemical structures of isolated naphthoquinones and their distribution in subfamily Bombacoideae.

Compound name	Structure				Plant names	Parts used	References
	R_1	R_2	R_3	R_4			
Isohemigossypolone (**85**)	OH	Me	OH	CHO	*Pachira aquatica* *Pachira glabra*	Root barks Root barks	(Shibatani et al. 1999a) (Paula et al. 2006a)
Bombaxquinone B (**86**) (2-*O*-Methyl-isohemigossypolone)	OMe	Me	OH	CHO	*Bombax anceps* *Bombax ceiba* *Ceiba pentandra* *Pachira aquatica*	Roots Heartwood Root barks Roots Root barks Root barks, Stems	(Sichaem et al. 2010) (Sankaram et al. 1981) (Reddy et al. 2003) (Zhang et al. 2007) (Rao et al. 1993) (Shibatani et al. 1999a), (Cheng et al. 2017)
11-nor-2-*O*-Methyl isohemigossypolone (**87**)	OMe	Me	OH	H	*Bombax ceiba* *Pachira aquatica*	Heartwood Root barks	(Sreeramulu et al. 2001) (Shibatani et al. 1999a)
2,7-Dimethoxy-8-formyl-5-isopropyl-3-methyl-1,4-naphthoquinone (**88**)	OMe	Me	OMe	CHO	*Bombax ceiba*	Root barks	(Sankaram et al. 1981)
Bombamalone D (**89**) (5,8-Dihydro-2-hydroxy-4-isopropyl-7-methoxy-6-methyl-5,8-dioxonaphthalene-1-carboxylic acid)	OMe	Me	OH	COOH	*Bombax ceiba*	Roots	(Zhang et al. 2007)

Table 15.12: Chemical structures of isolated sesquiterpenes and their distribution in subfamily Bombacoideae.

Compound name	R_1	R_2	R_3	R_4	R_5	Plant name	Parts used	References
Hemigossypol (**90**)	OH	H	OH	OH	CHO	*Bombax ceiba*	Roots	(Seshadri et al. 1973)
Hemigossypol-6-methylether (**91**)	OH	H	OMe	OH	CHO	*Bombax ceiba*	Roots	(Seshadri et al. 1973)
Hemigossypol-1,6,7-trimethylether (**92**)	OMe	H	OMe	OMe	CHO	*Bombax ceiba*	Roots	(Seshadri et al. 1973)
Isohemigossypol (**93**)	OH	OH	H	OH	CHO	*Bombax ceiba*	Roots	(Seshadri et al. 1976)
Isohemigossypol-1-methyl ether (**94**)	OMe	OH	H	OH	CHO	*Bombax ceiba*	Roots	(Seshadri et al. 1973, Sankaram et al. 1981)
						Bombax anceps	Roots	(Sichaem et al. 2010)
Isohemigossypol-2-methylether (**95**)	OH	OMe	H	OH	CHO	*Bombax ceiba*	Roots	(Sankaram et al. 1981)
						Bombax anceps	Roots	(Sichaem et al. 2010)
Isohemigossypol-1,2-methylether (**96**)	OMe	OMe	H	OH	CHO	*Bombax ceiba*	Roots	(Sankaram et al. 1981, Puckhaber and Stipanovic 2001)
2-Acetyl-isohemigossypol-1-methylester (**97**)	OMe	OCOMe	H	OH	CHO	*Bombax ceiba*	Root barks	(Sankaram et al. 1981)
7-Hydroxycadalene (**98**)	H	OH	H	OH	Me	*Bombax ceiba*	Roots	(Sankaram et al. 1981)
						Ceiba pentandra	Heartwood	(Sreeramulu et al. 2001)
							Root barks	(Rao et al. 1993)

Table 15.12 contd. ...

...Table 15.12 contd.

Compound name	Structure	Plant name	Parts used	References
Lacinilene C (**99**)		*Bombax ceiba*	Roots	(Zhang et al. 2007)
Bombamalone A (**100**)		*Bombax ceiba*	Roots	(Zhang et al. 2007)
Hibiscone D (**101**)		*Pachira aquatica*	Stems	(Cheng et al. 2017)

O-Methyl hibiscone D (**102**)	*Pachira aquatica*	Stems	(Cheng et al. 2017)
Bombamalone B (**103**)	*Bombax ceiba*	Roots	(Zhang et al. 2007)
Bombamalone C (**104**)	*Bombax ceiba*	Roots	(Zhang et al. 2007)

Table 15.12 contd. ...

...Table 15.12 contd.

Compound name	Structure	Plant name	Parts used	References
Hibiscone C (Gmelofuran) (**105**)		*Bombax ceiba* *Pachira aquatica*	Roots Stems	(Zhang et al. 2008) (Cheng et al. 2017)
Bombamaloside (**106**)		*Bombax ceiba*	Roots	(Zhang et al. 2007)
Bombaside (**107**)	 	R$_1$	R$_2$	R$_3$	*Bombax ceiba*	Roots	(Zhang et al. 2008)

For compound **107**:

R$_1$	R$_2$	R$_3$
H	H	Apiosyl-(1→6)-glucosyl

	R1	R2	R3	R4			
Bombaxone (**108**)	H	OH	H		*Bombax ceiba*	Roots	(Zhang et al. 2008)
7-*O*-β-Gluco-pyranosyl bmbaxone (**109**)	O-Glucosyl	OH	OH	H	*Bombax ceiba*	Roots	(Zhang et al. 2008)

	R1	R2	R3	R4			
Hibiscolactone A (**110**)	OH	H	H	OH	*Pachira aquatica*	Stems	(Cheng et al. 2017)
5-Isopropyl-3-methyl-2,4,7-trimethoxy-8,1-naphthalene carbolactone (**111**)	OMe	OMe	H	OMe	*Bombax ceiba*	Roots	(Reddy et al. 2003)
Isohemigossylic acid lactone 2-methyl ether (**112**)	OMe	H	H	OH	*Bombax ceiba*	Roots	(Puckhaber and Stipanovic 2001, Zhang et al. 2007)
Isohemigossylic acid lactone-7-methyl ether (**113**)	OH	H	H	OMe	*Bombax ceiba* *Ceiba pentandra* *Pachira aquatica*	Roots Root barks Stems	(Puckhaber and Stipanovic 2001) (Rao et al. 1993) (Cheng et al. 2017)
Isohemigossylic acid lactone-2,7-dimethyl ether (**114**)	OMe	H	H	OMe	*Bombax ceiba* *Ceiba pentandra*	Roots Root barks	(Puckhaber and Stipanovic 2001) (Rao et al. 1993)
11-Hydroxy-2-*O*-methylhibiscolactone A (**115**)					*Pachira aquatica*	Stems	(Cheng et al. 2017)

Table 15.13: Chemical structures of isolated triterpenes and their distribution in subfamily Bombacoideae.

Compound	Structure	Plant name	Parts used	References
Ursolic acid (116)	**R:** COOH	*Adansonia digitata*	Pulps	(Al-Qarawi et al. 2003)
α-Amyrin (117)	**R:** Me	*Adansonia digitata* *Bombax ceiba*	Pulps Flowers	(Al-Qarawi et al. 2003) (El-Hagrassi et al. 2011)
	R₁			
	R₂			

Lupeol (**118**)	OH	H	*Bombax ceiba*	Stem barks	(Saleem et al. 2003)
			Bombax anceps	Roots	(Jain and Verma 2012)
			Cavanillesia hylogeiton	Roots	(Sichaem et al. 2010)
			Ochroma pyramidale	Stem bark	(Bravo et al. 2002)
			Bombacopsis glabra	Leaves	(Vázquez et al. 2001)
			Pachira aquatica	Stem barks	(Paula et al. 2006a)
				Root barks	(Paula et al. 2006c)
Lupeol acetate (**119**)	OCOCH$_3$	H	*Bombax ceiba*	Flowers	(Jain and Verma 2012)
Lupenone (**120**)	=O	H	*Bombacopsis glabra*	Stem barks	(Paula et al. 2002)
			Pachira aquatica	Stems	(Cheng et al. 2017)
2α,3β-Dihydroxylup-20(29)-ene (**121**)	OH	OH	*Pachira aquatica*	Stems	(Cheng et al. 2017)

	R$_1$	R$_2$			
β-Amyrin (**122**)	OH	Me	*Bombax ceiba*	Flowers	(Jain and Verma 2012)
β-Amyrin palmitate (**123**)	β-*O*-palmitoyl	Me	*Adansonia digitata*	Pulps	(Al-Qarawi et al. 2003)
β-Amyrone (**124**)	=O	Me	*Chorisia crispiflora*	Flowers	(Hassan 2009)
Oleanolic acid (**125**)	OH	COOH	*Bombax ceiba*	Roots	(Qi et al. 1996)
			Ochroma pyramidale	Leaves	(Vázquez et al. 2001)

Table 15.13 contd. ...

...Table 15.13 contd.

Compound	Structure		Plant name	Parts used	References
(24R)-9,19-Cyclolanost-25-ene-3β,24-diol (**126**)	**R:** β-OH		*Bombacopsis glabra*	Stem barks	(Paula et al. 2002)
			Pachira aquatica	Stems	(Cheng et al. 2017)
(24S)-9,19-Cyclolanost-25-ene-3β,24-diol (**127**)	**R:** α-OH		*Bombacopsis glabra*	Stem barks	(Paula et al. 2002)
9,19-Cyclolanost-23-ene-3β, 25-diol (**128**)			*Bombacopsis glabra*	Stem barks	(Paula et al. 2002)
			Pachira aquatica	Stems	(Cheng et al. 2017)

Gossypol (**129**)

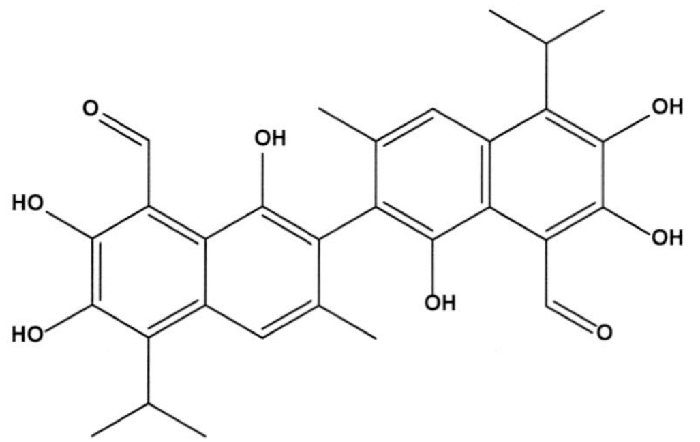

Bombax ceiba

Roots

(Seshadri et al. 1973)

Aquatidial (**130**)

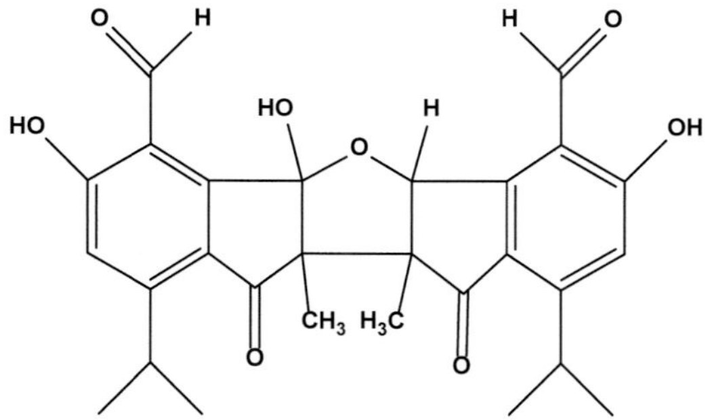

Pachira aquatica

Root barks

(Paula et al. 2006b)

O. lagopus, and *Pachira glabra* (Azab et al. 2013, Chauhan et al. 1980, Jain and Verma 2012, Ngounou et al. 2000, Paula et al. 1996, Paula et al. 2006a, Refaat et al. 2015a, Vázquez et al. 2001). Moreover, *β*-sitorstenone (141) was isolated from *Cavanillesia hylogeiton* stem bark, and cholestenone (142) was isolated from *Bombax anceps* roots (Bravo et al. 2002, Sichaem et al. 2010). Sterol glycosides were also reported, including 24-*β*-ethylcholest-5-en-3-*β*-yl-*α*-l-arabinosyl-(1→6)-*β*-*D*-glucoside (135) from *B. ceiba* flowers (Antil et al. 2013), stigmasterol-3-*O*-*β*-*D*-glucoside (138) from *Chorisia crispiflora* leaves (Azab et al. 2013), and *β*-sitosterol-3-*O*-*β*-D-glucoside (Daucosterol) (134) from *B. ceiba*, *C. pentandra*, *Chorisia crispiflora*, *Cavanillesia hylogeiton*, and *Ochroma pyramidale* (Antil et al. 2013, Azab et al. 2013, Bravo et al. 2002, Ngounou et al. 2000, Qi et al. 1996, Vázquez et al. 2001) (Table 15.14).

(xi) Fatty Acids

The oil content of several members of Bombacoideae were evaluated for their fatty acid composition, including *Adansonia digitata*, *A. fony*, *A. za*, *A. madagascariensis*, *A. suarezensis*, *A. grandidiera*, *Bombax costatum*, *Chorisia speciosa*, *Lagunaria patersonii*, *Pachira glabra*, *P. aquatica*, and *Ochroma lagopus*. Among normal fatty acids, palmitic acid (147), stearic acid (152), oleic acid (154), linoleic acid (155), linolenic acid (156), and sterculic acid (164) were observed in most species. Caproic (143), caprylic (144), arachidic (157), lignoceric (160), and vernolic (161) acids were found in *B. costatum* seed oil. Investigation of different *Adansonia* species demonstrated the existence of myristic (tetradecanoic acid) (145), pentadecanoic acid (146), palmitoleic acid (148), heptadecanoic acid (149), heptadecenoic acid (150), heptadecadienoic acid (151), octadec-7-enoic acid (153), arachidic acid (157), eicosenoic acid (158), and behenic acid (159) (Table 15.15).

(xii) Alkaloids

Alkaloids are not accumulated to any significant extent in Bombacoideae. Only three compounds were isolated from *Quararibea funebris* flowers, which are funberine (165), funberal (166), and funebradiol (167) (Table 15.16).

(xiii) Other Miscellaneous Compounds

Other compounds were reported, including bombaxoin (3-methyl-3,4-dihydrobenzo[c][1,2]dioxine-5-carbaldehyde) (168), a minor compound isolated from *Bombax anceps* roots. Also, benzophenone (169) and triacontyl *p*-coumarate (170) were isolated from *Bombacopsis glabra* and *Pachira aquatica* (Table 15.17).

Comprehensive phytochemical investigations of different Bombacoideae species resulted in the isolation and identification of almost 170 natural compounds belonging to different major phytochemical classes (Figure 15.1). Although the subfamily comprises of about 30 genera, only a few of them have been investigated for their secondary metabolites, including *B. ceiba*, Adansonia, *Ceiba pentandra*, Pachira, Chorisia, Pseudobombax, Ochroma, *Cavanillesia hylogeiton*, and Quararibea.

Biological Activities of Subfamily Bombacoideae

(i) Antioxidant Activity

The methanol extracts of *Adansonia digitata* demonstrated significantly high antioxidant activity comparable to orange juice activity when evaluated using DPPH and SO (super oxide) radical scavenging assays. The integral antioxidant capacity (IAC) of hydroalcoholic extracts obtained from *A. digitata* leaves and fruit pulp were evaluated by means of a photochemiluminescence method,

Table 15.14: Chemical structures of isolated steroids and their distribution in subfamily Bombacoideae.

Compound name (No.)	Structure		Plant name	Parts used	References
		R₁			
Cholesterol (**131**)	OH		*Adansonia digitata*	Seeds	(Bianchini et al. 1982)
			A. fony, A. za		
			A. madagascariensis		
			A. suarezensis		
			A.grandidiera		
			Bombax anceps	Roots	(Sichaem et al. 2010)
			Bombax ceiba	Flowers	(El-Hagrassi et al. 2011)
Campesterol (24-Methyl-cholesterol) (**132**)	OH		*Adansonia digitata*	Seeds	(Bianchini et al. 1982)
			A. fony, A. za		
			A. madagascariensis		
			A. suarezensis		
			A.grandidiera		
			Bombax ceiba	Flowers	(El-Hagrassi et al. 2011)
β-Sitosterol (24-ethyl-cholesterol) (**133**)	OH		*Adansonia digitata*	Seeds	(Bianchini et al. 1982)
			A. fony, A. za		
			A. madagascariensis		
			A. suarezensis		
			A.grandidiera		
			Bombax ceiba	Roots	(Chauhanm et al. 1980)
				Stem barks	(Jain and Verma 2012)
			Ceiba pentandra	Stem barks	(Ngounou et al. 2000)
			Chorisia chodatii	Flowers	(Refaat et al. 2015a)
			Chorisia crispiflora	Leaves	(Azab et al. 2013)
			Ochroma pyramidale	Leaves	(Vázquez et al. 2001)
			Ochroma lagopus	Heartwood	(Paula et al. 1996)
			Pachira glabra	Stem barks	(Paula et al. 2006a)
				Root barks	

Table 15.14 contd. ....

...Table 15.14 contd.

Compound	Structure	Plant name	Parts used	References
β-Sitosterol-3-O-β-D-glucopyranoside (Daucosterol) (**134**)	O-β-Glc.	*Bombax ceiba* *Bombax ceiba* *Ceiba pentandra* *Chorisia crispiflora* *Cavanillesia aff. hylogeiton*	Flowers Roots Stem barks Leaves Stem barks	(Antil et al. 2013) (Qi et al. 1996) (Ngounou et al. 2000) (Azab et al. 2013) (Bravo et al. 2002)
		Ochroma pyramidale	Leaves	(Vázquez et al. 2001)
24-β-Ethylcholest-5-en-3β-yl-α-l-arabincsyl-(1→6)-β-D-glucopyranoside (**135**)	O- β-Glc. (6→1)- α-Ara.	*Bombax ceiba*	Flowers	(Antil et al. 2013)
Isofucosterol (24-ethylidene-cholesterol) (**136**)	OH	*Adansonia digitata* *A. fony, A. za* *A. madagascariensis* *A. suarezensis* *A.grandidiera*	Seeds	(Bianchini et al. 1982)
Stigmasterol (24-Ethyl-5,22-cholestadien-3-β-ol) (**137**)	OH	*Adansonia digitata* *A. fony, A. za* *A. madagascariensis* *A. suarezensis* *A.grandidiera* *Bombax ceiba* *Ochroma pyramidale* *Ochroma lagopus* *Pachira glabra*	Seeds Flowers Leaves Heartwood Stem barks Root barks	(Bianchini et al. 1982) (El-Hagrassi et al. 2011) (Vázquez et al. 2001) (Paula et al. 1996) (Paula et al. 2006a)
Stigmasterol-3-O-β-D-glucopyranoside (**138**)	O- β-Glc.	*Chorisia crispiflora*	Leaves	(Azab et al. 2013)

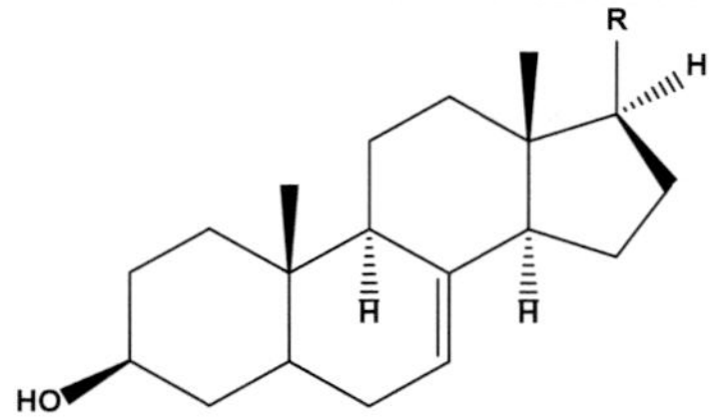

Δ⁷-Avenasterol
(24-Ethylidene-7-cholesten-3β-ol) **(139)**

R:

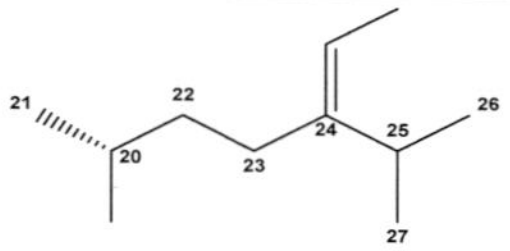

Adansonia digitata
A. fony, A. za
A. madagascariensis
A. suarezensis

Seeds   (Bianchini et al. 1982)

Δ⁷-Stigmasterol
(24-Ethyl-7-cholesten-3 β -ol)
(140)

R:

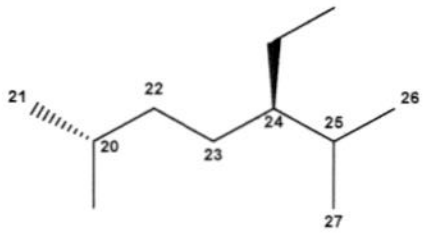

Adansonia digitata
A. fony, A. za
A. madagascariensis
A. suarezensis
A.grandidiera

Seeds   (Bianchini et al. 1982)

β-Sitostenone
(141)

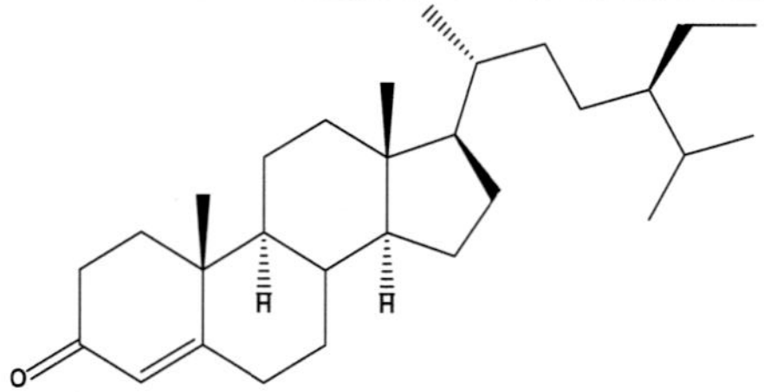

Cavanillesia aff. hylo-geiton

Stem barks   (Bravo et al. 2002)

Cholestenone **(142)**

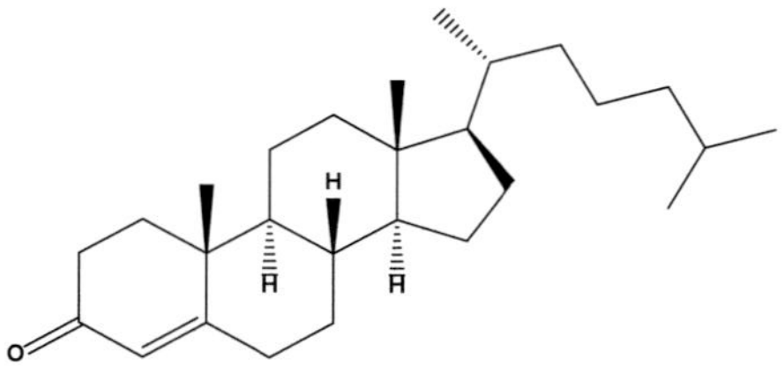

Bombax anceps

Roots   (Sichaem et al. 2010)

Table 15.15: List of identified fatty acids and their distribution in subfamily Bombacoideae.

Compound name (No.)	Lipid no.	Plant name	Parts used	References
Caproic (Hexanoic) **(143)**	6:07	*Bombax costatum*	Seeds	(Ogbobe et al. 1996)
Caprylic (Octanoic) **(144)**	8:0	*Bombax costatum*	Seeds	(Ogbobe et al. 1996)
Myristic (Tetradecanoic acid) **(145)**	14:0	*Adansonia digitata* *A. fony, A. za* *A. madagascariensis* *A. suarezensis* *A. grandidiera*	Seeds	(Ralaimanarivo et al. 1982)
Pentadecanoic acid **(146)**	15:0	*Adansonia digitata* *A. fony, A. za* *A. madagascariensis* *A. suarezensis* *A. grandidiera*	Seeds	(Ralaimanarivo et al. 1982)
Palmitic acid (Hexadecanoic acid) **(147)**	16:0	*Adansonia digitata* *A. fony, A. za* *A. madagascariensis* *A. suarezensis* *A.grandidiera*	Seeds	(Ogbobe et al. 1996) (Osman 2004), (Ralaimanarivo et al. 1982)
		Bombax ceiba	Seeds	(Dhar and Munjal 1976),
		Bombax costatum	Seeds	(Bohannon and Kleiman 1978)
		Chorisia speciosa	Seeds	(Bohannon and Kleiman 1978)
		Lagunaria patersonii	Seeds	(Rao et al. 1989)
		Pachira insignis	Seeds	(Yeboah et al. 2012)
		Pachira glabra	Seeds	(Cornelius et al. 1965)
		Pachira aquatica	Seeds	(Bohannon and Kleiman 1978)
		Ochroma lagopus	Heartwood	(Paula et al. 1996)
Palmitoleic acid **(148)**	16:1	*Adansonia digitata* *A. fony, A. za* *A. madagascariensis* *A. suarezensis* *A.grandidiera*	Seeds	(Ralaimanarivo et al. 1982)
Heptadecanoic acid **(149)**	17:0	*Adansonia digitata* *A. fony, A. za* *A. madagascariensis* *A. suarezensis* *A.grandidiera*	Seeds	(Ralaimanarivo et al. 1982)
Heptadecenoic acid **(150)**	17:1	*Adansonia digitata* *A. fony, A. za* *A. madagascariensis* *A. suarezensis* *A.grandidiera*	Seeds	(Ralaimanarivo et al. 1982)
Heptadecadienoic acid **(151)**	17:2	*Adansonia digitata* *A. fony, A. za* *A. madagascariensis* *A. suarezensis* *A.grandidiera*	Seeds	(Ralaimanarivo et al. 1982)
Stearic acid **(152)**	18:0	*Adansonia digitata* *A. fony, A. za* *A. madagascariensis* *A. suarezensis* *A. grandidiera*	Seeds	(Ralaimanarivo et al. 1982)
		Pachira glabra	Seeds	(Cornelius et al. 1965)
		Pachira aquatica	Seeds	(Bohannon and Kleiman 1978)
		Ochroma lagopus	Heartwood	(Paula et al. 1996)

Table 15.15 contd. ...

...Table 15.15 contd.

Compound name (No.)	Lipid no.	Plant name	Parts used	References
Octadec-7-enoic acid **(153)**	18:1 ω7	*Adansonia digitata* *A. fony, A. za* *A. madagascariensis* *A. suarezensis* *A. grandidiera*	Seeds	(Ralaimanarivo et al. 1982)
Oleic acid **(154)**	18:1 ω9	*Adansonia digitata* *A. fony, A. za* *A. madagascariensis* *A. suarezensis* *A. grandidiera*	Seeds	(Osman 2004, Ralaimanarivo et al. 1982)
		Bombax costatum	Seeds	(Ogbobe et al. 1996)
		Chorisia speciosa	Seeds	(Arafat et al. 2011)
		Lagunaria patersonii	Seeds	(Rao et al. 1989)
		Pachira glabra	Seeds	(Cornelius et al. 1965)
		Pachira aquatica	Seeds	(Bohannon and Kleiman 1978)
		Ochroma lagopus	Heartwood	(Paula et al. 1996)
Linoleic acid v**(155)**	18:2 ω6	*Adansonia digitata* *A. fony, A. za* *A. madagascariensis* *A. suarezensis* *A.grandidiera*	Seeds	(Osman 2004), (Ralaimanarivo et al. 1982)
		Bombax costatum	Seeds	(Ogbobe et al. 1996)
		Chorisia speciosa	Seeds	(Arafat et al. 2011)
		Lagunaria patersonii	Seeds	(Rao et al. 1989)
		Pachira glabra	Seeds	(Cornelius et al. 1965)
		Pachira aquatica	Seeds	(Bohannon and Kleiman 1978)
		Ochroma lagopus	Seeds	(Paula et al. 1996)
Linolenic acid **(156)**	18:3 ω3	*Adansonia digitata* *A. fony, A. za* *A. madagascariensis* *A. suarezensis* *A. grandidiera*	Seeds	(Ralaimanarivo et al. 1982)
		Bombax ceiba	Flowers	(Jain and Verma 2012, El-Hagrassi et al. 2011)
		Chorisia speciosa	Seeds	(Arafat et al. 2011)
		Ochroma lagopus	Heartwood	(Paula et al. 1996)
Arachidic acid **(157)**	20:0	*Adansonia digitata* *A. fony, A. za* *A. madagascariensis* *A. suarezensis* *A.grandidiera*	Seeds	(Ralaimanarivo et al. 1982)
		Bombax costatum	Seeds	(Ogbobe et al. 1996)
Eicosenoic acid **(158)**	20:1	*Adansonia digitata* *A. fony, A. za* *A. madagascariensis* *A. suarezensis* *A. grandidiera*	Seeds	(Ralaimanarivo et al. 1982)
Behenic acid **(159)**	22:0	*Adansonia digitata* *A. fony, A. za* *A. madagascariensis* *A. suarezensis* *A. grandidiera*	Seeds	(Ralaimanarivo et al. 1982)
		Bombax ceiba	Flowers	(Jain and Verma 2012, El-Hagrassi et al. 2011)

Table 15.15 contd. ...

...Table 15.15 contd.

Compound name (No.)	Lipid no.	Plant name	Parts used	References
Lignoceric acid (tetracosa-noic acid) **(160)**	24:0	*Bombax costatum*	Seeds	(Ogbobe et al. 1996)
Vernolic acid **(161)**	Lin-oleic acid 12:13-oxide	*Bombax costatum*	Seeds	(Ogbobe et al. 1996)
Malvalic acid **(162)**	8,9-Methy-lene-heptadec-8-enoic	*Adansonia digitata A. fony, A. za A. madagascariensis A. suarezensis A. grandidiera*	Seeds	(Ralaimanarivo et al. 1982)
		Bombax oleagineum	Seeds	(Bravo et al. 2002)
		Ceiba pentandra C. acuminata	Seeds	(Bohannon and Kleiman 1978)
Dihydromalvalic acid **(163)**	8,9-Methy-lene dihydro-heptadec-8-enoic	*Ceiba pentandra*	Seeds	(Kaimal and Lakshminarayana 1972)
Sterculic acid **(164)**	9,10-meth-ylene-octa-dec-9-enoic	*Adansonia digitata A. fony, A. za A. madagascariensis A. suarezensis A. grandidiera*	Seeds	(Ralaimanarivo et al. 1982)
		Bombax oleagineum	Seeds	(Bravo et al. 2002)
		Ceiba pentandra	Seeds	(Bohannon and Kleiman 1978)
		Ceiba acuminata	Seeds	
		Pachira aquatica	Seeds	(Yeboah et al. 2012)
		Pachira glabra	Seeds	(Cornelius et al. 1965)

and compared to those antioxidants derived from other natural sources, with particular consideration to ascorbic acid existing in orange, kiwi, apple, and strawberry. Results demonstrated an excellent antioxidant potency of *A. digitata* fruit pulp, showing an IAC value 10 times higher than that of orange pulp, being 11.1 and 0.3 mmol/g, respectively. This powerful antioxidant activity of *Adansonia* fruit pulp was attributed to its high content of vitamin C, ranging between 2.8–3 g/kg (Vertuani et al. 2002). Besides, the antioxidant activity of different fractions as well as the pure isolated compounds from *A. digitata* fruit pericarp were investigated by DPPH assay.

The ethyl acetate fraction in addition to the isolated proanthocyanidins compounds **(53-56)** showed high antioxidant activity with IC_{50} values between 2.40–9.60 µg/mL when compared to the reference Trolox that showed IC_{50} value of 12.18 µg/mL (Shahat et al. 2008). The methanol, hydroethanol, and dichloromethane extracts of *A. digitata* stem bark were examined for their *in vitro* antioxidant activities using 1, 1-diphenyl-2-picrylhydrazyl (DPPH) scavenging test. Both methanol and hydroethanol extracts showed promising free radical scavenging activities at a dose of 10 µg/mL with inhibition percentages of 79.81% and 77.39%, respectively, compared to the positive control (quercetin) that showed a value of 75.9% as an inhibition percentage (Lagnika et al. 2012). The antioxidant activities of the different parts of *A. digitata* represented by the leaves, stem, fruit pulp, seeds, and bark were also evaluated using DPPH scavenging test and revealed powerful antioxidant activities (Gahane and Kogje 2013). Moreover, the methanol extract of *A. digitata* leaf was additionally evaluated for its *in vitro* antioxidant activities, using both DPPH scavenging assay, as well as oxygen-radical-absorbance capacity (ORAC) assay. The extract demonstrated a strong ROS scavenging effect and showed higher potency, by about 10.2 times, comparable to vitamin C (Ayele et al. 2013).

Table 15.16: Chemical structures of isolated alkaloids and their distribution in subfamily Bombacoideae.

Compound name	Structure	Plant name	Parts used	References
Funberine **(165)**		*Quararibea funebris*	Flowers	(Raffauf et al. 1984)
Funberal **(166)**		*Quararibea funebris*	Flowers	(Zennie et al. 1986)
Funebradiol **(167)**		*Quararibea funebris*	Flowers	(Zennie and Cassady 1990)

The methanol extract of *Bombax ceiba* flowers was evaluated for its capacity to scavenge hydroxyl free radicals as well as DPPH. Its activity against lipid peroxidation induced by ascorbyl radicals and peroxynitrite was also evaluated using soybean phosphatidylcholine liposomes and rat liver microsomes, and the extract demonstrated promising antioxidant activity in all assays (Vieira et al. 2009). The aqueous and ethanol extracts of *B. ceiba* bark also demonstrated potent *in vitro* antioxidant activity using different antioxidant models, and subsequently compared to ascorbic acid that acts as a standard (Gandhare et al. 2010). The *n*-hexane and methanol extracts of *B. ceiba* flowers were evaluated for their ability to scavenge free radical DPPH assay. The effective antioxidant concentrations of the extracts were found to exist between 0.55–0.03 and 0.5–0.03 mg/mL regarding the *n*-hexane and the methanol extract, respectively (El-Hagrassi et al. 2011). Also, the aqueous 80% acetone and the 50% ethanol extracts of *B. ceiba* flowers were evaluated for their *in vitro* antioxidant activities using ORAC, DPPH, and prohibition of phosphatidylcholine liposome peroxidation, in addition to estimating the total phenolic content as well as the total flavonoid content.

Potent antioxidant capacities of the extracts were demonstrated compared to standard ascorbic and gallic acids, in which the 80% acetone extract revealed the highest total phenolic content (Yu et al. 2011). Besides, bioassay-guided fractionation of *B. ceiba* leaf methanol extract led to the isolation of the three known constituents, mangiferin, stigma-5-en-3-*O*-*β*-glucoside, and *β*-amyrin, in addition to the new xanthone *C*-glucoside, shamimoside. The activity was observed to increase with the polarity of the extracts and fractions, where mangiferin **(48)** demonstrated the most potent

Table 15.17: Chemical structures of other miscellaneous compounds and their distribution in the subfamily Bombacoideae.

Compound name	Structure	Plant name	Parts used	References
Bombaxoin (3-Methyl-3,4-dihydro-benzo[c][1,2]dioxine-5-carbaldehyde) **(168)**		*Bombax anceps*	Roots	(Sichaem et al. 2010)
Benzophenone **(169)**		*Pachira aquatica*	Stems	(Cheng et al. 2017)
Triacontyl *p*-coumarate **(170)**		*Bombacopsis glabra*	Stem barks Root barks	(Paula et al. 2006a)
		Pachira aquatica	Root barks	(Paula et al. 2006b)

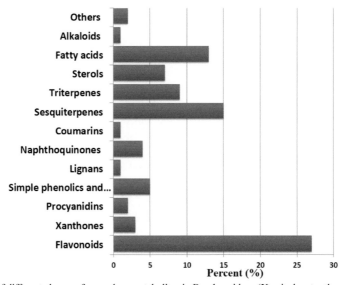

Class of secondary metabolites

Figure 15.1: Prevalence of different classes of secondary metabolites in Bombacoideae (X axis denotes the percent, Y axis denotes the class of secondary metabolites).

antioxidant activity adopting the DPPH assay (Faizi et al. 2012). A comparative study was conducted between the antioxidant activities of different solvent extracts of *B. ceiba* root grown in Bangladesh using DPPH radical scavenging assay. The methanol extract demonstrated the most potent DPPH radical scavenging activity, followed by dichloromethane (DCM) and petroleum ether extracts. The activity was correlated to the phenolic and flavonoid content, which was the highest in the methanol extract, representing (187.42 ± 3.77 mg/g, GAE) and (74.67 ± 4 mg/g, QE), respectively (Chauhan et al. 2017).

Different extracts of *Ceiba pentandra* spike and young fruit were also examined for their *in vitro* antioxidant activities, including the aqueous, methanol, chloroform, and ethyl acetate extracts. Among all the tested extracts evaluated using DPPH and TBARS models, the methanol extracts of *C. pentandra* spike and young fruit exhibited the maximum scavenging activities compared to ascorbic acid that is used as a positive control (Divya et al. 2012). On the contrary, the ethanol extract of *C. pentandra* root demonstrated weak antioxidant potential when evaluated using DPPH, FRAP, and ORAC models that is ultimately owing to the poverty of the extract with polyphenolic constituents (Bothon et al. 2012). In addition, the phenolic extract of *C. pentandra* seeds as well as its oil also demonstrated potent *in vitro* antioxidant activity using different systems in which the extract exhibited dose-dependent reducing power activity as well as dose-dependent DPPH• radical and hydroxyl radical scavenging activity (Ch et al. 2012, Loganayaki et al. 2013).

The aqueous and 70% alcohol extracts, together with the ethyl acetate and *n*-butanol fractions of *Chorisia insignis* leaves, demonstrated significant *in vivo* antioxidant activities, as indicated by the rise in blood glutathione levels in diabetic rats, as compared to vitamin E (El-Alfy et al. 2010). In a comparative study, the free radical scavenging potentials of the total ethanol extracts of *Chorisia chodatii* and *C. speciosa* leaves, flowers, fruits, and seeds, in addition to four main fractions of their leaf and flower extracts were investigated. Different extracts and fractions of both *Chorisia species* demonstrated concentration-dependent scavenging abilities by quenching DPPH radicals, except petroleum ether fractions. The aqueous, ethyl acetate, and chloroform fractions exhibited the highest scavenging activities. On the other side, the total ethanol extracts of seeds of both species showed the least scavenging properties among the tested total extracts of other plant parts (Refaat et al. 2015b).

Moreover, the soluble and insoluble phenolic fractions from *P. aquatica* seeds demonstrated high antioxidant activity measured by the oxygen radical absorption capacity (ORAC) and the trolox equivalent antioxidant capacity (TEAC) assays (Rodrigues et al. 2019).

(ii) Anti-inflammatory, Analgesic, and Antipyretic Activities

The aqueous extract of *Adansonia digitata* fruit pulp also displayed significant *in vivo* anti-inflammatory activity at 400 as well as 800 mg/kg against formalin-induced rat paw oedema at different intervals of time. The extract anti-inflammatory activity was 80–90% of the activity of standard phenylbutazone (15 mg/kg) after 12 and 24 hours, respectively. Besides, at a dose of 800 mg/kg, the extract displayed marked analgesic activity at the dose of 800 mg/kg when evaluated using hot plate method. The ability of the extract to induce analgesia was 90% that of standard acetyl salisylic acid analgesia at a dose of 50 mg/kg (Ramadan et al. 1994). Moreover, the aqueous extract produced significant antipyretic activity at the doses of 400 and 800 mg/kg by significantly decreasing the rectal temperature of hyperthermic rats at 1, 2, 3, and 4 hours (h) post treatment, when compared to the control group (Ramadan et al. 1994). Besides, a comparative study on cytokine modulatory activities of different sources of *A. digitata* was conducted. Aqueous, methanol, and DMSO extracts of commercial products of *A. digitata* leaves, fruits, and seeds, which were standardized were evaluated for their ability to secrete cytokine (IL-6 and ILr-8) in human epithelial cell cultures. The behavior of their bioactivities was variable for the three plant sources (Selvarani and Hudson 2009). The methanol extract of *A. digitata* leaf proved to possess potent *in vitro* anti-inflammatory activity. It significantly inhibited lipopolysaccharide (LPS)-induced iNOS expression within murine macrophage RAW264.7 cells, demonstrating an IC_{50} of 28.6 µg/mL. The extract failed to change the cell viability, as revealed from the MTT assay. Consequently, this indicated that the prohibition of NO synthesis by the extract was not simply attributed to its toxicity. The extract was proved to inhibit both NF-κB activation as well as IκBα degradation. So the potent anti-inflammatory effects of *A. digitata* methanol extract was assigned to its inhibitory effect on IκBα-mediated NF-κB signal transduction (Ayele et al. 2013).

Also, mangiferin isolated from *B. ceiba* leaf methanol extract had no detectable anti-inflammatory activity in carrageenan-induced rat paw edema, whereas, it exhibited significant *in vivo* analgesic activity in acetic acid-induced writhing and hot plate models. The induced analgesia was attributed to mangiferin interaction with opioid receptors at peripheral site with a minimal significance at the neuronal level (Dar et al. 2005). *Bombax ceiba* bark ethanol extract demonstrated significant *in vivo* analgesic activity at the dose of 500 mg/kg. The extract significantly inhibited acetic acid-induced writhing in mice when compared to diclofenac sodium as a standard at the dose of 25 mg/kg (Sharker 2009). The 70% methanol extract of *B. ceiba* flowers demonstrated both *in vivo* anti-inflammatory, as well as analgesic effects. The extract (at doses 25 and 50 mg/100 g body weight) significantly reduced acetic acid-induced writhing in mice relative to indomethacin, in addition to prolonging the reaction time in the hot-plate test compared to tramadol (2 mg/100 g), indicating both central and peripheral analgesic activities of the extract. Besides, the extract significantly reduced carrageenan-induced rat paw edema comparable to standard indomethacin (2 mg/100 g), indicating the extract's anti-inflammatory effect, which was attributed to its flavonoid content (Said et al. 2011). Methanol extract from *B. ceiba* leaves also proved significant *in vivo* dose-dependent antipyretic activity at the doses of 200 and 400 mg/kg. The extract significantly reduced rectal temperature in Baker's yeast-induced hyperthermic rats within 3 hours, compared to 6 hours for standard paracetamol (150 mg/kg) drug (Hossain et al. 2011a). Additionally, the extract proved to possess significant *in vivo* anti-inflammatory effects at the doses of 100, 200, and 400 mg/kg, decreasing carrageenan-induced rat paw edema, which was attributed to its dose-dependent inhibition of the production of NO (Hossain et al. 2013). Different extracts *of B. ceiba* bark (petroleum ether, ethanol, and aqueous) were also evaluated for their *in vitro* anti-inflammatory activity using Human Red Blood Corpuscles (HRBC) membrane stabilization method. The ethanol extract revealed the most significant ($p < 0.001$) anti-inflammatory activity, followed by the aqueous extract ($p < 0.01$), and finally petroleum ether extract ($p < 0.05$) (Anandarajagopal et al. 2013).

The methanol extract of *Bombax buonopozense* leaves also proved to possess potent *in vivo* analgesic, anti-inflammatory, and antipyretic activities. The extract at doses of 50 and 100 mg/kg caused late phase inhibition of the formalin pain test in rats, as well as a notable decline in acetic acid-induced writhing in mice, indicating analgesic activity superior to aspirin activity at 150 mg/kg. Besides, the extract significantly inhibited albumin-induced oedema over a period of 120 minutes (min) with 59.71% inhibition, compared to standard aspirin. Antipyretic activity of the extract was ascertained by its significant and dose dependant reduction in the yeast induced elevated rectal temperature (Akuodor et al. 2013).

Besides, the isoflavone compounds isolated from *Ceiba pentandra* stem bark- Vavain 3'-*O*-β-d-glucoside **(46)**, as well as its aglycone, vavain **(45)**, and (+)-catechin **(39)** demonstrated inhibitory effects against cyclooxygenase-2-catalyzed prostaglandin biosynthesis, with IC_{50} values of 381, 97, and 80 μM, respectively, compared to indomethacin activity (IC_{50} of 1.1 μM). Compounds **(45)** and **(46)** were found to be inactive against cyclooxygenase-2-catalyzed prostaglandin biosynthesis (Noreen et al. 1998). The petroleum ether and ethanol extracts of *C. pentandra* seeds significantly reduced carrageenan-induced paw edema in rats at 200 and 400 mg/kg, compared to aspirin activity (300 mg/kg) (Alagawadi and Shah 2011). *Ceiba pentandra* thorn extract demonstrated marked *in vivo* anti-inflammatory activity with doses of 50 and 500 mg/kg by inhibiting carrageenan-induced rat paw edema compared to dexamethasone at 5 mg/kg (Hashim et al. 2014). The aqueous extract of *C. pentandra* stem bark also demonstrated potent *in vivo* analgesic in addition to anti-inflammatory activities. The extract at the doses of 400 and 800 mg/kg reduced carrageenan-induced paw edema, inhibiting acute inflammation, but not comparable to standard diclofenac drug (5 mg/kg). On the other side, it significantly inhibited cotton pellet granuloma formation and inhibited edema progression to varying degrees. The mechanism of the extract's anti-inflammatory activity was postulated to be via the extract inhibition of the release of pro-inflammatory substances. The aqueous extract also exhibited significant analgesic activity evidenced by analgesymeter, Koster, and hot plate methods when compared to standard morphine drug (2 mg/kg), in which the central analagesic mechanism was considered (Itou et al. 2014). Also *C. pentandra* seed oil demonstrated excellent *in vitro* and *in vivo* anti-inflammatory activities evidenced by membrane stability assay in addition to C-reactive protein evaluation. The membrane stability percentage exerted by the oil was found to be concentration dependent and comparable to that of standard diclofenac. Besides, the oil significantly reduced the acute phase reactant C-reactive protein (Ravi Kiran and Raghava Rao 2014).

In a comparative study, the *in vivo* antipyretic activity of extracts of *C. pentandra* and *G. arboretum* leaves were evaluated in yeast-induced hyperthermic mice. Both extracts exhibited promising antipyretic activities. However, the activity of *C. pentandra* extract was better with effective dose of 189 mg/kg, compared to 1120 mg/kg of *G. arboretum* extract (Saptarini and Deswati 2015). The methanol extract of *C. pentandra* stem bark demonstrated significant *in vivo* anti-inflammatory and analgesic activities. The extract at doses of 100, 200, and 300 mg/kg significantly reduced carrageenan-induced paw edema and increased the reaction time in both Eddy's hot plate and tail-flick methods, when compared to standard indomethacin and pentazocine drugs (Kharat et al. 2015).

Different extracts of *Chorisia insignis* leaves, including petroleum ether, aqueous, and 70% alcohol extracts displayed significant *in vivo* anti-inflammatory activities at 100 mg/kg, versus carrageenan-induced rat paw oedema, as compared to indomethacin (20 mg/kg). Also, successive fractions of 70% alcohol extract were investigated with the ethyl acetate fraction, exhibiting the highest inhibitory activity at 100 mg/kg (El-Alfy et al. 2010). Besides, Rhoifolin isolated from different *Chorisia* s proved to possess potent anti-inflammatory activity against carrageenin-induced rat paw edema. At doses of 2.5, 25, and 250 mg/kg, rhoifolin showed both a time—as well as dose-dependent decline in rat paw edema volume estimated by 14, 25, and 45%, respectively, with respect to the control group (Eldahshan and Azab 2012).

(iii) Antimicrobial Activity

The petroleum ether, aqueous extracts, and ethanol of *Adansonia digitata* were evaluated for their antimicrobial activities against *Escherichia coli* that was isolated from both urine and water specimens using cup plate agar diffusion method. The ethanol extract showed prominent activity at concentrations of 100 and 75 mg/mL towards both urine and water isolates compared to different tested standard antibiotics. The aqueous extract showed greater antimicrobial activity against urine isolates. On the contrary, the petroleum ether extract showed no activity, indicating that the active phytoconstituents are not extracted by petroleum ether (Yagoub 2008). Crude ethanol and aqueous extracts of *A. digitata* stem, as well as root barks, demonstrated potent antimicrobial activities when evaluated using agar-well method.

The extracts exhibited Minimum Inhibitory Concentrations (MICs) ranging from 6 to 1.5 mg/mL, compared to standard gentamicin antibiotic. Although the stem bark extracts demonstrated broad spectrum antimicrobial activity, the root bark antimicrobial activity was restricted to Gram positive bacteria only (Masola et al. 2009). Comparing the antimicrobial activities of the ethanol and chloroform extracts of *A. digitata* stem bark, the ethanol extract proved to be more active against tested bacterial isolates of *Proteus mirabilis, Escherichia coli, Staphylococcus species*, and *Klebsiella pneumoniae*, irrespective of the extraction method employed (Yusha'u et al. 2010). The dichloromethane, methanol, and hydroalcoholic extracts of *A. digitata* stem bark also exhibited potent antimicrobial activity when evaluated against six Gram positive as well as Gram negative strains using the microplate dilution method. Besides, they demonstrated antifungal activities against different *Aspergillus* species with maximum activity exhibited by the hydroethanolic extract. Results also showed that the antifungal activities of the extracts were related to the inhibition of sporulation rather than inhibition of fungi mycelia development (Lagnika et al. 2012). The ethanol of *A. digitata* leaves also showed significant antimicrobial activity against *Aspergillus niger* and *Pseudomonas aeruginosa*, with moderate activity against *E. coli*, *Bacillus subtilis*, and *Candida albicans*, compared to different standard antibiotics and antifungals (Kabbashi et al. 2014). *Bombax ceiba* root methanol extract was successively fractionated using *n*-hexane, followed by carbon tetrachloride, and then chloroform. Then the factions were investigated for their antimicrobial activities against 16 microorganisms, comprising Gram-positive as well as Gram-negative bacteria and three fungi strains. Inhibition zones were measured using standard disc diffusion method at a concentration of 200 μg/disc, and compared to kanamycin standard (30 μg/disc). The *n*-hexane fraction showed significant activity *versus P. aeruginosa* and *Sarcina lutea*. The chloroform fraction showed promising activity, combating *Vibrio mimicus* with promising activity versus both *Bacillus megaterium* and *Vibrio parahemolyticus*. Carbon tetrachloride fraction showed potent activity versus almost all bacterial strains. On the other side, chloroform and carbon tetrachloride fractions showed prominent antifungal activities against *Aspergillus niger* and *Candida albicans* (Islam et al. 2011). Comparing the antimicrobial activities of the *n*-hexane and methanol extracts of *Bombax ceiba* flower against different bacterial, fungal, and yeast strains, the methanol extract exhibited prominent activity against *Bacillus subtilis*, *Staphylococcus aureus*, *S. faecalis*, and *Neisseria gonorrhoeae*, *Pseudomonas aeruginosa*, and *Candida albicans*.

On the other side, the *n*-hexane extract displayed moderate to weak activities against the same microorganisms. Also, methanol extract displayed weak to moderate activities against *Aspergillus niger* and *A. flavus*, although the hexane extract displayed no antifungal activity against them (El-Hagrassi et al. 2011). The aqueous extracts of *Bombax ceiba* bark also displayed significant antibacterial effects at the dose of 100 μg/mL. The extract activity was investigated using the Pour plate method against six medically important bacterial strains and compared to standard Gentamicin. The extract exhibited more significant activity against Gram-positive bacteria (*Bacillus subtilis*, *B. aureus*, and *Staphylococcus aureus*) than Gram-negative bacterial strains (*Escherichia coli*, *Klebsiella pneumoniae*, and *Pseudomonas aeruginosa*) (Kuthar et al. 2015). Biosynthesized silver nanoparticles (AgNPs) incorporating *Bombax ceiba* thorn extract also proved to exhibit remarkable antimicrobial activity against *S. aureus* with MIC of 25 μg/mL (Telrandhe et al. 2017). Moreover, a comparative study was performed on the antimicrobial activity of different solvents of *B. ceiba* root extracts using different solvents. The petroleum ether, dichloromethane, as well as the alcohol extracts demonstrated intermediate *in vitro* antibacterial activity against the tested Gram-positive as well as Gram-negative bacterial strains, showing an inhibition zone in the range of 7 mm to 13 mm (Hoque et al. 2018).

Bombax buonopozense leaf and root extracts were evaluated for their antimicrobial activities against *Bacillus subtilis*, *Staphylococcus aureus*, *Proteus* spp., *Klebsiella pneumoniae*, and *Escherichia coli* using agar diffusion method. The leaf extract demonstrated antimicrobial activity against *S. aureus* and *Bacillus subtilis* only while the root extract showed activity against all tested organisms (Akuodor et al. 2012a).

The methanol extract of *Ceiba aesculifolia* bark as well as its methanol and hexane fractions were investigated for their antibacterial as well as their antifungal activities. The hexane fraction exhibited no antibacterial activity. The methanol extract and methanol fraction were active against all tested Gram-positive bacteria. The extract was active against six Gram-negative bacteria, with the lowest MIC in *Staphylococcus aureus*, *S. epidermidis*, and *Vibrio cholerae* .The methanol fraction exhibited bactericidal activity against five Gram-negative bacteria. Neither the extract nor any of the fractions exhibited any antifungal activity against tested strains (Orozco et al. 2013).

The ethanol extracts of the leaves and stem bark of *Ceiba pentandra* and their combination were evaluated for their antibacterial activities against *Escherichia coli*, *Staphylococcus aureus*, *Klebsiella pneumoniae*, as well as *Pseudomonas aeruginosa*. The three extracts showed mean diameter of inhibition zone less than 12 mm at concentrations ranging between 30–50 mg/mL. The combined extract did not show any significant difference in activity with respect to that obtained from stem bark as well as the leaves. It was concluded that the combined extract lacked any synergistic or additive antibacterial activities on the test organisms (Asare and Adebayo 2012). Different extracts of *Ceiba petandra* stem bark, including *n*-hexane, acetone, and ethanol were investigated for their antimicrobial activities against *Staphylococcus aureus*, *Pseudomonas aeruginosa*, and *Klebsiella pneumoniae* compared to ampicillin standard. The acetone extract revealed the greatest antimicrobial activity at a concentration of 300 mg/mL, followed by the ethanol extract with lower activity at a concentration of 100 mg/mL. On the other side, the *n*-hexane extract exhibited no activity against all test organisms (Ezigbo et al. 2013). The methanol and dichloromethane extracts of *Ceiba pentandra* leaf and stem bark were evaluated for their antimycobacterial activities against *Mycobacterium fortuitum*, *M. smegmatis*, *M. abscessus*, and *M. phlei*. Different extract concentrations (10, 20, 100, and 200 mg/mL) were investigated using agar cup diffusion method; MIC and MBC were estimated for different extracts using agar dilution method. The most susceptible organism was *M. fortuitum*, while the most resistant one was *M. abscessus*. The stem bark methanol extract demonstrated the most potent antimycobacterial activity, producing the lowest MIC value of 20 mg/mL for some of the bacteria (Lawal et al. 2014). The ethyl acetate extract of *Ceiba pentandra* bark was evaluated for its antibacterial activity against Imipenem and Ceftazidime resistant *Pseudomonas aeruginosa* and resistant *Staphylococcus aureus*. The extract proved to be active against studied bacteria, with MIC ranging between 0.78 and 6.25 mg/mL, and MBC between 1.04 and 8.33 mg/mL. The study proved that purification of the ethyl acetate extract did not influence the activity against tested bacteria, since the most active fraction demonstrated MIC ranging from 0.52 to 6.25 mg/mL and MBC from 1.04 to 10.42 mg/mL (Julien et al. 2015). Investigation of antimicrobial activity of 70% alcohol extract of *Chorisia insignis* leaves as well as its successive fractions revealed that the ether and the ethyl acetate fractions have prominent antibacterial activity versus *Bacillus subtilis* and *B. cereus* (El Sawi et al. 2014).

Isohemigossypolone **(85)** isolated from *Pachira aquatica* root bark proved to possess antifungal activity against *Pythium ultimum* at a minimum dose of 10 ug/disk. It was suggested it had a defensive function protecting storage tissues of *Pachira aquatica* (Shibatani et al. 1999). Besides, the essential oils of many members in the plant kingdom in general (Ayoub et al. 2015, Youssef et al. 2014), and specifically belonging to Malvaceae, revealed a notable antimicrobial activity (Thabet et al. 2018a). Thus, a comparative study was performed comparing the antimicrobial activities of *P. aquatica* and *P. glabra* leaf essential oils. Only *P. aquatica* oil demonstrated effectiveness against both *Helicobacter pylori* and *Mycobacterium tuberculosis* infections, with MIC values of 20 and 50 µg/mL, respectively (Gamal El-Din et al. 2018).

(iv) Antiviral Activity

The methanol extracts of *Adansonia digitata* leaves and root bark demonstrated the most potent antiviral activities against *Herpes simplex*, Sindbis, and Poliovirus, among other evaluated plants grown in Togo. Besides their virucidal activity (direct inactivation of virus particles), they proved to have intracellular antiviral activities as well (Anani et al. 2000). In a comparative study, standardized preparations of *A. digitata* leaves, seeds, and fruit pulp extracted with water, methanol, and DMSO were evaluated for their antiviral activities against *H. simplex*, Influenza, and respiratory syncytial viruses. Leaf extracts demonstrated the most potent antiviral activity. Seeds and pulp extracts demonstrated significant but less active antiviral properties (Selvarani and Hudson 2009). The methanol extract of *A. digitata* root bark also demonstrated *in vivo* antiviral activity against Newcastle disease in poultry birds, especially at doses of 200 and 250 mg/mL (Sulaiman et al. 2013).

The ethyl acetate extract of *Bombax ceiba* flowers demonstrated significant inhibitory effects on activation of the early antigen Epstein-Barr virus at different concentrations (1, 10, and 100 µg/mL) (Said et al. 2011). Besides, the lignan compounds isolated from *B. ceiba* roots represented by (+)-pinoresinol **(66)**, matairesinol **(67)**, and 5,6-dihydroxymatairesinol **(68)** were evaluated for their anti-Hepatitis B Virus (HBV) activity. The examined samples revealed inhibitory activity against

HepG2 2.2.15 cell lines. Lignans isolated exhibited relative differences in their abilities to inhibit HBsAg secretion, with IC_{50} values of 123.7, 118.9, and 218.2 mM, respectively (Wang et al. 2013).

(v) Amoebicidal, Larvicidal, and Anthelmintic Activities

The 95% ethanol extracts of both *Adansonia digitata* leaves and *Ceiba pendantra* bark proved their anthelmintic activity against the nematode, *Haemonchus contortus* among 60 different plants species gathered in the Ivory Coast based on their ethnobotanical history. The extracts induced mortality of 80–94% of the larva supporting their traditional uses (Diehl et al. 2004).

Powdered leaves as well as leaf methanol extract of *Bombax ceiba* were evaluated for their larvicidal activity against different larval forms of the filarial vector *Culex quinquefasciatus*. Mortality rates, LC_{50} and LC_{90} were estimated at a different time interval. Different graded concentrations (0.1%, 0.2%, 0.3%, 0.4%, 0.5%) of powdered leaves exhibited significant ($p < 0.05$) larval mortality. The mortality rate was higher in 50 ppm doses of the methanol extract, demonstrating LC_{50} value of 6.97 ppm after 24 hours of exposure. The study suggested *B. ceiba* leaf methanol extract to be safely used in the aquatic ecosystem against larva of *C. quinquefasciatus* since no mortality occurred in the non-target organisms (Hossain et al. 2011b). Different extracts of *Adansonia digitata* leaves represented by benzene, hexane, chloroform, and methanol were investigated for their larvicidal and repellent activities against malarial vector, *Anopheles stephensi*. The larvae were exposed to various concentrations of extracts (30–180 mg/L) for 24 hours, and mortalities were subjected to log-probit analysis. Repellent activities of crude extracts at the dosages of 2, 4, and 6 mg/cm^2 were evaluated in a net cage containing 100 blood-starved female mosquitoes of *A. stephensi* using the protocol of (Organization 1996). The LC_{50} and LC_{90} values of different extracts were estimated against *A. stephensi* larvae in 24 h with the lowest LC_{50} and LC_{90} values demonstrated by the benzene extract (88.55, 78.18 mg/L, respectively). The methanol extract demonstrated the most effective repellent activity against *A. stephensi* (Krishnappa et al. 2012). The methanol extract of *Adansonia digitata* seeds was investigated for *in vivo* anti-trypanosomal activity at doses ranging from 50–500 mg/kg. The extract at the dose of 400 mg/kg exhibited significant antitrypanosomal activity in albino mice infected with *Trypanosoma brucei* (Ibrahim et al. 2013). The aqueous extract of *A. digitata* stem bark demonstrated promising antimalarial activity against *Plasmodium berghei* tested *in vivo*. The extract demonstrated the highest chemosuppression of parasitaemia, greater than 60% in *P. berghei* infected mice model (Musila et al. 2013).

Various concentrations represented by 500, 250, as well as 125 µg/mL of the ethanol extract of *Adansonia digitata* leaves were investigated for their amaebicidal activity against *Entamoeba histolytica*. The extract demonstrated 100% inhibition at a concentration of 500 µg/mL after 72 hours, which was comparable to Metronidazole standard (312.5 mg/kg) demonstrating 75% inhibition (Kabbashi et al. 2014). In a similar study, the aqueous and methanol stem bark extracts of *A. digitata* were evaluated for their antimalarial activities in mice infected with chloroquine sensitive *Plasmodium berghei*. Two different doses (200 and 400 mg/kg) of each extract were evaluated against chloroquine as positive control.

Significant dose dependent chemosuppressive effect was exhibited by the extracts in the two doses at different levels of infection. However, the dose of 400 mg/kg proved to be more effective in parasite clearance. Remarkable elevation in Packed Cell Volume (PCV) was also observed in the groups treated with extracts, compared to control ones (Adeoye and Bewaji 2015). Moreover, the ethyl acetate extract of *Adansonia digitata* seeds demonstrated significant ($P < 0.05$) *in vivo* antitrypanosomal activity at a dose of 400 mg/kg comparable to control treated group with berenil at dose of 3.5 mg/kg (Ibrahim et al. 2017).

The aqueous extract of *Bombax buonopozense* stem bark was determined for *in vivo* antiplasmodial effect in mice with infection triggered by chloroquine sensitive *Plasmodium berghei*. The extract at doses (100–400 mg/kg) exhibited dose-dependent activity with $p < 0.05$ versus the parasite in suppressive and curative tests (Iwuanyanwu et al. 2012). The methanol extract of *B. buonopozense* leaves also proved to possess significant *in vivo* antiplasmodial activity at the doses of 200–600 mg/kg. The activity was evaluated versus chloroquine sensitive *Plasmodium berghei* in mice through early and occurred infections. The established LD_{50} of the extract was found to be greater than 5,000 mg/kg (Akuodor et al. 2012b).

Bombax ceiba leaves methanol extract demonstrated highly significant anthelmintic activity against the trematode, *Paramphistomum explanatum* belonging to phylum Platyhelminthes, responsible for the acute parasitic gastroeneritits known as paramiphistomosis with elevated rates of morbidity as well as mortality. The extract at the doses of 10, 25, 50, and 100 mg/mL caused death of all trematodes within a short period of time (less than 45 minutes) compared to standard albendazole drug (10 mg/mL) (Hossain et al. 2012).

The hydroalcoholic extract of *Cavanillesia hylogeiton* was evaluated against chloroquine-resistant and sensitive strains of *Plasmodium falciparum* with an activity index of IC_{50} 0f 1 µg/mL. Bioassay guided fractionation of the extract was performed, where only two fractions of 25 were obtained, which showed a complete inhibition of parasitaemia of *P. falciparum* at less than 1 µg/mL (Bravo et al. 2002).

(vi) Antidiarrheal Activity

The ethanol extract of the fruit pulp of *Adansonia digitata* proved to possess a significant potency *in vivo* antidiarrheal activity at 500 mg/kg with respect to Loperamide reference drug (3 mg/kg). The extract significantly and dose-dependently prevented castor oil-induced diarrhea in rats by decreasing the frequent defecation and feces weight. The activity was related to the astringent action of *Adansonia* tannins as well as the inflammatory action of mucilage on the intestinal mucous membrane (Abdelrahim et al. 2013). In a similar experiment, the methanol extract of *A. digitata* fruit demonstrated potent dose-dependent *in vivo* antidiarrhoeal activity at doses of 300 and 700 mg/kg. The extract significantly decreased the intestinal transit time in mice and significantly prohibited diarrhea triggered by magnesium sulphate as well as castor oil in mice (Suleiman et al. 2014).

The methanol extract of *Bombax buonopozense* leaves also displayed a significant *in vivo* antidiarrheal effect at doses of 100–400 mg/kg. The extract was evaluated against castor oil-induced diarrhea, showing significant dose-dependent decrease in stooling frequency, enter-pooling, and intestinal motility in rats. The antidiarrheal activity of the extract was related to its anticholinergic effect (Akuodor et al. 2011).

Ceiba pentandra stem bark methanol extract showed significant *in vivo* antidiarrheal effect at the dose of 1,000 mg/kg. The extract displayed a promising protection versus castor oil-induced diarrhea without a pronounced delay in intestinal transit time (Sule et al. 2009).

The methanol extract of *Chiranthodendron pentadactylon* flowers was found to possess anti-secretory activity against *in vivo* toxin *Vibrio cholerae* in rat jejunal loops model. Bioassay-guided fractionation of the extract revealed three anti-secretory flavonoids. Epicatechin **(40)** displayed a potent anti-secretory action with ID_{50} of 8.3 mM/kg close to the activity of loperamide reference drug (ID_{50} = 6.1 mM/kg). Isoquercitrin **(25)** and catechin **(39)** showed moderate and weak activities, with ID_{50} of 19.2 mM/kg and 51.7 mM/kg, respectively (Velázquez et al. 2009, 2012). Additionally, the methanol extract of *Chiranthodendron pentadactylon* flowers as well as isolated fractions and major isolated flavonoids were evaluated for their *in vitro antiprotozoal* activities against *Entamoeba histolytica* and *Giardia lamblia*, and their antibacterial activities against nine bacterial enteropathogens.

In vivo antiarrheal activities were also evaluated using cholera toxin-induced diarrheal model in male Balb-c mice. Tiliroside **(20)** and epicatechin **(40)** proved to be the most potent isolated compounds responsible for the potent antidiarrheal, antiprotozoal, and antibacterial properties of *C. pentadactylon* flowers extract (Calzada et al. 2017).

(vii) Antidiabetic and Anti-hyperlipidemic Activities

The methanol extract of *Adansonia digitata* stem bark proved to possess *in vivo* antihyperglycaemic activity in streptozotocin-induced diabetic Wistar rats. The plant extract was intraperitoneally administered at 100, 200, and 400 mg/kg. Treatment with the extract significantly reduced the blood glucose levels in streptozotocin diabetes rats comparable to insulin. The highest antidiabetic activity was shown at 100 mg/kg with 51% percentage glycemic alteration after 7 hours of extract administration, while the other two doses of 200 and 400 mg/kg showed 39% and 31% glycemic change, respectively after 7 hours of extract administration (Tanko et al. 2008). The ethanol extract of *A. digitata* bark also demonstrated anti-hyperglycemic as well as hypolypidimic activity in alloxan-induced diabetic rats. The extract administered at 250 and 500 mg/kg significantly reduced plasma glucose levels by 26.7% and 35.9%, and stimulated glycogenesis by 11.3% and 32%, respectively. Plasma and hepatic lipid profiles were significantly reduced as well. Results were compared to standard Glipizide drug at the dose of 500 mg/kg (Bhargav et al. 2009). The methanol extract of *A. digitata* leaves also proved to possess strong antidiabetic and hypolipidaemic properties when evaluated in streptozotocin (STZ)-induced diabetic rats at doses of 200 mg/kg and 400 mg/kg. The extract administration caused a significant reduction in the blood glucose, glycosylated hemoglobin, cholesterol, triglycerides, low-density lipoprotein (LDL), interleukin 6 (IL-6), tumor necrosis factor-alpha (TNF-α), and malondialdehyde (MDA) levels after the sixth week of treatment compared to the diabetic group (Ebaid et al. 2019).

Bombax ceiba bark ethyl acetate extract also proved hypoglycemic and hypolipidemic activities in streptozotocin-induced diabetic rats when oral doses of 200, 400, 600 mg/kg were administered for 21 days. The most significant hypoglycemic and hypolipidemic activity was observed at the dose of 600 mg/kg, significantly lowering blood glucose level, total cholesterol, and triglyceride level (Bhavsar and Talele 2013).

Bombax ceiba methanol extract of stem bark proved a significant ameliorative potential against high fat diet induced obesity in rats (Gupta et al. 2013). *B. ceiba* bark and seed powders were evaluated for their *in vivo* antihyperlipidaemic activities on high fat high cholesterol (HFHC) fed rats group. Significant decrease in serum and tissue phospholipid, triglycerides, total cholesterol, free fatty acid, LDL-C levels were demonstrated on feeding rats with *B. ceiba* bark and seed powder at 200 mg/kg body weight of rats. Elevation of albumin, protein, and HDL-C level was also demonstrated in experimental groups (Singh et al. 2018).

The methylene chloride/methanol extract of *Ceiba pentandra* root bark proved its hypoglcaemic activity in diabetes induced by streptozotocin in rats. The effect of graded doses of the extract (40, 75, 150, and 300 mg/kg) was evaluated in fasted normal and diabetic groups. The extract at the two doses (40 and 75 mg/kg) caused a significant decrease in both blood as well as urine glucose levels relative to the initial values. The blood glucose level was reduced by 59.8 and 42.8% at 40 and 75 mg/kg, respectively, while the urine glucose level was reduced by 95.7 and 63.6%, respectively (Djomeni et al. 2006). In a similar study, the methylene chloride/methanol extract of *C. pentandra* root bark also proved its antihyperglycaemic activity in streptozotocin-induced type-2 diabetic rats at 40 and 75 mg/kg. The extract significantly reduced both water and food intake, enhanced glucose tolerance, and reduced the levels of serum cholesterol, triglyceride, creatinine, and urea, in addition to reducing blood glucose levels. *Ceiba pentandra* root bark extract proved to possess hypoglycaemic effect in normal and alloxan induced diabetic rats at the dose of 150 mg/kg. Significant reduction in blood glucose level was observed after seven weeks of treatment with the extract (Saif-ur-Rehman et al. 2010). Different fractions of *C. pentandra* methanol leaf extract, including petroleum ether, chloroform, ethyl acetate, and methanol fractions were investigated for their antihyperglycemic effect. Different doses were administered to normal in addition to alloxan-induced diabetic rats, using glibenclamide as a standard drug. The methanol fraction of *C. pentandra* extract at the dose of 200 mg/kg exhibited the maximal lowering of blood glucose level in diabetic rats (Dolui et al. 2011). The ethanol extract of *C. pentandra* leaf proved to have promising potential in obesity management.

Administration of 125 mg/kg of the extract significantly reduced body weight gain, weight of liver and fat pads, as well as Body Mass Index (BMI) in obese rats.

The intestinal activity of enzyme Alkaline Phosphatase (ALP) was found to be reduced in *C. pentandra* treated rats, suggesting the mechanism of the extract's anti-obesity activity to be mediated through inhibition of intestinal lipid absorption and thermogenesis (Patil et al. 2012). *C. pentandra* root hydroalcoholic extract was assessed for its hypoglycemic and anti-hyperlipidemic activity in both normal as well as alloxan-induced diabetic rats. At the dose of 300 mg/kg, the extract significantly decreased the elevated blood glucose level, glycosylated haemoglobin, as well as cholesterol, triglycerides, phospholipids, LDL, and VLDL. The extract also significantly elevated liver insulin, glycogen, and HDL levels, confirming its promising antidiabetic and antihyperlipidemic potential (Parameshwar et al. 2012). The hypoglycaemic as well as the antihyperglycaemic activities of *C. pentandra* bark ethanol extract were determined in both normal as well as streptozotocin-induced diabetic rats. In the single dose study of the extract at the doses of 200 and 400 mg/kg, significant elimination in blood glucose level was noticed in diabetic rats at the dose of 200 mg/kg, although insignificant effect was noticed in normal rats. In oral glucose tolerance test, significant decrease in glucose level was noticed in both normal and diabetic rats. Long term treatment with 200 mg/kg for 21 days significantly reduced blood glucose level, triglycerides, and total cholesterol. Levels of serum insulin and liver glycogen were significantly elevated. The study suggested the extract to be beneficial in the management of type I diabetes (Satyaprakash et al. 2014). In another study, *C. pentadra* stem bark extract and its combination with *Amaranthus viridis* aerial parts extract were evaluated for their antidiabetic and hypolipidaemic activities. Three test groups of albino rats received subcutaneous plant extracts of *C. pentandra*, *A. viridis*, as well as their mixture at 400, 450, and 450 mg/kg, respectively, after pretreatment with dexamethasone (10 mg/kg) for ten days. Significant reduction in serum glucose TG, LDL, VLD, TC, and significant elevation in body weight, HDL, tissue glycogen levels, and liver glycogen were noticed on the administration of both the extracts as well as their combination. They proved their antihyperglycaemic and antihyperlipidemic potentials without hypoglycaemic activity in normal individuals (Paramesha et al. 2014). The ethyl acetate fraction of *Ceiba pentandra* leaf extract was evaluated for its hypoglycaemic effect in alloxan-induced diabetic rats. Significant reduction in blood glucose was noticed in all the treated groups compared to standard drug (glibenclamide) with the highest hypoglycaemic activity observed at the dose of 200 mg/kg. Opposite to untreated groups body weight remaining stable, hematological abnormalities accompanied with diabetes mellitus were ameliorated, including red blood cells, platelet, and hemoglobin count, as well as packed cell volume (Lami et al. 2015).

Both the aqueous and methanol extracts of *Ceiba pentandra* trunk bark were evaluated for their antidiabetic properties on an experimental model of type 2 diabetes induced by the combination of a high-fat diet and a single dose of streptozotocin (40 mg/kg, intraperitoneal) on the seventh day of experimentation. Both extracts significantly reduced the hyperglycemia by up to 62%, and significantly improved the oral glucose tolerance test. The impaired levels of cholesterol and triglycerides registered in diabetic control were also significantly reversed by both extracts. The antidiabetic effects of the extracts could result from their ability to improve the peripheral use of glucose, lipid metabolism, or from their capacity to reduce oxidative stress (Fofie et al. 2019).

The aqueous and 70% alcohol extracts of *Chorisia insignis* leaves as well as the ethyl acetate fraction of the alcohol extract demonstrated significant *in vivo* antihyperglycemic activities in alloxan-induced diabetic rats. Results were compared to metformin reference drug (150 mg/kg) (El-Alfy et al. 2010). The ethanol extracts of various parts of *C. chodatii* and *C. speciosa* as well as their successive fractions were evaluated to study their effect on adipogenesis using the 3T3-L1 preadipocytes model. The extracts and their fractions demonstrated dose-dependent induction of 3T3-L1 preadipocytes differentiation, but with a remarkable reduction in the size of the lipid droplets at the lower concentrations of 5 and 10 μg/mL. The aqueous, ethyl acetate, and chloroform fractions of different plant parts demonstrated the greatest effects on adipogenesis, as well as the highest polyphenol contents. The study suggested the potential value of *Chorisia* in obesity-related disorders (Refaat et al. 2015b).

(viii) Hepatoprotective Activity

The aqueous extract of *Adansonia digitata* fruit pulp also displayed a pronounced *in vivo* hepatoprotective effect at the dose of 1 mg/kg. The activity was postulated to be due to the steroids and triterpenoids content of the fruit, and thus the anti-inflammatory, analgesic, immunostimulant, and antimicrobial activities of *A. digitata* fruit pulp play a role in hepatic protection (Al-Qarawi et al. 2003). The ethyl acetate fraction of *A. digatata* leaf ethanol extract exhibited potent *in vivo* liver protection versus hepatic toxicity triggered by carbon tetrachloride. Furthermore, at 100 mg/kg as well as 200 mg/kg, the extract significantly decreased the pronounced elevation in Alkaline Phosphate (ALP), Aspartate Aminotransferase (AST), Lactate Dehydrogenase (LDH), and alanine aminotransferase (ALT) levels, showing more pronounced effect at the lower dose (Oloyede et al. 2013).

Besides, mangiferin **(48)**, isolated from *Bombax ceiba* leaf methanol extract, proved to possess significant *in vivo* hepatoprotective activity at the doses of 0.1, 1, 10 mg/kg, combating carbon tetrachloride triggered hepatic damage, which consequently supported its free radical scavenging ability (Dar et al. 2005). The hepatoprotective activity of *B. ceiba* flower methanol extract was evaluated *in vivo* using hepatotoxicity model produced by combining the two anti-tubercular drugs isoniazid and rifampicin. Pretreatment with the methanol extract at the doses of 150, 300, and 450 mg/kg significantly reduced AST, ALT, alkaline phosphatase (ALP), TBARS, and total bilirubin levels, and elevated the level of total protein and GSH after anti-tubercular challenge when compared to silymarin control (2.5 mg/kg). Histopathological studies together with the obtained biochemical parameters suggested that although the extract was not able to completely resolve the antitubercular drugs induced hepatotoxicity, but it could limit their effect to necrosis extent (Ravi et al. 2010). In a similar study, the 70% methanol extract of *B. ceiba* flowers at 250 and 500 mg/kg significantly reduced the elevated ALT and AST levels caused by pracetamol-induced hepatotoxicity. The hepatoprotective activity of *B. ceiba* flower extract was assigned to its antioxidant activity related to its flavonoid content (Said et al. 2011). In another study, the aqueous methanol extract of *B. ceiba* flowers proved to ameliorate hepatosteatosis induced by ethanol and relatively moderate fat diet in rats at a dose of 200 mg/kg/d. Treatment with *Bombax ceiba* flower extract ameliorated the alcohol-induced increase of liver enzyme activities, and significantly increased the level of hepatic liver antioxidants and decreased malondialdehyde (MDA) level (Arafa et al. 2019).

Moreover, the ethyl acetate fraction of *Ceiba pentandra* stem bark methanol extract at the dose of 400 mg/kg also proved to possess potent *in vivo* hepatoprotective activity against paracetamol-induced liver injury. Histopathological screening as well as the significant reduction in serum enzymes ALT, AST, ALP, and total bilirubin content compared to standard silymarin drug (100 mg/kg) confirmed its hepatoprotective potential (Bairwa et al. 2010).

Besides, different extracts of *Chorisia insignis* leaves were evaluated for their *in vivo* hepatoprotective activities against CCl_4 induced liver damage. The aqueous, 70% alcohol extracts, as well as the ethyl acetate fraction at the doses of 100 mg/kg significantly declined the levels of ALT, AST, and ALP with respect to standard silymarin drug (25 mg/kg), proving their potent hepatoprotective activities (El-Alfy et al. 2010).

(ix) Cytotoxic Activity

Bombax ceiba stem bark methanol extract proved to possess a potent *in vitro* inhibitory effect on tube formation of human umbilical venous endothelial cells (HUVEC). Bioactivity-guided fractionation afforded lupeol as the active principle, showing marked inhibitory activity at doses of 50 and 30 µg/mL. However, it didn't show significant inhibition for growth of tumor cell lines represented by SK-MEL-2, A549, as well as B16-F10 melanoma (You et al. 2003). The antitumor potential of *B. ceiba* root methanol extract was measured using the lethality bioassay in brine shrimp. It exhibited pronounced cytotoxic activity with LC_{50} value of 3.90 μg/mL. Meanwhile, vincristine sulphate, the standard cytotoxic agent, has LC_{50} value of 0.625 μg/mL (Islam et al. 2011). The aqueous ethanol extract (80%) of *B. ceiba* flowers was evaluated for its cytotoxic activity on Ehrlich ascites carcinoma cells (EACC) using Trypan blue exclusion method in a dose-dependent manner. *B. ceiba* flower extract showed a mild inhibition of tumor volume as well as viable tumor cell count with concomitant elevation in the life span of the tumor-bearing mice. The study suggested that the activity of the extract is attributed to the reduction in the nutritional fluid volume in addition to arresting the tumor growth, resulting in elevating the life span of EACC bearing mice (El-Toumy et al. 2013). The diethyl ether and petroleum ether extracts of *B. ceiba* flowers were investigated for their antiproliferative responses against seven human cancer cell lines, including MCF-7, LNCaP, HeLa, ACHN, COR-L23, A375, and C32 and compared to human normal cell line. In a concentration-dependent manner, both extracts showed the highest activity against human renal adenocarcinoma (ACHN) (Tundis et al. 2014).

Besides, *Ceiba pentandra* stem methanol extract showed the most potent inhibitory effect upon the tube-like formation of HUVEC adopting angiogenesis *in vitro* assay of 58 Vietnamese medicinal plants with inhibition percentage of 87.5% at a dose of 100 μg/mL (Nam et al. 2003).

The *n*-hexane, ethyl acetate, butanol, and methanol extracts of *Chorisia crispiflora* leaves were investigated for their cytotoxic effect in MCF-7 breast cancer cells. The ethyl acetate extract was the most active extract, exhibiting IC_{50} of 5.2 and 4.2 μg/mL for 48 and 72 hours, comparable to standard doxorubicin. Further molecular characterization of the extract was performed, leading to the conclusion that down-regulation of NF-κB as well as up-regulation of p21 levels may be the underlying mechanism by which the extract inhibits MCF-7 proliferation (Ashmawy et al. 2012).

(x) Antiurolithiatic Activity

Bombax ceiba fruits aqueous and ethanol extracts were evaluated for their curative efficacy in calcium oxalate urolithiatic rats at 400 mg/kg. The extracts significantly reduced the increased urinary oxalate levels due to ethylene glycol, demonstrating a regulatory action on endogenous oxalate synthesis. Besides, the extracts significantly decreased the elevated precipitation of stone forming compounds in the kidneys of calculogenic rats when compared to standard cystone (750 mg/kg) as a reference antiurolithiatic drug.

They also proved to possess potent diuretic activity, evidenced by increasing total urine volume and electrolytes excretion at the doses of 200 and 400 mg/kg, comparable to hydrochlorothiazide and Frusemide (25 mg/kg each) as standard diuretic drugs (Gadge et al. 2009, Jalalpure and Gadge 2011, Gadge and Jalalpure 2012). The aqueous, *n*-butanol, and ethyl acetate extracts of *B. ceiba* leaves were assessed for their *in vivo* protective effects against gentamicin-induced renal toxicity. 200 mg/kg of the aqueous, *n*-butanol extracts reduced renal oxidative damage triggered by gentamicin induced in rats, showing a pronounced decline in serum levels of urea, uric acid, creatinine, and malondialdehyde (Vasita and Bhargava 2014).

The aqueous and alcohol extracts of *Ceiba pentandra* bark also proved to possess curative potential for calcium oxalate urolithiasis in albino rats. Extracts administration significantly decreased the increased urinary oxalate levels due to ethylene glycol and reduced the precipitation of stone forming agents enhanced by ammonium chloride administration (Choubey et al. 2010).

(xi) Anti-ulcer Activity

The aqueous extract of *Bombax buonopozense* leaves was assessed for its anti-ulcer activity at 100, 200, and 400 mg/kg *in vivo* using ethanol-induced ulcer model. It showed a pronounced dose-dependent effect comparable to ranitidine. Oral LD_{50} value was found to be 2828.42 mg/kg in mice

(Nwagba et al. 2013). Also, *B. ceiba* flower extract was investigated for its protective and curative effects against ethanol-induced gastric injury in rats. Treatment with the extract reduced the severity of ethanol gastric mucosal damage, the elevated ulcer index, and cell organelle marker enzymes, and suppressed gastric inflammation at a dose of 300 mg/kg (Barakat et al. 2019).

Ceiba pentandra stem bark methanol extract proved to possess *in vivo* protective properties against indomethacin and ethanol-induced gastric ulcers. At 100, 200, and 400 mg/kg, potent dose-dependent ulcer inhibition was observed and confirmed by the pronounced reduction in the ulcer index of the groups receiving the treatment. Moreover, histological examination of the gastric wall of rats pre-treated with the extract revealed a reduced ulcer area and sub-mucosal edema in addition to the absence of leucocytes infiltration owing to administration of 50 mg/kg indomethacin, as well as of 0.5 mL of 95% absolute ethanol (Anosike and Ofoegbu 2013). *Ceiba pentandra* leaves methanol extract also exhibited a protective activity against indomethacin and ethanol-induced gastric ulcer. It prohibited indomethacin-induced gastric ulcer at 100, 200, and 400 mg/kg in a dose-dependent manner by 70, 82, and 84%, and ethanol-induced gastric ulcer by 19, 53, and 58%, respectively (Anosike et al. 2014).

(xii) Immuno-modulatory Activity

The methanol extract of *Bombax ceiba* bark proved to possess promising *in vivo* immunostimulatory activity at 150 and 300 mg/kg. The extract activity was investigated both in normal and immunosuppressed mice by evaluating its efficacy on Hemagglutinating antibody (HA) titer, hematological profile (Hb, WBC, RBC), delayed type of hypersensitivity (DTH) response, lipid peroxidation (LPO), superoxide dismutase (SOD), reduced glutathione (GSH), catalase (CAT), and cytokine release. The methanol extract increased the antibody titer values, and produced a significant dose-related elevation in DTH reactivity in mice responding to cell-dependent antigen, which revealed the stimulation of T cells by the extract. Besides, the extract elevated the hematological profile, SOD, CAT, GSH activity, accompanied by a pronounced decline in LPO levels in immunosuppressed mice triggered by cyclophosphamide. Moreover, the animals administered the extract displayed a significant up regulation of cytokines represented by IL-6 and TNF-α relative to the control group. Results suggested that the methanol extract showed that the promising immunostimulatory principle is strongly correlated to its ability to stimulate humoral immunity via acting by different mechanisms (Wahab et al. 2014).

(xiii) Cardioprotective Activity

The aqueous extract of *Bombax ceiba* flower proved its *in vivo* protective activity against Adriamycin (Adr)-induced cardiotoxicity. The extract co-administered at the doses of 150, 300, and 450 mg/kg with vitamin E significantly elevated the level of cardiac antioxidant enzymes (myocardial superoxide dismutase), catalase as well as reduced glutathione with concomitant decline in the level of lipid peroxidation compared to Adr-treated animals. Besides, it significantly reduced the serum level of cardiac marker enzyme, LDH and AST enzyme, and this protective effect was further supported by the microscopic studies of the aqueous extract (Patel et al. 2011).

(xiv) Hypotensive Activity

Shamimin **(38)**, the *C*-flavonol glucoside isolated from the leaves of *Bombax ceiba* proved to possess prounonced potency as a hypotensive agent at 1, 3, and 15 mg/kg. Studies also showed that it causes mortality in rats at 500 mg/kg, which is considered as a lethal dose (Saleem et al. 1999).

(xv) Antiangiogenic Activity

Bombax ceiba stem bark methanol extract proved to possess a promising *in vitro* inhibitory activity on tube formation in HUVEC. Bioactivity-guided fractionation afforded lupeol as the active principle, showing marked inhibitory activity at doses of 50 and 30 µg/mL. However, it didn't show significant inhibition of A549, SK-MEL-2 in addition to B16-F10 melanoma cells (You et al. 2003).

(xvi) Antivenom

Ceiba pentandra leaves aqueous methanol extract was evaluated for its antivenom effect versus *Echis ocellatus* snake venom. The extract demonstrated potent *in vivo* snake venom-neutralizing capacity with an LD_{50} value of 0.280 ± 0.065 mg/kg. The extract administration significantly reduced haemolysis due to venom from 66% to 27.4%. It also inhibited venom-induced changes in packed cell volume, total protein, haemoglobin contents, and phospholipase A_2 activity (Sarkiyayi et al. 2010).

(xvii) Testicular Protection

The aqueous extract of *Adansonia digitata* leaf was tested *in vivo* for its activity versus carbon tetrachloride-induced testicular toxicity (2.5 mL/kg). The extract significantly ameliorated the low levels of follicle stimulating hormone, testosterone, and luteinizing hormone, as well as superoxide dismutase due to carbon tetrachloride toxicity. Cyto-architecture of testis also revealed minimum degree of testis distortion in treated animals relative to the control group. The study supported the therapeutic role of the extract in free radical mediated diseases (Oyetunji et al. 2015).

(xviii) Anti-osteoporotic Activity

Treatment with the petroleum ether and methanol extract of *Bombax ceiba* stem bark for 28 days significantly ameliorated the consequences of ovariectomy-induced bone porosity, and restored the normal architecture of bone at two doses: 100 and 200 mg/kg. The *in vitro* osteogenic activity was related to the presence of lupeol, gallic acid, and β-sitosterol constituents of the plant (Chauhan et al. 2018).

(xix) Entomototoxicity

Biological assays demonstrated the nectar toxicity of *Ochroma lagopus* flower, causing great mortality of bees and other insects. Chemical investigation of the nectar identified glucose, fructose, sucrose, and sixteen proteic amino acids. The toxicity proved not to be related to any sugar or nonproteic amino acids, although the toxic substance in the nectar was not yet identified (Paula et al. 1997).

(xx) Aphrodisiac Activity

Bombax malabaricum root extract showed potent aphrodisiac activity at the dose of 400 mg/kg/day evidenced by reduction of post-ejaculatory interval, intromission latency, mount latency, as well as ejaculation latency with concomitant elevation intromission frequency, mounting frequency, and ejaculation frequency in sexually active and inactive male mice (Chaudhary and Khadabadi 2012).

Mechanistic Interpretation of the Anti-inflammatory Activity of Reported Bombacoideae Members Possessing Anti-inflammatory Potential

After thoroughly reviewing the vast biological activities of Bombacoideae, it was observed that many extracts of various parts of the studied Bombacoideae plants demonstrated significant anti-inflammatory potential either *in vitro* or *in vivo*. It has been believed recently that inflammation is the cause of most diseases, being a common symptom in many disease conditions (Otimenyin 2018). Thus, it was encouraging to do further in-depth studies of the underlying mechanisms of the anti-inflammatory activities of different bombacoideae plants, as well as the phytoconstituents behind this potential.

Inflammation is a complex pathophysiological process mediated by a cascade of signaling molecules elicited by macrophages, leukocytes, and mast cells. Infiltration of leukocytes and extravasation of fluid and proteins at the inflammatory site by the activation of different complement factors results in the accompanying edema (Ashour et al. 2018). Natural products proved to be a promising reservoir of anti-inflammatory phytoconstituents, inhibiting inflammation through numerous mechanisms.

(i) Inhibition of Nitric Oxide Synthase (NOS)

Nitric oxide (NO) is a signaling molecule playing a crucial role in the pathogenesis of inflammation. NO is generated biochemically through the oxidation of the terminal guanidine nitrogen atom from L-arginine by nitric oxide synthetase (NOS). Having three isoforms, endothelial NOS (eNOS), neuronal NOS (nNOS), and inducible NOS (iNOS), nitric oxide synthase is an important cellular mediator of both physiological and pathological inflammatory processes. Endothelial NOS (eNOS) and neuronal NOS (nNOS) are constitutively expressed in the body under normal physiological conditions. However, inducible NOS (iNOS) is an inducible enzyme highly expressed by inflammatory stimuli. Overproduction of NO by inducible NOS occurs in response to different inflammatory mediators (e.g., tumor necrosis factor-α (TNF-α), interleukine-1β (IL-1β), and bacterial lipopolysaccharide (LPS)), aggravating the inflammatory process and acting synergistically with other inflammatory mediators. Many plants have recently proved to possess strong inhibitory potential of inducible NOS enzyme (iNOS), inhibiting overproduction of nitric oxide.

The methanol extract of *Adansonia digitata* leaf proved to possess significant anti-inflammatory activity through inhibiting iNOS expression in lipopolysaccharide (LPS)-stimulated Raw264.7 cells (Ayele et al. 2013). Besides, the anti-inflammatory activity of the methanol extract of *Bombax ceiba* leaves demonstrated significant anti-inflammatory activity, demonstrated in reducing carrageenan-induced rat paw edema. The activity was attributed to its dose-dependent inhibition of nitric oxide production (Hossain et al. 2013). Moreover, different studies reported the ability of flavonoids to inhibit the expression of inducible nitric oxide synthase isoform in several models. Meanwhile, differences may exist among several flavonoids in their inhibiting capacity to iNOS. Although kaempferol **(16)** and quercetin **(23)** demonstrate little differences in their inhibiting capacity of the expression of iNOS in RAW264.7 cells, the former demonstrated a greater extent of inhibition than quercetin in nitrite accumulation in culture medium of lipopolysaccharide (LPS)-stimulated J774.2 cells (González-Gallego et al. 2014).

(ii) Inhibition of Cyclooxygenases (COX)

The inflammatory processes may also activate some biomarkers, such as cyclooxygenases, which are enzymes that allow the body to produce prostaglandins from arachidonic acid. This type of enzymes can act as dioxygenase or peroxidase, as they are peripheral membrane proteins having two isoforms

(Salinas et al. 2007). COX1, present in most tissues, such as stomach, synthesize prostaglandins and perform maintenance of the gastric mucosa, which regulates the proliferation of normal cells and intervenes indirectly in physiological processes, such as protection and neutrophil migration to the epithelium. COX2, the other isoform, is formed from an increase of prostaglandins in tissues where an inflammatory response occurs. It is expressed after induction of inflammation caused by erosion of the mucosa (Díaz-Rivas et al. 2015). Inhibition of both COX-1 and COX-2 was found to be the basic mechanism of different anti-inflammatory plant extracts. However, selective COX-2 inhibitor mixes both the anti-inflammatory activity and minimum side effects usually related to COX-1 inhibition. The isoflavone compounds isolated from *Ceiba pentandra* stem bark were evaluated for their inhibition of cyclooxygenase-1-catalyzed prostaglandin biosynthesis together with the known flavan-3-ol, (+)-catechin **(39)**. Vavain **(45)**, vavain 3'-*O*-*β*-D-glucoside **(46)**, and (+)-catechin **(39)** demonstrated inhibitory effects with IC_{50} values of 97, 381, and 80 μM, respectively, compared to indomethacin activity (IC_{50} of 1.1 μM) (Noreen et al. 1998).

Besides, several flavonoids, such as apigenin, luteolin, kaempferol, and quercetin, as well as some of their glycosides, were repeatedly reported as inhibitors of COX (Ribeiro et al. 2015). This may explain the potent *in vivo* anti-inflammatory potential of *Bombax ceiba* flowers containing these flavonoids (Joshi et al. 2013, El-Hagrassi et al. 2011). Studying the structure-activity relationships (SAR) of the COX-2-inhibiting properties of various flavonoids, it was revealed that hydroxylation at the 4'-position and a free 5'-position are the only sufficient requirements for COX-2-inhibiting activity (Rosenkranz and Thampatty 2003).

(iii) Inhibition of Pro-inflammatory Cytokines

Cytokines are regulators of the human body, and react to infection, trauma, immune responses, and inflammation. Some cytokines act to increase the body's reaction to inflammatory stimulus, making it worse (pro-inflammatory), while others help to decrease inflammation, and induce the healing process (anti-inflammatory). The major pro-inflammatory cytokines include Interleukin-1alpha (IL-1α), IL1-beta (IL-1 β), IL-6, and TNF-alpha (TNFα). Inhibiting the expression of these pro-inflammatory cytokines has been demonstrated as the mechanism of action of many anti-inflammatory medicinal plants and their isolated compounds (Dinarello 2000). Conducting a comparative study on cytokine modulatory activities of different sources of *Adansonia digitata,* the aqueous, methanol, and DMSO extracts of commercial products of *A. digitata* leaves, fruits, and seeds were evaluated for their ability to secrete cytokine (IL-6 and ILr-8) in human epithelial cell cultures. Many of the extracts, especially leaf extracts, proved to be active as cytokine modulators, some of which were pro-inflammatory, and others anti-inflammatory. The overall results concluded the presence of multiple bioactive compounds in different parts of the plant, explaining the variable medical benefits in the treatment of infectious diseases and inflammatory conditions among the three plant sources (Selvarani and Hudson 2009).

The flavone glycoside, rhoifolin **(4)**, isolated from different *Chorisia* species, was reported to possess significant anti-inflammatory activity against carrageenin-induced rat paw edema. It demonstrated a dose-dependent decline in rat paw edema volume compared to the control group. Diminishing the TNF-*α* release was one of the mechanisms which proved to be behind the potent anti-inflammatory activity of rhoifolin (Eldahshan and Azab 2012).

Also, the potent anti-inflammatory activity of *Ceiba pentandra* stem bark aqueous extract against carrageenan and cotton pellet was attributed to its flavonoid content. The mechanism of the anti-inflammatory activity was postulated to be the inhibition of pro-inflammatory cytokines release (Itou et al. 2014). Moreover, the anti-inflammatory activity of *C. pentandra* seed oil was evaluated both *in vitro* and *in vivo* via membrane stability assay and C-reactive protein evaluation, respectively. The percentage of membrane stability demonstrated *in vitro* by the seed oil was concentration dependent and comparable to standard diclofenac. Meanwhile, the level of C-reactive protein was diminished significantly, reflecting systemic anti-inflammatory activity of the tested oil. The potent anti-inflammatory potential was attributed to the complex mixture of various fatty acids present in *C. pentandra* seed oil (Ravi Kiran and Raghava Rao 2014).

(iv) Antioxidant Activity

An additional mechanism which could participate in the anti-inflammatory properties of different Bombacoideae species is the modulation of the redox state by enhancing the different endogenous antioxidant defense mechanisms. Many extracts demonstrating significant anti-inflammatory activity proved to possess potent antioxidant activities as well. Examples of these extracts include *Adansonia digitata* fruit pulp extract, *Bombax ceiba* bark, leaf, and flower extracts, and *Chorisia insignis* leaf extracts (Eldahshan and Azab 2012, Yu et al. 2011, Faizi et al. 2012, Refaat et al. 2015b). Moreover, the potent anti-inflammatory rhoifolin **(4)** isolated from different *Chorisia* species proved to elevate the total antioxidant capacity in the inflammatory exudates when evaluated in carrageen-induced rat oedema model (Eldahshan and Azab 2012).

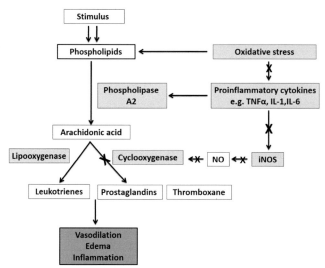

Figure 15.2: Mechanistic interpretation of the anti-inflammatory activity of reported Bombacoideae species with anti-inflammatory potential.

Conclusions

Bombacoideae, the worldwide distributed subfamily, comprises of a vast collection of economically *important* plants. Many of the Bombacoideae species were extensively used in traditional medicine for the management of various ailments. A chemical survey of Bombacoideae revealed great diversity in the secondary metabolites isolated from the different species. Classes of the isolated secondary metabolites included flavonoids, xanthones, procyanidins, lignans, naphthoquinones, sesquiterpenes, triterpenes, steroids, and alkaloids. Flavonoids constituted the most prevailing class of secondary metabolites, representing 27% of total isolated compounds, followed by sesquiterpenes and fatty acids. The biological survey illustrated a wide range of pharmacological and biological activities exhibited by different species of Bombacoideae.

Activities extended from the antioxidant, immunomodulatory, hepatoprotective, cardioprotective, anti-inflammatory, antimicrobial, antifungal activity, and antiviral activities to antidiarrheal, antidiabetic, antihyperlipidaemic, and many other pharmacological activities. The great diversity in the biological activities of Bombacoideae is mainly attributed to the wide variety of chemical classes present in the different species.

The anti-inflammatory potential of different Bombacoideae species was studied in-depth, revealing the various mechanisms beyond these activities. Inhibition of nitric oxide synthase, cyclooxygenases, and pro-inflammatory cytokines, together with the antioxidant potentials were the major mechanisms adopted by the most-studied Bombacoideae members and their isolated compounds. The anti-inflammatory potential was most commonly ascribed to the flavonoid content and to a lesser extent to other metabolites, such as fatty acids, despite the possibility of contribution of other reported secondary metabolites having anti-inflammatory potential, such as tannins and triterpenes (Mohammed et al. 2014).

Moreover, chemical investigations of the genus *Pachira* resulted in the identification of almost 20 secondary metabolites belonging to diverse classes. Major compound classes identified include naphthoquinone derivatives, flavonoids, coumarins, sesquiterpenes, and triterpenes. Only a few pharmacological activities have been evaluated, including antimicrobial, antifungal, and insecticidal activities. It's worth mentioning that *Pachira aquatica* is the most studied *Pachira* species, from which most compounds have been isolated. Meanwhile, studies in literature addressing the biological activities or the phytochemicals of *P. glabra* were very scarce. So, in-depth phytochemical investigation of the methanol extract of *P. glabra* leaves in an attempt to discover new molecules with promising pharmacological value is highly recommended.

Abbreviations

DPPH: 2,2-Diphenyl-1-picrylhydrazyl, ORAC: oxygen-radical-absorbance capacity, IAC: integral antioxidant capacity, TBARS: thiobarbituric acid reactive substances, FRAP: ferric reducing ability of plasma, HA: Hemagglutinating antibody, Hb: hemoglobin, WBC: White blood cells, RBC: red blood cells, DTH: delayed type of hypersensitivity, LPO: lipid peroxidation, SOD: superoxide dismutase, GSH: reduced glutathione, CAT: catalase, ALP: alkaline phosphate, AST: aspartate aminotransferase, LDH: lactate dehydrogenase, ALT: alanine aminotransferase, LPS: lipopolysaccharide, *MTT*: 3-[4,5-dimethylthiazol-2-yl]-2,5 diphenyl tetrazolium bromide, DMSO: Dimethyl sulfoxide, MeOH: methanol, LD_{50}: Median lethal dose, MIC: minimum inhibitory concentration, HBV: hepatitis B Virus, PCV: packed cell volume, TG: triglycerides, LDL: low density lipoprotein, VLDL: very-low-density lipoprotein, TC: total cholesterol, BMI: body mass index, TNF-α: tumor necrosis factor-α, IL-1β: interleukine-1β, LPS: bacterial lipopolysaccharide, COX: cyclooxygenases.

References

Abdelrahim, M.Y., Elamin, B.M., Khalil, D.J., El Badwi, S.M. 2013. Antidiarrhoeal activity of ethanolic extract of *Adansonia digitata* fruit pulp in rats. J. Phys. Pharm. Adv. 3: 172–178.

Adeoye, A., Bewaji, C. 2015. Therapeutic potentials of *Adansonia digitata* (Bombacaceae) stem bark in *plasmodium* berghei-infected mice. J. Biol. Sci. 15: 78–84.

Akuodor, G., Muazzam, I., Usman-Idris, M., Megwas, U., Akpan, J., Chilaka, K., Okoroafor, D., Osunkwo, U. 2011. Evaluation of the antidiarrheal activity of methanol leaf extract of *Bombax buonopozense* in rats. Ibnosina J. Med. Biol.S. 3: 15–20.

Akuodor, G., Usman, M.I., Ibrahim, J., Chilaka, K., Akpan, J., Dzarma, S., Muazzam, I., Osunkwo, U. 2013. Anti-nociceptive, anti-inflammatory and antipyretic effects of the methanolic extract of *Bombax buonopozense* leaves in rats and mice. Afr. J. Biotechnol. 10: 3191–3196.

Akuodor, G.C., Essien, A.D., Ibrahim, J.A., Bassey, A., Akpan, J.L., Ikoro, N.C., Onyewenjo, S.C. 2012a. Phytochemical and antimicrobial properties of the methanolic extracts of *Bombax buonopozense* leaf and root. Asian J. M. Biol. R. 2: 190–194.

Akuodor, G.C., Mbah, C.C., Megwas, U.A., Ikoro, N.C., Akpan, J.L., Okwuosa, B.O., Osunkwo, U.A. 2012b. *In vivo* antimalarial activity of methanol leaf extract of *Bombax buonopozense* in mice infected with *Plasmodium berghei.* Int. J. Biol. Chem. Sci. 5: 1790–1796.

Al-Qarawi, A., Al-Damegh, M., El-Mougy, S. 2003. Hepatoprotective influence of *Adansonia digitata* pulp. J. Herbs Spices Med. Plants 10: 1–6.

Alagawadi, K., Shah, A. 2011. Anti-inflammatory activity of *Ceiba pentandra* L. seed extracts. J. Cell Tissue Res. 11: 2781–2784.

Anandarajagopal, K., Sunilson, J.A.J., Ajaykumar, T., Ananth, R., Kamal, S. 2013. *In vitro* anti-inflammatory evaluation of crude *Bombax ceiba* extracts. Eur. J. Med. Plants 3: 99–104.

Anani, K., Hudson, J., De Souza, C., Akpagana, K., Tower, G., Arnason, J., Gbeassor, M. 2000. Investigation of medicinal plants of Togo for antiviral and antimicrobial activities. Pharm. Biol. 38: 40–45.

Anosike, C., CUgwu, J., Ojeli, P., Abugu, S. 2014. Anti-ulcerogenic effects and anti-oxidative properties of *Ceiba pentandra* leaves on alloxan-induced diabetic rats. Eur. J. Med. Plants 4: 458–472.

Anosike, C.A., Ofoegbu, R.E. 2013. Anti-ulcerogenic activity of the methanol extract of *Ceiba pentandra* stem bark on indomethacin and ethanol-induced ulcers in rats. Int. J. Pharm. Sci. 3: 223–228.

Antil V., Sinha, B.N., Pandey, A., Diwan, A.P.S. 2013. *Bombax malabaricum* DC: a salutary boon. Int. J. Pharm. Innovations 3(2): 27–28.

APG. 2009. An update of the Angiosperm Phylogeny Group classification for the orders and families of flowering plants: APG III. Botanical Journal of the Linnean Society 161(2): 105–121.

Arafa, A., Foda, D., Mahmoud, A., Metwally, N., Farrag, A. 2019. *Bombax ceiba* flowers extract ameliorates hepatosteatosis induced by ethanol and relatively moderate fat diet in rats. Toxicol. Rep. 6: 401–408.

Arafat, S.M., Abd El-Kader, E.M., Sayed, R. 2011. Fatty acids composition and quality assurance of Semal (*Bombax*) and Monsa *(Chorisia)* seed oils and use in deep-fat frying. Banat's J. of Biotechnol. 2: 66–75.

Asare, P., Adebayo, O.L. 2012. Comparative evaluation of *Ceiba pentandra* ethanolic leaf extract, stem bark extract and the combination thereof for *in vitro* bacterial growth inhibition. J. Nat. Sci. Res. 2: 44–49.

Ashmawy, A.M., Azab, S.S., Eldahshan, O.A. 2012. Effects of *Chorisia crispiflora* ethyl acetate extract on P21 and NF-κB in breast cancer cells. J. Am. Sci. 8: 965–972.

Ashour, M.L., Youssef, F.S., Gad, H.A., El-Readi, M.Z., Bouzabata, A., Abuzeid, R.M., Sobeh, M., Wink, M. 2018. Evidence for the anti-inflammatory activity of *Bupleurum marginatum* (Apiaceae) extracts using *in vitro* and *in vivo* experiments supported by virtual screening. J. Pharm. Pharmacol. 70(7): 952–963.

Ayele, Y., Kim, J.-A., Park, E., Kim, Y.-J., Retta, N., Dessie, G., Rhee, S.-K., Koh, K., Nam, K.-W., Kim, H.S. 2013. A methanol extract of *Adansonia digitata* L. leaves inhibits pro-inflammatory iNOS possibly via the inhibition of NF-κB activation. Biomol. Ther. 21: 146–152.

Ayoub, I.M., Youssef, F.S., El-Shazly, M., Ashour, M.L., Singab, A.N.B., Wink, M. 2015. Volatile constituents of *Dietes bicolor* (Iridaceae) and their antimicrobial activity. Z. Naturforsch. C. 70: 217–225.

Azab, S.S., Ashmawy, A.M., Eldahshan, O.A. 2013. Phytochemical investigation and molecular profiling by p21 and NF-[kappa] B of *Chorisia crispiflora* hexane extract in human breast cancer cells *in vitro*. Br. J. Pharm. Res. 3(1): 78–89.

Bairwa, N.K., Sethiya, N.K., Mishra, S. 2010. Protective effect of stem bark of *Ceiba pentandra* Linn. against paracetamol-induced hepatotoxicity in rats. Pharmacog. Res. 2: 26–30.

Barakat, M.M.A.-E., El-Boghdady, N.A., Farrag, E.K.E., Said, A.A., Shaker, S.E. 2019. Protective and curative effects of *Bombax ceiba* flower and *Ziziphus spina christi* fruit extracts on gastric ulcer. J. Biol. Sci. 19: 161–172.

Barwick, M., Schokman, L.M. 2004. Tropical & subtropical trees: An encyclopedia. Timber Press Portlande Oregon.

Baum, D.A., Oginuma, K. 1994. A review of chromosome numbers in Bombacaceae with new counts for *Adansonia*. Taxon, 43: 11–20.

Bayer, C., Kubitzki, K. 2003. Malvaceae, Flowering Plants Dicotyledons. Springer: 225–311.

Bhargav, B., Rupal, A., Reddy, A., Narasimhacharya, A. 2009. Antihyperglycemic and hypolipidemic effects of *Adansonia digitata* L. on alloxan induced diabetic rats. J. Cell & Tissue Res. 9(2): 1879–1882.

Bhavsar, C., Talele, G.S. 2013. Potential anti-diabetic activity of *Bombax ceiba*. Bangladesh J. Pharmacol. 8: 102–106.

Bianchini, J.P., Ralaimanarivo, A., Gaydou, E.M., Waegell, B. 1982. Hydrocarbons, sterols and tocopherols in the seeds of six *Adansonia* species. Phytochemistry 21: 1981–1987.

Bohannon, M.B., Kleiman, R. 1978. Cyclopropene fatty acids of selected seed oils from Bombacaceae, Malvaceae, and Sterculiaceae. Lipids 13: 270–273.

Bothon, F.T., Debiton, E., Yedomonhan, H., Avlessi, F., Teulade, J.-C., Sohounhloue, D.C. 2012. α-Glucosidase inhibition, antioxidant and cytotoxicity activities of semi-ethanolic extracts of *Bridellia ferruginea* Benth. and *Ceiba pentandra* L. Gaerth from Benin. Res. J. Chem. Sci. 2(12): 31, 36.

Bravo, J.A., Lavaud, C., Bourdy, G., Giménez, A., Sauvaine, M. 2002. First bioguided phytochemical approach to *Cavanillesia Aff. hylogeiton*. Rev. Bol. Quim. 19: 18–24.

Calzada, F., Juárez, T., García-Hernández, N., Valdes, M., Ávila, O., Mulia, L.Y., Velázquez, C. 2017. Antiprotozoal, antibacterial and antidiarrheal properties from the flowers of *Chiranthodendron pentadactylon* and isolated flavonoids. Pharmacog. Mag. 13(15): 240–244.

Ch, R.K., Madhavi, Y., Raghava Rao, T. 2012. Evaluation of phytochemicals and antioxidant activities of *Ceiba pentandra* (kapok) seed oil. J. Bioanal. Biomed. 4: 068–073.

Chaudhary, P.H., Khadabadi, S.S. 2012. *Bombax ceiba* Linn.: pharmacognosy, ethnobotany and phyto-pharmacology. Pharmacog. Comm. 2: 2–9.

Chauhan, S., Sharma, A., Upadhyay, N.K., Singh, G., Lal, U.R., Goyal, R. 2018. *In vitro* osteoblast proliferation and *in vivo* anti-osteoporotic activity of *Bombax ceiba* with quantification of Lupeol, gallic acid and β-sitosterol by HPTLC and HPLC. BMC Complem. Altern. M. 18(1): 233–245.

Chauhan, E.S., Singh, A., Tiwari, A. 2017. Comparative studies on nutritional analysis and phytochemical screening of *Bombax ceiba* bark and seeds powder. J. Med. Plants 5: 129–132.

Chauhan, J.S., Sultana, M., Srivastava, S.K. 1980. Constituents from *Salmalia malabaricum*. Canadian J. Chem. 58: 328–330.

Chauhan, J.S., Kumar, S., Chaturvedi, R. 1984. A new flavanonol glycoside from *Adansonia digitata* roots. Planta Med. 50(1): 113, doi: 10.1055/s-2007-969642.

Cheng, L.Y., Liao, H.R., Chen, L.C., Wang, S.W., Kuo, Y.H., Chung, M.I., Chen, J.J. 2017. Naphthofuranone derivatives and other constituents from *Pachira aquatica* with inhibitory activity on superoxide anion generation by neutrophils. Fitoterapia 117: 16–21.

Choubey, A., Choubey, A., Jain, P., Iyer, D., Patil, U. 2010. Assessment of *Ceiba pentandra* on calcium oxalate urolithiasis in rats. Der. Pharma. Chemica. 2: 144–156.

Cornelius, J., Hammonds, T., Shone, G. 1965. The composition of *Bombacopsis glabra* seed oil. J. Sci. Food Agri. 16: 170–172.

Coussio, J. 1964. Isolation of rhoifolin from *Chorisia species* (bombacaceae). Cell. Mol. Life Sci. 20: 562–562.

Cox, P.A. 1993. Saving the ethnopharmacological heritage of Samoa. J. Ethnopharmacol. 38: 177–180.

Dar, A., Faizi, S., Naqvi, S., Roome, T., Zikr-ur-Rehman, S., Ali, M., Firdous, S., Moin, S.T. 2005. Analgesic and antioxidant activity of mangiferin and its derivatives: the structure activity relationship. Biol. Pharm. Bull. 28: 596–600.

De Caluwé, E., Halamová, K., Van Damme, P. 2010. Adansonia digitata L.: a review of traditional uses, phytochemistry and pharmacology. Afrika focus 23: 11–51.

Dhar, D., Munjal, R. 1976. Chemical examination of the seeds of *Bombax malabaricum*. Plant Med. 29: 148–50.

Díaz-Rivas, J., Herrera-Carrera, E., Gallegos-Infante, J., Rocha-Guzmán, N., González-Laredo, R., Moreno-Jiménez, M., Ramos-Gómez, M., Reynoso-Camacho, R., Larrosa-Pérez, M., Gallegos-Corona, M. 2015. Gastroprotective potential of *Buddleja scordioides* Kunth Scrophulariaceae infusions; effects into the modulation of antioxidant enzymes and inflammation markers in an *in vivo* model. J. Ethnopharmacol. 169: 280–286.

Diehl, M., Atindehou, K.K., Téré, H., Betschart, B. 2004. Prospect for anthelminthic plants in the Ivory Coast using ethnobotanical criteria. J. Ethnopharmacol. 95: 277–284.

Dinarello, C.A. 2000. Proinflammatory cytokines. Chest 118: 503–508.

Divya, N., Nagamani, J., Suma, P. 2012. Antioxidant and antihemolytic activities of *Bombax ceiba, Petendra* spike and fruit extracts. Int. J. Pharm. Pharm. Sci. 4: 311–315.

Djomeni, P.D.D., Tédong, L., Asongalem, E.A., Dimo, T., Sokeng, S.D., Kamtchouing, P. 2006. Hypoglycaemic and antidiabetic effect of root extracts of *Ceiba pentandra* in normal and diabetic rats. Afr. J. Trad. CAM 3(1): 129–136.

Dolui, A., Das, S., Kharat, A. 2011. Antihyperglycemic effect of isolated fractions of *Ceiba pentandra* in alloxan induced diabetic rats. Asian J. Chem. 23: 2716–2718.

Dunstan, C.A., Noreen, Y., Serrano, G., Cox, P.A., Perera, P., Bohlin, L. 1997. Evaluation of some Samoan and Peruvian medicinal plants by prostaglandin biosynthesis and rat ear oedema assays. J. Ethnopharmacol. 57: 35–56.

Ebaid, H., Bashandy, S.A., Alhazza, I.M., Hassan, I., Al-Tamimi, J. 2019. Efficacy of a methanolic extract of *Adansonia digitata* leaf in alleviating hyperglycemia, hyperlipidemia, and oxidative stress of diabetic rats. BioMed. Res. Int. 19: 1–10.

El-Alfy, T., El-Sawi, S., Sleem, A., Moawad, D. 2010. Investigation of flavonoidal content and biological activities of *Chorisia insignis* Hbk. leaves. Austr. J. Basic & App. Sci. 4: 1334–1348.

El-Hagrassi, A.M., Ali, M.M., Osman, A.F., Shaaban, M. 2011. Phytochemical investigation and biological studies of Bombax *malabaricum* flowers. Nat. Prod. Res. 25: 141–151.

El-Toumy, S.A., Hawas, U.W., Taie, H.A. 2013. Xanthones and antitumor activity of *Bombax ceiba* against Ehrlich ascites carcinoma cells in mice. Chem. Nat. Comp. 49: 945–950.

El Sawi, S.A.M., Hanafy, D.M.M.M., El Alfy, T.S.M.A. 2014. Composition of the non-polar extracts and antimicrobial activity of *Chorisia insignis* HBK. leaves. Asian Pac. J. Trop. Dis. 4: 473–479.

Eldahshan, O.A., Azab, S.S. 2012. Anti-inflammatory effect of apigenin-7-neohesperidoside (rhoifolin) in carrageenin-induced rat oedema model. J. App. Pharm. Sci. 2(8): 74–79.

Etkin, N.L. 2000. Eating on the wild side: The pharmacologic, ecologic and social implications of using noncultigens. University of Arizona Press. P. 26.

Ezigbo, V., Odinma, S., Duruaku, I., Onyema, C. 2013. Preliminary phytochemical screening and antibacterial activity on stem bark extracts of *Ceiba Pentandra*. IOSR J. App. Chem. 6: 42–44.

Faizi, S., Ali, M. 1999. Shamimin. A new flavonol C-glycoside from leaves of *Bombax ceiba*. Planta Med. 65: 383–385.

Faizi, S., Zikr-ur-Rehman, S., Naz, A., Versiani, M.A., Dar, A., Naqvi, S. 2012. Bioassay-guided studies on *Bombax ceiba* leaf extract: isolation of shamimoside, a new antioxidant xanthone C-glucoside. Chem. Nat. Comp. 48: 774–779.

Fofie, C.K., Katekhaye, S., Borse, S., Sharma, V., Nivsarkar, M., Nguelefack-Mbuyo, E.P., Kamanyi, A., Singh, V., Nguelefack, T.B. 2019. Antidiabetic properties of aqueous and methanol extracts from the trunk bark of *Ceiba pentandra* in type 2 diabetic rat. J. Cell. Bioch. 120(7): 11573–11581.

Gadge, N., Jalalpure, S. 2012. Curative treatment with extracts of *Bombax ceiba* fruit reduces risk of calcium oxalate urolithiasis in rats. Pharm. Biol. 50: 310–317.

Gadge, N.B., Jalalpure, S.S., Pawase, K.B. 2009. Preliminary phytochemical screening and diuretic activity of *Bombax ceiba* L. fruit extracts. Pharmacol. Online 3: 188–194.

Gahane, R., Kogje, K. 2013. Antibacterial, antioxidant and phytochemical analysis of edible parts of potent nutraceutical plant–*Adansonia digitata.* Acta Horticulturae 972: 55–60.

Gamal El-Din, M.I., Youssef, F.S., Ashour, M.L., Eldahshan, O.A., Singab, A.N.B. 2018. Comparative analysis of volatile constituents of *Pachira aquatica* Aubl. and *Pachira glabra* Pasq., their anti-mycobacterial and anti-*Helicobacter pylori* activities and their metabolic discrimination using chemometrics. J. Ess. Oil Bear. Plants. 21: 1550–1567.

Gandhare, B., Soni, N., Dhongade, H.J. 2010. *In vitro* antioxidant activity of *Bombax ceiba*. Int. J. Biom. Res. 1: 31–36.

Gebauer, J., El-Siddig, K., Ebert, G. 2002. Baobab (*Adansonia digitata* L.): a Review on a Multipurpose Tree with Promising Future in the Sudan/Baobab (*Adansonia digitata* L.):. Gartenbauwissenschaft: 155–160.

González-Gallego, J., García-Mediavilla, M.V., Sánchez-Campos, S., Tuñón, M.J. 2014. Anti-inflammatory and immunomodulatory properties of dietary flavonoids, Polyphenols in Human Health and Disease. Elsevier: 435–452.

Gupta, P., Goyal, R., Chauhan, Y., Sharma, P.L. 2013. Possible modulation of FAS and PTP-1B signaling in ameliorative potential of *Bombax ceiba* against high fat diet induced obesity. BMC Complement. Altern. Med. 13: 281–289.

Hashim, R., Noordin, N., Zulkipli, F.H. 2014. Anti-inflammatory activity of kapok thorn (*Ceiba pentandra*) in acute inflammation. Asian Acad. Res. J. Multidis. 1: 278–287.

Hassan, A.A. 2009. Phytochemical and biological investigation of certain plants containing pigments. Mansoura University. Egypt. Ph.D Thesis.

Hoque, N., Rahman, S., Jahan, I., Shanta, M.A., Tithi, N.S., Nasrin, N. 2018. A comparative phytochemical and biological study between different solvent extracts of *Bombax ceiba* roots available in Bangladesh. Pharmacol. Pharm. 9: 53–66.

Hossain, E., Mandal, S.C., Gupta, J. 2011a. Phytochemical screening and *in vivo* antipyretic activity of the methanol leaf-extract of *Bombax malabaricum* DC (Bombacaceae). Tropical. J. Pharm. Res. 10: 55–60.

Hossain, E., Rawani, A., Chandra, G., Mandal, S.C., Gupta, J.K. 2011b. Larvicidal activity of *Dregea volubilis* and *Bombax malabaricum* leaf extracts against the filarial vector *Culex quinquefasciatus*. Asian Pacific J. Trop. Med. Dis. 4: 436–441.

Hossain, E., Chandra, G., Nandy, A.P., Mandal, S.C., Gupta, J.K. 2012. Anthelmintic effect of a methanol extract of *Bombax malabaricum* leaves on *Paramphistomum explanatum*. Parasitol. Res. 110: 1097–1102.

Hossain, E., Sarkar, D., Chatterjee, M., Chakraborty, S., Mandal, S.C., Gupta, J.K. 2013. Effect of methanol extract of *Bombax malabaricum* leaves on nitric oxide production during inflammation. Acta Poloniae Pharm. Drug Res. 70: 255–260.

http://www.webofknowledge.com.

http://www.ncbi.nlm.nih.gov/pubmed/.

https://scifinder.cas.org/scifinder/.

Ibrahim, H.M., Emmanuel, O.O., Kabiru, A.Y., Bello, M.U., Illumi, Y. 2013. Evaluation of antitrypanosomal activity of *Adansonia digitata* (methanol seed extract) in albino mice. Scien. J. Vet. Adv. 2: 1–6.

Ibrahim, H.M., Ogbadoyi, E.O., Adamu, K.Y. 2017. Evaluation of antitrypanosomal activity of ethyl acetate extract of *Adansonia digitata* seed extract in Tb brucei infected albino mice. Int. J. Drug Res. Tech. 2(7): 454–460.

Islam, M.K., Chowdhury, J., Eti, I. 2011. Biological activity study on a Malvaceae plant: *Bombax ceiba*. J. Sci. Res. 3: 445–450.

Itou, R.D.G.E., Sanogo, R., Ossibi, A.W.E., Ntandou, F.G.N., Ondelé, R., Pénemé, B.M., Andissa, N.O., Diallo, D., Ouamba, J.M., Abena, A.A. 2014. Anti-inflammatory and analgesic effects of aqueous extract of stem bark of *Ceiba pentandra* Gaertn. Pharmacol. Pharm. 5: 1113–1118.

Iwuanyanwu, T.C., Akuodor, G.C., Essien, A.D., Nwinyi, F.C., Akpan, J.L., Okoroafor, D. 2012. Evaluation of antimalarial potential of aqueous stem bark extract of *Bombax buonopozense* P. Beaur (Bombacaceae). Eastern J. Med. 17: 72–77.

Jain, V., Verma, S.K. 2012. Phytochemical Studies, Pharmacology of *Bombax ceiba* Linn. Springer 25–50.

Jalalpure, S., Gadge, N. 2011. Diuretic effects of young fruit extracts of *Bombax ceiba* L. in rats. Ind. J. Pharm. Sci. 73(3): 306–311.

Joshi, K.R., Devkota, H.P., Yahara, S. 2013. Chemical analysis of flowers of *Bombax ceiba* from Nepal. Nat. Prod. Comm. 8: 583–584.

Julien, G.K., Sorho, S., Yaya, S., Nathalie, G., Mireille, D., Joseph, D.A. 2015. Phytochemical study and antimicrobial activity of bark extracts of *Ceiba pentandra* (L.) Gaertn. (Bombacaceae) from Côte d'Ivoire on antibiotic resistant Staphylococcus aureus and Pseudomonas aeruginosa. Br. Microbiol. Res. J. 9, BMRJ.18103.

Kabbashi, A.S., Koko, W.S., Mohammed, S.E.A., Musa, N., Osman, E.E., Dahab, M.M., Allah, E.F.F., Mohammed, A.K. 2014. *In vitro* amoebicidal, antimicrobial and antioxidant activities of the plants *Adansonia digitata* and *Cucurbit maxima*. Advanc. Med. Plant Res. 2: 50–57.

Kaboré, D., Sawadogo-Lingani, H., Diawara, B., Compaoré, C., Dicko, M.H., Jacobsen, M. 2011. A review of baobab (*Adansonia digitata*) products: Effect of processing techniques, medicinal properties and uses. Afr. J. Food Sci. 5(16): 833–844.

Kaimal, T., Lakshminarayana, G. 1972. Changes in lipids of maturing *Ceiba pentandra* seeds. Phytochemistry 11: 1617–1622.

Kamatou, G., Vermaak, I., Viljoen, A. 2011. An updated review of *Adansonia digitata*: A commercially important African tree. S. Afr. J. Bot. 77: 908–919.

Khan, P.M.A., Hussain, S., Mohsin, M., Farooqui, M., Zaheer, A. 2012. *Bombax Ceiba* flower extract: biological screening and application as pH indicator. J. Chem. Biol. Phys. Sci. 2: 43–48.

Kharat, A., Ramteke, K., Kharat, K. 2015. Evaluation of anti-inflammatory and analgesic potential of methanolic extract of *Ceiba Pentandra*. Biopharm. J. 1: 22–26.

Krishnappa, K., Elumalai, K., Dhanasekaran, S., Gokulakrishnan, J. 2012. Larvicidal and repellent properties of *Adansonia digitata* against medically important human malarial vector mosquito *Anopheles stephensi* (Diptera: Culicidae). J. Vector Borne Dis. 49: 86–90.

Kuthar, S.S., Digge, V., Hogade, M.G., Poul, B., Jadge, D.R. 2015. Screening of Antibacterial activity of aqueous bark extract of *Bombax ceiba* against some gram positive and gram negative bacteria. Am. J. Phytomed. Clin. Ther. 3: 551–555.

Lagnika, L., Amoussa, M., Adjovi, Y., Sanni, A. 2012. Antifungal, antibacterial and antioxidant properties of *Adansonia digitata* and *Vitex doniana* from Bénin pharmacopeia. J. Pharmacog. Phytother. 4: 44–52.

Lami, M.H., Yusuf, K.A., Ndaman, S.A., Bola, B.M., Damilola, B.O., Siddique, A.A. 2015. Ameliorative properties of ethyl acetate fraction of *Ceiba pentandra* on serum glucose, hematological and biochemical parameters of diabetic rats. Asian Clin. Ther. Dis. 5: 737–742.

Lawal, T.O., Mbanu, A.E., Adeniyi, B.A. 2014. Inhibitory activities of *Ceiba pentandra* (L.) Gaertn. and *Cordia sebestena* Linn. on selected rapidly growing mycobacteria. Afr. J. Microbiol. Res. 8: 2387–2392.

Loganayaki, N., Siddhuraju, P., Manian, S. 2013. Antioxidant activity and free radical scavenging capacity of phenolic extracts from *Helicteres isora* L. and *Ceiba pentandra* L. J. Food Sci. Tech. 50: 687–695.

Masola, S., Mosha, R., Wambura, P. 2009. Assessment of antimicrobial activity of crude extracts of stem and root barks from *Adansonia digitata* (Bombacaceae) (African baobab). African J. Biotechnol. 8(19): 5076–5083.

Mohammed, M.S., Osman, W.J., Garelnabi, E.A., Osman, Z., Osman, B., Khalid, H.S., Mohamed, M.A. 2014. Secondary metabolites as anti-inflammatory agents. J. Phytopharmacol. 3: 275–285.

Musila, M., Dossaji, S., Nguta, J., Lukhoba, C., Munyao, J. 2013. *In vivo* antimalarial activity, toxicity and phytochemical screening of selected antimalarial plants. J. Ethnopharmacol. 146: 557–561.

Nam, N.H., Kim, H.M., Bae, K.H., Ahn, B.Z. 2003. Inhibitory effects of Vietnamese medicinal plants on tube-like formation of human umbilical venous cells. Phytother. Res. 17: 107–111.

Negi, J., Bisht, V., Singh, P., Rawat, M., Joshi, G. 2013. Naturally occurring xanthones: Chemistry and biology. J. App. Chem. 2013, 9, doi.org/10.1155/2013/621459.

Ngounou, F., Meli, A., Lontsi, D., Sondengam, B., Choudhary, M.I., Malik, S., Akhtar, F. 2000. New isoflavones from *Ceiba pentandra*. Phytochemistry 54: 107–110.

Noreen, Y., el-Seedi, H., Perera, P., Bohlin, L. 1998. Two new isoflavones from *Ceiba pentandra* and their effect on cyclooxygenase-catalyzed prostaglandin biosynthesis. J. Nat. Prod. 61: 8–12.

Nwagba, C., Ezugwu, C., Eze, C., Anowi, F., Ezea, S., Nwakile, C. 2013. Anti-ulcer activity of *Bombax buonopozense* P. Beauv. aqueous leaf extract (Fam: Bombacacea). J. App. Pharm. Sci. 3(2): 139–142.

Ogbobe, O., Ezeukwu, A., Ozoh, N. 1996. Physico-chemical properties of seed and fatty acid composition of *Bombax constantum* seed oil. Rivista Italiana delle Sostanze Grasse 73: 271–272.

Oloyede, G.K., Adaramoye, O.A., Oguntokun, O.J. 2013. Phytochemical and hepatotoxicity studies on *Adansonia digitata* leaf extracts. J. Exp. App. Anim. Sci. 1: 25–34.

Organization, W.H. 1996. Report of the WHO Informal Consultation on the Evaluation and Testing of Insecticides, WHO/HQ, Geneva, 7 to 11 October 1996. World Health Organization.

Orozco, J., Rodriguez-Monroy, M., Martínez, K., Flores, C., Jiménez-Estrada, M., Durán, A., Rosas-López, R., Hernández, L., Canales, M. 2013. Evaluation of some medicinal properties of *Ceiba aesculifolia* subsp. parvifolia. J. Med. Plants Res. 7: 309–314.

Osman, M.A. 2004. Chemical and nutrient analysis of baobab (*Adansonia digitata*) fruit and seed protein solubility. Plant Foods Hum. Nutr. 59: 29–33.

Otimenyin, S.O. 2018. Antiinflammatory medicinal plants: a remedy for most disease conditions? Natural Products and Drug Discovery. Elsevier: 411–431.

Oyetunji, O.A., Babatunde, I.R., Chia, S.L., Abraham, O.A., Adewale, F., Eweoya, O., Williams, F. 2015. Ameliorative Effects of *Adansonia digitata* Leaf extract on carbon tetrachloride (CCl4) induced testicular toxicity in adult male wistar Rats. Anat. J. Africa 4: 481–487.

Paramesha, B., Kumar, V.P., Bankala, R., Manasa, K., Tamilanban, T. 2014. Antidiabetic and hypolipidaemic activity of *Ceiba pentandra, Amaranthus viridis* and their combination on dexamethasone induced diabetic swiss albino rats. Int. J. Pharm. Pharm. Sci. 6(4): 242–246.

Parameshwar, P., Devi, D.A., Reddy, B.M., Reddy, K.S., Ramesh, E., Naik, G.N., Sravankumar, N. 2012. Hypoglycemic and anti-lipidemic effects of hydroethanolic extract of *Ceiba pentandra* Linn. Int. J. Pharm. Appl. 3: 315–323.

Patel, S.S., Verma, N.K., Rathore, B., Nayak, G., Singhai, A.K., Singh, P. 2011. Cardioprotective effect of *Bombax ceiba* flowers against acute adriamycin-induced myocardial infarction in rats. Rev. Bras. Farmacog. 21: 704–709.

Patil, A., Thakurdesai, P., Pawar, S., Soni, K. 2012. Evaluation of ethanolic leaf extract of *Ceiba pentandra* for anti-obesity and hypolipidaemic activity in cafeteria diet (CD) treated Wistar albino rats. Int. J. Pharm. Sci. Res. 3: 2664–2668.

Paula, V., Barbosa, L., Demuner, A., Campos, L., Pinheiro, A. 1997. Entomotoxicity of the nectar from *Ochroma lagopus* Swartz (Bombacaceae). Cienc. Cult. 49: 274–277.

Paula, V.F., Barbosa, L.C., Howarth, O.W., Demuner, A.J., Cass, Q.B., Vieira, I.J. 1995. Lignans from *Ochroma lagopus* Swartz. Tetrahedron 51: 12453–12462.

Paula, V.F., Barbosa, L., Demuner, A., Howarth, O., Piló-Veloso, D. 1996. Constituintes químicos de *Ochroma lagopus* Swartz. Quim. Nova 19: 225–229.

Paula, V.F., Barbosa, L.C.A., Errington, W., Howarth, O.W., Cruz, M.P. 2002. Chemical constituents from *Bombacopsis glabra* (Pasq.) A. Robyns: Complete 1H and 13C NMR assignments and X ray structure of 5-hydroxy-3,6,7,8,4'-pentamethoxyflavone. J. Braz. Chem. Soc. 13: 276–280.

Paula, V.F., Cruz, M.P., Barbosa, L.C.D.A. 2006a. Chemical constituents of *Bombacopsis glabra* (Bombacaceae). Quim. Nova 29: 213–215.

Paula, V.F., Rocha, M.E., Barbosa, L.C.d.A., Howarth, O.W. 2006b. Aquatidial, a new bis-norsesquiterpenoid from *Pachira aquatica* Aubl. J. Braz. Chem. Soc. 17: 1443–1446.

Puckhaber, L.S., Stipanovic, R.D. 2001. Revised structure for a sesquiterpene lactone from *Bombax malbaricum.* J. Nat. Prod. 64: 260–261.

Qi, Y., Guo, S., Xia, Z., Xie, D. 1996. Chemical constituents of *Gossampinus malabarica* (L.) Merr.(II). China J. Chinese Materia Med. 21: 234–235.

Raffauf, R.F., Zennie, T.M., Onan, K.D., Le Quesne, P.W. 1984. Funebrine, a structurally novel pyrrole alkaloid, and other. gamma.-hydroxyisoleucine-related metabolites of *Quararibea funebris* (Llave) Vischer (Bombacaceae). J. Org. Chem. 49: 2714–2718.

Ralaimanarivo, A., Gaydou, E.M., Bianchini, J.-P. 1982. Fatty acid composition of seed oils from six *Adansonia s* with particular reference to cyclopropane and cyclopropene acids. Lipids 17: 1–10.

Ramadan, A., Harraz, F., El-Mougy, S. 1994. Anti-inflammatory, analgesic and antipyretic effects of the fruit pulp of *Adansonia digitata*. Fitoterapia 65: 418–418.

Rao, K.S., Jones, G.P., Rivett, D.E., Tucker, D.J. 1989. Cyclopropene fatty acids of six seed oils from Malvaceae. J. Am. Oil Chem. Soc. 66: 360–361.

Rao, K.V., Sreeramulu, K., Gunasekar, D., Ramesh, D. 1993. Two new sesquiterpene lactones from *Ceiba pentandra*. J. Nat. Prod. 56: 2041–2045.

Ravi Kiran, C., Raghava Rao, T. 2014. Lipid profiling by GC-MS and anti-inflammatory activities of *Ceiba pentandra* Seed Oil. J. Biol. Act. Prod. Nat. 4: 62–70.

Ravi, V., Patel, S., Verma, N., Datta, D., Saleem, T.M. 2010. Hepatoprotective activity of *Bombax ceiba* Linn against isoniazid and rifampicin-induced toxicity in experimental rats. Int. J. App. Res. Nat. Prod. 3: 19–26.

Rates, S.M.K. 2001. Plants as source of drugs. Toxicon 39(5): 603–613.

Reddy, M.V.B., Reddy, M.K., Gunasekar, D., Murthy, M.M., Caux, C., Bodo, B. 2003. A new sesquiterpene lactone from *Bombax malabaricum*. Chem. Pharm. Bull. 51: 458–459.

Refaat, J., Samy, M.N., Desoukey, S.Y., Ramadan, M.A., Sugimoto, S., Matsunami, K., Kamel, M.S. 2015a. Chemical constituents from *Chorisia chodatii* flowers and their biological activities. Med. Chem. Res. 1–11.

Refaat, J., Yehia Desoukey, S., Ramadan, M.A., Kamel, M.S., Han, J., Isoda, H. 2015b. Comparative polyphenol contents, free radical scavenging properties and effects on adipogenesis of *Chorisia chodatii* and *Chorisia speciosa*. J. Herb. Drugs 5: 193–207.

Ribeiro, D., Freitas, M., Tomé, S.M., Silva, A.M., Laufer, S., Lima, J.L., Fernandes, E. 2015. Flavonoids inhibit COX-1 and COX-2 enzymes and cytokine/chemokine production in human whole blood. Inflammation 38: 858–870.

Rishton, G.M. 2008. Natural products as a robust source of new drugs and drug leads: past successes and present day issues. Am. J. Cardiol. 101: S43–S49.

Robyns, A. 1964. Flora of Panama. Part VI. Family 116. Bombacaceae. Annals of the Missouri Botanical Garden 51: 37-68.

Rodrigues, A.P., Pereira, G.A., Tomé, P.H.F., Arruda, H.S., Eberlin, M.N., Pastore, G.M. 2019. Chemical composition and antioxidant activity of monguba (*Pachira aquatica*) seeds. Food Res. Int. 121: 880–887.

Rosenkranz, H.S., Thampatty, B.P. 2003. ISAR: Flavonoids and COX-2 Inhibition. Oncol. Res. Featuring Preclin. Clin. Canc. Ther. 13: 529–535.

Said, A., Aboutabl, E.A., Nofal, S.M., Tokuda, H., Raslan, M. 2011. Phytoconstituents and bioctivity evaluation of *Bombax ceiba* L. flowers. J. Trad. Med. 28: 55–62.

Saif-ur-Rehman, S.A.J., Ahmed, I., Shakoor, A., Iqbal, H.M., Ahmad, B.M., Tipu, I. 2010. Investigation of hypoglycemic effect of *Ceiba pentandra* root bark extract in normal and alloxan induced diabetic albino rats. Int. J. Agro Vet. Med. Sci. 4: 88–95.

Saleem, R., Ahmad, M., Hussain, S.A., Qazi, A.M., Ahmad, S.I., Qazi, M.H., Ali, M., Faizi, S., Akhtar, S., Husnain, S.N. 1999. Hypotensive, hypoglycaemic and toxicological studies on the flavonol C-glycoside shamimin from *Bombax ceiba*. Planta Med. 65: 331–334.

Saleem, R., Ahmad, S.I., Ahmed, M., Faizi, Z., Zikr-ur-Rehman, S., Ali, M., Faizi, S. 2003. Hypotensive activity and toxicology of constituents from *Bombax ceiba* stem bark. Biol. Pharm. Bull. 26: 41–46.

Salinas, G., Rangasetty, U.C., Uretsky, B.F., Birnbaum, Y. 2007. The cycloxygenase 2 (COX-2) story: It's time to explain, not inflame. J. Cardiovas. Pharmacol. & Ther. 12: 98–111.

Sankaram, A.V.B., Reddy, N.S., Shoolery, J.N. 1981. New sesquiterpenoids of *Bombax malabaricum*. Phytochemistry 20: 1877–1881.

Saptarini, N.M., Deswati, D.A. 2015. The antipyretic activity of leaves extract of Ceiba pentandra better than *Gossypium arboreum*. J. of App. Pharm. Sci. 5: 118–121.

Sarkiyayi, S., Ibrahim, S., Abubakar, M., Shehu, S. 2010. Studies on antivenom activity of *Ceiba pentandra* leaves' aqueous methanol extract against Echis ocellatus' snake venom. Res. J. App. Sci. Eng. Tech. 2: 687–694.

Sati, S.C., Sati, M.D., Sharma, A. 2011. Isolation and characterization of flavone di-glucoside and acetoxyxanthone from the flowers of *Bombex ceiba*. J. Appl. Nat. Sci. 3: 128–130.

Satyaprakash, R., Rajesh, M., Bhanumathy, M., Harish, M., Shivananda, T., Shivaprasad, H., Sushma, G. 2014. Hypoglycemic and antihyperglycemic effect of *Ceiba pentandra* L. Gaertn in normal and streptozotocininduced diabetic rats. Ghana Med. J. 47: 121–127.

Selvarani, V., Hudson James, B. 2009. Multiple inflammatory and antiviral activities in *Adansonia digitata* (Baobab) leaves, fruits and seeds. J. Med. Plants Res. 3: 576–582.

Seshadri, V., Batta, A.K., Rangaswami, S. 1973. Phenolic components of *Bombax malabaricum*. Indian J. Chem. 11, 825.

Seshadri, V., Batta, A.K., Rangaswami, S. 1976. A new crystalline lactone from *Bombax malabaricum*. Indian J. Chem. 14: 616–617.

Shahat, A.A. 2006. Procyanidins from *Adansonia digitata*. Pharm. Biol. 44: 445–450.

Shahat, A.A., Ahmed, H.H., Hassan, R.A., Hussein, A.A. 2008. Antioxidant activity of proanthocyanidins from *Adansonia digitata* fruit. Asian Pacific J. Trop. Med. 1: 55–59.

Shahat, A.A., Hassan, R.A., Nazif, N.M., Van Miert, S., Pieters, L., Hammuda, F.M., Vlietinck, A.J. 2003. Isolation of mangiferin from *Bombax malabaricum* and structure revision of shamimin. Planta Med. 69: 1068–1070.

Sharker, S.M. 2009. Antinociceptive activity of crude ethanolic extract of *Paederia foetida, Butea monosperma, Bombex ceiba*. Pharmacologyonline 2: 862–866.

Shibatani, M., Hashidoko, Y., Tahara, S. 1999a. Accumulation of isohemigossypolone and its related compounds in the inner bark and heartwood of diseased *Pachira aquatica*. Biosci. Biotechnol. Biochem. 63: 1777–1780.

Shibatani, M., Hashidoko, Y., Tahara, S. 1999b. A major fungitoxin from *Pachira aquatica* and its accumulation in outer bark. J. Chem. Ecol. 25: 347–353.

Sichaem, J., Siripong, P., Khumkratok, S., Tip-Pyang, S. 2010. Chemical constituents from the roots of *Bombax anceps*. J. Chil. Chem. Soc. 55: 325–327.

Singh, A., Chauhan, E.S., Singh, O.P. 2018. Anti-hyperlipidaemic effect of *Bombax ceiba* bark and seeds powder on albino wistar rats. World J. Pharm. Pharm. Sci. 7(4): 1259–1276.

Singh, M.P., Panda, H. 2005. Medicinal herbs with their formulations. Daya Books.

Sreeramulu, K., Rao, K.V., Rao, C.V., Gunasekar, D. 2001. A new naphthoquinone from *Bombax malabaricum*. J. Asian Nat. Prod. Res. 3: 261–265.

Sulaiman, L.K., Oladele, O.A., Shittu, I.A., Emikpe, B.O., Oladokun, A.T., Meseko, C.A. 2013. In-ovo evaluation of the antiviral activity of methanolic root-bark extract of the African Baobab (*Adansonia digitata* Lin). Afr. J. Biotechnol. 10: 4256–4258.

Sule, M., Njinga, N., Musa, A., Magaji, M., Abdullahi, A. 2009. Phytochemical and anti-diarrhoeal studies of the stem bark of *Ceiba pentandra* (Bombacaceae). Nig. J. Pharm. Sci. 8: 143–148.

Suleiman, M., Mamman, M., Hassan, I., Garba, S., Kawu, M., Kobo, P., Suleiman, M.M., Mamman, M., Hassan, I., Garba, S. 2014. Antidiarrhoeal effect of the crude methanol extract of the dried fruit of L.(Malvaceae) *Adansonia digitata*. Vet. World 7: 495–500.

Tanko, Y., Yerima, M., Mahdi, M., Yaro, A., Musa, K., Mohammed, A. 2008. Hypoglycemic activity of methanolic stem bark of *Adansonnia digitata* extract on blood glucose levels of streptozocin-induced diabetic wistar rats. Int. J. App. Res. Nat. Prod. 1: 32–36.

Telrandhe, R., Mahapatra, D., Kamble, M. 2017. *Bombax ceiba* thorn extract mediated synthesis of silver nanoparticles: Evaluation of anti-Staphylococcus aures activity. Int. J. Pharm. Drug Anal. 5: 376–379.

Thabet, A.A., Youssef, F.S., El-Shazly, M., Singab, A.N.B. 2018a. GC-MS and GC-FID analyses of the volatile constituents of *Brachychiton rupestris* and *Brachychiton discolor*, their biological activities and their differentiation using multivariate data analysis. Nat. Prod. Res. 1–5, doi.org/10.1080/14786419.2018.1490908.

Thabet, A.A., Youssef, F.S., El-Shazly, M., El-Beshbishy, H.A., Singab, A.N.B. 2018b. Validation of the antihyperglycaemic and hepatoprotective activity of the flavonoid rich fraction of Brachychiton rupestris using *in vivo* experimental models and molecular modelling. Food Chem. Toxicol. 114: 302–310.

Tundis, R., Rashed, K., Said, A., Menichini, F., Loizzo, M.R. 2014. *In vitro* cancer cell growth inhibition and antioxidant activity of *Bombax ceiba* (Bombacaceae) flower extracts. Nat. Prod. Comm. 9: 691–694.

Ueda, H., Kaneda, N., Kawanishi, K., Alves, S.M., Moriyasu, M. 2002. A new isoflavone glycoside from *Ceiba pentandra* (L.) Gaertner. Chem. Pharm. Bull. 50: 403–404.

Vasita, A., Bhargava, S. 2014. Effect of different leaf extracts of *Bombax ceiba* on gentamicin induced nephrotoxicity in albino rats. Int. J. Adv. Res. Pharm. Bio. Sci. 4: 1–7.

Vázquez, E., Martínez, E.M., Cogordán, J.A., Delgado, G. 2001. Triterpenes, phenols, and other constituents from the leaves of *Ochroma pyramidale* (Balsa Wood, Bombacaceae). Preferred Conformations of 8-C-β-D-Glucopyranosyl-apigenin (vitexin). J. Mex. Chem. Soc. 45(4): 254–258.

Veeresham, C. 2012. Natural products derived from plants as a source of drugs. J. Adv. Pharm. Technol. Res. 3(4): 200–201.

Velázquez, C., Calzada, F., Esquivel, B., Barbosa, E., Calzada, S. 2009. Antisecretory activity from the flowers of Chiranthodendron pentadactylon and its flavonoids on intestinal fluid accumulation induced by *Vibrio cholerae* toxin in rats. J. Ethnopharmacol. 126: 455–458.

Velázquez, C., Correa-Basurto, J., Garcia-Hernandez, N., Barbosa, E., Tesoro-Cruz, E., Calzada, S., Calzada, F. 2012. Anti-diarrheal activity of (–)-Epicatechin from *Chiranthodendron pentadactylon* Larreat: Experimental and computational studies. J. Ethnopharmacol. 143: 716–719.

Versiani, M.A. 2004. Studies in the chemical constituents of *Bombax ceiba* and *Cuscuta reflexa*. University of Karachi, Karachi. Ph.D Thesis.

Vertuani, S., Braccioli, E., Buzzoni, V., Manfredini, S. 2002. Antioxidant capacity of *Adansonia digitata* fruit pulp and leaves. Acta Phytother. 2: 2–7.

Vieira, T.O., Said, A., Aboutabl, E., Azzam, M., Creczynski-Pasa, T.B. 2009. Antioxidant activity of methanolic extract of *Bombax ceiba*. Redox Rep. 14: 41–46.

Wahab, S., Hussain, A., Farooqui, A.H.A., Ahmad, M.P., Hussain, M.S., Rizvi, A., Ahmad, M.F., Ansari, N.H. 2014. *In vivo* Antioxidant and Immunomodulatory Activity of *Bombax ceiba* Bark-Focusing on its Invigorating Effects. Am. J. Adv. Drug Del. 2: 01–13.

Wang, G.K., Lin, B.B., Rao, R., Zhu, K., Qin, X.Y., Xie, G.Y., Qin, M.J. 2013. A new lignan with anti-HBV activity from the roots of *Bombax ceiba*. Nat. Prod. Res. 27: 1348 1352.

Wickens, G. 1980. The uses of the baobab (*Adansonia digitata* L.) in Africa. Browse in Africa. ILCA/FAO, Addis-Ababa, Ethiopie: 151–154.

Williamson, E.M. 2002. Major herbs of Ayurveda. Churchill Livingstone.

Wu, J., Zhang, X.H., Zhang, S.W., Xuan, L.J. 2008. Three novel compounds from the flowers of *Bombax malabaricum*. Helv. Chim. Acta. 91: 136–143.

Yagoub, S.O. 2008. Antimicrobial activity of *Tamarindus indica* and *Adansonia digitata* extracts against *E. coli* isolated from urine and water specimens. Res. J. Microbiol. 3: 193–197.

Yeboah, S.O., Mitei, Y.C., Ngila, J.C., Wessjohann, L., Schmidt, J. 2012. Compositional and structural studies of the oils from two edible seeds: Tiger nut, *Cyperus esculentum*, and asiato, *Pachira insignis*, from Ghana. Food Res. Int. 47: 259–266.

You, Y.J., Nam, N.H., Kim, Y., Bae, K.H., Ahn, B.Z. 2003. Antiangiogenic activity of lupeol from *Bombax ceiba*. Phytother. Res. 17: 341–344.

Youssef, F.S., Hamoud, R., Ashour, M.L., Singab, A.N. and Wink, M. 2014. Volatile oils from the aerial parts of *Eremophila maculata* and their antimicrobial activity. Chem. Biodivers. 11: 831–841.

Yu, Y.-g., He, Q.-t., Yuan, K., Xiao, X.-l., Li, X.-f., Liu, D.-m., Wu, H. 2011. *In vitro* antioxidant activity of *Bombax malabaricum* flower extracts. Pharm. Biol. 49: 569–576.

Yusha'u, M., Hamza, M., Abdullahi, N. 2010. Antibacterial activity of *Adansonia digitata* stem bark extracts on some clinical bacterial isolates. Int. J. Biomed. Health Sci. 6: 129–135.

Zennie, T.M., Cassady, J.M., Raffauf, R.F. 1986. Funebral, a new pyrrole lactone alkaloid from *Quararibea funebris*. J. Nat. Prod. 49: 695–698.

Zennie, T.M., Cassady, J.M. 1990. Funebradiol, a new pyrrole lactone alkaloid from *Quararibea funebris* flowers. J. Nat. Prod. 53: 1611–1614.

Zhang, X., Zhu, H., Zhang, S., Yu, Q., Xuan, L. 2007. Sesquiterpenoids from *Bombax malabaricum*. J. Nat. Prod. 70: 1526–1528.

Zhang, X., Zhang, S., Xuan, L. 2008. Three new furanosesquiterpenoids from *Bombax malabaricum* and revised NMR assignment of hibiscone C. Heterocycles 75: 661–668.

Zoghbi, M.d.G.B., Andrade, E.H.A., Maia, J.G.S. 2003. Volatiles from flowers of *Pachira aquatica* Aubl. J. Ess. Oil-Bear. Plants 6: 116–119.

16

Ayahuasca
Inherent Dangers in Its Consumption

Raquel Consul,[1,] Flávia Lucas[2,3] and Maria Graça Campos[1,4]*

Introduction

The interest in plants and their healing properties has always followed human evolution, and in the case of psychoactive plants, has been widely used in magical-religious rituals (Camargo 2014), as a vehicle for understanding existence, contact with the supernatural, and understanding the outside world (Albuquerque et al. 2005). In this context of searching for physical and spiritual cures, the Ayahuasca drink emerged in the Amazon (Zanela et al. 2018).

The Amazon rainforest is a true example of biculturalism,[a] due to its immensity, botanical wealth, its people, and *suis generis* customs, which give this region of the globe the fullest symbolism of what life in communion with nature is like, whilst exclusively relying on it.

Ayahuasca is a drink made by the decoction of *Psychotria viridis* Ruiz and Pav leaves, with the woody parts of the *Banisteriopsis caapi* (Spruce ex Griseb) Morton vine. It has been used by indigenous people to communicate with their ancestors, for prophetic purposes, divination, witchcraft, and healing. This state of soul liberation, which makes it possible to "travel to the world of the dead", is at the root of the term "Ayahuasca", which in Quechua means "wine of souls", and may also be called "Yagê", "Hoasca", "Caapi", "Dápa", "Mihi", "Kahí", "Natema", or "Pindé" (Schultes et al. 1992).

The altered state of consciousness reached by those who consume this drink, commonly called "tea", is due to its chemical composition, which highlights the metabolite N, N-dimethyl tryptamine (DMT), and the β-carbolines: harmine, harmaline, and tetrahydro harmine (THH) compounds, extracted from *Psychotria viridis* and *Banisteriopsis caapi*, respectively (Lorenzi and Matos 2008).

[1] Observatory of Drug-Herb Interactions/Faculty of Pharmacy, University of Coimbra, Health Sciences Campus, Azinhaga de Santa Comba, Coimbra, Portugal.
[2] Universidade do Estado do Pará, Belém, Pará, Brasil.
[3] Herbário MFS, Universidade do Estado do Pará, Belém, Pará, Brasil, Rua do Una, n°156, Telégrafo.
[4] Coimbra Chemistry Centre (CQC, FCT Unit 313) (FCTUC) University of Coimbra, Rua Larga, Coimbra, Portugal.
* Corresponding author: quel.consul@hotmail.com

[a] Term that combines "culture" and "diversity" and refers to the area of knowledge that interdisciplinarily articulates knowledge, languages, customs, and values derived from the relationship of man in a cultural and social context and nature, consisting of ecosystems and miscellaneous features (Salick et al. 2014).

By inhibiting the enzyme monoamine oxidase (MAO), β-carbolines enable the action of DMT in the Central Nervous System (CNS), which induces both mental and physical manifestations. The effects are mainly reflected in mood swings, synaesthesia, distortion of perception of space and time, as well as certain immobility, which occurs sometimes, uncoordinated movement, and more often nausea, diarrhea, and vomiting (Camargo 2014).

Although of a tribal origin, Ayahuasca is no longer unique to the forest, but has been integrated into peri-urban religious ceremonies, which have sprung up in northern Brazil and have spread throughout the world (Lorenzi and Matos 2008). It is estimated that over 20,000 people currently participate in these syncretic movements (UDV 2018).

The neo-psychedelic revival occurred in the 1980s, which brought to societies around the world the renewed desire to seek out the most varied ways to reach altered states of consciousness (Labate and Goulart 2005), a trend which seems to continue today. Analyzing the results of the research on the most used drugs in the United States of America (USA), there was a relative increase of 273% in use of tryptamines between 2007/2008 and 2013/2014, DMT being one of the most preferred substances. This is perhaps a result of its wide availability and ease of purchase online, allied to the anecdotal evidence presented by the media (Palamar and Le 2018).

At the same time, and possibly related, there is growing popularity of Ayahuasca, as it is a subject which has been increasingly referenced by the media, and heavily influenced by celebrity endorsement. This mainstreaming and prevalent exposure to Ayahuasca, given its composition, constitutes a danger to the health of those who seek it (Stiffler 2018). Although indigenous religions bind care and control to the use of Ayahuasca, the non-compliance with the traditional context can culminate in detrimental situations (Balick and Cox 1996).

The imprudent use of Ayahuasca has increased in Amazonian cities, especially by the tens of thousands of tourists who travel to Ecuador, Peru, and Colombia, motivated to experience shamanism, ignorant of the contraindicated combination of MAO inhibitors with certain drugs, food, and even natural herbal products, and their adverse effects (Bauer 2018).

In this text, in order to assess the inherent dangers of this drink, after a distinction of its ritualistic and religious use from recreational use has been made, followed by a detailed discussion of Ayahuasca pharmacology and its possible interactions, evidence from different tests performed *in vivo* in humans and cases reported in the literature were collected, with the purpose of reviewing the consequences of acute and chronic exposure of the beverage in question.

Although the main objectives are to evaluate its most harmful aspects, the potential therapeutic effects, already published in several studies (Domínguez-Clavé et al. 2016), will also be presented here.

Ritualistic and Religious Use: An Ethnopharmacobotanical Reflection

The Shamanism

Archaeological evidence of Ayahuasca ritual use, dating back to 500 B.C., is perfectly contestable, and may belong to any other ritualistic plant. In fact, there is no botanical evidence to accurately confirm when it was discovered, and that the combination of *B. caapi* and *P. viridis* resulted in an entheogen feeling (Bianchi 2005).

The "entheogen" concept was proposed by ethnologists Gordon Wasson, Karl Ruck, and Schultes in 1978 to characterize and encompass the complexity of the psychophysiological effects which invoking plants of divine entities can provide. It is not a theological or pharmacological term, but a cultural one, which means "internalizing God". This would be impossible to express through the term "hallucinogenic", as much by its negative connotation as by its limiting character, dispensable in the ethnopharmacobotanical understanding of a shamanic ritual (Camargo 2014).

Shamanism corresponds to the belief system of hunter and collector societies, and where the shaman, a charismatic figure, is the leader and is responsible for conducting the main community rituals, especially those to heal and protect the group. From an anthropological perspective, "cure" and "disease" are universal expressions of religiosity. The belief is that disease is caused by spiritual aggression and healing is achieved through religious practices. The shamanic ritual is unique and is also the most relevant activity in communities (Winkelman and Baker 2016).

Notwithstanding its lost temporal genesis, the use of Ayahuasca originated in the upper Amazon basin (Luna 2005), ingested by the shaman for healing purposes. By interpreting the visions, the shaman can identify the cause of the disease and wage a symbolic fight, which is described to the sick person through chanting until the evil spirit is released. The visions may involve their ancestors, mystical figures, or animals, especially snakes and jaguars (Camargo 2014).

The plants themselves enable communication with the spirits and entities which determine the fate of the human being because they also have their own spirit (Cruz 2016). They are "Teacher Plants" or "Power Plants" because they grant spiritual revelation to the shaman when they are ingested and teach others to be the best version of themselves, healing them and guiding them in the best path towards spiritual evolution. This religious implication is evident in shamanic traditions (Winkelman and Baker 2016, Labate and Goulart 2005). It reaches dimensions which exceed material life and leads to a "gateway" for the sacred purposes, perceived and interpreted in the human sphere.

With power plants, other types of diseases are treated as cultural diseases (Amorozo 2002). These, although not recognised by biomedicine, are characterized by presenting a consensus among individuals in the same community regarding their signs and symptoms, their origins, denominations, and their own sense and ways to cure them. These diseases have psychological, symbolic, and moral meaning, and have a direct and indirect impact on the suffering experienced by the individual.

The Preparation of the Ayahuasca Beverage

Despite not knowing when its use was transformed since pre-Columbian times (Bianchi 2005), Schultes et al. (1992) note that there are different ways to prepare Ayahuasca. Typically, in East Amazon regions, they use about two meters of *B. Caapi,* cut into smaller pieces, which are then placed in a container with plenty of water, together with the leaves of *P. viridis*, until a drink with a syrupy consistency is obtained. This is then ingested in small doses. In other regions, the *B. caapi* stems can be pulverised, layered with *P. viridis* leaves, and boiled for one hour. The preparation is finished after it has cooled. Due to its more liquid consistency, larger amounts are ingested, contrary to what was previously stated (Camargo 2014).

Other botanical species can also be added, especially those belonging to the *Rubiaceae* and *Solanaceae* families, in order to enhance the effects of the drink or provide the desired cure for specific diseases (Cruz et al. 2017, Camargo 2014). For example, *Brugmansia* spp. is used when the disease is caused by magic arrows or incantations or *Brunfelsia* spp. in the case of fever, rheumatism, and arthritis (Schultes et al. 1992), by producing the sensation of cold, through shivering, drives away evil spirits (Camargo 2014).

Ayahuasca's Effects

The various ways of preparation, taking into account the plants which are mixed and the different dosages, influence the manifestations of Ayahuasca (Balick and Cox 1996). Spruce, the driver of ethnobotanical studies in the Amazon, was a leading English botanist who devoted part of his life to researching the flora and the culture of this region (Spruce Project 2005). Also responsible for identifying the species *B. caapi*, he reported that the effects of the beverage are manifested two minutes after ingestion: initially observable by pallor and shaking, followed by perspiration, agitation, and intense delirium. After, approximately ten minutes, the shaman returns to calm and lull the person to sleep (Camargo 2014). The description outlined by Schultes et al. (1992) in "Plants of the Gods", states that, initially, symptoms are characterized by dizziness, nervousness, and nausea, as well as perspiration and intense tremors associated with increased heart rate and mydriasis.

The moment fatigue is manifested, the most impetuous chromatic and luminous experience begins, visions occur with the eyes closed, going from white to a bluish mist which gradually becomes more intense. However, deep sleep is interrupted by dream-like states, and occasionally feverish ones can also occur. At no time are movements negatively affected, with the only noticeable unpleasant effect being the severe diarrhea, which remains even after re-establishing a normal state of consciousness (Schultes et al. 1992).

Ayahuasca is one of the most powerful tools of Amazonian shamans, but its use can also be extended to other members who wish to see their gods or ancestors, humans, or animals (Balick and Cox 1996). Although in the scope of healing, the shaman is the vehicle of the "therapeutic" properties of this drink, the others are invited to consume Ayahuasca when the group goes through crisis situations that last longer than usual. In these cases, they take advantage of the emesis and purging which often occurs, which is a sign of soul purification and cleansing (Camargo 2014).

The Value of Indigenous Knowledge

Indigenous knowledge is often considered folkloric for its excessive superstition, but whenever an Indian uses a plant to cure a particular disease, the effectiveness of this treatment is empirically verified, as occurs in modern science, which acquires knowledge based on careful observation and is then subsequently tested. Obviously, the context of study and the experiments performed are different. However, epistemologically, they are very similar, empiricism being reflected in both (Albuquerque et al. 2005, Balick and Cox 1996). For example, traditional communities have developed plant classification systems based on trial-and-error processes, distinguishing the different types of "vine" (*B. caapi*) based on the different colors and visions they provide in the hallucination (Balick and Cox 1996).

In all medical systems, the disease underlies an imbalance. From a shamanic perspective, disturbances to homeostasis may be caused naturally, by excessive exposure to environmental elements or personally, caused by both mythical beings and the intervention of witches. However, diagnosis and treatment are always made by a specialist, in this case, by a shaman. Therefore, traditional medical knowledge is a coherent system of notions, concepts, models, and rituals, with a tendency towards health conservation (Cruz 2016).

Ethnobotany is dedicated to the study of the relationship between people and plants (Balick and Cox 1996), and ethnopharmacobotany is a more specific branch of this field of study, focusing on plants with therapeutic properties, based on ethnopharmacology, which scientifically and interdisciplinarily explores the bioactive compounds used in the medical systems of pre-literate people (Camargo 2014).

One of the biggest obstacles is correlating the aetiology of disease and its corresponding therapy, characteristic of healing rituals in the light of conventional medicine. It is utopian to draw a clear line between the magical and the rational, the spiritual and the scientific when the great emphasis of traditional medical systems associates illness with psychological and sociopathological causes. Therefore, the set of disorders caused by extra physical forces, of a magical-supernatural character, recognised in the culture of an ethnic group is referred to as "cultural disease" (Gruca et al. 2014).

Banisteriopsis caapi, for example, is used against the evil eye (Cruz 2016), also called the "cultural evil eye syndrome" (Camargo 2014). This plant is used as a purifier when prepared by bark infusion or to treat diabetes when stalk decoction is carried out (Cruz et al. 2017). Due to its alkaloid composition, the vine also provides an anthelmintic action, which has been demonstrated by a harmine inhibitory effect of 70% of the epimastigote, *Trypanosoma cruzi*, after 96 hours, due to the antagonism exerted on the parasite's neuromuscular system. Considering this gastrointestinal infection is endemic to tropical regions, the medicinal value of Ayahuasca is unquestionable, and it is through the physiological manifestations that the shaman takes advantage of this antiparasitic property, the psychoactivity serving as an effective dose marker (Pomilioet al. 1999).

Shamanism has existed since time immemorial and is currently practised in communities which still seek to live in harmony with nature (Camargo 2014). However, with the increasing westernization of these people, many of the traditions have begun to disappear because there ceases to be a transmission of knowledge, which was passed on exclusively to some members of the community, a fact determined by the cultural erosion, which,together with the loss of biodiversity, has become a real threat (Aswani et al. 2018, Balick and Cox 1996). Therefore, ethnobotany plays a key role in the preservation of indigenous traditions, taking into account the scientific and cultural value of ethnobotanical collections and the educational content they contain. These are a source of awareness of the importance of flora and its diversity, which also constitute a way to achieve the goals proposed by the 2011–2020 Global Plant Conservation Strategy (Melo et al. 2019).

The traditions and knowledge of each community are unique and the interaction with the environment influences healing, so it is essential that national laws and international conventions consider and respect the rights related to traditional plant knowledge and its use, and intellectual property (Cruz et al. 2017).

Today, there is a renewed interest in lifestyles that favor ecological and ancestral use of plants, which reflects the growth of alternative and traditional medicine (Cruz 2016). Although these societies have been transposed into more modern contexts, traditional medicine is still often the primary health care available, due to the limited access of the rural indigenous population to biomedicine, because of the cultural resistance which still exists, as well as the costs and insufficient coverage of public health (Cruz et al. 2017). Thus, and according to the World Health Organization strategy on traditional medicine (2014–2023), member states should make a careful assessment, based on sound scientific evidence, ensuring adequate protection of intellectual property, safeguarding against the inappropriate use of ancestral practices (such as Ayahuasca) (WHO 2013).

From the Forest to the City

Ayahuasca was used by at least 72 indigenous communities throughout Brazil, Bolivia, Colombia, Ecuador, Peru, and Venezuela by the end of the 19th and mid-20th century, a period in which the East Amazon region was occupied because of the exploration of rubber tree latex (Zanela et al. 2018).

Especially in Brazil, the ancestral use of Ayahuasca was transposed into a non-indigenous context through rubber tappers, giving rise to syncretic religious movements, such as "*Santo Daime*" and "*União do Vegetal*" (UDV) (Labate and Goulart 2005, Zanela et al. 2018).

In these churches, communion with Ayahuasca, also known as "*daime*" or "vegetable", takes place collectively, sometimes with the participation of hundreds of people. It occurs at least twice a month (Mello et al. 2019) and administered doses are relatively low, but the celebration is rich in sensory stimuli, including music and many dances (Ott 2013). Although the stages in the obtainment of Ayahuasca may differ slightly between the "*Ayahuasqueiros*" movements (Zanela et al. 2018, Albuquerque 2007), it is common to exclusively use "*mariri*" (*B. caapi*) and "*chacrona*" (*P. viridis*); by decoction and subsequent reduction, a thick/viscous, oily, brownish liquid is obtained from these plants (Figure 16.1) (Lanaro et al. 2015).

It is impossible to separate belief systems in Brazil from popular medical practices, since there is a large indigenous influence. Therefore, in the same way, the high point of these religions is working with mediumship, focusing on individual healing. Emphasizing the influence of the sociocultural context, the migration process of this drink to urban centers implies different hallucinatory responses from those found in the forest (Camargo 2014), since the mind plays a high-level function in the brain, as it receives previously filtered, structured, and organized information. With the constant acquisition of information about the outside world and inside our body, in connection with personal and cultural ideas, the mind produces a global scenario of reality (Winkelman and Baker 2016).

Figure 16.1: Ayahuasca – photo by Raquel Cônsul Lourenço.

The psychoactive substances which comprise Ayahuasca give it an illicit character, but in Brazil its use in a religious context is protected by Resolution No 01, of 25th January 2010, by the National Council on Drug Policy (CONAD), published in the Official Government Gazette of Brazil (CONAD 2010). The report presented by the CONAD Multidisciplinary Work Group, which establishes the ethics of Ayahuasca use as the means of preventing its misuse, highlights the distinction of the term "therapy". This term is based on ethical and scientific principles aimed at treatment, health maintenance or development—which, at no time, should be confused with the act of faith which takes place in a religious context, considering all the cures and personal problem-solving, without a necessary relation of cause and effect with Ayahuasca (GMT 2006).

Psychoactive substances, when interacting with serotonergic receptors, induce spiritual experiences and altered states of consciousness. They function as psychointegrators due to the overstimulation of crucial brain regions in the management of processes related to fundamental aspects of self, emotions, memories, and relationships. A feature of these substances is the enhancement of theta brain waves, which create feelings of healing, fullness, and cosmic awareness, when synchronised along the cerebrospinal axis (Winkelman and Baker 2016).

In a study aimed to understand the impact of Ayahuasca's repeated use, in a religious context, on psychological well-being, mental health, and cognitive processes, a psychopathological assessment, neuropsychological performance, and attitude to life in members of "Ayahuasca" religion, compared to a control group, was carried out for a year. The results showed better neuropsychological performance, high levels of spirituality, and better psychosocial adaptation in the study group (Bouso et al. 2012).

The results of a comparative evaluation carried out on the state of health, psychosocial well-being, and lifestyle of individuals who make religious use of Ayahuasca is consistent with this data. It showed that participants in the study had, in general, a better perception of health and a healthier lifestyle, taking into account public health indicators (Ona et al. 2019).

Considering the religion-healing interconnection, characteristic of these syncretic religions, the healing effects must be understood in their universe, as a way to restore the balance and re-integration of the individual in a social environment, taking advantage of these benefits (Zanela et al. 2018).

Ayahuasca, which has been transposed from the forest to Brazil's urban environment, has expanded through these religions to other countries, including Portugal, and it is in continuous transition, marked by a certain decontextualization, as much due to the cultivation of plants in exotic phytogeographies as by non-religious use, and even by demographic and ethnic differences (Zanela et al. 2018). Due to this "internationalization", there is an interest in studying its mechanism of action and understanding the neuronal correlation which exists between the altered states of consciousness induced by Ayahuasca (Schenberg et al. 2015).

Ayahuasca Pharmacology

At this point, a brief description will be given of the plants used in Ayahuasca preparation, with the main focus being the pharmacokinetic and pharmacodynamic behavior of the most abundant phytochemical compounds in the beverage.

Psychotria viridis Ruiz and Pav.

Psychotria viridis is a plant of the *Rubiaceae* family, commonly known as "*chacrona*" or "*rainha*" (Figure 16.2). It is a non-endemic native shrub of the Amazon and Atlantic Forest, characteristic of clayey and sandy soils, with abundant water (Taylor et al. 2015). It has dark green leaves, flexible, with a circular petiole in the proximal part and flat-convex in the distal part, with lateral projections. The limbus is lanceolate, narrow at the base, and acute at the apex (Quinteiro et al. 2006).

DMT is the alkaloid found in the highest concentration in *P. viridis* leaves. When present, it represents about 99% of the total alkaloids, showing trace amounts of N-monomethyl tryptamine (NMT) and 2-methyl-1,2,3,4-tetrahydro-β-carboline (MTHC). It was found that, when DMT is not present, the NMT and MTHC become the dominant compounds (Rivier and Lindgren 1972, Parra-Estrella-Parra et al. 2009).

The amount of DMT in leaves of *P. viridis* can vary between zero to 17.75 mg/g of dehydrated materials, and it appears that such production suffers circadian fluctuations, being reduced in the warmer periods of the day (Callaway et al. 2005). In addition to temperature, soil characteristics and season also determine the biosynthesis of alkaloids, a condition which also occurs in *B. caapi*. Considering the quantitative variations in phytochemical content, the portions used of each raw material and the different forms of obtaining them, make each Ayahuasca preparation distinct, triggering psychotropic responses of varying intensities (Lanaro et al. 2015).

Figure 16.2: *Exsicata* of *P. viridis* from MFS Herbarium – photo by Raquel Cônsul Lourenço.

DMT

Endowed with ubiquity, DMT is present in several plant and animal species, and can also be chemically synthesized. In humans, it results from tryptophan metabolism, but its function is not yet well understood. The biosynthetic mechanism seems to be common to the plant kingdom (Cameron and Olson 2018).

DMT is a classic hallucinogenic compound considered illicit by the 1971 United Nations Convention on Psychotropic Substances (UN 1971). It belongs to the class of indolamines due to its bicyclic structure, through the combining of a benzene group with a pyrrole group, which forms the indole nucleus, common to serotonin and melatonin (Araújo et al. 2015, Cameron and Olson 2018).

Pharmacokinetics

When ingested, it does not cause any activity in the body due to the extensive first-pass effect, essentially by the action of the monoamine oxidase (MAO) enzymes present in the liver and intestine, which convert DMT into inactive metabolites before it can permeate the blood-brain barrier to interact with target receptors. DMT is only orally active if co-administered with substances that inhibit the enzyme system in question, a condition guaranteed by β-carbolines in Ayahuasca (Barker 2018).

To cause the hallucinogenic effects, sufficient Ayahuasca must be ingested to achieve plasma concentrations in the range of 12–90 µg/L in an apparent distribution volume of 36–55 L/kg body weight (bw), which corresponds approximately to 0.06–0.50 µM of DMT in the plasma. If 0.1–0.4 mg/kg bw of only DMT is administered intravenously, the peak of the psychedelic effect occurs after five minutes and lasts roughly half an hour. In contrast, when MAO inhibitors are present, the half-life ($t_{1/2}$) increases, and consequently, the hallucinatory state is longer-lasting, because co-administration with β-carbolines can result in increased levels of DMT in the bloodstream (Carbonaro and Gatch 2016).

According to the pharmacokinetic evaluation of Ayahuasca, performed in a clinical trial, it was found that after equivalent administration of 0.6 mg/kg bw and 0.85 mg/kg bw of DMT, half an hour elapsed until the maximum plasma concentration (C_{max}) was reached (12 ng/mL and 17.44 ng/mL, respectively), coinciding with the peak of subjective effects, which began at 30 minutes and lasted up to six hours (Riba et al. 2003).

Oxidative deamination by MAO is the main detoxification pathway of DMT, but it is not exclusive, so it can be quantified in urine, in addition to the metabolite major indole-3-acetic acid (IAA), N-oxide-DMT (NO-DMT), NMT, and MTHC, resulting from N-oxidation, N-demethylation, and cyclization, respectively (Araújo et al. 2015). When co-administered with β-carbolines, there is a significant increase in NO-DMT quantified in urine. Therefore, it can be concluded that N-oxidation also becomes a major pathway.

However, the degree to which the shift of the MAO metabolic pathway to cytochrome P450 (CYP450) occurs, induced by the alkaloids in question, is not yet well understood (Riba et al. 2012).

Interestingly, at a high dose, DMT itself can have an inhibitory action of MAO-A for which it is selective, albeit of short duration. The maximum effect is found at 50 mg/kg bw and is reflected in decreased oxidative deamination of serotonin and dopamine (Carbonaro and Gatch 2016).

From the administration of Ayahuasca corresponding to 1 mg/kg bw DMT, less than 1% is recovered intact in the urine after 24 hours, and 95–97% of the quantified total is excreted within the first eight hours after ingestion (Riba et al. 2012).

Vitale et al. (2011) found that, after intravenous administration of radiolabelled DMT in rabbits, blood-brain barrier permeation and receptor binding occurs within 10 seconds. Despite being excreted in the urine in less than 24 hours, it is still detectable in the brain after 7 days, which proves the storage mechanism at the cerebral level. The release of stored DMT is triggered by specific stimuli, in this case, by stimulation of the rabbit's olfactory receptors.

The aforementioned transport and storage mechanism can be summarized in three steps: first DMT is actively transported by the blood-brain barrier by an ATP-dependent Mg^{2+} uptake mechanism through the endothelial membrane. It is then internalized in neuronal cells by serotonin transporters

(SERT) on the surface of neurons. Finally, DMT is sequestered in synaptic vesicles by the action of the monoamine vesicular transporter 2 (VMAT2). An inhibitory effect of serotonin uptake via SERT and VMAT2 by DMT was demonstrated (Carbonaro and Gatch 2016). Given these data, it appears unquestionable that this mechanism also reflects the influence of DMT in the CNS (Frecska et al. 2013).

Pharmacodynamics

Due to its structural simplicity, DMT has high affinity for many receptors.

Serotonergic system

The 5-HT$_{2A}$ receptor agonism is the most well-understood mechanism and explains the introspective and hallucinogenic effect, by increasing the frequency and amplitude of postsynaptic excitatory signals of the V pyramidal layer of the cortex, in addition to stimulating other regions, such as the amygdala, hippocampus, striated body, where the receptor in question has high expression. Stimulation of this receptor is also responsible for the neuroplasticity of DMT, which causes increased dendritic tree complexity of cortical neurons and promotes an increase in dendritic backbone density (Cameron and Olson 2018).

In the 5-HT$_{1A}$ somatodendritic receptors, which mediate inhibitory neurotransmission, DMT agonist action can result in a decreased release of serotonin in other brain regions because of the acute inhibition of the Raphe dorsal nucleus (Cameron and Olson 2018). Moreover, this mechanism also appears to be related to the visual effects (Carbonaro and Gatch 2016).

DMT also has affinity at the 5-HT$_{2C}$ receptor, although to a lesser extent compared to binding to a 5-HT$_{2A}$ receptor. However, it is believed that this mechanism has no implication in interoceptive effects due to the desensitisation phenomenon which occurs (Cameron and Olson 2018).

DMT also has an affinity for 5-HT$_{1D}$, 5-HT$_5$, 5-HT$_6$, and 5-HT$_7$ receptors, but little is known about the consequences of these interactions (Barker 2018), so further study is essential, in this sense. For example, 5-HT$_7$ receptors are implicated in learning and memory processes, so it is essential to understand the behavioral effects of DMT (Cameron and Olson 2018) as a result of Ayahuasca.

Sigma-1 receptor

The sigma-1 receptor is distributed throughout the body, including the CNS, lungs, heart, adrenal gland, spleen, and pancreas, and it is located between the endoplasmic reticulum and mitochondria. It is a target receptor in the treatment of depression, in addition to the numerous effects which result from its agonism (Carbonaro and Gatch 2016, Cameron and Olson 2018).

DMT agonises the sigma-1 receptor, but the relationship of this mechanism to hallucinogenic activity is not yet clear. The effects are more physiological. It was found that the DMT, to activate the sigma-1 receptor, influences the regulation of intracellular calcium and the expression of pro-apoptotic genes, which may result in a neuroprotective effect (Frecska et al. 2013).

The cells of the immune system also express the sigma-1 receptor. In an *in vitro* study with monocytes derived from human dendritic cells, it was shown that DMT can interfere with the innate and adaptive immune response via the sigma-1 receptor. Pro-inflammatory cytokines decreased, namely Interleukin (IL)-1b, IL-6, IL-8, and Tumor Necrosis Factor-α, and anti-inflammatory cytokine IL-10 increased. In the adaptive immune response, monocytes previously treated with DMT decreased the ability of Th$_1$ (T helper) and Th$_{17}$ effector T cells to differentiate, with an immunomodulatory effect of DMT upon interaction with the sigma-1 receptor (Szabo et al. 2014).

Trace amine-associated receptor-1 (TAAR-1)

TAAR-1 belongs to a recently discovered class of receptors which can mediate the effects of DMT. As most research has focused on the action of DMT in the 5-HT receptor$_{2A}$, there is little information on the role of TAAR-1 in the triggered effects. It is only known that there is a high affinity for the binding of DMT to this receptor and that it activates adenyl cyclase, leading to the accumulation of AMP$_c$ (Carbonaro and Gatch 2016, Barker 2018, Cameron and Olson 2018).

Dopaminergic system

The action of DMT on the dopaminergic system is controversial. On the one hand, the affinity of DMT for dopamine receptors is low, and in addition there is no stimulation of the dopamine-sensitive adenyl cyclase system. On the other hand, one study reported that DMT was able to reverse dopamine system damage, which led to increased dopamine concentrations. Distinctly, other authors found that DMT can interact with dopamine receptors, blocking overactive DMT action, after treatment with antagonists. However, these studies were performed long before the pharmacological understanding of such substances, and the interruption of DMT action could be explained by antipsychotic substances also used in those tests, which also have an affinity for serotonergic receptors, particularly for 5-HT$_{2A}$ (Cameron and Olson 2018).

Cholinergic system

Little has been studied about the action on the cholinergic system, but the administration of DMT provoked substantially lower levels of acetylcholine in the corpus striatum region. However, there was no change in the levels of acetylcholine in the cortex. It is likely that stimulation of the serotonergic system by DMT has an influence in this regard (Cameron and Olson 2018).

Metabotropic Glutamate Receptor II (mGlu2/3) and N-methyl-D-Aspartate Receptor (NMDA)

The interest in understanding the interactions between the functions of serotonin and glutamate, as mediators of DMT effects, has been increasing over the last decade, where the mGlu2/3 and NMDA receptors can be highlighted. The mGlu2/3 receptor agonism appears to suppress the release of glutamate at the presynaptic level, while the antagonism results in increased levels of glutamate in the synaptic cleft, enhancing hallucinogenic effects. However, the possible influence of this mechanism on the effects of DMT has not been systematically studied (Carbonaro and Gatch 2016).

With regard to the NMDA receptor, there is evidence that it plays a role in the effects triggered by DMT, since the receptor in question may be potentiated by the activation of sigma-1 receptor (Carbonaro and Gatch 2016).

Banisteriopsis caapi (Spruce Ex Griseb.) Morton

Also known as "*mariri*", "*caapi*", "*jagube*", or even "*ayahuasca*", *B.caapi* is a woody, robust vine with thick, sinuous stems, native to solid ground forests of the Amazon region, belonging to the *Malpighiaceae* family (Figure 16.3) (Lorenzi and Matos 2008).

It has quite a different alkaloid profile in composition and concentration, ranging from 0.5% to 1.95% of dry weight (Mckenna et al. 1998), but the three most prominent compounds exist in the leaves, stem, and roots. They are harmine, harmaline, and THH (Pomilio et al. 1999). The harmine is the major component, followed by THH, harmaline being the compound found in the least amount.

Through the phytochemical analysis of 33 samples of dry material of *B. caapi* stems, it was observed that the harmine composition can vary between 0.31 and 8.43 mg/g, harmaline presents values

Figure 16.3: Example of *B. caapi* from MFS Herbarium – photo by Raquel Cônsul Lourenço.

between 0.03 and 0.83 mg/g, and THH can vary between 0.05 and 2.94 mg/g. Harmine and harmaline composition is uniformly proportional in a 1:10 ratio. THH has a more variable distribution and is not clearly related to the other two β-carbolines. It was also found that the lowest values presented correspond to older plants, whereby plant age is also a factor which influences alkaloid biosynthesis, as already referenced (Callaway et al. 2005).

B-carbolines: harmine, harmaline, and THH

These compounds are related by the structure they share of the tricyclic indole alkaloids, for which reason they are called "β-carbolines" or also "harmala alkaloids", because they were first discovered in the *Peganum harmala* L. plant (Domínguez-Clavé et al. 2016). Their MAO inhibitory action defines the hallucinatory response and physiological effects of Ayahuasca, which may contain a varied range of β-carbolinic alkaloids, and this amount can vary substantially for the reasons already mentioned. In a quantitative evaluation of Ayahuasca samples from different times of the year, the range of harmine, harmaline, and THH concentrations was recorded, ranging from 294.5 to 2893.8 µg/mL, 27.5 to 181.3 µg/mL, and 849.5 to 2052 g/ml of Ayahuasca, respectively (Lanaro et al. 2015).

Pharmacokinetics

After ingestion of a dose of Ayahuasca equivalent to 3.4 mg/kg bw of harmine, 0.4 mg/kg bw of harmaline, and 2.14 mg/kg bw of THH, it was shown that the compound with the highest $t_{1/2}$ is THH, with 532 minutes, reaching $C_{max.}$ (91 ng/ml) at 174 minutes. The C_{max} reached by harmine and harmaline were 114.8 ng/mL and 6.3 ng/mL at 102 and 145 minutes, respectively. The $t_{1/2}$ of the harmine is substantially short compared to THH, about 115 minutes. For harmaline, this parameter could not be accounted for (Mckenna et al. 1998). The presence of harmine appears to be of a short duration, since THH plasma levels appear to have a dose-dependent relationship, with reduced elimination. However, in general, the peak plasma concentrations of β-carbolines are temporally offset from the acute vision effects. Therefore, DMT plays the greatest role in the pharmacology of this complex alkaloid combination (Riba et al. 2003).

From a metabolic point of view, β-carbolines essentially suffer *O*-demethylation, and it is possible to identify harmol, harmalol, and tetrahydroharmolin urine, as glucuronide and sulphate conjugates, but the existence of alternate pathways cannot be excluded (Riba et al. 2012).

It is important to emphasise the role of the 2D6 isoenzyme of the P450 cytochrome (CYP 2D6) in the harmine and harmaline biotransformation, and the inherent polymorphic degree of this isoenzyme in the population. In this sense, Callaway (2005) wanted to verify what the implication was of slow or fast metabolization of harmine and harmaline in the metabolism of DMT and THH, but the differences were not significant, highlighting the existence of an alternative route for THH. However, the similarities between DMT and harmine profiles clearly illustrate the dependence of DMT on plasma harmine levels.

In order to compensate for metabolic changes, Callaway (2005) describes that in religious worship, it is the "master", the ceremony officiant, who decides the amount of "*vegetable*" each person should ingest, taking into account the Ayahuasca's "strength". It is an empirical method, hardly concrete, but it enables the neutralization of the effects caused by inter-individual metabolic differences.

Pharmacodynamics

The action of Ayahuasca β-carbolines relates to their ability to reversibly and preferentially inhibit MAO-A, although it is recognised that at high concentrations their action can extend equally to MAO-B. However, there are no Ayahuasca preparations with those amounts (Santos 2007).

The inhibitory power of these alkaloids decreases with the increase of the saturation of the pyridine ring, but THH still has some blocking activity. In addition, the β-carboline effects are additive, not synergistic or antagonistic, and the mixture of only THH and harmine together can explain Ayahuasca activity. Although harmaline's inhibitory power is similar to harmine, its action contributes little to the overall action of Ayahuasca, given its prevalence (Pomilio et al. 1999).

Additionally, it was also found that β-carbolines have a direct effect on serotonin levels in the synaptic cleft by the action of THH, which acts as a selective serotonin reuptake inhibitor (Domínguez-Clavé et al. 2016, Estrella-Parra et al. 2019).

As MAO is responsible for endogenous neurotransmitter degradation, its inhibition alters, albeit indirectly, the homeostasis of the dopaminergic, adrenergic, and serotonergic systems, which is reflected in greater dopamine circulation, for example (Alsuntangled 2017). In addition to other less-studied mechanisms, the dopamine transport inhibition at high concentrations, specific inhibition of tyrosine-1A-phosphorylation regulatory kinase, as well as affinity for the imidazoline binding site contributes to this. Harmine, by way of example, also acts in the regulation of the excitatory amino acid transporter-2, as a primary mechanism of synaptic glutamate inactivation, as well as causing an inverse agonist effect at the benzodiazepine binding site on the $GABA_A$ (γ-aminobutiric acid) receptor (Hamill et al. 2018).

Some β-carbolines are known to have affinity for the $5-HT_{2A}$ and $5-HT_{2C}$ receptors, as is the case with harmine. However, it does not bind to the dopamine receptor, suggesting that dopamine efflux in the Accumbens nucleus is mediated by the $5-HT_{2A}$ mechanism (Hamill et al. 2018).

Ayahuasca's Biological Effects

Although the isolated understanding of the biological activity of Ayahuasca components is essential, its effects should be considered in light of the dynamic combination of β-carbolines with DMT. By measuring the cerebral electrical activity, using the electroencephalogram, it was possible to confirm a biphasic effect of the brain waves, which is characterized by the reduction of alpha waves 50 minutes after ingestion, and the potential increase of fast and slow gamma waves between 75 and 125 minutes. Correlating with the pharmacokinetics of the substances, it was possible to understand that DMT and harmine participate in the initial phase of the hallucinatory experience, with harmaline and THH being responsible for the later phase (Schenberg et al. 2015).

It was verified that Ayahuasca ingestion promotes the activation of several brain regions, including the insula, the parahippocampal and the inferior frontal gyrus, the anterior and frontomedial cingulate cortex and amygdala, which are involved in the modulation of feelings and emotions, in perception and self-awareness (Dos Santos et al. 2017b). Also, interestingly, the neuronal activation of the primary visual cortex area, produced by Ayahuasca whilst the eyes are closed, is comparable to the activation levels of natural image receiving whilst the eyes are open (De Araujo et al. 2012).

The induced altered state of consciousness is always difficult to compare and describe because of its abstract character. However, the most referenced subjective effects are introspection, serenity, almost biographical memories of experiences, sensation of well-being, hallucinations and synaesthesia, more specifically, visual and auditory, as well as mystical and religious experiences (Dos Santos et al. 2017b).

From a somatic perspective, Ayahuasca causes mydriasis, tachycardia, and increased blood pressure, tingling sensation, muscle contraction, increased body temperature, and increased secretion of prolactin, cortisol, and growth hormone (Domínguez-Clavé et al. 2016). It also causes vomiting due to vagus nerve stimulation (Lanaro et al. 2015), and severe diarrhea by stimulation of the peripheral serotonergic receptors (Hamill et al. 2018).

From a psychological perspective, profound mood swings may occur, during which a depressive state quickly changes to a euphoric state or vice versa, as well as feelings of panic, apathy, fear, depersonalization, and insomnia (Pires et al. 2010).

Addiction and tolerance

Ayahuasca, like almost all hallucinogenic substances, does not seem to cause physical dependence, which can be reflected in the pattern of use at the religious level, which is characterized by intermittent consumption without an increase of the dose (Santos 2007).

However, repeated exposure demonstrates some signs of tolerance observable due to the slight increase in growth hormone and prolactin secretion after a second administration, as well as a reduced effect on heart rate and diastolic pressure (Hamill et al. 2018).

Possible interactions with other bioactive molecules

Concomitant use of Ayahuasca with medicinal products is always risky and should be avoided, especially with those medications which may potentiate the serotonergic pathway, namely: selective serotonin reuptake inhibitors, MAO-A inhibitor drugs, tricyclic antidepressants, opioids, central antitussives or drugs used to treat migraines, such as triptans. This occurs because excessive accumulation of serotonin in nerve endings can result in a set of adverse effects called "serotonergic syndrome", which is characterized by intense CNS and peripheral serotonergic activity. Typical symptoms begin with an initial state of euphoria, followed by tremors and seizures, loss of consciousness, and may even occasionally result in death (Lanaro et al. 2015, Hamill et al. 2018). Although these are the most evident interactions, the same applies for herbal medicines containing *Panax ginseng* CA Mey or *Hypericum perforatum* L. (Hamill et al. 2018).

Pharmacokinetic interactions should also be taken into consideration. The β-carbolines are CYP2D6 substrates and competitively inhibit this isoenzyme. More specifically, harmine and its harmol metabolite are capable of competitively inhibiting the 3A4 isozyme of cytochrome P450. Therefore, special care must be taken in the administration of drugs which use the aforementioned pathways, either because of potential intoxication triggered by the drug and/or Ayahuasca, or because of the possibility of therapeutic inefficacy (Hamill et al. 2018). For example, *Valeriana officinalis* L, the root of *Glycyrrhiza glabra* L., *Curcuma longa* L., *Actaea racemosa* L., or *Camellia sinensis* (L.) Kuntze contain compounds which inhibit CYP2D6, and therefore, their simultaneous consumption with Ayahuasca should be avoided (Campos et al. 2018).

Tyramine-rich foods, usually fermented foods, such as cheese, beer, and meat, are also potential sources of interaction. Tyramine is degraded by MAO and the impediment of this pathway by Ayahuasca can lead to increased sympathetic stimulation, and consequently blood pressure increases, and the risk of stroke and intracranial haemorrhaging increases (Alsuntangled 2017).

Ayahuasca's Potential Therapeutic Effects

Preclinical and observational studies state that Ayahuasca has anxiolytic, antidepressant, and antiaddictive effects. The anxiolytic and antidepressant effects have already been tested in controlled trials, where a single dose of Ayahuasca has been shown to reduce the feeling of panic. A link has also been verified between Ayahuasca and a rapid reduction of depressive symptoms after seven days in patients with treatment-resistant major depression (Dos Santos et al. 2017b).

Regarding the potentiality of addiction treatment, based on animal models, Ayahuasca seems to modulate the dopaminergic system (Nunes et al. 2016), but there are no controlled human studies yet. There are only observational studies in religious contexts which suggest such an effect (Dos Santos et al. 2017b). Given the low prevalence of drug use by individuals belonging to religious groups and the reduced tendency of illicit drug use after their integration in the group, it remains to be seen whether it is due to Ayahuasca's mechanism of action or the result of the support found in religion (Cameron and Olson 2018).

Acute and Chronic Exposure Effects

In order to understand the potential dangers of Ayahuasca use, a systematic review of the literature was performed with the data-based resource, "PubMed", of *in vivo* studies, clinical studies, and reported cases, from ten years ago to the present day, using the terms "ayahuasca" and "chronic", "acute", "intoxication", "long-term" or "risks". From the 63 results, systematic reviews, articles which directly studied potential therapeutic effects, or trials whose Ayahuasca preparation included plants other than *Banisteriopsis caapi* and *Psychotria viridis*, were excluded, and literature not written in English, Portuguese, or Spanish was also excluded. The following is the results analysis and discussion of eight trials, three reported cases, and a descriptive analysis of the notifications sent to Poison Control Centers (PCC) in the USA.

Table 16.1: Methods and main results of citations included in the literature review.

Study Focus	Methods and main results	Ref.
Evaluation of acute toxicity in *Danio rerio* (zebrafish) embryos and the implications of exposure on their development	The assay of embryo toxicity consisted of subjecting the eggs, immediately after fertilization up to 96 hours, at concentrations ranging from 0 to 1000 mg of lyophilised Ayahuasca per liter of culture water, and thus assess the mortality incidence of pericardial oedema, defects, hatching balance, and developmental delays. Concentrations that did not induce any abnormalities or mortality (between 0.0064 to 20 mg /L) were selected for behavioral assessment by tracking the locomotor activity of zebrafish larvae. The lethal concentration (LC50 = 236.3 mg/L) corresponding to 0.02 mg/mL of DMT, 0.017 mg/mL of harmaline, and 0.22 mg/mL of harmine was quantified. All embryos exposed to 1000 mL/L died up to 48 hours and at 200 mg/ml fatal cases occurred only after hatching, reaching 90% mortality 96 hours after fertilization. Occurrence of oedema and clots was observed at higher concentrations. Embryonic development was affected, with early hatching at lower concentrations (0.3 mg/mL and 1.6 mg/mL) and the inability to complete this process at higher concentrations, showing a decreased dose-dependent hatching rate. Reduction in locomotion was observed, with shorter total swimming distance at higher doses. *Observations:* Ayahuasca sample provided by UDV with the following composition: DMT: 0.141 mg/mL; Harmine: 1.56 mg/mL; Harmaline: 0.122 mg/mL; THH: no information (ni). Analytical results cited from Pic-Taylor et al. 2015.	Andrade et al. 2018
Investigation of Ayahuasca high-dose acute intoxication in female Wistar rats by determining lethal dose, behavioral impact and neurotoxic potential	Acute oral toxicity of Ayahuasca at an initial dose which was 30 times higher than the religious dose (30x) was evaluated and gradually tested at higher concentrations. Rats were treated at 15x (4.5 mg/kg bw DMT) and 30x (9 mg/kg bw) higher concentrations to assess behavior in forced swim, open field, and elevated maze tests. Neuronal activity and neurotoxicity were assessed after exposure to 30x dosage by brain dissection and evaluation of the Raphe dorsal nucleus, amygdaloid nucleus, and hippocampal formations (dentate gyrus and 1, 2, and 3 areas from *Cornuammonis*). It was not possible to determine the lethal dose for technical reasons, but it was estimated to be greater than 15 mg/kg bw DMT, that is 50 times the dose used in a ritual context. In the behavioral assessment, decreased exploratory and locomotor activity was observed in the open field test and higher number of entrances with longer stay in the open arms in the elevated maze test, compared to the control group, which indicates low levels of anxiety. In the forced swim test, rats exposed to Ayahuasca showed less immobility and more swimming activity compared to the control group. The involvement of serotonergic neurotransmission in the observed brain structures was confirmed, but without apparent brain damage. *Observations:* Ayahuasca sample provided by UDV with the following composition: DMT: 0.141 mg/mL; Harmine: 1.56 mg/mL; Harmaline: 0.122 mg/mL; THH: ni. The authors consider that the amount of tea ingested in a religious context is 150 mL, equivalent to 0.302 mg/kg bw of DMT, 3.34 mg/kg bw of harmine and 0.261 mg/kg bw of harmaline.	Pic-Taylor et al. 2015

Table 16.1 contd. ...

..Table 16.1 contd.

Study Focus	Methods and main results	Ref.
Assessment of maternal toxicity and developmental consequences in Wistar rats	Three different concentrations of Ayahuasca, the typical dose used in the religious context, and doses 5 and 10 times higher were tested during the gestational period, namely in the period of organogenesis and foetal development. It evaluated the reproductive capacity of pregnant rats and possible changes of the organs and tissues. Foetuses were also evaluated and all results compared to the control group. Maternal toxicity was observed with decreased food intake and consequently lower weight gain in the groups exposed to the highest concentration, which was also reflected in the lower weight of the corresponding foetuses. Skeletal changes were found in all foetuses of the groups exposed to the highest and intermediate concentrations, demonstrating a risk of dose-dependent toxicity. The group treated with the highest dose had higher liver mass. *Observations:* Ayahuasca sample provided by an unspecified religious group with the following composition: DMT: 0.42 mg/mL; Harmine: 1.37 mg/mL; Harmaline: 0.62 mg/mL; THH: 0.35 mg/mL. Analytical results cited by another author. The authors consider that the amount of tea ingested in a religious context is 100 mL per 70 kg bw, equivalent to 0.6 mg/kg bw of DMT, 1.95 mg/kg bw of harmine, 0.88 mg/kg bw of harmaline and 0.5 mg/kg of THH.	Oliveira et al. 2010
Effect of intermittent consumption on liver function	Biochemical parameters were evaluated: alanine aminotransferase, aspartate aminotransferase, creatinine, bilirubin, lactate dehydrogenase, alkaline phosphatase, and gamma glutamyl transpeptidase in 22 subjects taking Ayahuasca more than two times a month for at least one year. The tests were performed at 0, 4, 24, and 168 hours after ingestion. No changes were observed in any of the parameters, even during the acute effects of the drink, which corresponds to the first 4 hours after ingestion. *Observations:* Ayahuasca sample provided by Luz de Vegetal Integrated Development Centre with the following composition: DMT: 2.07 mg/mL; Harmine: 2.89 mg/mL; Harmaline: 0.15 mg/mL; THH: 1.89 mg/mL quoted from Lanaro et al. 2015 corresponding to sample 7. The equivalent of 3.10 mg/kg bw of DMT, 4.33 mg/kg bw of harmine, 0.22 mg/kg bw of harmaline and 2.83 mg/kg of THH.	Mello et al. 2019
Ontogenic study of the behavioral effect of intermittent exposure to Ayahuasca in mice	Mice of different age groups were exposed to 1.5 mL/kg bw of Ayahuasca twice a week to mimic exposure in ritual context. In order to evaluate the effect of Ayahuasca on memory and anxiety, the mice were subjected to the Morris water maze test, open field test, and high cross maze test. The results suggest that locomotor activity is not affected in any of the development phases; anxiogenic effects and memory impairment, respectively, were confirmed in the representative mice from infancy in the high cross maze test results and in the results of the Morris water maze test from the adolescent group. These effects are not maintained in subsequent age groups. *Observations:* Ayahuasca sample provided by Luz de Vegetal Integrated Development Centre with the following composition: DMT: 2.07 mg/mL; Harmine: 2.89 mg/mL; Harmaline: 0.15 mg/mL; THH: 1.89 mg/mL quoted from Lanaro et al. 2015 corresponding to sample 7. The equivalent of 3.10 mg/kg bw of DMT, 4.33 mg/kg bw of harmine, 0.22 mg/kg bw of harmaline and 2.83 mg/kg of THH.	Correa-Netto et al. 2017
Effect of acute and chronic exposure on aorta artery of Wistar rats	2 mL or 4 mL of Ayahuasca was administered once and for 14 days, thus forming four treated groups. The rats were killed and the aorta was removed for qualitative and quantitative morphometric analysis. Stretching and flattening of vascular smooth muscle cells and alteration in the arrangement and distribution of collagen fibers and elastic fibers were observed in all treated groups compared to the control group. In chronic treatment, at the highest dose, a significant increase in wall thickness was observed, as well as an increase in the average wall thickness: lumen diameter ratio. *Observations:* Ayahuasca sample provided by Santo Daime with the following composition: DMT: 0.24 mg/mL; Harmine: ni mg/mL; Harmaline: ni mg/mL; THH: ni mg/mL. Two different doses, equivalent to 0.48 mg/kg bw and 0.96 mg/kg bw of DMT were administered.	Pitolet al. 2015

Table 16.1 contd. ...

..Table 16.1 contd.

Study Focus	Methods and main results	Ref.
Interference of acute and chronic exposure in the zebrafish learning process	Zebrafish were subjected to two different concentrations of Ayahuasca: 0.1 mL and 0.5 mL per liter of aquarium water for one hour over a period of 13 days to assess chronic exposure. In the acute exposure assessment, this procedure was performed only once. Discrimination tests were used to assess the process of memorization and learning. . Acute exposure to low concentrations did not seem to interfere, but prolonged use even at low concentrations negatively affected memory and learning processes. *Observations:* Ayahuasca sample provided by Igreja da Barquinha with the following composition: DMT: 0.36 mg/mL; Harmine: 1.86 mg/mL; Harmaline: 0.24 mg/mL; THH: 1.20 mg/mL. Cited analytical composition of another author.	Lobao-Soares et al. 2018
Effect of long-term Ayahuasca exposure on memory and anxiety in Wistar rats	Male Wistar rats were subjected to 120, 240, and 480 mg/kg bw of freeze-dried Ayahuasca for 30 days and were evaluated 48 hours after discontinuation of treatment. The tests were performed in the elevated cross maze and the Morris water maze. Results show that long-term exposure did not affect task performance in the two specified tests. There was also a higher contextual conditioned fear response which remained for several weeks compared to the control group. *Observations:* Ayahuasca sample provided by an unspecified religious group with the following composition: DMT: 0.26 mg/mL; Harmine: 0.56 mg/mL; Harmaline: 0.17 mg/mL; THH: 0.44 mg/mL.	Favaro et al. 2015
Reported case	A 25-year-old man was admitted to a hospital in the United States, with a clearly altered state of consciousness and seriously disturbed after ingestion of Ayahuasca acquired over the internet. He suffered from schizophrenia and had a history of drug abuse; however, he did not take any medication. He was given I.V. solutions for fluid restoration and was discharged on the fourth day of hospitalization, accompanied by a primary care provider.	Bilhimer et al. 2018
Reported case	In Australia, a 40-year-old man was admitted to a hospital for suffering persecutory delusions, becoming physically and verbally threatening after drinking Ayahuasca and alcohol. He had a previous history of substance abuse and was diagnosed with acute mania and psychosis. The situation was resolved within one day after receiving olanzapine and valproate treatment.	Pope et al. 2019
Reported case	In Spain, a 36-year-old man went to emergency services with behavioral changes after recreational consumption of Ayahuasca. He had hyperactivity, euphoria, reduced sleep, and feeling of well-being, being partially aware of his condition. He had a history of manic disease and psychosis, having been under the care of a psychiatrist since the age of 15 due to cannabis use. The patient also reported occasional cocaine and LSD use. He was diagnosed with organic psychosis, because of drug use, and was treated with antipsychotics until his emotional state was stabilized.	Márquez and Gómez-Luengo 2017
Descriptive analysis of notifications sent to US PCCs between 2005 and 2015	A total of 538 cases of exposure to Ayahuasca were reported, of which 83% reported health care providers as notifiers. Most of the reported exposures were acute situations, affecting mainly men between 18 and 29 years of age. Major clinical manifestations occurred in 7% of the cases, with 12 cases of seizures, 7 cases of respiratory arrest, and 4 cases of cardiac arrest, and 55% of cases involved moderate manifestations. The most common clinical manifestations were hallucinations, agitation, confusion, mydriasis, hypertension, and tachycardia. Three fatal cases were reported, two of which were indirectly related to Ayahuasca. Regarding the duration of the effects, it was not always possible to obtain this information, but in 27 cases, the manifestations lasted less than 2 hours, in 144 cases they lasted from 2 hours to 8 hours, in 105 cases they lasted from 8 hours to 24 hours, in 57 cases there was a duration of between 24 hours and 3 days, in 25 cases between 3 days and 1 week, 6 cases lasted more than 1 week but less than 1 month, 2 cases lasted more than 1 month, and there was 1 case with permanent effects. About 48% of cases were treated and discharged from the institution; however, 17% were admitted to the critical care unit and 11% to the non-critical care unit. Hospitalization in a psychiatric unit occurred in 6% of the situations, and the remaining 18% were not screened. In 35% of cases, treatment was by intravenous fluid restoration, and 30% by benzodiazepines, in addition to 28 cases requiring tracheobronchial intubation.	Heise and Brooks 2017

Reported cases

In the descriptive analysis of the notifications sent to the US PCCs, from 1st September 2005 to 1st September 2015, it was possible to identify 538 cases of Ayahuasca exposure. Most of the cases resulted from acute exposure, encompassing mostly male individuals, aged between 18 and 29 years old. Moderate to severe clinical manifestations occurred in 63% of the cases, including twelve cases resulting in seizures, four cases in cardiac arrest, and seven cases in respiratory arrest. The most common clinical manifestations observed were hallucinations, agitation, confusion, mydriasis, hypertension, and tachycardia. Three fatal cases and one which resulted in permanent damage were reported (Heise and Brooks 2017).

However, the context of its use, whether recreational or religious, has not been reported, as the doses administered may vary substantially. In a religious context, about 120 mL of Ayahuasca (Estrella-Parra et al. 2019) is normally ingested, although the alkaloid composition may vary, as previously described. According to the results of compositional analysis of Ayahuasca samples, those seized due to illicit use, compared to those used in religious contexts, have been shown to contain more than four times the DMT content (Lanaro et al. 2015).

There is also no evidence of the actual qualitative composition of Ayahuasca, which may be tampered with by the addition of other more toxic psychotropic molecules (Dos Santos et al. 2017a). Nor has the concomitant use of other substances been described, nor the pre-existence of pathologies.

In the reported case in Spain in 2017, and in the USA in 2018, the individuals involved needed medical assistance due to their psychological situation. Drug abuse history and the existence of underlying psychiatric illness, such as manic disorder and schizophrenia respectively, is common in both situations (Márquez and Gómez-Luengo 2017, Bilhimer et al. 2018). This seems to corroborate the reviewed data by Dos Santos et al. (2017a) on the incidence of psychosis in Ayahuasca users, who concluded that individuals with a personal or family history of any psychotic or manic pathology are predisposed to the onset of such adverse reactions and should avoid ingestion of psychomimetic substances.

In Australia, an individual, who went to hospital in a state of persecutory delirium after drinking Ayahuasca and alcohol, was diagnosed with acute mania and psychosis. Since it was not possible to detect DMT in the urine sample, due to its low $T_{1/2}$ and lower percent recovery, a simpler way to confirm its consumption was explored: through the detection of β-carbolines. Thus, and with strong suspicions that the reported drink would be Ayahuasca, harmine, harmaline, and THH were detected and included in the drug screening test database, thus facilitating detection in future situations (Pope et al. 2019).

Studies

In order to assess acute toxicity, in a study of zebrafish embryos, it was possible to determine the lethal concentration in these organisms corresponding to 0.02 mg/mL of DMT, 0.017 mg/mL of harmaline, and 0.22 mg/mL of harmine. Exposure to Ayahuasca has also been found to induce developmental and behavioral changes in the embryonic and larval phase, visible due to hatching delay, oedema, and erythrocyte accumulation, as well as decreased locomotor activity, which is characteristic of zebrafish when an accumulation of serotonin in the synaptic clefts occurs (Andrade et al. 2018). However, the results were not extrapolated to humans, and can therefore only be interpreted as an indicator of possible toxicity. In adulthood, according to the results of another test also conducted with zebrafish, it has been shown that acute exposure to low concentrations does not interfere with memory processing. However, at higher doses the locomotor activity is affected. In prolonged exposures, cognitive performance, namely memory and learning, may be negatively affected (Lobao-Soares et al. 2018).

In the same context, a study in female Wistar rats concludes that the lethal dose of Ayahuasca is more than 50 times the dose used in religious contexts. It was also observed that exposure to concentrations 30 times higher than those usually administered in a religious context, triggers a strong serotonergic activation of several brain regions involved in emotional information processing and behavior modulation, and that, despite causing neurodegenerative signs, there were no permanent changes in the brain structures concerned. In the behavioral assessment of the rats, an antidepressant effect similar to fluoxetine and lower levels of anxiety in mice treated with Ayahuasca was observed (Pic-Taylor et al. 2015).

Despite the anxiolytic power so often claimed in literature and contrary to what was reported by Pic-Taylor et al. (2015), there was no evidence of changed anxiety levels in male Wistar rats exposed to Ayahuasca for 30 days. Furthermore, in the study, it was found that long-term exposure can interfere with contextual association of emotions, probably resulting from biological plasticity in certain areas of the brain, which appears to enhance the process of contextual learning and may even endure a few weeks after treatment ends. Once again, these results cannot be directly extrapolated to humans, but they emphasize the importance of including emotional memory tests for a better understanding of Ayahuasca's role at the cerebral level (Favaro et al. 2015).

The results of the ontogenic study in mice intermittently exposed to Ayahuasca are consistent with the results of the study described above, as no changes in anxiety parameters have been reported in adult mice. However, it has been shown that infancy seems to be a sensitive period to the possible anxiogenic effect of Ayahuasca. Likewise, in adolescence, there was also memory impairment, which reveals, once again, its action in the hippocampus. Both effects were transient, suggesting that neuroplasticity induced by Ayahuasca is probably reversible (Correa-Netto et al. 2017). This study is very important, because although the religious use of Ayahuasca is considered a safe practice, according to CONAD's 2010 Resolution No 1 of 25th January, its consumption by children and pregnant women in the aforementioned context, when properly authorised, by their guardians, or by personal decision, in the respective cases, is allowed (CONAD 2010).

In this sense, the assay performed in pregnant Wistar rats, proved Ayahuasca's possible teratogenic effect, namely, in the organogenesis and foetal development period. Decreased food intake, and consequent decrease in weight gain by pregnant rats, seems to be at the origin of this possible toxicity, as dietary restrictions may cause malformations, delays in growth, and development of the foetus, as noted. The incomplete ossification of the hyoid and nasal bones and asymmetry of the sternum were the most frequently observed skeletal variations, as well as third ventricle and lateral brain ventricle defects, the incidence of which is dose dependent (Oliveira et al. 2010).

Toxic effects on the foetus at lower doses appear to be indirect, resulting from maternal toxicity. At higher doses, the results suggest an accumulative effect of indirect and direct toxicity (Oliveira et al. 2010), which is consistent with what is reported in literature about the ability of β-carbolines to cross the placental barrier (Estrella-Parra et al. 2019). However, despite being pioneering, this study does not reflect the actual pattern of Ayahuasca religious-context consumption, either due to the frequency of administration, or the concentration tested, so no deductions can be made about the existence of actual risks. However, considering the relevance of this issue, the probable toxicity in pregnant women should be further studied (Dos Santos 2010).

In the studies by Oliveira et al. (2010), it was found that exposure to the highest dose of Ayahuasca led to increased liver mass in rats. Despite there being no changes in biochemical parameters, the authors consider this information as a sign of hepatotoxicity due to the induction of microsomal enzymes, which act on the DMT metabolism. However, in the evaluation of liver function of Ayahuasca users in a religious context, no changes in biochemical markers were found. It was considered that the pattern of consumption in question does not appear to compromise the liver. However, the sample size of the study in question is very limited (Mello et al. 2019).

In the cardiovascular system, it was demonstrated that in mice, acute and chronic exposure produces changes in the smooth muscle cells in collagen and elastic fibers, suggesting a vascular remodelling effect. In addition, acute exposure to higher doses has been found to produce arterial hypertrophy—a classic structural change produced by hypertension (Pitol et al. 2015). Even if, at lower doses, the hypertensive potential is more modest, long-term exposure must be better clarified (Dos Santos et al. 2017b).

Conclusion

From the results presented, it was possible to infer that there are risks in the acute and chronic use of Ayahuasca, as there are several factors which may influence the triggering of harmful health situations.

In the studies conducted in animals, behavioral changes, teratogenic effects, memory and learning impairment, as well as morphological changes in the cardiovascular system were observed. Yet, these data are always difficult to extrapolate to humans, so further research in this regard is essential.

It was also verified that there is no uniformity in composition and administered dose of tested Ayahuasca. Additionally, there are even studies which cite the Ayahuasca chemical composition from other studies. This practice undermines the strength of the experiments, jeopardizing the true dose-effect relationship.

Another important aspect is the clear distinction between use in a religious context, inherent to conduct, and recreational use, often associated with marginalization. In the religious use of Ayahuasca, there seems to be an awareness of the behaviors to be adopted regarding interactions and underlying diseases, as can be seen in the interview with Master "João" from an Ayahuasca religious center, which was held in Belém-Brazil—on March of 2019, transcribed below, which has been attached. Nevertheless, increasing globalization has made Ayahuasca an easily accessible drug, and therefore, it is important for the public to be aware of the consequences of its use.

Attachment
Interview with Master "João"—Transcript of Audio Recording

When did you start using Ayahuasca and why?

"I started using it 1ˢᵗ January of 2002, around 3:30h in the morning. There was no reason to start, I went to a place that served the vegetable without knowing what it was and someone offered it to me. And so, that's why I started!"

What are the main differences in your daily life?

"I can't talk about the differences in my current daily life, only in the differences I noticed when I started using it. Now it is my daily life but when I started, I felt great differences. My biggest priorities no longer persisted and others came up. For example, I had completely abandoned the spiritual part of me and church matters were no longer important, even though I had a lot of knowledge and interest in it. When I first drank Ayahuasca, I felt the need to seek God, with a free thinker spirit, instead of thinker free spirit. I think because I wanted to, not because I was imposed to. I smoked for 46 years, around 6 packs a day (120 cigarettes) and suddenly, I stopped without any side effects, like shaking. Not only smoking, but I also stopped drinking sodas. Another thing I did was stop eating meat for one year, which allowed me to detox my body (cigarettes, meat, and sugar). I felt an almost immediate physical improvement and some people didn't even recognize me after two months."

Is Ayahuasca a cure? Because of the "vegetable" itself or for the monitoring provided?

"Ayahuasca is not a cure, not a curative medicine, or a medication. If you have a heart condition, you have to take the right medication. Ayahuasca works as a conditioning of the human being with its own ideology. As I seek to be a good person, Ayahuasca gives me these good thoughts. I want to be a person who is dedicated to helping others and it gives me the strength to dedicate and sacrifice for them. There is no monitoring, you drink whenever you can or want, as I told you (Belém meeting), it is not addictive, there"s no need for continuous use. When I go to São Paulo-Brazil (when I don't have access to Ayahuasca), I don't drink for 2 or 3 months and I don't feel the need to do it. It is not an addictive psychoactive drug that requires you to use it continuously. It's an entheogenic drink that puts you in touch with a divine entity. It's like in Portugal, everyone believes in "Our Lady" and the "Fátima miracle". There isn't a single Portuguese person who doesn't believe, unless you don't believe in anything. So how can you say that it didn't happen? Then I have to question the 3 little shepherds. Lúcia could hear and see "Our Lady" but the other girl could only see and the boy could only hear Her or vice versa. Each of them had a process, let's say, I wouldn't say clairvoyance or maybe I shouldn't say it because science puts it in another way, right? A contact with a divine entity has occurred but it stopped there. The vegetable allows you to keep in touch with a divine entity and reach an epiphany, that's the right word to describe it, when you get involved with something you can't explain it, but it is related with the personal "me". It happened to me; I don't know whether it will happen to others."

What makes people seek Ayahuasca?

"I ask you the same question, what was your motivation to study Ayahuasca? "Because everybody talks about it". So, if everybody talks about it, it's because lots of people search for it. "Because I read an article that says it's addictive and a drug". Is that a motive to seek it out?" No, but you did it anyway, even knowing you were going to study the negative effects of Ayahuasca.

We believe that when someone has the need to search for Ayahuasca, that person has received the call. Ayahuasca is for everyone, but not everybody is for Ayahuasca. There's one thing that explains it really well. There aren't enough plants to make Ayahuasca for 8 billion people, because there aren't enough plants in the world. It's only for some, the chosen ones."

What are the hallucinations like? Do they have specific a color? Does it happen with your eyes open or closed?

"It is a very interesting phenomenon. Once, before I knew Ayahuasca, I tried to impress on my students the effects of trying new things. If I tried to teach about Portuguese grammar exposition, I had to give examples of other grammatical cycles. If you don't have any foundation in Latin, you won't be able to understand some grammatical issues of Neo Latin languages. So, I posed a question to my students. You know Brazil is a very rich country (minerals, food production), but the money is in the hands of a few people. Most of the people are very poor and live in those neighborhoods called "favelas". I asked my students "If you go to a favela and ask the people if they like lobster, they won't even know what a lobster is because they never tried it, but they will say that they did and that they didn't like it. It's called truth distortion. The hallucinations can only be understood if you try Ayahuasca. It can't be explained, it's like a miracle! Why didn't anybody believe in Lúcia, except for the priest and her mother and father? Because it was really difficult for someone who didn't experience it to believe in such a miracle. Only after the "Miracle of the Sun" did people believe in Lúcia. It is impossible to describe the hallucinations, unless you drink Ayahuasca and feel it. It has specific colors, but not only one. Thousands of colors at the same time, it's a really divine event. Eyes open or closed? It can occur both ways, it's a strange, mysterious force, beyond our capacity of understanding."

How long do the hallucinations last?

"The hallucinations can last from an hour, an hour and a half, to two days if necessary. There is no correct time. It depends on the need that the spiritual plan defines for you. But they usually last 2 hours"

After communion, do you feel changes in your psychological state in the following days? If so, how long exactly?

"Let's see, feeling a change in the psychological state is thinking, "I feel regret", "I shouldn't have done it" or "I did it and want to do it again…". It is not exactly those types of questions you ask after a day or two. You feel a deep peace you cannot explain. Even if you're feeling boredom, even if you're having difficulties, this peace totally envelopes you and your thinking is not focused on that act, but on what you can do because of that. For example, if you feel "I have to work on the dissertation, I'll wait for next week to start", Ayahuasca makes you feel that at that moment what you need is to start working. It is natural and spontaneous. This time of sensation is not specific. Once a change occurs, that change can be for life. I will give a practical example: how does love occur? How do you fall in love? How long does it last? How do you feel in the following days? If you can answer these questions, you will answer the question you asked me."

What type of selection is made? Who can or who cannot take drink the "vegetable"?

"There is no selection. This is not a joke, it is not a place that you buy a ticket, or go to try on some clothes. That's not it. First, you feel the urge to know and get in touch with a nearby place where they serve it, then they will ask what your intentions are and you have to answer that your intentions are the best and you want to know how it works. From the moment you demonstrate this interest, you will surely be well accepted within the group.

Who can and cannot drink it? This is an interesting question. Imagine that a person has neurological problems. People who are undergoing psychiatric treatment, they take very strong medicines. So, if the plant activates the mind, bringing a sense of heightened awareness, a sense of responsibility, but the person is not right, they may perhaps have, not a psychological outbreak, but enter a state they do not need at that moment. We avoid giving the "vegetable" to those who take these medicines. What is done is they are asked to stop taking the medicine by medical order for a week, if the doctor thinks it is possible, then the person can drink the "vegetable"."

What is your opinion regarding the marginalized use of Ayahuasca? Do you consider there are risks in it use?

"I will not give my opinion because it is the worst possible. I think that something so serious might fall so low because of some individuals, but there is everything in this world. Just as there still is trafficking of white female slaves, organ trafficking, especially in richer countries that take advantage of poverty. Children sacrificed to take their little hearts and kidneys to favor children who were born in a more golden cradle! So, when we speak of marginalization, not only of Ayahuasca, but in all things, today, even medicines are falsified. There are a lot of pills you pay their weight in gold for from Swiss industries which are nothing more than flour. You have to be careful about that. If you enter this side of marginalization, it is better not to even talk about the "vegetable" anymore. Forget it because it's not for that kind of conversation. I will not consider the risks, as they are the worst possible! I don't know what you believe about what might happen to us in the future, if you believe in the afterlife, in karma, I don't know, because we didn't have the time to get to know each other, but if you have a materialistic point of view, that life is only this, forget Ayahuasca, it's not for that kind of thinking. I'm sorry to be this way, frank, but it's something that we consider divine. It's a serious thing, not child's play."

What diet should be followed before ingestion? And after? What is the preparation?

"In addition to diet, there is behavior. To drink Ayahuasca, you have to go three days without sex, three days without smoking, and no drinking alcohol. Preferably on the day you drink it, you should go at least 4 hours before without eating meat. Out of respect for the drink, for it is a sanctified drink. If your body is dirty, it will not have the effect it should."

List of Abbreviations

bw	–	Body weight
CCE	–	*Centros de Controlo de Envenenamento* (Poison Control Centers)
$C_{máx}$	–	Maximum plasma concentration
CONAD	–	*Conselho Nacional de Políticassobre Drogas* (National Council on Drug Policy)
CYP 2D6	–	Isoenzyme 2D6 from Cytochrome P450
CYP 450	–	Cytochrome P450
DA	–	Dopamine
DMT	–	N, N-dimethyltryptamine
EUA	–	*Estados Unidos da Amé*rica (United States of America)
5-HT receptor	–	5-hydroxytryptamine receptor (serotonin receptor)
IAA	–	Indoleacetic acid
I.V.	–	intravenously
IL	–	Interleucine
MAO	–	Monoamine Oxidase
MFS	–	Dr. Marlene Freitas da Silva Herbarium
mGlu2/3	–	Metabotropic glutamate receptor II
MTHC	–	2-Methyl-1,2,3,4-tetrahydro-β-carboline
NMDA	–	N-methyl-D-aspartate
NMT	–	N-monomethyl-tryptamine
SERT	–	Serotonin transporter
CNS	–	Central Nervous System
$t_{1/2}$	–	Half-life time
TAAR-1	–	"Trace Amine-Associated"-1 Receptor
Th	–	Helper T cells
THH	–	Tetra-hydro-harmine
UDV	–	*União do Vegetal* (Vegetable Union)
VMAT-2	–	Vesicular monoamine transporter 2

References

Albuquerque, M.B.B. 2007. Abc Do Santo Daime. 1ª Ed. Belém: Eduepa. ISBN:978-85-88375-18-5.

Albuquerque, U.P., Almeida, C.F.C.B.R. Martins, J.F.A. 2005. Tópicos Em Conservação, Etnobotânica e Etnofarmacologia de Plantas Medicinais e Mágicas. 1ª Ed. Recife: Nupeea / Sociedade Brasileira de Etnobiologia e Etnoecologia. ISBN: 85-7716-005-X.

Amorozo, M.C.M. 2002. Uso e diversidade de plantas medicinais em Santo Antonio do Leverger, MT, Brasil. Acta Bot. Bras. 16(2): 189–203.

Andrade, T.S., De Olveira, R., Lopes, M., Von Zuben, M.V., Grisolia C.K., Domingues, I., Caldas, E.D., Pic-Taylor, A. 2018. Exposure to ayahuasca induces to developmental and behavioral alterations on early life stages of zebrafish. Chem. Biol. Interact. 293: 133–140.

Araújo, A.M., Carvalho, F., Bastos, M.L., Guedes De Pinho, P., Carvalho, M. 2015. The hallucinogenic world of tryptamines: an updated review. Arch. Toxicol. 89(8): 1151–1173.

Aswani, S., Lemahieu, A., Sauer, W.H.H. 2018. Global trends of local ecological knowledge and future implications. PLoS One13(4): 1–19.

Balick, M.J., Cox, P.A.—Plants, People and Culture. 1996. 1ª Ed. New York: Scientific American Library. ISBN: 0-615-12953-6.

Barker, S.A. 2018. N, N-Dimethyltryptamine (DMT), an endogenous hallucinogen: past, present, and future research to determine its role and function. Front Neurosci. 12: 536.

Bauer, I.L. 2018. Ayahuasca: A Risk for Travellers? Travel. Med. and Infect. Di. 21: 74–76.

Bianchi, A. 2005. Ayahuasca e o XamanismoIndígenananaSelva Peruana: O Lento Caminho Da Conquista. *In*: Labate B.C., Goulart, S.L. (eds.). O Uso Ritual das Plantas de Poder. Campinas Sp Brasil: Mercado De Letras. P. 319-328. ISBN: 978-8575910498.

Bilhimer, M.H., Schult, R.F., Higgs, K.V., Wiegand, T.J., Gorodetsky, R.M., Acquisto, N.M. 2018. Acute intoxication following dimethyltryptamine ingestion. Case Rep. Emerg. Med. 1–3.

Bouso, J.C., González, D., Fondevila, S., Cutchet, M., Fernández, X., Barbosa, P.C.R., Alcázar-Córcoles, M.Á., Araújo, W.S., Barbanoj, M.J., Fábregas, J.M., Riba, J. 2012. Personality, psychopathology, life attitudes and neuro psychological performance among ritual users of ayahuasca: a longitudinal study. PloS One. 7, 8.

Callaway, J.C. 2005. Fast and slow metabolizers of hoasca. Journal of Psychoactive Drugs 37(2): 157–161.

Callaway, J.C., Brito, G.S., Neves, E.S. 2005. Phytochemical analyses of banisteriopsiscaapi and psychotriaviridis. J. Psychoactive Drugs 37(2): 145–150.

Camargo, M.T.L.A. 2014. As PlantasMedicinais e o Sagrado: EtnofarmacologiaemumaRevisãoHistoriográfica da Medicina Popular no Brasil. 1ª Ed. São Paulo: Câmara Brasileira Do Livro, Brasil. ISBN: 136-1582014.

Cameron, L.P., Olson, D.E. 2018. Dark classics in chemical neuroscience: N, N-Dimethyltryptamine (DMT). Acs Chemical Neuroscience 9(10): 2344–2357.

Campos, M.G., Cupido, M., Tavares, R., Cônsul, R. 2018. Clinical outcomes from tamoxifen drug-herb interactions. Int. J. Clin. Pharmac. & Pharmacotherapy 3: 140.

Carbonaro, T.M., Gatch, M.B. 2016. Neuropharmacology of N,N-Dimethyl Tryptamine. Brain Res. Bull. 126: 74–88.

Conselho Nacional De PolíticasSobre Drogas. 2010. Resolução Nº 1 De 25 de Janeiro de 2010 Sobre O Uso Religioso da Ayahuasca. Diário Oficial do Brasil.

Correa-Netto, N.F., Masukawa, M.Y., Nishide, F., Galfano, G.S., Tamura, F., Shimizo, M.K., Marcato, M.P., Junior, J.G.S., Linardi, A. 2017. An ontogenic study of the behavioral effects of chronic intermittent exposure to ayahuasca in mice. Braz. J. Med. Biol. Res. 50(7): 1–11.

Cruz, O.V. 2016. AlgunasEspécies y Usos de las PlantasÚtilesen la MedicinaTradicional de los Kichwa del Napo. Yachay-Kusunchi: Saber, Coñocer y Aprender. ISSN: 2346-402. Vol. 4, Nº 1: 25–31.

Cruz, O.V., Medina, D., Iñiguez, J., Navarrete, H. 2017. Los Kichwas del Alto Napo y Sus PlantasMedicinales. 1ª Ed. Quito: Facultat de Ciencias Exactas y Naturales, Escuela de CienciasBiológicas y Herbario CQA de la Pontificia Universidad Católica del Ecuador. ISBN: 978-9978-77-312-3.

De Araujo, D.B., Ribeiro, S., Cecchi, G.A., Carvalho, F.M., Sanchez, T.A., Pinto, J.P., De Martinis B.S., Crippa, J.A., Hallak, J.E.C., Santos, A.C. 2012. Seeing with the eyes shut: neural basis of enhanced imagery following ayahuasca ingestion. Hum. Brain Mapp. 33(11): 2550–2560.

Domínguez-Clavé, E., Soler, J., Elices, M., Pascual, J.C., Álvarez, E., Revenga, M.F., Friedlander, P., Fielding, A., Riba, J. 2016. Ayahuasca: pharmacology, neuroscience and therapeutic potential. Brain Res. Bull. 126: 89–101.

Dos Santos, R.G. 2010. Toxicity of chronic ayahuasca administration to the pregnant rat: how relevant it is regarding the human, ritual use of ayahuasca? Birth Defects Res. B Dev. Reprod. Toxicol. 89(6): 533–535.

Dos Santos, R.G., Bouso, J.C., Hallak, J.E.C. 2017a. Ayahuasca, dimethyltryptamine, and psychosis: a systematic review of human studies. Ther. Adv. Psychopharm. 7(4): 131–157.

Dos Santos, R.G., Bouso, J.C., Hallak, J.E.C. 2017b. Ayahuasca: what mental health professionals need to know. Rev. Psiquiatr. Clin. 44(4): 103–109.

Estrella-Parra, E.A., Almanza-Pérez, J.C., Alarcón-Aguilar, F.J. 2019. Ayahuasca: uses, phytochemical and biological activities. Nat. Prod. Bioprospect. 9(4): 251–265.

Favaro, V.M., Yonamine, M., Soares, J.C.K., Oliveira, M.G.M. 2015. Effects of long-term ayahuasca administration on memory and anxiety in rats. PloS One. 10(12): 1–10.

Frecska, E., Szabo, A., Winkelman, M.J., Luna, L.E., Mckenna, D.J. 2013. A possibly sigma-1 receptor mediated role of dimethyltryptamine in tissue protection, regeneration, and immunity. J. Neural. Transm. 120(9): 1295–1303.

Gruca, M., Cámara-Leret, R., Macía, M.J., Balslev, H. 2014. New categories for traditional medicine in the economic botany data collection standard. J. Ethnopharmacol. 155(2): 1388–1392.

Grupo Multidisciplinar de Trabalho (GMT). 2006. Ayahuasca: Relatório Final. [Consult. August 18th, 2019]. In: Http://www.mp.go.gov.br/portalweb /hp/41/docs/relatorio _final_-_grupo_multidisciplinar_ de_trabalho_-_ gmt_-_ayahuasca.pdf.

Hamill, J., Hallak, J., Dursun, S.M., Baker, G. 2018. Ayahuasca: psychological and physiologic effects, pharmacology and potential uses in addiction and mental illness. Curr. Neuropharmacol. 17(2): 108–128.

Heise, C.W., Brooks, D.E. 2017. Ayahuasca exposure: descriptive analysis of calls to us poison control centers from 2005 to 2015. J. Med. Toxicol. 13(3): 245–248.

Labate, B.C., Goulart, S.L. 2005. O Uso Ritual das Plantas de Poder. 1ª Ed. Campinas, SP Brasil: Mercado de Letras. ISBN: 978-8575910498.

Lanaro, R., Calemi, D.B.A., Togni, L.R., Costa, J.L., Yonamine, M., Cazenave, S.O.S., Linardi, A. 2015. Ritualistic use of ayahuasca versus street use of similar substances seized by the police: a key factor involved in the potential for intoxications and overdose? J. Psychoactive Drugs 47(2): 132–139.

Lobao-Soares, B., Eduardo-Da-Silva, P., Amarilha, H., Pinheiro-Da-Silva, J., Silva, P.F., Luchiari, A.C. 2018. It's tea time: interference of ayahuasca brew on discriminative learning in zebrafish. Front Behav. Neurosci. 12: 1–10.

Lorenzi, H., Matos, F.J.A. 2008. PlantasMedicinais no Brasil: Nativas e Exóticas. 2ª Ed. Nova Odessa SP: Instituto Plantarum de Estudos da Flora. ISBN: 615.3210981,

Luna, L.E. 2005. Narrativas de Alteridade: A Ayahuasca e o Motivo de Transformaçãoem Animal. In: Labate B.C., Goulart, S.L. (eds.). O Uso Ritual das Plantas de Poder. Campinas SP Brasil: Mercado de Letras. ISBN: 978-8575910498. 333-340.

Márquez, B.P., Gómez-Luengo, B.D. 2017. Intoxicación por Ayahuasca. Med. Clin. 149(3): 136–137.

Mckenna, D.J., Callaway, J.C., Grob, C.S. 1998. The scientific investigation of ayahuasca: a review of past and current research. The Heffter Review of Psychedelic. Research. 1: 65–76.

Mello, S.M., Soubhia, P.C., Silveira, G., Corrêa-Neto, N.F., Lanaro, R., Costa, J.L., Linardi, A. 2019. Effect of ritualistic consumption of ayahuasca on hepatic function in chronic users. J. Psychoactive Drugs. 51(1): 3–11.

Melo, P.M.C.O., Fonseca-Kruel, V.S., Lucas, F.C.A., Coelho-Ferreira, M. 2019. Coleções Etnobotânicas no Brasil frente à Estratégia Global para a Conservação de Plantas. Boletim do Museu Paraense Emílio Goeldi. Ciências Humanas. 14(2).

Nunes, A.A., Dos Santos, R.G., Osório, F.L., Sanches, R.F., Crippa, J.A.S., Hallak, J.E.C. 2016. Effects of ayahuasca and its alkaloids on drug dependence: a systematic literature review of quantitative studies in animals and humans. J. Psychoactive Drugs. 48(3): 195–205.

Oliveira, C.D.R., Moreira, C.Q., De Sá, L.R.M., Spinosa, H.D.S., Yonamine, M. 2010. Maternal and developmental toxicity of ayahuasca in wistar rats. Birth Defects Res. B Dev. Reprod. Toxicol. 89(3): 207–212.

Ona, G., Kohek, M., Massaguer, T., Gomariz, A., Jiménez, D.F., Dos Santos, R.G., Hallak, J.E.C., Alcázar-Córcoles, M.A., Bouso, J.C. 2019. Ayahuasca and public health: health status, psychosocial well-being, lifestyle, and coping strategies in a large sample of ritual ayahuasca users. J. Psychoactive Drugs 51(2): 135–145.

Organização das NaçõesUnidas (ONU). 1971. Convention on Psychotropic Substances. Final Act of The United Conference for the Adoption if a Protocol on Psychotropic Substances 1–28.

Ott, J. 2013. YajéChamánico: Ni Sacramento Religioso, Ni TampocoRemedio Contra la "DependenciaQuímica". pp, 433–457. In: Labate, B.C. and Bouso, J.C. (eds.). Ayahuasca Y Salud. Barcelona: La Liebre De Marzo, S.L. ISBN: 978-84-92470-25-9.

Palamar, J.J., Le, A. 2018. Trends in DMT and other tryptamine use among young adults in the United States. Am. J. Addict. 27(7): 578–585.

Pic-Taylor, A., Da Motta, L.G., De Morais, J.A., Junior, W.M., Santos, A.F.A., Campos, L.A., Mortari, M.R., Von Zuben, M.V., Caldas, E.D. 2015. Behavioural and neurotoxic effects of ayahuasca infusion (*BanisteriopsisCaapi* and *PsychotriaViridis*) in female wistar rat. Behav. Process. 118: 102–110.

Pires, A.P.S., Oliveira, C.D.R., Yonamine, M. 2010. Ayahuasca: a review of pharmacological and toxicological aspects. Rev. Ciênc. Farm. Básica Apl. 31(1): 15–30.

Pitol, D.L., Siéssere, S., Dos Santos, R., Rosa, M.L.N., Hallak, J.E., Scalize, P., Pereira, B., Iyomasa, M., Semprini, M., Regalo, S.C. 2015. Ayahuasca alters structural parameters of the rat aorta. J. Cardiovasc. Pharmacol. 66(1): 58–62.

Pomilio, A.B., Vitale, A.A., Ciprian-Ollivier, J., Cetkovich-Bakmas, M., Gómez, R., Vázquez, G. 1999. Ayahuasca: an experimental psychosis that mirrors the transmethylation hypothesis of schizophrenia. J. Ethnopharmacol. 65(1): 29–51.

Pope, J.D., Choy, K.W., Drummer, O.H., Schneider, H.G. 2019. Harmala alkaloids identify ayahuasca intoxication in an urine drug screen. J. Anal. Toxicol. 43(4): 23–27.

Quinteiro, M.M.C., Silva, J.G., Moraes, M.G., Teixeira, D.C. 2006. Anatomia Foliar de *PsychotriaViridis* Ruiz & Pav. (Rubiaceae). Revista Universidade Rural. Série Ciências Da Vida. 26(2): 30–41.

Riba, J., Mcilhenny, E.H., Valle, M., Bouso, J.C., Barker, S.A. 2012. Metabolism and disposition of N,N-dimethyltryptamine and harmala alkaloids after oral administration of ayahuasca. Drug Test Anal. 4(7-8): 610–616.

Riba, J., Valle, M., Urbano, G., Yritia, M., Morte, A., Barbanoj, M.J. 2003. Human pharmacology of ayahuasca: subjective and cardiovascular effects, monoamine metabolite excretion, and pharmacokinetics. J. Pharmacol. Exp. Ther. 306(1): 73–83.

Rivier, L., Lindgren, J.E. 1972 "Ayahuasca," the south american hallucinogenic drink: an ethnobotanical and chemical investigation. Econ. Bot. 26(2): 101–129.

Salick, J., Konchar, K., Nesbitt, M. 2014. Biocultural collections: needs, ethics and goals. pp. 1–13. *In*: Salick, J., Konchar, K., Nesbitt, M. (eds.). Curating Biocultural Collections: A Handbook. London: Royal Botanic Gardens Kew. ISBN: 9781842464984.

Santos, R.G. 2007. Ayahuasca: Neuroquímica e Farmacologia. SMAD: Rev. Eletrônica Saúde Mental Álcool Drog (Edição em Português). 3(1): 01–11.

Schenberg, E.E., Alexandre, J.F.M., Filev, R., Cravo, A.M., Sato, J.R., Muthukumaraswamy, S.D., Yonamine, M., Waguespack, M., Lomnicka, I., Barker, S.A., Da Silveira, D.X. 2015. Acute biphasic effects of ayahuasca. PloS One 10(9): 1–27.

Schultes, R.E., Hoffman, A., Ratsch, C. 1992. The magic drink of the amazon. pp. 124–135. *In*: Schultes, R.E., Hoffman, A., Ratsch, C. (eds.). Plants of The Gods: Their Sacred, Healing, and Hallucinogenic Powers. Rochester, Vermont: Healing Arts Press. ISBN: 0-89281-979-0.

Spruce Project, 2005. Natural History Museum. [Consult. August 15th, 2019]. In: https://www.nhm.ac.uk/ourscience/data/spruce/introduction/introduction_spruce.dsml.

Stiffler, J.D. 2018. Ayahuasca: from the Amazon to a City Near You. Am. J. Addict. 27(8): 648–649.

Szabo, A., Kovacs, A., Frecska, E., Rajnavolgyi, E. 2014. Psychedelic N,N-Dimethyltryptamine and 5-Methoxy-N,N-dimethyltryptamine modulate innate and adaptive inflammatory responses through the sigma-1 receptor of human monocyte-derived dendritic cells. PloS One 9(8): 1–12.

Taylor, C., Gomes, M., Zappi, D. 2015. Lista de Espécies da Flora do Brasil. [Consult. August 18th, 2019]. In: http://floradobrasil.jbrj.gov.br/jabot/floradobrasil /fb24581.

The Alsuntangled Group. 2017. Alsuntangled 40: Ayahuasca. Amyotrophic Lateral Sclerosis and Frontotemporal Degeneration 18(7-8): 627–631.

União Do Vegetal. 2018. UDV temSócios de 60 Nacionalidadesem 11 Países. [Consult. August 12th, 2019]. In: http://udv.org.br/blog/udv-tem-socios-de-60-nacionalidadesem -11-paises/.

Vitale, A.A., Pomilio, A.B., Cañellas, C.O., Vitale, M.G., Putz, E.M., Ciprian-Ollivier, J. 2011. *In vivo* long-term kinetics of radiolabeled N,N-dimethyltryptamine and tryptamine. J. Nucl. Med. 52(6): 970–977.

Winkelman, M., Baker, J.R. 2016. Supernatural as Natural: a Biocultural Approach to Religion. 2º Ed. New York: Routledge. ISBN: 978-0131893030.

World Health Organization (WHO). 2013. Who Global Report on Traditional And Complementary Medicine 2019. Geneva: World Health Organization. ISBN: 978-92-4-151543-6.

Zanela, J.P.P., Junior, M.R.D.M., Lucas, F.C.A., Nery, M.I.D.S. 2018. Práticas de Cura no Centro de Unificação Rosa Azul - SaberesTradicionais e SaúdeHolísticanaBeberagem da Ayahuasca. *In*: Dos Santos, S.F., Lucas, F.C.A., Junior, M.R.M., Santos, A.S.S. Bioculturalidade, Conservação E Biotecnologia Na Amazônia Oriental. Curitiba Brasil: CVR. ISBN: 978-85-444-2575-6. P. 19-34.

17

Exploring the Plant Kingdom for Sources of Skincare Cosmeceuticals

From Indigenous Knowledge to the Nanotechnology Era

Mayuri Napagoda[1],* and *Sanjeeva Witharana*[2]

Introduction

The skin is the outer covering of the human body and is considered as the largest organ in a human. It is comprised of three major structural layers viz., epidermis, dermis, and hypodermis (Tabassum and Hamdani 2014). The epidermis is the outermost layer of the skin which encompasses five sub-layers/strata; stratum corneum, stratum lucidum, stratum granulosum, stratum spinosum, and stratum basale. The dermis lies beneath the epidermis and is attached to an underlying hypodermis or subcutaneous connective tissue (Tabassum and Hamdani 2014).

As the outermost barrier of the body, the skin is constantly being challenged by invading microorganisms, parasites, as well as environmental factors, such as solar ultraviolet radiation (UV), humidity, airborne allergens, and pollutants, etc. For example, the exposure of the skin to solar UV radiation leads to photochemical generation of reactive oxygen species (ROS). When the production of ROS overwhelms the cellular intrinsic antioxidant capacity, a phenomenon known as oxidative stress occurs (Varma et al. 2011, Nita and Grzybowski 2016). The oxidative stress is usually associated with inflammation, immunosuppression, impaired wound healing, DNA damage, and activation of signaling pathways that affect gene transcription, cell cycle, proliferation, and apoptosis. These alterations ultimately promote carcinogenesis, and are also responsible for the induction of several chronic and degenerative disease conditions (Amaro-Ortiz et al. 2014, Pizzino et al. 2017, Dunaway et al. 2018). Moreover, environmental factors, such as air pollutants, significantly contribute to the premature/extrinsic skin aging process by inducing the generation of ROS, telomere-based DNA damage, and activation of aryl hydrocarbon receptor (AhR) signaling (Vierkötter and Krutmann 2012). Exogenous factors, such as dry climate, colder temperatures, repeated washing, and exposure to various alkali and detergents could deteriorate the ability of the skin to maintain moisture, thus impairing the normal functions of the skin, such as thermoregulation, gaseous exchange, protection against pathogens, and maintenance of proper hydration (Wan et al. 2014).

[1] Department of Biochemistry, Faculty of Medicine, University of Ruhuna, 80 000, Sri Lanka
[2] Faculty of Engineering, Higher College of Technology, PO Box 4793, United Arab Emirates
* Corresponding author: mayurinapagoda@yahoo.com

In order to protect the skin from numerous harmful factors and to improve skin regeneration, elasticity, and smoothness, and to reduce the degradation of primary structural constituents of skin (collagen, elastin, etc.), a diverse array of products are available in the market. These range from skin moisturizers to skin rejuvenation products, sunscreens, anti-wrinkle and antiaging agents, anti-acne products, and skin whitening agents, which act by different mechanisms, i.e., as protectives from ultraviolet light, maintaining healthy skin, as well as by enhancing attractiveness of person by improving skin radiance (Goh 2009, Yadav and Chaudhary 2015). The aforementioned products could be better described as "cosmeceuticals", rather than using the conventional term "cosmetics".

The aim of this chapter is to describe the significance of various herbal ingredients in cosmeceutical preparations based on the scientific evidence of their biological activities, and to emphasize modern day approaches to develop more consumer-friendly and biologically-enhanced products.

Cosmeceuticals: The Hybrids between Cosmetics and Pharmaceuticals

The term "cosmetics" was supposed to be originated from the word "cosmetae", which was used for Roman slaves whose role was to bathe royal men and women in perfume (Chaudhri and Jain 2009, Ahsan 2018). However, some believe that it is derived from the Greek "Kosmetos", which means "adornment" or "ornament" (González-Minero and Bravo-Díaz 2018). According to the United States Federal Food, Drug, and Cosmetic Act (FFDC), cosmetics are defined as "articles intended to be rubbed, poured, sprinkled, or sprayed on or introduced into, or otherwise applied to the human body or any other part thereof for cleansing, beautifying, promoting attractiveness, or altering the appearance without affecting structure or function" (Schneider et al. 2001, Pandey and Sonthalia 2019). This signifies that cosmetic agents are meant for beautification and decorative purposes. Nevertheless, since the onset of the concept of "cosmeceuticals" in the late 1970s, more efforts have been dedicated to developing cosmeceutical products that contain functional ingredients with pharmaceutical benefits to the skin (i.e., therapeutic, disease-fighting, or healing properties), thus capable of providing goodness beyond the typical decorative function of the conventional cosmetic agents. As a result, cosmeceuticals are considered to be cosmetic-pharmaceutical hybrids intended to enhance the health as well as the beauty of skin (Wanjari and Waghmare 2015). Moreover, the active ingredients in cosmeceutical products can penetrate the stratum corneum, and act at the cellular level of the skin, whereas cosmetic products could only deliver their ingredients at a very superficial level into the skin (Verma et al. 2016).

The field of cosmeceuticals is an interdisciplinary integration of physics, chemistry, and biology that involves the application of techniques, such as chromatography and spectrometry along with the use of active ingredients in *in vitro* and *in vivo* tests (González-Minero and Bravo-Díaz 2018). Active ingredients in cosmeceutical products could be of synthetic or natural origin. However, with the increasing awareness of the negative side-effects of chemicals used in synthetic products, there has been a rise in demand for natural cosmeceutical preparations that are mainly based on traditional herbal formulations for skincare (Ahmad et al. 2008, Rekha and Gokila 2015, Bodeker et al. 2017). The present-day herbal skincare cosmeceuticals are formulated with the incorporation of one or more herbal ingredients, particularly those that had been used since time immemorial for beautification purposes, or for the treatment of various ailments of the skin. These products are available in the market in the form of ointments, creams, emulsions, powder solutions, compacts, etc. (Sagbo and Mbeng 2018).

Herbal Preparations in Ancient Cosmetology

The use of natural ingredients in skincare dates back to the ancient civilizations (Figure 17.1). For example, castor oil (*Ricinus communis*), anise (*Illicium verum*), belladonna (*Atropa belladonna*),

Figure 17.1: The concept of beauty in ancient Greece- Evidence from the National Archaeological Museum in Athens, Greece.
(a) The Judgment of Paris: A story on a beauty contest in Greek mythology
(b) Mycenaean perfumes: The scent of antiquity

cinnamon (*Cinnamomum*), cardamom (*Elettaria cardamomum*), myrrh (*Boswellia sacra*), and mustard (*Sinapis alba*) were used by Sumerians, Assyrians, and Babylonians to remove skin devils. Cosmetics played a prominent role in ancient Egyptian skincare. From the tombs of Egyptian pharaohs, vessels filled with rare species and exquisite oils and portraits were discovered, showing female faces enhanced by beauty products (Oumeish 2001). The Egyptian queen Cleopatra was known to bathe in goat's milk, almond, and honey to soften her skin. Scented oils and ointments were used by Egyptians to clean and soften their skin and mask body odor. Oils and creams containing myrrh, thyme (*Thymus*), chamomile (*Matricaria*), lavender (*Lavandula*), peppermint (*Mentha*), rosemary (*Rosmarinus officinalis*), cedar (*Cedrus libani*), rose (*Rosa*), aloe (*Aloe barbadensis*), olive (*Olea europaea*), sesame (*Sesamum indicum*), and almond (*Prunus dulcis*) were used by them to protect their skins from desert sun and winds (Chaudhri and Jain 2009, González-Minero and Bravo-Díaz 2018). Clay and herbal masks were popular among Egyptian women in the time as far back as 69 BC (Oumeish. 2001). Malachite (copper) and "kohl" (made of lead and antimony) were applied to decorate eyelids. A mixture of beeswax, virgin olive oil, cypress resin (*Cupressus*), and milk were applied as a face mask (González-Minero and Bravo-Díaz 2018). The famous "Papyrus Ebers" records the use of aloe in various cosmetic preparations, in addition to its usage as a medication for burns, cut wounds, and skin rashes (Oumeish 2001).

The Egyptian traditions were transmitted to Greece and Italy over time, triggering the development of various ointments, perfumes, and facial creams of herbal origin. The Greeks were considered as the first Europeans to use aromatic oils of Eastern origin (from India and Arabia through the opening of trade routes) and ointments as a form of make-up. Greeks and the Romans applied ointments composed of cypress, cedar, and incense resins at night. Olive oil was reputed as a cleanser and also employed to combat wrinkles, while rose water was used to extract perfumes. A customary facial cream of Romans was made up of figs (*Ficus carica*), banana (*Musa*), oats (*Avena*), and rose water, while the roots of *Asparagus*, wild anise, wild lily bulbs in goat's milk, and manure were filtered and applied with soft bread on the face (González-Minero and Bravo-Díaz 2018).

Historical records further reveal that the Arabs learned cosmetology through their interaction with the Egyptian, Roman, Persian, and Medieval European cultures. Arabic "aromatherapy" used the oils extracted from flowers and pine trees. Creams made of fruit acids obtained from sugar cane, mangos, avocados, apples, etc., were used as skin peelers. The extracts of roses and lemon flowers diluted in water or mixed with glycerine were commonplace as cleansers and moisturizers for the face, neck, and hands (Oumeish 2001). The prominence of cosmetology in Arab culture is further evidenced in the medical encyclopedia compiled by Doctor Abu'al-Qasim Al Zahrawi (936–1013). He dedicated one full chapter for this cause, and considered cosmetics as a branch of medicine, and referred to it as Adwiyat al-Zinah (Medicine of Beauty) (González-Minero and Bravo-Díaz 2018).

In the Middle Ages, "Schola Medica Salernitana" was founded in the south Italian city of Salerno, where the first written work about cosmetics, i.e., "De Ornatu Mulierum" was published. This work was aimed at teaching women about conserving and improving the beauty and treating skin diseases through a series of precepts, advice, and natural remedies. It depicted 96 plant species of cosmetic value, some of which are still in use in this 21st century (Cavallo et al. 2008, González-Minero and Bravo-Díaz 2018). Furthermore, a kind of "phytotherapy research" had taken place within monasteries that were converted into knowledge hubs during the Middle Ages, with the custom of using plants and minerals for medicinal and cosmetic purposes by the monks (Cavallo et al. 2008).

Ancient Indian literature provides a wealth of knowledge for numerous cosmeceutical preparations and self-beautification concepts in ancient India. Influenced by Indian traditional medical schools such as Ayurveda, and social and cultural factors, these practices were targeting both the external and internal beauty and happiness. Lip balms, deodorants, skin lightening and exfoliating scrubs, anti-dandruff preparations, breast developers, face packs, anti-acne preparations, mouth fresheners, and hair care products were known to be used in ancient India. As far as medical plants are concerned, *Emblica officinalis*, *Euphorbia nivulia*, *Acorus calamus*, *Pongamia pinnata*, *Berberis aristata*, *Coriandrum sativum, Cinnamomum camphora*, and *Punica granatum* were common in them (Patkar 2008). The women in India used a turmeric cream, and the formulation composed of gram flour or wheat husk mixed with milk, instead of soap. Even today, cosmetic preparations comprised of turmeric, almonds, sandal, etc., are popular among Indians (González-Minero and Bravo-Díaz 2018).

As much as in Europe and Middle-East, herbal cosmetics were an integral part of the traditional Chinese and Japanese cultures. The historical data reveals that China imported sesame oil scented with *Murraya paniculata* and rose water from Persia along the Silk Road, and also aromatic compounds, such as cloves (*Syzygium aromaticum*), ginger (*Zingiber officinale*), and nutmeg (*Myristica fragrans*) from Indonesia. The Japanese used crushed safflower petals (*Carthamus tinctorius*) to paint the eyebrows, the edges of eyes, and the lips, while rice powder (*Oryza sativa*) was employed to whiten the face and back (González-Minero and Bravo-Díaz 2018).

After the renaissance, there was a rapid development in the field of cosmetics. Yet, the indigenous knowledge of natural skincare agents plays a vital role in expanding the horizons of modern cosmetics and cosmeceutical industry. In this respect, ethnobotanical surveys could contribute significantly to gather folklore knowledge and rationalize the traditional claims in the limelight of modern science.

Ethnobotanical Surveys on Cosmetics and Skincare Agents

Gathering and preservation of knowledge on traditional uses of natural ingredients such as plants and minerals for cosmetic purposes, and harnessing their potential for body care products is referred to as "cosmetopoeia" (Ansel et al. 2016, Jost et al. 2016). Preservation of traditional knowledge through ethnobotanical studies is an effective approach to reveal the hidden potential as well as the sustainable use of medicinal plants (Napagoda et al. 2019). It has been reported that 90% of the medicinal species are used by the people who are native to a particular geographic area (Baquar 1989). This undocumented vast indigenous knowledge is fast diminishing due to cultural changes, migration, urbanization, modernization, and the descent of western medicinal practices (Rashid and Arshad 2002). Therefore, it is imperative that a concerted effort should be made to document and preserve this residual knowledge. Catering to this need, a large number of ethnobotanical studies have emerged over the recent years, particularly in plant-rich African and Asian regions (Ekpendu et al. 1998, Singh and Singh 2001). Although the information on medicinal plants is being documented this way, the local knowledge on cosmetic and cosmeceutical plants is still large. The handful of studies reported in the literature are summarized below.

The use of plant materials by tribal women of Kashmir Himalayas for cosmetic purposes assessed by Shaheen et al. (2014) revealed the traditional use of 39 plants species belonging to 20 families and the practice of 70 different herbal recipes for conditions, such as treating acne, boosting hair growth, treating facial spots and wrinkles, enhancing fairness, and for eye and lip care. *Citrus limon, Lycopersicum esculentum, Mentha longifolia, Raphanus sativus, Rosa indica, Allium sativum*, and *Allium cepa* were listed among the major plant species having cosmetic applications. Cosmetic ethnobotany was the only choice the women in that region had, due to the geographic remoteness, poverty, and their faith in folklore herbal remedies (Shaheen et al. 2014).

An ethnobotanical study of herbal cosmetics conducted in the Northern Province of Sri Lanka enabled the identification of 62 plant species belonging to 36 families. These plants were used for acne treatment, fairness, antiaging, pigmentation, allergy, hair loss, dandruff, hair cleansing, hair coloring, nail care, eye care, lip care, and body odor. Either fresh paste of these plants were used or they were dried for cosmetics preparations. Out of the recorded plant species, *Curcuma longa, Coscinium fenestratum, Mentha arvensis, Azardiracta indica, and Cinnamomum zeylanicum* were widely present in acne treatments, while *Santalum album, Cassia auriculata, Carica papaya*, and *Daucus carota* were used as fairness enhancers. *Cocos nucifera, Coffea arabica*, and *Hemidesmus indicus* were found in antiaging formulations, whereas *Colocynthis citrullus* and *Trigonella foenum-graecum* were present in the formulations against pigmentation problems (Nirmalan 2017).

A recent study on the use of phytocosmetics in three districts of North-eastern Algeria revealed that the soap and a fixed oil were popular preparation forms. *Aloe vera, Matricaria recutita, Lavandula angustifolia, Citrus limon*, and *Ricinus communis* were reported as common ingredients in these preparations, and the highest relative citation frequency (RFC) was attributed to *A. vera* (Bouzabata 2017).

A phytocosmetics study conducted in Marquesas Islands (French Polynesia) recorded over 500 cosmetic recipes. Plant species, such as *Calophyllum inophyllum, Cananga odorata, Citrus aurantifolia, Cocos nucifera, Curcuma longa, Gardenia taitensis, Mentha* spp., *Ocimum basilicum, Rauvolfia nukuhivensis*, and *Santalum insulare* var. marchionense were identified with high use-values. Coconut (*Cocos nucifera*) water and coconut oil were the main excipients used in the preparation of cosmetic recipes. These plant species were used for perfumes, skin hydration, and medicinal care. Most of these preparations were applied on skin, hair, and on genital organs (Jost et al. 2016).

For the purpose of documenting the plant species used as natural-based cosmetics and cosmeceuticals by the Vhavenda women in Vhembe district municipality, Limpopo province, South Africa, an ethnobotanical study was conducted by Ndhlovu et al. (2019). This study led to the identification of 49 plant species from 31 plant families. *Dicerocaryum zanguebaricum* and *Ricinus communis* were the most commonly cited plants. The cultural importance index (CI) was highest in *Dicerocaryum zanguebaricum*, followed by *Ricinus communis*, and *Helinus integrifolius*. In terms of the plant families, Fabaceae had the highest number of plants (04), while Meliaceae and Rhamnaceae were reported with three plants. Leaves and barks of the plants were widely used to prepare herbal cosmetics and cosmeceuticals, and the preparation methods included infusions, decoctions, poultice, or juice from fresh plants. The majority of plant preparations were applied topically (Ndhlovu et al. 2019).

A similar study was conducted in four states in the South West region of Nigeria to investigate and to prepare an inventory of plants used in traditional cosmetics recipes. It permitted the identification of 80 plant species belonging to 39 families, which were used mainly in dried form or after extracting the juice of the plants. High use values were reported for *Lawsonia inermis, Cocos nucifera, Butyrospermum paradoxum*, and *Pterocarpus osun*. Some of the species identified in the aforementioned study as a part of cosmetics recipes—*Achyranthes aspera, Allium sativum, C. nucifera, Elaeis guineensis, Ageratum conyzoides, Chromolaena odorata, Vernonia amygdalina, Baphia nitida, Azadirachta indica, Aloe vera, Curcuma longa*, etc., have also been identified as being useful in the treatment of skin diseases in Nigeria in the previous studies. Thus, it is suggested that there is no clear demarcation between phytocosmetics and plants used for skin diseases in folklore medicine (Fred-Jaiyesimi et al. 2015). Therefore, ethnobotanical studies focused on the utility of medicinal plants against skin diseases would also provide indispensable information on cosmetic and cosmeceutical herbal preparations. Nevertheless, the number of such investigations conducted so far is also scarce.

A study conducted in northern Maputaland, South Africa by De Wet et al. (2013) reported the utility of 47 plant species from 35 families for the treatment of 11 different skin disorders, including abscesses, acne, burns, boils, incisions, rashes, shingles, sores, wounds, and warts. Out of the above-documented plant species, nine species, i.e., *Acacia burkei, Brachylaena discolor, Ozoroa engleri, Parinari capensis* subsp. *capensis, Portulacaria afra, Sida pseudocordifolia, Solanum rigescens, Strychnos madagascariensis,* and *Drimia delagoensis* were recorded for the first time globally as therapeutics for skin disorders. The most frequently cited species was *Senecio serratuloides*, a medicinal plant that is well-reputed in South Africa for the treatment of various skin disorders and wounds (De Wet et al. 2013).

A review carried out using the ethnobotanical literature of 105 plant species utilized by the people of Eastern Cape Province in South Africa for various cosmetic purposes indicated that the majority of those plants were used for skincare. Those plant materials were applied topically or used as a paste or infusion by the local communities. Although bioactivities that are directly associated with skincare, such as wound healing, antioxidant, antityrosinase, and anti-inflammatory activities have been reported for some of the documented plant species, further scientific explorations are warranted in order to rationalize the traditional claims (Sagbo and Mbeng 2018).

Several studies were carried out in different regions in India to gather information about the utilization of herbal remedies for the treatment of skin diseases by local communities. Twenty-one plant species belonging to 15 different plant families were recorded by Balaraju et al. (2015) in their study conducted in Mahabubnagar district. According to their findings, the paste of the entire plant of *Aloe vera* was used by the tribal communities to cure burns, wounds, wrinkles, and for fairness, while plant species such as *Citrus niman, Cocos nucifera*, and *Tagetes erecta* were used to cure dark spots (Balaraju et al. 2015). The knowledge of medicinal plants used for the treatment of skin diseases in Balod District, Chhattisgarh was collected through an ethnobotanical survey, and it recorded 75 important medicinal plant species belonging to 42 families (Gupta and Gupta 2018). In another study conducted in India, a total of 57 medicinal plants representing 34 families were reported for their therapeutic use against skin ailments and for skincare. The most preferred species were identified as *Andrographis paniculata, Annona squamosa, Azadirachta indica, Calophyllum inophyllum, Cissampelos pareira, Croton sparsiflorus, Glinus oppositifolius, Lantana camara, Ocimum sanctum, Pongamia pinnata*, and *Tridax procumbens* (Panda et al. 2016).

In summary, ethnobotanical observations play a crucial role in the selection of plants for pharmacological screening, as well as means of documenting and preserving local knowledge, which is obtained by trial and error and transferred over generations. Some of the scientific investigations conducted on pharmacological aspects of popular phyocosmetic and phytocosmeceutical ingredients are described in the following section.

Bioactivity Studies on Herbal Cosmeceutical Ingredients

Since most of the skincare cosmeceuticals are used as sunscreens, skin lightening, and antiaging agents, more emphasis is given in this section to elaborate the scientific evidence related to these activities in the context of herbal extracts and secondary metabolites thereof.

Photoprotective Potential

Ultraviolet (UV) component of the solar electromagnetic radiation is capable of causing skin damage either via direct absorption or photosensitization mechanisms (Saewan and Jimtaisong 2015). In the direct absorption mechanism, cellular chromophores, such as nucleic acids, amino acids, quinines, flavins, porphyrins, etc., absorb UV radiation that gets transformed into a biochemical signal, initiating biological responses. On the other hand, photosensitization mechanisms involve endogenous and/or exogenous sensitizers that change from a ground state to an excited state upon the absorption of

UV radiation, resulting in further reactions that lead to the formation of reactive oxygen species (ROS), such as hydroxyl radical, superoxide anion, peroxyl radicals, and their active precursors-singlet oxygen, hydrogen peroxide, and ozone. Reactive nitrogen species (RNS), such as nitric oxide and nitric dioxide are also produced due to the photosensitization mechanisms. Although ROS are constantly generated in keratinocytes and fibroblasts, the nonenzymatic and enzymatic antioxidants are involved in the removal of these ROS, thereby maintaining a balance between prooxidant and antioxidant levels, ensuing cell structure stabilization. However, the excessive generation of free radicals due to prolonged and repeated exposure of the skin to UV light overwhelm these defense mechanisms, leading to modifications of DNA and abnormal expression of cellular genes, ultimately resulting in a loss of cellular integrity (Ziegler et al. 1994, Goihman-Yahr 1996, Schwarz et al. 2005, Saewan and Jimtaisong 2015). The UV-generated ROS affect the mitogen-activated protein kinase (MAPK) pathway and initiate the activation of transcription factor families, nuclear factor kappa B (NF-κB), and activator protein 1 (AP-1) that are involved in the processes of cell proliferation, cell differentiation, and cell survival, hence having significant roles in tumorigenesis. The activation of NF-κB and AP-1 may contribute to the induction of heme oxygenase-1 and matrix metalloproteinases in the skin (Cooper and Bowden 2007). Induction of heme oxygenase-1 could elevate cellular iron levels and promote further ROS generation, while the increase in matrix metalloproteinases results in the extracellular matrix protein degradation, causing wrinkle formation and metastases (Cooper and Bowden 2007, Saewan and Jimtaisong 2015). In addition, excessive generation of ROS/RNS will lead to oxidative damage that may trigger inflammation, and local and systemic immunosuppression (Saewan and Jimtaisong 2015).

In order to overcome the harmful effects caused by the UV radiation, a novel concept known as "photochemoprevention/photoprotection" was introduced. This involves the use of various photochemopreventive/photoprotective agents. These agents are capable of preventing the damage caused by UV radiation and/or manipulating different cellular responses to UV radiation to prevent, stop, or correct tumor promotion and progression (Napagoda et al. 2016). In this respect, a number of plant extracts and secondary metabolites thereof have been subjected to extensive investigations, and proven to be beneficial as photochemopreventive agents by inhibiting the expression of UV-induced AP-1, NF-κB, and several other pathways. For example, epigallocatechin-3-gallate (EGCG) (Figure 17.2a), one of the major polyphenolic compound in green tea, is found to be effective in inhibiting UVB-induced nitric oxide, NF-kB activation, and also the expression of IL-6, a cytokine which plays pathological roles in chronic inflammatory condition (Xia et al. 2005, Song et al. 2006, Cooper and Bowden 2007). Resveratrol (Figure 17.2b), which is found at high levels in red wine and grapes, has been reported with strong anti-inflammatory and antiproliferative properties, suggesting the necessity of developing resveratrol-containing emollient, sunscreens, and other skincare products for the prevention of skin cancer and other conditions caused by UV radiation (Aziz et al. 2005). Silibinin (Figure 17.2c), a naturally occurring flavonoid, has also displayed chemopreventive effects on UVB-induced skin carcinogenesis by interfering several cellular mechanisms, including the modulation of cell cycle regulators and mitogen-activated protein kinases (Mallikarjuna et al. 2004).

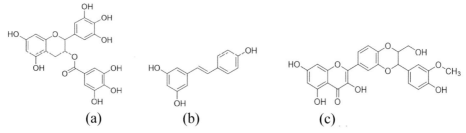

(a) (b) (c)

Figure 17.2: Some examples of natural photochemopreventive agents capable of modulating cellular responses. (a) Epigallocatechin 3-gallate (b) Resveratrol (c) Silibinin

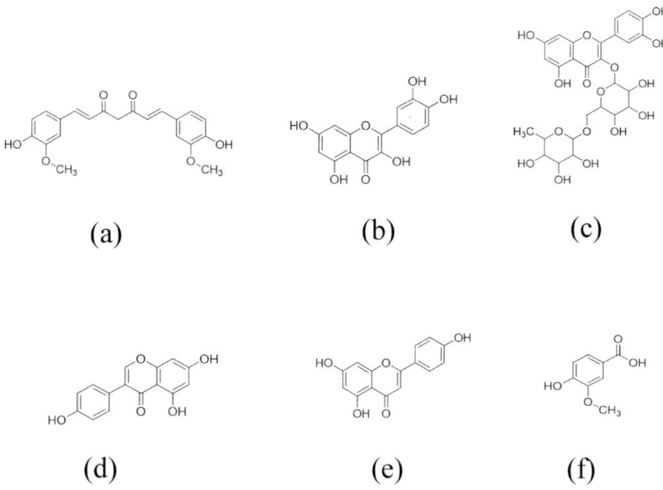

Figure 17.3: Examples of some secondary metabolites effective as UV-blockers. (a) Curcumin (b) Quercetin (c) Rutin (d) Genistein (e) Apigenin (f) Vanillic acid

Furthermore, a wide array of plant secondary metabolites have been reported as effective UV blockers. This includes phenolic acids, flavonoids, lichen polyphenols, terpenoids, and mycosporine-like amino acids (Saewan and Jimtaisong 2015). The structures of some well-known UV blockers are presented in Figure 17.3. These compounds are capable of preventing the penetration of UV radiation into the skin, thus reducing inflammation, oxidative stress, and DNA damaging effects (Nichols and Katiyar 2010).

In addition, a large number of plant extracts have exerted photoprotection potential by interfering with different cellular mechanisms (F'guyer et al. 2003), and some examples are summarized in Table 17.1.

The effectiveness of a sunscreen is usually expressed by its sun protective factor (SPF), and the sunscreens with SPF value of 15 or greater are highly recommended (Napagoda et al. 2016). Several plant species used in traditional medicine in Sri Lanka as dermatological remedies were reported with high sunscreen potential. The aqueous-methanolic extracts prepared from *Atalantia ceylanica*, *Hibiscus furcatus*, *Leucas zeylanica*, *Mollugo cerviana*, *Olax zeylanica*, and *Ophiorrhiza mungos* were identified as strong photoprotectants with high UV-filtering and antioxidant activities (Napagoda et al. 2016). Further, the *in vitro* SPF analysis of hydroalcoholic extracts of 12 commonly used vegetables revealed that extracts prepared from beetroot, green pea, drumstick, and sweet potato possess high SPF values (Mazumder et al. 2018), while a sunscreen formulated using a mixture of *Pongamia pinnata* and *Punica granatum* in 3:2 ratio also showed effective sunscreen potential (Patil et al. 2015).

Skin Lightening Activity

Skin color is primarily determined by the amount of melanin present in the skin. Melanin is a pigment produced by melanocytes through a process known as melanogenesis, from which the amino acid L-tyrosine gets converted by the enzyme tyrosinase into dopaquinone (Cooksey et al. 1997). Although melanogenesis and skin pigmentation are considered as natural photoprotective approaches in response to UV-induced skin photocarcinogenesis, the increased melanin synthesis and accumulation of these pigments give rise to many aesthetic and dermatological problems, such as melasma, periorbital hyperpigmentation, freckles, or lentigines (Smit et al. 2009, Zolghadri et al. 2019). Pigmentation is either dependent on the number, size, composition, and distribution of melanocytes, or activity of melanogenic enzymes. Furthermore, cutaneous pigmentation is resulted from melanin synthesis by melanocytes and transfer of melanosome to keratinocytes (Lin et al. 2008).

Table 17.1: Multiple mechanisms of photoprotection by various plant extracts.

Plant extract	Mechanism of action	References
Green tea, Black tea (*Camellia sinensis*)	Inhibit, reverse, or retard the process of the skin photodamage via sunscreen and antioxidant properties, regulation of signal transduction pathway and gene expression, alleviation of DNA damage, and modulation immunological function	Li et al. 2014
Aloe vera	Decrease UVA-induced redox imbalance, decrease UVA associated lipid membrane oxidation and increase overall cell survival	Rodrigues et al. 2016 Dunaway et al. 2018
Walnut extract (*Juglans regia*)	Prevent ROS generation and lipid peroxidation as well as UVB-activated inflammatory markers	Muzaffer et al. 2018
Turmeric (*Curcuma longa*)	Exert anti-inflammatory effects by inhibiting NFkB and MAPK signaling pathways with reduction of the expression of inducible nitric oxide (iNOS) and COX2 Inhibit UVB-induced TNF-α at the mRNA level and reduce the expression of matrix metalloproteinase-1 (MMP-1) expression in keratinocytes and fibroblasts	Guo et al. 2008 Jang et al. 2012 Dunaway et al. 2018
Garlic (*Allium sativum*)	Decrease UVB and *cis*-urocanic acid- induced immunosuppression	Reeve et al. 1993
Red clove (*Trifolium pretense*)	Reduce UV-induced erythema and edema	Widyarini et al. 2000
Capparis spinosa	Protect phospholipidic biomembranes from UV light-induced peroxidation and protect against UVB-induced erythema	Bonina et al. 2002 Saija et al. 2000
Culcitium reflexum	Inhibit UV light-induced peroxidation in phosphatidylcholine multilamellar vesicles and protect against UVB-induced erythema	Aquino et al. 2002
Ginseng (*Panax ginseng*)	Exert anti-inflammatory activity by reducing nitric oxide production and iNOS mRNA synthesis, Inhibit the UVB-induced COX2 expression and TNF-α transcription	Lee et al. 2012 Dunaway et al. 2018

Although becoming tan is a desirable feature in Western culture, a light complexion is considered to be equivalent to youth and beauty in Eastern countries. Consequently, interest in skin whitening has grown tremendously over recent years, and more attention has been paid on the identification of tyrosinase inhibitors from natural sources (Napagoda et al. 2018). There are six different classes of tyrosinase inhibitors—reducing agents, O-dopaquinone scavengers, alternative enzyme substrates, nonspecific enzyme inactivators, specific tyrosinase inactivators, and specific tyrosinase inhibitors. Among these six types, only specific tyrosinase inactivators and specific tyrosinase inhibitors actually bind to the enzyme and inhibit its activity, hence only they are regarded as "true inhibitors" (Chang 2009).

Several plant species commonly found in the Indian subcontinent, such as *Aloe vera, Carica papaya, Cinnamomum zeylanicum, Curcuma longa, Rosa alba, Syzygium aromaticum*, and *Cassia auriculata* (Adhikari et al. 2008, Vaibhav and Lakshaman 2012, Vardhan and Pandey 2014, Gupta and Masakapalli 2013, Napagoda et al. 2018) have been already identified with anti-tyrosinase activity. Out of 299 parts of 263 plant species collected from Jeju Island of the Korean Peninsula, *Cornus walteri, Maackia fauriei, Toxicodendron succedaneum*, and *Sophora flavescens* have shown potent tyrosinase inhibition (Moon et al. 2010). Meanwhile, a study conducted using Amazonian plants indicated that the extracts obtained from the leaves and stem of *Ruprechtia* sp. and from the aerial organs of *Rapanea parviflora* were most active in inhibiting the tyrosinase activity (Macrini et al. 2009). Also, a study conducted on tyrosinase inhibitory activity of 91 native plants from central Argentina revealed significant inhibition of tyrosinase by the extracts prepared from *Dalea elegans, Lepechinia meyenii,* and *Lithrea molleoides* with IC_{50} values of 0.48, 10.43, and 3.77 μg/mL, respectively (Chiari et al. 2010).

Moreover, tyrosinase inhibitory potential has been evaluated in numerous plant secondary metabolites by various research groups. Phenolic compounds and their derivatives, as well as some terpenoids, coumarins and quinones, have displayed strong tyrosinase inhibitory action. For example, an apigenin flavone glucoside vitexin and a C-glycosylflavone isovitexin isolated from *Vigna radiata* extracts inhibited the enzyme with IC_{50} values of 6.3 and 5.6 mg/mL, respectively (Yao et al. 2012). Also, significant inhibition of tyrosinase was observed in five flavones isolated from the stem barks of *Morus lhou*-viz., mormin (IC_{50}=0.088 mM), cyclomorusin (IC_{50}=0.092 mM), morusin (IC_{50}=0.250 mM), kuwanon C (IC_{50}=0.135 mM), and norartocarpetin (IC_{50}=1.2 µM) (Ryu et al. 2008). Glabridin (IC_{50}=0.43 µM), isolated from the root of *Glycyrrhiza glabra* has exhibited excellent inhibitory effects on tyrosinase (Chen et al. 2016), while Glyasperin C isolated from *Glycyrrhiza uralensis* also showed a strong tyrosinase inhibitory activity with an IC_{50} value of 0.13 µg/mL (Kim et al. 2005). The tyrosinase inhibitory studies on several other isoflavones, such as formononetin, genistein, daidzein, texasin, tectorigenin, odoratin, and mirkoin isolated from the stems of *Maackia fauriei*, revealed that, out of the above-isolated compounds, mirkoin (IC_{50}=5 µM) possessed a stronger tyrosinase inhibition than the positive control kojic acid, and it could inhibit the enzyme reversibly in a competitive manner (Kim et al. 2010). In a recent study aimed at the screening of natural products for the development of cosmetic ingredients, two major compounds in *Humulus japonicus*, *trans*-N-coumaroyltyramine (IC_{50}=40.6 µM) and *cis*-N-coumaroyltyramine (IC_{50}=36.4 µM) displayed potent tyrosinase inhibition (Yang et al. 2018).

In addition, several studies were focused on the evaluation of the synergistic strategy for tyrosinase inhibitors for the improvement of their inhibitory activities. In this respect, the mixtures of 4-methyl catechol:catechol, aloesin:arbutin, glabridin:resveratrol, glabridin:oxyresveratrol, and resveratrol:oxyresveratrol have shown a synergistic effect on tyrosinase inhibition (Schved and Kahn 1992, Jin et al. 1999, Wang et al. 2018).

Antiaging Activity

The causes of skin aging are hitherto uncertain. Current theories are assigned to the damage concept, whereby the accumulation of damage within the cell may cause biological systems to fail, or to the programmed aging concept, whereby internal processes may cause aging (Gems et al. 2009, Jin 2010). In the DNA damage concept, one of the theories explains the damage caused by the free radicals that are originated from exogenous sources, such as UV and ionizing radiations, and from several intracellular sources (Beckman and Ames 1998). Ultimately, these damages lead to skin aging with wrinkles, dryness, spots, discoloration, and sagging.

Skin health is considered as one of the principal factors of beauty. Hence, the antiaging treatments have come up with major cosmetic consideration and several antiaging treatments have become more prominent during the past years. These include non-invasive and even invasive procedures (Ganceviciene et al. 2012). The extracellular matrix provides a structural framework essential for the growth and elasticity of the skin, and contains fibroblasts and proteins, including collagen and elastin. Degradation of the extracellular matrix has directly been linked to skin aging, and is correlated with increased activity of several enzymes-elastase, collagenase, and hyaluronidase (Ndlovu et al. 2013).

Eighty percent of the dry weight of skin is considered to be collagen and is responsible for the tensile strength of the skin. Collagenases are a type of metalloproteinase that can cleave molecules in the extracellular matrix. Elastase is a proteolytic enzyme involved in the degradation of the extracellular matrix that contains elastin. Elastin provides much of the elastic recoil properties of skin, arteries, lungs, and ligaments. Loss of elastin is a major part of what causes visible signs of aging in the skin. Hyaluronic acid has a role in retaining the moisture, structure, and elasticity of the skin while facilitating rapid tissue proliferation, regeneration, and repair. The levels of collagen, elastin, and hyaluronic acid would decrease with aging, and this could lead to a loss of strength and flexibility in the skin, causing the emergence of wrinkles (Ndlovu et al. 2013). Moreover, the high levels of ROS

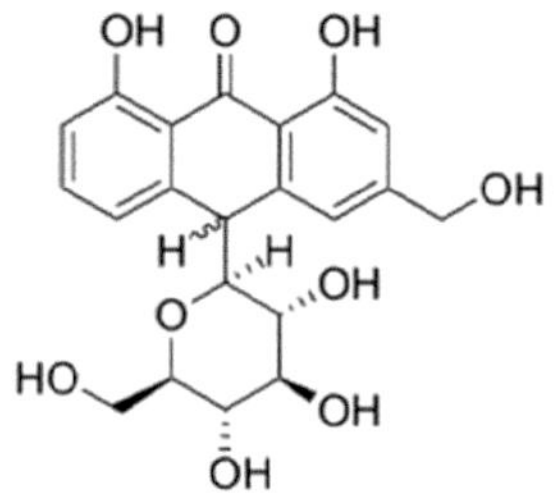

Figure 17.4: Aloin, a major constituent in Aloe vera.

induce the action of collagenase, elastase, and hyaluronidase, which can further contribute to skin aging (Labat-Robert et al. 2000, Ndlovu et al. 2013). However, natural materials with anti-collagenase, anti-elastase, and anti-hyaluronidase properties can help to prevent the undesirable age-associated destruction of collagen, elastin, and hyaluronic acid (Thring et al. 2009, Ndlovu et al. 2013).

The anti-aging properties of four southern African medicinal plants—*Clerodendrum glabrum*, *Schotia brachypetala*, *Psychotria capensis*, and *Peltophorum africanum*, were investigated by Ndlovu et al. (2013). Their study revealed high anti-elastase activity in ethyl acetate extract of bark of *S. brachypetala* and leaves of *P. capensis*. Moreover, the methanol extract of *S. brachypetala* bark displayed the highest anti-hyaluronidase activity, whilst the ethyl acetate extract of *P. africanum* bark exhibited the highest antioxidant activity (Ndlovu et al. 2013). In another study, high anti-elastase activity was observed in the extracts prepared from *Aesculus turbinata*, *Taxillus yadoriki*, and *Cornus walteri* (Moon et al. 2010). The anti-elastase activity of 150 medicinal plants was investigated by Lee et al. Out of these plants, six plant extracts, i.e., *Areca catechu*, *Cinnamomum cassia*, *Myristica fragrans*, *Curcuma longa*, *Alpinia katsumadai*, and *Dryopteris cassirhizoma* exhibited more than 65% of inhibition of elastase activity at a concentration of 1 mg/mL. Only *Areca catechu* showed a high inhibitory effect on hyaluronidase activity (Lee et al. 1999).

The flowers of *Tagetes erecta* are traditionally used to treat skin diseases, such as sores, burns, wounds, ulcers, eczema, and several other skin ailments. Hyaluronidase, elastase, and matrix metalloproteinase (MMP-1) inhibitory activity of this flower extract was investigated to determine its anti-wrinkle potential. The methanol extract showed significant hyaluronidase and elastase inhibition with IC_{50} of 11.70 g/mL and 4.13 g/mL, respectively, along with a moderate inhibition of MMP-1. Syringic acid and β-amyrin isolated from this extract were also capable of inhibiting the above enzymes, rationalizing the traditional uses of the plant (Maity et al. 2011). Furthermore, procyanidins extracted from *Vitis vinifera*, curcumin present in *Curcuma* longa, as well as phenolic compounds, such as epicatechin, resveratrol, galangin, kaempferol, quercetin, and myricetin had also exhibited potential elastase inhibition (Maffei Facino et al. 1994, Chainani-Wu 2003, Hrenn et al. 2006, Kanashiro et al. 2007). Aloin (Figure 17.4) in *Aloe vera* plant inhibited *Clostridium histolyticum* collagenase reversibly and noncompetitively. Aloe gel and aloin were also proved to be effective inhibitors of stimulated granulocyte matrix metalloproteinases (Barrantes and Guinea 2003)

Other Important Bioactivities for the Purpose of Skincare

Apart from the abovementioned bioactivities, antimicrobial activity is also important for natural skincare cosmeceuticals, particularly, the antimicrobial activity against acne-causing bacterial species. Acne vulgaris is considered as the most abundant dermatologic condition which affects late adolescents, and is characterized by follicular hyperkeratinization, seborrhea, microbial colonization, and inflammation (Lynn et al. 2016). This condition is triggered by the activity of some bacterial species, such as *Propionibacterium acnes*, *Staphylococcus aureus*, and *Staphylococcus epidermidis*. Although antibiotics are prescribed to treat acne vulgaris, more emphasis is paid on the application of natural remedies as an alternative strategy to treat acne vulgaris. In the search for natural anti-acne agents, a number of plant extracts and phytochemicals thereof have been investigated for antibacterial activity against acne-causing bacterial species. The extracts, such as *Punica granatum, Morus alba*, and *Angelica anomala* have exhibited potent antibacterial activity against *P. acnes* and *S. epidermidis*. Similarly, the essential oils of *Citrus obovoides, Citrus natsudaidai, Cryptomeria japonica*, and *Cymbopogon nardus*, as well as phytochemicals, such as pulsaquinone, hydropulsaquinone, rhodomyrtone, and rhinacanthin-C, were found to possess strong antimicrobial activity against *P. acnes* (Sinha et al. 2014). Moreover, a recent study revealed a strong antibacterial activity in novel topical gel formulations comprising of *N. sativa*. These formulations did not exhibit undesirable side effects on human subjects. The results suggested the suitability of the prepared gel formulations to be developed for commercial products (Nawarathne et al. 2019).

Another important aspect of cosmeceuticals is its moisturizing ability. The balance between the water content of the stratum corneum and skin surface lipids is detrimental to the appearance and function of the skin. When this balance gets disrupted, a dermatological condition known as dry skin ensues. This is a phenomenon commonly observed in atopic dermatitis patients. Under these circumstances, effective cosmetic products must be used to improve skin hydration (Dal'Belo et al. 2006). In this regard, the polysaccharide-rich *Aloe vera* extracts, which are often used in cosmetic formulations, were investigated using skin bioengineering techniques, and the results confirmed its effectiveness in improving skin hydration (Dal'Belo et al. 2006). In another study, *Corchorus olitorius* leaves that are rich in mucilaginous polysaccharide were proved to be effective in increasing skin hydration (Yokoyama et al. 2014). Similarly, the polysaccharide gel extracted from the fruit hulls of durian (*Durio zibethinus*) had a significant effect on skin capacitance in human subjects, and exerted a positive but moderate effect on skin firmness (Futrakul et al. 2009).

The above-discussed bioactivities in herbal extracts and the secondary metabolites provide clear justification for the traditional utility of those plants as skincare agents while highlighting their potential in cosmeceutical industry. The potency of these natural materials could be further improved with nanotechnology, and some examples for novel nano-based approaches in herbal cosmeceutical field are discussed below.

Nanotechnology, the Paradigm in Herbal Cosmeceuticals

Nanotechnology has opened up new avenues in the field of herbal cosmeceuticals with the introduction of nanoemulsions, nanocapsules, nano pigments, liposome formulations, fullerenes, niosomes, nanocrystals, and solid lipid nanoparticles, etc. This new technology has provided solutions to the long-lasting issues in the development of herbal cosmetics, where the lower penetration and high compound instability had hindered the sustained and enhanced delivery of the active phytoconstituents. The nano-sized delivery systems are beneficial in many aspects—enhance the encapsulation and stability of active ingredients, increase the penetration of cosmeceuticals through the epidermis, target the active ingredients to the desired site, and controlled release of those ingredients for a prolonged effect. Moreover, these nano-systems can increase the aesthetics of the products (Lohani et al. 2014).

Antioxidants are able to delay the aging of the skin and to protect it. However achieving the stability of these compounds remains a challenge (Vinardell and Mitjans 2015). To address this need, types of nanomaterials have been developed to encapsulate antioxidant molecules. For example, niosomes prepared with polyglyceryl-3 dioleate or glycerol monooleate and cholesterol resulted in high cutaneous accumulation of resveratrol (Pando et al. 2013), while the simultaneous encapsulation of resveratrol with curcumin in lipid-core nanocapsules, resulted in increased delivery of resveratrol into deeper skin layers. This was attributed to the interaction of curcumin with the lipid bilayers of the stratum corneum that would have facilitated the penetration of less lipophilic resveratrol across the skin barrier into the epidermis and dermis (Friedrich et al. 2015). Apart from this, the co-encapsulation of resveratrol and curcumin in niosomal systems enhanced the ability to reduce free radicals due to a synergic antioxidant action (Tavano et al. 2014). On the other hand, tocopherol has been incorporated into nanostructured lipid carriers to produce a non-irritant, stable, and cosmetically appealing aqueous formulation, which is capable of inducing a high release of tocopherol (Mahamongkol et al. 2005, Ben-Shabat et al. 2013). Furthermore, a nanoemulsion system formulated with prenylated flavanones isolated from *Eysenhardtia platycarpa* was capable of enhancing antiaging activity (Domínguez-Villegas et al. 2014). Encapsulation of curcumin in photo-stable nanospheres was able to protect curcumin from photodegradation, and hence prolong its antioxidant activity (Suwannateep et al. 2012). In another study, curcumin along with lauric acid was delivered to the skin for the inhibition of *P. acnes* via niosomes. The antimicrobial activity of curcumin and lauric acid against this acne-causing bacteria was significantly enhanced due to the development of nano-sized vehicles (Liu and Huang 2013).

The bioavailability and skincare properties of *Aloe vera* leaf gel extract were enhanced by liposome encapsulation, and this has significantly increased the collagen synthesis, in comparison to the *Aloe vera* leaf gel extract alone (Takahashi et al. 2009). Similarly, the bioavailability and antioxidant activity of catechin have been significantly enhanced by the preparation of nanoformulation using biodegradable polymer Eudragit L 100 (Monika et al. 2017). Furthermore, a novel herbal nano-emulsion containing lemon juice and/or rose water has been developed for topical treatment of acne and other skin disorders, with increased stability and percutaneous penetration, low skin irritation, and with reservoir effect that enabled the controlled delivery of the active therapeutic agents (Chaudhary and Naithani 2011).

The efficacy of some of the photoprotective herbal extracts could be markedly increased by developing into nano-preparations. This could be attributable to the improved solubility, permeability, and stability of the nanoformulations. Advanced lipid nanocarriers based on rice bran oil and raspberry seed oil were developed and incorporated into creams containing synthetic UV-B and UV-A filters, and these formulations have exhibited improved antioxidant activity and photoprotection due to the synergistic effect between rice bran and raspberry seed oil and organic filters (Niculae et al. 2014). Similarly, a significant enhancement in the photoprotection was observed for green coffee oil in combination with the synthetic sunscreen ethylhexyl methoxycinnamate, which is also due to the synergistic effect between plant oils and organic filters (Chiari et al. 2014). Another example of the synergistic photoprotective action was the incorporation of inorganic filters to oil-in-water emulsions containing flavonoid compounds quercetin and rutin. Interestingly, the addition of TiO_2 significantly enhanced the photoprotective capability, while a moderate effect was observed with the incorporation of ZnO, signifying that ZnO has exerted purely an additive effect (Choquenet et al. 2008). On the contrary, the combination of zinc oxide nanoparticles into lyophilized methanolic extract of the top flowerings of *Teucrium polium* has considerably increased the SPF value. Thus, the addition of nanoparticles into herbal sunscreens has plausible outcomes, such as reducing the photodecomposition of the compounds in the sunscreen and controlled release of the UV absorbents into the skin (Sharififar et al. 2013). In addition, the novel sunscreen formulations comprised of polymeric nanoparticles of morin displayed high SPF values and antioxidant activities without any cytotoxic effects (Shetty et al. 2015).

Based on these observations, it is obvious that nanotechnology-based herbal cosmeceuticals could offer the advantages of diversified products, increased bioavailability of active constituents, and increased aesthetic appeal with prolonged effects.

Conclusions

In summary, the field of herbal skincare cosmeceuticals is exponentially growing, and the ethnobotanical studies on herbal cosmetics and skincare agents would be imperative not only for documenting and preserving the local knowledge, but also for conducting scientific investigations on the specific biological activities in the herbal extracts that would bring about the desired effects. Once the herbal extracts with potent biological activities are identified, nanotechnological approaches could be employed to enhance the efficacy, bioavailability, and safety of those herbal skincare formulations.

References

Adhikari, A., Devkota, H.P., Takano, A., Masuda, K., Nakane, T., Basnet, P., Skalko-Basnet, N. 2008. Screening of Nepalese crude drugs traditionally used to treat hyperpigmentation: *in vitro* tyrosinase inhibition. Int. J. Cosmet. Sci. 30: 353–360.

Ahmad, M., Khan, M.A., Zafar, M. 2008. Traditional herbal cosmetics used by local women communities in district Attock of Northern Pakistan. Indian J. Tradit Know. 7(3): 421–424.

Ahsan, H. 2018. The biomolecules of beauty: biochemical pharmacology and immunotoxicology of cosmeceuticals. J. Immunoass. Immunoch. 40(1): 1–18.

Amaro-Ortiz, A., Yan, B., D'Orazio, J.A. 2014. Ultraviolet radiation, aging and the skin: prevention of damage by topical cAMP manipulation. Molecules 19(5): 6202–6219.

Ansel, J.-L., Morett, C., Raharivelomanana, P., Hano, C. 2016. Cosmetopoeia. CR Chim. 19: 1033–1034.

Aquino, R., Morelli, S., Tomaino, A., Pellegrino, M., Saija, A., Grumetto, L., Puglia, C., Ventura, D., Bonina, F. 2002. Antioxidant and photoprotective activity of a crude extract of *Culcitium reflexum* H.B.K. leaves and their major flavonoids. J. Ethnopharmacol. 79(2): 183–191.

Aziz, M.H., Reagan-Shaw, S., Wu, J., Longley, B.J., Ahmad, N. 2005. Chemoprevention of skin cancer by grape constituent resveratrol: relevance to human disease? FASEB J. 19(9): 1193–1195.

Balaraju, S., Ramamurthy, N., Konkala, A., Suresh, S. 2015. Ethnomedicinal plants used to cure skin diseases by tribals of Mahabubnagar district, Telangana state. IOSR-JPBS. 10(6): 25–27.

Baquar, S.R. 1989. Medicinal and poisonous plants of Pakistan. Printas. Karachi, Pakistan.

Barrantes, E., Guinea, M. 2003. Inhibition of collagenase and metalloproteinases by aloins and Aloe gel. Life Sci. 72: 843–850.

Beckman, K.B., Ames, B.N. 1998. The free radical theory of aging matures. Physiol Rev. 78(2): 547–581.

Ben-Shabat, S., Kazdan, Y., Beit-Yannai, E., Sintov, A.C. 2013. Use of alpha-tocopherol esters for topical vitamin E treatment: Evaluation of their skin permeation and metabolism. J. Pharm. Pharmacol. 65: 652–658.

Bodeker, G., Ryan, T.J., Volk, A., Harris, J., Burford, G. 2017. Integrative skin care: Dermatology and traditional and complementary medicine. J. Altern. Complement Med. 23(6): 479–486.

Bonina, F., Puglia, C., Ventura, D., Aquino, R., Tortora, S., Sacchi, A., Saija, A., Tomaino, A., Pellegrino, M.L., de Caprariis, P. 2002. *In vitro* antioxidant and *in vivo* photoprotective effects of a lyophilized extract of *Capparis spinosa* L buds. J. Cosmet. Sci. 53(6): 321–335.

Bouzabata, A. 2017. Contemporary use of phytocosmetics in three districts from North-Eastern Algeria. Pharmacog. J. 9(6): 762–766.

Cavallo, P., Proto, M.C., Patruno, C., Del Sorbo, A., Bifulco, M. 2008. The first cometic treatise of history; A female point of view. Int. J. Cosmet. Sci. 30: 79–86.

Chainani-Wu, N. 2003. Safety and anti-inflammatory activity of curcumin: A component of tumeric (*Curcuma longa*). J. Altern. Complement Med. 9: 161–168.

Chang, T.S. 2009. An updated review of tyrosinase inhibitors. Int. J. Mol. Sci. 10: 2440–2475.

Chaudhary, M., Naithani, V. 2011. Topical herbal formulation for treatment of acne and skin disorder. US Patent 20110262499 A1.

Chaudhri, S.K., Jain, N.K. 2009. History of Cosmetics. Asian J. Pharm. 3: 164–167.

Chen, J., Yu, X., Huang, Y. 2016. Inhibitory mechanisms of glabridin on tyrosinase. Spectrochim Acta A Mol. Biomol. Spectrosc. 168: 111–117.

Chiari, M.E., Joray, M.B., Ruiz, G., Palacios, S.M., Carpinella, M.C. 2010. Tyrosinase inhibitory activity of native plants from central Argentina: isolation of an active principle from *Lithrea molleoides*. Food Chem. 120: 10–14.

Chiari, B.G., Trovatti, E., Pecoraro, É., Corrêa, M.A., Cicarelli, R.M.B., Ribeiro, S.J.L., Isaac, V.L.B. 2014. Synergistic effect of green coffee oil and synthetic sunscreen for health care application. Ind. Crop Prod. 52: 389–393.

Choquenet, B., Couteau, C., Paparis, E., Coiffard, L.J.M. 2008. Quercetin and rutin as potential sunscreen agents: Determination of efficacy by an *in vitro* method. J. Nat. Prod. 71(6): 1117–1118.

Cooksey, C.J., Garratt, P.J., Land, E.J., Pavel, S., Ramsden, C.A., Riley, P.A., Smit, N.P. 1997. Evidence of the indirect formation of the catecholic intermediate substrate responsible for the autoactivation kinetics of tyrosinase. J. Biol. Chem. 272(42): 26226–262235.

Cooper, S.J., Bowden, G.T. 2007. Ultraviolet B regulation of transcription factor families: roles of nuclear factor-kappa B (NF-ΚB) and activator protein-1 (AP-1) in UVB-induced skin carcinogenesis. Curr. Cancer Drug Tar. 7: 325–334.

Dal'Belo, S.E., Gaspar, L.R., Maia Campos, P.M. 2006. Moisturizing effect of cosmetic formulations containing *Aloe vera* extract in different concentrations assessed by skin bioengineering techniques. Skin Res. Technol. 12(4): 241–246.

De Wet, H., Nciki, S., van Vuuren, S.F. 2013. Medicinal plants used for the treatment of various skin disorders by a rural community in northern Maputaland, South Africa. J. Ethnobiol. Ethnomed. 9: 51. DOI:10.1186/1746-4269-9-51.

Domínguez-Villegas, V., Clares-Naveros, B., García-López, M.L., Calpena-Campmany, A.C., Bustos-Zagal, P., Garduno-Ramirez, M.L. 2014. Development and characterization of two nano-structured systems for topical application of flavanones isolated from Eysenhardtia platycarpa. Colloids Surf B Biointerfaces 116: 183–192.

Dunaway, S., Odin, R., Zhou, L., Ji, L., Zhang, Y., Kadekaro, A.L. 2018. Natural antioxidants: Multiple mechanisms to protect skin from solar radiation. Front. Pharmacol. 9: 392. DOI: 10.3389/fphar.2018.00392.

Ekpendu, T., Obande, O.E., Anayogo, P.U., Attah, A.D. 1998. Nigerian ethnomedicine and medicinal plant flora—the Benue experience. Part I. J. Pharm. Res. And Dev. 3(1): 37–46.

F'guyer, S., Afaq, F., Mukhtar, H. 2003. Photochemoprevention of skin cancer by botanical agents. Photodermatol. Photoimmunol. Photomed. 19(2): 56–72.

Fred-Jaiyesimi, A., Ajibesin, K.K., Tolulope, O., Gbemisola, O. 2015. Ethnobotanical studies of folklore phytocosmetics of South West Nigeria. Pharm. Biol. 53(3): 313–338.

Friedrich, R.B., Kann, B., Coradini, K., Offerhaus, H.L., Beck, R.C., Windbergs, M. 2015. Skin penetration behavior of lipid-core nanocapsules for simultaneous delivery of resveratrol and curcumin. Eur. J. Pharm. Sci. 78: 204–213.

Futrakul, B., Kanlayavattanakul, M., Krisdaphong, P. 2009. Biophysic evaluation of polysaccharide gel from durian's fruit hulls for skin moisturizer. Int. J. Cosmetic. Sci. 32(3): 211–215.

Ganceviciene, R., Liakou, A.I., Theodoridis, A., Makrantonaki, E., Zouboulis, C.C. 2012. Skin anti-aging strategies. Dermatoendocrinol. 4(3): 308–319.

Gems, D., Doonan, R. 2009. Antioxidant defense and aging in *C. elegans*: is the oxidative damage theory of aging wrong? Cell Cycle. 8(11): 1681–1687.

Goh, C. 2009. The need for evidence-based aesthetic dermatology practice. J. Cutan Aesthet. Surg. 2(2): 65–71.

Goihman-Yahr, M. 1996. Skin aging and photoaging: an outlook. Clin. Dermatol. 14: 153–160.

González-Minero, F.J., Bravo-Díaz, L. 2018. The use of plants in skin-care products, cosmetics and fragrances: Past and present. Cosmetics 5(3): 50. DOI:10.3390/cosmetics5030050.

Guo, L.Y., Cai, X.F., Lee, J.J., Kang, S.S., Shin, E.M., Zhou, H.Y., Jung, J.W., Kim, Y.S. 2008. Comparison of suppressive effects of demethoxycurcumin and bisdemethoxycurcumin on expressions of inflammatory mediators *in vitro* and *in vivo*. Arch. Pharm. Res. 31: 490–496.

Gupta, D.K., Gupta, S.G. 2018. Endemic use of medicinal plants for the treatment of skin diseases in the Balod District. IOSR J. Pharm. 8(2): 18–24.

Gupta, S.D., Masakapalli, S.K. 2013. Mushroom tyrosinase inhibition activity of *Aloe vera* L. gel from different germplasms, Chin. J. Nat. Med. 11: 616–620.

Hrenn, A., Steinbrecher, T., Labahn, A., Schwager, J., Schempp, C.M., Merfort, I. 2006. Plant phenolics inhibit neutrophil elastase. Planta Med. 72: 1127–1131.

Jang, S., Chun, J., Shin, E.M., Kim, H., Kim, Y.S. 2012. Inhibitory effects of curcuminoids from *Curcuma longa* on matrix metalloproteinase-1 expression in keratinocytes and fibroblasts. J. Pharm. Investig. 42: 33–39.

Jin, Y.H., Lee, S.J., Chung, M.H., Park, J.H., Park, Y.I., Cho, T.H., Lee, S.K. 1999. Aloesin and arbutin inhibit tyrosinase activity in a synergistic manner via a different action mechanism. Arch. Pharm. Res. 22(3): 232–236.

Jin, K. 2010. Modern biological theories of aging. Aging Dis. 1(2): 72–74.

Jost, X., Ansel, J.-L., Lecellier, G., Raharivelomanana, P., Butaud, J.-F. 2016. Ethnobotanical survey of cosmetic plants used in Marquesas Islands (French Polynesia). J. Ethnobiol. Ethnomed. 12: 55. DOI: https://doi.org/10.1186/s13002-016-0128-5.

Kanashiro, A., Souza, J.G., Kabeya, L.M., Azzolini, A.E., Lucisano-Valim, Y.M. 2007. Elastase release by stimulated neutrophils inhibited by flavonoids: Importance of the catechol group. Z Naturforsch C 62: 357–361.

Kim, H.J., Seo, S.H., Lee, B.G., Lee, Y.S. 2005. Identification of tyrosinase inhibitors from *Glycyrrhiza uralensis*. Planta Med. 71: 785–787.

Kim, J.M., Ko, R.K., Jung, D.S., Kim, S.S., Lee, N.H. 2010. Tyrosinase inhibitory constituents from the stems of *Maackia fauriei*. Phytother Res. 24: 70–75.

Labat-Robert, J., Fourtanier, A., Boyer-Lafargue, B., Robert, L. 2000. Age dependent increase of elastase type protease activity in mouse skin effect of UV-irradiation. J. Photochem. Photobiol. B. 57: 113–118.

Lee, K.K., Kim, J.H., Cho, J.J., Choi, J.D. 1999. Inhibitory effects of 150 plant extracts on elastase activity, and their anti-inflammatory effects. Int. J. Cosmetic. Sci. 21(2): 71–82.

Lee, H., Lee, J.Y., Song, K.C., Kim, J., Park, J.H., Chun, K.H., Hwang, G.S. 2012. Protective effect of processed Panax ginseng, sun ginseng on UVB-irradiated human skin keratinocyte and human dermal fibroblast. J. Ginseng Res. 36: 68–77.

Lin, J.W., Chiang, H.M., Lin, Y.C., Wen, K.C. 2008. Natural products with skin-whitening effects. J. Food Drug Anal. 16: 1–10.

Li, N.-N., Deng, L., Xiang, L.-P., Liang, Y.-R. 2014. Photoprotective effect of tea and its extracts against ultraviolet radiation-induced skin disorders. Trop. J. Pharm. Res. 13(3): 475–483.

Liu, C.H., Huang, H.Y. 2013. *In vitro* anti-propionibacterium activity by curcumin containing vesicle system. Chem. Pharm. Bull. 61(4): 419–425.

Lohani, A., Verma, A., Joshi, H., Yadav, N., Karki, N. 2014. Nanotechnology-based cosmeceuticals. ISRN Dermatol. 2014: 843687. DOI:10.1155/2014/843687.

Lynn, D.D., Umari, T., Dunnick, C.A., Dellavalle, R.P. 2016. The epidemiology of acne vulgaris in late adolescence. Adolesc Health Med. Ther. 7: 13–25.

Macrini, D.J., Suffredini, I.B., Varella, A.D., Younes, R.N., Ohara, M.T. 2009. Extracts from Amazonian plants have inhibitory activity against tyrosinase: An *in vitro* evaluation. Braz. J. Pharm. Sci. 45(4): 715–721.

Maffei Facino, R., Carini, M., Aldini, G., Bombardelli, E., Morazzoni, P., Morelli, R. 1994. Free radicals scavenging action and anti-enzyme activities of procyanidines from *Vitis vinifera*. A mechanism for their capillary protective action. Arzneimittelforschung 44: 592–601.

Mahamongkol, H., Bellantone, R.A., Stagni, G., Plakogiannis, F.M. 2005. Permeation study of five formulations of alpha-tocopherol acetate through human cadaver skin. J. Cosmet. Sci. 56: 91–103.

Maity, N., Nema, N.K., Abedy, M.K., Sarkar, B.K., Mukherjee, P.K. 2011. Exploring *Tagetes erecta* Linn flower for the elastase, hyaluronidase and MMP-1 inhibitory activity. J. Ethnopharmacol. 137(3): 1300–1305.

Mallikarjuna, G., Dhanalakshmi, S., Singh, R.P., Agarwal, C., Agarwal, R. 2004. Silibinin protects against photocarcinogenesis via modulation of cell cycle regulators, mitogen-activated protein kinases, and Akt signaling. Cancer Res. 64(17): 6349–6356.

Mazumder, M.U., Das, K., Choudhury, A.D., Khazeo, P. 2018. Determination of sun protection factor (SPF) number of some hydroalcoholic vegetable extracts. PharmaTutor. 6(12): 41–45.

Monika, P., Basava, R.B.V., Chidambara Murthy, K.N., Ahalya, N., Gurudev, K. 2017. Nanocapsules of catechin rich extract for enhanced antioxidant potential and *in vitro* bioavailability. J. App. Pharm. Sci. 7(1): 184–188.

Moon, J.Y., Yim, E.Y., Song, G., Lee, N.H., Hyun, C.G. 2010. Screening of elastase and tyrosinase inhibitory activity from Jeju Island plants. Eurasia J. Biosci. 4(1): 41–53.

Muzaffer, U., Paul, V., Prasad, N.R., Karthikeyan, R., Agilan, B. 2018. Protective effect of *Juglans regia* L. against ultraviolet B radiation induced inflammatory responses in human epidermal keratinocytes. Phytomedicine 42: 100–111.

Napagoda, M.T., Malkanthi, B.M., Abayawardana, S.A., Qader, M.M., Jayasinghe, L. 2016. Photoprotective potential in some medicinal plants used to treat skin diseases in Sri Lanka. BMC Complement Altern Med. 16(1): 479. DOI: 10.1186/s12906-016-1455-8.

Napagoda, M.T., Kumari, M., Qader, M.M., De Soyza, S.G., Jayasinghe, L. 2018. Evaluation of tyrosinase inhibitory potential in flowers of *Cassia auriculata* L. for the development of natural skin whitening formulation. Eur. J. Integr. Med. 21: 39–42.

Napagoda, M.T., Sundarapperuma, T., Fonseka, D., Amarasiri, S., Gunaratna, P. 2019. Traditional uses of medicinal plants in Polonnaruwa district in North Central Province of Sri Lanka. Scientifica. 2019: Article ID 9737302, 11 pages. DOI: https://doi.org/10.1155/2019/9737302.

Nawarathne, N.W., Wijesekera, K., Wijayaratne, W.M.D.G.B., Napagoda, M. 2019. Development of novel topical cosmeceutical formulations from *Nigella sativa* L. with antimicrobial activity against acne-causing microorganisms. Sci. World J. 2019: Article ID 5985207, 7 pages, https://doi.org/10.1155/2019/5985207.

Ndhlovu, P.T., Mooki, O., Otang Mbeng, W., Aremu, A.O. 2019. Plant species used for cosmetic and cosmeceutical purposes by the Vhavenda women in Vhembe District Municipality, Limpopo, South Africa. S. Afr. J. Bot. 122: 422–431.

Ndlovu, G., Fouche, G., Tselanyane, M., Cordier, W., Steenkamp, V. 2013. *In vitro* determination of the anti-aging potential of four southern African medicinal plants. BMC Complement Altern Med. 13: 304. DOI:10.1016/j.jep.2011.07.064.

Nichols, J.A., Katiyar, S.K. 2010. Skin photoprotection by natural polyphenols: anti-inflammatory, antioxidant and DNA repair mechanisms. Arch. Dermatol. Res. 302: 71–83.

Nirmalan, T.E. 2017. Cosmetic perspectives of ethno-botany in Northern part of Sri Lanka. J. Cosmo. Trichol. 3: 126. DOI: 10.4172/2471-9323.1000126.

Niculae, G., Lacatusu, I., Badea, N., Stan, R., Vasile, B.S., Meghea, A. 2014. Rice bran and raspberry seed oil-based nanocarriers with self-antioxidative properties as safe photoprotective formulations. Photochem. Photobiol. Sci. 13(4): 703–716.

Nita, M., Grzybowski, A. 2016. The role of the reactive oxygen species and oxidative stress in the pathomechanism of the age-related ocular diseases and other pathologies of the anterior and posterior eye segments in adults. Oxid. Med. Cell Longev. 2016: Article ID 3164734, 23 pages, 2016. https://doi.org/10.1155/2016/3164734.

Oumeish, O.Y. 2001. The cultural and philosophical concepts of cosmetics in beauty and art through the medical history of mankind. Clin. Dermatol. 19(4): 375–86.

Panda, T., Mishra, N., Pradhan, B.K. 2016. Folk knowledge on medicinal plants used for the treatment of skin diseases in Bhadrak district of Odisha, India. Med. Aromat .Plants 5: 262. DOI:10.4172/2167-0412.1000262.

Pando, D., Caddeo, C., Manconib, M., Fadda, A.M., Pazos, C. 2013. Nanodesign of olein vesicles for the topical delivery of the antioxidant resveratrol. J. Pharm. Pharmacol. 65: 1158–1167.

Pandey, A., Sonthalia, S. 2019. Cosmeceuticals. In: StatPearls [Internet]. Treasure Island (FL): StatPearls Publishing; 2019 Jan-. Available from: https://www.ncbi.nlm.nih.gov/books/NBK544223/.

Patil, S., Fegade, B., Zamindar, U., Bhaskar, V.H. 2015. Determination of sun protection effect of herbal sunscreen cream. WJPPS. 4(8): 1554–1565.

Patkar, K.B. 2008. Herbal cosmetics in ancient India. Indian J. Plast Surg. 41(Suppl): S134–S137.

Pizzino, G., Irrera, N., Cucinotta, M., Pallio, G., Mannino, F., Arcoraci, V., Squadrito, F., Altavilla, D., Bitto, A. 2017. Oxidative Stress: Harms and Benefits for Human Health. Oxid. Med. Cell Longev. 2017: 8416763. DOI: 10.1155/2017/8416763.

Rashid, A., Arshad, M. 2002. Medicinal plant diversity, threat imposition and interaction of a mountain people community. In Proceeding of Workshop on Curriculum Development in Applied Ethnobotany. Published by the Ethnobotany Project, WWF Pakistan, 34-D/2, Sahibzada Abdul Qayuum Road Peshawar, Pakistan. 84–90.

Reeve, V.E., Bosnic, M., Rosinova, E., Boehm-Wilcox, C. 1993. A garlic extract protects from ultraviolet B (280–320 nm) radiation induced suppression of contact hypersensitivity. Photochem. Photobiol. 58: 813–817.

Rekha, M.B., Gokila, K. 2015. A study on consumer awareness, attitude and preference towards herbal cosmetic products with special reference to Coimbatore city. Int. J. Interdiscip. Multidiscip. Stud. 2(4): 96–100.

Rodrigues, D., Viotto, A.C., Checchia, R., Gomide, A., Severino, D., Itri, R., Baptista, M.S., Martins, W.K. 2016. Mechanism of *Aloe vera* extract protection against UVA: shelter of lysosomal membrane avoids photodamage. Photochem. Photobiol. Sci. 15: 334–350.

Ryu, Y.B., Ha, T.J., Curtis-Long, M.J., Ryu, H.W., Gal, S.W., Park, K.H. 2008. Inhibitory effects on mushroom tyrosinase by flavones from the stem barks of *Morus lhou* (S.) Koidz. J. Enzyme Inhib. Med. Chem. 223(6): 922–930.

Saewan, N., Jimtaisong, A. 2015. Natural products as photoprotection. Journal of Cosmetic Dermatology, 14(1): 47–63.

Sagbo, I.J., Mbeng, W.O. 2018. Plants used for cosmetics in the Eastern Cape povince of South Africa: A case study of skin care. Phcog. Rev. 12: 139–156.

Saija, A., Tomaino, A., Trombetta, D., De Pasquale, A., Uccella, N., Barbuzzi, T., Paolino, D., Bonina, F. 2000. *In vitro* and *in vivo* evaluation of caffeic and ferulic acids as topical photoprotective agents. Int. J. Pharm. 199(1): 39–47.

Schneider, G., Gohla, S., Schreiber, J., Kaden, W., Schönrock, U., Schmidt-Lewerkühne, H., Kuschel, A., Petsitis, X., Pape, W., Ippen, H., Diembeck, W. 2001. Skin Cosmetics. Ullmann's Encyclopedia of Industrial Chemistry. Ullmanns Encyclopedia of Industrial Chemistry. DOI: 10.1002/14356007.a24_219.

Schved, F., Kahn, V. 1992. Synergism exerted by 4-methyl catechol, catechol, and their respective quinones on the rate of DL-DOPA oxidation by mushroom tyrosinase. Pigment Cell Res. 5(1): 41–48.

Schwarz, A., Maeda, A., Kernebeck, K., van Steeg, H., Beissert, S., Schwarz, T. 2005. Prevention of UV radiation induced immunosuppression by IL-12 is dependent on DNA repair. J. Exp. Med. 201: 173–179.

Shaheen, H., Nazir, J., Firdous, S.S., Khalid, A.U. 2014. Cosmetic ethnobotany practiced by tribal women of Kashmir Himalayas. Avicenna J. Phytomed. 4(4): 239–250.

Sharififar, F., Ansari, M., Kazemipour, M., Mahdavi, H., Sarhadinejad, Z. 2013. *Teucrium polium* L. extract adsorbed on zinc oxide nanoparticles as a fortified sunscreen. Int. J. Pharm. Investig. 3(4): 188–193.

Shetty, P. K., Venuvanka, V., Jagani, H.V., Chethan, G.H., Ligade, V.S., Musmade, P.B., Nayak, U.Y., Reddy, M.S., Kalthur, G., Udupa, N., Rao, C.M., Mutalik, S. 2015. Development and evaluation of sunscreen creams containing morin-encapsulated nanoparticles for enhanced UV radiation protection and antioxidant activity. Int. J. Nanomedicine 10: 6477–6491.

Singh, N.K., Singh, D.P. 2001. Ethnobotanical survey of Balrampur. Flora-fauna 7(2): 59–66.

Sinha, P., Srivastava, S., Mishra, N., Yadav, N.P. 2014. New perspectives on antiacne plant drugs: contribution to modern therapeutics. Biomed. Res. Int. 2014, Article ID 301304, 19 pages. DOI: http://dx.doi.org/10.1155/2014/301304.

Smit, N., Vicanova, J., Pavel, S. 2009. The hunt for natural skin whitening agents. Int. J. Mol. Sci. 10(12): 5326–5349.

Song, X.Z., Bi, Z.G., Xu, A.E. 2006. Green tea polyphenol epigallocatechin- 3-gallate inhibits the expression of nitric oxide synthase and generation of nitric oxide induced by ultraviolet B in HaCaT cells. Chin. Med. J. (Engl). 119(4): 282–287.

Suwannateep, N., Wanichwecharungruang, S., Haag, S.F., Devahastin, S., Groth, N., Fluhr, J.W., Lademann, J., Meinke, M.C. 2012. Encapsulated curcumin results in prolonged curcumin activity *in vitro* and radical scavenging activity *ex vivo* on skin after UVB-irradiation. Eur. J. Pharm. Biopharm. 82: 485–490.

Tabassum, N., Hamdani, M. 2014. Plants used to treat skin diseases. Pharmacogn. Rev. 8(15): 52–60.

Takahashi, M., Kitamoto, D., Asikin, Y., Takara, K., Wada, K. 2009. Liposomes encapsulating *Aloe vera* leaf gel extract significantly enhance proliferation and collagen synthesis in human skin cell lines. J. Oleo Sci. 58(12): 643–650.

Tavano, L., Muzzalupo, R., Picci, N., de Cindio, B. 2014. Co-encapsulation of lipophilic antioxidants into niosomal carriers: Percutaneous permeation studies for cosmeceutical applications. Colloids Surf. B Biointerfaces. 114: 144–149.

Thring, T.S., Hili, P., Naughton, D.P. 2009. Anti-collagenase, anti-elastase and anti-oxidant activities of extracts from 21 plants. BMC Complement Altern. Med. 9: 27. DOI: 10.1186/1472-6882-9-27.

Vaibhav, S., Lakshaman, K. 2012. Tyrosinase enzyme inhibitory activity of selected Indian herbs. Int. J. Res. Pharm. Biomed. Sci. 3: 977–982.

Vardhan, A.K., Pandey, B. 2014. Screening of plant parts for anti-tyrosinase activity by tyrosinase assay using mushroom tyrosinase. Indian J. Sci. Res. 4: 134–139.

Varma, S.D., Kovtun, S., Hegde, K.R. 2011. Role of ultraviolet irradiation and oxidative stress in cataract formation-medical prevention by nutritional antioxidants and metabolic agonists. Eye Contact Lens. 37(4): 233–245.

Verma, A., Gautam, S.P., Devi, R., Singh, N., Harjaskaran, L.S. 2016. Cosmeceuticals: acclaiming its most fascinating position in personal care industry. Ind. Res. J. Pharm. & Sci. 3(1): 506–518.

Vinardell, M.P., Mitjans, M. 2015. Nanocarriers for delivery of antioxidants on the skin. Cosmetics, 2: 342–354.

Vierkötter, A., Krutmann, J. 2012. Environmental influences on skin aging and ethnic-specific manifestations. Dermatoendocrinol. 4(3): 227–231.

Wan, D.C., Wong, V.W., Longaker, M.T., Yang, G.P., Wei, F.C. 2014. Moisturizing different racial skin types. J. Clin. Aesthet. Dermatol. (6): 25–32.

Wang, Y., Hao, M.M., Sun, Y., Wang, L.F., Wang, H., Zhang, Y.J., Li, H.Y., Zhuang, P.W., Yang, Z. 2018. Synergistic promotion on tyrosinase inhibition by antioxidants. Molecules. 23(1): 106. DOI:10.3390/molecules23010106.

Wanjari, N., Waghmare, J. 2015. Review on Latest Trend of Cosmetics-Cosmeceuticals. IJPRR. 4(5): 45–51.

Widyarini, S., Spinks, N., Reeve, V.E. 2000. Protective effect of isoflavone derivative against photocarcinogenesis in a mouse model. Redox Rep. 5(2-3): 156–158.

Xia, J., Song, X., Bi, Z., Chu, W., Wan, Y. 2005. UV-induced NF-kappaB activation and expression of IL-6 is attenuated by (-)-epigallocatechin-3-gallate in cultured human keratinocytes *in vitro*. Int. J. Mol. Med. 16 (5): 943–950.

Yadav, K.D., Chaudhary, A.K. 2015. Cosmeceutical assets of ancient and contemporary ayurvedic astuteness. Int. J. Green Pharm. 9(4): S1–S6.

Yang, H.H., Oh, K.E., Jo, Y.H,. Ahn, J.H., Liu, Q., Turk, A., Jang, J.Y., Hwang, B.Y., Lee, K.Y., Lee, M.K. 2018. Characterization of tyrosinase inhibitory constituents from the aerial parts of *Humulus japonicus* using LC-MS/MS coupled online assay. Bioorg. Med. Chem. 26(2): 509–515.

Yao, Y., Cheng, X., Wang, L., Wang, S., Ren, G. 2012. Mushroom tyrosinase inhibitors from mung bean (Vigna radiatae L.) extracts. Int. J. Food Sci. Nutr. 63: 358–361.

Yokoyama, S., Hiramoto, K., Fujikawa, T., Kondo, H., Konishi, N., Sudo, S., Iwashima, M., Ooi, K. 2014. Topical application of *Corchorus olitorius* leaf extract ameliorates atopic dermatitis in NC/Nga mice. Dermatol Aspects 2:3. DOI: http://dx.doi.org/10.7243/2053-5309-2-3.

Ziegler, A., Jonason, A.S., Leffell, D.J., Simon, J.A., Sharma, H.W., Kimmelman, J., Remington, L., Jacks, T., Brash, D.E. 1994. Sunburn and p53 in the onset of skin cancer. Nature. 372: 773–776.

Zolghadri, S., Bahrami, A., Khan, M.T.H., Munoz-Munoz, J., Garcia-Molina, F., Garcia-Canovas, F., Saboury, A.A. 2019. A comprehensive review on tyrosinase inhibitors. J. Enzyme Inhib. Med. Chem. 34(1): 279–309.

18

Ethnomedicinal and Pharmacological Importance of *Glycyrrhiza glabra* L.

Ashish K. Bhattarai* and Sanjaya M. Dixit

Introduction

Glycyrrhiza glabra is a medicinal plant that belongs to Fabaceae/Leguminosae family (Sharma et al. 2018). It is commonly known as jethimadhu in Nepali, and liquorice or licorice in English. The name Glycyrrhiza was derived from the Greek words "glykys" meaning sweet and "rhiza" meaning root. The species name glabra was derived from the Latin "glaber", which means smooth or bald and refers to the smooth husks or pod-like fruit. Its common names are licorice, licorice-root, liquorice in English; réglisse in French; Lakritze, Süßholz in German; Jethimadhu in Nepali; Mulhatti, Jethimadhu, Mithilakdi in Hindi; gan-cao in Chinese; Sus, IrikSus, rib el-sus in Arabic; liquirizia in Italian; alcaçuz, pau-doce in Portuguese; alcazuz, licorice, orozuz, regaliz in Spanish; lakritsrot in Swedish (Esmail 2018).

In ancient times, plants were the major source of drugs- the word drug itself comes from the French word "drogue" which means "dry herb" (Wadud et al. 2007, Gootz 2010). In the traditional system of medicine, various indigenous plants were used for diagnosis, prevention, and treatment of both acute and chronic diseases. *G. glabra* is used in traditional medicine across the world for its ethnopharmacological value (Thakur and Raj 2017). The roots and rhizomes are the main medicinal parts of liquorice. Liquorice is essentially the dried rhizome and root of *G. glabra*, commercially known as Spanish licorice, or of *G. glabra* var. *glandulifera*, commercially known as Russian licorice, or of other varieties of *G. glabra* that yield a yellow and sweet wood (Tyler et al. 1998). Glycyrrhiza contains a saponin-like glycoside, glycyrrhizin (glycyrrhizic acid), which is 50 times as sweet as sugar (Omar et al. 2012). Glycyrrhizin is the major active constituent obtained from liquorice roots. Glycyrrhiza is considered to possess ulcer-protective, demulcent, antitussive, expectorant, and laxative properties (Kaur et al. 2013). It is also used as a flavoring agent to mask the taste of bitter drugs, such as aloe, quinine, and others (Zheng et al. 2018).

Department of Pharmacology, Kathmandu Medical College, Kathmandu University, Nepal
* Corresponding author: ashishakb33@gmail.com

Distribution

The genus *Glycyrrhiza* consists of about 30 species and is distributed all over the world (Thakur and Raj 2017). It is native to central and south western Asia, Europe, North and South America, Australia, and the Mediterranean region. Among these species, *G. uralensis, G. inflata,* and *G. glabra* are the only species mentioned in the Chinese Pharmacopoeia (Marisa et al. 2013). A monograph on *G. glabra* has also been mentioned in British Herbal Pharmacopoeia (Marisa et al. 2013).

Cultivation

Glycyrrhiza glabra is a perennial herb. It grows to about 1 m in height. Its roots are stoloniferous and fruit is oblong pod. In the Indian subcontinent it is popularly known as JethiMadhu or Mulhatti. In many parts of this region, it is also cultivated for medicinal purposes (Badkhane et al. 2014).

Morphology

The root is approximately 1.5 cm long and subdivides into subsidiary roots, about 1.25 cm long, from which the horizontal woody stolons arise. They may reach 8 m and when dried and cut, together with the root, constitute commercial licorice. It may be found peeled or unpeeled. The pieces of root break with a fibrous fracture, revealing the yellowish interior with a characteristic odor and sweet taste. The leaves are compound, imparipinnate, alternate, and have 4–7 pairs of oblong, elliptical, or lanceolate leaflets covered with soft hairs on the abaxial side. The flowers are narrow, typically papilionaceous, borne in axillary spikes, and violet in color. The calyx is short, campanulate, with lanceolate tips, and bearing glandular hairs. The fruit is a compressed legume or pod, upto 1.5 cm long, erect, glabrous, somewhat reticulately pitted, and usually contains 3–5 brown, reniform seeds (Pandey et al. 2017).

Physiochemical Properties

Glycyrrhiza glabra roots reveal that extractive values are (petroleum ether $4.67 \pm 0.23\%$, chloroform $10.56 \pm 1.53\%$, n-butanol, $6.54 \pm 0.84\%$, and methanol $13.89 \pm 2.42\%$); ash values are (total ash $4.67 \pm 0.35\%$, acid-insoluble ash $0.56 \pm 0.34\%$, and water-soluble ash $6.54 \pm 0.22\%$); loss on drying $5.87 \pm 0.65\%$, moisture contents $0.56 \pm 0.054\%$, pH of the extract (1% solution) 5.04 ± 0.65, pH of the extract (10% solution) 6.26 ± 0.54 (Esmail 2018).

Chemical constituents

Glycyrrhiza glabra roots contain alkaloids, glycosides, carbohydrates, starches, phenolic compounds, flavonoids, proteins, pectin, mucilage, saponins, lipids, tannins, sterols, and steroids. Liquorice root contains triterpenoidsaponins (4–20%), mostly glycyrrhizin. A mixture of potassium and calcium salts of 18β-glycyrrhizic acid (also known as glycyrrhizic or glycyrrhizinic acid and a glycoside of glycyrrhetinic acid) is 50 times sweeter than sugar (Omar et al. 2012).

Other triterpenes present are liquiritic acid, glycyrretol, glabrolide, isoglaborlide, and liquorice acid (Isbrucker and Burdock 2006). The root also contains an isoflavane known as Glabridin. EMA (2013) was the first to isolate 18β-glycyrrhizic acid from the roots of *Glycyrrhiza glabra* and he called 18β-glycyrrhizic acid as glycyrrhizin (EMA 2013). Glycyrrhizin is the major bioactive compound in the underground parts of *Glycyrrhiza* plants, which possess a wide range of pharmacological properties, and are used worldwide as a natural sweetener. Due to its economic value, the biosynthesis of glycyrrhizin has received substantial importance in many parts of the world. The percentage of Glycyrrhizin present in the root as potassium and calcium salts depends on plant species, geographic, and climatic conditions (Sabbioni et al. 2005). It is the flavonoids and chalcones which impart the

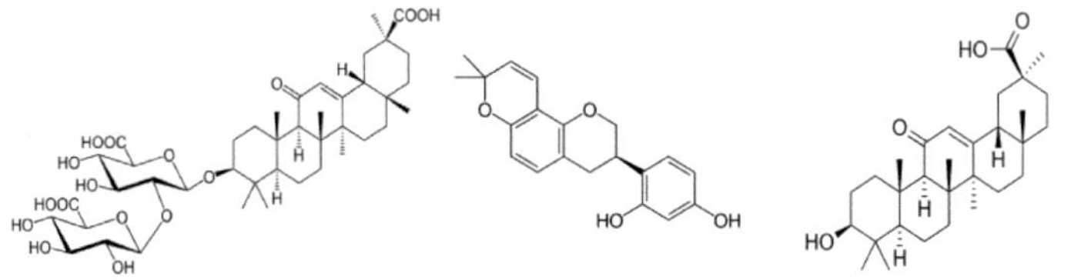

Glycyrrhizic Acid   Glabridin   Glycyrrhetinic Acid

Figure 18.1: Structures of important constituents of Glycyrrhiza.

yellow color to liquorice (Damle 2014). Examples of such flavonoids and chalcones include liquiritin, liquiritigenin, rhamnoliquiritin, neoliquiritin, isoliquiritin, isoliquiritigenin, neoisoliquiritin, licuraside, glabrolide, and licoflavonol (Damle 2014). Structures of important constituents of Glycyrrhiza have been given in Figure 18.1 (Pandey et al. 2017).

Traditional Uses

It had extensible uses in traditional Ayurveda and Chinese medicine for different liver and skin diseases for hundreds of years (EMA 2013). The earliest evidence of the use of liquorice comes from the ancient catacombs of Egyptian rulers. One of the earliest record of its use in medicine is found in "code Hummurabi" (2100 BC). It is also one of the important plants mentioned in Assyrian herbal studies (2000 BC) (Kaur et al. 2013). The people in ancient Greece and Rome commonly used liquorice as a tonic and as a cold remedy (Marisa et al. 2013). Theophrastus is known to have suggested liquorice as a remedy to combat infertility, to heal wounds and ulcerations of the mouth, and to treat malaises of the throat. Among the ancient Hindus, it was believed that liquorice, administered as a mixture with milk and sugar, increased sexual potency (Marisa et al. 2013). The ancient Chinese believed that liquorice root gave them strength and endurance, and they prepared it most often as tea for its tonic, expectorant, rejuvenating, laxative, and nutritive properties (Marisa et al. 2013).

During the middle ages, Arabic medical scientists such as IbnSinna wrote about licorice. English physician Nicholas Culpeper documented numerous uses of licorice in his work, the "Complete Herbal" (Marisa et al. 2013). In the 19th century, American Samuel Stearns and John Monroe proclaimed that the liquorice root has soothing, demulcent, expectorant, detergent, and diuretic property (Marisa et al. 2013). In India, licorice was believed to ease thirst, and has antitussive and demulcent activity. It was also thought to serve as a treatment for nausea, influenza, and urinary tract diseases (Marisa et al. 2013). In the Chinese subcontinent, licorice was used as a guide to enhance the effectiveness of the other ingredients, reduce toxicity, provide flavor, and improve the taste (Marisa et al. 2013). Licorice in present days also continues to serve as a flavoring agent, sweetening the bitter taste of many drugs (Zheng et al. 2018). It is still used as a filler for pills, as an essential ingredient in ointments for treating different skin diseases, for prolonging the effects of strong tonic medicines, and to potentiate glucocorticoid actions (Marisa et al. 2013). In 1949, Costello and Lynn suggested that this plant can be used for the medicinal purposes for the hormonal imbalances associated with menstruation. He extracted the estrogenic constituents from *G. glabra* (Marisa et al. 2013).

At present, licorice extracts have been commonly used in many European countries to relieve gastric and duodenal ulcers. Carbenexolone sodium, an antipeptic ulcer drug, which is a succinate derivative of 18β-glycyrrhetinic acid, has been extensively employed for the purpose of alleviating ulcers (Yano et al. 1989).

Licorice is the most-used crude drug in Kampo medicines (traditional Chinese medicines modified in Japan) (Fukai et al. 2002).

Anemia: A decoction of licorice powder is generally prescribed with honey to treat anemia.

Aphrodisiac: A mixture of licorice powder with honey taken with milk is used as an aphrodisiac and as an intellect-promoting tonic.

Burns and Bruishes: Warm clarified butter mixed with licorice, is used topically on wounds, bruises, and burns.

Cardiotonic: A paste of licorice and Picirrhizakurroa with sugar water is used as a cardiotonic.

Edema: In edema, a paste of licorice and Sesamumindicum milk mixed with butter is used.

Greying of hair: A decoction of the root is considered a good wash for falling and greying of hair.

Haematemesis: A mixture of licorice and Santalum album, powdered with milk is used to treat haematemesis.

Hoarseness: A confection of rice milk, prepared with licorice, is used for the treatment of hoarseness of voice.

Lactation: After mixing with cow's milk, it is used for promoting lactation.

Menmetrorrhagia: Equal parts of root powder and sugar are pounded with rice water and are commonly prescribed in menometrorrhagia.

Clinical Studies and Therapeutic Implications

The main constituents with pharmacological value of licorice are glycyrrhizin and its aglycone, namely 18β-glycyrrhetinic acid. These two compounds were found to show wide biological activities. They exhibit anti-ulceric, anti-inflammatory, anti-allergic, antioxidative, antiviral, anticarcinogenic, antithrombotic, hepatoprotective, neuroprotective, and antidiabetic activities (Zhang and Ye 2009). The glycyrrhizin has also been used as a potential therapeutic agent for different viral diseases such as chronic hepatitis B and C, human immune deficiency virus (Zhang and Ye 2009). It has traditional applications in stimulating digestive system functions, eliminating phlegm, relieving coughing, nourishing qi, and alleviating pain (Yang et al. 2017). Other studies have elaborated its other effects, such as nootropic action (Dhingra et al. 2004), anticariogenic action (Ajagannanavar et al. 2014), anti-tussive action (Kuang et al. 2018), and hair growth (Roy et al. 2014).

Anti-inflammatory Effects

The effect of glycyrrhizin on inflammatory mediators, such as neutrophil functions, including reactive oxygen species (ROS) generation was examined. The finding was that glycyrrhizin is not an ROS scavenger, but exerts an anti-inflammatory action by inhibiting the generation of ROS by neutrophils, which is a potent inflammatory mediator (Akamatsu et al. 1991).

The alcoholic and petroleum ether extracts of licorice roots can be safely used as anti-inflammatory agents determined in carrageenan-induced edema in experimental rats (Anmar et al. 1997). Topical application of extracts of raw licorice obtained by ethanol (LE) or roasted licorice obtained by ethanol (rLE) onto the mouse ear prior to 12-*O*-tetradecanoylphorbol-13-acetate (TPA) treatment inhibited TPA-induced acute inflammation, and oral administration of LE or rLE effectively suppressed the inflammatory response and tissue damage in collagen-induced arthritis (CIA) mouse model. rLE exhibited a more potent inhibition on TPA-induced acute inflammation than LE, but the antiarthritic effect of rLE was similar to that of LE in the CIA model (Kim et al. 2010). Overall, these data suggest that supplementation with LE and rLE may be beneficial in preventing and treating both acute and chronic inflammatory conditions, including rheumatoid arthritis. Furthermore, LE and rLE treatment also prevented oxidative damages in liver and kidney tissues of CIA mice (Kim et al. 2010).

The *G. glabra* extract has a significant anti-inflammatory action when compared to aspirin when the anti-inflammatory activity of it was screened by protein denaturation assay using aspirin as control (Jitesh 2017).

Antiulcer Effects

The pathogenesis of peptic ulcer disease includes an imbalance between gastric offensive factors such as acid, pepsin secretion, *Helicobacter pylori*, bile salts, ethanol, some medications such as NSAIDS, lipid peroxidation, nitric oxide (NO), and defensive mucosal factors such as prostaglandins (PG's), gastric mucus, cellular renovation, blood flow, mucosal cell shedding, glycoproteins, mucin secretion, proliferation, and antioxidant enzymes such as catalase (CAT), Superoxide dismutase (SOD), and glutathione level (Kaur et al. 2012).

Licorice causes the inhibition of 15-hydroxyprostaglandin dehydrogenase and delta13-prostaglandin reductase. 15-hydroxyprostaglandin dehydrogenase converts prostaglandins E2 and F2α to 15-ketoprostaglandins, which are inactive. This prevents the concentration of prostaglandin E2 and F2α towards degradation into inactive compounds (Baker 1994). Different prostaglandins have gastro shielding role. They help to maintain the mucosal integrity, and PGE2 is basically more important for this action. Increase in concentration of local prostaglandins promote the mucous secretion and cell proliferation in the stomach that ultimately promotes the healing of ulcers (Takeuchi and Amagase 2018).

The saponins and flavonoids are both considered to be the major bioactive constituents of licorice. The protective effect of licorice extract against gastric ulcer was recognized to glycyrrhizic acid-free fractions. And the role of flavonoids was also identified as part of the pharmacological activities of licorice (Zhang and Ye 2009).

The antioxidative mechanism of GutGard (a standardized extract of *G. glabra*) against gastric mucosal lesions was supported by its *in vitro* antioxidant potency, as evidenced by its high oxygen radical absorbance capacity assay (ORAC) value. These results support the ethnomedical uses of licorice in the treatment of ulcers (Mukherjee et al. 2010).

The different chemical constituents of the plant, glabridin and glabrene, which are the components of *G. glabra*, exhibited inhibitory activity against the growth of *Helicobacter pylori in vitro*. These flavonoids also showed anti-*H. pylori* activity against a Clarithromycin and Amoxicillin resistant strain (Fukai et al. 2002). A study even suggested licorice could be promoted as a replacement in the treatment for quadruple therapy when this regimen is not available as licorice has a low-cost, is highly tolerable, and has minimal side-effects (Rahnama et al. 2013).

In a study where anti-ulcer activity of DGL (Deglycyrrhizinated licorice) was evaluated in some common etiologies of ulcer, its extract was found to lower the frequency of ulcers, reduce the severity and inflammation of ulcers in all the three ulcer models, i.e., ethanol-induced ulcer model, aspirin-induced ulcer model, and stress-induced ulcer model (Kulkarni 2017).

Carbenoxolone

Carbenoxolone is a glycyrrhetinic acid derived from the root of licorice plant. It is one among few ulcer healing drugs used for treatment of esophageal, peptic, oropharyngeal inflammations, and ulceration (Pinder et al. 1976).

Carbenoxolone sodium accelerates the rate of healing of both gastric and duodenal ulcers. Carbenoxolone may act by affecting both the proliferative activity of gastric epithelium and the differentiation of the epithelial cells to produce mucus, as well as favorably altering the physicochemical properties of mucus and by reducing peptic activity. These factors may be useful for the prevention of acute gastric ulcers. Optimum therapeutic effect in gastric ulcer with the least side-effects is achieved with a dosage of 100 mg Carbenoxolone tablets three times daily for the first week, followed by 50 mg three times daily thereafter, best taken before meals. A lower dosage is recommended in the elderly and in patients with other medical conditions, such as those with renal, cardiac, or liver disease (Pinder et al. 1976).

The exact mechanism of the action of Carbenoloxone is still undecided. However, its benefits in treatment of gastric and duodenal ulcers are well established. Carbenoxolone is a compound developed as glycyrrhizate analog, and has shown to be effective in clinical trials in the treatment of gastric ulcer at the medium dose of 100 mg three times a day (Horwich and Galloway 1965, Turpie and Thomson 1965, Fraser et al. 1972, Langman et al. 1973) and duodenal ulcers (Brown et al. 1972, Doll et al. 1968, Montgomery et al. 1968).

Antimicrobial and Immune-stimulatory Effects

The presence of secondary metabolites such as saponins, alkaloids, flavonoids in hydro-methanolic root extract of *Glycyrrhiza glabra*, exhibits potent antibacterial activity (Sharma et al. 2013). *In vitro* studies have proved that aqueous and ethanolic extracts of licorice show inhibitory activity on cultures of *Staphylococcus aureus* and *Streptococcus pyogenes*. Isoflavonoids such as glabridin, glabrol, and their derivatives are responsible for *in vivo* inhibition of *Mycobacterium smegmatis* and *Candida albicans* (Alonso 2004). Methanolic extract of licorice was reported to have fungicidal activity against *Arthrinium sacchari* M001 and *Chaetomium funicola* M002. Glabridin was found to be the active compound giving antifungal activity (Hojo and Sato 2002).

A constituent of licorice "Licochalcone A" was reported to possess very good antimalarial action. *In vivo* studies against *P. yoelii* in mice with oral doses of 1000 mg kg^{-1} have shown to eradicate the malarial parasite completely. The toxicity was also not reported with this (Sianne and Fanie 2002).

Different studies have confirmed that the *Glycyrrhiza glabra* derived compound glycyrrhizin and its derivatives have antiviral activities. Animal studies demonstrated a reduction of mortality and viral activity in herpes simplex virus encephalitis and influenza. *In vitro* studies revealed antiviral activity against HIV 1, Severe Acute Respiratory Syndrome (SARS) related corona virus, respiratory syncytial virus, arboviruses, vaccinia virus, and vesicular stomatitis virus. Mechanisms for antiviral activity of *Glycyrrhiza* spp. include reduced transport to the membrane and sialylation of hepatitis B virus surface antigen, reduction of membrane fluidity leading to inhibition of fusion of the viral membrane of HIV 1 with the cell, induction of interferon-gamma in T cells, inhibition of phosphorylating enzymes in vesicular stomatitis virus infection, and reduction of viral latency (Fiore 2008).

Glycyrrhizin has been reported as the most active in inhibiting replication of the SARS associated corona virus. This study suggests that glycyrrhizin should be assessed for treatment of SARS (Cinati et al. 2003). The observations done by Wolkerstorfer et al. (2009) lead to the conclusion, that the antiviral activity of GL is mediated by an interaction with the cell membrane, which most likely results in reduced endocytosis activity, hence reducing the virus uptake. These understandings can help in the development of structurally related anti-influenza compounds.

Future research needs to explore the potency of compounds derived from licorice in the prevention and treatment of influenza A virus pneumonia and as an adjuvant treatment in patients infected with HIV resistant to antiretroviral drugs (Fiore et al. 2008).

The *Glycyrrhiza glabra* at 100 µg/ml concentration, showed increased production of lymphocytes and macrophages from human granulocytes in *in vitro* studies. In *in vivo* studies, licorice root extract was found to prevent the rise in the number of immune-complexes in different auto-immune diseases (Alonso 2004).

Antioxidant Activity

Powdered dry roots of licorice were extracted with methanol. Licorice extract was tested for antioxidative activity in comparison to antioxidants sodium metabisulfite and Butylatedhydroxytoluene (BHT) at 0.1%, 0.5%, 1.0%, and 2.0% w/w in 2% w/w hydroquinone cream. The extract demonstrated more antioxidant activity than two other commercial antioxidants at all concentrations, suggesting the possibility of using a licorice extract as an effective natural antioxidant for substances that are oxidation-susceptible (Morteza-Semnani et al. 2003).

A group of neolignan lipid estersand phenolic compounds isolated from the roots and stolons of licorice *Glycyrrhiza glabra* were found to have chemo preventive properties. Of these compounds, hispaglabridin B isoliquiritigenin, and paratocarpin B were found to be the most potent antioxidant agents (Chin et al. 2007).

The phytochemical composition, antioxidant, cytotoxic, and antimicrobial activities of a methanol extract from *Glycyrrhiza glabra* (Ge) was investigated. The antioxidant activity was evaluated by scavenging 2, 2-diphenyl-1-picrylhydrazyl (DPPH) and 2, 20-azino-bis (3-ethylbenzothiazoline-6-sulphonic acid) (ABTS) radicals, and reducing ferric complexes, and the total phenolic content was tested with the Folin–Ciocalteu method. According to this study, Ge has no antioxidant potential by this method, but suggested to be further studied for their potential to be developed as antioxidant. Instead, Ge showed moderate antibacterial activity against the 5 bacterial strains (Zhou et al. 2019).

Antiatherogenic Effects

The root extract of *G. glabra* was described to have antilipidemic and antihyperglycemic activity at low doses when its extract was studied on serum lipid profiles and liver enzymes in albino mice (Revers 1956).

Supplementation of licorice root extract (0.1 g/d) to patients for 1 month resulted in moderate hypercholesterolemia patients. It reduced the plasma LDL cholesterol level, plasma triacylglycerol levels, and after consumption of placebo for one month, the parameter reversed towards the baseline. Licorice extract supplementation also reduced systolic blood pressure by 10%, which was sustained during the placebo consumption (Fuhrman et al. 2002).

The antioxidant property of *Glycyrrhiza glabra* root extracts using *in vitro* models was evaluated. The dose-dependent aqueous and ethanolic extracts demonstrated the scavenging activity against nitric oxide, superoxide, hydroxyl radicals. Further, both extracts showed strong reducing power and iron-chelating capacities. The ethanolic extract of *G. glabra* possesses considerable antioxidant activity and protective effect against the human lipoprotein oxidative system (Visavadia et al. 2009).

Hepatoprotective effects

The *in vivo* protection of glycyrrhizin against CCl_4-induced hepatotoxicity was illustrated by Jeong et al. (2002). Glycyrrhizin showed a significant reduction in the release of CCl_4 induced AST and LDH at the concentration of 25–200 ug/ml. It has been speculated that this function was due to an alteration of membrane fluidity by the glycyrrhizin, or an inhibition of CCl_4-induced membrane lipid peroxidation. 18β-glycyrrhetic acid (an aglycone of glycyrrhizic acid) shows hepatoprotective activity by inhibiting both free radical generation and lipid peroxidation. The depletion of hepatic glutathione was also reduced in a dose-dependent manner by glycyrrhizin treatment. Besides this, glycyrrhizin also showed efficacy in reducing different drug-induced toxicities. It has shown usefulness in treating acetaminophen-induced hepatotoxicity (Xu ying et al. 2009) and diclofenac-induced hepatotoxicity in rats (Alaaeldin 2007).

Alpha-naphthylisothiocyanate (ANIT) is a common hepato-toxicant experimentally used to reproduce the pathologies of drug-induced liver injury in humans, but the mechanism of its toxicity remains unclear. Pre-treatment of glycyrrhizin (GL) and glycyrrhetinic acid (GA) prevented ANIT-induced liver damage and reversed the alteration of bile acid metabolites. These results suggested that GL/GA could prevent drug-induced liver injury and ensuing disruption of bile acid metabolism in humans (Wang et al. 2017).

Anticarcinogenic activity

Constituents of licorice include triterpenoids, such as glycyrrhizin and its aglycone glycyrrhizic acid, various polyphenols, and polysaccharides. A number of pharmaceutical effects of licorice are known or suspected (anti-inflammatory, antivirus, anti-ulcer, anticarcinogenesis, and others). Licorice and its derivatives may protect against carcinogen-induced DNA damage and may be suppressive agents as well. Glycyrrhizic acid is an inhibitor of lipoxygenase and cyclooxygenase, inhibits protein kinase C, and downregulates the epidermal growth factor receptor. Licorice polyphenols induce apoptosis in cancer cells (Wang and Nixon 2001).

When normal serum-free mouse embryo (SFME), tumorigenic human c-Ha-ras and mouse c-myccotransfected highly metastatic serum-free mouse embryo-1 (r/m HM-SFME-1) cells were treated with various concentrations of clinically available antitumor agents or glycyrrhetinic acid (GA), the anti-proliferative effects of these compounds were determined by the MTT assay. Western blotting analysis, RT-PCR, fluorescence staining, and confocal laser scanning microscopic observation were adopted to analyze H-Ras regulation. GA exhibited the tumor cell-selective toxicity through H-Ras downregulation, and its selectivity was superior to those of all the clinically available antitumor agents examined (Tao et al. 2010). For the selective toxicity of tumor cells, GA was most effective at 10 μM. Amusingly, this concentration was the same as the previously reported maximum plasma GA level reached in humans ingesting licorice. The result suggests that GA with its cytotoxic effects could be utilized as a suitable antitumor chemo therapeutic agent (Tao et al. 2010, Yamaguchi et al. 2010).

GA induces actin disruption and has tumor cell-selective toxic properties, and its selectivity is superior to those of all the clinically available antitumor agents tested in this study. The cytotoxic activity of GA and the tested antitumor agents showed a better correlation with the partition coefficient (log *P*) values rather than the polar surface area (PSA) values. For selective toxicity against tumor cells, GA was most effective at 10 μM (Yamaguchi et al. 2010).

Licorice and its derivatives may protect against carcinogen-induced DNA damage and may be suppressive agents as well. Glycyrrhizic acid is an inhibitor of lipoxygenase and cyclooxygenase, inhibits protein kinase C, and downregulates the epidermal growth factor receptor. Licorice polyphenols induce apoptosis in cancer cells (Wang 2001).

Persistent Hepatitis C Virus (HCV) infection and necro-inflammatory changes in chronic hepatitis C accelerate the development of liver cirrhosis and can promote in Hepatocellular carcinoma (HCC). When intravenous injection with Stronger Neo-Minophagen C (SNMC) was started in patients with chronic hepatitis or liver cirrhosis, most of them turned out to be infected with hepatitis viruses (Kumada 2002). In a multicenter double-blind study, Alanine AminoTransferase (ALT) levels decreased in the patients who received 40 ml/day of SNMC for four weeks at a rate significantly higher (p < 0.001) than controls receiving placebo. Furthermore, 100 ml/day of SNMC for eight weeks improved liver histology in 40 patients with chronic hepatitis, in correlation with improved ALT levels in serum. Liver cirrhosis occurred less frequently in 178 patients on long-term SNMC than in 100 controls (28 vs 40% at year 13, p < 0.002) (Kumada 2002). HCC developed less frequently in the 84 patients on long-term SNMC than in the 109 controls (13 vs 25% at year 15, p < 0.002). These results indicate that long-term treatment with SNMC prevents the development of HCC in patients with chronic hepatitis. SNMC is also helpful in patients with chronic hepatitis C who fail to respond to interferon, and in cases where interferon cannot be given due to many reasons (Kumada 2002).

Glycyrrhetinic acid (GA) and some of its derivatives may offer a role in combating cancer types having bad prognosis. Some GA derivatives are indeed able to target both the proteasome and Peroxisome Proliferator-Activated Receptors (PPARs), two proteins that play major roles in cancer cell biology, but are not related to Multi-Drug Resistant (MDR) and/or apoptosis-related resistance phenotypes (Lallemand et al. 2011).

Anticariogenic

The effect of licorice and its active sweet component glycyrrhizin was tested on the growth and adherence to glass of the cariogenic *Streptococcus mutans*. Neither licorice nor glycyrrhizin promoted growth or induced plaque formation. In the presence of sucrose, glycyrrhizin did not affect bacterial growth, but instead the plaque formation was markedly repressed. At 0.5–1% glycyrrhizin, inhibition was almost complete. These results support that glycyrrhizin might serve as an efficient vehicle for topical oral medications (Segal et al. 1985).

The effect of Aqueous and Alcoholic Licorice root extract against *Streptococcus mutans* and *Lactobacillus acidophilus* in comparison to Chlorhexidine was studied. At the end of 48 hours, statistically significant antimicrobial activity was demonstrated by all the test specimens used in this study. The inhibitory effect shown by alcoholic licorice root extract against *S. mutans* and *L. acidophilus* was found superior when compared to that of Chlorhexidine (CHX) and aqueous licorice (Ajagannanavar et al. 2014).

According to (Jatav et al. 2011) in the presence of sucrose, 0.5–1% glycyrrhizin had no effect on growth, but significantly inhibited bacterial adherence to glass by nearly 100% at the highest concentration tested. This also implies its anticariogenic role.

Antitussive and Expectorant Activity

Glycyrrhiza roots are useful for treating cough because of their demulcent and expectorant properties. The licorice powder and extract were found to be effective in treatment of sore throat, cough, and bronchial catarrh. It decreases irritation and produces expectorant effects. Licorice extract may stimulate tracheal mucous secretions producing demulcent and expectorant effects, although exact mechanism is not known. It efficiency is compared to be equivalent to that of codeine in sore throat (Murray 1998).

Licorice has been used as an antitussive and expectorant herbal medicine for a long time. Cough is produced in different medical conditions, such as common cold, bronchitis, and other respiratory illnesses. Expectorants help to bring up mucus and other materials from the lung, bronchi, and trachea. The activities of 14 major compounds and crude extracts of licorice, using the classical ammonia-induced cough model and phenol red secretion model in mice was evaluated. Liquiritinapioside, liquiritin, and liquiritigenin at 50 mg/kg (i.g.) could significantly decrease cough frequency by 30–78% (p < .01). The compounds Liquiritinapioside, liquiritin, and liquiritigenin showed potent expectorant activities after 3 days of treatment (p < .05). The water and ethanol extracts of licorice, which contain abundant Liquiritinapioside and liquiritin, could decrease cough frequency at 200 mg/kg by 25–59% (p < .05). The result indicates liquiritinapioside and liquiritin are the major antitussive and expectorant compounds of licorice. Their antitussive effects depend on both peripheral and central mechanisms (Kuang et al. 2018). Glycyrrhizin is responsible for demulcent action of licorice. Liquiritinapioside, an active compound present in the methanolic extract of licorice, is found to inhibit capsaicin-induced cough (Kamei et al. 2003).

Hair growth

The hydro-alcoholic extract of 2% licorice showed better hair growth activity than 2% of the standard drug Minoxidil. Thus, its further study for alopecia after efficacy and safety analysis is suggested (Roy et al. 2014).

Anti Coagulant

Glycyrrhizin is one of the first plant-based inhibitors of thrombin. It is found to prolong the thrombin and fibrinogen clotting time (Mendes-Silva et al. 2003). Glycyrrhizin causes inhibition in thrombin-induced platelet aggregation (Mauricio et al. 1997).

Nootropic action

Nootropics are drugs, supplements, and other substances that may improve cognitive functions, such as memory, creativity, etc., in healthy individuals. Significant improvement in learning and memory of mice was reported at the dose of 150 mg/kg when investigated in mice. Three doses of aqueous extract of licorice were administered (75, 150, and 300 mg/kg p.o) for seven successive days in separate groups of animals. Elevated plus-maze and passive avoidance paradigm were used as experimental setup to test learning and memory (Dhingra et al. 2004).

Uses in Skin

In vitro tyrosinase enzyme inhibition studies have showed that 21.2 μg/ml of methanolic extract of licorice caused 50% tyrosinase enzyme inhibition. The inhibition of tyrosinase enzyme and reduction in enzyme activity is caused due to modification of action site of the enzyme. Due to good tyrosinase inhibition activity, licorice extract can be used for depigmenting activity (Zuidhoff and Rijsbergen 2001).

Some other active compounds in licorice extract, such as glabrene, Licochalcone A, Isoliquiritin are also responsible for inhibition of tyrosinase activity. Liquiritin present in licorice extract disperses melanin, thereby inducing skin lightening (Nohata et al. 2005).

Safety and Side-effects of Licorice and its Constituents

Different genotoxic studies have shown that glycyrrhizin is neither teratogenic nor mutagenic, and may possess antigenotoxic properties under certain conditions. Based on the *in vivo* and clinical evidence, it is proposed that a daily intake of 0.015–0.229 mg glycyrrhizin/kg body weight/day is acceptable (Isbrucker and Burdock 2006).

High intake of licorice can cause hyper mineralocorticoidism with sodium retention and potassium loss, oedema, increased blood pressure, and depression of the renin-angiotensin-aldosterone system. As a result, the number of related clinical symptoms are reported. There is increased cortisol level in the kidneys and other mineralocorticoid selective tissues because of the inhibition of enzymes involved in the metabolism of corticosteroids. Glycyrrhetic acid inhibits the enzyme 11β-hydroxysteroid dehydrogenase involved in the metabolism of corticosteroids which is produced after glycyrrhizic acid is hydrolyzed in the intestine. This cortisol results in hyper mineralocorticoid effect. The compensatory physiological mechanisms following hyper mineralocorticoids, which is depression of the renin-angiotensin system, can last several months. The inhibitory effect on 11β-hydroxysteroid dehydrogenase is reversible. So after the withdrawal of consumption of licorice, there is physiological reversal of hyper mineralocorticoids, but it takes several months (Størmer et al. 1993).

The continuous, high-level exposure to glycyrrhizin compounds can produce hyper mineralocorticoid-like effects in both animals and humans. The glycyrrhizinates inhibit 11beta-hydroxysteroid dehydrogenase, the enzyme responsible for inactivating cortisol. And these effects are reversible upon withdrawal of glycyrrhizin (Isbrucker and Burdock 2006, Yang et al. 2017).

Severe mineralocorticoid-like toxic effects, such as sodium and water retention and hypokalemia appear, most frequently in those receiving excessive doses of Carbenoxolone. It should be used in patients with careful and regular observation of serum electrolytes, especially potassium. If the signs of the toxicity appears, it should be stopped and the complication is treated (Pinder et al. 1976).

Short-term use of less than 4–6 weeks of licorice preparations is regarded safe. The most common serious side-effects reported following chronic use of high dose of licorice root are hypokalemia and hypertension. Especially in vulnerable people, prolonged daily intake even of low doses of licorice, corresponding to 80–100 mg of glycyrrhizic acid, may provoke severe hypertension. More rarely, cardiac rhythm disorders can occur. And there is insufficient data to support the safety of licorice root during pregnancy and lactation in children and adolescents under 18 years, therefore the use is not recommended for these patient groups (EMA 2013).

Conclusion

Glycyrrhiza glabra is a plant with huge ethno-pharmacological importance. Its role has been identified in many clinical conditions, such as anti-inflammatory, anti-ulcer, antiviral, antimicrobial, antioxidants, antiatherogenic, anticarcinogenic, antimutagenic, antitussive expectorant, and hepatoprotective, etc. Carbenoxolone is a glycyrrhetinic acid derived from the root of this plant. It is one of the established commercial drugs found to have good clinical efficacy for esophageal, peptic, oropharyngeal inflammations, and ulceration.

In medieval times, *Glycyrrhiza glabra* was used for soothing, demulcent, expectorant, detergent, and diuretic properties. In the middle ages, it was used for maintaining hormonal balance in menstrual irregularities, to combat infertility, and increase the sexual potency. As a traditional herbal use, it has been considered for the dyspepsia and expectorants as herbal infusion in boiling water or as a decoction. Its use as soft extract indicated to improve the gastric function, and dry extract in combination with other expectorants is also available in markets around the world.

It has been suggested for depigmenting and skin lightening property. Thus, licorice extract has a possibility to be used solely or as an ingredient for making different cosmetic preparations. It is claimed to have better hair re-growing activity than Standard Minoxidil preparation in some studies. It is implied to have anticaries role as well.

Short term and appropriate use of licorice preparations is regarded as safe. Still, serious side effects, such as hypokalemia and hypertension are reported following chronic use of high dose of licorice root. And there is insufficient data to support the safety of licorice root during pregnancy, lactation, and in children. Therefore, it requires more extensive study before recommending the use in those age groups. The constituents of licorice are found to have anticarcinogenic activity. However, there is still a need of ample tests on reproductive toxicity, teratogenicity, and carcinogenicity before its actual benefits can be acclaimed.

Licorice in present days also continues to serve as a flavoring agent, sweetening the bitter taste of many drugs. This review was focused on its general introduction, traditional uses, and pharmacological activities. This could help in further studies on *Glycyrrhiza glabra* for exploring its potential in preventing and treating diseases and other commercial uses.

References

Ajagannanavar, S.L., Battur, H., Shamarao, S., Sivakumar, V., Patil, P.U., Shanavas, P. 2014. Effect of aqueous and alcoholic licorice (*Glycyrrhiza glabra*) root extract against *Streptococcus mutans* and *Lactobacillus acidophilus* in comparison to chlorhexidine: an *in-vitro* study. J. Int. Oral Hea. 6(4): 29–34.

Akamatsu, H., Komura, J., Asada, Y., Niwa, Y. 1991. Mechanism of anti-inflammatory action of glycyrrhizin: effect on neutrophil functions including reactive oxygen species generation. Planta Med. 57(2): 119–121.

Alaaeldin, A.H. 2007. *Curcuma longa*, *Glycyrrhiza glabra* Linn and *Moringa oleifera* Ameliorate Diclofenac-induced Hepatotoxicity in Rats. Am. J. Pharmacol. Toxicol. 2(2): 80–88.

Alonso, J. 2004. Tratado de Fitofármacos y Nutracéuticos. *In*: www.fitoterapia.net. Barcelona: Corpus, 905–911.

Anmar, N.M., Okbi, S.Y.A., Mohamed, D.A. 1997. Study of the inflamatory activity of some medicinal edible plantys growing in Egypt. J. Islam. Acad. Sci. 10(4): 113–122.

Badkhane, Y., Yadav, A.S., Bajaj, A., Sharma, A.K., Raghuwanshi, D.K. 2014. *Glycyrrhiza glabra* L., a miracle medicinal herb. Indo. Amer. J. Pharm. Res. 4(12).

Baker, M.E. 1994. Licorice and enzymes other than 11β-hydroxysteroid dehydrogenase: An evolutionary perspective. Stero. 59(2): 136–141.

Brown, R.C., Langman, M.J., Lambert, P.M. 1976. Hospital admissions for peptic ulcer during 1958–72. Br. Med. J. 1(6000): 35–37.

Chin, Y.W., Jung, H.A., Liu, Y., Su, B.N., Castoro, J.A., Keller, W.J., Pereira, M.A., Kinghorn, A.D. 2007. Anti-oxidant constituents of the roots and stolons of licorice (*Glycyrrhiza glabra*). J. Agric. Food Chem. 55(12): 4691–4697.

Cinatl, J., Morgenstern, B., Bauer, G., Chandra, P., Rabenau, H., Doerr, H.W. 2003. Glycyrrhizin, an active component of liquorice roots, and replication of SARS-associated coronavirus. Lancet 361(9374): 2045–2046.

Damle, M. 2014. *Glycyrrhiza glabra* (Liquorice)—a potent medicinal herb. Int. J. Herb. Med. 2(2): 132–136.

Dhingra, D., Parle, M., Kulkarni, S.K. 2004. Memory enhancing activity of *Glycyrrhiza glabra* Linn mice. J. Ethnopharma. 91(2-3): 361–365.

EMA. 2013. Assessment report on Glycyrrhizaglabra L. and/or GlycyrrhizainflataBat.and/orGlycyrrhizauralensisFisch., radix EMA/HMPC/571122/2010.

Esmail, Al-Snafi, A. 2018. *Glycyrrhiza glabra*: A phytochemical and pharmacological review. Journal of Pharmacy 8(6): 1–17.

Fiore, C., Eisenhut, M., Krausse, R., Ragazzi, E., Pellati, D., Armanini, D., Bielenberg, J. 2008. Antiviral effects of *Glycyrrhiza* species. Phytother. Res. 22(2): 141–148.

Fuhrman, B., Volkova, N., Kaplan, M., Presser, D., Attias, J., Hayek, T., Aviram, M. 2002. Antiatherosclerotic effects of licorice extract supplementation on hypercholesterolemic patients: increased resistance of LDL to atherogenic modifications, reduced plasma lipid levels, and decreased systolic blood pressure. Nutrit. 18(3): 268–273.

Fukai, T., Marumo, A., Kaitou, K., Kanda, T., Terada, S., Nomura, T. 2002. Anti-Helicobacter pylori flavonoids from licorice extract. Life Sci. 71(12): 1449–1463.

Gootz, T.D. 2010. The global problem of antibiotic resistance. Crit. Rev. Immunol. 30(1): 79–93.

Hojo, H., Sato, J. 2002. Antifungal Activity of Licorice (*Glycyrrhiza glabra* Linn) and Potential Applications in Beverage Foods. J. Food Ingred. 203.

Isbrucker, R.A., Burdock, G.A. 2006. Risk and safety assessment on the consumption of Licorice root (*Glycyrrhiza* sp.), its extract and powder as a food ingredient, with emphasis on the pharmacology and toxicology of glycyrrhizin. Regul. Toxicol. Pharmacol. 46(3): 167–192.

Jatav, V.S.S.S.K., Khatri, P., Sharma, A.K. 2011. Recent Pharmacological trends of *Glycyrrhiza glabra* Linn. International Journal of Pharmaceutical Frontier Research 1(1): 170–185.

Jeong, H.G., You, H.J., Park, S.J., Moon, A.R., Chung, Y.C., Kang, S.K., Chun, H.K. 2002. Hepatoprotective effects of 18β-glycyrrhetinic acid on carbon tetrachloride-induced liver injury: inhibition of cytochrome P450 2E1 expression.

Jitesh, S. 2017. Anti-inflammatory activity of *Glycrrhiza glabra* extract—An *in-vitro* study. J. Pharm. Sci. & Re. 9(4): 451–452.

Jun-Xian Zhou, M.S.B., PilleWetterauer, Bernhard Wetterauer, Michael, W. 2019. Antioxidant, Cytotoxic, and Antimicrobial Activities of *Glycyrrhiza glabra* L., *Paeonia lactiflora* Pall., and *Eriobotrya japonica* (Thunb.) Lindl. Extracts. Med., 6(43).

Kamei, J., Nakamura, R., Ichiki, H., Kubo, M. 2003. Antitussive principles of Glycyrrhiza radix, a main component of Kampo preparations Bakumondo-to (Mai-men-dong-tang). E. J. Pharm. 69: 159–163.

Kaur, D.J., Rana, A.C., Sharma, N.G. 2012. Herbal Drugs with AntiulcerActivity. J. Appl. Pharma. Sci. 2(3): 160–165.

Kaur, R., Kaur, H., Dhindsa, A.S. 2013. *Glycyrrhiza glabra*: a phytopharmacological review. Int. J. Pharm. Sci. Res. 4(7): 2470–2477.

Kim, K.R., Jeong, C.K., Park, K.K., Choi, J.H., Park, J.H.Y., Lim, S.S., Chung, W.Y. 2010. Anti-inflammatory effects of licorice and roasted licorice extracts on TPA-induced acute inflammation and collagen-induced arthritis in mice. J. Biomed. Biotechnol. 709378.

Kuang, Y., Li, B., Fan, J., Qiao, X., Ye, M. 2018. Antitussive and expectorant activities of licorice and its major compounds. Bioorg. Med. Chem. 26(1): 278–284.

Kulkarni, K. 2017. Evaluation of Antiulcer Activity of DGL (Deglycyrrhizinate dliquorice). J. Pharm. 9(11): 2019–2022.

Kumada, H. 2002. Long-term treatment of chronic hepatitis C with glycyrrhizin [stronger neo-minophagen C (SNMC)] for preventing liver cirrhosis and hepatocellular carcinoma. Oncology, 62(Suppl 1): 94–100.

Lallemand, B.G.M., Dubois, J., Prevost, M., Jabin, I., Kiss, R. 2011. Structure-Activity Relationship Analyses of Glycyrrhetinic Acid Derivatives as Anticancer Agents. Mini-Rev. Med. Chem. 11(10): 881–887.

Langman, M.J.S., Knapp, D.R., Wakley, E.J. 1973. Treatment of Chronic Gastric Ulcer with Carbenoxolone and Gefarnate: A Comparative Trial. British Medical Journal 3: 84–86.

Licorice. 101. Traditional Medicinals—Wellness Teas. Accessed from www.traditionalmedicinals.com.

Marisa, R., Assessor, D., Calapai, G., Delbò, M. Date. Assessment report on Glycyrrhizaglabra L. and/or Glycyrrhizainflata Bat. and/orGlycyrrhizauralensisFisch., radix,.

Mauricio, I., Francischett, B., Monterio, R.Q., Guimaraeas, J.A. 1997. Identification of Glycyrrhizin as thrombin inhibitor, Biochim. Biophys. Res. Commun., 235: 259–263.

Mendes-Silva, W., Assafim, M., Ruta, B., Monteiro, R.Q., Guimarães, J.A., Zingali, R.B. 2003. Antithrombotic effect of Glycyrrhizin, a plant-derived thrombin inhibitor. Thromb, Res. 112: 93–98.

Morteza-Semnani, K., Saeedi, M., Shahnavaz, B. 2003. Comparison of antioxidant activity of extract from roots of licorice (*Glycyrrhiza glabra* L.) to commercial antioxidants in 2% hydroquinone cream. J. Cosmet. Sci. 54(6): 551–8.

Moumita Mukherjee, N.B., Srinath, R., Shivaprasad, H.N., Joshua Allan, J., Shekhar, D., Agarwal, A. 2010. Anti-ulcer and antioxidant activity of GutGard TM. Ind. J. Exp. Biol., 48: 269–274.

Murray, W.J. 1998. Herbal Medications for Gastrointestinal Problems, Herbal Medicinals-A Clinician's Guide. Pharmaceutical Products Press, New York, pp. 79–93.

Nohata, Yasuhiro, K, Tomomi. 2005. Japan Patent 2002029929.

Omar, H.R., Komarova, I., El-Ghonemi, M., Fathy, A., Rashad, R., Abdelmalak, H.D., Yerramadha, M.R., Ali, Y., Helal, E., Camporesi, E.M. 2012. Licorice abuse: Time to send a warning message. Ther. Adv. Endocrinol. Metab. 3(4): 125–138.

Pan, C., Chen, Y.G., Ma, X.Y., Jiang, J.H., He, F., Zhang, Y. 2011. Phytochemical constituents and pharmacological activities of plants from the genus Adiantum: A review. Trop. J. Pharm. Res. 10(5): 681–692.

Pandey, S., Verma, B., Arya, P. 2017. A review on constituents, pharmacological activities and medicinal uses of *Glycyrrhiza glabra*. Univ. J. Pharma. Res. 2(2): 6–11.

Pinder, R.M., Brogden, R.N., Sawyer, P.R., Speight, T.M., Spencer, R., Avery, G.S. 1976. Carbenoxolone: a review of its pharmacological properties and therapeutic efficacy in peptic ulcer disease. Drugs 11(4): 245–307.

Rahnama, M., Mehrabani, D., Japoni, S., Edjtehadi, M., Saberi Firoozi, M. 2013. The healing effect of licorice (*Glycyrrhiza glabra*) on Helicobacter pylori infected peptic ulcers. J. Res. Med. Sci. 18(6): 532–533.

Revers, F.E. 1956. Clinical and pharmacological investigations on extract of licorice. Acta Med. Scand. 154: 749–751.

Roy, S.D., Karmakar, P.R., Dash, S., Chakraborty, J., Das, B. 2014. Hair growth stimulating effect and phytochemical evaluation of hydro-alcoholic extract of *Glycyrrhiza glabra*. Glob. J. Res. Med. Plants Indigen. 3(2): 40–47.

Sabbionia, C., Mandriolia, R., Ferrantia, A., Bugamellia, F., Saracinoa, M.A., Fortib, G.S., Fanalic, S., Raggia, M.A. 2005. Separation and analysis of glycyrrhizin, 18_-glycyrrhetic acid and 18_-glycyrrhetic acid in liquorice roots by means of capillary zoneelectrophoresis. Journal of Chromatography A 1081: 65–71.

Segal, R., Pisanty, S., Wormser, R., Azaz, E., Sela, M.N. 1985. Anticariogenic activity of licorice and glycyrrhizine I: Inhibition of *in-vitro* plaque formation by *Streptococcus mutans*. J. Pharm. Sci. 74(1): 79–81.

Sharma, V., Agrawal, R.C., Pandey, S. 2013. Phytochemical screening and determination of anti-bacterial and anti-oxidant potential of *Glycyrrhiza glabra* root extracts. J. Environ. Res. Develop. 7(4A): 1552–1558.

Sharma, V., Katiyar, A., Agrawal, R.C. 2018. *Glycyrrhiza glabra*: chemistry and pharmacological activity. pp. 87–100. *In*: Mérillon, J.M., Ramawat, K. (eds.). Sweeteners. Reference Series in Phytochemistry. Springer, Cham.

Sianne, S., Fanie, R.V.H. 2002. Antimalarial activity of plant metabolites. Nat. Prod. Rep. 19: 675–692.

Størmer, F.C., Reistad, R., Alexander, J. 1993. Glycyrrhizic acid in liquorice-Evaluation of health hazard. Food Chem. Toxicol. 31(4): 303–312.

Takeuchi, K., Amagase, K. 2018. Roles of Cyclooxygenase, Prostaglandin E2 and EP Receptors in Mucosal Protection and Ulcer Healing in the Gastrointestinal Tract. Curr. Pharm. Des. 24(18): 2002–2011.

Tao, Y., Toshiro Noshita, H.Y., YumiKidachi, Hironori Umetsu, Kazuo Ryoyama. 2010. Selective cytotoxicity of glycyrrhetinic acid against tumorigenic r/m HM-SFME-1 cells: Potential involvement of H-Ras down regulation. Toxicol. 192(3): 425–430.

Thakur, A.K., Raj, P. 2017. Pharmacological Perspective of *Glycyrrhiza glabra* Linn.: a Mini-Review. J. Anal. Pharm. Res. 5(5).

Tyler, V., Brady, L., Robbers, J. 1998. *Pharmacognosy*. 9th ed. Philadelphia: Lea & Febiger, pp. 69.

Visavadiya, N.P., Soni, B., Dalwadi, N. 2009. Evaluation of antioxidant and anti-atherogenic properties of *Glycyrrhiza glabra* root using *in vitro* models. Int. J. Food Sci. Nutr., 60(sup2): 135–149.

Wadud, A., Prasad, P.V.V., Rao, M.M., Narayana, A. 2007. Evolution of drug: a historical perspective. Bull. Indian Inst. Hist. Med. 37(1): 69–80.

Wang, H., Fang, Z.Z., Meng, R., Cao, Y.F., Tanaka, N., Krausz, K.W., Gonzalez, F.J. 2017. Glycyrrhizin and glycyrrhetinic acid inhibits alpha-naphthyl isothiocyanate-induced liver injury and bile acid cycle disruption. Toxicol. 386: 133–142.

Wang, Z.Y., Nixon, D.W. 2001. Licorice and Cancer. Nutr. Can. 39(1): 1–11.

Wolkerstorfer, A., Kurz, H., Bachhofner, N., Szolar, O.H. 2009. Glycyrrhizin inhibits influenza A virus uptake into the cell. Antiv. Res. 83(2): 171–178.

Xu-ying, W., Ming, L., Xiao-dong, L., Ping, H. 2009. Hepatoprotective and anti hepatocarcinogenic effects of glycyrrhizin and matrine. Chem. Biol. Interact. 181(1): 15–19.

Yamaguchi, H., Noshita, T., Yu, T., Kidachi, Y., Kamiie, K., Umetsu, H., Ryoyama, K. 2010. Novel effects of glycyrrhetinic acid on the central nervous system tumorigenic progenitor cells: induction of actin disruption and tumor cell-selective toxicity. Eur. J. Med. Chem. 45(7): 2943–2948.

Yang, R., Yuan, B.C., Ma, Y.S., Zhou, S., Liu, Y. 2017. The anti-inflammatory activity of licorice, a widely used Chinese herb. Pharm. Biol. 55(1): 5–18.

Yano, S., Harada, M., Watanabe, K., Nakamaru, K., Hatakeyama, Y., Shibata, S., Takahashi, K., Mori, T., Hirabayashi, K., Takeda, M., Nagata, N. 1989. Antiulcer Activities of Glycyrrhetinic Acid Derivatives in Experimental Gastric Lesion Models. Chem. Pharm. Bull. 37(9): 2500–2504.

Zhang, Q., Ye, M. 2009. Chemical analysis of the Chinese herbal medicine Gan-Cao (licorice). J. Chromatogr. A 1216(11): 1954–1969.

Zheng, X., Wu, F., Hong, Y., Shen, L., Lin, X., Feng, Y. 2018. Developments in taste-masking techniques for traditional Chinese medicines. Pharmaceu., 10(3).

Zuidhoff, H.W., Rijsbergen, J.M.V. 2001. Whitening Efficacy of Frequently Used Whitening Ingredients, C&T. 116(1): 53–59.

Index

About the Editors

Dr. Mahendra Rai is a Senior Professor and UGC-Basic Science Research Faculty at the Department of Biotechnology, Sant Gadge Baba Amravati University, Maharashtra, India. He was a visiting scientist at the University of Geneva, Debrecen University, Hungary; University of Campinas, Brazil; Nicolaus Copernicus University, Poland; VSB Technical University of Ostrava, Czech Republic, and National University of Rosario, Argentina. He has published more than 400 research papers in national and international journals. In addition, he has edited/authored more than 55 books and 6 patents.

Dr. Shandesh Bhattarai is a Senior Scientific Officer at Nepal Academy of Science and Technology, Khumaltar, Lalitpur, Nepal. He worked in the collaborative research projects funded by NUFU-Norway, VW Foundation-Germany, Darwin Initiative-UK, DFID-UK, IUCN-Bangkok, UGC-Nepal, etc. He is a coordinator of the Flora of Nepal Project, an international initiative of the institutions of Nepal, UK, and Japan. He has published more than 35 research papers in national and international journals and has authored one book.

Dr. Chistiane M. Feitosa is a Professor at the Department of Chemistry, Federal University of Piaui, Teresina-Piaui, Brazil. Her research area includes natural products with the potential for the treatment of neurodegenerative diseases. She has published more than 50 research papers in national and international journals. In addition, she has edited/authored more than 10 books/chapters and 5 patents.